FINE PARTICLES

Aerosol Generation, Measurement, Sampling, and Analysis

ACADEMIC PRESS RAPID MANUSCRIPT REPRODUCTION

Proceedings of a symposium by the
U.S. Environmental Protection Agency
held in Minneapolis, Minnesota, May 28-30, 1975

FINE PARTICLES

Aerosol Generation, Measurement, Sampling, and Analysis

Edited by

BENJAMIN Y. H. LIU

Particle Technology Laboratory
Mechanical Engineering Department
University of Minnesota
Minneapolis, Minnesota

Academic Press, Inc.

New York San Francisco London 1976

A subsidiary of Harcourt Brace Jovanovich, Publishers

ACADEMIC PRESS, INC.
111 Fifth Avenue, New York, New York 10003

United Kingdom Edition published by
ACADEMIC PRESS, INC. (LONDON) LTD.
24/28 Oval Road, London NW1

Library of Congress Cataloging in Publication Data

Symposium on Fine Particles, Minneapolis, 1975.
Fine particles.

Includes bibliographies and index.
1. Aerosols--Congresses. I. Liu, Benjamin Y. H.,
(date) II. Title.
TP244.A3S9 1975 660.2'9'4515 75-45096
ISBN 0-12-452950-X

PRINTED IN THE UNITED STATES OF AMERICA

CONTENTS

CONTENTS

PREFACE

This volume contains the technical papers presented at the Symposium on Fine Particles held in Minneapolis, Minnesota, May 28-30, 1975. In addition, several papers that were not presented at the symposium, due to the lack of time, are also included.

The purpose of the symposium was to review the state of art and recent developments in instrumentation and experimental techniques for aerosols studies. The focus was on fine particles below about 3.5 μm in diameter. The topics covered include aerosol generation, measurement, sampling, and analysis.

The importance of fine particles to air pollution control is well known: Many of the undesirable effects of particulate air pollutants, such as those on human health and atmospheric visibility, are due to fine particles. However, the control of fine particles is considerably more difficult than the control of coarse particles. In industrial hygiene, mining safety, fire detection, and other related areas, fine particles have also been found to play an important role.

In a rapidly growing field of science where there is a diversity of workers, there is a need for a periodic review of the subject and a convenient source of reference. It is hoped that this symposium volume will help in part to meet that need.

I wish to thank Dr. Dennis C. Drehmel of the Industrial Environmental Research Laboratory of the Environmental Protection Agency for helping to bring the symposium into being. I am grateful also to Mr. Joe Kroll of the Department of Continuing Education and Extension of the University of Minnesota for helping to organize the symposium. In addition, the help of the Program Committee–J.H. Abbott (EPA), J.A. Dorsey (EPA), D.C. Drehmel (EPA), and K.T. Whitby (UM)–and that of the Session Chairmen–A.B. Craig (EPA), D.C. Drehmel (EPA), J. Nader (EPA), K. Willeke (UM), and D.B. Kittelson (UM)–are also gratefully acknowledged. I would like to thank the staff and the students of the Particle Technology Laboratory of the University of Minnesota for their help during the symposium. A special word of thanks goes to David Y.H. Pui for his help in bringing the manuscripts to a form suitable for printing. Last, but not least, the contributions of the authors and the financial support of EPA are gratefully acknowledged.

B.Y.H. Liu

FIGURE AND TABLE CREDITS

Figure 2, P. 41: Reprinted with permission from R.N. Berglund and B.Y.H. Liu, "Generation of Monodisperse Aerosol Standards", Environ. Sci. Technol. 7:147-153, 1973. Copyright by the American Chemical Society.

Figure 3, P. 44: Reprinted with permission from R.N. Berglund and B.Y.H. Liu, "Generation of Monodisperse Aerosol Standards", Environ. Sci. Technol. 7:147-153, 1973. Copyright by the American Chemical Society.

Figure 4, P. 44: Reprinted with permission from R.N. Berglund and B.Y.H. Liu, "Generation of Monodisperse Aerosol Standards", Environ. Sci. Technol. 7:147-153, 1973. Copyright by the American Chemical Society.

Figure 6, P. 45: Reprinted with permission from B.Y.H. Liu, R.N. Berglund and J.K. Agarwal, "Experimental Studies of Optical Particle Counters", Atmos. Environ. 8:717-732, 1974.

Figure 7, P. 48: Reprinted with permission from B.Y.H. Liu and D.Y.H. Pui, "A Submicron Aerosol Standard and the Primary, Absolute Calibration of the Condensation Nuclei Counter", J. Colloid Interface Sci. 47:155-171, 1974.

Figure 8, P. 48: Reprinted with permission from B.Y.H. Liu and D.Y.H. Pui, "A Submicron Aerosol Standard and the Primary, Absolute Calibration of the Condensation Nuclei Counter", J. Colloid Interface Sci. 47:155-171, 1974.

Figure 9, P. 50: Reprinted with permission from B.Y.H. Liu, D.Y.H. Pui, A.W. Hogan and T.A. Rich, "Calibration of the Pollak Counter with Monodisperse Aerosols", J. Appl. Meteorol. 14:46-51, 1975. Copyright by the American Meteorological Society.

Figure 10, P. 50: Reprinted with permission from B.Y.H. Liu and D.Y.H. Pui, "A Submicron Aerosol Standard and the Primary, Absolute Calibration of the Condensation Nuclei Counter", J. Colloid Interface Sci. 47:155-171, 1974.

Figure 3, P. 67: Reprinted with permission from K.E. Lauterbach, A.D. Hayes and M.A. Coelho, "An Improved Aerosol Generator", A M A Arch. Ind. Health 13:156-160, 1956. Copyright 1956, American Medical Association.

FIGURE AND TABLE CREDITS

Figure 7, P. 71: Reprinted with permission from T.T. Mercer, M.I. Tillery and H.Y. Chow, "Operating Characteristics of Some Compressed Air Nebulizers", Am. Ind. Hyg. Assoc. J. 29: 66-78, 1968.

Figure 12, P. 77: Reprinted with permission from O.G. Raabe, G.M. Kanapilly and G. J. Newton, "New Methods for the Generation of Aerosols of Insoluble Particles for Use in Inhalation Studies", Inhaled Particles III, pp. 3-17, Unwin Brothers, Surrey, England, 1971.

Figure 15, P. 84: Reprinted with permission from O.G. Raabe, "The Dilution of Monodisperse Suspensions for Aerosolization", Am. Ind. Hyg. Assoc. J. 29:439-443, 1968.

Figure 1, P.293: Reprinted with permission from J.R. Lynch, "Evaluation of Size-Selective Presamplers", Am. Ind. Hyg. Assoc. J. 31:548-551, 1970.

Figure 5, P.304: Reprinted with permission from M. Lippmann and T.L. Chan, "Calibration of Dual-Inlet Cyclones for respirable Mass Sampling", Am. Ind. Hyg. Assoc. J. 35:189-200, 1974.

Figure 6, P.359: Reprinted with permission from M.I. Tillery, " A Concentric Aerosol Spectrometer", Am. Ind. Hyg. Assoc. J. 35:62-74, 1974.

Figure 8, P.361: Reprinted with permission from O.R. Moss, H.J. Ettinger and J.R. Coulter, "Aerosol Density Measurements Using a Modified Spiral Centrifuge Aerosol Spectrometer", Environ. Sci. Technol. 6:614-617, 1972. Copyright by the American Chemical Society.

Figure 20, P.374: Reprinted with permission from J. Kops, L. Hermans and J.F. van de Vate, "Calibration of a Stober Centrifugal Aerosol Spectrometer", J. Aerosol Sci. 5:379-386, 1974.

Figure 21, P.375: Reprinted with permission from J. Kops, L. Hermans and J.F. van de Vate, "Calibration of a Stober Centrifugal Aerosol Spectrometer", J. Aerosol Sci. 5:379-386, 1974.

Figure 1, P.413: Reprinted with permission from Marple, V.A., B.Y.H. Liu and K.T. Whitby, "On the Flow Fields of Inertial Impactors", J. Fluids Eng. 96:394-400, 1974.

Figure 3, P.417: Reprinted with permission from V.A. Marple and B.Y.H. Liu, "Characteristics of Laminar Jet Impactors", Environ. Sci. Technol. 8:648-654, 1974. Copyright by the American Chemical Society.

FIGURE AND TABLE CREDITS

Figure 5, P.420: Reprinted with permission from V.A. Marple and B.Y.H. Liu, "Characteristics of Laminar Jet Impactors", Environ. Sci. Technol. 8:648-654, 1974. Copyright by the American Chemical Society.

Figure 1, P.537: Reprinted with permission from K.T. Whitby, R.B. Husar and B.Y.H. Liu, "The Aerosol Size Distribution of the Los Angeles Smog", J. Colloid Interface Sci. 39:117, 1972.

Figure 1, P.609: Reprinted with permission from B.Y.H. Liu and D.Y.H. Pui, "On the Performance of the Electrical Aerosol Analyzer", J. Aerosol Sci. 6:249-264, 1975.

Figure 16, P.614: Reprinted with permission from B.Y.H. Liu, K.T. Whitby and D.Y.H. Pui, "A Portable Electrical Analyzer for Size Distribution Measurement of Submicron Aerosol", J. Air Pollut. Control Assoc. 24:1067-1072, 1974.

Figure 4, P.707: Reprinted with permission from B.Y.H. Liu, R.N. Berglund and J.K. Agarwal, "Experimental Studies of Optical Particle Counters", Atmos. Environ. 8:717-732, 1974.

Figure 5,6 P.708: Reprinted with permission from B.Y.H. Liu, R.N. Berglund and J.K. Agarwal, "Experimental Studies of Optical Particle Counters", Atmos. Environ. 8:717-732, 1974.

Figure 7, P.710: Reprinted with permission from B.Y.H. Liu, R.N. Berglund and J.K. Agarwal, "Experimental Studies of Optical Particle Counters", Atmos. Environ. 8:717-732, 1974.

Figure 8, P.711: Reprinted with permission from B.Y.H. Liu, R.N. Berglund and J.K. Agarwal, "Experimental Studies of Optical Particle Counters", Atmos. Environ. 8:717-732, 1974.

Figure 9, P.712: Reprinted with permission from B.Y.H. Liu, R.N. Berglund and J.K. Agarwal, "Experimental Studies of Optical Particle Counters", Atmos. Environ. 8:717-732, 1974.

Figure 12, P.718: Reprinted with permission from B.Y.H. Liu, R.N. Berglund and J.K. Agarwal, "Experimental Studies of Optical Particle Counters", Atmos. Environ. 8:717-732, 1974.

EUROPEAN AEROSOL STUDIES

C. N. Davies
Department of Chemistry, University of Essex
Wivenhoe Park, Colchester, Essex
England

ABSTRACT

Over the last 5 years about 60% of publications on aerosol research originated in Europe. Experimental work can be traced back to Tyndall's use, in 1869, of an intense beam of light to show the presence of particles, a method which is still of basic importance.

Recent work is reviewed and documented, the main subjects being filtration, particle size measurement and the atmospheric aerosol; there is also discussion of papers on diffusion, light scattering, sampling, gas-particle conversion, condensation, coagulation, evaporation, electric charging, inhalation, fluid mechanics, and so on.

EUROPEAN AEROSOL STUDIES

by

C. N. Davies

Department of Chemistry, University of Essex,

Wivenhoe Park, Colchester, Essex, England

The Journal of Aerosol Science displays on the inside back cover its policy to "publish research work and occasional reviews of recent work of basic scientific value. The intention is also to encourage the presentation of results in forms which can be generalized as far as possible, rather than the restricted presentation of most applied research". In this context it is now an important medium of publication. The first five volumes (1970-74) have been analyzed with the following results.

Countries of origin	Number of papers published
U.S.A.	66
Britain	50
Germany	25
France	19
U.S.S.R.	14
Canada	8
Czechoslovakia	7
Sweden	7
Italy	5
Holland, Hungary, India 3 each	9
Austria, Israel, Japan Norway; 2 each	8
Australia	1
TOTAL	219

Of the papers submitted for the 1974 volume, 65% were accepted, mostly with changes suggested by referees. From the table it will be seen that 63% of the published papers originated in Europe and 30% in the U.S.A.

The cover design of the Journal recalls the first experimenter with aerosols, John Tyndall, who in 1869[1] showed how a powerful beam of light could be used to reveal the presence of minute particles in air. With this useful tool he demonstrated gas to particle conversion; destruction of organic particles by radiant heat; formation of a fine aerosol by heating platinum; filtration by cotton wool; putrefaction in ordinary air, which did not take place if the air was free of particles; the correctness of the germ theory of disease and the principle of sterilization used so successfully by Lister; the blue color of light, scattered by fine particles, which gradually becomes white, this being due to coagulation which he did not understand. It is rather extraordinary that he wrote about "Heat - a mode of motion" but never associated Brownian motion with thermal energy.

Experimental aerosol science therefore started in Europe with the Tyndall beam. Applications date from Aitken's work on condensation nuclei in 1880 and his subsequent particle counters[2]. Aerosols became an integral part of fundamental science with the cloud-chamber of C.T.R. Wilson (1895)[3], Townsend's studies of electricity in gases[4] and on clouds (1897) which led to the first aerosol experiments in the U.S.A. by Millikan[5] and his determination of e.

Theoretical aerosol science could be said to start with the speculation of Leonardo da Vinci that the blue color of the sky was due to "the most minute imperceptible particles"[6]. The real take-off was Einstein's relationship between diffusion and thermal motion in 1905[7] followed by Smoluchowski's coagulation theory (1916)[8] which was applied to aerosols by Whytlaw-Gray who made the first accurate measurements of particle concentration by number (1922)[9]; Lomax and he constructed the first thermal precipitator which, as developed by Green and Watson (1935),[10] is still the best instrument for sampling many kinds of aerosol.

Improvised filters against dust and smoke long preceded applications of aerosol science which started with World War I. The Tyndall beam was used extensively to measure smoke and filtration efficiency. Interest in the optical properties of the atmosphere and in smoke screens developed. Aerosols of organic arsenicals appeared, the differentation of nose and chest symptoms being correctly associated with particle size. Applications to industrial hygiene also started about this time. The Kotze konimeter (1916) was introduced into the South African gold mines to assess the risk of silicosis according to concentration by number of particles instead of by weight, thus emphasising the small particles which were shown to be important in silicotic lungs.

The experimental study of aerosols was stimulated in a big way by military interests and occupational health. Particulate air pollution, due to the action of sunshine on atmospheric SO_2, was studied by Aitken in 1911[2], although his idea of the wavelength concerned was perhaps erroneous; the stimulus to aerosol science did not come until after the Donora incident, in 1948, and the British Clean Air Act in 1956 which followed the London fog of 1952.

It is a peculiarity of aerosol science that, like biochemistry sixty years ago, it exists because it is needed. The fine early advances were the product of the enthusiasm, energy and sometimes the money of a few individuals; then the practical problems took over and it has since been difficult to pursue consistent lines of experimental research over the necessarily long periods. There is very little of the university teaching and fundamental research to which, in other fields, applied science looks repeatedly for basic knowledge and for trained staff. This is reflected in the subject matter of this symposium - generation, sampling, measurement and analysis - and in the skills of some of the delegates which have to extend beyond aerosol science to an ability to invent convincing one-to-three year projects and to fill in grant application forms, the questions on which having been answered, there seems little need to carry out the research.

Looking back over the last three years of the Journal of Aerosol Science, the most popular subjects in Europe were filtration theory (8 papers), particle size (7) and the atmospheric aerosol (6). The latter subject is overlapped by 5 papers dealing with the growth of atmospheric particles with rising atmospheric humidity and its effect on impactor sampling. Five papers deal with diffusion, 4 with light scattering and 3 each with inhalation, non-spherical particles, sedimentation, spectrometers and sprays, 2 each with coagulation, diffusiophoresis, sampling and sulphur dioxide. There are also papers on condensation nuclei, deposition from moving air, electric charging, evaporation, fluid mechanics, ions, impaction and visibility.

It is rather noticeable how the early interests persist - filtration light scattering, atmospheric particles and the effects of Brownian motion. Because of the nature of its support, experimental aerosol science has tended to limited objectives and the early interests are evidently not yet terminated. There are too few fundamental experimental studies. Theory, like the cactus, can flourish in an unfavourable environment so there is too much theory, doubly so because the intrinsic difficulty of handling aerosols encourages theoretical speculation. At the same time, the theoretical back-up of difficult or necessarily incomplete experiments is very important.

Work on filtration by Fuchs and his colleagues[12] has been backed with experiments over a period of 15 years. It was shown that the Kuwabara flow field gave an accurate description of the velocity round individual fibers and that the theoretical difficulties were not important for particles of the most penetrating size. Work with model filters showed that the fan model, consisting of a series of parallel sheets of parallel fibers, but with each sheet turned radomly in its own plane, represented ordinary manufactured filters of the same volume fraction quite well. Filter resistance has been studied at reduced pressure up to $Kn < 0.5$ and the effects of fiber arrangement and inhomogeneity explored. His latest paper deals with resistance and efficiency of filters made of polydisperse fibers of 0.2 μm mean radius. Zebel[13], following Smutek and Pich[54], contributed to the theory of membrane filters. The writer[73] analyzed the results of failure of particle adhesion. Gebhart[14] and his colleagues have made an exhaustive experimental study of glass bead filters for particles between 0.1 and 2 μm diameter. An entirely new means of realizing inertial deposition, the rotary impaction filter, is due to Soole and colleagues.[15]

Naturally a lot of thought has gone into particle size determination. There has been doubt about polystyrene latices. Fuchs[16] advises the use of only highly diluted suspension which avoids the formation of multiplet particles in the aerosol and of "empties" consisting of particles of dried, non-volatile stabilizer which are the bane of photoelectric counters; stabilizer shells on polystyrene particles may increase the size. Direct size measurements at latex spheres were reported by Bierhuizen and Ferron[17] who compared Stokes diameters with geometrical ones using Perrin's method of averaging from the number in a straight chain. Bexon and Ogden[18] give useful tables for finding the original spherical droplet size from the measured diameter of a droplet spread to a spherical cap with given angle of contact on a treated glass surface. This is an accurate and useful way of dealing with sprays provided that evaporation can be checked. Other methods of measuring particle size include a laser spectrometer[26] which measures diameters between 0.1 and 0.7 μm and is claimed to measure the equivalent spherical volume diameter, irrespective of the shape of the particle; a parallel plate mobility analyzer, described by Maltoni et al.[19], with 16000 volts between the plates, spreads the particles out over 50 cm according to size and charge and, at 4 l/min, covers diameters from 0.3 to 3 μm. Owing to the alternating rise and and fall with particle size of the intensity of scattered light in a given direction, ambiguities arise in interpreting measurements of aerosols made with photoelectric particle counters in the range 0.5 - 5 μm diameter; particles of different sizes may scatter the same amount of light. Bakhanova and Ivanchenko[20] describe a method of deriving the particle size distribution from the distribution of scattered light intensities, which are recorded as a series of pulses from the particles, using a calibration curve obtained for a series of homogeneous aerosols. Sutugin and colleagues[22] suggest that aerosols with bimodal size distributions may be formed from condensing vapours. The reason is that collision efficiency between a vapour molecule and a cluster of molecules decreases rapidly as the number of molecules in the cluster decreases. This means that condensation is delayed so that supersaturation is very high and molecule-with-molecule collisons result first in homogeneous dimerisation. Larger particles may form heterogeneously at the same time, depending on the vapour concentration and whether it is maintained. Each mechanism has its own peak on the size distribution. A computer simulation showed these ideas to be plausible and they were verified experimentally with highly dispersed aerosols ($r < .05$ μm) of NaCl, Ag and AgI.

Storebo[21] has made a series of calculations of condensation in the fireball of a nuclear explosion using the classical equations for nucleation, condensation and coagulation. The fireball reaches maximal size at a temperature of 2000 K and is supposed to contain Fe vapour which condenses to liquid FeO and freezes at 1641 K. The radioactivity of a particle is taken to be proportional to its mass. The initial process is one of homogeneous nucleation but this switches to vapor condensation upon the primary particles. In the case of ground bursts there are also particles of contaminent, mainly above 1 μm in radius. The particles in air bursts are too small for sedimentation under gravity to be appreciable; there is no fall-out.

E. Mészáros[23] has taken filter samples in Budapest and shown correlations of the sulphate/SO_2 ratio with the near ultraviolet of sunlight and with temperature. However, there was no significant difference between summer and winter levels of sulphate (~ 10 μg/m^3) so that some mechanism other than gas-phase photochemical conversion seems necessary; it is suggested that sulphate is formed in winter on the surface of atmospheric aerosol particles since positive correlation was found with the concentration of aerosol by mass and also by number of particles exceeding 0.15 μm radius.

A. Mészáros[24] carried out a membrane filter study of the air over the ocean of the southern hemisphere which is likely to be less affected by human contamination than the northern. The particles consisted mainly of ammonium sulphate, sea salt and mixtures of the two; some sulphuric acid was found. A Gardner condensation nucleus counter was also used. Aitken particles numbered 300-450/cm^3, compared with 600/cm^3 in the North Atlantic; above 0.03 μm radius there were up to 55/cm^3. The peak size was near to 0.1 μm radius. A case for sampling during summer in Greenland, where there is little exchange of air with lower latitudes, is made out by Megaw and Flyger[25] since long term trends would become evident. A Danish expedition obtained useful data in 1974 and another is going again this summer. The size distribution of particles in city air is complex in form - broadly bimodal by mass but showing numerous irregularities which are presumably due to individual sources; between 0.1 and 1 μm radius the aerosol is stable. The writer[28] gave a simple technique for fitting distributions of atmospheric particles by number or mass with compatible sets of lognormal components.

Cascade impactors, invented by K. R. May, have been used for 25 years for sampling the atmospheric aerosol. It is interesting that Jaenicke and Blifford[29] now conclude that the cut-off of each stage is sharper than has hitherto been believed from the results of experimental calibrations and, in fact, is closer to theory. This may be caused by errors in calibrations with aerosols of broad size distribution or by the presence of multiplets in nominally monodisperse aerosols. A new theory has been presented by Pelassy.[27]

Another problem with impactor sampling is the effect of humidity. Winkler[30] shows that in atmospheric relative humidities above 75% (NaCl crystals liquefy) adhesion of particles to the plates is good but in drier air there is often a loss of particles. He also studied[31] the growth of various particles with rising humidity, demonstrating the very important point that particles of mixed constitution as found in the atmopshere, increase in weight gradually unlike pure crystals which grow suddenly at the liquefaction value. Another complication is the effect on cascade impactor performance of the growth of atmospheric particles. Hänel and Gravenhorst[32] study this theoretically and show that the cut-off radius can nearly double over the whole range of humidity. The number of particles per stage is not so much affected and there is no problem below 70% RH.

Mechanisms of sulphur dioxide oxidation have received a good deal of attention, both theoretically and in the laboratory. Takahashi at al.[33] deal with the sequence: - photo-oxidation to sulphur trioxide, combination with water to give sulphuric acid vapor , nucleation by the combination of several molecules of sulphuric acid with several water molecules to form critical embryos, growth of the embryo by condensation of both vapors , coagulation. The critical size of embryo is about 10^{-7} cm and contains 5 - 20 molecules of sulphuric acid. The nucleation rate is strongly dependent on relative humidity.

Stauffer and colleagues[34] start with the assumption that the concentration of sulphuric acid which is required to nucleate water vapor is very low and that, during growth by condensation of the vapors , the droplets receive one molecule of acid for every ten of water. Heat of condensation is neglected and the condensation coefficient is taken to be one. The results are compared with experimental data obtained by Cox[35] who irradiated SO_2 - N_2 mixtures at 290-400 nm and at 185 nm, the former giving very slow oxidation. The measured concentration of particles of acid changed much less with relative humidity than the calculated figures; there was rough agreement at about 20% R.H., enough to suggest the general plausibility of the heteromolecular condensation theory.

Work on light scattering continues to be mainly theoretical but there are two useful experimental papers by Proctor et al.[36] dealing with the extinction coefficients of ground dust particles, and one by Seaney[37] on the forwards scattering by limestone and coal dusts. On the theoretical side Bradshaw[38] gives a program and results for total and angular scattering for various refractive indices and values of α up to 60; Tonna[39] has worked out the correction for finite acceptance angle (0.2-1°) up to α = 200 for spheres (m = 1.33). The correction amounts to 6% on the extinction efficiency at 0.2° and 42% at 1°, when α= 100, thus underlining the problem of designing apparatus for particles up to 50 μm diameter.

Diffusion is employed for measuring the size of fine particles. The writer[40] reviewed the solution of the diffusion equations for flow through tubes, giving criteria for avoiding errors due to short or curved tubes and for the employment of long tubes for transporting aerosols without appreciable loss by diffusion or sedimentation. The mathematical solution becomes difficult to evaluate and compute for large particles in short tubes at high velocities: Ingham[41] gives an empirical formula which works well under these conditions and is also valid under the opposite conditions. A solution is also deduced with allowance for slip at the tube wall. Diffusion theory is applied to the experimental problem of evaluating the results of flow through diffusion batteries, systems of parallel tubes; the difficulty then arises of interpreting the penetration when the aerosol is polydisperse. Maigné et al.[42] propose a new method of finding the size distribution from diffusion battery data, by measuring penetration at a series of rates of flow, and present an experimental check against the electron microscope. Another experimental method of using the diffusion equation is described by Matteson, Sandlin and Preining[44]; they used commercially available collimated hole structures with holes 15.5 μm mean diameter and a length to diameter ratio of 32.7. Experimental results at room temperature agreed with theory but at -16 and -72°C the rate of diffusion of the particles was much increased. It is suggested that this might be due to gas molecules adsorbing on the particles. Many cases of evaporation and condensation are diffusion controlled; Gallily[43] discusses the change of shape of a particle which these processes may cause, so complicating the mathematical analogy with electrostatic capacity. He shows that an ellipsoidal particle in still air will maintain its shape with constant axial ratios.

Two papers are concerned with diffusiophoresis. Yalamov and Gaidukov[45] deal with particles in the slip-flow regime using the Onsager principle; account is taken of diffusion in the surrounding gas mixture, thermophoresis and evaporation of the particle. Hochrainer and Zebel[46] made an experimental test of the possible significance of diffusiophoresis on the deposition of aerosol particles in the lungs during breathing. A single two channel model was made with a membrane filter separating the channels which contained aerosols dispersed in different gas phases. It is concluded that diffusiophoresis has little effect due to the counter-flow of oxygen and carbon dioxide.

Coagulation is very rapid in the early stages of formation of aerosols. Walter[47] has obtained solutions of the integro-differential equation and shows how sensitive they are to the gas-kinetic correction of Fuchs to the coagulation constant, which occurs in the kernel of the equation. The importance of previously existing large particles in mopping up the newly formed small ones is demonstrated and the distributions obtained suggest that homogeneous gas reactions are far more important in forming atmospheric aerosols than ion-molecular reactions. Commenting on this paper, Storebø[48] points out that the early, non-gaseous nuclei of radioactive decay or gas phase chemical reaction in the atmosphere are larger than can be accounted for in terms of their birth mechanism. He thinks that other gas molecules are collected on them to create embryos and subsequently evaporated after coagulation of the embryos. The low coagulation rate due to the shell adsorbed to nuclei is formulated.

Observations of the mobility of small ions near the ground are reported by Ungethüm[49] and confirm the existence of discrete mobility groups with break-points in between. Mobilities up to 11 $cm^2\ V^{-1} sec^{-1}$ were measured. The small ions decay to larger ones by several routes and the sharpness of the break-points depends on air conditions. It is considered that no conclusions can yet be drawn about the structure of the ions in different groups.

New calibrations of two condensation nuceli counters by O'Connor[50], a portable one for use in mines and an automatic one, correlated well with a Nolan-Pollak instrument using both natural and artificial nuclei over the range 5000 to 300,000 cm^{-3}.

A good deal of work has been carried out with Stöber spiral centrifuges. Porstendörfer[51] obtained size distributions of various polydisperse aerosols by allowing radom decay products to deposit on the particles, sampling the tagged aerosol with the centrifugal spectrometer and scanning the deposit for α-activity. An important factor is the deposition coefficient of the decay atoms upon the aerosol particles which depends upon their size. Kops and his colleagues[52] have calibrated one of these centrifuges over the range of aerodynamic diameter from 0.06 to 1.5 μm. The aerosols were of polystyrene and important practical points for users of these centrifuges are discussed.

A very much simpler aerosol centrifugal spectrometer was made by Matteson et al.[53] and calibrated from 0.5 to 1.5 μm diameter, the agreement with theory, based on air flow and sedimentation, was reasonable.

Sedimentation under gravity is an important mechanism of aerosol deposition. Pich[54] developed a comprehensive theory for deposition on cylinders, with filters particularly in mind, and Heyder[55], thinking of the airways in the human lungs, dealt with settlement upon the walls of tubes, inclined at any angles, from aerosol flowing through the tubes.

The rotating drum as a container for aerosols is as near as we are ever likely to come to switching off gravity. Such systems have been important for studying the viability of bacterial aerosols over long periods. Frostling[56] describes splendid equipment which he tested on particles up to 3 μm diameter. The advantage over an ordinary chamber is quantified. He had previously used the rotating drum for determining the vapor pressure of high boiling organics dispersed as aerosols[57].

Deposition of particles on surfaces in the open air has been studied for many years because of radioactive aerosols. Clough[58] described new measurements of the deposition rate in a large wind tunnel. Particles over a range of diameters from 0.08 to 30 μm were used with wind speeds up to 6 m/sec. Grass acts as a filter for fine particles up to 0.5 μm and is also a good collector of coarse particles which bounce off smooth surfaces.

Sampling in the open air and some industrial situations involves orifice inertial effect with large particles. A detailed analysis has been made by Belyaev and Levin[59], who consider the shape of the nozzle, effects of departure from isokinetic conditions and the effect of fluctuating wind velocity. Inertial effect is used in a 35 l/min sampler by Stevens and Churchill[60] to segregate fine particles and coarse particles. All are collected on a single filter paper disc but a direct stream, carrying mainly particles above 3 μm diameters, passes through the periphery of the filter and the rest of the air, with only fine particles, is directed round sharp bends into the centre of the disc.

Problems concerned with aerodynamic drag of aerosol particles and filters often extend to high values of the Kundsen number, Kn. There is little experimental data apart from spheres, but Dahneke[62] has in three papers described procedures for correcting the drag of discs, cylinders and spheroids for the hydrodynamic slip regime , free molecular flow, and in between.

Of course, sprays are used a great deal as a source of aerosols and recent work has involved several varieties. May[64] has published a full account of the Collison atomizer which has long been a favorite source of fine particles. Topp[65] obtained some excellent photographs of the ultrasonic nebulizer which throw a lot of light on how it works. Philipson[66] has illustrated the formation of droplets by the spinning disc atomizer and presented useful performance data. Dombrowski and colleagues[67,68] present some excellent photographs of spray production and discuss the correct procedure for analysis of photographs for drop size distribution. They also studied fan spray nozzles and showed how vibrating the nozzle could result in the production of uniform droplets.

Low Reynolds number hydrodynamics is represented by a single paper by Gallily and Mahrer[69] who derive expressions for the trajectory of a sphere approaching a wall and for the slowing down of a sphere projected towards a wall.

Experiments on human inhalation of aerosols continue to be important for a variety of applications. The results of early work were widely scattered and considerable overestimates of alveolar deposition have been incorporated in safety criteria in health physics and occupational health. One reason for overestimating alveolar deposition has been a large theoretical underestimate of the efficiency of impaction in the airways of the dead space. This error, dating from 1950, has been explained by Johnston and Muir[70].

A large number of experiments on the human inhalation of aerosols by Heyder et al.[71] are in progress. Particle size was measured by laser spectrometry from 0.2 to 0.8 μm diameter and by higher order Tyndall spectra above 0.8 μm. The particles of di-2-ethyl hexyl sebacate showed serious loss by evaporation below 0.2 μm. The results show little dependence of deposition on particle size between 0.3 and 1.0 μm diameter with tidal volumes near to 500 cm^3 and 15 breaths/min.

Fry and Black[72], using γ -tagged particles from 2 - 10 μm diameter and collimated crystal counters, studied deposition in the human nose and clearance of deposited particles. They demonstrated the indefinite holding of particles in the anterior passages which indicates that the nose can be a critical organ.

An exhaustive analysis, supported by experimental checks, of the evaporation of saline droplets (radius 1 - 8 μm) has been made of El Golli et al[74]. Allowance is made for the change in activity with concentration, as the solvent evaporates.

Horvath[75] has made a number of experimental observations of visual range in the laboratory, using aqueous suspensions of homogeneous latex spheres and other particles between 0.1 and 5 μm diameter. Using white light from above and taking the wavelength as 550 nm, where eye sensitivity is maximal, he showed that the visibility was inversely proportional to the extinction efficiency of the particles and that the aqueous turbid medium simulated the atmopshere satisfactorily. The writer[76] has studied experimental data on visual range in fogs and show that the measured ranges are sometimes only about half the calculated values because the latter were derived from measured concentrations of water droplets taken with samplers which could not collect particles below 1 μm diameter. It is shown that the visual range can be calculated from the sum of the reciprocals of the ranges due to each size fraction. Since inferences can be made about particle size from observations of visual range, it is considered that these should be made routinely when sampling for atmospheric particles by weight.

CURRENT WORK

Work in progress in Europe, or not yet published can be briefly mentioned. Interest in the performance of filters at high altitudes resulted in laboratory experiments being carried out in the U.S. over a range of pressures; interpretation of these results and their correlation with theory is being examined. Filters of plastic fibers seem to perform better than mechanical theory indicates, perhaps because of electric charge on the fibers. Filtration and analysis of aerosols containing fibers of asbestos is being studied.

Polishing up diffusion mathematics, particularly with reference to radon decay products is in progress. Turbulent diffusion in jets of aerosol is being studied theoretically. Development of the Stöber centrifuge continues with a piezo electric method of weighing the deposit at intervals along its length. This instrument is in use for sorting out the aerodynamic diameters of various types of aggregated particles; studies with DOP aerosols indicate difficulties due to coagulation and evaporation as well as loss by deposition in the intake of the centrifuge. The use of two lognormal distributions is suggested, one for singlet and one for doublet particles each, of course, being spherical.

K. R. May will shortly be publishing a review of his work on aerosol impaction jets which has extended over many years.

New generators are being tried out, one using a welding torch to evaporate refractory substances and another, developed by Tarroni and his colleagues at Bologna, using a 5 megaherz oscillator of up to 12 kilowatts to actuate an argon plasma torch at over 10,000°C through which the aerosol material is blown. Everything vaporizes at this temperature and aerosols have been generated down to 5 nm diameter with $\sigma_g = 1.3$.

Theoretical work on La Mer-Sinclair generators is in progress. A device for recording short term output fluctuations of powder and liquid aerosol dispensers is in use; it works by tagging the particles with a fluorescent substance and impacting them on to a rotating drum.

The wide size range of atmospheric particles makes it necessary to use simultaneously several different kinds of sampler; hence there is interest in techniques for combining the measurements to produce a single size distribution. One of the most recent ideas is an inversion method which minimizes the deviation between the measured data and the ordinates of a specified type of size distribution. The data come from an integrating nephelometer, a condensation nucleus counter and a Royco counter; this work is being carried out by J. Heintzenberg at Mainz.

Atmospheric pollution stimulates research on SO_2, NH_3 and other gas-particle reactions. Botanists are studying the role of vegetation in producing aerosols as well as adverse effects on plant growth of lead and salt aerosols. Interest in the atmospheric aerosol is associated with work on light-scattering methods of size analysis, including laser holography (above 1 μm). Interesting new methods of presenting the results of air pollution surveys are being developed by M. Benarie of Vert-le-Petit, France. In France, also, comprehensive surveys of Aitken nuclei of human and natural origin are in progress. A balloon-carried nucleus counter has been used by Käselau, Lindan, Harz up to a height of 27 km[77].

Aerosols are frequently connected with corrosion. Impingement of salt containing droplets in boiler superheater tubes is enhanced by a force of complicated origin (evaporation, local surface tension variation on the droplet, motion of adjacent liquid and vapor phases, force on droplet opposite to the temperature gradient) which is being studied by Gardner at the Central Electricity Research Laboratories, Leatherhead. Corrosion, also, has motivated current studies of the size distribution and concentration of salt particles over the North Atlantic which are being conducted from weather ships and reduced at the Heriot-Watt University, Edinburgh.

Current meteorological work overlaps with air pollution studies in being concerned with the physical and chemical nature of atmospheric particles, their sources and sinks. There is much work on water droplet clouds, including the extent to which they are influenced by radiation from the sky and from the ground. The nature of freezing nuclei continues to offer a challenge.

More accurate measurements of the deposition of aerosols in the human respiratory tract are being obtained at the Gesellschaft für Strahlen- und Umweltforschung, Frankfurt, and at the University of Essex. Ranges of particle size and breathing pattern are being covered, nasal deposition is being evaluated and the causes of subject-to-subject variations are being thought about. At Saarbrücken, Stauffer is comparing different mammalian species by a dimensional scaling argument.

There is much interest in aerosol therapy which is gradually acquiring the sound scientific basis which was formerly lacking. Drugs administered as aerosols by pressure pack generators depend for their effectiveness on particle size. In some cases the Freon propellant may not fully evaporate with two serious possibilities - the dose of propellant droplets absorbed through the mucous membranes of the upper respiratory tract may be dangerous, and the drug will fail to reach the alveolated regions of the lungs and so be ineffective. Various drugs are used for enhancing mucociliary clearance and for bronchodilation, in the treatment of bronchitis and asthma. The effects can be gauged by scanning the subjects chest with crystal counters after depositing in his airways a 5 μm diameter γ-active insoluble aerosol.

Some research on aerosols has industrial applications. This includes the generation of metal particles by flash heating a metal in a cooler atmosphere of argon. Buckle, at Sheffield University, has produced unusual shapes of particles, including dendrites, in this way[78]. Graham and Homer (Shell Research, Chester)[79] have worked on the coagulation of lead aerosols on a millisecond time scale. The light scattered per particle is proportional to the square of the particle volume; assuming the size distributions of the coagulating aerosol to be self-preserving enables the whole scattered light from all particles to be taken as proportional to the mean particle volume. The aerosols were produced by shock heating of tetramethyl lead vapor in argon at about 940 K.

Studies of radiation from flames (A. R. Jones, Dept. of Chemical Engineering, Imperial College, Longon)[80] give information about the nature of soot particles which are present and their effect on the emissive properties of the flame. Optical properties of metal oxides in flames have also been studied[82]. Particles in flames are charged electrically so that they can be manipulated by electric fields at any stage of growth; this holds out the possibility of improving combustion efficiency and diminishing pollution.[81]

A powerful instrument for measuring rapidly changing distributions of particle size has been developed by Carabine and Moore.[83] Light scattered from a laser beam by the aerosol particles is collected at angles from 8° to 172° in 3.75° steps by a system of multiple fixed and a single rotating mirror which flashes each signal in turn to a photomultiplier. Angular intensity distributions were calculated for lognormal aerosols having values of σ_g between 1.1 and 1.65 and the measured angular distribution of light intensity was inverted by a computer program to the representative sum of the weighted log-normal distributions. The system should operate in the size range 0.1 to 1.0 μm radius and has been used to demonstrate coagulation. There are obvious problems related to the effects on the scattered light of aggregates of particles.

In 1972 Hinds and Reist[84] described the measurement of particle size by Laser Doppler Spectrometry based on the Brownian motion of the particles for diameters between 0.3 and 3 μm. An alternative method, based on particles moving across a grid of light fringes at a constant velocity has been discussed by Fristrom et al.[85] Such systems have been used for liquid suspensions with some success but the setting up of the apparatus is very difficult. Hopefully, developments in the aerosol field cannot be too far away. Such a tool for studying particle motion in fluid mechanical problems would have many applications, apart from measuring particle size. Laser holography, also, has a long way to go, particularly in the concise handling of the information which can be obtained.

Finally, continued interest is maintained in the evaporation of aerosol particles. The writer[86] described an easy and direct method of calculating the lifetime of particles in which both the Kelvin effect and the defect of vapor concentration near the surface were of importance. It was based on an evaporation coefficient of unity; in the discussion this was questioned and a brief review of the literature was given. New measurements are in progress. The evaporation of particles in mono and heterodisperse aerosols of infinite extent is also being studied, including the effects of small scale variations of particle concentration due to turbulence.

Clearly, some current projects must have been missed from this review of research on aerosols which is now going on in Europe. It is hoped that those who are engaged in them will accept my apologies, and that readers of this article, both in Europe and elsewhere, will find some new information about work which is related to their own interests. Perhaps, too, some valuable personal contacts may be encouraged.

REFERENCES

1. J. Tyndall. Essays on the floating matter of the air. Longmans Green, London, 1881.

2. J. Aitken. Collected Scientific Papers. Ed. by C. G. Knott. Cambridge, 1923.

3. C. T. R. Wilson, Nature 52, 144, 1895.

4. J. S. Townsend. Proc. Camb. Philosoph. Soc. 9, 244, 1897.

5. R. A. Millikan. The electron. Chicago, 1917.

6. The notebooks of Leonardo da Vinci. Ed. by E. MacCurdy 1, 418. Cape, London, 1948.

7. A. Einstein. Ann. d. Physik. 17, 549, 1905.

8. M. von Smoluchowski. Phys. Zeits. 17 557, 583, 1916.

9. R. Whytlaw Gray & H. S. Patterson. Smoke. Arnold, London 1932.

10. H. L. Green & W. H. Watson. Med. Res. council Spec. Rep. Ser. No. 199, HMSO, London, 1935.

11. N. A. Fuchs, A. A. Kirsch & I. B. Stechkina. Symp. of Faraday Soc. No. 7, 143, 1973.

12. A. A. Kirsch, I. B. Stechkina & N. A. Fuchs. J. Aerosol Sci. 6, 119. 1975; 5, 39, 1974; 4, 287, 1973.

13. G. Zebel, ib. 5, 473, 1975.

14. J. Gebhart, C. Roth & W. Stahlhofen, ib. 4, 355, 1973.

15. B. W. Soole, H. C. W. Meyer & K. Pannell. ib. 4, 59, 1973; B. W. Soole, ib. 4, 171, 1973.

16. N. A. Fuchs. ib. 4, 405, 1973.

17. H. W. J. Bierhuizen & G. A. Ferron ib. 6, 19, 1975.

18. R. Bexon &. T. L. Ogden. ib. 5, 509, 1974.

19. G. G. Maltoni, et al. ib. 4, 447, 1973.

20. R. A. Bakhanova & L. V. Ivanchenko,ib. 4, 485, 1973.

21. P. B. Storebo, ib 5, 557, 1974.

22. A. G. Sutugin, A. A. Lushnikov & G. A. Chernyaeva, ib. 4, 295, 1973.

23. E. Mészáros, ib. 4, 429, 1973.

24. A. Mészáros & K. Vissy. ib. 5, 101, 1974.

25. W. J. Megaw & H. Flyger. ib. 4, 179, 1973.

26. J. Heyder & J. Porstendörfer. ib. 5, 387, 1974.

27. P. Pelassy ib. 5, 531, 1974.

28. C. N. Davies. ib. 5, 293, 1974.

29. R. Jaenicke & I. H. Blifford. ib. 5, 457, 1974.

30. P. Winkler, ib. 5, 235, 1974.

31. P. Winkler, ib. 4, 373, 1973.

32. G. Hänel, & G. Gravenhorst. ib. 5, 47, 1974.

33. K. Takahashi, M. Kasabara & M. Itoh. ib. 6, 45, 1975.

34. D. Stauffer, V. A. Mohnen & C. S. Kiang, ib. 4, 461, 1973.

35. R. A. Cox. ib. 4, 473, 1973.

36. T. D. Proctor & G. W. Harris. ib. 5, 81, 91, 1974.

37. R. J. Seaney. ib. 4, 245, 1973.

38. P. Bradshaw. ib. 6, 147, 1975.

39. G. Tonna. ib. 5, 579, 1974.

40. C. N. Davies. ib. 4, 317, 1973.

41. D. B. Ingham. ib. 6, 125, 1975.

42. J. P. Maigné, P. -Y. Turpin, G. Madelaine & J. Bricard. ib. 5, 339, 1974.

43. I. Gallily, ib. 4, 457, 1973.

44. M. J. Matteson, I. W. Sandlin, & O. Preining. ib. 4, 307, 1973.

45. Y. I. Yalamov & M. N. Gaidukov. ib. 4, 65, 1973.

46. D. Hochrainer & G. Zebel. ib. 5, 525, 1974.

47. W. Walter. ib. 4, 1, 1973.

48. P. B. Storebø, ib. 4, 337, 1973.

49. E. Ungethüm. ib. 5, 25, 1974.

50. D. T. O'Connor. ib. 6, 23, 1975.

51. J. Porstendörfer. ib. 4, 345, 1973.

52. J. Kops, L. Hermans & J. F. Van de Vate. ib. 5, 379, 1974.

53. M. J. Matteson, G. F. Boscoe & O. Preining. ib. 5, 71, 1974.

54. J. Pich. ib. 4, 217, 1973.

55. J. Heyder, ib. 6, 133, 1975.

56. H. Frostling. ib. 4, 411, 1973.

57. H. Frostling. ib. 1, 341, 1970.

58. W. S. Clough, ib. 4, 227, 1973.

59. S. P. Belyaev & L. M. Levin, ib. 5, 325, 1974.

60. D. C. Stevens & W. L. Churchill. ib. 4, 85, 1973.

61. M. Smutek & J. Pich. ib. 5, 17, 1974.

62. B. E. Dahneke. ib. 4, 139, 147, 163, 1973.

63. D. Hochrainer &. G. Hänel. ib. 6, 97, 1975.

64. K. R. May. ib. 4, 235, 1973.

65. M. N. Topp. ib. 4, 17, 1973.

66. K. Philipson. ib. 4, 51, 1973.

67. C. J. Clarke & N. Dombrowski. ib. 4, 27, 1973.

68. N. Dombrowski & N. D. Neale. ib. 5, 551, 1974.

69. I. Gallily & Mahrer. ib. 4, 253, 1973.

70. J. R. Johnston & D. C. F. Muir. ib. 4, 269, 1973.

71. J. Heyder, J. Gebhart, G. Heigwer, C. Roth & W. Stahlhofen. ib. 4, 191, 1973.

72. F. A. Fry & A. Black. ib. 4, 113. 1973.

73. C. N. Davies. ib. 5, 487, 1974.

74. S. El Golli, J. Bricard, P. -Y. Turbin & C. Treiner. ib. 5, 273, 1974.

75. H. Horvath. ib. 6, 73, 1975.

76. C. N. Davies. ib. 6, 1975.

77. Report of 2nd Annyal Meeting of Ges. fur Aerosol Forschung. ib. 6 (4) 1975.

78. E. R. Buckle. Symp. Faraday Soc. No. 7, 17, 1973.

79. S. C. Graham & J. B. Homer. ib. 85, 1973.

80. P. A. Tesner, ib. 104, 1973.

81. F. J. Weinberg. ib. 120, 1973.

82. A. R. Jones, J. Phy. D. Appl Physics. 6, 417, 1973.
7, 1369, 1974.

83. M. D. Carabine & A. P. Moore. Symp. Faraday. Soc. No. 7, 176, 1973.

84. W. Hinds & P. C. Reist. J. Aerosol Sci. 3, 501, 515, 1972.

85. R. M. Fristrom, A. R. Jones, M. J. R. Schwar and F. J. Weinberg. Symp. Fraday Soc. No. 7, 183, 1973.

86. C. N. Davies. Symp. Faraday Soc. No. 7, 34, 1973.

RESEARCH AND DEVELOPMENT IN JAPAN ON FINE PARTICLE MEASUREMENT AND NEW CONTROL DEVICES

Koichi Iinoya
Kyoto University
Kyoto, Japan

ABSTRACT

Fundamental research on the measurement and sampling of fine dust particulates has been conducted both analytically and experimentally in several laboratories of Japanese universities and institutes. The research efforts include work on anisokinetic sampling errors, particle settling in a sampling line, light scattering calculations, cascade impactors, and special analytical methods for the determination of particle size distributions. Some of the work is introduced in detail along with the titles of the research papers which have appeared in Japanese journals. Several new instruments and apparatuses, which have been developed in Japan for dust generation and for the measurement of particle size, dust concentration and other particle characteristics, are also introduced along with the name of their manufacturer. A few new control devices for fine particulates are also referred to briefly.

RESEARCH AND DEVELOPMENT IN JAPAN ON FINE PARTICLE MEASUREMENT AND NEW CONTROL DEVICES

Koichi Iinoya

Kyoto University, Kyoto, Japan

INTRODUCTION

This paper introduces some of the recent fundamental and applied work which has been done in Japan on the development of instruments and new control techniques for fine dust. Fundamental research has been conducted both analytically and experimentally in Japanese universities and institutes. A few industrial developments have also taken place in Japanese industry.

FUNDAMENTAL RESEARCH

Several laboratories in Japanese institutes are devoted to the study of techniques for the measurement of fine dust particles. This includes, e.g., anisokinetic sampling error, particle loss in a sampling line, light-scattering phenomena, cascade impactors, and various analytical methods for the determination of particle size distribution. A few examples are given as follows:

Figure 1 shows a typical calculated result of the anisokinetic sampling error for fine particles, and Figure 2 is the comparison between the experimental values and theoretical curves for the sampling error. The inertia theory for fine particle sampling has been confirmed through comparison with experiments. However, the amount of particle adhesion in a sampling probe and a sampling tube is quite significant and leads to erroneous results if neglected.

Yuu and Iinoya[1-5] have studied the separation mechanism of a cascade impactor for several years. The experimental fractional efficiency curves are in agreement with their analytical solution as shown in Figure 3, when the velocity distribution caused by the boundary layer at a nozzle outlet is taken into account in the gas flow model. Iinoya and Makino are trying to develop an electric dust-dislodging technique for fabric filter cleaning. They have found the optimal arrangement of electrodes, and the optimum frequency and phase of the applied electricity.

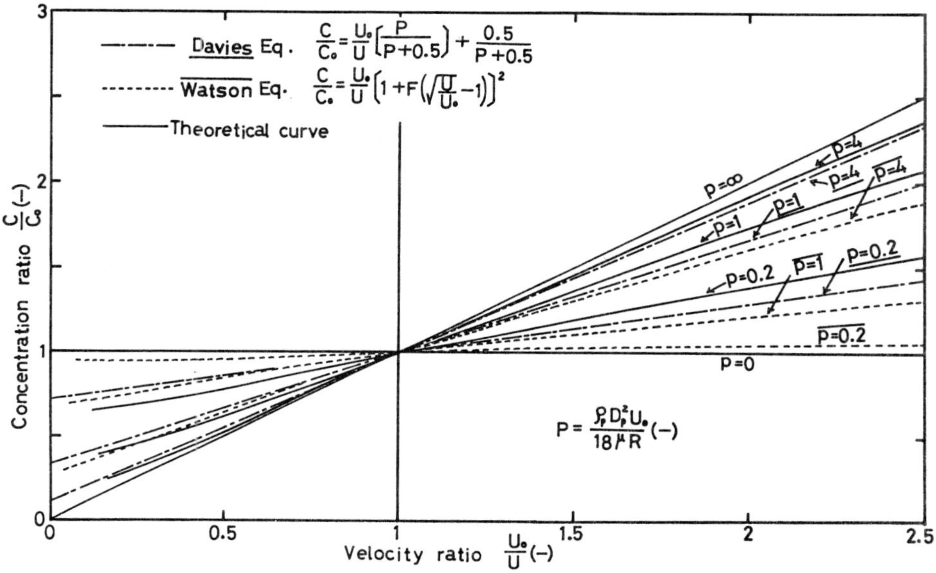

Fig. 1 Calculated results of anisokinetic sampling errors for particle concentration

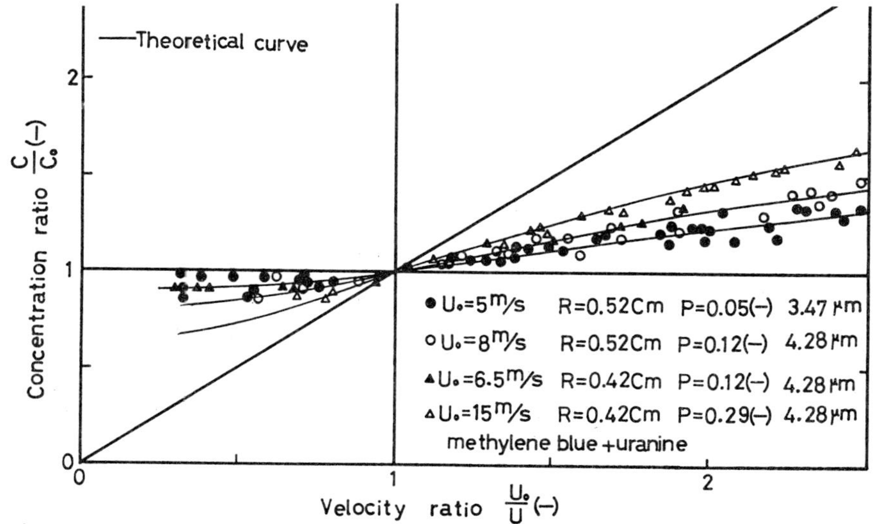

Fig. 2 Errors in particle concentration due to anisokinetic sampling

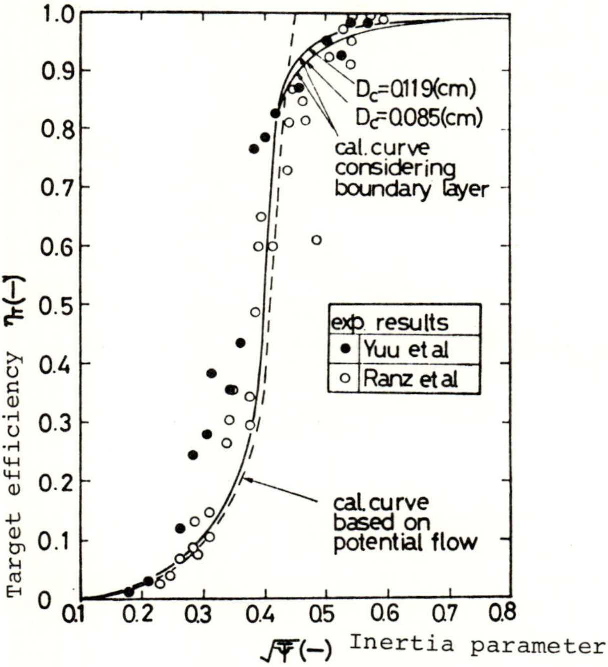

Fig. 3 Comparison of experimental and calculated target collection efficiency in an impactor

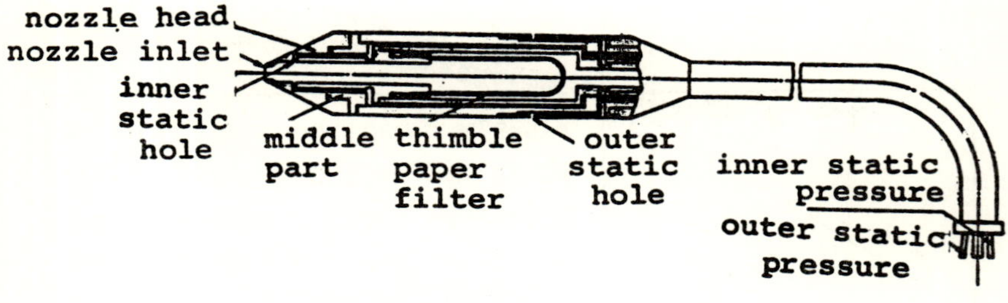

Fig. 4 New equilibrium type of dust sampling probe

Takahashi[9-22] of Kyoto University, has studied the aerosol size distribution and the electrical charge on particles by use of numerical calculations and various experimental methods. Kanagawa[23-30] of Nagoya University, has analytically studied the light scattering method for particle size analysis and for the measurement of aerosol concentration. He recommends a scattering angle of 60 degrees from the incident rays. Yoshida[31-33] of Osaka Prefectural University, has developed a new technique for aerosol size analysis based on the measurement of settling velocities using an ultra-microscope and video system.

Kitani[34-35] of the Japan Atomic Energy Research Institute, also has developed a method for the direct measurement of aerosol concentration in a fast reactor chamber using a light scattering photometer. He has also studied the thermal precipitation of aerosol particles.

Ikebe[36] of Nagoya University, has also used a unique method of diffusion response for submicron particle size analysis. Suganuma[37-38] of Tokyo University, has studied the effect of humidity on airborne dust generation.

Tamori[39-43] of the National Research Institute for Pollution and Resources has studied a new equilibrium type of dust sampling probe which has an enlarged inner inlet, as shown in Figure 4. Isokinetic sampling can be achieved by using this type of probe, when the inner static pressure is controlled so as to be equal to the outer static pressure.

Hashimoto[44] of Keio University is using a neutron activation analysis for the study of aerosol size.

Masuda[45-46] of Tokyo University has developed the electric curtain for dust particles and a new type of electric precipitator.

Fourteen kinds of industrial test dust have been authorized by Japanese Industrial Standard JIS Z 8901. Their size distributions are fixed as shown in Figure 5. These test dusts are sold through the Association of Powder Process Industry and Engineering, Japan. (See Table I). The quoted Japanese research papers reflect only some of the recent research work in Japan.

INDUSTRIAL DEVELOPMENT

I would like to introduce several recent develpments in the field of fine particle measurement and new control devices.

Hosokawa Micromeritics Laboratory has manufactured a powder characteristics tester which is based on Carr's research work. The powder

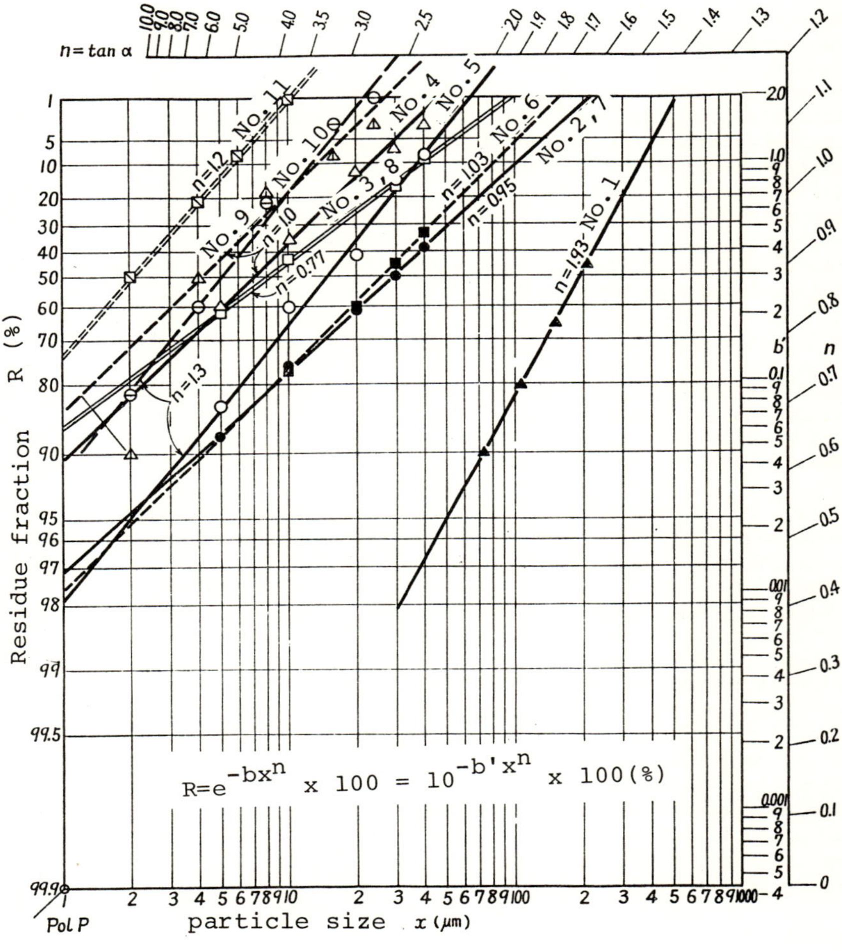

Fig. 5 Industrial test dusts in Japan (JIS Z 8901)

tester can give flowability and floodability indices for any fine dust. Figure 6 shows an overall schematic of the device.

Shimazu Seisakusho Ltd. has developed many kinds of dust measuring instruments such as: two types of cascade impactor for the size analysis of stack and ambient dust particles, a dust concentration meter for the measurement of suspended particle mass in a building and in the environment, and a smoke meter measuring the opacity of exhaust gas from a diesel engine. The company is also manufacturing a powder disperser for test purposes, which is shown in Figure 7, and a micro multicyclone as a dust control device in which the unti cyclone is one inch in diameter as shown in Figure 8.

Kony Company Ltd. has developed an automatic, continuous, dust concentration meter, Konytest, under a technical agreement with Dr. Prochazka of West Germany. It is based on the contact electrification principle. Figure 9 shows a general view of it.

Shibata Chemical Apparatus Manufacturing Company manufactures a light scattering type of dust concentration meter, Digital Dust Counter, which has a 45 degree scattering angle. A multishelves type of gravitational classifier is attached to the meter in order to remove the fraction above 10 micrometers by environmental regulation.

Aichi Watch Company has also developed an air volume sampler with a centrifugal classifier for 10 micrometers separation.

Dan Industry Company, under Professor Kanagawa's instruction, has developed an aerosol size photo-counter based on a 60 degree scattering angle.

Ishikawajima-Harima Heavy Industry Company has developed a new two-stage type electrostatic precipitator, which comprises a conventional one-stage type ESP of fairly small size in its charging zone and a newly developed channel electrode system called "electrostatic screen" in its particle collection zone.

Onoda Cement Manufacturing Company has developed a hybrid type electrostatic precipitator consisting of a dry and wet stage combined inside a single casing as an integral unit and a slurry processing unit coupled to the casing.

Sumitomo Heavy Machinery Company has manufactured an electrostatic precipitator mounted directly on the roof of a workshop, as shown in Figure 10. This precipitator is light in weight because the dust collecting electrodes are made from reinforced plastic or synthetic resin. A suction blower is not used.

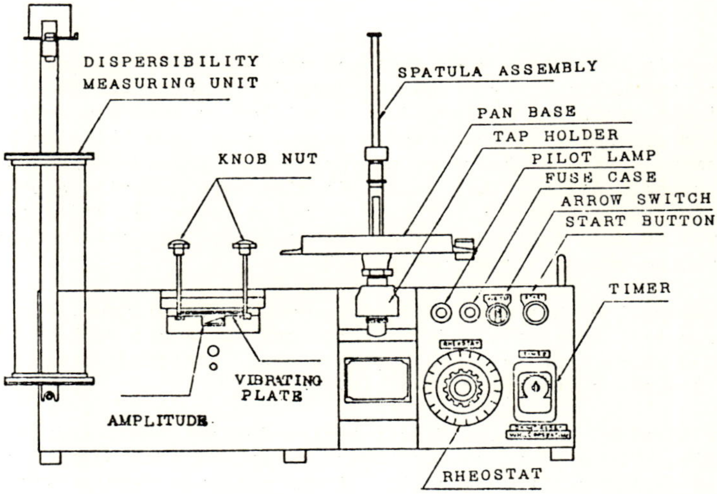

Fig. 6 Powder tester (Hosokawa)

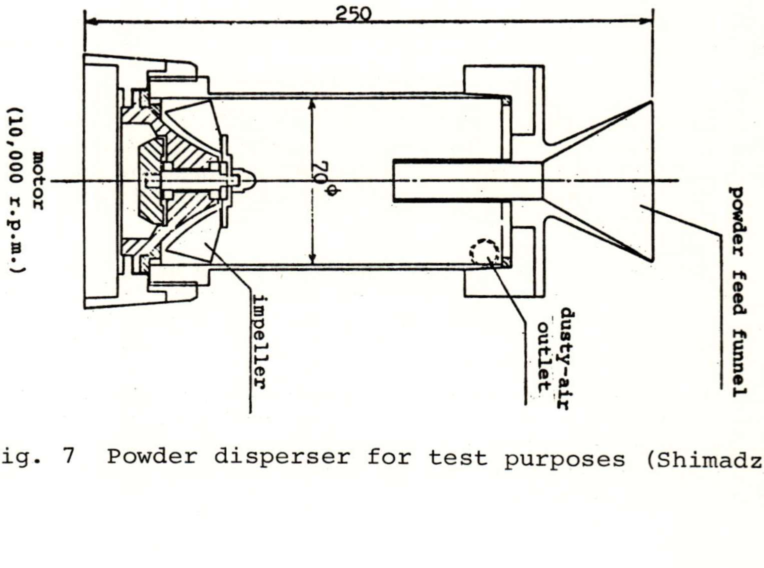

Fig. 7 Powder disperser for test purposes (Shimadzu)

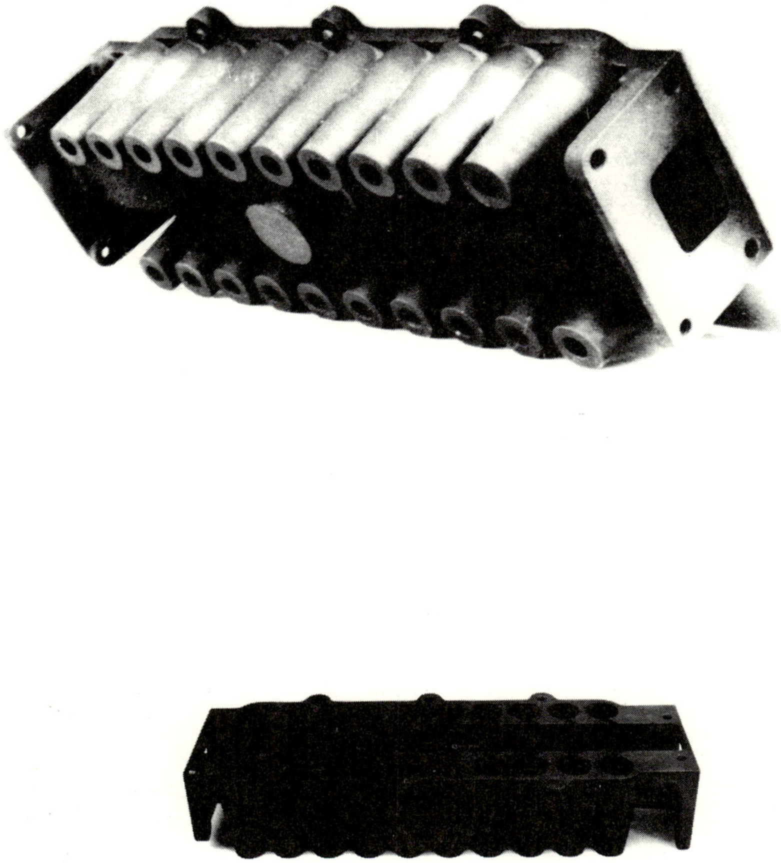

Fig. 8 Micro multicyclone (Shimadzu)

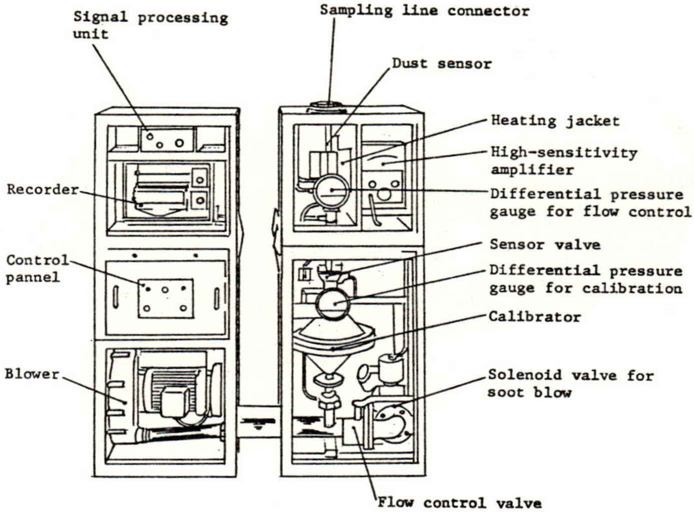

Fig. 9 Kon ytest (Electrification dust concentration meter, Kony)

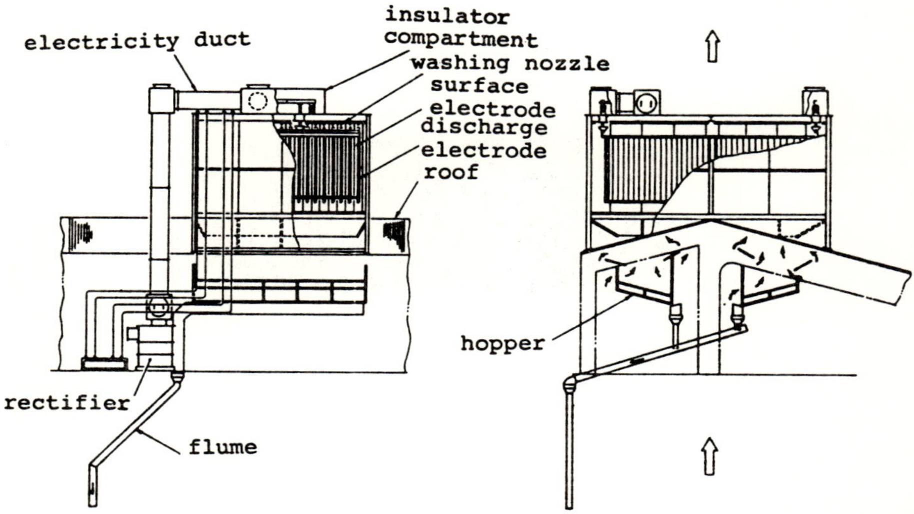

Fig.10 Natural convection type of electrostatic precipitator mounted on roof

Kimoto Electronics Company has also developed a four-stage annular slit nozzle type of cascade impactor.

Table I

Dusts and Aerosols for Industrial Testing (1974)

(JIS Z 8901)

No	Materials	mmd (μm)	geo. st. dev. σ_g
1	Silica sand	195	2.14
2	"	30	4.00
3	"	8	3.63
4	Talc	7	2.43
5	Flyash	15	2.24
6	Portland cement	26	3.65
7	Kanto Loam	30	4.00
8	"	8	3.63
9	Talc	4.2	2.95
10	Flyash	5.1	2.40
11	Kanto Loam	2	2.15
12	Carbon black	-	-
13	Aerosol	0.3	1.0
14	"	0.8	1.8
15	Mixed dust	(72% No. 8+25% No. 12+cotton linter)	

REFERENCES

1 Iinoya K., S. Yuu, K. Makino and K. Nakano: On Measurement of Particle Size Distribution by Cascade Impactors - In Case of Setting the Clearance Ratio Three for Round Nozzle, Kagaku Kogaku 33, 689 (1969).

2 Yuu, S. and K. Iinoya: On Separation Mechanism of Two Dimensional Cascade Impactor, Kagaku Kogaku, 33, 1265 (1969).

3 Yuu, S. and K. Iinoya: Particle Precipitation on Two Dimensional Nozzle Wall, Kagaku Kogaku, 35, 1251 (1971).

4 Yuu, S., T. Yukawa and K. Iinoya: Effect of Gravitation on a Round Nozzle Cascade Impactor, J. Chem. Eng. Japan, 5, 285 (1972).

5 Yuu, S., N. Miyake and K. Iinoya: Effect of the Velocity Boundary Layer at the Nozzle Outlet on the Target Collection Efficiency of an Impactor, Kagaku Kogaku Ronbunshu, 1, 115 (1975).

6 Iinoya, K. and H. Yamanaka: Experiments on Anisokinetic Sampling Errors for Solid-Liquid Two Phase Flow, Kagaku Kogaku 34, 69 (1970).

7 Iinoya, K., Z. Tanaka and H. Takai: Particle Size Analysis With a Gas Centrifuge at Reduced Pressures, Kagaku Kogaku 35, 1041 (1971).

8 Iinoya, K., K. Makino, S. Toyama and A. Goto: Development of Mist Size Analyser, Kagaku Kogaku, 37, 858 (1973).

9 Takahashi, K. and S. Iwai: Estimation of Size Distribution of Small Aerosol Particles by Light Scattering Measurement, J. Colloid Interf. Sci. 23, 113 (1967).

10 Takahashi, K.: Determination of Number Concentration of Polydispersed Small Aerosol Particles by Turbidity Measurement, J. Colloid Interf. Sci. 24, 159 (1967).

11 Takahashi, K.: Application of Diffusion Tube Method to Characterization of Gas-Particle Mixture, Hoken Butsuri (J. Japan Health Phys. Soc.) 2, 115 (1967)

12 Takahashi, K. and M. Kasahara: A Theoretical Study of the Equilibrium Particle Size Distribution of Aerosols, Atmos. Environ. 2, 441 (1968).

13 Takahashi, K. and M. Kasahara: Numerical Calculation of Light Scat-

tering from Polydispersed Small Aerosol Particles, Techn. Rept. Eng. Res. Inst., Kyoto Univ., No. 143 (1968).

14 Kudo, A. and K. Takahashi: Numerical Calculation for Electrical Charge on Aerosol Particles, Pt. 1, Techn. Rept. Eng. Res. Inst., Kyoto Univ., No. 147 (1969).

15 Takahashi, K.: Changes in Particle Size Distribution of Aerosols Flowing Through Vessels, Techn. Rept. Eng. Res. Inst., Kyoto Univ., No. 149 (1970).

16 Kudo, A., K. Takahashi and T. Kojima: Numerical Calculation for Electrical Charge on Aerosol Particles, Pt. 2 Estimation Method of Particle Size Distribution, Techn. Rept. Eng. Res. Inst., Kyoto Univ., No. 151 (1971).

17 Takahashi, K.: Numerical Verification of Boltsmann's Distribution for Electrical Charge of Aerosol Particles, J. Colloid Interf. Sci. 35, 508 (1971).

18 Takahashi, K.: Concentration Change of Aerosols in Closed Vessel, Nihon-Genshiryoku-Gakkaishi (J. Atomic Energy Soc. Japan), 13, 632 (1971).

19 Kudo, A. and K. Takahashi: A Method Determining Aerosol Particle Size Distribution Applying Boltzmann's Law, Atmos. Environ. 6, 543 (1972).

20 Takahashi, K. and T. Tamachi: Monitoring of Atmospheric Aerosol Particles in Uji-City, Kukiseijo (J. Japan Air Cleaning Assoc.) 10, 72 (1972).

21 Takahashi, K. and A. Kudo: Electrical Charging of Aerosol Particles by Bipolar Ions in Flow Type Charging Vessels, J. Aerosol Sci. 4, 209 (1973).

22 Kasahara, M. and K. Takahashi: Numerical Calculation of Aerosol Particle Concentration Passing Through Diffusion Tube, Techn. Rept. Inst. Atomic Energy, Kyoto Univ., No. 165 (1974).

23 Kanagawa, A.: Response Calculations for Sideways Light Scattering Aerosol Counters, Kagaku Kogaku, 34, 521, (1970).

24 Kanagawa, A.: Response Calculations for Forward Light Scattering Aerosol Particle Counters, ibid., 34, 991 (1970).

25 Yokochi, A. and A. Kanagawa: Improvement in Sideways Light Scat-

tering Particle Counter-Rearrangement of Optical Lens Geometry, ibid, 34, 997, (1970).

26 Kanagawa, A.: Size Determination of Aerosol Particles by Owl Spectrometer-Measurable Particle Size and Dependency on Refractive Index, ibid, 36, 91 (1972).

27 Kanagawa, A.: Size Determination of Aerosol Particles by Owl Spectrometer-Influence on Dispersion of Aerosol Particles, ibid, 36, 97 (1972).

28 Kanagawa, A.: Size Determination of Aerosol Particles by Owl-Type Photometer-Effect of Wavelength Ratio, ibid, 36, 647 (1972).

29 Yokochi, A. and A. Kanagawa: Problems in Generation of Monodisperse Latex Aerosol, ibid, 36, 676 (1972).

30 Kanagawa, A.: Response Calculation of Light Scattering Photometer for Measurement of Aerosol Concentration, ibid., 38, 513 (1974).

31 Yoshida, T., Y. Kousaka and K. Okuyama: A New Technique of Particle Size Analysis of Aerosols and Fine Powders Using an Ultramicroscope, I. & E.C. Fund. 14, 47 (1975).

32 Yoshida, T., Y. Kousaka, K. Okuyama and S. Nishio: Effect of Brownian Coagulation and Brownian Diffusion on Gravitational Settling of Polydisperse Aerosols, J. Chem. Eng. Japan, 8, 137 (1975).

33 Yoshida, T., Y. Kousaka, S. Inage, and S. Nakai: Pressure Drop and Collection Efficiency of an Irrigated Bag Filter, I. & E.C. Process Design and Develop., 14, (2)(1975).

34 Kitani, S., S. Uno and J. Takada: Direct Measurement of Variable Aerosol Concentration in a Fast Reactor Chamber,Kuki Seijo (J. Japan Air Cleaning Association) 10, (2) 34 (1972).

35 Nishio, G., S. Kitani and K. Takahashi: Thermophoretic Deposition of Aerosol Particles in a Heat Exchanger Pipe, I. & E.C. Process Design Develop., 13, 408 (1974).

36 Ikebe, Y.: Determination of the Size Distribution of Heterogeneous Submicron Aerosols by Response Matrix Method, Pure Appl. Geophs., 98, 197 (1972).

37 Mori, Y., A. Suganuma and K. Ishibashi: Influence of Air Humidity on the Degree of Airborne Dust Agglomeration, J. Res. Assoc. Powder Tech. Japan, 7, 29 (1970).

38 Mori, Y., A. Suganuma, M. Oka and Y. Kaida: Influence of Air Humidity on Airborne Dust Generation Phenomena, Kagaku Kogaku, 34, 198 (1970).

39 Tamori, I. and Tadao Shirasawa: On Collection Efficiency of Dust Tube Sampler, J. Res. Assoc. Powder Tech. 8, 8 (1971).

40 Tamori, I., N. Kogure and K. Imagami: Proposal of Cylindrical Filter Paper Method for Stack Dust Sampling, J. Environ. Pollut. Control, 7, 435 (1971).

41 Tamori, I., N. Kogure and K. Imagami: Measuring of Stack Dust Contents by Cylindrical Filter Paper Method, Kogai 6, 289 (1971).

42 Imagami, K., N. Kogure and I. Tamori: Development of New Isokinetic Sampler by Comparison between Two Kinetic Pressures, Kogai, 8, 362 (1973).

43 Shirasawa, T., A. Ito, I. Tamori and T. Ohyanagi: Filtration of Combustion Gas by Coal Packed Bed, Kogai, 5, 10 (1970).

44 Fujimura, M. and Y. Hashimoto: Analysis of Size Distribution Data of Particulate Matters by Andersen Sampler, Bunseki Kagaku, 24, 36 (1975).

45 Shibuya, A. and S. Masuda: EP-ES Type Electrostatic Precipitator, Proc. 1975-General Conf. Institute of Electr. Engrs. Japan, No. 920.

46 Masuda, S., S. Ago, T. Itoh and H. Saito: Hybrid-Type Electrostatic Precipitator, Proc. 1975-General Conf. Inst. Electr. Engrs. Japan, No. 921.

STANDARDIZATION AND CALIBRATION OF AEROSOL INSTRUMENTS

BENJAMIN Y.H. LIU
Particle Technology Laboratory
Mechanical Engineering Department
University of Minnesota
Minneapolis, Minnesota 55455

ABSTRACT

This paper reviews the aerosol standards development work at the Particle Technology Laboratory, University of Minnesota. The operating principle and the performance characteristics of the vibrating orifice monodisperse aerosol generator and those of the electrical aerosol generator are described. It is shown that with the use of these generators, monodisperse aerosols of a known particle size and concentration can be generated from about 0.01 μm to 50 μm in particle diameter. In addition, calibration studies on condensation nuclei counters and the diffusion battery have also been briefly reviewed.

STANDARDIZATION AND CALIBRATION OF AEROSOL INSTRUMENTS

BENJAMIN Y.H. LIU
Particle Technology Laboratory
Mechanical Engineering Department
University of Minnesota
Minneapolis, Minnesota 55455

INTRODUCTION

The calibration of aerosol measuring and sampling devices is facilitated by the use of monodisperse aerosols. When the size or concentration of the aerosol, or both, are known to a sufficiently high degree of accuracy, the aerosol can be referred to as an aerosol standard.

During the past few years, the Particle Technology Laboratory, University of Minnesota has devoted much effort to the development of apparatus and procedure for generating monodisperse aerosols of a primary standard quality. Monodisperse aerosols can now be generated from 0.01 μm to over 50 μm in diameter at concentration levels up to 10^6 particles /cc in certain size ranges. In addition, both solid and liquid particles can be generated and the particle size can be calculated from the operating conditions of the aerosol generators to a high degree of accuracy (better than 1%). And the need to measure particle size by the tedious and often inaccurate microscopic method has largely been eliminated.

In this paper we shall briefly review the aerosol standards development work at the Particle Technology Laboratory and the application of these standards to instrument evaluation and standardization.

GENERATION OF MONODISPERSE AEROSOL STANDARDS

Vibrating Orifice Monodisperse Aerosol Generator

For generating monodisperse aerosols in the 0.5 to 50 μm diameter range, the vibrating orifice monodisperse aerosol generator[1] has been developed. The generator, shown schematically in Figure 1, is comprised of a droplet generation and dispersion system and an aerosol dilution and transport system. In addition, a radioactive source of Krypton 85 (10 mCi activity) is placed in the generator to neutralize the particle electrostatic charge incurred during the droplet generation process.

Briefly, the operation of the generator is as follows. The material to be aerosolized is first dissolved in a suitable solvent, such as water or alcohol. The solution is then forced by a syringe pump through a small (5 to 20 μm diameter) orifice, which is vibrated by a piezoelectric ceramic.

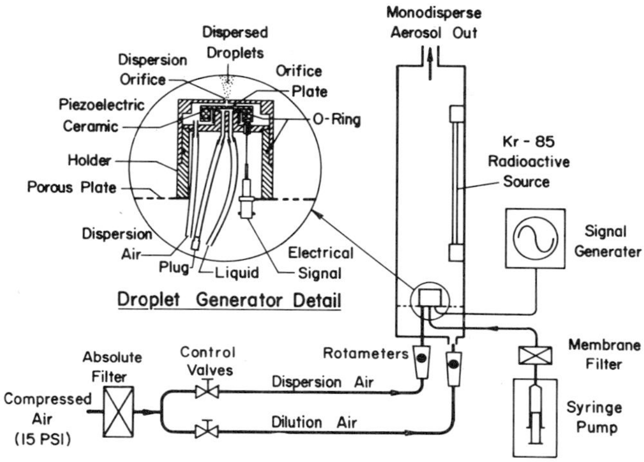

Figure 1 Schematic diagram of the vibrating orifice monodisperse aerosol generator.

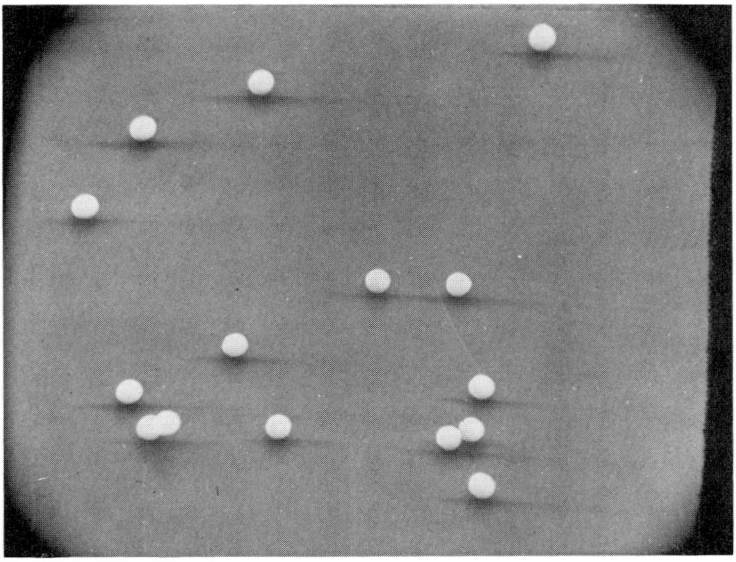

Figure 2 Electron micrograph of methylene blue particles (3.7 μm diameter) produced by the vibrating orifice generator.

The liquid jet is broken up by this vibration into uniform droplets, which are then injected along the center of a turbulent air jet to disperse the droplets. The dispersed droplets are then mixed with a much larger volume of clean dry air to evaporate the solvent and to transport the particles to where they are needed. Table 1 summarized the operating conditions of the generator and Figures 2 through 5 show some typical particles generated by the device.

The fact that particles of a known size can be generated by the vibrating orifice principle is based on the fact that in the region of uniform droplet production, the rate of droplet production is equal to the frequency of oscillation of the piezoelectric ceramic. Consequently, the volumetric rate of flow of the liquid, Q_1 (cc/sec), through the orifice is given by

$$Q_1 = V_d \ f \tag{1}$$

where V_d (cc) is the volume of the individual droplets and f (Hz) is the vibrating frequency. It follows from Equation (1) that the droplet diameter is given by

$$D_d = (6 \ Q_1 \ / \ \pi \ f \ )^{1/3} \tag{2}$$

Further, if C is the volumetric concentration of the non-volatile solute in the solution, then following the evaporation of the solvent from the solution droplets, an aerosol of the diameter, D_p, is obtained where

$$D_p = (\ C \ )^{1/3} \ D_d \tag{3}$$

By the use of Equations (2) and (3), the particle diameter, D_p, can be calculated if the quantities Q_1, f and C are known.

Assuming a 1% error in the measurement of Q_1, negligible error in the measurement of f and a 2% error in the measurement of C, the overall error in the calculated particle size is about 1%. However, to achieve such a small error, the following conditions must be met:

(1) There must be no leak in the liquid flow system.
(2) The solvent used must contain a negligible amount of non-volatile impurity.
(3) The solvent must be completely evaporated from the solution droplets and
(4) the density of particles must be the same as the intrinsic density of the material.

Condition (1) stated above can be easily met through careful operation of the droplet generator.

Table I Charateristics of the vibrating orifice monodisperse aerosol generator (Thermo-Systems Inc., 2500 Cleveland, St. Paul, Minnesota 55113)

Diameter of Liquid Orifice, μm	Nominal Frequency kHz	Droplet Diameter μm*	Particle Diameter Range, μm**	Nominal Concentration Particles/cc***
5	450	15	0.6 - 15	273
10	225	25	1.0 - 25	137
20	60	40	1.8 - 40	36

* Continuously adjustable over an approximate 25% range by varying the frequency

** Obtainable by the solvent evaporation technique

*** Theoretical concentration based on the nominal aerosol output of 100 liters per minute.

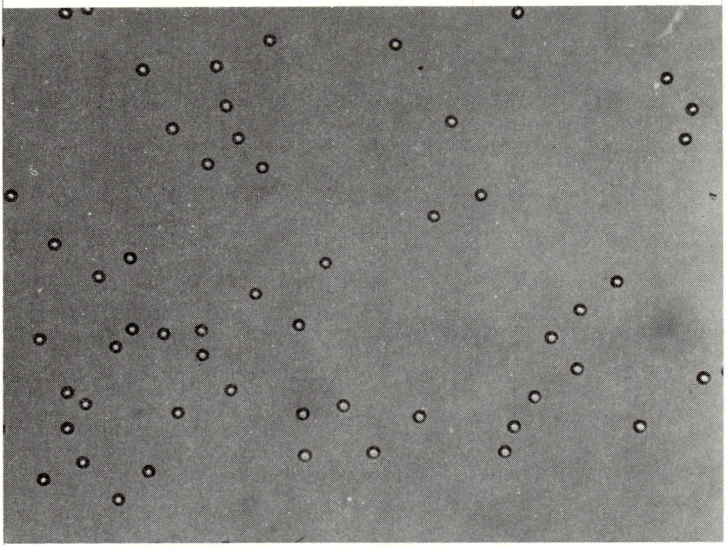

Figure 3 Optical micrograph of monodisperse DOP particles (9.5 μm diameter) produced by the vibrating orifice generator.

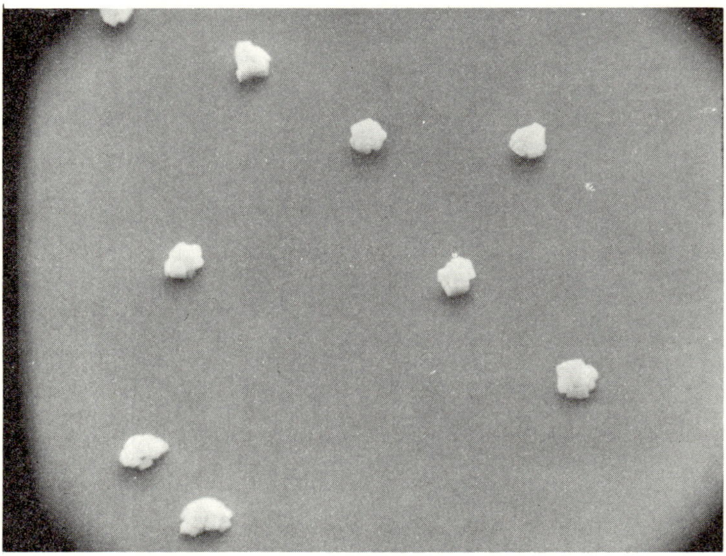

Figure 4 Electron micrograph of sodium chloride particles of 27.4 μm volume produced by the vibrating orifice generator.

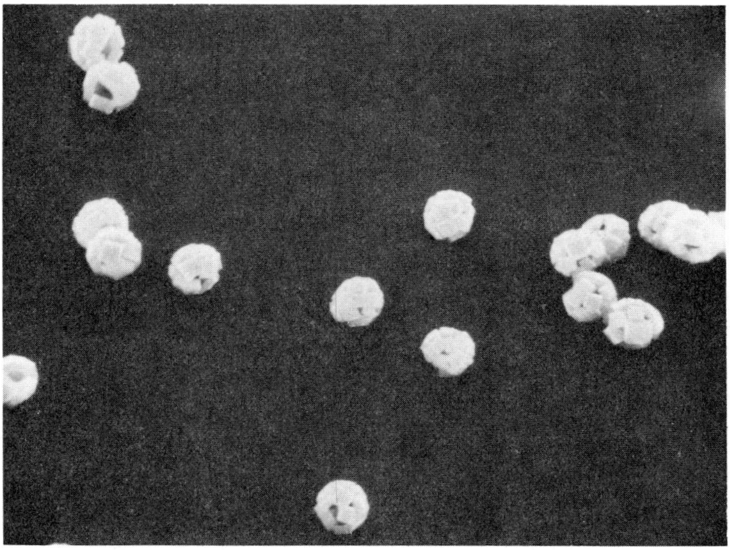

Figure 5 Electron micrograph of hollow NaCl particles (125 μm^3 total solid volume) produced by the vibrating orifice generator.

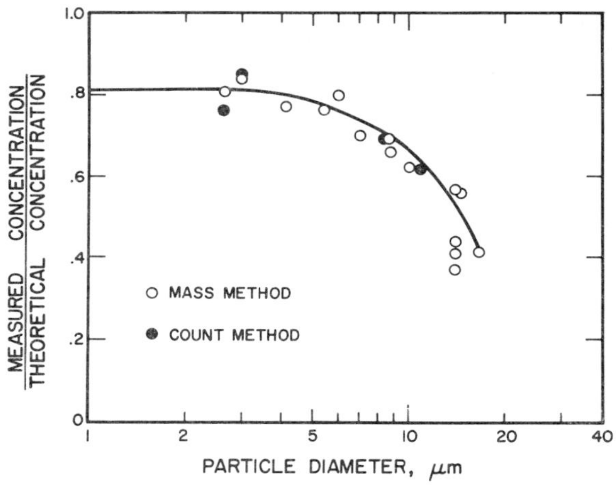

Figure 6 Output aerosol concentration of the vibrating orifice generator as a function of particle diameter.

To insure that condition (2) is obtained, it is necessary that C, the solution concentration used, be large compared to the non-volatile impurity content of the solvent. For instance, for a diameter reduction of 20, i.e. $D_d/D_p = 20$, the solution concentration is $C = (1/20)^3 = 125 \times 10^{-6}$ or 125 ppm. If the non-volatile impurity concentration, I, of the solvent is 10 ppm, this impurity will contribute an error of 10/125 = 8% to the value of C. This corresponds to a particle diameter error of about 2.6%. However, by measuring I and calculating the particle diameter as follows

$$D_p = (C + I)^{1/3} D_p \qquad (4)$$

the impurity contribution to the particle diameter error can be corrected for. One of the simplest methods of measuring I is to spray the solvent directly through the droplet generator and measuring the residue particle size with an optical particle counter.

To insure that condition (3) is obtained, the dilution air used must be dry and sufficient time must be allowed for the solvent to completely evaporate from the solution droplets.

In the case of condition (4), there is generally no difficulty when the aerosol material is a liquid. The droplets are spherical (see Figure 3) and the particle density is the same as the bulk liquid density. In the case of solids, experience shows that amorphous solids, such as methylene blue, will ususally dry from a solution state to form spherical solid particles (see Figure 2) and the particle density will be the same as that of the parent material. However, the behavior of crystaline solids, such as NaCl, is considerably more complex and the solution droplets may crystalize to form particles such as those shown in Figures 4 and 5. Consequently, Equation (4) must be used with caution under such circumstances.

The ability of the vibrating orifice aerosol generator to produce aerosols of a known particle concentration is based on the fact that the rate of droplet production is equal to the vibrating frequency of the piezoelectric ceramic. When the droplets are mixed with a total air flow of Q_a (cc/sec), this will give rise to an aerosol having the theoretical concentration of

$$N_{th} = f / Q_a \qquad (5)$$

Unfortunately, there are losses in the system and the actual aerosol concentration at the generator output is somewhat less than the theoretical concentration calculated above. Figure 6[2] shows the measured output of the generator, expressed as a percentage of the theoretical output, as a function of the particle size. With the use of Equation (5) and Figure 6 the generator can be used as a secondary standard for aerosol concentration in the particle diameter range below 20 μm.

Generation of Monodisperse Aerosols by the Electrical Mobility Classifier

For generating monodisperse aerosols below 0.5 μm, the system shown in Figure 7 has been developed[3]. The system is comprised of a Collison atomizer, a diffusion dryer, a Kr-85 bipolar charger and a mobility classifier. The operation of the system is based on the fact that for particle diameters smaller than about 0.1 μm an aerosol in charge equilibrium with bipolar ions will acquire a bipolar charge distribution with most of the particles either electrically neutral or carrying only ±1 elementary unit of charge. Because of the small particle size, only a small fraction of the particles will be multiply charged. Consequently, by passing this aerosol through a mobility classifier, a monodisperse fraction can be extracted according to electrical mobility. Since the particles are singly charged, the electrical mobility is related to particle size as follows,

$$Z_p = e\, C \,/\, 3\, \pi\, \mu\, D_p\, 300, \ \mathrm{cm^2/volt\text{-}sec} \qquad (6)$$

where $e = 4.8 \times 10^{-10}$ esu is the elementary unit of charge, C is the slip correction, μ (poise) is the viscosity of the gas and D_p (cm) is the particle diameter.

In the system shown in Figure 7, a polydisperse aerosol is first produced by spraying a liquid solution with the Collison atomizer and drying the particles with the diffusion dryer. The aerosol is then passed through the Kr-85 bipolar charger in which the particles are brought to an equilibrium state with the bipolar ions produced by the radioactive source. This charge-equilibrated, polydisperse aerosol is then classified electrostatically.

In the mobility classifier shown in Figure 7, the aerosol is introduced along with the clean air in two concentric, laminar streams in the annular gap region between the concentric metal cylinders. Because of the voltage applied on the inner cylinder, charged particles are deflected through the clean air stream. Those that have the appropriate electrical mobility would arrive at the slit near the lower end of the electrode and be swept out by the small airstream flowing through the slit. The mobility of these particles can be calculated by the following equation,

$$Z_p = [\, q_c + (1/2)(q_s - q_a)\,]\, \ln (r_2/r_1) \,/\, 2\, \pi\, V\, L \qquad (7)$$

where q_o (cc/sec) is the clean air flow in the mobility classifier, q_a (cc/sec) is the aerosol flow at the inlet, q_s (cc/sec) is the aerosol flow through the slit, r_1 and r_2 (cm) are the inner and outer radii of the annular gap region between the cylindrical electrodes, L is the length of the precipitating region and V (volt) is the applied voltage on the inner

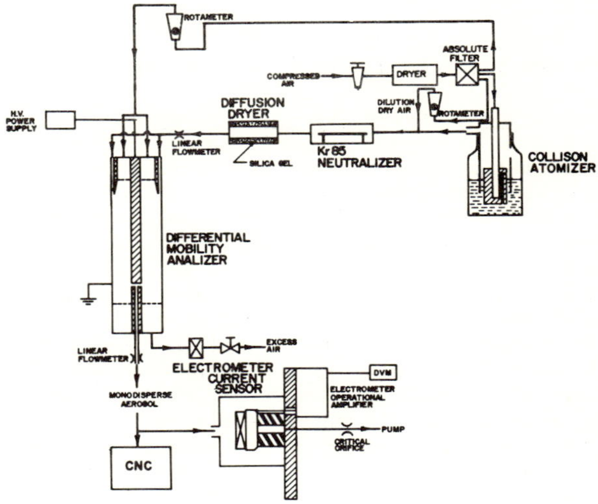

Figure 7 Schematic diagram of apparatus for generating submicron aerosol standards by the electrical classifier.

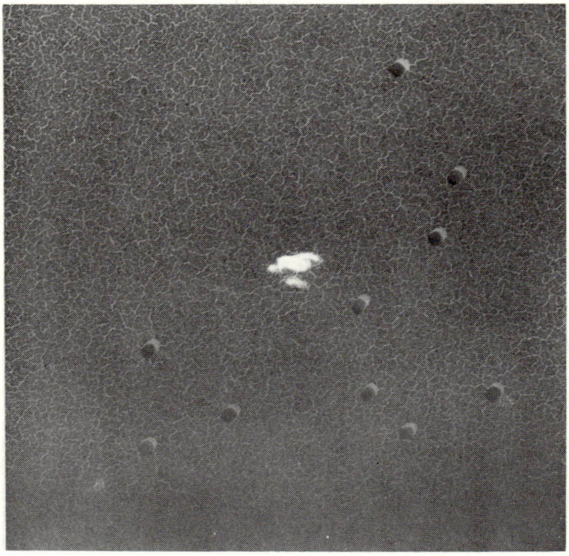

Figure 8 Electron micrograph of monodisperse NaCl particles produced by the electrical aerosol generator.

electrode, the outer electrode being grounded. By the use of Equations (6) and (7), the size of the particles can be calculated when the flow rates, the applied voltage and the dimensions of the device are known. In the experiment reported by Liu and Pui[3], the uncertainty in the calculated particle size is estimated to be ±2%, based on the measurement accuracies for the various quantities that enter into the calculation. Figure 8 is an electronmicroscope picture of the particles generated by this method.

An additional advantage of the electrical classification method of monodisperse aerosol generation is the fact that the aerosol concentration can be easily measured to a high degree of accuracy. This can be accomplished by collecting the aerosol onto an absolute filter in a Faraday cage and measuring the collected particle charge with an electrometer. Since the particles are singly charged, the electrometer current, I (amp) is related to the aerosol concentration N (particles/cc) as follows,

$$I = q_e \, e \, N \tag{8}$$

where $e = 1.61 \times 10^{-19}$ coulomb is the elementary unit of charge and q_e (cc/sec) is the sampling flow rate into the electrometer current sensor. More details concerning the method including the procedure for correcting for the effect of the small percentage of doubly charged particles in the output aerosol stream is given in reference 3.

INSTRUMENT CALIBRATION AND STANDARDIZATION

The monodisperse aerosol generators described above have been used to calibrate a variety of aerosol measuring and sampling devices. The work on optical particle counters has been described by Liu, Berglund and Agarwal[2] and is reviewed elsewhere in this symposium volume (see the paper on optical particle counters by Willeke and Liu). The work on the electrical aerosol analyzer has been reported by Liu and Pui[4] and reviewed by Whitby in the paper on electrical measurements, also in this volume. The following review is limited to the calibration study on condensation nuclei counter and the diffusion battery, two devices that have found widespread use in aerosol studies.

The condensation nuclei counter is an aerosol concentration measuring device based on the growth of particles to a visible size by condensing vapor on the particles. Particles as small as 20 Å can theoretically be detected by this method. Such particles are usually referred to as condensation nuclei.

Figure 9 shows the result of calibration studies reported by Liu, Pui, Hogan and Rich[5] on the Pollak counter, a manually operated condensation nuclei counter. The Pollak counter has been carefully calibrated

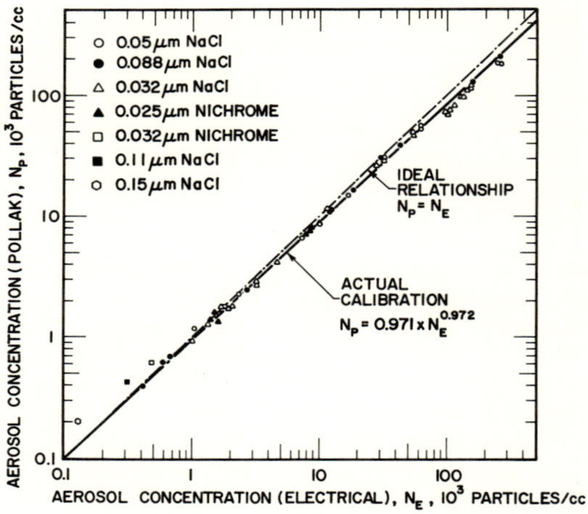

Figure 9. Calibration of the Pollak counter with the electrical aerosol generator. generator. (Reprinted courtesy of the American Meteorological Society.)

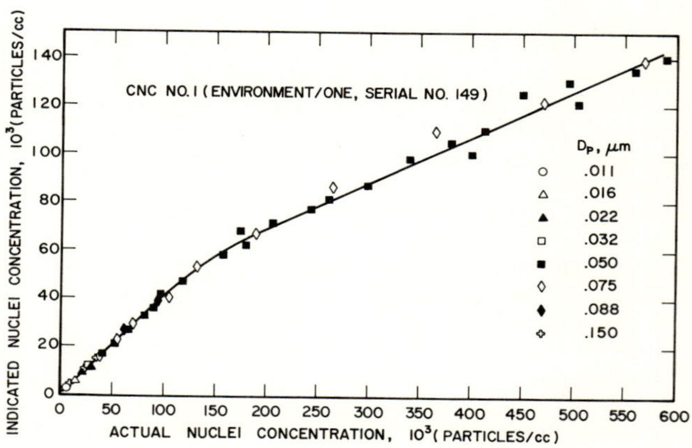

Figure 10 Calibration of the Environment/One, Rich 100 condensation nuclei counter with the electrical aerosol generator.

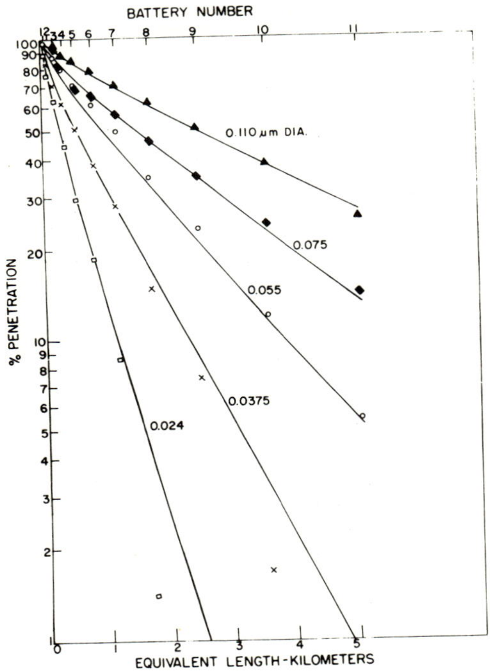

Figure 11 Penetration of monodisperse NaCl aerosol through a portable diffusion battery measured with a GE CN counter at 6 lpm, ▲ 0.110 μm diam., ◆ 0.075 μm, ○ 0.055 μm, × 0.0375 μm, □ 0.024 μm, solid lines (———) calculated from the theory.

by Pollak and Metnieks[6] and has been used as a secondary standard for condensation nuclei measurement. The calibration study of Liu, Pui, Hogan and Rich[5], made with monodisperse aerosol generated by the electrical mobility classifier described in the preceeding section, shows that the original calibration of the Pollak counter is in good agreement with the present calibration, and that the maximum difference of these two calibration is only about 18%. Further, the study shows that within the particle size range studied, viz. from 0.025 to 0.5 μm diameter, the counter response is essentially independent of the particle size. However, a similar calibration study of a commercially available automatic condensation nuclei counter[3] shows that there is a substantial discrepancy between the factory calibration and the present calibration (see Figure 10).

The calibration study on the diffusion battery was made with the portable diffusion battery described by Sinclair[7]. The diffusion battery is comprised of a series of porous metal plates with approximately 50% porosity and holes having diameters of approximately 230 μm. In the calibration study, which was recently reported by Sinclair, Countess, Liu and Pui[8], monodisperse aerosols of various diameters were generated by mobility classification as previously described and the penetration of the aerosols through the diffusion battery was measured with a condensation nuclei counter. The measured penetration was compared with the theoretically predicted penetration. Some sample results are shown in Figure 11. The agreement is seen to be very good. Such good agreement between theory and experiment constitutes a direct verification of the theory of diffusion battery.

CONCLUSION

In this paper we have briefly reviewed the aerosol standards development and instrument calibration studies made at the Particle Technology Laboratory, University of Minnesota. We have shown that the use of the vibrating orifice monodisperse aerosol generator and the electrical aerosol generator has led to the development of monodisperse aerosol standards in the 0.01 to 50 μm diameter range. Through the use of these aerosol standards we have performed accurate calibration studies on such devices as the condensation nuclei counter, the electrical aerosol analyzer, the optical particle counter and the diffusion battery. Such instrument standardization work should lead to improved measurement accuracies for aerosols, both in the laboratory and in the open atmosphere.

REFERENCES

1. Berglund, R.N. and B.Y.H. Liu. Generation of Monodisperse Aerosol Standards. Env. Sci. Tech. 7:147-153, 1973.

2. Liu, B.Y.H., R.N. Berglund and J.K. Agarwal. Experimental Studies of Optical Particle Counters. Atm. Env. 8:717-732, 1974.

3. Liu, B.Y.H. and D.Y.H. Pui. A Submicron Aerosol Standard and the Primary Absolute Calibration of the Condensation Nuclei Counter. J. Colloid and Interface Sci. 47:155-171, 1974.

4. Liu, B.Y.H. and D.Y.H. Pui. On the Performance of the Electrical Aerosol Analyzer. Aerosol Sci. 6:249-264, 1975.

5. Liu, B.Y.H., D.Y.H. Pui, A.W. Hogan and T.A. Rich. Calibration of the Pollak Counter with Monodisperse Aerosols. J. Appl. Meteor. 14:46-51, 1975.

6. Pollak, L.W. and A.L. Metnieks. Intrinsic Calibration of the Photoelectric Nucleus Counter Model 1957, with Convergent Light Beams. Tech. Note No. 9, Contract AF61 (052)-26, School of Cosmic Physics, Dublin Institute for Advanced Studies, 1960.

7. Sinclair, D. A Portable Diffusion Battery. Am. Ind. Hyg. J. 33: 729, 1972.

8. Sinclair, D., R.J. Countess, B.Y.H. Liu and D.Y.H. Pui. Experimental Verification of Diffusion Battery Theory. Presented at the 68th Annual Meeting of the Air Pollution Control Association, Boston, Mass. June 15-20, 1975.

PART II

AEROSOL GENERATION

THE GENERATION OF AEROSOLS OF FINE PARTICLES*

Otto G. Raabe
Inhalation Toxicology Research Institute
P. O. Box 5890
Albuquerque, New Mexico 87115

ABSTRACT

Aerosols of fine particles required for experimental applications vary widely in necessary characteristics and require means of aerosol production that are consistent with the scientific goals of an investigation. In this report, the common techniques for generation of aerosols in the respirable size range are reviewed with emphasis upon experimental, non-therapeutic applications. Methods described include dispersion of dry powders and the atomization of solutions and suspensions with various compressed air and ultrasonic nebulizers. Special emphasis is placed on the generation of aerosols of insoluble forms under controlled and reproducible conditions. Heat treatment is described as an ancillary step to speed the drying of droplets or to change the chemical nature of aerosols after they are produced. Particular attention is devoted to the various methods for producing monodisperse aerosols including nebulization of suspensions of monodisperse particles, spinning disk droplet generation, electrostatic dispersal, controlled condensation and periodic dispersion of liquid jets. Reduction of the electrostatic charge of aerosols as an important part of aerosol generation is discussed.

NOMENCLATURE

A = liquid aerosol concentration at time of droplet formation expressed as volume of liquid aerosolized per unit volume of air at ambient conditions; a measure of nebulizer efficiency (μl/s, ml/s, μl/min or ml/min, depending on equation units).

c = solute concentration in solution being nebulized (g/ml or g/cm^3).

c_o = initial solute concentration in solution to be nebulized (g/ml or g/cm^3).

D = geometric diameter of a spherical particle (cm or μm).

D_d = geometric diameter of a spherical droplet (cm or μm).

* Prepared under U. S. Energy Research and Development Administration Contract E(29-2)-1013.

D_g = geometric diameter of a cylindrical liquid stream (cm or μm) emitted from a circular orifice.

D_{opt} = geometric diameter of droplets generated at the optimum frequency for periodic dispersion of liquid jets.

e = either the base of Naperian logarithms equal to about 2.718 (unitless) or the basic electrostatic unit of charge equal in absolute value to the charge of an electron.

F = fraction by volume of monodisperse particles in a stock suspension (unitless).

f = frequency of periodic vibrations or disturbance for dispersal of cylindrical liquid streams (Hz).

f_{opt} = optimum value of f (Hz).

f_{max} = maximum value of f defined as the lowest frequency required to produce the smallest possible droplets by periodic dispersal of a cylindrical liquid stream (Hz).

L = length of a separated segment of a cylindrical stream of liquid (cm or μm).

L_{min} = minimum value of L.

ln = natural logarithm.

L_{opt} = optimum value of L given by f_{opt}.

MMD = mass median diameter of an aerosol size distribution (cm or μm).

MNC^* = maximum number concentration of particle per milliter of water for nebulization of an aqueous suspension of monodisperse particle to produce an aerosol with no less than 95% single particles.

MNC = MNC^* expressed in units of radioactivity concentration in nebulizer suspension (Ci/ml) for radioactive particles.

n = integer number.

P_a = ambient barometric pressure (dynes/cm^2, cm Hg, psi).

P_m = minimum pressure in nebulizer jet (dynes/cm^2, cm Hg, psi).

Q = volumetric flow rate (cm^3/s, ml/s or l/min).

R = the ratio of droplets with single particles produced by a nebulizer from a suspension of monodisperse particles to all droplets produced that contain particles, i.e., the fraction of singlets in an aerosol produced by nebulization of a monodisperse suspension.

r = droplet radius (cm or μm).

r_o = initial droplet radius (cm or μm).

s = specific activity of radioactive material (Ci/g).

t = time (s or min).

VMD = volume median diameter of a droplet distribution (cm or μm).

V_o = initial volume of solution to be nebulized (ml or cm^3).

W = volume of water evaporated from a nebulizer per unit volume of aerosol-containing air generated at ambient conditions (μl/s, ml/s, μl/min or ml/min, depending on equation units).

y = liquid dilution ratio required for preparing monodisperse suspensions (unitless).

π = ratio of circumference to diameter for a circle $\simeq$ 3.14159 (unitless).

ρ = physical density of solid material (g/cm^3).

σ_g = geometric standard deviation of log-normal distribution (unitless); minimum value is unity.

τ = surface tension (dynes/cm).

ω = speed of angular rotation (radians/s).

ΔP = gage pressure above P_a (dynes/cm^2, cm Hg, psig).

THE GENERATION OF AEROSOLS OF FINE PARTICLES

Otto G. Raabe
Inhalation Toxicology Research Institute
Lovelace Foundation
P. O. Box 5890
Albuquerque, New Mexico 87115

INTRODUCTION

Aerosols of fine particles can be defined as relatively stable suspensions in a gaseous medium of liquid droplets or solid particles with settling speeds under the influence of gravity less than the settling speed of a unit density sphere of 3.5 μm geometric diameter. Such aerosols are encountered in all phases of life. In fact, it appears to be a fundamental property of the earth's atmosphere to contain aerosols. Even in thoroughly filtered air, aerosols will spontaneously form from trace gases under the influence of radiant energy.[1-6] There is great scientific interest in aerosols of fine particles not only because of their use in medicine, industry and research and their collection by the respiratory tree during inhalation, but also because of their intimate involvement in the workings of nature. It is because of natural aerosols that the sky appears blue and the sunset red.

Fundamental to the use, study or understanding of aerosols of fine particles is the ability to produce suitable forms for experimental or applied purposes. The extensive use of aerosols in medical therapy has led to the design of many types of aerosol generators which fall into three basic groups: (1) compressed air nebulizers, (2) ultrasonic nebulizers and (3) dust blowers. In addition to medical nebulizers and dust blowers, other specialized nebulizers, dust blowers, spray nozzles, spray cans and aerosol generation devices are widely used in experimental research. Among these specialized instruments are some designed to produce uniform droplets or particles. Aerosols of fine particles can also be produced by photochemical, radiolytic and condensation processes.

The characteristics of the ultimate form of aerosols produced by various means depend not only upon the generation method but also upon the physical and chemical nature of the source material and the temperature and humidity conditions or treatment received. On this basis, aerosols can be produced that are heterodisperse, polydisperse, or monodisperse. The term heterodisperse is used to describe aerosols of particles which vary widely not only in size but also in physical and chemical characteristics. Polydisperse usually implies homogeneity of basic physical and chemical characteristics but with widely different particle sizes and differences in those aerodynamic properties which depend in part upon particle size. Monodisperse is a term which implies both homogeneity of basic chemical and physical characteristics including composition and physical density and relative uniformity of particle size and aerodynamic properties. Monodisperse is a relative term, since a distribution of particles which is satisfactorily monodisperse in one application may be considered polydisperse in another. However, the definition suggested by Fuchs and Sutugin[7] provides a guideline for defining "practical monodispersity". Their definition of monodisperse particle size distribution is one having a coefficient of variation less than 0.2. Nevertheless, a distribution of particles which extends from 1 to 5 μm and has a coefficient of variation much bigger than 0.2 may be clearly polydisperse to most, but a distribution that extends from 0.01 to 0.05 μm may be considered essentially monodisperse by some investigators even though the coefficient of variation may be much greater than 0.2. In addition, the phenomenon being studied may affect the definition. Particles may be numerically monodisperse with respect to geometric size, but because of variations in physical density, they may be polydisperse with respect to aerodynamic properties such as settling speed.

This review summarizes a variety of methods commonly used to produce aerosols of fine particles from the perspective of the investigator who is interested in aerosol behavior or the biological effects of aerosols inhaled by experimental animals or man.

POLYDISPERSE AEROSOLS

NEBULIZERS

A nebulizer is an atomizer used to produce aerosols of fine particles by atomization of liquids to produce droplets, the majority of which are less than 10 μm in diameter, as opposed to those from a spray atomizer which may range up to 100 μm or larger. Both compressed air and ultrasonic nebulizers produce aerosols from liquids. Often, the particles of residue remaining after the droplets evaporate form the desired aerosol of fine particles. Either solutions or suspensions may be aerosolized with nebulizers.

Important factors in describing the operation of nebulizers include (a) the output rate and concentration of usable aerosol, (b) the volumetric rate of air, (c) the evaporation losses of liquid which are independent of usable aerosol, (d) the droplet size distribution, (e) the volume of liquid required for proper operation and (f) the maximum unattended operating time.[8] Cognizance of these factors in relation to a particular application determines the choice of a nebulizer.

The output rate of usable aerosol is usually described as volume of liquid at initial formation associated with droplets that leave the generator and may be expressed as ml/min or μl/min. The concentration of droplets from most nebulizers is usually between 10^6 and $10^7/cm^3$ of air; however, since the volumetric rate of air may be several liters per minute, the droplet output rate may be well in excess of 10^9/min.

Evaporation losses increase the concentration of the solute in solution or particles suspended in the aerosolized liquid. This can be described as given by Mercer, et al.[9]

$$\frac{c}{c_0} = \left[\frac{V_0}{V_0-(A+W)Qt}\right]^{\frac{W}{A+W}} \quad (1)$$

with the c_0 and c solute concentrations at time t = 0 and t(min), respectively, and W and A are the evaporated and aerosolized liquid output concentrations in air (ml/l), respectively, V_0 is the initial volume of solutions and Q is the volumetric output of the nebulizer (l/min). This evaporation occurs primarily from the surface of the liquid and from the droplets which evaporate slightly and then hit the wall of the nebulizer to be returned to the reservoir. This change in concentration causes an increase in the sizes of the particles that are formed when the droplets dry. Since this is undesirable, it is advantageous to keep the ratio W/(A + W) as small as possible. This is accomplished by high aerosolization efficiency (as given by A) and by keeping the nebulizer solution cool. Using a nebulizer with a small reservoir reduces W by the natural cooling from evaporation and air expansion in compressed air operated devices. An effective way to reduce evaporation is to operate the nebulizer with the reservoir submerged in an ice-water bath.[10] Other methods of maintaining a more uniform concentration of liquid to be nebulized that have been used include continuous feed of liquid[11] and humidification pretreatment of the compressed air.[12]

Nebulizers produce droplets of many sizes and resultant aerosol particles after evaporation are concomitantly polydisperse.[13] The droplet distributions described for nebulizers are the initial distributions at the instant of formation; droplet evaporation begins immediately even at saturation humidity since the vapor pressure on a curved surface is elevated.[14] The rate of evaporation depends upon many factors including surface tension, energy availability, degree of saturation of the air, the solute concentration,[15] the hygroscopicity of the solute,[16-18] the presence of immiscible liquids or evaporation inhibitors[19-20] and the size of the droplets (smaller droplets have higher vapor pressures and dry faster[14]).

The distribution of droplets produced by nebulizers has been described satisfactorily in many ways.[13,21-23] Most useful is the assumption that the logarithms of droplet size are normally distributed. This log-normal distribution of sizes allows for simple mathematical transformations[24-25] and usually satisfactorily describes droplet volume distributions.[23]

The characteristic parameters of a log-normal distribution are the median (or geometric mean) and the geometric standard deviation (σ_g). The median of a distribution of droplet diameters is called the droplet count median diameter (CMD); the median of the mass or volume distribution of the droplets is called either the mass median diameter (MMD) or the volume median diameter (VMD). These are related by:

$$\ln(\mathrm{MMD}) = \ln(\mathrm{CMD}) + 3(\ln \sigma_g)^2 \tag{2}$$

in which ln refers to the natural logarithm.[25]

Aerosols produced from aqueous solutions (and some other methods) are electrostatically charged by the imbalance of ions in the droplets as they form.[26-27] After evaporation, aerosol particles can be relatively highly charged; this may cause a small evaporating droplet to break up if the Rayleigh limit[28] is reached due to the repelling forces of the electrostatic charges overcoming the liquid surface tension.[29-30] The Rayleigh limit, the minimum stable diameter, D_d, of a droplet with n electronic units of charge, e, can be expressed as:

$$D_d = \sqrt[3]{\left[\frac{n^2 e^2}{2\pi\tau}\right]} \tag{3}$$

with τ the surface tension. In some cases the net charge on a particle may be tens or even hundreds of electronic charge units, which may affect both aerosol stability and behavior. Therefore, a reduction in the net charge of aerosol particles produced by nebulization either by mixing with bipolar ions[31] or by passing through a highly ionized volume near a radioactive source[32-33] is desirable and may be imperative in some experiments.

Compressed air nebulizers generate droplets by shattering a liquid stream in fast moving air.[9,34-37] The liquid is usually drawn into the air flow by the natural reduction in pressure that occurs at right angles to the fast moving air stream (Venturi effect given by Bernoulli's Theorem).

When such a nebulizer is operated at gage pressures greater than the ambient barometric pressure (which is usually the case), the air jet operates under sonic conditions, that is, the maximum velocity of the air jet is the speed of sound at the temperature of the air stream at a given location. (Supersonic atomizers require special configurations.[38]) The flow rate, Q, at a given barometric pressure, P_a, and gage pressure, ΔP, is:[39]

$$Q = Q' \left[\frac{P_a'}{P_a}\right]\left[\frac{P_a + \Delta P}{P_a' + \Delta P'}\right] \tag{4}$$

with Q' the known flow rate at some ambient barometric pressure P_a' when operated at gage pressure $\Delta P'$.

Adiabatic expansion occurs at the appropriate sonic velocities until the sum of the pressure head and velocity head equal the ambient pressure, P_a, at which point the pressure is a minimum P_m given by:[39]

$$P_m = 0.83 P_a \tag{5}$$

If it is assumed that liquid is aerosolized at a rate proportional to $P_a - P_m$ and that the total flow rate is given by equation 4, the nebulizer output concentration, A, is given by:

$$A = A' \left[\frac{P_a}{P_a'}\right]^2 \left[\frac{P_a' + \Delta P'}{P_a + \Delta P}\right] \tag{6}$$

with A', $\Delta P'$ and P_a' a set of known operating conditions.

This equation can be used to correct reported or measured nebulizer characteristics for the effect of change in barometric pressure since these characteristics are customarily reported in terms of the gage pressures (ΔP) at which they are operated under local conditions. Much of the available nebulizer data were collected by Mercer and his colleagues at a barometric pressure of 62 cm Hg.[9,23,26,40-42] An attempt has been made in this report to correct nebulizer operation data to 21 °C and barometric pressure 76 cm Hg.

Auxiliary Air Flow

The air flow exiting from a nebulizer creates a negative pressure at the back side of the nebulizer jet. An opening placed nearby will allow auxiliary air to be drawn into the device and mixed with the outflow. Such auxiliary flow can not only increase the effective volumetric output of a nebulizer, but also markedly increase aerosol droplet output frequently without markedly affecting aerosol output concentration. Many atomizers are designed to incorporate auxiliary air flow into normal operation.[42]

DeVilbiss No. 40

The DeVilbiss No. 40 (DeVilbiss Co., Somerset, Pa.) is one of the simplest compressed air nebulizers. This glass nebulizer (Fig. 1) has a vertical jet and a separate capillary through which the liquid to be aerosolized is drawn into the air stream. Its reservoir holds only about 10 ml. About 10 liters of air and 0.22 ml of liquid droplets are released per minute at an operating pressure of 20 psig with VMD of 3.2 μm and σ_g 1.8.[9]

Dautrebande D-30

The Dautrebande D-30 nebulizer (Fig. 2) also uses a vertical jet with separate feed capillary.[43-44] The aerosol droplets, however, are required to follow tortuous paths through baffle holes so that most of the larger drops are unable to negotiate this scrubbing action and are collected and returned to the liquid reservoir. Consequently, droplets which leave the Dautrebande generator as useful aerosol are much smaller than those produced by other nebulizers and the aerosol output concentration is low. The D-30 releases about 25 liters of air and 60 μl of liquid droplets per minute at an operating pressure of 20 psig with VMD of 1.4 μm and σ_g 1.7.[9] It is usually constructed of plastic.

Lauterbach Nebulizer

The Lauterbach nebulizer[45] uses a jet consisting of a metal tube sealed at one end with a small hole drilled near the sealed end. This jet is operated by attaching a compressed air line to the open end of the tube with the orifice positioned very near the surface of the liquid so that the air stream is emitted parallel to the surface. The liquid touching the tip of the metal tube is drawn directly into the air stream because of the reduction in pressure at right angles to the stream. The Lauterbach improved nebulization by using a recirculating reservoir system, consisting of the generator reservoir and a larger volume supply reservoir. Liquid from a 200 ml supply reservoir continually flows into the generation reservoir, which is maintained at a constant level with a

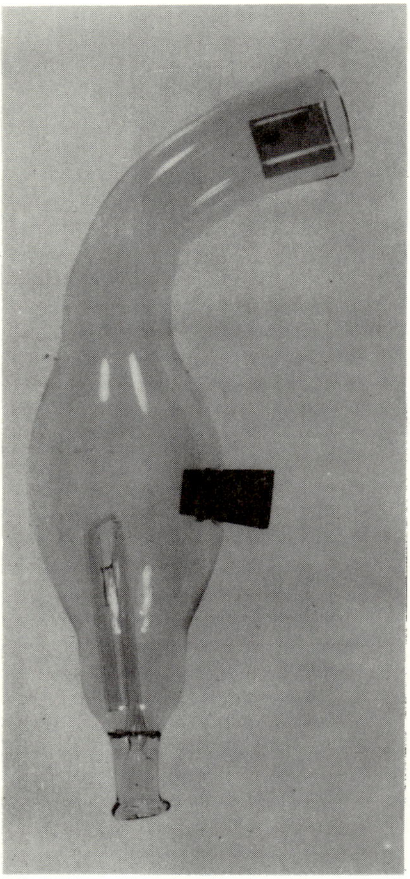

Figure 1. The DeVilbiss No. 40 glass nebulizer; the compressed air inlet is at the bottom and the aerosol outlet is at the upper right. The auxiliary air vent is corked.

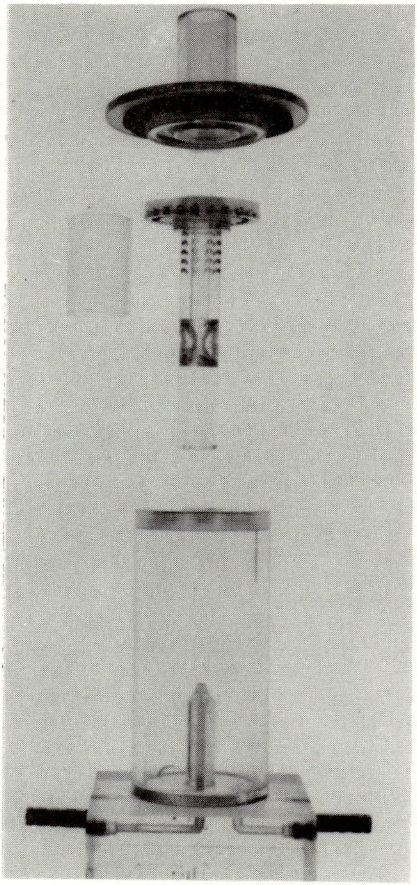

Figure 2. Exploded view of a Dautrebande D-30 nebulizer showing the vertical jet, orifice, neck and baffle holes. This unit has a separate liquid feed to allow the addition of liquid during operation.

fixed overflow tube that allows excess liquid to be pumped back to the main reservoir. Since the recirculation system works independently of the jet operation, the liquid is continually mixed and concentration changes are minimized. The reported output is about 2.4 liters of air and 17 μl of liquid droplets per minute at an operating pressure of 20 psig, with VMD of 2.4 μm and σ_g 2.0.[45]

A schematic of the glass Lauterbach nebulizer is shown in Figure 3. Mercer et al[9] report a plastic modification[40] of the Lauterbach generator that is somewhat different in design.

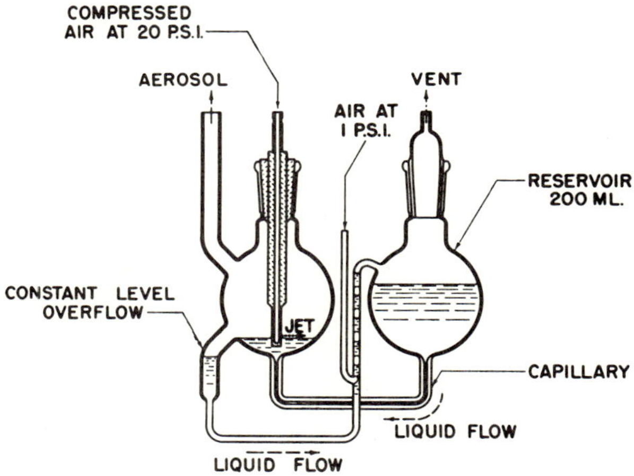

Figure 3. Schematic drawing of the Lauterbach glass nebulizer. [From Lauterbach, K.E., Hayes, A. D., and Coelho, M. A. The Archives of Industrial Health 13:156--160, 1956. Copyright 1956 by the American Medical Association.]

Collison Nebulizer

One of the early medical nebulizer modifications, which is still in use, is the nebulizer introduced in 1932 by W. E. Collison.[46] It has been widely used in Great Britian for both medical and experimental purposes.[47] The Collison jet consists of one or more orifices of various diameters, e.g. 0.035 cm, set at right angles to a vertical liquid feed channel within a cylindrical metallic rod. A simple baffle arrangement consisting of a hollow cylinder surrounding the jet is used to collect the larger droplets by impaction. Output concentrations are low with 55 μl/min of aerosol in 7.1 l/min at 20 psig; water evaporation exceeds the aerosol output and is equal to 90 μl/min at 20 psig with VMD about 2.0 μm and σ_g 2.0.[47]

Laskin Nebulizer

The Laskin nozzle (designed by Dr. Sidney Laskin, Department of Environmental Medicine, New York University Medical Center, New York, New York) consists of an orifice in a sealed metal tube. The basic arrangement is similar to the Collison nebulizer. The orifice is positioned at the top of a metal capillary through which the liquid is drawn into the air stream. The bottom of the feed tube can be submerged into the liquid at various levels without markedly affecting the aerosol output. Although the efficiency of the Laskin nozzle is not remarkable, it is useful for producing large quantities of aerosol when large volumes of air are either desirable or not prohibitive. Operating characteristics are similar to those of the Collison nebulizer.

Wright Nebulizer

B. M. Wright introduced the principle of the jet baffle in 1958 with the design of the Wright Nebulizer.[48] By placing a collection surface close to the point of disintegration of a nebulizer jet, the output concentration of usable aerosol is markedly increased. The basic arrangement is shown in Figure 4. The outlet of the liquid feed tube is placed near the point of minimum static pressure of the air jet emitted from a small orifice (0.074 cm ID) so that liquid is drawn into the jet flow within a small compartment and sprayed through a larger hole (0.16 cm) and impinged upon an open-sided flat collector at short distance (0.116 cm). Only the smaller droplets can negotiate the turn at the baffle. The liquid from the large droplets that collects on the baffle is forced to the edges by the airstream where it is extensively aerosolized enhancing the output of small droplets. If the jet baffle is too close to the point of disintegration of the liquid stream, it will interrupt the formation of small droplets and if it is too far away, the secondary aerosolization of small droplets will be less because of the lower speed of the air passing the baffle. By optimizing

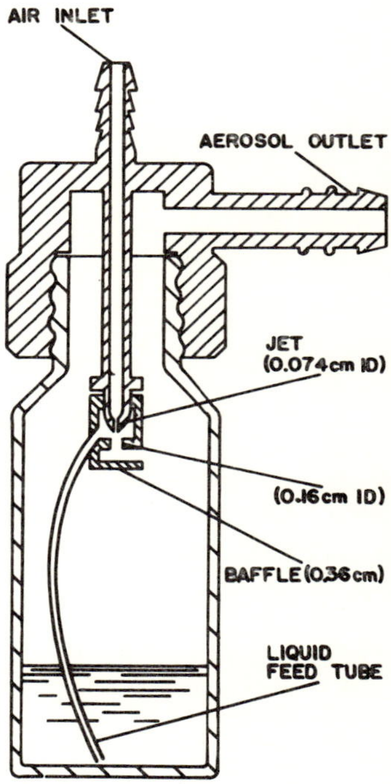

Figure 4. Schematic drawing of the Wright Nebulizer showing the air jet, liquid feed tube, secondary flow hole and jet baffle.

the position of the baffle, the concentration of small droplets (< 10 μm in diameter) in the output is greatly enhanced. Use of the Wright jet baffle principle is a common feature of modern nebulizers. The Wright nebulizer jet assembly is usually constructed of plastic. Typically, the reservoir holds about 500 ml. About 12 liters of air and 350 μl of water droplets are released per minute at operating pressure of 24 psig. The VMD has been reported to be 2.6 μm with a σ_g about 2.0.[49-50]

Lovelace Nebulizer

The "Lovelace" (Lovelace Foundation for Medical Education and Research, Albuquerque, New Mexico) Nebulizer (Figures 5 and 6) is a miniature generator which uses the Wright jet baffle principle but with a small rounded baffle rather than a flat one. A small adjustable nylon or teflon screw with rounded end mounted near and in direct opposition to the air jet serves as this primary baffle in the Lovelace design. For its size and simplicity, it has an outstanding efficiency in the generation of usable aerosols at a small volumetric rate. It operates with a small liquid volume of about 4 ml and an air flow of about 1.4 l/min at 20 psig jet pressure. This device can provide up to 70 μl/min of liquid droplets with VMD about 5.8 μm and σ_g about 1.8. Newton et al.[51] describe the basic design of the Lovelace nebulizer. The construction is of lucite for the liquid cup, epoxy for the top, stainless steel for the outlet tube and stainless steel for the tube supplying compressed air to the lucite jet assembly. The jet assembly contains a small orifice (0.25 mm ID), a cylindrical capillary through which liquid is fed to the jet and a support for the baffle screw. Mercer et al[9] demonstrated the importance of positioning the jet baffle by measuring the output for various settings of distances between the end of the screw and the orifice (Figure 7); they also report a very favorable output-to-evaporation ratio.

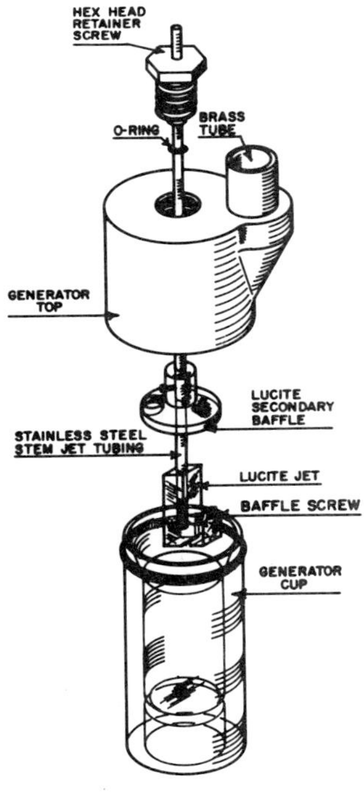

Figure 5. Exploded schematic of the Lovelace Nebulizer which uses a small screw as the jet baffle.

Retec X-70/N Nebulizer

The Retec X-70/N Nebulizer (Retec Development Laboratory, Portland, Oregon) is another small, efficient nebulizer using the Wright jet-baffle principle (Figure 6). The jet orifice is 0.5 mm ID; the secondary orifice is 0.75 mm. Clever design of the components allows all parts to be molded of plastic for easy assembly and disassembly while the operating dimensions are maintained within narrow limits. The jet baffle is a plastic sphere 6.0 mm in diameter. A double orifice system as with the Wright Nebulizer (Figure 4) allows use of a liquid flow tube. By recessing the secondary orifice, the speed of airflow passing the jet baffle increases so that the secondary aerosolization is enhanced. The reservoir volume is normally 10 ml, but larger reservoirs can be used. Burns[52] determined the aerosol output to be 280 ml of liquid per minute with a flow rate of 5 l/min when operated at 20 psig.

Babington Nebulizer

A new class of compressed air nebulizers is possible based upon the Babington principle.[53-55] Liquid is allowed to flow over an orifice, usually a slit at a right angle to the flowing liquid, at a rate which

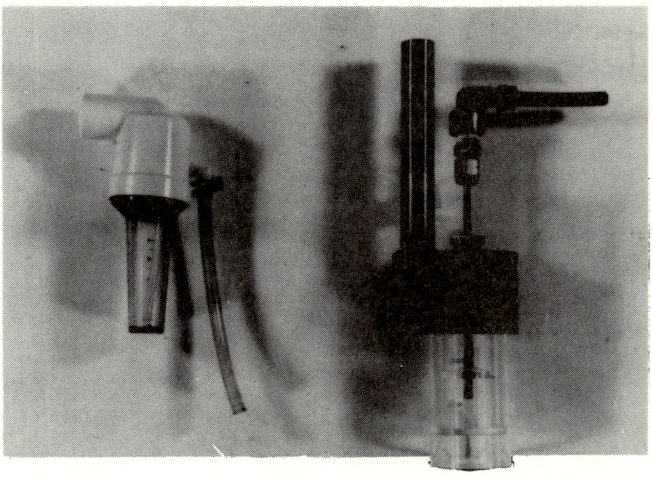

Figure 6. Photograph of the Retec X-70/N Nebulizer (left) and the Lovelace Nebulizer (right). The height of the Retec Nebulizer is 11.5 cm.

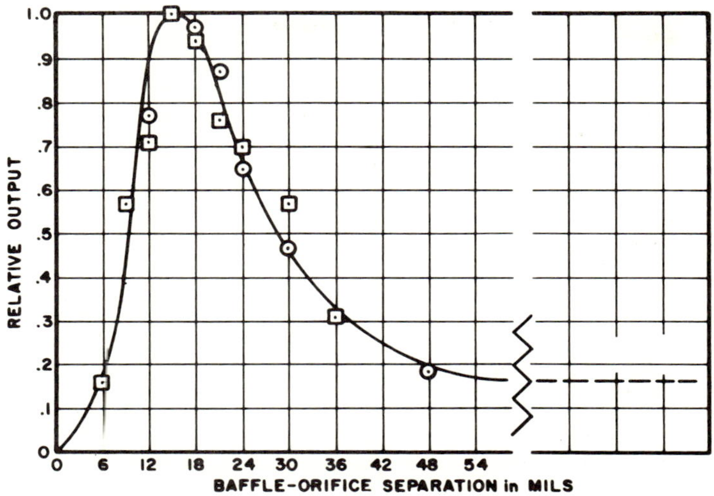

Figure 7. Effect of baffle screw position on output of useful aerosol from the Lovelace Nebulizer showing the Wright jet-baffle effect which provides enhanced output by proper jet-to-baffle spacing. The output at optimum position is about six times the output of the unbaffled jet. The data squares are based on mass output measurements and the data circles are based on light-scattering measurements. (From Mercer et al[9]).

maintains an untorn film over the orifice even as compressed air is forced through the liquid film. Small droplets are generated at the surface of the liquid and carried away in the air stream without the intermediate Rayleigh stream disintegration which is characteristic of conventional nebulizers. The usual design is to have several small slit orifices on the surface of a hollow sphere. The inside of the sphere is pressurized and liquid to be aerosolized is pumped from the reservoir to the outside top of the sphere so that it flows over the surface of the sphere at a volumetric rate sufficient to maintain a closed film over each orifice.

The Hydrosphere Nebulizer (Owens-Illinois, Toledo, Ohio) is a commercial nebulizer using the Babington principle. The jets are slits about 0.2 mm by 1.2 mm cut in two hollow glass spheres 1.5 cm in diameter. Two slits in one sphere and one in the other provide three orifices. The unit is designed for auxiliary air flow on demand to enhance output. It also incorporates jet baffles based on the Wright principle; these are metal spheres 4 mm in diameter. Large auxiliary reservoirs are required with the Hydrosphere Nebulizer. At 20 psig with auxiliary air, the volumetric output rate is about 110 l/min with about 3.3 ml of liquid aerosolized per minute. Of the 3.3 ml of liquid aerosolized per minute about 2.7 ml represents a usable droplet aerosol with VMD equal to about 4 μm and σ_g about 2.3.[56]

Ultrasonic Nebulizers

Ultrasonic nebulizers operate differently from compressed air nebulizers.[57-62] Although a flow of air is used to carry off the aerosol droplets, the air is not involved in the initial formation of these droplets which are formed by ultrasonic vibrations. A sectional view of an experimental ultrasonic generator (designed by Mr. G. J. Newton, Lovelace Foundation, Albuquerque, New Mexico) is shown in Figure 8. A 110-volt, 60-cycle line current is converted to a high-frequency signal (in this model approximately 800 kHz) and transmitted to the transducer A through the shielded cable B. The transducer, a cylindrical piezoelectric device, transforms the high-frequency electrical current to mechanical oscillations. Because of the shape and nature of the crystal, these mechanical vibrations are highly directional and create an intense acoustic field in the coupling fluid C. The energy is carried through

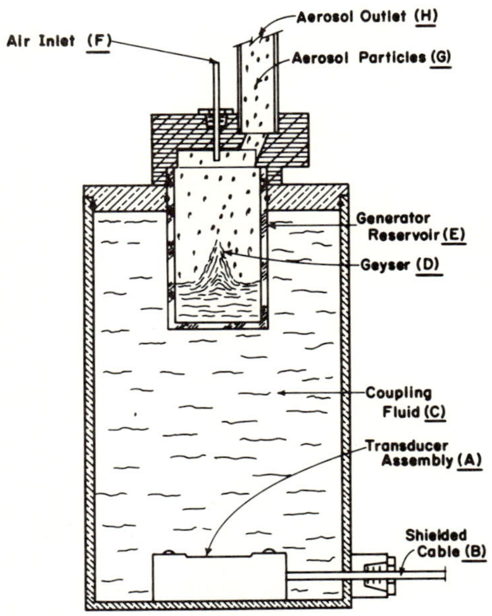

Figure 8. Sectional schematic view of an operating Ultrasonic Aerosol Generator showing transducer assembly (A) receiving power through shielded cable (B) generating an acoustic field in the coupling fluid (C) creating an ultrasonic geyser (D) in the generator reservoir (E) and air entering at (F) carrying away aerosol (G) through the aerosol outlet (H) (from Raabe[8]).

the liquid by the motion of the molecules along the direction of propagation. Intense adiabatic compressions and rarefactions occur with corresponding density and temperature changes. The actual amplitude of the motion is only about 0.2 μm in an 800 kHz field, but the acceleration attained is in excess of 500,000 g. This turbulence creates a pressure gradient along the cylindrical axis of the transducer and results in a water geyser D. The high-frequency turbulence in the geyser produces the aerosol droplets.

The sizes of the droplets formed depend on the frequency of the acoustical field and the physical-chemical character of the liquid. The sizes of the droplets carried by the air stream out of the generator depend upon the rate at which the droplets are carried away from the geyser since coagulation is rapid at the high concentrations of droplets initially formed. For example, with an air flow of 1 l/min, the droplet distribution had a VMD of 10 μm while at a flow of 10 l/min it was only 3 μm.[63]

Ultrasonic nebulizers impart intense energy to the liquids being nebulized which causes heating with an increase in evaporation. Also, materials in solution or suspension can be denatured, degraded or deformed. Ultrasonic nebulizers are also sensitive to the chemical properties of the aqueous solutions or suspension to be nebulized and may malfunction because of this sensitivity. Operating characteristics of several ultrasonic nebulizers have been reported.[41,64-65]

Summary

The operating characteristics with water of several compressed air and ultrasonic nebulizers are summarized in Table 1. The output concentration, A(μl/l), is a measure of aerosolization efficiency. Only selected operating pressures have been described for each compressed air nebulizer, although most may be operated at various pressures with associated variations in operating characteristics. Other conditions being equal, compressed air nebulizers which utilize the Wright jet baffle principle are about 5 times as efficient as those that do not.

AEROSOL GENERATION WITH NEBULIZERS

Fine aerosols of both soluble and insoluble* materials may be produced by nebulization of solutions or suspensions. For example, spherical insoluble particles of plastics may be made from solutions with suitable organic solvents,[66] or soluble forms may be made from aqueous solutions of electrolytes.[67] Changes may be made in the chemical state of the particles formed after solvent evaporation by heating, and in the size distribution by selective collection of a portion of the particles.

* The terms "soluble" and "insoluble" are relative terms especially when used to describe the character of inhaled aerosols.

Table 1

Typical Operating Characteristics of Selected Nebulizers With Reference, Gage Pressure (ΔP), Output Concentration (A), Evaporation Output (W), Volumetric Flow Rate (Q) and Droplet Distribution (VMD and σ_g) Corrected, as Required, to Operation at 76 cm Hg and 21 °C; Values in Parentheses are my Estimates.

Nebulizer	ΔP (psig)	A (μl/l)	W (μl/l)	Q (l/min)	VMD (μm)	σ_g
Dautrebande D-30	10	1.6	9.6	17.9	1.7	1.7
(9)	20	2.3	8.6	25.4	1.4	1.7
	30	2.4	8.2	32.7	1.3	1.7
Lauterbach	10	3.9		(1.7)	3.8	2.0
(9)	20	5.7	(12)	(2.4)	2.4	2.0
	30	5.9		(3.2)	2.4	2.0
Collison	20	7.7	12.7	7.1	(2.0)	(2.0)
(3 jet model)	25	6.7	12.6	8.2	2.0	2.0
(47)	30	5.9	12.6	9.4		
	40	5.0	12.6	11.4		
DeVilbiss #40	10	16	(10)	10.8	4.2	1.8
(9)	15	15.5	8.6	13.5	3.5	1.8
	20	14	7.0	15.8	3.2	1.8
	30	12	7.2	20.5	2.8	1.8
Lovelace	15	27	(10)	1.3		
(Baffle screw set	20	40	10	1.5	5.8	1.8
for optimum operation	30	31	11	1.6	4.7	1.9
at 20 psig)	40	21	9	2.0	3.1	2.2
(63, 10)	50	27	11	2.3	2.6	2.3
Retec X-70/N	20	56	20	5.0	5.1	2.0
(52, 10)	20	53	12	5.4	5.7	1.8
	30	54	11	7.4	3.6	2.0
	40	53	7	8.6	3.7	2.1
	50	49	9	10.1	3.2	2.2
DeVilbiss Ultrasonic	--	150	33	41	6.9	1.6
(setting #4)						
(Somerset, Pa.) (41)						

Both crystalline and amorphous forms of aerosol particles may be produced from aqueous solutions. Often the type of particles produced will depend upon the conditions of drying as well as the chemical nature of the materials. If drying is too rapid low density particles, which are essentially shells, may be formed by the encrustation of the surfaces of the drying droplets.[67] More often, drying is seriously hindered by the hygroscopicity of the solute,[16,17] the presence of immiscible liquids or evaporation inhibitors,[19] the insufficiency of energy available to the droplet, the saturation of the air, and other factors. Adequate drying usually requires mixing the primary droplet aerosol with filtered, dry air. Warming of an aerosol to speed drying, by passing it though a heated tube, may be necessary in some experiments. Reduction of electrostatic charge to Boltzmann equilibrium is usually essential.[33]

When droplets evaporate, the residue particles become the aerosol. Since nebulizers produce droplets of many sizes, the aerosols formed after evaporation are polydisperse with geometric standard deviation about equal to the value for the droplet distribution produced by the nebulizer. The size of a given particle depends upon the solution concentration, c, and the droplet diameter, D_d. These are related for spherical particles by:

$$D^3\rho = D_d{}^3 c \tag{7}$$

with D the geometric diameter of the resultant spherical particle of density, ρ. Since aerosol particles vary in density and shape, it is useful to relate various size particles in terms of their dynamic characteristics. For example, the aerodynamic equivalent diameter for a particle may be described as the diameter of a unit density sphere with the same falling speed as the particle.[18] The aerodynamic equivalent diameters of solid spheres of various densities produced from a 5-μm droplet for various solution concentrations are shown in Figure 9.

Nebulization of an aqueous colloidal suspension forms insoluble particles of the aggregates of the colloidal micelles. This method has the advantage of requiring no organic solvents. If the micelles are small and in high concentration,the resultant particles will be nearly spherical in shape and their size distribution may be predicted by assuming that the suspension behaves as a solution. For example, a colloidal suspension of ferric hydroxide may be aerosolized to produce relatively insoluble and spherical particles of ferric oxide for inhalation experiments.[68-69] Water physically trapped or chemically bonded in such particles may be removed by heating. If the colloidal micelles are comparable in size to the droplets produced by the nebulizer,[70] or if the concentration is small, physical factors and the statistics of the random pickup of micelles by droplets will determine the size distribution

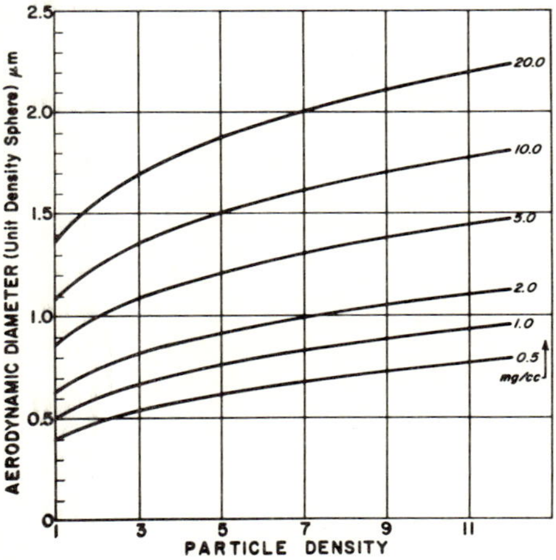

Figure 9. The aerodynamic (equivalent) diameter of solid spherical particles formed by the evaporation of a liquid droplet of 5 μm geometric diameter for various solute concentrations in the droplet (milligrams per cubic centimeter or per milliliter) plotted with respect to the physical density (grams per cubic centimeter) of the resultant solid particles.

of the resultant aerosol. Unfortunately, aerosols produced from colloids sometimes have inherent porosity[71] because of the interstices between micellular components. Aerosol of bacteria and viruses can also be produced by nebulization of suitable suspensions.[72]

Changing the chemical nature of aerosols by heat treatment after they are produced is a useful way of creating particles with desirable chemical and physical characteristics.[73-74] Kanapilly, et al[67,75] describe the generation of spherical particles of insoluble oxides from aqueous solutions with heat treatment of the aerosols with temperatures up to 1400 °C. This procedure involves (a) nebulizing an appropriate aqueous preparation containing the desired metal cation, (b) drying the droplets, (c) passing the aerosol through a high temperature heating column to produce the spherical oxide particles and (d) cooling the aerosol with the addition of diluting air. Figure 10 shows a ZrO_2 aerosol produced by nebulizing zirconium oxalate with heat treatment at 1150 °C. Thermal degradation of aerosols to produce desired chemical forms has also been used by Horstman, et al[77] and by Arvik and Zimdahl.[78] Another example of aerosol alteration is the production of spherical aluminosilicate particles with entrapped radionuclides by heat fusion of clay aerosols.[76] This method involves (a) ion exchange of the desired radionuclide cation into clay in aqueous suspension and washing away of the unexchanged fraction, (b) nebulization of the clay suspension yielding a clay aerosol as shown in Figure 11 and (c) heat fusion of clay aerosol removing water and forming an aerosol of smooth solid spheres of density 2.3 g/cm^3 as shown in Figure 12. Freeze-drying of droplets has also been reported as a useful technique for preparing certain types of aerosols.[79]

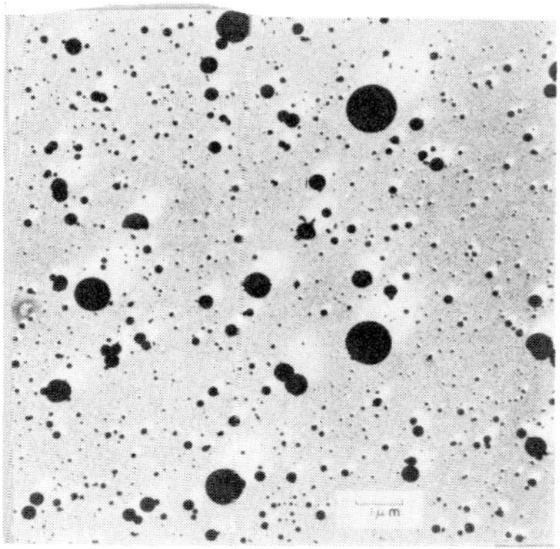

Figure 10. Electronmicrograph of an aerosol sample of zirconium dioxide (shadowed with chromium vapor) produced by nebulization of an aqueous zirconium oxalate mixture and degraded at 1100 °C. (from Kanapilly et al[75]).

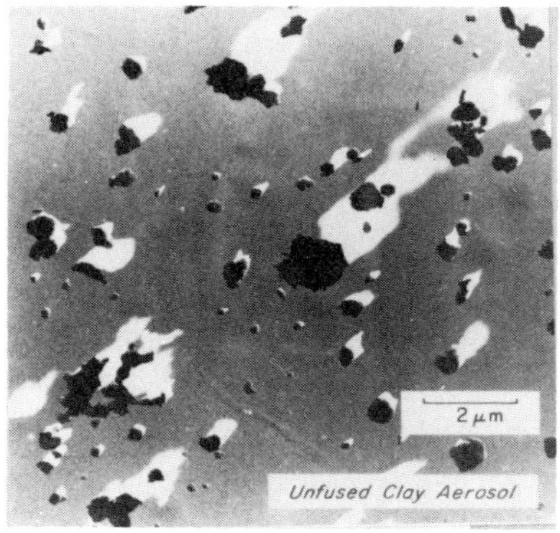

Figure 11. Electronmicrograph of an aerosol sample of montmorillonite clay (shadowed with chromium vapor) generated by nebulization of an aqueous clay suspension.

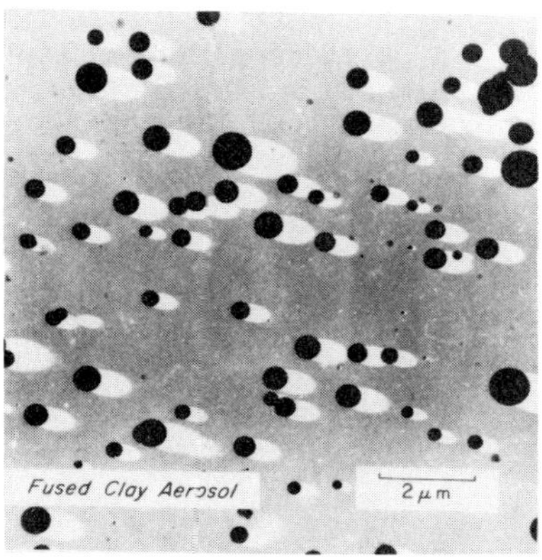

Figure 12. Electronmicrograph of a sample of an aerosol of aluminosilicate spheres (shadowed with chromium vapor) generated by heat fusion at 1150 °C of the clay aerosol shown in Figure 11. (from Raabe et al[76]).

SPRAY CANS

Commercial "aerosol" containers operate on a very different principle than nebulizers.[80-81] They use various mixtures of the liquid to be atomized with a suitable volatile liquid (usually a fluorinated hydrocarbon such as dichloro-difluoro-methane (freon)). Numerous formulations have been tested to evaluate the applicability to different purposes.[82] The pressure in the sealed container caused by the volatile liquid forces the fluid mixture through a feed tube to the nozzle orifice when the nozzle is depressed (Figure 13). The rapid evaporation of the volatile liquid causes numerous bubbles to form in the liquid mixture as it approaches the orifice and shatters the released liquid stream into droplets which usually have a broad range of sizes, often up to 100 μm in diameter. However, the aerosols produced by these pressure containers can be widely varied by orifice nozzle design and chemical formulation.[83]

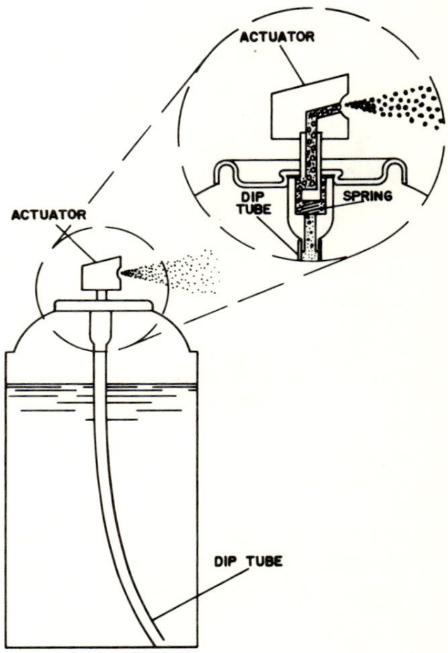

Figure 13. Schematic illustration of the basic components of a spray can which uses liquid fluorinated hydrocarbons or other hydrocarbons as propellant.

A major portion of the aerosols produced by spray cans may, in many cases, be smaller than 3.5 μm in aerodynamic equivalent size, and, therefore, qualify as fine particles. Spray cans have been in use to deliver respirable aerosols of drugs.[84] They also may produce respirable aerosols of fine particles as undesirable side-products. Vos and Thompson determined an average of 57% by weight of the nonvolatile aerosols produced by commercial deodorant spray cans were in the fine particle range.[85] In another study, they observed mass fractions in the fine particle range of 6.6% for a hair spray, 11.3% for a frying pan spray, 9.3% for disinfectants, 7.5% for an insect killer, 31.4% for an air freshener and 3.2% for another deodorant.[86]

Some investigators may object to the use of the term aerosol in describing spray products.[87] Many books and articles that contain the word aerosol in their titles are devoid of aerosol information in the scientific sense. The word is certainly misused when applied to cans that dispense such commodities as shaving cream, tooth paste or whipped potatoes.

DRY PARTICLE AEROSOLIZATION

Dust generators such as the DeVilbiss dust blower #175 (Figure 14) (DeVilbiss Co., Somerset, Pa.) are designed to turbulently suspend dry dusts and carry the resultant aerosol into an air stream. The DeVilbiss unit uses turbulent air flow in the glass dust container to stir the dust. Many methods may be used to generate dry dust: however, all involve two factors: getting the dust in motion by shaking, grinding, stirring, placing in fluidized bed, etc., and moving the proper amount into the desired air volume. Some methods may not yield reproducible or unvariable concentrations or size distributions. One unusual approach to dust generation involves the loading of dry powder into a capsule and propelling it with an air gun through a sharp cutting device into an aerosol chamber.[88] This method has even been used for animal exposures.[89-90]

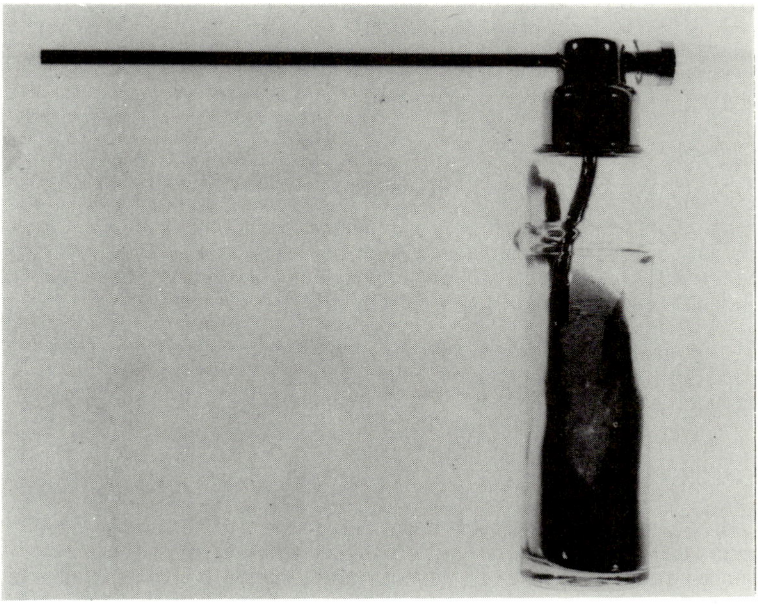

Figure 14. The DeVilbiss No. 175 dry-dust blower used to aerosolize powders.

One of the most popular dust generating devices has been the B. M. Wright Dust Feed[91] which allows dust to be ground off of a packed plug. A timing and gearing mechanism provides a constant rate for this dust feeding device. Another somewhat similar device, providing a narrower size range of particles, has been described by Dimmick.[92] Variations on dry dust feed systems have been described by Ebens and Vos[93] and by Fuchs and Murashkevich.[94]

An elaboration of a kitchen blender for producing aerosols of dry powders and fibers has been described by Drew and Laskin.[95] In this system a preground dust is mixed in a container having a four-bladed, high-speed fan at the bottom. The air-dust mixture is aspirated into a separate baffling chamber mounted at the top of the fluidizing container. From the baffling chamber, the dry aerosol is conducted to its intended destination. Drew and Laskin showed the fluidizing dust chamber to be of practical value in inhalation studies with fiberglass dust with concentrations as high as 200 mg/m^3 and coefficient of variation of concentration only 6.7% for 12 samples taken at half-hour intervals.

Fluidized beds of solid particles such as bronze spheres kept in motion by upward air flow have been used for aerosolization of dry dust.[96-97] Dusts are injected into the fluidized bed which serves to break up the dust particles and readily releases them as an aerosol.

PHOTOCHEMICAL AND RADIOLYTIC AEROSOLS

Chemical reaction can occur at the molecular level in air under the influence of light or other radiation to produce a variety of aerosols of ultrafine and fine particles. The production of condensation nuclei by the irradiation of filtered nuclei-free air by X-rays, alpha particles and beta particles has been described by Megaw and Wiffen,[1] Madelaine[3] and Mercer and Tillery.[99] Aerosols produced by alpha, beta and gamma (X-ray) radiation are usually called radiolytic aerosols.

The topic of photochemical production of aerosols of fine particles is beyond the scope of this report. Needless to say, it has and will continue to receive considerable attention with respect to its role in air pollution.

MISCELLANEOUS

Useful aerosols of dry particles of metal and metal oxides have also been produced with electrically heated[100-104] or exploded wires.[105-107] These techniques have some disadvantages because of the broad size distributions of the resulting particles and because of the tendency of particles to coalesce. There are applications, however, for this type of aerosol since they tend to be in the ultrafine range (less than 0.1 μm). It is possible with a technique such as Couchman's[100] wire heating method to produce spherical particles of many different metals or their oxides. Aerosols of ultrafine particles have also been produced by arc vaporization.[108-110]

Spurny and Lodge have described a method for producing radiolabeled aerosols of ultrafine particles by vaporization of radiolabeled metals and metallic compounds.[111] Lasers have also been employed to provide heat for vaporization and combustion of metals under conditions which lead to condensation aerosols of ultrafine particles or chain aggregates.[112-113] Sinclair and Henchliffe used an induction furnace to vaporize silver to yield ultrafine spherical particles of metallic silver.[114] Many other vaporization and combustion methods can be designed on the basis of these techniques.[115-118]

MONODISPERSE AEROSOLS

GENERAL

The generation of monodisperse aerosols with particles of similar size and physical-chemical characteristics is desirable in many experimental applications since the effects of particle size can be evaluated. Many of the methods for producing monodisperse aerosols have been reviewed by Fuchs and Sutugin,[7] Raabe,[8] Liu[119] and Mercer.[120] These methods include the growth of uniform aerosol particles or droplets by controlled condensation, the formation of uniform droplets (which may dry to solid particles) by controlled dispersion of liquids electrostatically, by dispersion of liquid jets with periodic vibration, or with a spinning disk or top. Another popular method for producing monodisperse aerosols is the nebulization of a suspension of monodisperse particles, but involves some inherent difficulties. It is not a simple matter to produce suitable monodisperse aerosols for particular applications. The variety of physical-chemical types of such aerosols is still somewhat limited and the specialized equipment usually requires careful operation.

The definition for monodispersity suggested by Fuchs and Sutugin[7] provides a guide for evaluation of monodispersity of particle size distributions. If the coefficient of variation (ratio of standard deviations to the mean) of the distribution of sizes is less than 0.2 (20%), the aerosol may be satisfactorily described as having "practical monodispersity". For a log-normal distribution, this is about equivalent to a $\sigma_g < 1.2$. In general this criterion is not very stringent, and many investigators may prefer to achieve greater size uniformity. However, even with very effective devices for producing uniform particles or droplets, it is sometimes necessary to tolerate a small fraction of odd-sized particles or doublets (two primary particles which have coalesced).

NEBULIZATION OF MONODISPERSE SUSPENSIONS

General

The nebulization of suspensions of uniform particles which have been chemically grown or separated from polydisperse particles has been a simple and useful means of generating monodisperse aerosols. Suspensions of polystyrene latex spheres of fairly uniform size grown by emulsion polymerization as developed by Bradford and Vanderhoff[121-122] have been commonly used for this purpose.[123-126] Other types of monodisperse hydrosols have also been made.[127-128] Polystyrene latex spheres[129-132] and others[133] can be labeled with radionuclides.

It is not possible at reasonable dilutions to guarantee that only individual particles will be aerosolized and droplets containing more than one of the suspended particles will become undesirable aggregates upon evaporation. Raabe[126] has calculated the dilution, y, for polystyrene latex suspensions required to generate an aerosol with singlet ratio R defined as the ratio of droplets with but one monodisperse particle to all droplets that contain one or more monodisperse particle:

$$y = \frac{F(\mathrm{VMD})^3 e^{4.5(\ln \sigma_g)^2} \left[1 - 0.5 e^{(\ln \sigma_g)^2}\right]}{(1-R) D^3} \tag{8}$$

for $R > 0.9$ and $\sigma_g < 2.1$ with F the fraction by volume of the particles in the original stock suspension, D the diameter of the monodisperse spheres and with VMD and σ_g the volume median diameter and geometric standard deviation of the droplet distribution from the nebulizer.

Equation 8 can be used to calculate the maximum number concentration, MNC* of monodisperse particles of any type in a liquid suspension to be nebulized so that the singlet ratio is not less than R:

$$\mathrm{MNC}^* = \frac{6(1-R)\, e^{-4.5(\ln \sigma_g)^2}}{\pi(\mathrm{VMD})^3 \left[1 - 0.5\, e^{(\ln \sigma_g)^2}\right]} \tag{9}$$

When VMD is the nebulizer droplet median diameter in centimeter units, MNC* is number of monodisperse particles per milliliter of the suspension.

Raabe also showed that more than 90% and probably more than 99% of the droplets produced at these low suspension concentrations contain no spheres when the singlet ratio, R, is taken at 0.95. These empty droplets dry to form ultrafine particles of residue of the impurities in the liquid; these secondary aerosols can be undesirable since they are considerably more numerous than the monodisperse particles themselves.[126]

Polystyrene and Polyvinyl Toluene Latex

Monodisperse suspensions of polystyrene and polyvinyl toluene latex spheres (Dow Diagnostics, Indianapolis, Indiana) have been available in sizes from 0.088 μm to 3.5 μm and have been widely used for preparing monodisperse aerosols.[123-126] Their monodispersity of size is excellent with $\sigma_g < 1.05$. However, many problems have been encountered in their use.[134-136] The density of polystyrene latex has been measured to average 1.05 g/cm^3 and the density of polyvinyl toluene latex at 1.027 g/cm^3.[137] These particles are grown by emulsion polymerization and are stabilized in aqueous suspensions with an anionic surfactant having an active sulfonate radical. These particles carry a negative charge in aqueous suspensions, which contributes to their stability. The compositions of some typical latex suspensions are given in Table 2. Solids normally represent 10% of the aqueous suspension as supplied by the manufacturer. Using equation 8, the dilutions of 10% by volume stock suspensions of monodisperse spheres required to generate 95% single particles are shown in Figure 15 for various log-normal droplet distributions and sphere diameters.

Table 2

Relative Concentrations of Solids in Latex Suspensions[138]

Reference Number	Particle Diameter	Composition of Solids,%		
		Polymer	Emulsifier	Inorganics
LS-040-A	0.088	92.59	5.56	1.85
LS-052-A	0.126	97.09	2.43	0.48
LS-057-A	0.264	98.23	1.13	0.64
LS-63-A	0.557	98.78	0.26	0.95
LS-449-E	0.796	99.04	0.25	0.70
LS-464-E	1.305	99.15	0.29	0.56

If the 0.264 μm particles, (batch LS-057-A in Table 2) were generated with a Lovelace nebulizer (VMD = 5.8 μm, σ_g = 1.8) with no dilution, the 0.177% emulsifier and inorganics in the suspension would increase the diameter of a single particle which happens to be aerosolized in a 5.8 μm droplet from 0.264 μm to 0.714 μm. Proper dilution to 30,000 to 1 to generate an aerosol with 95% singles (Equation 8 and Figure 15)

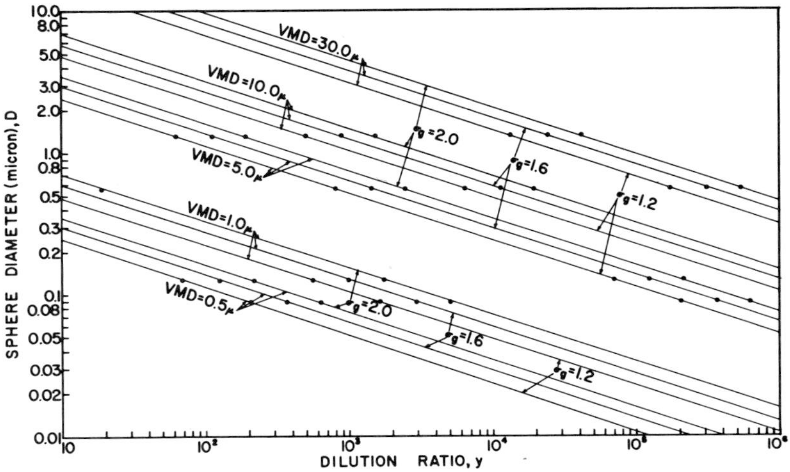

Figure 15. The dilution ratio, y, required to generate a singlet ratio (R) equal to 0.95 vs. the sphere diameter from stock of 10% spheres by volume for various values of the volume median diameter (VMD) and geometric standard deviation (σ_g) of the droplet distribution. The lines were derived by the empirical equations and the points calculated numerically by a theoretical equation (from Raabe[126]).

makes a 0.264 μm particle aerosolized in a 5.8 μm droplet yield a latex particle essentially unchanged in diameter. For the larger particles, it is convenient to separate the particles from the liquid by centrifugation and resuspend them in purified and filtered water before use.

Monodisperse aerosols of polystyrene and polyvinyl toluene particles are highly charged when generated by nebulization.[139-140] A reduction of charge to Boltzmann equilibrium with a suitable aerosol neutralizer is absolutely essential in studies using these aerosols.[33] The presence of a background aerosol produced from the numerous empty droplets must be given appropriate consideration as well.[134-136]

Monodisperse Particles Separated by Centrifugation

Another method for obtaining a suspension of monodisperse particles for nebulization is to generate a suitable polydisperse aerosol of insoluble particles and aerodynamically separate the particles into monodisperse size groups using a spiral-duct aerosol centrifuge. Kotrappa and Moss[141] introduced this technique and Kotrappa et al[142] used it to prepare monodisperse particles of $^{239}PuO_2$. Raabe et al[143] produced $^{238}PuO_2$ with this method and described an elaborate system for separating monodisperse particles of PuO_2 which simultaneously used four Lovelace Aerosol Particle Separators.[144] Particles are collected on stainless-steel foils which are cut into small segments for suspension in water. Dilute solutions of various surface active agents and dilute NH_4OH have been used to stabilize aqueous suspensions. Samples of the polydisperse primary aerosol and three monodisperse size groups of PuO_2 are shown in Figure 16. A monodisperse aerosol of PuO_2 generated by nebulization of an aqueous suspension of monodisperse particles is shown in Figure 17. Note the presence of some doublets and the ultrafine background aerosols; cascade impactor samples demonstrate that these background particles do not contain plutonium and that the alpha activity distribution is in fact monodisperse with regard to aerodynamic size.

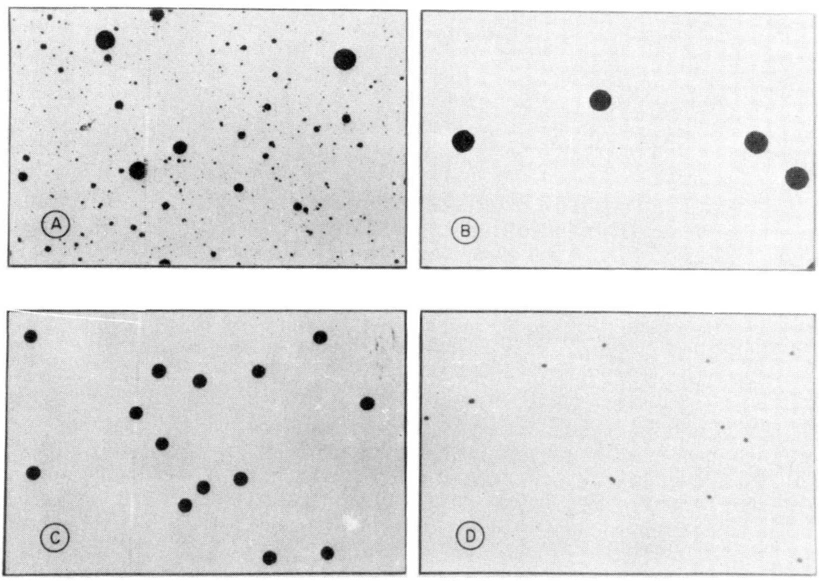

Figure 16. Composite of electronmicrographs of four aerosol samples of PuO_2 particles: (A) primary polydisperse aerosol of PuO_2 collected with an electrostatic precipitator and (B), (C) and (D) monodisperse particles collected at three positions on the foil of a spiral-duct centrifuge (Lovelace Aerosol Particle Separator)(from Raabe et al[143]).

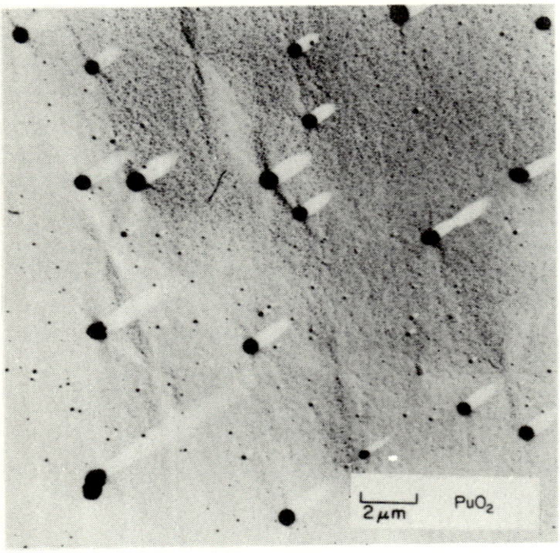

Figure 17. Electronmicrograph of a sample of an aerosol of PuO_2 (shadowed with chromium vapor) generated by nebulization of an aqueous suspension of separated monodisperse particles as shown in Figure 16. Note the presence of some doublets and the background aerosol of ultrafine particles which are not PuO_2 but are formed from trace impurities in "empty" water droplets (from Raabe et al[145]).

CONTROLLED CONDENSATION

The isothermal growth of droplets under supersaturated conditions by vapor diffusion and condensation causes the droplet surface area to increase linearly with time as given by Wilson and LaMer.[146]

$$r^2 = r_o^2 + bt \tag{10}$$

with r the droplet radius at the time t, r_o the initial radius and b may be maintained nearly constant under controlled conditions. Hence, if droplets are grown by controlled condensation onto two small nuclei of widely different size, say radii 0.02 μm and 0.1 μm, with $bt \simeq 0.164$ μm^2 the droplets formed will be almost of identical size with radii of 0.40 μm and 0.41 μm, respectively. Sinclair and LaMer[147] used this principle to design an apparatus for generating monodisperse droplet aerosols of such materials as oleic acid, stearic acid, lubricating oils, menthol, dibutyl phthalate, dioctyl phthalate and tri-o-cresyl phosphate. Many researchers have used and improved on the basic Sinclair-LaMer generator[148-159] and the basic technique has also been used to produce solid aerosols by sublimation.[160-161] Heyder et al,[162] Muir and Davies[163] and Davies et al[164] have used this method to produce monodisperse aerosols of di-2-ethylhexyl sebacate for inhalation studies with humans. Giacomelli-Maltoni et al have made monodisperse carnauba wax particles for human deposition studies.[165]

GENERATION OF UNIFORM DROPLETS

General

Monodisperse aerosols of both soluble and insoluble forms can be produced from solutions or suspensions if small uniform droplets can be dispersed. These droplets can then be dried and, if desired, altered in character by heat treatment to yield the required aerosols. Many devices have been designed to produce small uniform droplets. Two of these, the vibrating reed[166] and vibrating thread,[167] have created only moderate interest because of the low concentrations available. Liu and associates have developed a method for separating a chosen monodisperse aerosol fraction of a polydisperse aerosol using an electrostatic mobility analyzer.[168-170]

Spinning Disk

The most popular device for dispersing uniform droplets has been the spinning disk aerosol generator first described by Walton and Prewett.[171] They observed that primary droplets thrown off at the perimeter of a spinning disk were of uniform size. Liquid is fed to the center of the disk and flows to the edge by centrifugal forces where it accumulates until the centrifugal force, which increases with increasing liquid at the edge, overcomes the surface tension and disperses the liquid. This dispersion also produces some secondary fragments (satellite droplets) which are easily separated dynamically from the larger primary droplets. This is usually done by a separate flow of air near the disk, into which the satellites move but beyond which the larger primary droplets are thrown. A spinning disk generator, shown schematically in Figure 18, can attain disk speeds up to 100,000 rpm. The drop diameter, D_d, produced by a spinning disk is given theoretically by:

$$D_d = K\sqrt{\left[\frac{\tau}{\rho\omega^2 d}\right]} \tag{11}$$

with τ the surface tension, ρ the fluid density, d the disk diameter, ω the speed of angular rotation of the disk and K a constant given theoretically by $\sqrt{12}$ which varies in practice from 2 to 7. Many investigators have developed and successfully used spinning disk monodisperse aerosol generators for a variety of experimental applications including aerosol studies and inhalation experiments.[172-179] The spinning top generator introduced by K. R. May,[180-181] works on essentially the same principle as the Walton and Prewitt spinning disk. It, too, has been successfully used by many investigators.[182-194]

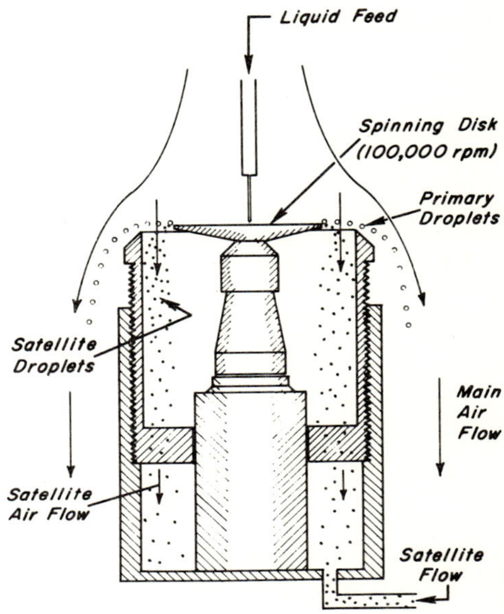

Figure 18. Schematic drawing of a Spinning Disk Generator used to produce monodisperse aerosols of both soluble and insoluble forms from solutions of suspensions. Air flow into the satellite collector is adjusted so that the inertia of the primary particles allows them to enter the main air flow.

Although the spinning disk and top have probably been the most generally successful methods for producing monodisperse aerosols, the production of particles smaller than 0.5 μm geometric diameter is not practical because droplets produced with these devices are generally in the 20 to 30 μm diameter range and even purified water has trace impurities which may yield resultant particles as big as 0.5 μm. Monodisperse aerosols produced with spinning disk and spinning top devices are usually near or larger than 3.5 μm in aerodynamic equivalent diameter. As with other methods, the resulting aerosols are highly charged and must be passed through a suitable neutralizer soon after being produced.

Electrostatic Dispersion of Liquids

Electrostatic dispersion[195] provides another method for generating uniform droplets for certain solutions. Vonnegut and Neubauer[196] and others[197-198] have studied this approach using a filament of electrically charged liquid released from a small capilllary. The filament breaks up into fragments, forms a conical spray and further breaks up into small droplets of almost uniform charge and size. These droplets must be discharged soon after generation and are difficult to control because of electrostatic behavior.

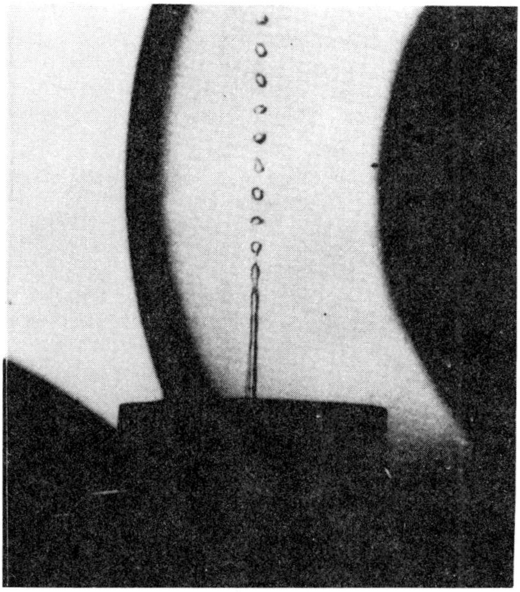

Figure 19. Early observation of the principle of periodic dispersion of a liquid jet into uniform droplets as photographed by Castleman[35] in 1930 using a spark illumination technique (from Castleman[35]).

Periodic Dispersion of Liquid Jets

If a thin liquid stream is emitted from an orifice under pressure, this stream is by nature unstable and will soon disintegrate into droplets by the action of any external forces as described by Rayleigh.[34] The collapse of such a stream into very uniform droplets is easily attainable with the application to the stream of a periodic vibration of suitable amplitude and frequency. In 1931 Castleman[35] demonstrated this principle experimentally using stroboscopic and spark pictures (Figure 19). Castleman described the principle that the collapse of a cylindrical stream of liquid can "be made so beautifully regular that it may be readily viewed by stroboscopic means if the chamber from which the jet issues be influenced by a periodic vibration of proper amplitude and frequency".[35] Although sometimes called a vibrating orifice generator, this description is too specific since the disintegration occurs in the same manner irrespective of where the oscillatory vibrations are imparted to the liquid, and may even be accomplished at a distant reservoir.[199] Although the rate of disruption of a liquid stream depends upon both the viscosity and surface tension of the liquid, the characteristics of the droplets formed depend only upon the diameter of the stream and the frequency of the vibrations. This is true because the minimum length of such a stream that can be separated into a separate droplet is equal to the circumference of the stream; a shorter length cannot be separated since the surface tension along the circumference of the stream will tend to elongate a shorter segment and prevent separation. Hence the smallest segment of a stream which can be separated has a length, L_{min}, given by:[200-201]

$$L_{min} = \pi D_g \tag{12}$$

with D_g the diameter of the stream. For most liquids, it is satisfactory to assume that the stream diameter is equivalent to the orifice diameter. If the liquid is emitted with a flow rate, Q, and the frequency of periodic disturbance is f, then a spherical droplet formed of each separated cylindrical segment will have a diameter D_d given by:

$$D_d = \sqrt[3]{\frac{6Q}{\pi f}} \tag{13}$$

The frequency required to produce the minimum droplet size is given by a maximum frequency, f_{max}, for disruption of the stream:

$$f_{max} = \frac{4Q}{\pi^2 D_g{}^3} \tag{14}$$

Any frequency of disturbance less than f_{max} will disrupt the stream into uniform droplets. However, Castleman[35] showed that the optimum disturbance cuts off a length of stream given by:

$$L_{opt} = \pi D_g \sqrt{2} \tag{15}$$

and therefore the optimum frequency of disruption is:

$$f_{opt} = \frac{4Q}{\pi^2 D_g{}^2 \sqrt{2}} \tag{16}$$

and the minimum diameter droplet is given by:

$$D_{min} = D_g \sqrt[3]{\frac{3\pi}{2}} \approx 1.77 D_g \tag{17}$$

and the optimum diameter is given by:

$$D_{opt} = D_g \sqrt[3]{\left[\frac{3\pi\sqrt{2}}{2}\right]} \approx 1.88\,D_g \tag{18}$$

Hence the minimum or optimum diameters of droplets that can be generated in practice depend only on the stream diameter (orifice diameter). Measurement of volumetric flow rate from the orifice, Q, can be made for any combination of orifice diameter and reservoir pressure, so that droplet sizes can be calculated.

Since Castleman's work was reported in 1931, this principle has been rediscovered and studied by many investigators.[200-212] Fulwyler [202] reported the use of a device of this type as a cell separator. Fulwyler and Raabe,[200] Raabe[8] and Raabe and Newton[201] described the application of the Fulwyler droplet generator for producing monodisperse aerosols. In the Fulwyler device an audio oscillator and amplifier provide a high frequency power signal which is converted to mechanical vibrations by an ultrasonic transducer linked by a coupling rod to a small liquid reservoir. The reservoir is pressurized (≃ 30 psig) to emit the liquid through a small orifice (≃ 10 μm) as a fine stream. This stream is uniformly disrupted by the ultrasonic vibrations into droplets that vary less than 1% in volume. To produce an aerosol of these uniform droplets, they must be disturbed from their forward direction of motion so that their forward speed is reduced without coalescence. This can be accomplished by concentric mixing or cross flow of air streams.

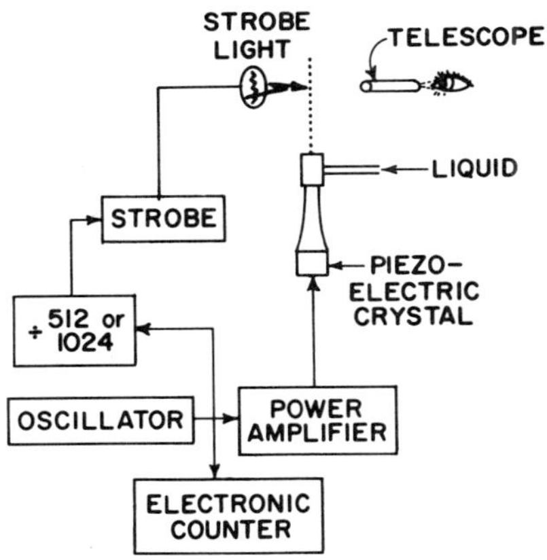

Figure 20. Block diagram of equipment used for operation of the Fulwyler Droplet Generator for producing monodisperse droplets by periodic dispersion of aliquid jet (from Fulwyler and Raabe[200]).

A block diagram of the operation of the Fulwyler droplet generator is shown in Figure 20 and the experimental arrangement used by Raabe and Fulwyler[200] and Raabe and Newton[201] for dispersing the aerosol is shown in Figure 21. Monodisperse particles of ZrO_2 produced from monodisperse droplets containing Zr oxalate and heat treated at 1150 °C are shown in Figure 22.[76]

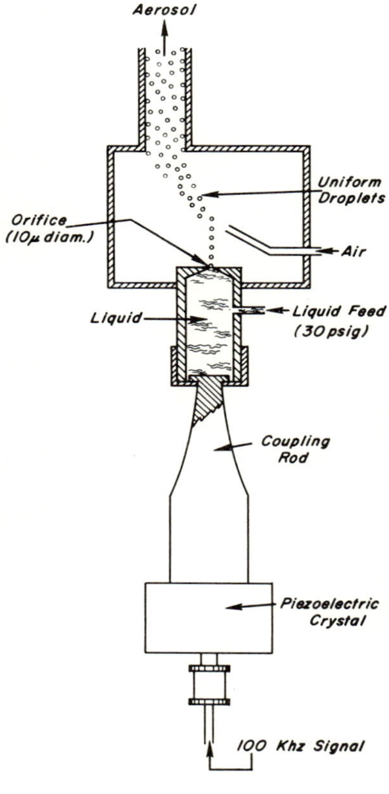

Figure 21. Schematic drawing of the Fulwyler Droplet Generator used to produce monodisperse droplets to yield monodisperse aerosols of both soluble and insoluble materials. Air is directed at the stream of droplets to perturb them out of alignment and create the droplet cloud (from Fulwyler and Raabe[200]).

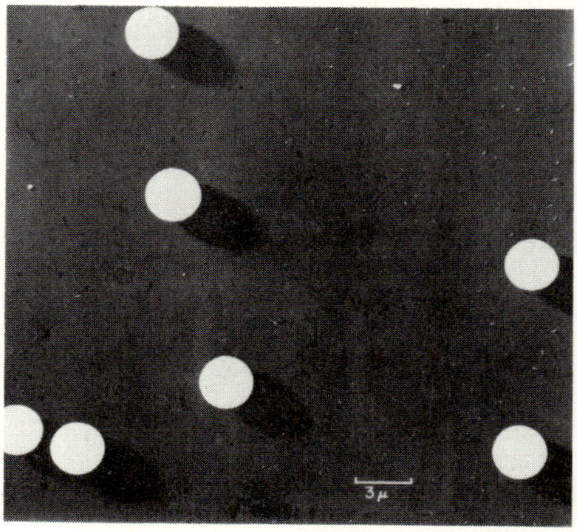

Figure 22. Uniform spherical aerosol particles of ZrO_2 produced with the Fulwyler Droplet Generator from solutions of zirconium oxalate degraded to the oxide with a heating column by the method of Kanapilly et al[67] (from Fulwyler and Raabe[200]).

Generation of monodisperse aerosols with similar devices using electro-constructive elements around the orifice rather than an ultrasonic transducer has been described by Strom[204] and by Berglund and Liu.[211]

One operational problem that exists with the periodic dispersion method of producing aerosols of fine particles is that the small orifices required an subject to clogging if small particles are present in the liquid; membrane filtration of the liquid is essential but doesn't com-Since the minimum practical orifice diameter is in the range 8 μm to 15 μm, the smallest droplets that can be produced by periodic dispersion of liquid jets are in the range 14 μm to 27 μm, which is about the same range of droplet sizes as produced with spinning disks and spinning tops. Hence, as with the spinning disk, it is difficult to produce solid particles smaller than 0.5 μm geometric diameter with periodic dispersion devices, and most useful aerosols will be at the high end and above the upper limit of the definition of fine particles.

ACKNOWLEDGEMENT

The author is indebted to Dr. Brian Mokler and Mr. George Newton for beneficial advice, to Mr. Emerson Goff and Mr. Michael Rios for preparation of illustrations, to Mr. Fred Rupprecht for editorial and composition assistance, to Mr. David Velasquez for proofreading and to Mrs. Judith Miller for typing the manuscript. This report was prepared under U. S. Energy Research and Development Administration Contract E(29-2)-1013.

REFERENCES

1. Megaw, W. J. and R. D. Wiffen. The Generation of Condensation Nuclei by Ionising Radiation. Geofis Pura Appl 50:118-128, 1961.

2. Renzetti, N. A. and G. J. Doyle. Photochemical Aerosol Formation in Sulfur Dioxide-Hydrocarbon Systems. Intern J Air Water Pollution. 2:327-345, 1960.

3. Madelaine, G. J. Formation et e´volution des aerosols dans l'air filtre´ et dans l'air naturel Action de la radioactivite´, Doctoral Thesis, Commissariat a l'Energie Atomique (France) CEA-R-3614, 1968.

4. Bricard, J., F. Billard and G. Madelaine. Formation and Evolution of Nuclei of Condensation that Appear in Air Initially Free of Aerosols. J Geophys Res. 73:4487-4496, 1968.

5. Knott, M. J. and M. Malanchuk. Analysis of Foreign Aerosol Produced in NO_2-Rich Atmospheres of Animal Exposure Chambers. Am Ind Hyg Assoc J. 30:147-152, 1969.

6. Pfefferkorn, G. Photochemische Bildung Schwerflüchtiger Tröpfchen aus organischen Dämpfen in Luft. Staub Reinhaltung Luft. 27:138-140, 1967.

7. Fuchs, N. A. and A. G. Sutugin. Generation and Use of Monodisperse Aerosols. In: Aerosol Science, Davies, C. N. (ed.). New York, Academic Press, Inc., 1966, pp. 1-30.

8. Raabe, Otto G. Generation and Characterization of Aerosols. Inhalation Carcinogenesis, pp. 123-172, U.S.A.E.C. Division of Technical Information, Oak Ridge, Tennessee, April 1970.

9. Mercer, T. T., M. I. Tillery and H. Y. Chow. Operating Characteristics of Some Compressed Air Nebulizers. Am Ind Hyg Assoc J. 29:66-78, 1968.

10. Raabe, O. G. Operating Characteristics of Two Compressed Air Nebulizers Used in Inhalation Experiments. Fission Product Inhalation Program Annual Report, 1971-1972, LF-45, Lovelace Foundation, Albuquerque, New Mexico, pp. 1-6, November 1972.

11. Liu, Benjamin Y. H. and K. W. Lee. An Aerosol Generator of High Stability. Amer Ind Hyg Assoc J. In press, 1976.

12. Berke, Harry L. and T. E. Hull. An Improved Aerosol Generator. Amer Ind Hyg Assoc J. 36:43-48, 1975.

13. Mugele, R. A. and H. D. Evans. Droplet Size Distribution in Sprays. Ind Eng Chem. 43:1317-1324, 1951.

14. LaMer, V. K. and R. Gruen. A Direct Test of Kelvin's Equation Connecting Vapour Pressure and Radius of Curvature. Trans Faraday Soc. 48:410-415, 1952.

15. Straubel, Harold. Condensation, Evaporation, Crystallization and Charge Changes of Solution Droplets. Staub Reinhaltung der Luft in English. 33:179-182, April 1973.

16. Orr, C., F. K. Hurt and W. J. Corbett. Aerosol Size and Relative Humidity. J Colloid Sci. 13:472-482, 1958.

17. Pilat, M. J. and R. J. Charlson. Theoretical and Optical Studies of Humidity Effects on the Size Distribution of Hygroscopic Aerosol. J Rech Atmospheriques. 2:165-170, 1966.

18. Morrow, P. E. (Chairman, Task Group on Lung Dynamics). Deposition and Retention Models for Internal Dosimetry of the Human Respiratory Tract. Health Phys. 12:173-208, 1966.

19. Snead, C. C. and J. T. Zung. The Effects of Insoluble Films Upon the Evaporation Kinetic of Liquid Droplets. J Colloid Interface Sci. 27:25-31, 1968.

20. Anshus, Byron E. The Effect of Surfactants on the Breakup of Cylinders and Jets. J Colloid and Interface Sci. 43:113-121, April 1973.

21. Nelson, P. A. and W. F. Stevens. Size Distribution of Droplets from Centrifugal Spray Nozzles. A I Ch E Journal. 7:80-86, 1961.

22. Lewis, H. C. D. G. Edwards, M. J. Goglia, R. I. Rice and L. W. Smith. Atomization of Liquids in High Velocity Gas Streams. Ind Eng Chem. 40:67-74, 1948.

23. Mercer, T. T., R. F. Goddard and R. L. Flores. Output Characteristics of Several Commercial Nebulizers. Ann Allergy. 23:314-326, 1965.

24. Hatch, T. and S. P. Choate. Statistical Description of the Size Properties of Non-Uniform Particulate Substances. J Franklin Inst. 207:369-387, 1929.

25. Raabe, Otto G. Particle Size Analysis Utilizing Grouped Data and the Log-Normal Distribution. Aerosol Sci 2:289-303, 1971.

26. Mercer, T. T. Aerosol Production and Characterization - Some Considerations for Improving Correlation of Field and Laboratory Derived Data. Health Phys. 10:873-887, 1964.

27. Matteson, Michael J. The Separation of Charge at the Gas-Liquid Interface by Dispersion of Various Electrolyte Solutions. J Colloid and Interface Sci. 37:879-890, December 1970.

28. Rayleigh, Lord. On the Equilibrium of Liquid Conducting Masses Charged with Electricity. Phil Mag. 14:184-186, 1882.

29. Whitby, K. T. and B. Y. H. Liu. The Electrical Behavior of Aerosols. In: Aerosol Science, Davies, C. N. (ed.). New York, Academic Press, Inc., 1966, pp. 59-86.

30. Doyle, Arnold, D. Read Moffett and Bernard Vonnegut. Behavior of Evaporating Electrically Charged Droplets. J Colloid Sci. 19:136-143, 1964.

31. Whitby, K. T. Generator for Producing High Concentrations of Small Ions. Rev Sci Instr. 32:361-355, 1961.

32. Soong, A-L. The Charge on Latex Particles Aerosolized from Suspensions and Their Neutralization in a Tritium De-Ionizer. M. S. Thesis, University of Rochester, Rochester, New York, 1968.

33. Liu, Benjamin Y. H. and David Y. H. Pui. Electrical Neutralization of Aerosols. Aerosol Sci. 5:465-472, 1974.

34. Rayleigh, Lord. On the Instability of Jets. London Math Soc Proc. 10:4-13, 1878-1879.

35. Castleman, R. A. The Mechanism of the Atomization of Liquids. Bur Std J Res. 6:369-376, 1931.

36. Schweitzer, P. H. Mechanism of Disintegration of Liquid Jets. J Appl Phys. 8:513-521, 1937.

37. Merrington, A. C. and E. G. Richardson. The Break-Up of Liquid Jets. Proc Phys Soc (London). 59:1-13, 1947.

38. Bitron, Mosche D. Atomization of Liquids by Supersonic Air Jets. Ind and Eng Chem. 47:23-28, 1955.

39. Raabe, Otto G. Analysis of the Operation and Behavior of Compressed Air Nebulizers. Unpublished, 1975.

40. Mercer, T. T., M. I. Tillery and M. A. Flores. Operating Characteristics of the Lauterbach and Dautrebande Aerosol Generators. AEC Research and Development Report LF-6, Lovelace Foundation for Medical Education and Research, Albuquerque, New Mexico, 1963.

41. Mercer, T. T., R. F. Goddard and R. L. Flores. Output Characteristics of Three Ultrasonic Nebulizers. Ann Allergy. 26:18-27, 1968.

42. Mercer. T. T., R. F. Goddard and R. L. Flores. Effect of Auxiliary Air Flow on the Output Characteristics of Compressed-Air Nebulizers. Ann Allergy. 27:211-217, 1969.

43. Dautrebande, L. Microaerosols. New York, Academic Press, Inc., 1962. pp. 7-22.

44. Dautrebande, L., H. Beckmann and W. Walkenhorst. New Studies on Aerosols. III. Production of Solid, Small-Sized Aerosols. Arch Int Pharmacodyn. 66:170-186, 1958.

45. Lauterbach, K. E., A. D. Hayes and M. A. Coelho. An Improved Aerosol Generator. A M A Arch Ind Health. 13;156-160, 1956.

46. Collison, W. E. Inhalation Therapy Techniques, Heinemann, London, 1935.

47. May, K. R. The Collison Nebulizer: Description, Performance and Application. J Aerosol Sci. 4:235-243, 1973.

48. Wright, B. M. A New Nebulizer. Lancet. 2:24-25, July 5, 1958.

49. Geoffrion, Louis A. Evaluation of Sodium Chloride Aerosols Generated for Filter and Respirator Studies. American Industrial Hygiene Conference, Miami Beach, Florida, May 12-17, 1974.

50. Williams, Truman. Personal Communication from Frontier Enterprises, Albuquerque, New Mexico.

51. Newton, G. J., J. E. Bennick and S. Posner. A Miniature High Output Aerosol Generator. In: Selected Summary of Studies on the Fission Product Inhalation Program from July 1965 Through June 1966. AEC Research and Development Report LF-33, Lovelace Foundation for Medical Education and Research, Albuquerque, New Mexico. 1966. pp. 29-35.

52. Burns, H. L. Nebulizer Evaluation by the Dew Point Method. Respiratory Therapy. Sept./Oct. 1972.

53. Babington, R. S., W. R. Slivka and A. A. Yetman. Method of Atomizing Liquids in a Mono-Dispersed Spray. U. S. Patent 3421692. January 14, 1969.

54. Babington, R. S., W. R. Slivka and A. A. Yetman. Apparatus for Spraying Liquids in Monodispersed Form. U. S. Patent 3421699. June 14, 1969.

55. Babington, R. S. Fuel Burner. U. S. Patent 3425058. January 28, 1969.

56. Raabe, Otto G. Report on Prototype Model of Owens-Illinois Hydrosphere Nebulizer. Unpublished Report to Owens-Illinois, March 1973.

57. Sollner, K. The Mechanism of the Formation of Fogs by Ultrasonic Waves. Trans Faraday Soc. 32:1532, 1936.

58. Antonevich, J. N. Ultrasonic Atomization of Liquids. In: Proceedings of the National Electronics Conference. Chicago, Illinois, October 7-9, 1957, pp. 798-807.

59. Muromstev, S. N. and V. P. Nenashev. The Study of Aerosols III. An Ultrasonic Aerosol Atomizer. J Microbiol Epidemiol Immunobiol (USSR) (English Transl.). 31:1840-1846, 1960.

60. Il'in, B. I. and O. K. Eknadiosyants. Nature of the Atomization of Liquids in an Ultrasonic Fountain. Soviet Phys Acoust (English Transl.). 12:269-275, 1967.

61. Boucher, R. M. G. and J. Kreuter. The Fundamentals of the Ultrasonic Atomization of Medicated Solutions. Ann Allergy. 26:591-600, 1968.

62. Denton, M. B. and D. B. Swartz. An Improved Ultrasonic Nebulizer System for the Generation of High Density Aerosol Dispersion. Rev Sci Instrum. 45:81-83, January 1974.

63. Newton, G. J. Lovelace Foundation for Medical Education and Research Unpublished Observation. 1968.

64. Goddard, Roy F., T. T. Mercer, Peter X. F. O'Neill, Richard L. Flores and Ray Sanchez. Output Characteristics and Clinical Efficacy of Ultrasonic Nebulizers. J Asthma Research. 5:355-368, June 1968.

65. Reif, Arnold E. and Margot Holcomb. Operating Characteristics of Commercial Nebulizers and Their Adaptation to Produce Closely Sized Aerosols. Ann Allergy. 16:626-638, 1958.

66. Raabe, Otto G. The Adsorption of Radon Daughters to Some Polydisperse Submicron Polystyrene Aerosols. Health Phys. 14:397-416, 1968.

67. Kanapilly, G. M., O. G. Raabe and G. J. Newton. A New Method for the Generation of Aerosols of Insoluble Particles. Aerosol Sci. 1:313-323, 1970.

68. Gibb, F. R. and P. E. Morrow. Alveolar Clearance in Dogs After Inhalation of an Iron-59 Oxide Aerosol. J. Appl Physiol. 17:429-432, 1962.

69. Ramsden, D. and D. A. Waite. Inhalation of Insoluble Iron-Oxide Particles in the Submicron Range, Assessment of Radioactive Contamination in Man. International Atomic Energy Agency, Vienna, 1972, pp. 65-81.

70. McKnight, M. E. and R. W. E. Norgan. A Study of the Exchange Characteristics of Montmorillonite Clay for Fission Product Cations for

82. Sciarra, John J. and Leonard Stoller. The Science and Technology of Aerosol Packaging. New York, John Wiley and Sons, 1974. 710 p.

83. Polli, Gerald P., Wayne M. Grim, Frederick A. Bacher and Martin H. Yanker. Influence of Formulation on Aerosol Particle Size. J Pharm Sci. 58:484-486, 1969.

84. Nelson, Sidney W. and Anthimos J. Christoforidis. An Automatic Inhalation-Actuated Aerosol Anesthesia Unit: A New Method of Applying Topical Anesthesia to the Oropharynx and Tracheobronchial Tree. Radiology. 82:226-293, 1964.

85. Vos, Kenneth D. and David B. Thompson. Particle Size Measurement of Three Commercial Pressurized Products. Aerosol Age. 20:31-32, 34, 52, 54, 56, 58, 1975.

86. Vos, Kenneth D. and David B. Thompson. Particle Size Measurement of Eight Commercial Pressurized Products. Powder Technology. 10: 103-109, 1974.

87. Davies, C. N. Book Review of Principles of Aerosol Technology. Aerosol Science. 2:125, 1971.

88. Gordieveff, V. A. Studies on Dispersion of Solids as Dust Aerosols. A M A Arch Ind Health. 15:510-515, 1957.

89. Bair, W. J. and B. J. McClanahan. Plutonium Inhalation Studies. Arch Environ Health. 2:48-55, 1961.

90. Bair, W. J., D. H. Willard, J. P. Herring and L. A. George II. Retention, Translocation and Excretion of Inhaled $^{239}PuO_2$. Health Phys. 8:639-649, 1962.

91. Wright, B. M. A New Dust-Feed Mechanism. J Sci Instrum. 27:12-15, 1950.

92. Dimmick, R. L. Jet Disperser for Compacted Powders in the One-to-Ten Micron Range. A M A Arch Ind Health. 20:8-14, 1959.

93. Ebens, Rinze and Maria Vos. A Device for the Continuous Metering of Small Dust Quantities. Staub-Reinhalt Luft in English. 28:24-25, 1968.

94. Fuchs, N. A. and F. I. Murashkevich. Laboratory Powder Dispenser (Dust Generator). Staub-Reinhalt Luft in English. 30:1-3, 1970.

95. Drew, Robert T. and Sidney Laskin. A New Dust-Generated System for Inhalation Studies. Amer Ind Hyg Assoc J. 32:327-330, 1971.

96. Willeke, K., C. S. K. Lo and K. T. Whitby. Dispersion Characteristics of a Fluidized Bed. Aerosol Sci 5:449-455, 1974.

97. Guichard, J. C. and J. L. Magne. Application du lit Fluidise en Milieu Gazeux a la Generation des Aerosols. Note Interieure No. 51. IRCHA, VERT-le-Petit, France, 1967.

98. Pfefferkorn, Gerhard. Photochemische Bildung Schwerflüchtiger Tröpfchen aus Organischen Dämpfen in Luft. Staub-Reinhalt. Luft. 27:138-140, 1967.

99. Mercer, T. T. and M. I. Tillery. Coagulation Rates of Particles Produced in Air by Thoron. J Colloid Interface Sci. 37:785-792, 1971.

100. Couchman, J. C. Metallic Microsphere Generation, EG&G., Inc. Santa Barbara, California, 1966.

101. Goldsmith, P., A. C. Wells and L. C. Cox. Aerosol Emission from Chromium Alloys when Heated in Air and in Carbon Dioxide/1% Carbon Monoxide. AERE-R5426, Atomic Energy Research Establishment, Harwell, U.K., 1968. 20 p.

102. Goldsmith, P., F. G. May and R. D. Wiffen. Chromium Trioxide Aerosol from Heated 80:20 Nickel Chromium Wire. Nature. 210:475-477, April 30, 1966.

103. Maiwald, Elenor. Ein Wolframoxid - Testaerosol. Staub. 25:535-537, 1965.

104. Spurny, Kvetoslav and James P. Lodge, Jr. Production of Highly Dispersed Model Aerosols. Staub-Reinhalt Luft in English. 33:171-173, 1973.

105. Karioris and B. R. Fish. An Exploding Wire Aerosol Generator. J Colloid Sci. 17:155-161, 1962.

106. Phalen, Robert F. Evaluation of an Exploded-Wire Aerosol Generator for Use in Inhalation Studies. Aerosol Sci. 3:395-406, 1972.

107. Chace, W. G. and H. K. Moore (eds). Exploding Wires. New York, Plenum Press, (2 volumes), 1962.

108. Holmgren, J. D., J. O. Gibson and C. Sheer. Some Characteristics of Arc Vaporized Submicron Particulates. J Electrochem Soc. III: 362-369, 1964.

109. Pfender, E. and C. V. Boffa. Generation of Ultrafine Aerosols With a Transpiration Labeled Anode in a High Intensity Arc. Rev Sci Instrum. 4:655-657, 1970.

110. Pfefferkorn, Gerhard and Hans Desler. Investigating Smokes Generated in Electric-Arc Welding. Staub-Reinhalt Luft in English. 27:1-5, 1967.

111. Spurny, Kvetoslav R. and James P. Lodge, Jr. Radioactively Labeled Aerosols. Atmos Environ. 2:429-440, 1968.

112. Nelson, L. S. Explosion of Burning Zirconium Droplets Caused by Nitrogen. Science. 148:1594-1595, 1965.

113. Nelson, Lloyd S., Herman S. Levine, Daniel E. Rosner and Shelby C. Kurzius. The Combustion of Zirconium Droplets in Oxygen/Rare Gas Mixtures, High Temp. Sci. 2:343-375, 1970.

114. Sinclair, David and Lawrence Hinchliffe. Production and Measurement of Submicron Aerosols. In: Assessment of Airborne Particles, Mercer, T. T., P. E. Morrow and W. Stöber (eds.). Springfield, Ill., Charles C. Thomas, 1972, pp. 182-199.

115. Chatfield, E. J. Some Studies of the Aerosols Produced by the Combustion or Vaporization of Plutonium-Alkali Metal Mixtures (I and II). J Nuclear Materials. 32:228-267, 1969.

116. Polishchuk, D. I., V. L. Velikanova and I. N. Nechitailo. Combustion of Multicomponent Aluminum-Combustion Systems. In: Advances in Aerosol Physics No. 4, Fedoseev, V. A. (ed.). New York, John Wiley and Sons, 1973. pp. 89-93.

117. Selivanov, S. E. Oxidation of Magnesium in Nitrous Oxide. In: Advances in Aerosol Physics No. 4, Fedoseev, V. A. (ed.). New York, John Wiley and Sons, 1973. pp. 68-73.

118. Fedoseev, V. A. Combustion of Aerogels. In: Advances in Aerosol Physics No. 4, Fedoseev, V. A. (ed.). New York, John Wiley and Sons, 1973. pp. 55-67.

119. Liu, B. Y. H. Methods of Generation of Monodisperse Aerosols. COO-1248-10, Department of Mechanical Engineering, University of Minnesota, Minneapolis, Minnesota, 1967.

120. Mercer, T. T. Aerosol Technology in Hazard Evaluation. New York, Academic Press, Inc., 1973. 394 p.

121. Bradford, E. B., J. W. Vanderhoff and T. Alfrey, Jr. The Use of Monodisperse Latexes in an Electron Microscope Investigation of the Mechanism of Emulsion Polymerization. J Colloid Sci. 11:135-149, 1956.

122. Bradford, E. B. and J. W. Vanderhoff. Electron Microscopy of Monodisperse Latexes. J Appl Phys. 26:864-871, 1955.

123. Reist, P. C. and W. A. Burgess. Atomization of Aqueous Suspensions of Polystyrene Latex Particles. J Colloid Interface Sci. 24:271-273, 1967.

124. Langer, G. and J. M. Pierrad. Anomalous Behavior of Aerosols Produced by Atomization of Monodisperse Polystyrene Latex. J Colloid Sci. 18:95-97, 1963.

125. Lichtenbelt, J. W. Th., C. Pathmamanoharah and P. H. Wiersema. Rapid Coagulation of Polystyrene Latex in a Stopped-Flow Spectrophotometer. J Colloid Interface Sci. 49:281-285, 1974.

126. Raabe, Otto G. The Dilution of Monodisperse Suspensions for Aerosolization. Am Ind Hyg Assoc J. 29:439-443, 1968.

127. Ottewill, R. H. and R. F. Woodbridge. Studies on the Electrokinetic Behavior of Monodisperse Silver Halide Sols. J Colloid Sci. 19: 606-620, 1964.

128. Stöber, Werner and Arthur Fink. Controlled Growth of Monodisperse Silica Spheres in the Micron Size Range. J Colloid Interface Sci. 26:62-69, 1968.

129. Singer, Jacques, M., Carel J. Van Oss and John W. Vanderhoff. Radioiodination of Latex Particles. J Reticuloendothelial Soc. 6: 281-286, 1969.

130. Wilkins, D. J. Fluorescent Labelling of Polystyrene Latex for Tracing in Biological Systems. Nature. 202:798-799, 1964.

131. Black, A. and M. Walsh. The Preparation of Bromine-82 and Iodine-131 Labeled Polystyrene Microspheres with Diameters From 0.1 to 30 Microns. Ann Occup Hyg. 13;87-100, 1970.

132. Bogen, Donald D. Preparation of RadioactiveLabeled Polystyrene Latex Monodisperse Submicron Aerosols. Am Ind Hyg Assoc J. 31: 349-352, 1970.

133. Flachsbart, H. and W. Stöber. Preparation of Radioactively Labeled Monodisperse Silica Spheres of Colloidal Size. J Colloid Interface Sci. 30:568-573, 1969.

134. Fuchs, N. A. Latex Aerosols - Caution! Aerosol Sci. 4:405-410, 1973.

135. Langer, G. and A. Lieberman. Anomalous Behavior of Aerosols Produced by Atomization of Monodisperse Polystyrene Latex. J Colloid Sci. 15:357-360, 1960.

136. Rimberg, D., R. Knuth and L. Hinchliffe. Detection of Emulsifien Particles in Latex Aerosols. J Colloid Interface Sci. 40:315-316, 1972.

137. Pugh, Thomas L. and Wilfried Heller. Density of Polystyrene and Polyvinyltoluene Latex Particles. J Colloid Sci. 12:173-180, 1957.

138. Lippie, L. J. Letter to Aerosol Group at University of Rochester, New York. The Dow Chemical Company, Midland, Michigan, January 17, 1966.

139. Whitby, Kenneth T. and Benjamin Y. H. Liu. Polystyrene Aerosols - Electrical Charge and Residue Size Distribution. Atmos Environ. 2:103-116, 1968.

140. Thomas, Jess W. and David Rimberg. Ein einfache Methode zur Messung der Mittleren elektrischen Ladung eins Monodispersen Aerosol. Staub-Reinhalt Luft. 27:354-357, 1967.

141. Kotrappa, P. and Owen R. Moss. Production of Relatively Monodisperse Aerosols for Inhalation Experiments by Aerosol Centrifugation. Health Phys. 21:531-535, 1971.

142. Kotrappa, P., C. J. Wilkinson and H. A. Boyd. Technology for the Production of Monodisperse Aerosols of Oxides of Transuranic Elements for Inhalation Experiments. Health Phys. 22:837-843, 1972.

143. Raabe, Otto G., Howard A. Boyd, George M. Kanapilly, Charles J. Wilkinson and George J. Newton. Development and Use of a System for Routine Production of Monodisperse Particles of $^{238}PuO_2$ and Evaluation of Gamma Emitting Labels. Health Phys. In Press, 1975.

144. Kotrappa, P. and M. E. Light. Design and Performance of the Lovelace Aerosol Particle Separator. Rev Sci Instrum. 43:1106-1112, 1972.

145. Raabe, O. G., H. A. Boyd, J. A. Mewhinney, J. J. Miglio and C. J. Wilkinson. Production of Monodisperse Aerosols of $^{239}PuO_2$ Labeled with ^{51}Cr. Fission Product Inhalation Program Annual Report, 1971-1972, LF-45, Lovelace Foundation for Medical Education and Research, Albuquerque, New Mexico, pp. 21-28, 1972.

146. Wilson, Irwin B. and Victor K. LaMer. The Retention of Aerosol Particles in the Human Respiratory Tract as a Function of Particle Radius. J Ind Hyg Toxicol. 30:265-280, 1948.

147. Sinclair, David and Victor K. LaMer. Light Scattering as a Measure of Particle Size in Aerosols. Chem Rev. 44:245-267, 1949.

148. LaMer, Victor K., Edward C. Y. Inn and Irwin B. Wilson. The Methods of Forming, Detecting and Measuring the Size and Concentration of Liquid Aerosols in the Size Range of 0.01 to 0.25 Microns Diameter. J Colloid Sci. 5:471-497, 1950.

149. Rapaport, E. and S. E. Weinstock. A Generator for Homogeneous Aerosols. Experientia. 11:363-364, 1955.

150. Lassen, Lars. A Simple Generator for the Production of Monodisperse Aerosols in the Size Range of 0.15 to 0.70 Microns (Particle Radius). Z Angew Phys. 4:157-159, 1960.

151. Movilliat, P. Production et observation d'aerosols Monodisperses Application a l'etude de quelques proprietes physiques d'etats disperses. Ann Occupational Hyg. 4:275-294, 1962.

152. Matijevic, E., S. Kitani and M. Kerker. Aerosol Studies by Light Scattering. II. Preparation of Octanoic Acid Aerosols of Narrow and Reproducible Size Distributions. J Colloid Sci. 19:223-237, 1964.

153. Swift, D. L. A Study of the Size and Monodispersity of Aerosols Produced in a Sinclair-LaMer Generator. Ann Occupational Hyg. 10:337-348, 1967.

154. Liu, Benjamin, Kenneth T. Whitby and Henry H. S. Yu. A Condensation Aerosol Generator for Producing Monodispersed Aerosols in the Size Range, 0.036 μ to 1.3 μ. J Rech Atmospheriques. 397-406, 1966.

155. Edwards, J. and B. J. Brinkworth. A Device for the Control fo Particle Size in the LaMer Aerosol Generator. J Sci Instrum, Ser 2. 1:636-637, 1968.

156. Sutugin, A. G. A Simple Monodisperse Aerosol Generator. In: Advances in Aerosol Physics No. 4, Fedoseev, V. A. (ed.). New York, John Wiley and Sons, 1973. pp. 36-41.

157. Nicolaon, G., D. D. Cooke, M. Kerker and E. Matijevic. A New Liquid Aerosol Generator. J Colloid Interface Sci. 34:534-544, 1970.

158. Nicolaon, G., D. D. Cooke, E. J. Davis, M. Kerker and E. Matijevic. A New Liquid Aerosol Generator, II. The Effect of Reheating and Studies on the Condensation Zone. J Colloid Interface Sci. 35: 490-501, 1971.

159. Davis, E. James and G. Nicolaon. A New Liquid Aerosol Generator, III. Operating Characteristics and Theoretical Analysis. J Colloid Interface Sci. 37:768-778, 1971.

160. Matijevic, E., W. F. Espenscheid and M. Kerker. Aerosols Consisting of Spherical Particles of Sodium Chloride. J Colloid Sci. 18:91-93, 1963.

161. Kitani, Susumu and Sohei Ouchi. Preparation of Monodisperse Aerosols of Sodium Chloride. J Colloid Interface Sci. 23:200-202, 1967.

162. Heyder, J., J. Gebhart, G. Heigwer, C. Roth and W. Stahlhofen. Experimental Studies of the Total Deposition of Aerosol Particles in the Human Respiratory Tract. Aerosol Sci. 4:191-208, 1973.

163. Muir, D. C. F. and C. N. Davies. The Deposition of 0.5 μ Diameter Aerosols in the Lungs of Man. Ann Occup Hyg. 10:161-174, 1967.

164. Davies, C. N., J. Heyder and M. C. Subba Ramu. Breathing of Half-Micron Aerosols, I. Experimental. J Applied Physiol. 32:591-611, 1972.

165. Giacomelli-Maltoni, Carlo Melandri, Vittorio Prodi and Giuseppe Tarroni. Deposition Efficiency of Monodisperse Particles in Human Respiratory Tract. Amer Ind Hyg Assoc J. 33:603-610, 1972.

166. Wolf, W. R. Study of the Vibrating Reed in the Production of Small Droplets and Solid Particles of Uniform Size. Rev Sci Instrum. 32:1124-1129, 1961.

167. Binek, Bedrich and Blanka Dohnalova. Ein Generator zur Herstellung monodisperser Aerosole aus der flüssigen Phase. Staub Reinhaltung Luft. 27:492-494, 1967.

168. Liu, Benjamin Y. H. and David Y. H. Pui. A Submicron Aerosol Standard and the Primary Absolute Calibration of the Condensation Nuclei Counter. J Colloid Interface Sci. 47:155-171, 1974.

169. Liu, B. Y. H., V. A. Marple, K. T. Whitby and N .J. Barsie. Size Distribution Measurement of Airborne Coal Dust by Optical Particle Counters. Amer Ind Hyg Assoc J. 35:443-451, 1974.

170. Liu, Benjamin Y. H., David Y. H. Pui, Austin W. Hogan and Ted A. Rich. Calibration of the Pollak Counter with Monodisperse Aerosols. J Applied Meteor. 14:46-51, 1975.

171. Walton, W. H. and W. C. Prewett. The Production of Sprays and Mists of Uniform Drop Size by Means of Spinning Disc Type Sprayers. Proc Phys Soc (London). 62:341-350, 1949.

172. Schwendiman, L. C., A. K. Postma and L. F. Coleman. A Spinning Disc Aerosol Generator. Halth Phys. 10:947-953, 1964.

173. Albert, R. E., H. G. Petrow, A. S. Salam and J. R. Spiegelman. Fabrication of Monodisperse Lucite and Iron Oxide Particles with a Spinning Disk Generator. Health Phys. 10:933-940, 1964.

174. Sehmel, G. A. The Density of Uranine Particles Produced by a Spinning Disc Aerosol Generator. Am Ind Hyg Assoc J. 28:491-492, 1967.

175. Lippmann, Morton and Roy E. Albert. A Compact Electric-Motor Driven Spinning Disc Aerosol Generator. Am Indus Hyg Assoc J. 28:501-506, 1967.

176. Partridge, Jennings E. and Harry J. Ettinger. Calibration of a Spinning-Disc Aerosol Generator and Two-Stage Air Samplers. AEC Research and Development Report LA-4066, Los Alamos Scientific Laboratory, Los Alamos, New Mexico, 1969.

177. Albert, Roy E., Morton Lippmann, Jack Spiegelman, Anthony Liuzzi and Norton Nelson. The Deposition and Clearance of Radioactive Particles in the Human Lung. Arch Environ Health. 14:10-15, 1967.

178. Tarroni, G. Emploi et caracteristiques D'un Generateur D'Aerosols Monodisperse a Disque Tournant. Aerosol Sci. 2:257-260, 1971.

179. Lourenco, Ray V., Mary F. Klimek and Claudia J. Borowski. Deposition and Clearance of 2 μ Particles in the Tracheobronchial Tree of Normal Subjects - Smokers and Non Smokers. J Clinical Invest. 50:1411-1420, 1971.

180. May, K. R. An Improved Spinning Top Homogeneous Spray Apparatus. J Appl Phys. 20:932-938, 1949.

181. May, K. R. Spinning-Top Homogeneous Aerosol Generator with Shock-Proof Mounting. J Sci Instrum. 43:841-842, 1966.

182. Ellison, John McK. Adaptation of the Spinning Top Generator to Provide Aerosols in the Respirable Range. Ann Occup. Hyg. 10:363-367, 1967.

183. Kajland, A., M. Edford, L. Friberg and B. Holma. Radioactive Monodisperse Test Aerosols and Lung Clearance Studies. Health Phys. 10:941-945, 1964.

184. Booker, D. V., A. C. Chamberlain, J. Rundo, D. C. F. Muir and M. L. Thompson. Elimination of 5μ Particles from the Human Lung. Nature. 215:30-33, 1967.

185. Muir, D. C. F. and C. N. Davies. The Deposition of 0.5μ Diameter Aerosols in the Lungs of Man. Ann Occupational Hyg. 10:161-174, 1967.

186. Holma, Bo. Lung Clearance of Mono- and Di-Disperse Aerosols Determined by Profile Scanning and Whole-Body Counting. A Study on Normal and SO_2 Exposed Rabbits. From the Department of Hygiene, Karolinska Institute and the Department of Environmental Hygiene, National Institute of Public Health, Stockholm, Sweden, 1967.

187. Holma, Bo. The Initial Lung clearance of Di-Disperse (2μ and 7μ) Polystyrene Particles in Rabbits. Arch Environ Health. 17:871-873, 1968.

188. Black, A. and M. Walsh. Production of Uniform Radioactive Particles for Aerosol Inhalation Experiments. Part I. Particles in the Size Range 1-10 microns. United Kingdom Atomic Energy Authority Research Group Report, AERE-R 5716, Health Physics and Medical Division, Atomic Energy Research Establishment, Harwell, Berkshire, 1968.

189. Philipson, Klas, Per Camner and John Svedberg. Tagging 7μ Monodisperse Polystyrene Particles with ^{11}C. J Appl Rad and Isotopes. 21:639-642, 1970.

190. Camner, P. and K. Philipson. Production of 7μ Monodisperse Fluorocarbon Resin Particles Tagged with ^{18}F. J Applied Rad and Isotopes. 22:349-353, 1971.

191. Camner, P., K. Philipson and L. Linnman. A Simple Method for Nuclidic Tagging of Monodisperse Fluorocarbon Resin Particles. J Applied Rad and Isotopes. 22:731-734, 1971.

192. Camner, P. Per-Aki Hellström and Margot Lundborg. Coating 5μ Particles with Carbon and Metals for Lung Clearance Studies. Arch Environ Health. 27:331-333, 1973.

193. Philipson, Klas. On the Production of Monodisperse Particles With a Spinning Disc. Aerosol Sci. 4:51-57, 1973.

194. Aldrich, John E. and J. R. Johnston. Use of the Spinning Disk Technique to Product Monodisperse Microspheres of Human Serum Albumin for Labeling with Radionuclides. J Applied Rad and Isotopes. 25:15-18, 1974.

195. Drozin, Vadim G. The Electrical Dispersion of Liquids as Aerosols. J Colloid Sci. 10:158-164, 1955.

196. Vonnegut, B. and R. Neubauer. Production of Monodisperse Liquid Particles by Electrical Atomization. J Colloid Sci. 7:616-621, 1952.

197. Yurkstas, Edward P. and Charles J. Meisenzahl. Solid Homogeneous Aerosol Production by Electrical Atomization. AEC Research and Development Report UR-652, University of Rochester Atomic Energy Project, Rochester, New York, 1964.

198. Huebner, A. L. Disintegration of Charged Liquid Jets: Results With Isopropyl Alcohol. Science. 168:118-119, April 3, 1970.

199. Fulwyler, M. J., J. D. Perrings and L. S. Cram. Production of Uniform Microspheres. Rev Sci Instrum. 44:204-206, 1973.

200. Fulwyler, M. J. and O. G. Raabe. The Ultrasonic Generation of Monodisperse Aerosols. Presented at the American Industrial Hygiene Conference, Detroit, Michigan, May 11-15, 1970.

201. Raabe, O. G. and G. J. Newton. Development of Techniques for Generating Monodisperse Aerosols with the Fulwyler Droplet Generator. Fission Product Inhalation Program Annual Report, 1969-1970, LF-43, Lovelace Foundation for Medical Education and Research, Albuquerque, New Mexico, pp. 13-17, 1970.

202. Fulwyler, M. J. Electronic Separation of Biological Cells by Volume. Science. 150:910-911, 1965.

203. Fulwyler, M. J., R. B. Glascock and R. B. Hiebert. Device Which Separates Minute Particles According to Electronically Sensed Volume. Rev Sci Instrum. 40:42-48, 1969.

204. Strom, L. The Generation of Monodisperse Aerosols by a Means of a Disintegrated Jet of Liquid. Rev Sci Instrum. 40:778-782, 1969.

205. Crane, L., S. Birch and P. D. McCormack. The Effect of Mechanical Vibration on the Break-Up of a Cylindrical Water Jet in Air. Brit J Applied Phys. 15:743-750, 1964.

206. Schneider, J. M., N. R. Lindblad and C. D. Hendricks. An Apparatus to Study the Collision and Coalescence of Liquid Aerosols. J Colloid Sci. 20:610-616, 1965.

207. Lindblad, N. R. and J. M. Schneider. Production of Uniform-Sized Liquid Droplets. J Sci Instrum. 42:635-638, 1965.

208. McCormack, P. D., L. Crane and S. Brich. An Experimental and Theoretical Analysis of Cylindrical Liquid Jets Subjected to Vibration. Brit J Appl Phys. 16:395-408, 1965.

209. Dabora, E. K. Production of Monodisperse Sprays. Rev Sci Instrum. 38:502-506, 1967.

210. Lindblad, N. R. and J. M. Schneider. Method of Producing and Measuring Charged Single Droplets. Rev Sci Instrum. 38:325-327, 1967.

211. Berglund, Richard N. and Benjamin Y. H. Liu. Generation of Monodisperse Aerosol Standards. Environ. Sci and Technol. 7:147-153, 1973.

212. Liu, Benjamin Y. H. Laboratory Generation of Particulates With Emphasis on Submicron Aerosols. APCAJ. 24:1170-1172, 1974.

GENERATION OF MONODISPERSE SUBMICRON AEROSOLS BY ABLATION FROM TRANSPIRATION-COOLED POROUS MATRICES°°

by

Cesare V.Boffa,^ Augusto Mazza,^^ Delfino Maria Rosso ^^^

Istituto di Fisica Tecnica - Politecnico di Torino - Italy

ABSTRACT

In this paper the generation is described of monodisperse submicron particles of controlled size by ablation from transpiration-cooled porous matrices.

A new experimental apparatus is illustrated which has been realized in the Istituto di Fisica Tecnica - Politecnico di Torino - Italy, and which is based on previous works in cooperation with the University of Minnesota, USA and the Sonderforschungsbereich 80, Universität Karlsruhe, West Germany.

In the new version of the apparatus an ewternal heat source (a plasma jet) heats a porous matrix up to ablation temperatures and particles are generated by ablation of the porous material itself. The temperature of the ablated surface of the porous matrix is controlled by a flow of gas transpiring through the matrix, which also quenches and dilutes the particles formed in the ablation process. Particles can be generated from any porous material of sufficiently fine structure and uniform porosity, in particular from graphite, tungsten, ceramics, etc.

(°°) This work has been sponsored by the Italian National Research Council, Centro Nazionale delle Ricerche.

(^) Cesare V.Boffa, Professor.

(^^) Augusto Mazza, Graduate student.

(^^^) Delfino Maria Rosso, Laboratory technician.

The experimental apparatus is described as well as the particle measuring techniques, based on the utilization of a calibrated T.S.I. Model 3030 Whitby Electrical Aerosol Analyzer.
Experimental results are given referring to the generation of submicron graphite particles. The measured number, surface, and volume distributions of these particles are given, together with the logarithmic standard deviation σg. The results show that, for the generated particles $1.13 < \sigma g < 1.32$.

GENERATION OF MONODISPERSE SUBMICRON AEROSOLS BY ABLATION FROM TRANSPIRATION-COOLED POROUS MATRICES °

Cesare V.Boffa,^ Augusto Mazza,^^ Delfino Maria Rosso ^^^

Istituto di Fisica Tecnica - Politecnico di Torino - Italy

INTRODUCTION

This report describes a new approach for the generation of monodisperse aerosols which are presently of interest for all kinds of medical and air pollution research, including evaluating aerosol sampling and measuring equipment, air cleaner evaluation, inhalation studies, and for interrelated studies. Although numerous research-type aerosol generators utilising electric arcs have been developed or proposed for generating monodisperse aerosols ranging from 0.02 to 20 μm, there is at present no reliable method available for generating ultrafine, monodisperse aerosols for a wide spectrum of materials, and in particular of refractory materials, based on the utilisation of electric arcs.

Aerosol generation by means of electric arcs between solid electrodes relies on the high temperature of the attachment spots at the electrodes which causes an evaporation of some electrode material. This metal vapor, if cooled rapidly, condenses into an aerosol. The properties of the aerosol particles, including their size, depend on a number of parameters, particularly on the temperature distribution at

---

(°) This work has been sponsored by the Italian National Research Council, C.N.R.

(^) Cesare V.Boffa, Professor.

(^^) Augusto Mazza, Graduate student.

(^^^) Delfino Maria Rosso, Laboratory technician.

and in the vicinity of the arc attachment spot. It has not been possible to generate particles with a nearly uniform size or monodisperse particles under these conditions because of the temperature variations and the varying conditions at the location at which individual particles are generated. The time needed to collect a sufficient number of particles is large compared to the characteristic time during which changes on the surface occur, i.e., the relevant parameters cannot be held constant during this period. The main reason for these varying conditions lies in the nature of the attachment spot of the arc which moves more or less quickly over the electrodes, even if current and voltage do not fluctuate appreciably.

But these variations, in addition to the explosive character of the vaporization process, change the conditions for the aerosol production so rapidly that for any useful length of time the produced aerosol has a wide distribution of its particle size.

Various experimental configurations have been suggested for the production of aerosols with these techniques (Chace et als[1], Reichelt[2], Holmgren et als[3], Kuhn[4]) employing both high and low intensity arcs. Solid electrodes have been used by all investigators. The electrodes have frequently been cooled by a conventional water cooling system, especially in conjunction with high intensity arcs. With low intensity arcs coagulated and chainlike particles are produced. The size distribution of these particles covers a wide range, from approximately 100 to 1600 Å (Reichelt[2]) and is poorly controlled. Better results have been obtained by using high intensity arcs (Holmgren et als[3]). The feed material is incorporated in the body of the anode which consumes a large fraction of the arc power. In most cases the solid electrodes are water-cooled and dilution gas can be blown through the arc. By varying the cooling rate of the electrodes and the dilution flow rate through the plasma in the region of maximum condensation, the particle size may to a certain degree be controlled. Monodisperse spheroidal shaped particles with a diameter ranging from approximately 100 to 1000 Å have been obtained.

The results of the above-discussed investigations indicate that electric arcs are a poor source for generating aerosols. It seems to be impossible to obtain small monodisperse particles of uniform size because coagulation occurs as soon as the particles are produced. In addition, the particle generation process is characterized by strong non-uniformities which are difficult to control in conventional arc arrangements with solid electrodes.

During the past two years, a new method for the generation of ultrafine, monodisperse aerosol has been successfully explored (Boffa et als[5]). This method makes use of a transpiration-cooled anode in a high intensity arc. In this situation, the aerosol is generated continuously rather than in an explosive manner at the steady attachment spot which is the hottest part of the anode surface. The transpiring gas passing through the anode provides quenching of the vapor and an immediate dilution of the generated particles in the vicinity of the anode surface avoiding in this way coagulation and formation of agglomerations.

The method as such has been very successful, but the particle generator itself is a rather complicated device which requires skilled personnel for its proper operation. The knowledge acquired during previous work has been, however, instrumental for the success of the new approach described in this report.

In the new approach here described the arc plasma source is entirely separated from the body from which the aerosol is generated; but the basic features for the aerosol production process inherent to the previously-developed device are retained. Instead of a transpiration cooled anode which is an integral part of the previously-developed plasma torch, a porous, transpiration-cooled body will be utilized heated by the plasma jet emanating from a conventional arc plasma torch. Since transpiration cooling fullfills two tasks simultaneously; namely, quenching of the generated vapor and dilution of the particles formed in the quenching process, utilization of this mechanism is considered as the basic principle which also governs this approach.

The separation of the heat source from the particle providing surface has at least two important advantages over the previous method:

I) The resulting particle generator is simple and straight forward to operate, and easily movable, as required by the experiments which will be performed in this research program.

II) The useful parameter range for operation of this device and the associated flexibility is substantially enhanced.

EXPERIMENTAL APPARATUS

The experimental apparatus consists essentially of a commercial power supply and control unit, which enable the operation of a plasma torch, which in turn represents the heat source, necessary for the operation of the aerosol generator.
The plasma torch and the aerosol generator are schematically represented in fig. 1.
The plasma torch consists essentially of a water cooled, thoriated tungsten cathode, with a conical tip, and a nozzle shaped water cooled copper anode, coaxial with the cathode. Argon is injected tangentially at the base of the catode, thus providing a swirl flow around the cathode tip, which stabilizes the arc. This gas, passing through the arc, emerges from the anode nozzle in the form of a highly energized plasma jet.
The cooling of anode and cathode is provided by a water cooling system, as indicated in fig. 2.
For an overall energy balance the cooling water flow is measured by means of flowrators, and the temperature increase of the anode and cathode cooling water is measured by means of four thermocouples located in the water circuit immediately upstream and downstream from the electrodes.
In this way the energy losses to the electrodes, that is the heat transferred from the arc to the anode and cathode, can be determined. The power in the emerging plasma jet can be

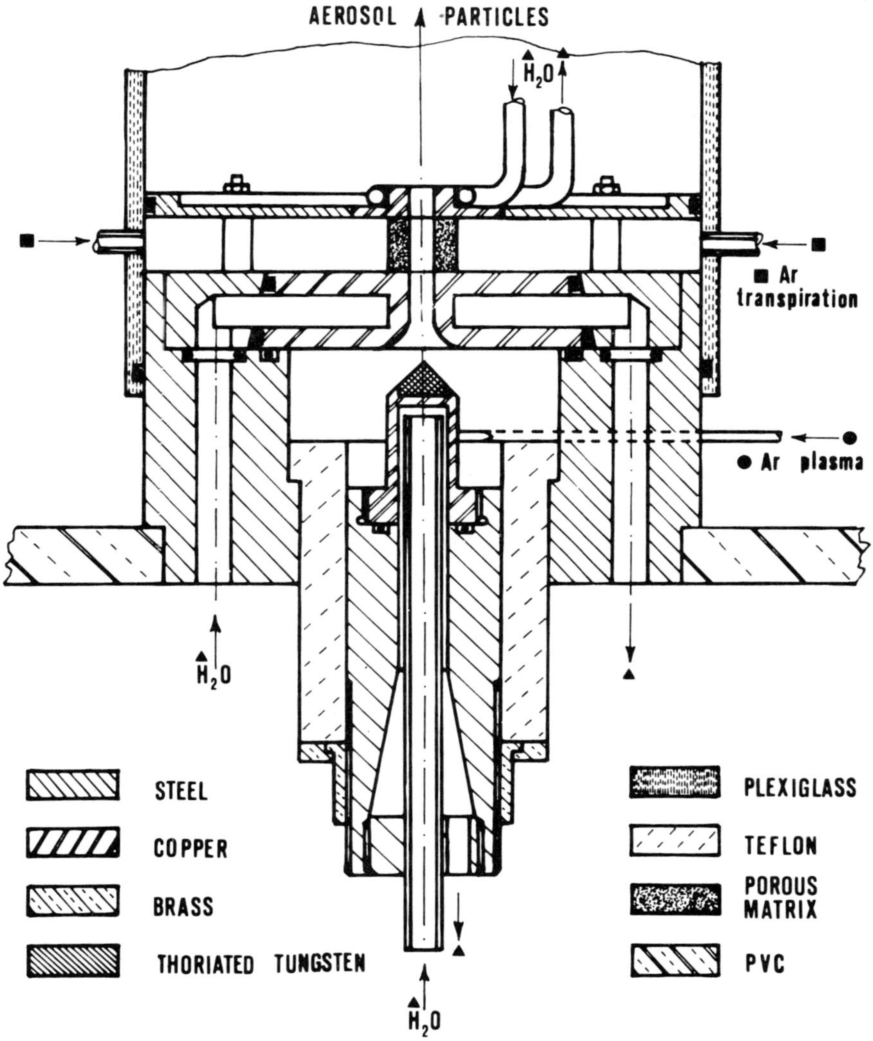

Fig. 1 Schematic of the plasma torch with aerosol generator.

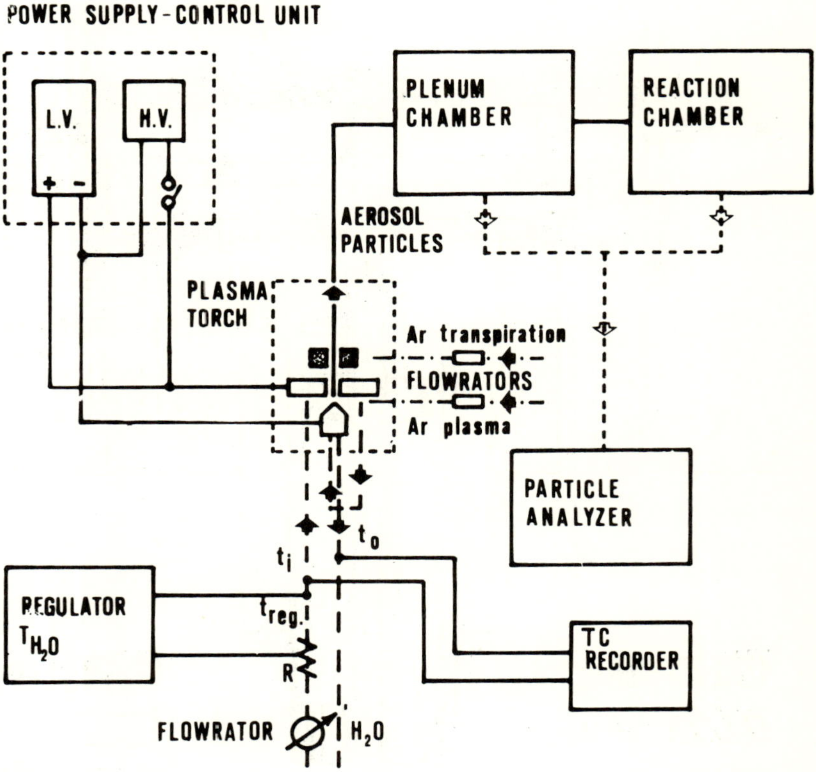

Fig. 2 Schematic of the experimental set-up.

calculated as the difference between the electrical power input and the power losses to the electrodes, carried away by the cooling water.

As before mentioned,the plasma jet emerging from the anode nozzle enters the aerosol generator located on top of the plasma torch and coaxial with it.

The aerosol generator itself consists essentially of a transpiration cooled porous carbon nozzle, concentrical with the plasma jet, inbedded in a steel and plexiglass structure which supports the porous nozzle and acts as plenum chamber feeding the cooling gas to the nozzle itself. The cooling gas transpires through the nozzle and emerges into the plasma jet, diluting the aerosol particles, which are continuously produced by ablation from the porous carbon nozzle inside surface facing the plasma.

A cylindrical plexiglass window surrounds the plenum chamber and seals it, by means of "O" rings. Through this window pyrometric readings of the outer surface temperature of the porous nozzle are possible. The plenum chamber is equipped with an inlet for the cooling gas and with a pressure tap. The steel portion of the structure supporting the porous nozzle is water cooled.

The aerosol emerging from the generator, in suspension in the argon gas, is collected into a 2 m^3 collection chamber, where it can be additionally diluted.

A block diagram of the experimental setup is shown in fig. 2.

The aerosol size distribution measuring system consists mainly of a calibrated T.S.I. Model 3030 Whitby Electrical Aerosol Analizer described elsewhere (Liu et als[6]).

EXPERIMENTAL RESULTS AND CONCLUSIONS

The results referring to carbon particles are summarized in fig. 3 to 7.

In fig. 3 the normalized number distributions of the produced aerosols are depicted, corresponding to different values of the argon flowrate transpiring through the ablat-

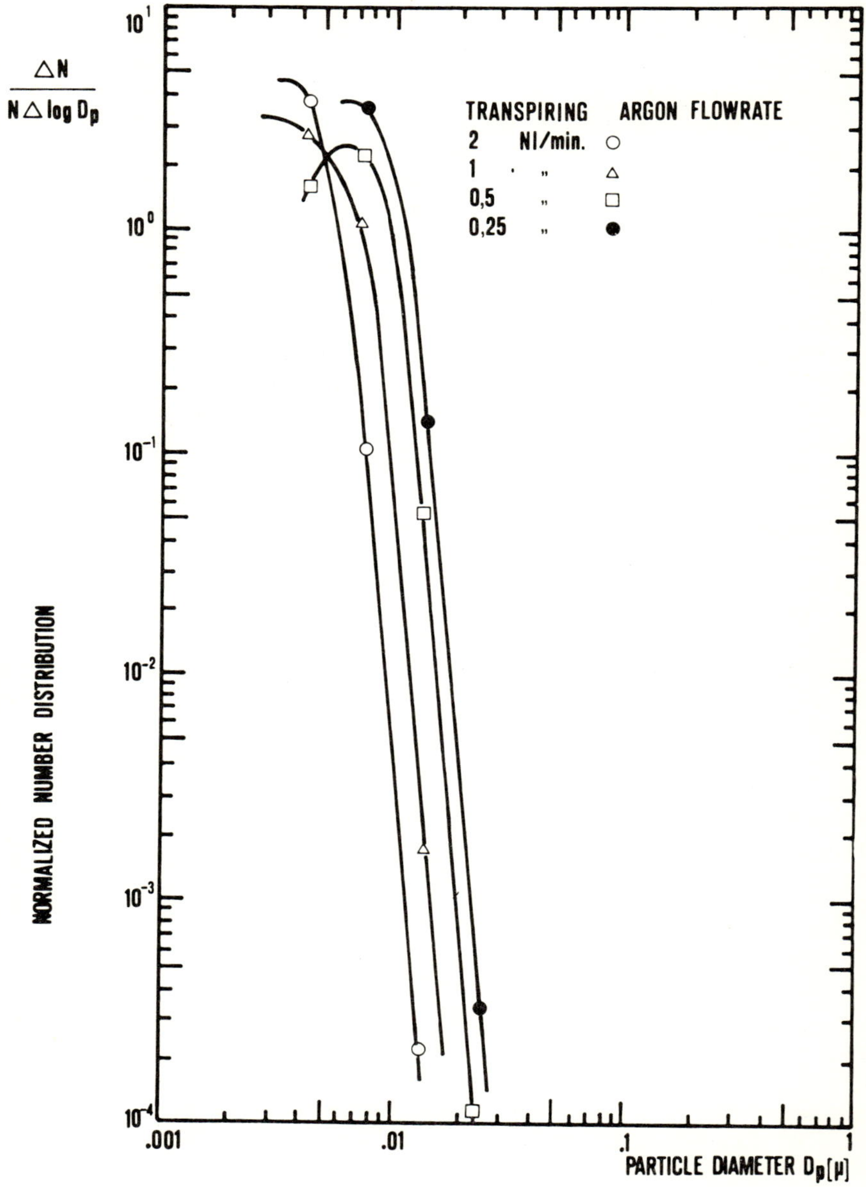

Fig. 3 Normalized number distributions of the generated aerosols.

ing porous matrix. Figures 4 and 5 show the corresponding normalized surface and volume distributions.

All the data are taken with an arc current of 100 Amp.

The results show that, for a given arc current value, the mean diameter of the generated aerosol decreases with increasing of the argon flowrate transpiring through the ablating matrix.

This finding can be qualitatively understood considering the influence of the transpiring gas flow-rate on the temperature of the ablating surface of the porous nozzle. By increasing the transpiring mass flow-rate, the temperature of the ablating surface and the associated rate of ablation decreases, yielding smaller particle densities, lower coagulation rates and, therefore, smaller particle diameters. At the same time, an increase in transpiring mass flow-rate enhances the dilution of the ablated particles in front of the surface, which decreases the coagulation rate and, consequently, the particle size also.

In fig. 6 the logaritmic standard deviation is depicted as a function of the transpiring argon flowrate for an arc current of 100 Amp.

It can be seen that for the generated particles $1.13 < \sigma_g < 1.32$.

The logarithmic standard deviation, characterizing the ablated aerosol size distribution, is a decreasing function of the transpiring mass flow-rate. It is speculated that the gas transpiring through the ablating surface has a quenching effect on the particles immediately after they are produced, preventing them from further coagulation. The higher the gas flow-rate, the smaller the number of particles which can undergo coagulation over an extended time period and randomly increase their size, resulting in a wider size distribution. On the basis of these results, it is obvious that, keeping the arc current at the fixed value of 100 A, the logarithmic standard deviation decreases with increasing transpiring argon flowrate, as shown in fig. 6.

In fig. 7 the generated aerosol particles total number density is plotted as a function of the transpiring argon flowrate for an arc current of 100 Amp.

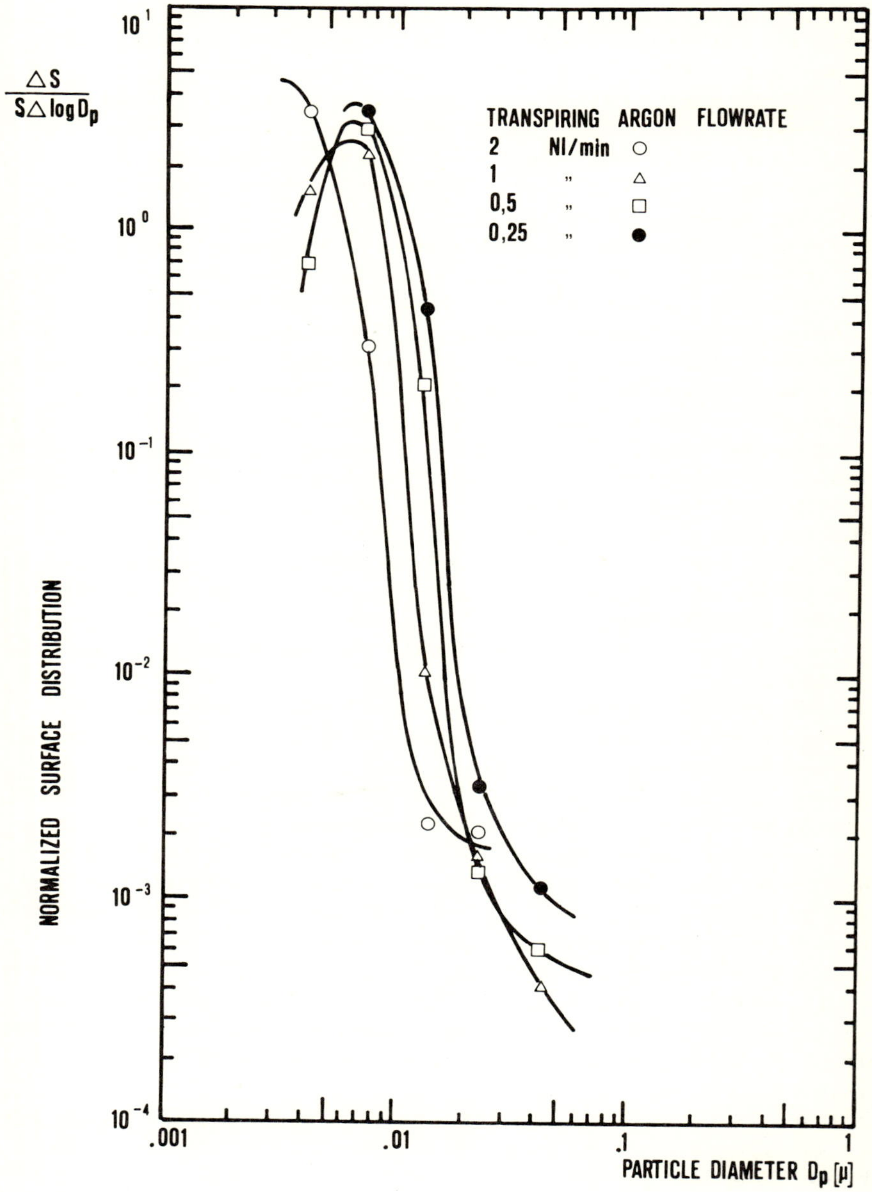

Fig.4 Normalized surface distributions of the generated aerosols.

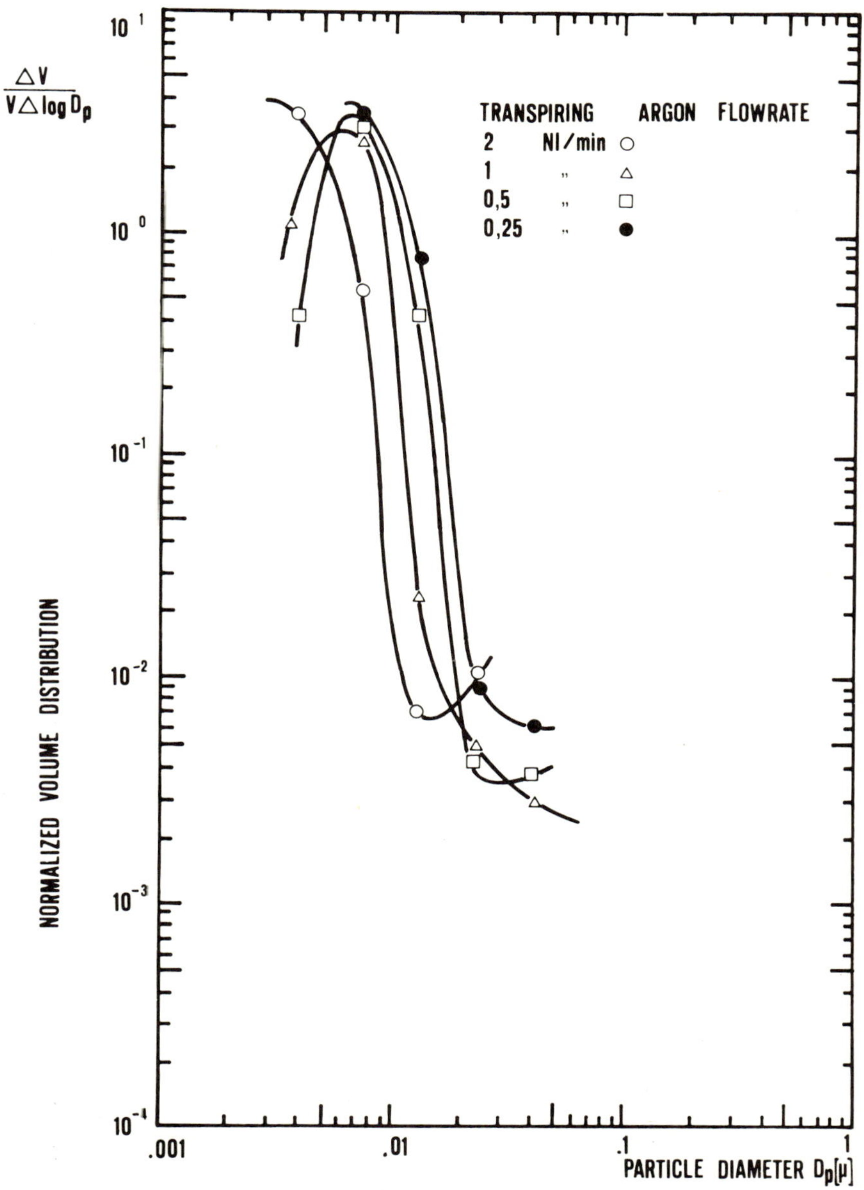

Fig. 5 Normalized volume distributions of the generated aerosols.

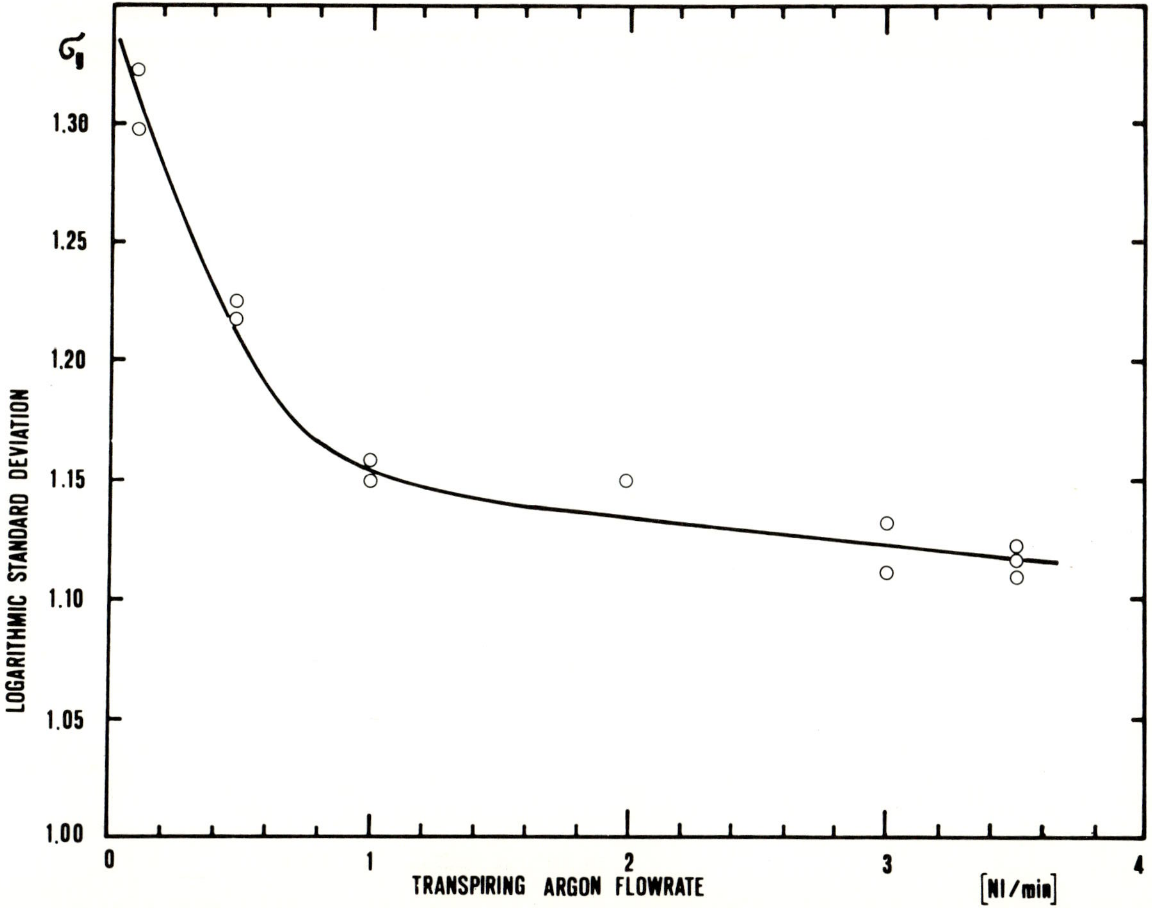

Fig. 6 Logarithmic standard deviation σ_g as a function of the argon flowrate transpiring through the porous matrix.

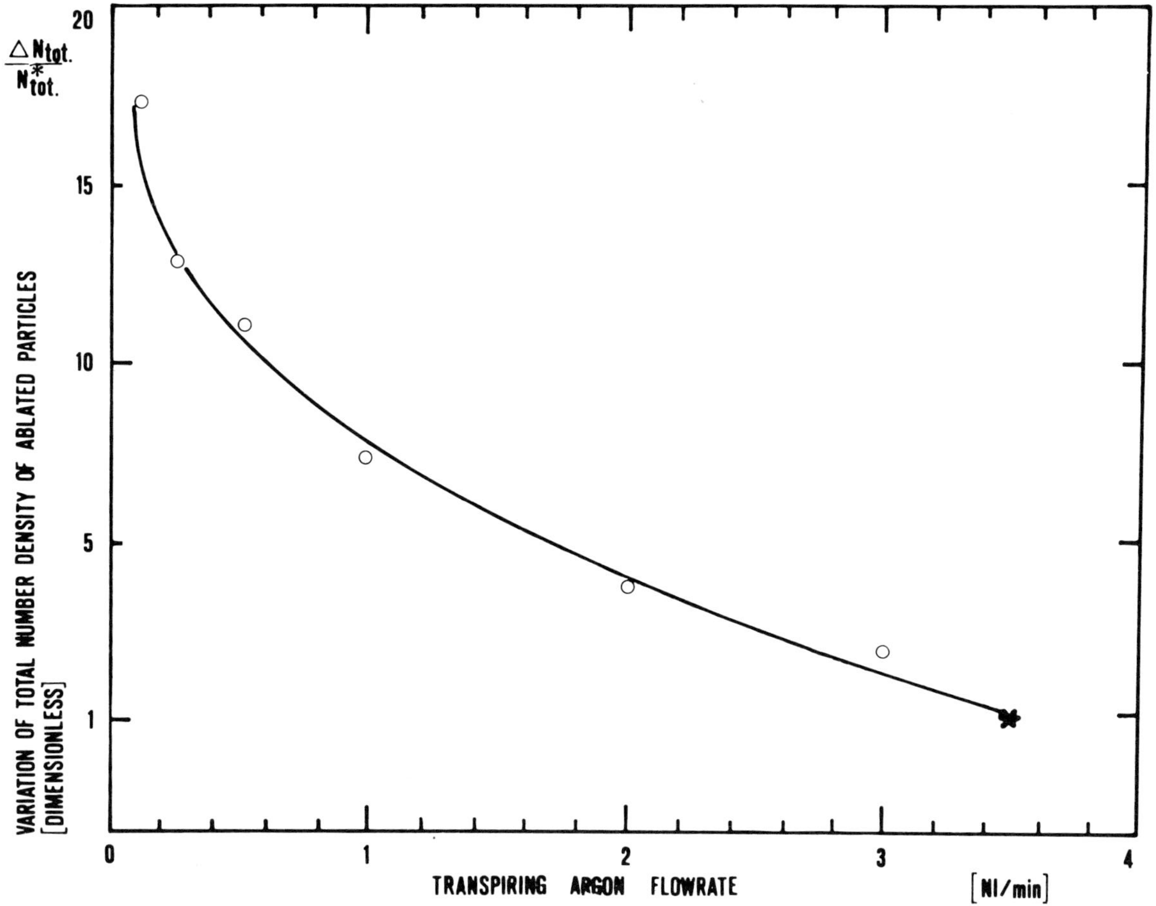

Fig. 7 Variation of total number density of the ablated particles as a function of the argon flowrate transpiring through the porous matrix.

The total concentration values N_{tot}, corresponding to a given mass flowrate, are plotted in dimensionless form according to the following relation:

$$\frac{N_{tot} - N^*_{tot}}{N^*_{tot}} = \frac{\Delta N_{tot}}{N^*_{tot}}$$

where N^*_{tot} = 8.736.000 [particle/cm^3] is the reference value and represents the total concentration corresponding to a transpiring argon flowrate of 3.5 Nl/min.

The results given in fig. 7 are in agreement with the qualitative explanation of the aerosol production mechanism given above.

During the experiments consumption of the matrix takes place, and hence, for a given value of the arc current and of the pressure in the plenum chamber feeding the transpiring gas through the matrix, the temperature of the matrix ablating surface decreases. This is due to the lower temperature attained by the ablating surface which can be attributed to two main causes. One is related to the ablation of the matrix which enlarges the inner diametre of the matrix nozzle, decreasing the heat transfer rate from the plasma to the unit area of the nozzle walls. The other one is related to the increase of the transpiring argon flowrate through the matrix.

For a given value of the relative pressure in the chamber and a given matrix temperature distribution the argon flow rate transpiring through the matrix is inversely proportional to the matrix thickness.

The ablation of the matrix reduces its thickness and therefore the mass flowrate increases. This effect is enhanced by the lower temperature attained by the matrix, caused by the decrease in the heat transfer rate from the plasma per unit area of the matrix nozzle, related to the increase in the inner diameter of the ablating nozzle, as above mentioned.

Therefore during a long operation of the particle generator the mean diameter tends to decrease. This can be easily

compensated for by increasing the arc current and decreasing the relative pressure in the plenum chamber for the feeding of the transpiring gas through the matrix.
Further experiments will be performed with arc current up to 300 Amp; various materials will be investigated such as tungsten, ceramic and stainless steel, which are available as porous matrices. Particles from other materials, which are not available as porous matrices, will be generated by mixing the material to be studied with graphite. Since the evaporation temperature of graphite is in most cases higher than that of other elements, it is expected that the admixture predominately evaporates, producing aerosols of the desired material.

BIBLIOGRAPHIC REFERENCES

1) Chace, W.G., Moore, H.K., Exploding Wires, Vol.1 (1959), Vol. 2 (1962), Vol.3 (1964), Plenum Press, New York.

2) H. Reichelt, "An Aerosol Generated in Electric Arcs Between Metal Electrodes", Aerosol Research at the First Physics Institute, Wien (1968).

3) J.G. Holmgren, J.O. Gibson, C. Sheer, "Some Characteristics of Arc Vaporized Submicron Particulates", Ultrafine Particles, John Wiley & Sons, New York (1963).

4) W.E. Kuhn, "The Formation of Silicon Carbide in the Electric Arc", Ultrafine Particles, John Wiley & Sons, New York (1963).

5) Boffa, C., Pfender, E., "Controlled generation of monodisperse aerosols in the submicron range", Journal of Aerosol Science, vol. 4 (1973).

6) Liu, B.Y.H., Whitby, K.T., Pui, D.Y.H., "A portable electrical analyzer for size distribution measurement of submicron aerosols" Journal of the Air Pollution Control Association, vol.24, n.11 (1974).

GENERATION OF MONODISPERSE AEROSOLS OF ^{67}Ga-LABELED ALUMINOSILICATE AND ^{198}Au-LABELED GOLD SPHERES*

George J. Newton, O. G. Raabe, R. L. Yarwood and G. M. Kanapilly
Inhalation Toxicology Research Institute
Lovelace Foundation
P. O. Box 5890
Albuquerque, New Mexico 87115

ABSTRACT

A system has been designed, constructed, tested and used for producing monodisperse ($\sigma_g < 1.2$) submicrometer particles of both ^{198}Au-labeled gold and ^{67}Ga-labeled aluminosilicate spheres for inhalation deposition studies in man, animals and in other types of experiments where short-lived radioactive, monodisperse aerosols are useful. Aerosol generation and particle separation equipment housed in a system of five stainless steel glove boxes produces a primary polydisperse aerosol which is passed via a quartz tube through a high temperature furnace (1200 °C) which converts the clay into insoluble aluminosilicate spheres entrapping the ^{67}Ga in the particles. Metallic gold particles are produced by nebulizing an aqueous suspension of "fulminating gold", a mixture of gold-ammonium-acetate compounds, which is reduced to metallic gold in the high temperature heating column. Two Lovelace Aerosol Particle Separators are used to separate the polydisperse aerosol into monodisperse fractions onto stainless steel foils. Gold particles are removed from the collection foil and subjected to neutron irradiation to form the ^{198}Au label. Monodisperse particles are then suspended in water and nebulized to provide monodisperse aerosols in the respirable size range between 0.6 μm and 3.5 μm aerodynamic diameter. The particle densities are 2.3 g/cm^3 for aluminosilicate and about 19 g/cm^3 for gold.

* Research supported by the National Institute of Environmental Health Sciences via ERDA Contract E(29-2)-1013.

GENERATION OF MONODISPERSE AEROSOLS OF ^{67}Ga-LABELED ALUMINOSILICATE AND ^{198}Au-LABELED GOLD SPHERES

George J. Newton, O. G. Raabe, R. L. Yarwood and G. M. Kanapilly
Inhalation Toxicology Research Institute
Lovelace Foundation
P. O. Box 5890
Albuquerque, New Mexico 87115

INTRODUCTION

Data from inhalation deposition and particle clearance studies in laboratory animals are used to help evaluate the inhalation toxicity of a variety of potentially toxic aerosols to which man could be exposed. Since the extrapolation of experimental animal data to man is not precise, a useful comparison is made in studies of the same aerosol in laboratory animals and man. Ideally, an aerosol for human deposition studies should be monodisperse, insoluble, easily detected, of low toxicity and reproducible in the size ranges of major interest. To this end, we have developed systems for generating monodisperse, insoluble aerosols of aluminosilicate spheres labeled with ^{67}Ga and gold spheres labeled with ^{198}Au. The aluminosilicate particles have biological retention half-times in the lung on the order of one year or more attesting to their insolubility in body fluids. Gallium-67 decays with a 78.1 hr. half-life by electron capture to stable ^{67}Zn and has no particulate emissions (excepting Auger electrons). Gallium-67 decay is accompanied by emission of three major gamma photon groups at 92, 182 and 300 keV with 69, 24 and 22% yields respectively. These gamma photons have an average energy of 150.5 keV and could deposit a maximum dose to a 1000 g human lung of 36 mrad/μCi. Gallium-67 is made by a ^{68}Zn(p, 2n) ^{67}Ga reaction in an accelerator.

The ^{198}Au-gold aerosol offers several desirable features not found in other standard aerosols. The radiolabel, ^{198}Au, decays by beta emission accompanied by 412 KeV gamma photons. Its half-life of 64.75 hours is ideal for deposition and early clearance studies in animals and allows for repeated exposure of the same animals to a variety of particle sizes. The density of gold (ρ = 19.3 g/cm^3) is another desirable feature of this aerosol. Many of the environmental aerosols of interest such as the oxides of the actinides and other heavy metals are dense, and since the important deposition mechanism of Brownian diffusion is independent of density, a need for a high density aerosol exists. As an example, consider two particles both having 0.5 μm aerodynamic diameters with one particle having a density of 1.0 g/cm^3 and the other a density of 11.0 g/cm^3. These two particles would have geometric diameters of 0.40 μm and 0.081 μm, respectively; however, the smaller particle would have a diffusion coefficient ten times greater than the larger particle. Gold aerosols are also insoluble, a requisite for conducting inhalation deposition and particle clearance studies.

The production of monodisperse particles involves collecting the polydisperse aerosol with the Lovelace Aerosol Particle Separator (LAPS)[1] which separates the particles with respect to aerodynamic size and deposits the particles on a 46 cm long by 0.015 cm thick stainless steel foil. The foil is then cut into narrow segments, each containing a narrow size distribution of particles ($\sigma_g < 1.1$). Particles are then resuspended from the foil segments and aerosolized.

METHODS

PREPARATION OF GALLIUM-67 LABELED ALUMINOSILICATE SPHERES

Using procedures developed at the Inhalation Toxicology Research Institute, montmorillonite clay was prepared as described by McKnight and Norgon.[2] This procedure is illustrated by the flow chart in Figure 1. Basically, the procedure consists of treating a raw clay sample with concentrated H_2O_2 until no reaction occurs, decanting the finely divided clay suspension, "packing" the exchange sites with sodium, dialysis with running water to remove the excess sodium, allowing the cation of interest to exchange with the sodium ions, filtering and washing the labeled clay. The ^{67}Ga-labeled clay suspension is then aerosolized and heat treated to fuse the clay particles as described by Raabe, et al.[3] Preliminary tests using 10 mCi $^{67}GaCl_3$ (Amersham Searle, Chicago, Illionis) yielded a 3% activity exchange into 200 mg clay. Since the exchange capacity of the clay is 1.3 mEq/g, 200 mg of clay should have a capacity of 5.81 mg of ^{67}Ga. The specific activity was listed as 10^4 Ci/g of Ga(III), therefore other cations competing for the clay exchange sites were assumed to be present indicating the necessity for separating ^{67}Ga(III) from the zinc target material and other possible impurities.

Gallium-67 was extracted from 6 M HCl solution into isopropyl ether as the chlorocomplex $H^+GaCl_4^-$ as described by Morrison and Freiser.[4] Under these conditions, Zn(II) was not extracted.

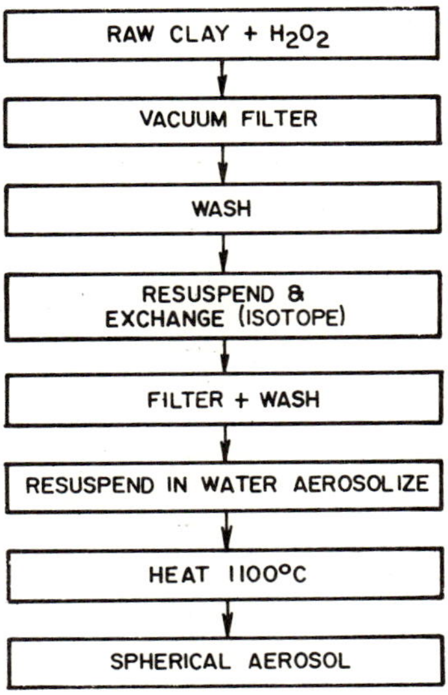

Figure 1. Flow Chart of Clay Preparation and Exchange.

The ^{67}Ga from the organic phase was stripped by shaking with 0.04 M HNO_3. The ^{67}Ga obtained was carrier free and was used directly for exchange into clay.

GENERATION OF ^{67}Ga-LABELED CLAY AEROSOL

After preparation of the labeled clay, the clay suspension is placed in a Lovelace nebulizer in an aerosol production system housed in a series of five connected stainless steel glove boxes described by Raabe, et al[5] and illustrated in Figure 2. The first glove box in the assembly line houses the generator and a low temperature (≃ 350 °C) heating column used to enhance drying of the aerosol droplets. The second box contains a high temperature tube furnace (1200 °C) to fuse the clay aerosol into aluminosilicate spheres entrapping the ^{67}Ga. The third glove box contains the sampling chamber from which the LAPS samples and other samples are drawn as needed. Routine samples taken include point-to-plane electrostatic precipitator samples for electron microscopy, cascade impactor samples for aerodynamic size determination and a continuous filter sample for activity concentration assays. A lead-shielded conconcentric electrostatic precipitator collects all particles not sampled. The fourth glove box houses two LAPS units that sample the resultant polydisperse aerosol and collect particles on stainless steel foils according to their aerodynamic properties. The fifth box is used for cutting the collection foils into segments and preparing the segments for individual storage and is not shown in the schematic.

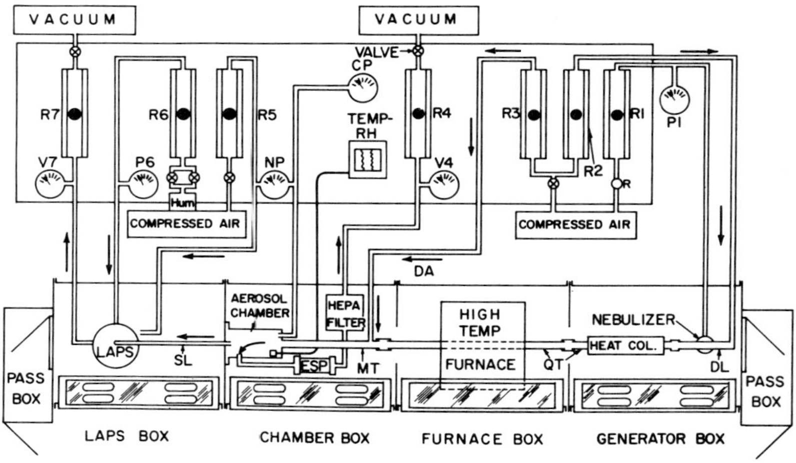

Figure 2. Schematic of LAPS Glove Box Line.

PRODUCTION OF GOLD AEROSOL

Aerosol Generator Solution

It was assumed that solutions of Au(III) could be aerosolized and subjected to thermal degradation at 1200 °C which should reduce the Au(III) to metallic gold since Au(III) is relatively unstable and gold melts at 1063 °C. All chemical systems using Au(III) in the presence of chlorine failed to yield spherical particles. One system using chlorine free Au(III) yielded spherical metallic gold particles at usable concentrations and is shown in the flow chart, Figure 3. Massive gold (^{197}Au) was dissolved in aqua regia with gentle (50 °C) heat. This solution was then continuously boiled and diluted with 1 M HCl for several hours to remove nitrate. The resultant chloroauric acid, $HAuCl_4$, contained 190 mg Au per ml and pH of 1-2. This stock solution was diluted with filtered deionized water to yield a chloroauric solution of ≃ 1.0 mg Au per ml and 4 < pH < 6. NH_4OH (1 M) was then added until 8.0 < pH < 8.5 is obtained. This precipitated an ammonium complex, "fulminating gold",[6] a mixture of two compounds: $Au_2O_3 \cdot NH_3$ and $NH(AuNH_2Cl)_2$ in an average ratio of 1:2.

$$4\ AuCl_4^- + 4\ NH_4OH + H_2O \rightarrow Au_2O_3 \cdot NH_3 + NH\ (AuNH_2Cl)_2 + H_2O + HCl \quad (1)$$

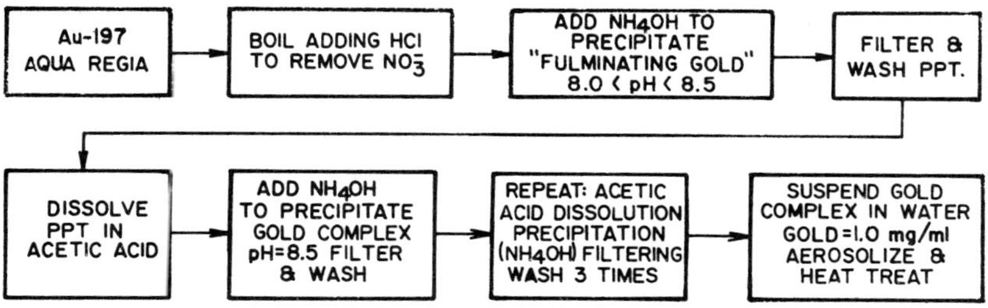

Figure 3. Gold Aerosol Production Flow Chart.

This precipitate was vacuum filtered using 0.45 µm pore sized membrane filters and washed with deionized water. The wet precipitate was then dissolved in 25 ml of 1 M CH_3COOH. The gold complex was reprecipitated with 1 M NH_4OH at pH = 8.5, filtered and washed with deionized water. This procedure of dissolution, precipitation, filtering and washing was repeated three times. The final precipitate was resuspended in deionized, filtered water using ultrasonic agitation to yield a concentration of ≃ 1.0 mg Au/ml. This suspension was a gold-ammonium-acetate complex free of chlorine ions.

Generation of Gold Aerosol

The acetate washed fulminating gold suspension was aerosolized using a Lovelace type nebulizer[7] aspirated with compressed air. The same aerosol production system described for the ^{67}Ga-labeled aluminosilicate aerosol was used to separate the polydisperse gold particles into relatively monodisperse fractions according to aerodynamic size.

Gold particles collected on a LAPS foil were resuspended from a segment of the foil, $1.02 < \sigma_g < 1.2$, in aqueous ammonia, pH = 10.1 ± 0.1), centrifuged, placed in quartz vials, vacuum dried, sealed and neutron irradiated to produce the radiolabel ^{198}Au. The production of the ^{198}Au-labeled gold spheres is illustrated schematically in Figure 4.

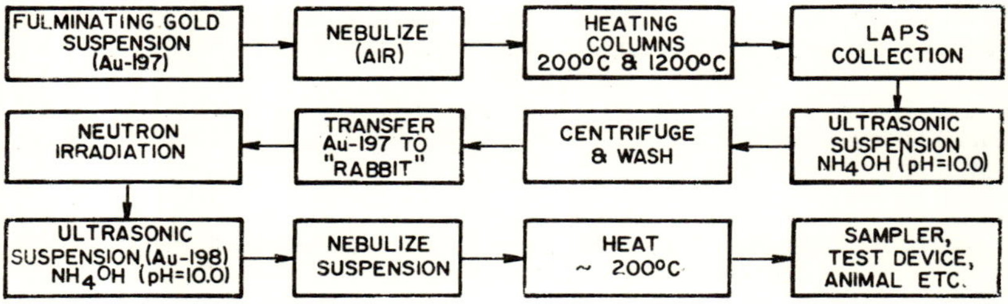

Figure 4. Monodisperse Gold Particle Preparation and Neutron Activation Flow Chart.

Monodisperse Aerosol Generation

Generator suspensions to produce monodisperse aerosols were prepared by suspending a monodisperse fraction of the aerosol collected on a foil segment in 0.001 M NH_4OH using ultrasonic agitation. The maximum number concentration (MNC) of the particle suspension was adjusted to yield 95% singlets using a Lovelace nebulizer as previously described.[8] The aerosol was passed through a ^{85}Kr discharger heated to 70 °C to reduce the charge on the particles and to enhance drying of the droplets. The dry aerosol was then mixed with clean, dry, diluting air and passed into an exposure apparatus.

RESULTS AND DISCUSSION

GALLIUM-67 ALUMINOSILICATE

Since the production of usable quantities of ^{67}Ga-clay monodisperse particles would require handling 100 mCi to 200 mCi of ^{67}Ga, the system of solvent extraction and clay exchange was modified to permit manipulator cell operations. Figure 5 is a schematic diagram of the procedures. The entire production procedure was tested using 10 mCi ^{67}Ga plus 0.5 mg carrier, Ga(III), to simulate a 100 mCi to 200 mCi production run. The following list indicates the relative losses in the solvent extraction and clay exchange processes:

Aqueous phase after organic solvent extraction	= 0.6%
Organic phase after back extraction into HNO_3	= 0.2%
Filtrate after cation exchange and filtration	= 16.7%
Amount exchanged into clay	= 82.6%

The total slightly exceeds 100% due to counting statistics.

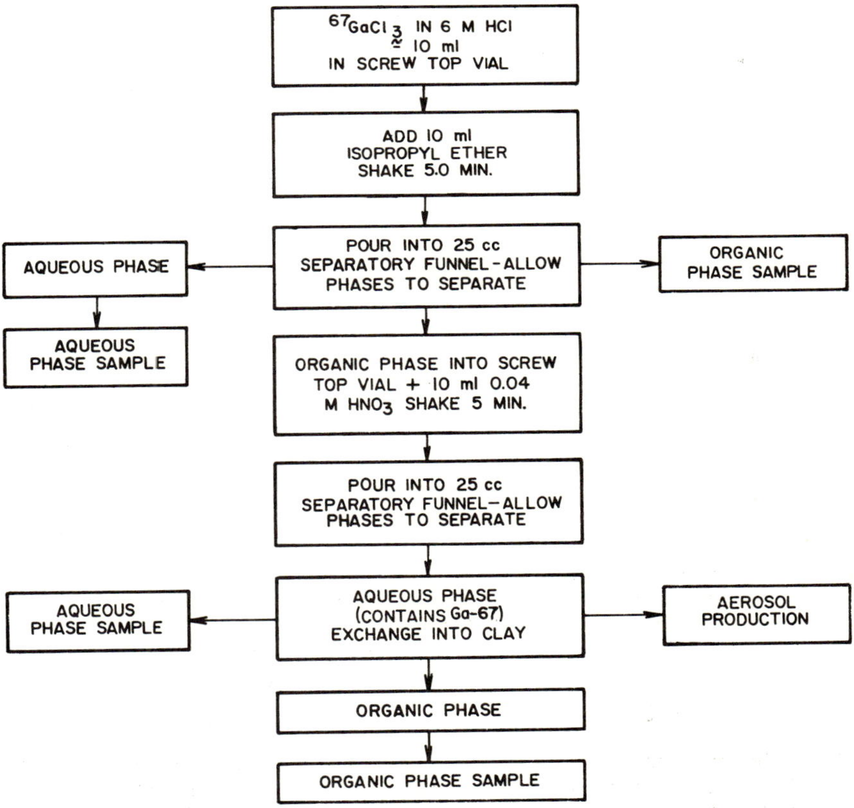

Figure 5. Flow Chart, Preparation of Monodisperse ^{67}Ga-Aluminosilicate Particles.

Preliminary tests indicate that 100 to 200 mCi ^{67}Ga can be exchanged into 40 mg of clay. The clay suspension could be aerosolized in 1 to 4 hours and the subsequent resuspension would require approximately 2 hours. Total production of the monodisperse particles would require 6 to 12 hours. Figure 6 is a composite electron micrograph of four sizes of monodisperse aluminosilicate particles labeled with Ga-67. The geometric standard deviation of these particles ranged from 1.02 to 1.10.

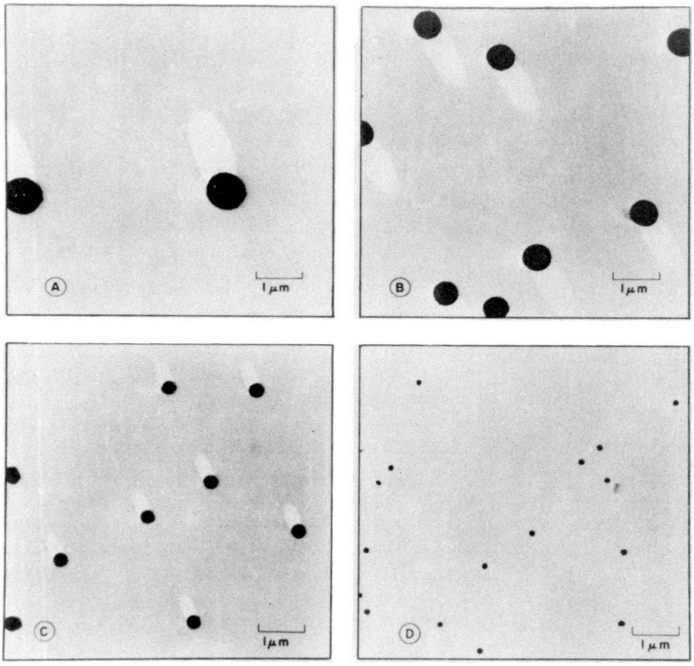

Figure 6. Composite Electron Micrograph of Four Monodisperse Sizes of ^{67}Ga-Labeled Aluminosilicate Particles from Collection Foil: (a) 0.86 μm; (b) 0.64 μm; (c) 0.32 μm and (d) 0.12 μm.

INHALATION EXPOSURES

Consider the mass and activity concentration of ^{67}Ga-labeled aluminosilicate spheres of 1 μm geometric diameter with a specific activity of 0.5 Ci/g clay and a maximum number concentration of 3.5×10^8 particles/ml in an aqueous suspension.

with

$$V = \text{volume of each sphere}$$

$$\rho = \text{density of the aluminosilicate} = 2.3 \text{ g/cm}^3$$

$$\text{mass/ml} = V \times \rho \times 3.5 \times 10^8 \text{ particles/ml}$$

$$= \frac{\pi D^3 \times 2.3 \times 3.5 \times 10^8}{6}$$

$$= 4.215 \times 10^{-4} \text{ g/ml}$$

$$\text{Activity/ml} = \frac{\text{mass}}{\text{ml}} \times \frac{\text{activity}}{\text{mass}} = 4.215 \times 10^{-4} \text{ g/ml} \times 0.5 \text{ Ci/g}$$

$$= 2.107 \times 10^{-4} \text{Ci/ml}$$

Using the above calculations, mass and/or activity concentrations for any particle size of interest can be obtained by multiplying the results by the cube of the diameter in micrometer units. Table 1 lists parameters of interest for various particle sizes for both dog and human exposures.

Table 1

^{67}Ga-Labeled, Monodisperse Aluminosilicate Spheres for Inhalation Studies With Dogs and Humans

Generator Concentration μCi/ml at 3.5 x 10^8 Particle/ml	Aerodynamic Diameter[a] (μm)	Geometric (Real) Diameter (μm)	Expected Pulmonary Deposition in a Dog μCi/min	Expected Pulmonary Deposition in a Human μCi/min
1400	3.00	1.88	2.5	5.3
383	2.00	1.22	0.67	1.4
211	1.66	1.00	0.37	0.79
18	0.80	0.44	0.03	0.07
6	0.60	0.31	0.01	0.02

[a]Lovelace Aerodynamic Diameter as discussed by Raabe, et al[5] where $D_{aer} = D_{real} \sqrt{(\rho C)}$

^{198}Au-GOLD AEROSOLS

Neutron activation of stable gold-197 was the method used to produce the ^{198}Au tag on the particle. Consider an 8-hr irradiation time for 1 g of ^{197}Au. Then the specific radioactivity (A) is given by

$$A = \frac{\sigma\ f\ n(1 - e^{-\lambda t_1})e^{-\lambda t_2}}{3.7 \times 10^{10}} = 36.94\ \text{Ci/g} \qquad (2)$$

where

σ = thermal neutron cross section activation area

= 9.8×10^{-23} cm^3

f = neutron flux = 8.0×10^{13} n/cm^2/s (Omega West reactor at LASL)

n = number of target nuclei per gram of gold = 3.06×10^{21}

t_1 = irradiation time = 8 hr

t_2 = time from removal from reactor to use ($\simeq$ 34 hr)

$\lambda = \frac{\ln 2}{T_{1/2}} = \frac{0.693}{64.75\ h}$ = gold-198 decay constant

Consider the mass and activity concentration of 1.0 μm diameter ^{198}Au-gold particles at the specific activity of 36.94 Ci/g and a concentration of 3.5×10^8 particles/ml in an aqueous suspension.

$$\text{Mass/particle} = \text{volume} \times \rho \qquad (3)$$

where ρ = density of Au = 19.3 g/cm^3

$$\text{Total mass/ml} = \frac{V}{\text{particle}} \times \rho \times 3.5 \times 10^8\ \text{particles/ml}$$

$$= \frac{\pi D^3}{6} \times 19.3 \times 3.5 \times 10^8 = 3.54 \times 10^{-3}\ \text{g/ml} \qquad (4)$$

$$\text{Activity/ml} = \frac{\text{mass}}{\text{ml}} \times \frac{\text{activity}}{\text{mass}} = 3.54 \times 10^{-3}\ \text{g/ml} \times 36.94\ \text{Ci/g}$$

$$= 0.131\ \text{Ci/ml} \qquad (5)$$

Using the calculation above, the mass or activity for any particle size of interest can be computed by multiplying the results obtained by the cube of the diameter in micrometer units.

Consider the following hypothetical dog exposure. Wanted: the activity/mass concentration required to yield an aerosol of ^{198}Au-gold that will deposit in the lungs of a Beagle dog ≃ 1.0 μCi initial lung burden (ILB) in 10 min.

Using a dog exposure apparatus previously described[9] and assuming 25% pulmonary deposition and a respiratory minute volume of 2.8 liters, the activity concentration of the particle suspension in a Lovelace nebulizer (0.05 ml/min output) with 20 l/min total aerosol volume was:

$$\frac{1\ \mu\text{Ci} \times 20\ \text{l/min}}{10\ \text{min} \times 0.7\ \text{l/min} \times 0.05\ \text{ml/min}} = 57.1\ \mu\text{Ci/ml} \qquad (6)$$

Table 2 lists parameters of interest for various particle sizes for both dog and human inhalation exposures.

Table 2

Monodisperse ^{198}Au-Gold Aerosols for Inhalation Studies With Dogs and Humans

Generator Concentration μCi/ml at 3.5 x 10^8 Particle/ml	Aerodynamic Diameter[a] (μm)	Geometric (Real) Diameter (μm)	Expected Pulmonary Deposition in a Dog μCi/min	Expected Pulmonary Deposition in a Human μCi/min
131000	4.81	1.00	≃ 230	≃ 478
6636	2.00	0.37	11.6	24.2
2302	1.50	0.26	4.0	8.4
442	1.00	0.15	0.8	1.7
174	0.80	0.11	0.3	0.6
28	0.55	0.06	0.05	0.1

[a]Lovelace Aerodynamic Diameter = $D_{real} \sqrt{(\rho C)}$ [5]

Figures 7 and 8 show electron micrographs of monodisperse gold particles collected on electron microscope grids placed on the LAPS sampling foil. The grids contain a replica of a diffraction grating with a line spacing of 0.83 μm. Figure 8 contains monodisperse gold particles and reference particles of polystyrene latex, ρ = 1.05. Both populations of particle sizes have the same aerodynamic diameter.

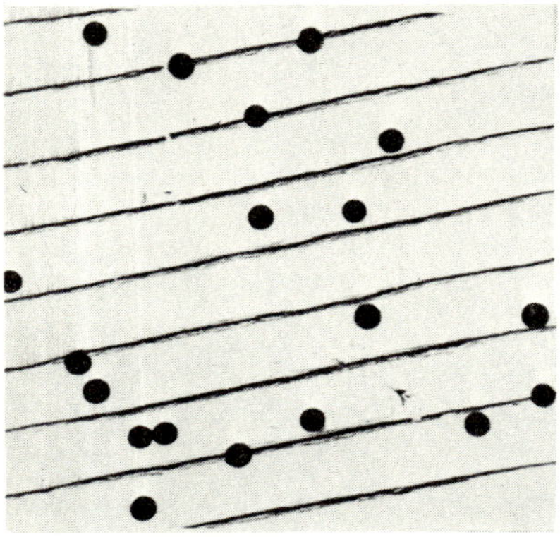

Figure 7. Electron Micrograph of Monodisperse Gold Particles from Collection Foil (0.83 μm Ruling)

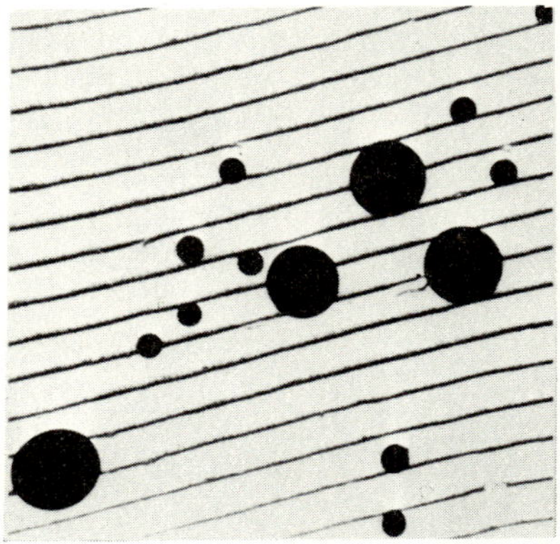

Figure 8. Electron Micrograph of Monodisperse Gold Particles and Reference Polystyrene Latex Particles (0.83 μm Ruling).

Estimate of LAPS Sampling Time Necessary to Obtain Useful Numbers of Monodisperse Gold Particles:

Wanted: Estimated run time utilizing two LAPS units to obtain enough gold particles for one generator loading for a single size range. ILB for a Beagle dog = 1.0 μCi.

Assume: Lovelace generator (0.05 ml/min output), 4 l/min diluting air and 300 ml/min/LAPS sampling rate, and a generator suspension concentration of 1 mg Au/ml and 5% of LAPS sample per segment. Then the sampling rate is given by:

$$1 \text{ mg/ml} \times 0.05 \text{ ml/min} \frac{2 \times 0.3 \text{ l/min/foil}}{4 \text{ l/min}} \times 0.05 \text{ seg sample/foil}$$

$$= 3.75 \times 10^{-7} \text{ g/min/foil segment} \qquad (7)$$

By assuming a generator load for exposure of 10 ml at a concentration of 60 μCi/ml, the LAPS sampling time required per generator load is given by

$$\frac{60 \ \mu\text{Ci/ml} \times 10 \text{ ml}}{3.75 \times 10^{-7} \text{ g/min} \times 3.694 \times 10^{7} \ \mu\text{Ci/g}} = 43.3 \text{ min/generator loading} \qquad (8)$$

A 10 ml generator solution was adequate for 120 min of aerosol generation time.

RESULTS AND DISCUSSION

The yield of ^{198}Au in monodisperse gold particles neutron irradiated in the Omega West reactor at the Los Alamos Scientific Laboratory was comparable to the calculated values and the difference was due to lack of precision in estimating the mass of gold. Numerous small animal inhalation studies[10] have been conducted using monodisperse ^{198}Au-gold aerosols to radioactivity levels between 0.1 μCi and 0.8 μCi per liter of air. The most significant problem encountered was difficulty in breaking up of agglomerates in the smaller particles.

CHARGE DISTRIBUTION

Both monodisperse aerosols, ^{67}Ga-aluminosilicate and ^{198}Au-gold, were studied for charge effects using a miniature charge spectrometer by Yeh et al.[11] After aerosolization of the particle suspension from 0.001 M NH_4OH, the aerosol was passed through a charge neutralizer containing 2 mCi of ^{85}Kr and then into the charge spectrometer. The configuration of the aerosol train resulted in about a 0.5 minute time lag from the particles exiting the ^{85}Kr neutralizer until entering the charge spectrometer. Preliminary tests of ^{67}Ga labeled spheres indicated a charge distribution close to Boltzman equilibrium. Yeh et al[11] found that the ^{198}Au-gold spheres exhibited a net positive charge that was proportional to the specific activity of ^{198}Au. This net positive charge approached Boltzman equilibrium as the radioactivity per particle decreased with time as seen in Figure 9.

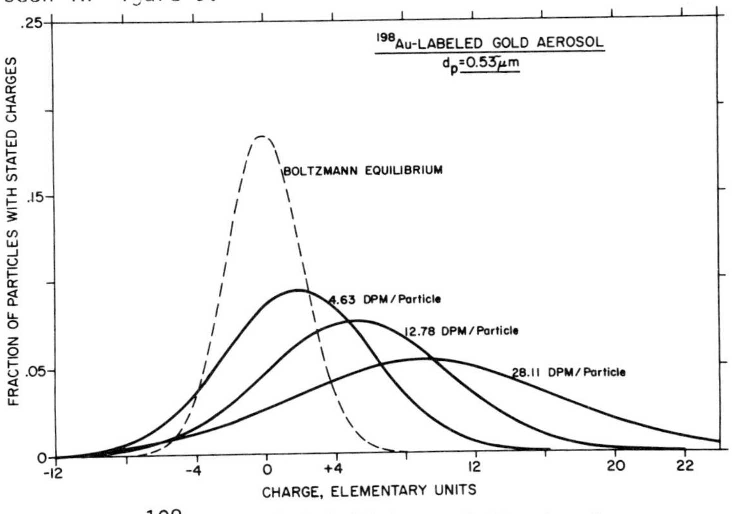

Figure 9. ^{198}Au-Labeled Gold Aerosol Showing Decrease in Net Positive Charge With Decreasing Specific Activity.

SUMMARY

The production of monodisperse spherical particles of ^{67}Ga-labeled aluminosilicate and ^{198}Au-labeled gold has been described. Gallium-67 labeled aluminosilicate particles ($\rho = 2.3$) and a specific radioactivity of $\simeq 0.5$ Ci/gm can be produced to yield aerosols ranging in size from 3.0 to 0.60 μm aerodynamic diameter. These particles are suitable for human inhalation studies involving initial deposition and early clearance. The maximum time required to achieve a 0.5 μCi initial lung deposit is 25 to 30 min for the smallest particle size. Similarily, monodisperse radioactive gold aerosols can be used for similar studies in Beagle dogs, rodents and man. Gold as a basic laboratory aerosol is desirable because it is stable under the electron beam of an electron microscope, insoluble, chemically inert, non-hygroscopic, non-toxic and available in high purity. Radioactive gold aerosols could also find applications where a unipolar charge distribution is desirable. Gold aerosols also have a high density ($\rho = 19.3$) which might be important in evaluating inertial sampling devices designed to sample atmospheres contaminated with compounds of actinides.

ACKNOWLEDGEMENTS

The authors are indebted to Mr. Emerson Goff and Mr. Michael Rios for preparation of illustrations, to Mr. Fred Rupprecht for editorial assistance and to Mrs. Judith Miller for typing the manuscript. This report was prepared under research supported by the National Institute of Environmental Health Sciences via ERDA Contract E(29-2)-1013.

REFERENCES

1. Kotrappa, P. and M. E. Light. Design and Performance of the Lovelace Aerosol Particle Separator. Rev Sci Instrum. 43:1106-1112, 1972.

2. McKnight, M. E. and R. W. E. Norgon. A Study of the Exchange Characteristics of Montmorillonite Clay for Fission Product Cations for Use in the Generation of Insoluble Aerosols. USAEC Research and Development Report, LF-37, Lovelace Foundation for Medical Education and Research, Albuquerque, New Mexico, October 1967.

3. Raabe, Otto G., George M. Kanapilly and George J. Newton. New Methods for the Generation of Aerosols of Insoluble Particles for Use in Inhalation Studies. In: Inhaled Particles III, Unwin Brothers Limited, Surrey, England, 1971. p. 3-18.

4. Morrison, G. H. and H. Freiser. Solvent Extraction in Analytical Chemistry. John Wiley and Sons, New York, 1957.

5. Raabe, O. G., H. A. Boyd, G. M. Kanapilly, C. J. Wilkinson and G. J. Newton. Development and Use of a System for Routine Production of Monodisperse Particles of $^{238}PuO_2$ and Evaluation of Gamma Emitting Labels. Health Phys. (in press)

6. Sneed, M. C., J. L. Maynard and R. C. Brasted (eds.). Comprehensive Inorganic Chemistry, Vol. II, Cooper, Silver and Gold, pp. 222, Van Norstrand Co., Inc., New York, 1954.

7. Raabe, O. G. Aerosol Generation and Characterization. In: Inhalation Carcincgenesis, Hanna, M. G., P. Nettesheim and J. R. Gilbert (eds.). Oak Ridge, Tennessee, U. S. Atomic Energy Commission Division of Technical Information. 1970. pp. 123-172.

8. Raabe, O. G. The Dilution of Monodisperse Suspensions for Aerosolization. Amer Ind Hyg Assoc J. 29:439-443, 1968.

9. Boecker, B. B., F. L. Aguilar and T. T. Mercer. A Canine Exposure Apparatus Utilizing a Whole Body Plethysmograph. Health Phys. 10:1077, 1964.

10. Raabe, O. G., Hsu-Chi Yeh, G. J. Newton, R. F. Phalen and D. J. Velasquez. Deposition of Inhaled Monodisperse Aerosols in Small Rodents. To appear in Proceedings of Fourth International Symposium on Inhaled Particles and Vapours, Edinburgh, Scotland, September 22-26, 1975.

11. Yeh, Hsu-Chi, G. J. Newton, O. G. Raabe and D. R. Boor. Self Charging of ^{198}Au-Labeled Monodisperse Gold Aerosols Studied With a Miniature Electrical Mobility Spectrometer. Aerosol Science (submitted for publication).

AEROSOL GENERATION FOR INDUSTRIAL RESEARCH AND PRODUCT TESTING

by

EUGENE E. GRASSEL

SR. RESEARCH CHEMIST, DONALDSON COMPANY, INC.

ABSTRACT

Aerosol generation plays an important role in developing reliable air cleaning systems. Aerosols needed for product testing, where the emphasis is on life performance, are different and consequently require a different method of generation than those normally used in research and theoretical studies. A number of aerosol generators and dust feeders have been used during the past 20 years to supply the broad range of aerosol needs for research, quality control and performance testing. The aerosol needs range in type from aerosols containing uniform particle size to aerosols having broad distributions in a specific size range. They range in capacity from small volumes at low concentrations to large volumes at high concentrations. The systems reviewed include commercially available units and specially adapted equipment. They are described in terms of aerosol needs, method of generation, kind of aerosol produced, methods of particle sizing and method of measuring performance. Experience with these general methods is discussed in terms of equipment limitations and problems in producing fine particle aerosols for life testing air cleaning systems. Also discussed are methods of fine particle aerosol generation that have been adapted for specific applications in lieu of more general systems.

AEROSOL GENERATION FOR

INDUSTRIAL RESEARCH AND PRODUCT TESTING

by

EUGENE E. GRASSEL

SR. RESEARCH CHEMIST, DONALDSON COMPANY, INC.

INTRODUCTION

Aerosol generation is an important element in developing reliable air cleaning systems. It is industry's function to translate understanding, materials, ideas and manpower into reliable products with a competitive edge to a profitable market. This competitive edge is defined in terms of size, price or performance or, in some cases, as simply a solution to an unsolved problem. This goal is accomplished by developing new materials and concepts, evaluating prototypes and providing a method of measuring quality. Each of these steps may require aerosol testing.

The purpose of this paper is to review twenty years experience with aerosols and aerosol generators used to satisfy an air cleaner company's needs for research and product testing needs. The life parameters such as life efficiency, dust capacity and cleanability are as important to product design as particle size efficiencies determined by initial efficiency tests. Life parameters are problem-system interactions not readily obtained by theoretical or semi-empirical calculations. They are usually measured by a comparative test designed to examine how well a solution satisfies the problems needs. Many of these tests have become standards for company performance of products from different manufacturers. Life testing of air cleaning systems designed to remove fine particles from industrial effluents will require a new type aerosol test. Each of the generators described here is examined in terms of this need as well as their present application, operation, capability and limitations. An effort is made to define the type aerosol and method of generation needed to measure performance of products designed for fine particle removal.

AEROSOL NEEDS: RESEARCH VERSUS PRODUCT EVALUATION

Many of the aerosol generators used by industry may seem crude and antiquated when compared with generators used in academic research. In most cases, an industrial aerosol is a simple dispersion of a collected sample or a standard dust. Such generators, however, should not be judged on the academic standard of aerosol size uniformity, but on how well the aerosol meets the industrial needs.

Aerosols used in research to examine natural phenomenon are generally different in kind and quantity from those used for product design and qualification testing. Experimental conditions in research can be selected to isolate and quantitavely measure a single mechanism, phenomenon or property as a function of variables. The emphasis is in quality not quantity. The quantity can be kept small by keeping the test sample small but the test particle should be round with a very small geometric standard deviation to simplify mathematical manipulation of data. Effects of system-aerosol interactions, that might change the experimental condition's with time, are minimized by using low aerosol concentration and keeping exposure times as short as is consistent with the sensitivity of the measuring method. A generator that provides a few liters per minute of monodispersed aerosols at a relatively low concentration is usually adequate.

Most of the aerosol generators used in industry are for improving products and customer qualification testing. The emphasis is on quantity and reproducibility. The aerosol must provide an estimate as to how well large systems as conceptual models, prototypes or production systems will perform in the field. This requires measuring the aerosol-system interactions as life parameters that make up a cleaning system's total performance.

The complexity of life performance parameters make comparative testing as useful as absolute measurements in evaluating the merits of a system modification or change. For such testing, idealized aerosols are neither practical nor desirable, as they usually shed little light on overall performance under field conditions. For such studies, broad distributions are desirable; and, in many cases, attempts are made to approximate application conditions in the laboratory by dispersing a typical sample of the material to be encountered. Because heavy dust loadings are required to collect life data and because it is desirable to minimize testing time, the concentrations of aerosol generated are maintained as high as practical. This requires generators capable of providing high concentrations at high airflows. For comparative tests to have meaning, these generators must provide a reproducible aerosol cloud with long term stability.

Donaldson Company furnishes filtration products to four market areas as shown in Table 1. Products developed for each of these markets require a different qualification test to evaluate performance in their application area. The need to retain a method of testing for the market life of a product and the range of airflows, concentrations, and kinds of aero-

Table 1. Comparison of Filter Requirements in Four Market Areas

	ENGINE MARKET	INDUSTRIAL EFFLUENT	SPACE CLEANING	HEPA FILTERS
Placement	Intake Air	Effluent Air	Intake and Recirculation	Intake and Effluent
Loaded Δp (in. wg)	25	3-5	0.5	1
Renewal Procedure	Replace	Clean	Replace	Replace
Material to be Removed by Filter	Atmospheric Dust	Dust from Grinding, Process and Conveying	Lint, Atmospheric dust, Atmospheric Fume	Toxic or Radioactive Fume
Size Range of Interest (microns)	>1	>1	all	<1
Design, Face Velocity for Media (m/min)	.9 to 6.1	.3 to 4.6	15.2 to 91.4	1.5
Test Concentration in (gm/m^3)	.88	2.29 to 91.5	Atmospheric	.080
Efficiency or Concentration Downstream	99.9% on AC Fine	.00046 After Steady State	25% Removal of Staining Power	99.97% on .3 DOP

sols required has resulted in a number of aerosol generators to meet the day to day testing needs. These generators have been grouped according to a market area. The generators consist of two types: 1) Those that generate aerosol by dry dispersion of a standard dust or a collected sample, and 2) those that generate aerosols either by condensation or atomization of a liquid. Most generators used for life testing are large particle generators and fall into the first category. The research and fine particle generators usually fall into the second category. The fine particle generator group will receive more industrial attention as the markets for air cleaning systems with increased efficiency on fine particles grows.

INDUSTRIAL GENERATORS

ENGINE AIR CLEANING MARKET

The range of feed rates and airflows covered by coarse particle generators used in the engine market are summarized in Table 2. To better understand industrial aerosol generation, it is necessary to examine the history of air cleaning at the time a product was introduced to a market and a test method was selected for product qualification.

Industrial practices make it difficult to document dates at which some test procedures were first adopted. Even today, equipment or materials will diffuse in and be well established before they are recorded. The early history of the air-floated dust feeder and the evolution of standardized test dust are good examples.

The need to remove airborne silica dust in the intake air to minimize wear was recognized as an air cleaner market prior to 1920. At that time there was little worry about filtration theory or optimizing design of filter materials. Trial-and-error methods provided an adequate basis for solving most problems and the technique of modify-and-test provided a method for refining the system.

Standardized Arizona Road Dust

Early engine air cleaning designs were tested on equipment at an Arizona test track. There is no record as to when field testing was first supplemented by hand dribbling dust from a scoop into an engine air cleaner inlet as an expedient in design. The dust used was collected near the Arizona engine proving ground and shipped to the laboratory in 55-gallon drums. At first, it was prepared by screening through a 200 mesh sieve and analyzed by a sieve analysis to divide the mass into two size ranges. The test code adopted by the SAE in 1941 recommended either a sedimentation analysis or a roller analysis for evaluating the fines, but left the size distribution problem to the air cleaner manufacturer[1].

Table 2. Aerosol Generators used for the Engine Air Cleaning Market

Dust Feeder	Feeder Capacity (gm)	Feed Rate Range (gm/min)	Test Velocity Range (c/m at .025/cfm)
Wright Dust Feeder	65	.0125 - .125	.5 - 5
Air-Floated Dust Feeder	700	manual adjustment	up to 400
Sonic Feeder	3,000	.25 - 250	10 - 10,000
Automatic Feeder	Reload during operation	5 - 230	200 - 9,200
Barrel Feeder	136,000	5 - 1135	200 - 45,400
Novadel Feeder	Reload during operation	1 - 1250	40 - 50,000

It was necessary to increase the percentage of fines in order to measure improvements oil bath air cleaner designs in the test dust and an air elutriator method was adopted to do this. During the war years, the air elutriator method was replaced by a ball milling operation[2]. Shortly thereafter, coarse and fine roller analyzed dust became commercially available. The commercial availability of standardized Air Cleaner Coarse and Fine dust was not recognized by the SAE until the 1960 change in the SAE code[3].

With the introduction of dry air cleaning systems to the engine air cleaning market, it has been necessary to supplement standardized coarse and fine testing with other dusts to provide adequate design information. Standard AC Coarse dust was found to be unsuited for loading studies used for improving the dust capacity of paper. Total cake depth relative to dust paper interaction depth made the test insensitive to the paper structure. Use of fine test dust improves the definition of the paper structure but provides little information about fine particle loading that causes high pressure drops with light loadings.

Using dust-carbon mixtures to simulate high pressure drops at light loads was unsuccessful, although the dust-carbon mixtures were of similar composition to those determined from problem situations in the field. This is undoubtedly due to the inability to deagglomerate the carbon-dust mix and feed each particle as its discrete size. Standardized coarse or fine dust does not provide adequate information about cake support. Dust cakes formed by most dusts on fiber beds will become unstable and migrate if either the pressure drop and/or velocity become too high. The values at which this takes place are dependent on distribution of the dust and the structure of the fiber bed supporting the cake. For papers with 90 micron maximum pore and dust with a mixture containing a large portion of 5-15 micron particles, the values are 20 ft/min face velocity and less than 20 in. wg pressure drop. A dust containing such a mixture is used to develop filter papers with proper fiber spacings to support these poor cake formers.

Air-Floated Dust Feeder

The air-floated feeder shown in Figure 1 is a fabricated item and consists of a 2.5 in. o.d. tube 21-1/2 in. long with a rounded bottom containing a 1/8 in. dent. A 5/16 in. o.d. copper tube brings air into the system. The operation of the air-floated dust feeder depends on the range of sizes in the distribution of the dust to be fed. It feeds dust by classifying the floating out the fines first. The rate is maintained by increasing the air pressure to float out larger sizes. The range of sizes in the distribution of both coarse and fine Air Cleaner dust permits a reasonable uniform feed. The feeder is operated manually and the feed rate is dependent on the particle size distribution of the dust and

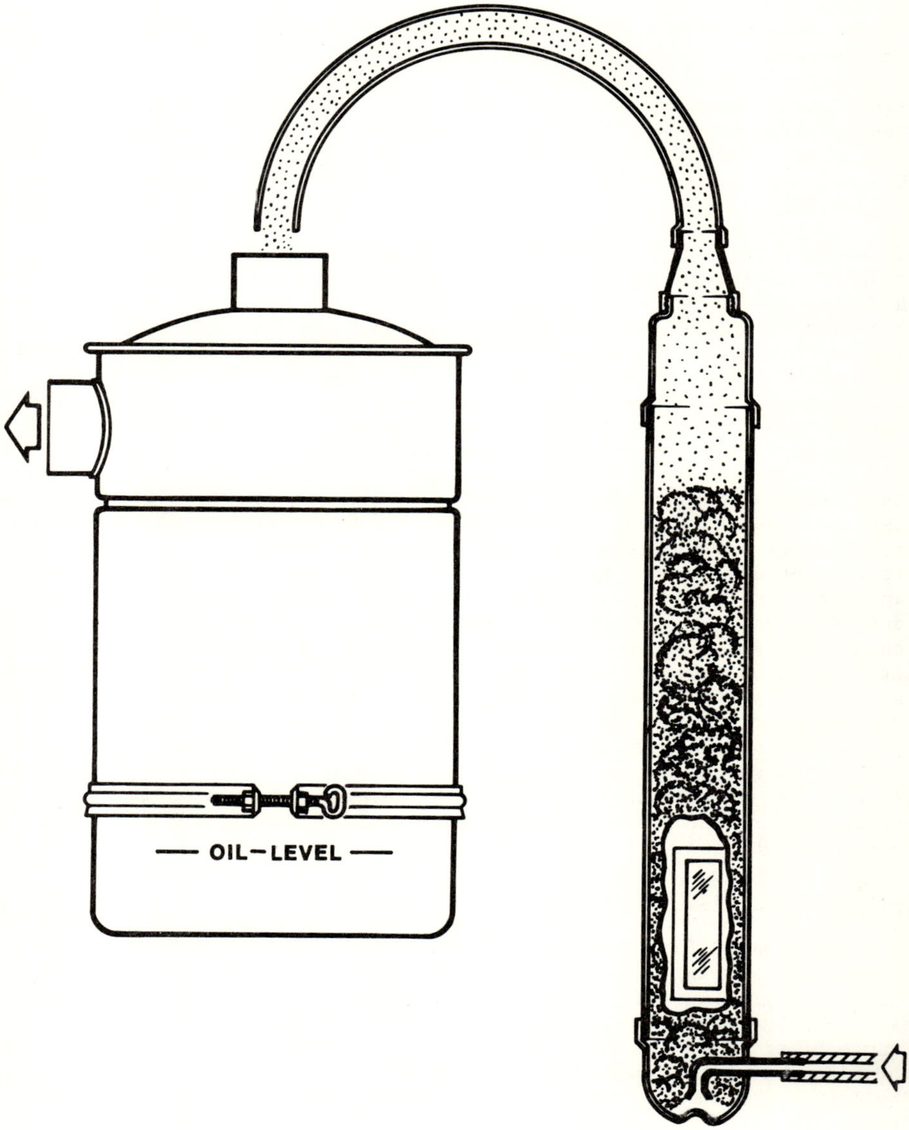

Figure 1 Air-Floated Dust Feeder

operator judgement. It is unsuited for dusts having a distribution with a small σ_g.

The air-floated dust feeder was adopted to test the oil bath air clean-ers (a type of oil wetted scrubber) sometime prior to 1940. The life efficiency of oil bath air cleaners, like many steady state air clean-ing systems, is a function of the size distribution of the particles or agglomerates fed and is only slightly influenced by particle classifi-cation during feed. This feeder represented an advance and, in the hands of only a few experienced operators, the disadvantages of the air-floated feeder were not apparent until it was well established as a standard method.

In 1940 this feeder[4] was proposed to the Society of Automotive Engineers. The proposed feeder and test procedure were adopted in 1941[1]. 1953 changes in the Code recommended improvements for increasing test repro-ducibility along with a new type dust feeder called the Ordnance feeder[5]. Although the new feeder was motor driven and did not classify dust, it was more difficult to use. Action of the vibrators and piston compacted the loose dust in the feeder, sometimes forming a cake that did not move readily in relation to the cylinder wall causing the feeder to jam.

Until its recent displacement by the automatic feeder, the air-floated feeders simplicity, portability, ease of set-up and use, and our exten-sive experience of interpreting results made it the preferred feeder for most testing. It was used as a single unit for airflows less than 400 cfm and in multiples necessary to meet flow requirements on higher air-flows. It is still specified for qualification testing of military air cleaners.

Wright Dust Feeder

The smallest dust feeder and the standard for paper research in our laboratories is the Wright dust feeder. It is a purchased item normally employed by others for toxicological studies of animals. Its operation depends on air picking up the small fraction of a dust cake scraped into a channel as the compacted dust cake rotates past a fixed scraper blade. The usable range with silica dust is not as great as the gear combinations furnished with the instrument. At zero visibility, 0.025 gm/ft^3, the lowest feed rate is limited by the amount of air pressure and resulting airflow (0.5 cfm) needed to pick up the separated dust. The highest feed rate is limited by the available motor power and difficulty of cutting the compacted dust cake. Uniformity of feed is dependent upon the way the dust is packed in the dust cup. When the dust cup is properly loaded with silica dust, the average dust feed over both reasonably long and short intervals is constant, although the in-stantaneous feed may vary greatly. To obtain good feeds, the cake must be built very slowly in extremely thin layers packed under pressure. The feeder cannot be used for every kind of dust, especially soft materials with high cohesive forces (for example, coal dust). The size distribution of the airborne coal fed from the Wright feeder may have no relationship to the distribution packed in the dust cup.

The Wright feeder has the advantage of being light and portable and easy to set-up and use. Except for the care necessary in packing the dust cups, results are operator independent. The degree of agglomeration is subject to the concentration of water vapor in the pick-up of air. Loading results on filter materials show winter-summer variations that can be modified by humidity control. To improve reproducibility between tests, the feeder should be sealed in a chamber to minimize leaks that develop intermittently around threads and machined surfaces. The low flow rate of the feeder limits its use to filter media development studies and a very few low flow tests. Its primary use in our laboratory is on small flow benches for life testing flat samples of papers and felts. It has been used in filter media studies to examine such factors as: 1) Effect of fiber bed structure on loading characteristics of media, 2) effects of fiber shape, length, diameters and bonding on sheet properties relative to efficiency and loading, and 3) the effect of average fiber spacing on dust cake support.

Sonic Feeder

In 1959 the sonic dust feeder, shown in Figure 2 and 3, was designed and fabricated. It was based on the same principle as a smaller feeder designed at the University of Minnesota by Dr. Whitby's group[19]. Its operation depends on a pressurized chamber containing a screw driven platform entering from the bottom and a pick-up foot mounted on a stem coming down from the top. The stem is connected to the feed head. Dust to be fed is loosely loaded in a dust cup. The cup is placed on the platform and the platform raised until the pick-up touches the dust surface. The chamber pressure is adjusted so the airflow picks up only that dust pushed over to the pick-up opening. Dust arriving at the head is deagglomerated by a sonic shock due to high velocity air passing through the head at 90° to the incoming dust. By adjusting the pressure of this added air, the degree of agglomeration in the dust fed can be controlled. Agglomeration for road dust can be varied from almost completely dispersed to heavily flocked aerosol.

This feeder was offered as an alternate to the Ordnance feeder in 1959 and accepted as part of the SAE Standard in 1963[6].

The sonic feeder is capable of feeding both coarse and fine AC test dust at rates ranging from 0.25 to 250 gm/min. At zero visibility, this represents airflows of 10 to 10,000 cfm. This versatility is obtained

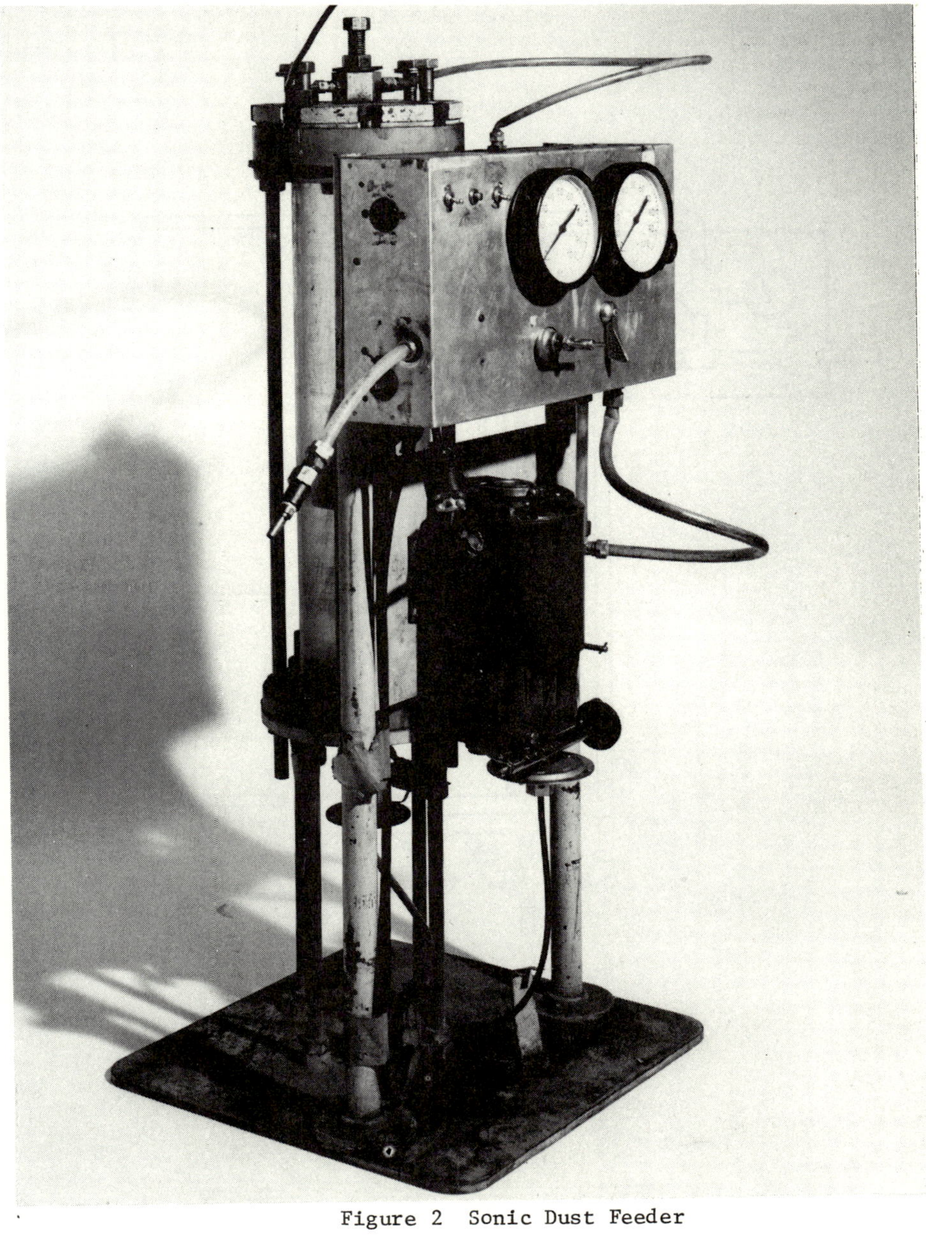

Figure 2 Sonic Dust Feeder

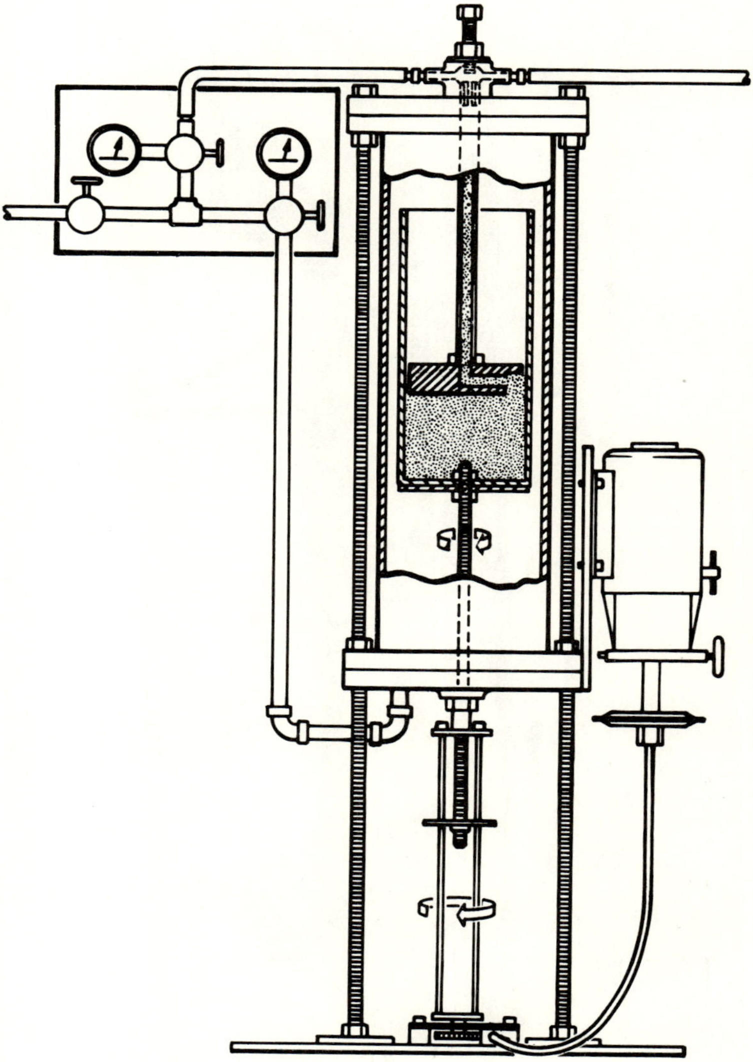

Figure 3 Functional Drawing of the Sonic Dust Feeder

by using different size dust cups and a variable speed drive.

The sonic feeder is heavy and difficult to transport from one point in the lab to another. Ease of over-pressurizing the chamber to pick up too much dust or under-pressurizing causing compaction of the dust by the pick-up foot, alters the bulk density and makes the feed rate difficult to estimate for critical testing.

This feeder found its major use in selecting candidate materials for self-cleaning air cleaners. Its present use is product qualification testing.

Barrel Feeder

The barrel feeder is a scaled-up version of the sonic dust feeder. It is limited to a single dust container and pick-up. Its variable drive allows feed rates from 5 to 1135 gm/min or to supply dust at zero visibility to airflows of 200 to 45,400 cfm. Its major use has been in examining prototypes of self-cleaning air cleaners. Its 300 lb dust capacity permits unattended testing of such devices for extended periods of time. This feeder, however, presents the same problems as the sonic dust feeder of compacting the dust and uneven feeds. An even greater disadvantage is its size and weight.

Novadel Feeder

The purchase of a NOVADEL VARI-FEEDER® coincided with interest in higher airflows for larger engines and gas turbines. It was selected to feed both coarse and fine road dust and was intended to replace the barrel dust feeder.

The system depends on mechanical agitation to fluidize the bed sufficiently to keep the auger channel full. The dust is delivered to an aspirator by a hollow spring auger. It provides as good a feed rate control and a wider range of feed rates than any other dust feeder listed in this report. It can feed over a range of 1 to 1250 gm/min to cover flow ranges of 40 to 50,000 cfm at zero visibility. It meters out dust with a minimum of classification and can be adjusted to a consistent rate. It can be loaded during operation and therefore provide continuous aerosols for extended periods of time.

® NOVADEL VARI-FEEDER is a registered tradename of Novadel Ltd., London, England

Automatic Feeder

The automatic feeder, shown in Figures 4 and 5, is the preferred feeder for most of our present engine air cleaner test purposes. It was designed and fabricated to supplement a flow controller which automatically compensates for pressure drop increases across an air cleaner as it loads with dust. The feeder consists of a straight-sided hopper with a screen false bottom separated from and mounted above the true bottom containing a spring auger. A low velocity scraping device is mounted above the screen and a second identical unit scrapes the material falling between the screen and bottom into the offset auger channel. Such an arrangement minimizes bridging when feeding very fine dust and keeps the auger channel filled with dust which has a reasonably constant bulk density. The auger speed is set by digital controls. The feeder is mounted on a 20 Kg balance so the total dust fed is easily obtained at the end of tests or at intermediate points. The dust feeder, balance, and all its controls are mounted on a cart for ease of transportation to any test site.

This feeder is compact, mechanically reliable, and easy to operate. It delivers a consistent particle size over reasonably long and short intervals. The feed rate can be adjusted for feeds between 5 and 230 gm/min to accommodate testing at zero visibility for airflows between 200 and 9,200 cfm. The major disadvantages are tight packing and auger clogging when feeding soft or very fine dusts.

Methods of Evaluating Efficiency and Particle Size

Life efficiencies are determined from measuring the weight of total dust fed and material collected on an absolute downstream of the test device. Almost all of the information about particle size effects, from both the laboratory and the field, are generated from collected samples. Size distributions are determined on a mass basis using either the Whitby MSA Sedimentation method or a Coulter Counter. Both methods provide the most information about the larger sizes and the Coulter has a lower practical limit of 0.7 microns. With these methods, however, particle size efficiency measurements made by feeding a polydispersed sample and analyzing collected samples may tend to show a better fines efficiency than should be expected because there is no way to estimate the portion of efficiency that is due to agglomerates presented to the system. As the test dust to be dispersed moves toward the finer sizes, it becomes harder to disperse by sonic shock. As the dispersed dust approaches these finer distributions, it will be necessary to determine the airborne nature and the extent of flocculation of the dust fed to provide reliable information about efficiency and cleanability for design. Equipment to provide this information is being considered.

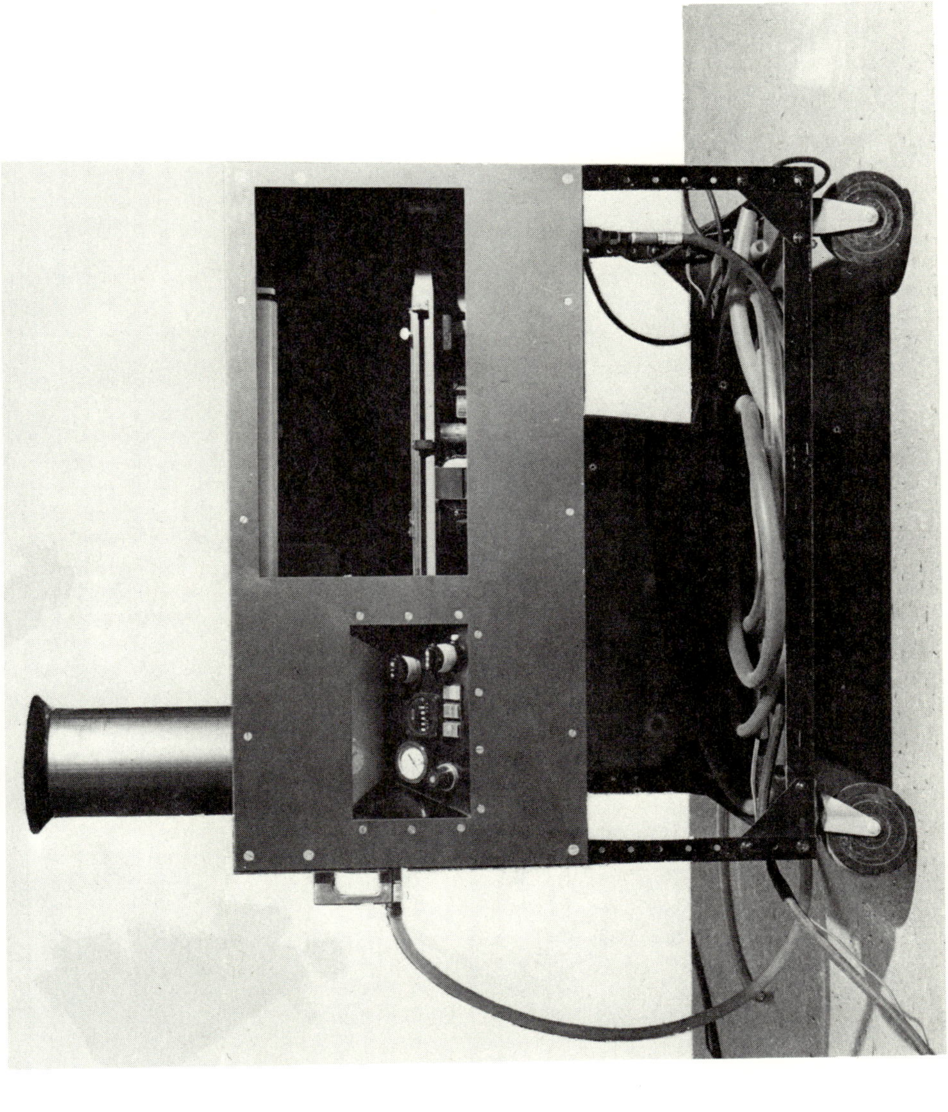

Figure 4 Automatic Dust Feeder

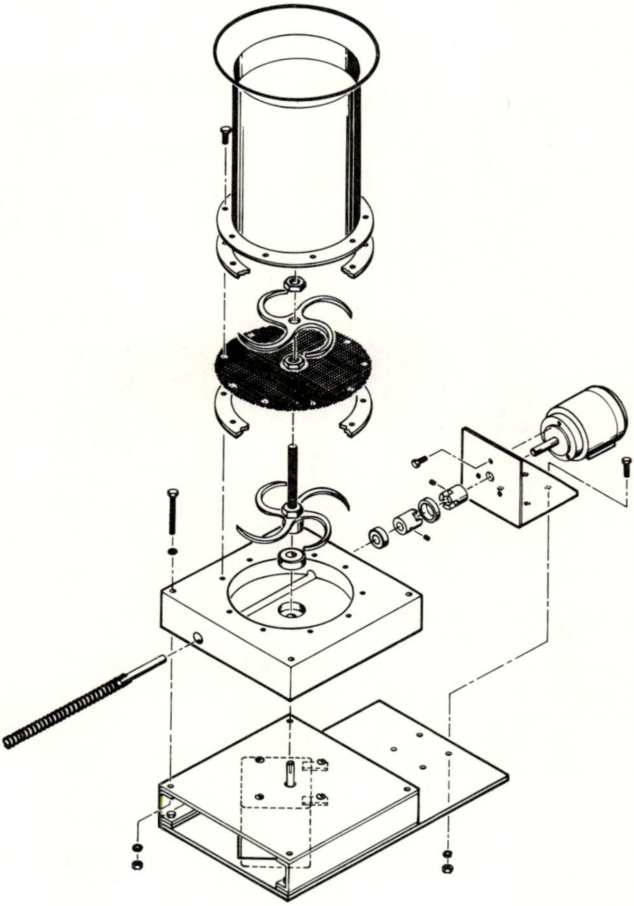

Figure 5 Exploded Drawing of the Automatic Dust Feeder Dust Metering Section

INDUSTRIAL EFFLUENT MARKET

Dust collection systems such as cyclones, electrostatic precipitators, scrubbers, and baghouse filters for the industrial effluent market also pre-date the extensive interest in submicron particles, refined air cleaning theory, or legal needs to clean air. At the time of their introduction, they were sold on the basis of either economics or public relations. In all cases, the emphasis was on the removal of coarse particles and mass efficiency.

The portion of the market covered here is baghouses and cabinet cleaners with airflows from 200 to 30,000 cfm. Conventionally, these cleaners are physically larger for the same airflows and require a different kind of filter material than is used in engine air cleaners. This is because they are intended to operate in a steady state condition, i. e., a constant pressure drop. They require constant cleaning by reverse air jets in order to operate in high dust concentrations for extended periods of time. Materials used in filter bags are purchased to specifications. The filtration characteristics for selecting materials are determined by testing on a small flow bench with a back flushing capability. Dusts are applied to the surface using a spring type auger transport system and a strong air jet to aspirate and deagglomerate the sample to form the aerosol cloud. After a number of loading-unloading cycles have been run and a steady state has been reached, penetration of the materials, normally expressed in grains/ft^3 air, is determined by sampling downstream of the test sample using a millipore filter. If size efficiency is determined, it is from Coulter results run on samples of feed dust and the dust on the millipore downstream of the test sample.

Tests are run on full size units using either a straight through or a looped system. There are no standard test procedures, dusts, or feed rates in this industry as provided by the SAE code in the automotive industry. The method of testing has been left to the discretion of the manufacturer. Duramite (a calcium carbonate dust), fly ash and iron oxide are representative of the test particles used. Test criterion requires a 24-hour continuous test without a change in pressure drop, called steady state conditions. After a steady state condition is reached, the penetration should not exceed 0.0046 gm/m^3 using a straight through test.

Two types of feeders, SYNTRON® and VIBRA SCREW®, are used interchangeably. The preferred feeder for life testing is the VIBRA SCREW feeder.

® SYNTRON is a registered tradename of Syntron Co., Homer City, PA.

® VIBRA SCREW is a registered tradename of Vibra Screw, Inc., Totowa,NJ

This is a purchased item capable of feeding 3 to 32,000 gms/min. The feed rate selected for testing may be varied from 1 to 40 grains/ft^3 (2.29 to 91.5 gm/m^3). Assuming a feed of 10 grains/ft^3 (22.9 gm/m^3) as normal, this feeder has a capacity of testing systems with airflows of 5 to 50,000 cfm. Under a 24-hour life test, a rotary valve is installed in the hopper outlet. The dust collected in the hopper is returned to the feeder which recycles it back to the cleaner inlet. Penetration is determined after steady state conditions are reached by sampling downstream of the bag, using either an absolute or an EPA probe.

SYNTRON feeders can also be purchased in a number of sizes and feed rates. They are best used in a straight through mode, although they can be looped for recycle. For longer runs under recycle operation, the SYNTRON tends to give ever increasing feed rates with respect to time as the viscosity decreases with time as powder is recycled. This increase is accentuated with certain powders if a method of dispersing, such as a materials handling fan, is used to break up the agglomerates. In such cases, Coulter measurements indicate a loss of the measurable fines. This fines loss appears to stabilize the flow characteristics of the bulk sample. The size distribution in the feed and the method of feeding dispersed dust cannot be expected to present either the total fines or fines at the same rate as can be expected in some industrial applications even though the total concentration of airborne dust may be the same. In applications where a large percentage of dust is fines, such a test method may not be a good indicator of either penetration or rate of blinding.

Sizing baghouses to an application is usually dependent on the experience and judgement of an applications engineer and, at best, supplemented by a Coulter analysis of a collected sample. If there is doubt about the percentage of fines, it is the practice to recommend a larger unit that provides more media area per cfm of airflow. Within the limits of available filter media, design improvement of baghouses is a packaging problem. Applying engineering practice normally reserved for the engine air cleaning market to the baghouse market has reduced the unit size and increased the media area without changing the throughput.

SPACE CLEANING MARKET

The ASHRAE Standard

The space cleaning market presents an entirely different set of parameters for filtration than those in the engine or industrial effluents markets. The application requires moving large quantities of air with blowers having low static pressures. For this application, filter media face velocities may reach 300 ft/min; but pressure drops, even in the loaded condition, cannot exceed 1 in. of water and may be limited to

0.5 inches of water.

In the home comfort market, the major purpose of deep bed filters containing 50 microns and larger fibers is to reduce lint collection on heat exchangers and in hot air ducts. Higher performance filters are specified by efficiencies measured according to the American Society of Heating, Refrigeration and Air Conditioning Engineers (ASHRAE) standard[7]. Filters with efficiencies less than 20 percent or greater than 98 percent fall outside the scope of this standard.

All ASHRAE standard testing is performed in a large duct with direct access to the outside atmosphere. The test involves a specific procedure and depends on the universal nature of aerosols found in the atmosphere. A test requires an initial estimate of the filter's performance. After its performance is estimated, the filter is placed in the test duct and the flow rates of the system adjusted. Sampling tips are selected from one of two sizes that best matches the duct velocity to provide semi-isokinetic sampling of the duct. Flows through the sampling tips are controlled by critical orifices and measured with dry gas meters. The downstream sampler is run continuously and the upstream sampler is set for intermittent operation, so the difference in darkening between the two sampling filters is no greater than 20 percent. The test length must be adjusted so the staining opacity of the samples shall not be greater than 40 percent or less than 10 percent. The opacity of staining is determined by an opacity meter calibrated to read relative light transmission.

Although the atmosphere provides a convenient source of fine aerosol for filter testing, it probably is far from a standard. To be a valid standard, it must be assumed that atmospheric aerosol is of universal composition, although experimentally if it is found that the length of the test varies from day to day and day to night. In three consecutive tests on the same filter, one test may be higher or lower than the other two. Information published by others indicates the particulate matter in the atmosphere is not universally uniformly distributed[8]. The second assumption is that the darkening power in the aerosol is equally distributed among all particle sizes in proportion to their mass. Without a knowledge of the relationship between staining power and particle size at the time of the test, it is difficult to correlate this standard to any other initial efficiency test. Products manufactured to this specification are best tested at the customer's site.

It takes one to four hours or longer to run an initial efficiency test on atmospheric aerosol. In tests to determine life-loading characteristics using atmospheric aerosols, the filter of interest is compared to a standard by running the two in parallel in a common test duct. Such life tests usually average three to six months, which is too long to be of much value in product improvement research, furthermore, the competitive nature of many segments of this market limits the type of improvements that can be justified.

AFI Feeder

The AFI feeder provides a method of feeding materials which are difficult to feed uniformly by other means. These include materials that contain fibers, those that might deform and fuse on packing, and those that are too free flowing because of their particle size distribution. The AFI feeder was until recently commercially available. The feeder consists of a support frame, a tray to hold the dust, a gear train to move the tray and metering gear, a motor to drive the gear train, a gear that meters the dust from the channel to the pick-up area, and a pick-up tube and aspirator for dispersing the dust. The original feeders have been modified to provide a wide range of feed rates. A full tray holds about 150 gms AC Fine or 50 gms coal. Feed time can be varied from 2-2/3 minutes to 146 minutes. It is the feeder specified in the ASHRAE code for feeding ASHRAE dust consisting of 72 percent standardized air cleaner fine dust, 23 percent Molocco Black and 5 percent #7 cotton linters. Experience with this feeder has been in feeding AC Fine for military qualification testing and coal dusts to study performance of air cleaning devices for the mining industry.

Turntable Feeder

The turntable feeder, shown in Figure 6, is similar in operation but not design to the AFI feeder. It was developed at Stanford Research Institute to feed small quantities of ground materials at a precise rate to a classifier system.

The feeder consists of a turntable rotated by a variable speed motor. The table contains a groove to accept dust from a hopper which holds the supply. The hopper contains agitating bars to keep the dust moving toward the groove. A scraper bar removes excess dust from the table to the groove and keeps the dust in the channel level with the table top. The system can use either of two types of pick-ups. The dust can be aspirated out of the channel as in the AFI feeder or the system can be pressurized to force the dust out the feed tube, as in the sonic feeder.

The turntable feeder has been built to provide a number of feed rate ranges. The practical limit on both the upper and lower ends of feed rate depends on the ability of the bulk powder to flow. On the upper end, the maximum depth and width channel are limited to a size that permits the powder to flow so freely it floods the table. The lower end is limited by the groove size which is so small that agglomerates from the hopper will bridge and prevent the channel from filling evenly. For

Figure 6 Turn Table Feeder

AC Fine, the lower practical feeding limit is between 0.1 to 0.5 gm/min.

The turntable feeder supplements the automatic feeder and is the preferred feeder for low feed rates. It provides a consistent feed and can be equipped with large hoppers for unattended runs for extended periods. Its compact size make it a convenient feeder for many tests.

HEPA MARKET AND MONODISPERSED AEROSOL GENERATORS

Interest in fine particles, filtration theory, and HEPA filters grew out of a World War 11 military need[9]. Pre-World War II gas mask filters did not provide the needed protection against toxic war gas when these gasses were presented as finely divided particles. The application required a filter having a small size, a low pressure drop at flow rates below 85 liters/min, and an efficiency of 99.97 percent minimum for the removal of all airborne particles. HEPA filters, however, have low dust capacities and a fragility that prevents cleaning. This prevents their use in solving most industrial effluent problems, limiting their use to space cleaning for contamination sensitive applications and industrial effluents containing highly toxic or radioactive materials.

Academic interest in filtration theory has provided a number of monodispersed aerosol generators which we have purchased or copied for research, calibration, or quality control. These can be split into drop generators and thermal dispersion methods. As the design, operation and application have been adequately described by others in the literature, only applications of industrial interest will be covered.

Spinning Disc Generators

Experience with drop generators includes both a high speed and a low speed spinning disc generator. The high speed generator is a purchased item and is used mainly to provide particles to calibrate equipment. The low speed spinning disc generator[10] was fabricated and used to provide 8 micron and larger particles to study velocities in fiber beds at which collection by inertial deposition decreased with increasing velocity.

Thermal Dispersion Generators for Research

Several generators based on the thermal dispersion of a liquid have been built and used. The earliest was a generator built under a government contract in 1955 to study inertial deposition in fiber pads of very fine particles (0.3 micron) using very high velocities (up to 100 ft/sec)[11].

This generator was capable of producing 50 cfm of aerosol and delivering it under pressure to a test manifold. Particle size was determined using a four position jet impactor with the jets operating in parallel. The jets covered the 0.1 to 1 micron range in six overlapping steps. The dye mass for both this and filter studies was determined colorimetrically. Dust loading studies were made using an oil orange aerosol; however, there is some doubt as to the practical use of this aerosol in fiber bed loading for comparing performance parameters of systems with improved loading, as the particle is deposited as an oil drop and then solidified. It is doubtful if thermally generated particles can be used to simulate solid particles in industrial effluents for loading and cleaning cycles in baghouse fine particle studies.

The oil orange generator was followed by a LeMer type generator[12] for laboratory studies of filter materials. It is the smallest of the available thermal dispersion generators, producing only 2 cfm of low concentration aerosol. It is the only one which employs a nuclei source and the only one that can be easily adjusted to produce particles in a wide range of sizes from DOP, steric acid, or oil orange. Tediousness in measuring both the particle size and concentration limited the extensive use of these earlier generators.

Thermal Dispersion Generators for Quality Control

Early use of DOP as a standard test for HEPA filters was under a military contract using a standard DOP machine of World War II vintage. Particle size adjustment of such generators is limited to a narrow band of sizes near 0.3 microns. Particle size control was sensitive to temperature changes in the laboratory. A similar more modern version of equal capacity has replaced this earlier DOP machine and is used routinely for quality control and customer acceptance testing.

The largest of the available thermal dispersion generators is a 1200 cfm DOP machine[13]. It is larger and self-nucleating but functions in the same manner as the smaller machines.

POTENTIAL NEW MARKET

Recent scrutiny of fine particles in the atmosphere is generating a new market. The market requires cleaning efficiencies approaching HEPA filters with capacities of baghouse filters. This is a new market but not a new problem. Until recent years industrial effluents passing cleaning devices could be, and were, exhausted to the atmosphere. Their relatively small mass made them uneconomical to recover and their effects difficult to trace and measure. This is fast becoming a real potential market area with a problem to be solved. The method of aerosol generation needed to measure achievement toward solving this problem will require one capable of producing aerosols in sufficient quantity to test a 1000 cfm system. The aerosol should reflect the problems of both penetration and cleanability observed in the field. Dust dispersion methods normally used in the engine, baghouse and space cleaning markets are not capable of deagglomerating the fines to provide a measure of the problems of field efficiency and cleanability. Thermally dispersed aerosols provide a measure of efficiency but are either a liquid or a solidifying liquid which may not duplicate most industrial effluents for filter loading and cleanability. Preliminary studies to meet our needs in this area have involved the following generators.

A Carbon Generator

Premature shortening of filter life in air cleaners exposed to diesel exhaust has generated an interest in filtration problems associated with fine particles.

The laboratory carbon generator shown in Figure 7 is based on an acetylene burner. It is built in an enclosure so the flow of both the air and gas released to the system can be controlled. The test system uses pressure regulators, rotometers and metering valves to measure, adjust and control the flow of both gas and air. Two types of burner heads have been tested. The first head was fabricated from a pipe cap and capillary tubing. This head showed decreasing aerosol output as a spiralling rim of carbon would build around the point at which the gas emerged, causing short term changes in aerosol output. The second head design was simply an orifice drilled in a pipe cup. This appeared to have a better long-term stability, but day-to-day reproducibility of filter media loadings could not be held closer than a factor of 10 by using repetitive settings on the controls. Measurements of the flame height at which smoking became visible was found to vary from day to day. These variances in flame height were related to atmospheric conditions, primarily changes in atmospheric pressure. The variability of outputs was determined by flat sheet loading tests to a constant pressure drop at a constant velocity. The generator requires monitoring instrumentation if it is to provide the reproducibility needed for industrial studies.

A particle even finer than that produced by the acetylene generator can be produced by burning commercial methane or propane. The particles may not be carbon, since they come from neither the propane nor methane. When either gas was filtered through a charcoal bed, neither formed particles, even when the gas to air ratio was varied over a wide range. The investigations were conducted using a Nuclei Counter.

Portable Air Operated Generators

Atomization of solutions appears to hold the most promise for solving industry's needs of large quantities of fine particles for fiber-bed loading studies. One such generation is the Model II NRL air operated

Figure 7 Carbon Generator

generator. It consists of six modified Laskin jets suspended in a lug top five-gallon can. This is connected to a high pressure air nozzle through a pressure regulator. Its design and operation was reported by NRL in 1963[14]. A modified version is available commercially.

The system is reasonably portable and easily set up and adjusted. When set at 30 psi upstream pressure, it provides a reasonably stable, reproducible output as determined with a light scattering photometer. The six jets are reported to produce 512 L/min with a mass concentration of 4.95 mg/L[14]. Because of contract restrictions, no tests have been made on liquids other than DOP. Until an instrument free of restrictions is obtained, we can only speculate as to its ability to provide an aerosol for industrial testing. If it is assumed soluable salts can be substituted for DOP, then the methylene blue-fluorescein[15] or sodium chloride techniques[16, 17, 18] could be applied. Echols and Young[14] measured the mass medium diameter for DOP as 0.8 micron with a geometric standard of 1.37. Assuming a saturated solution of sodium chloride, the generator should produce 0.9 gms of smoke with a MMD of 0.43. Each nozzle operates independently of the others, which should permit scaline from a single nozzle to multiples of six to cover a range of outputs to provide the aerosol needed for most industrial testing.

The primary contract use of this generator has been to examine 400 cfm HEPA elements for leaks. It provides a field test means of examining the integrity of installed banks of HEPA filters.

CONCLUSION

The performance of a number of generators providing aerosol for industrial testing have been explored. For most industrial purposes, comparative testing on a mass efficiency basis is adequate for both product improvement and sales. It has been shown that the largest number of aerosol generators to be found in the air cleaning industry are for generating coarse aerosols from standard dusts or collected samples. These generators produce the aerosols that support product improvement and customer acceptance testing. Because the largest dollar volume is in markets where mass recovery is the problem, only a few fine particle generators will be found. Most of the available fine particle generators do not provide a solid particle adequate for life testing products. The use of fine particle generators in industry has been slow to develop because of the small size of the market for fine particle air cleaning devices.

A basis exists for generators and techniques of generating fine particles to meet industrial needs for evaluating products for the fine particle market. For such generators to receive extensive industrial acceptance, they should provide a reproducible aerosol cloud that can relate to market problems and be easily ported, setup, and operated.

REFERENCES

1. 1941 SAE Handbook, Society of Automotive Engineers, New York, N. Y., 1941.

2. W. W. Lowther, "Dust and its Effects on Air Cleaner Design", SAE Preprint 365, SAE National Tractor Meeting, Milwaukee, Wisc., Sept. 13-15, 1949.

3. 1960 SAE Handbook , Society of Automotive Engineers, New York, N. Y., 1960.

4. W. H. Worthington, "Proposed Air Cleaner Test Code", SAE Journal (Transactions), 47 No. 1, pp. 294-99.

5. 1953 SAE Handbook, Society of Automotive Engineers, New York, N. Y., 1953.

6. 1963 SAE Handbook, Society of Automotive Engineers, New York, N. Y., 1963.

7. "Method of Testing Air Cleaning Devices used in General Ventilation for Removing Particulate Matter", ASHRAE Standard 52-68.

8. Robert E. Lee, "The Size of Suspended Particulate Matter in Air", Science, Vol. 178, No. 4061, pp. 567-75, 10 Nov. 1972.

9. H. F. Johnstone, Preface, AEC Handbook on Aerosols, Wash. D. C., 1950.

10. K. T. Whitby and R. C. Jordan, Progress Report Research on Air Cleaning, Aerosol Generation and Ionization, U. of Minn., Jan. 1962.

11. T. Wright, R. J. Stasny and C. E. Lappel, "High Velocity Air Filters", WADC Technical Report, pp. 55-457, 1957.

12. E. E. Grassel, "Construction of a LaMer Type Thermal Aerosol Generator for Radioactive Compounds", ORNL 54-3-46, April 1954.

13. Edgewood Arsenal Penetrometer, filter testing, DOP, Q107, B76-2-639 (Chemical Corps Drawing Number).

14. W. H. Echols and J. A. Young, "Studies of Portable Air Operated Aerosol Generation", NRL Report 5929, July 1963.

REFERENCES (continued)

15. R. T. Whitby, D. A. Lundgren and R. C. Jordan, "Homogeneous Aerosol Generators", Technical Report 13, U. of Minn., Jan. 1961.

16. B. I. Ferber, F. J. Brenenborg and A. Rhode, "Respirator Filter Penetration Using Sodium Chloride Aerosol", Report of Investigations 7403, Bureau of Mines, June 1970.

17. R. G. Dorman and L. E. J. Yeates, "A Comparison of the Methylene Blue and Sodium Flame Methods of Measuring Particulate Filter Penetration", Filtration & Separation, September/October 1966.

18. R. G. Dorman, L. E. J. Yeates and P. F. Sergison, "An Apparatus for Testing High-Efficiency Particulate Filters", Porton Technical Paper No. 873, September, 1963.

19. A. B. Aigren and K. T. Whitby, "Tentative Recommendation for a Method of Evaluating Air Cleaning Devices", Tech. Rpt. No. 7, U. of Minn., June 1957.

AEROSOL GENERATION USING FLUIDIZED BEDS

J.C GUICHARD

IRCHA BP N°1 91710 VERT-le-PETIT

FRANCE

Abstract

A well-know method to obtain airborne solid particles for industrial or laboratory purposes is to disperse pneumatically the corresponding powder. The classical way which uses an air ejector is not entirely satisfactory neither from the point of view of the break-up and separation of particles and nor from that of the constancy of concentration. Due to these shortcomings, these last years, we tried to build new powder dispersal device which makes use of fluidized beds. These are of two kinds.

1) Some of them uses the flow properties of the fluidized powder which is then fed to a separate dispersion system (which may be an air ejector or else) at a controlled rate. A system called "pulverotron" and a fluidized bed nebulizer are described.

2) Other devices use the elutriation properties of fluidized beds. In this case the fluidized bed by itself constitutes the dispersion device. The most simple case is that of a bed of monodisperse fluidized particles operating at a flow rate which is greater than the elutriation velocity of the particles. An alternate system is a normal fluidized bed where elutriation is locally forced. But the most useful system is that we called "puldoulit". In this case the powder to be dispersed is a added to a fluidized bed of larger particles (for example glass beads). When the system is in operation, the fine particles are elutriated from the bed.

AEROSOL GENERATION USING FLUIDIZED BEDS

J.C. GUICHARD

IRCHA -France

During the past ten years, we developed new aerosol generators in the field of fundamental research and for semi-industrial applications. They use some properties of fluidized beds and are able to disperse fine powders. This text explains some of these new aerosol generators without describing all possible systems but only a choice among those which were most pratical in use. For each of them we shall give the basic principles, some numerical examples of their results and the limits of their pratical applications but without details as those will be published elsewhere.

In a first part we shall consider some special properties of the fluidized bed that we studied because they are of current use in our generators. Then we shall describe devices which take advantage of the well-know liquid-like properties of the fluidized bed. Finally we shall speak about more original devices where the fluidized bed itself is the dispersing system.

I. SOME PROPERTIES OF THE GAS FLUIDIZED BED.

I.1-The fluidization of very fine powders.

It is not easy to fluidize fine particles because their natural tendancy to agglomerate. We understand that it is difficult, if not impossible, to put them in the quasi-browniam motion which is typical of the fluidized bed. If we try to fluidize a fixed bed of fine powder, we currently observe the formation of channels through which the air flows. In view of this experiment, we understand that it is necessary to fight this channeling and also to promote some mobility of the particles or of the agglomerate against each other.

A first measure is to use, as a porous plate, some microporous membrane (reinforced nylon) because it is advantageous to have a high pressure drop associated with fine pores. But that is not sufficient to promote fluidization. A well known although approximative solution is to use a tapered bed build with some electro magnetic vibrator fastened to the walls of the fluidization cylynder[1]. That gives good result for powder which are not too sticky (carbon black for example). This device although easy to operate is unsatisfactory, thus we developed another solution[2]. We mounted under the microporous membrane a Cagniard Latour Siren (a device use by firemen) which is essentially a rotating perforated disk facing a identical fixed disk. With an air flow around fifty liter/minute, we obtain a powerful audible sound (see figure 1). The vibrating energy generated promotes vibrations in the microporous membrane and is gradually absorbed by the fine powder in such a way that no sound is heard outside. The effect of the sound is to avoïd the channeling phenomena and to help the dispersion of fine particles. This system works very well for fluidized layer not thicker than twenty centimeters.

I.2 - Powder flowing from a fluidized bed through a hole in a thin wall

In some of the systems using liquid-like properties of fluidized bed, there is a flow of matter out of the bed through some orifice and this raises the question of the relationships governing the phenomena. It is a problem similar to that of the flow of a liquid through a hole in a thin wall. (for this latter case, the solution is found by application of Bernouilli's theorem)[3]. We studied the problem, only from an experimental point of view, using a variety of different powders and we came to the following formule

$$Q = K\, S^{\alpha}\, p^{\beta} \qquad (1)$$

where
- Q is the flow of matter out of the hole in g/s
- S the surface of the hole in cm^2
- p the pressure in the fluidized bed at the level of the hole expressed in baryes
- K α and β are parameters function of the nature of the powder (cohesion for example)

It is useful to notice that Q is independent of the air flow through the fluidized bed (p is also independent). We give some numerical values of the parameters in the following table.

Reference of the powder	K	α	β
Lycopode powder	0.0568	0.743	0.614
Glass beads smaller than 63µ with a median diameter 40µ	0.910	1.13	0.548
Glass beads of 800µ diameter	0.885	1.17	0.472

We remark that the numerical values are little dependent on the particle size distribution, except if the powder is a very fine one.

I.3 - Introduction of powder in a fluidized bed - fluidized bed with a constant level

For most of our generators, it is necessary to supply the fluidized bed with powder. There is some solutions available but our choice was limited by the necessity of getting simple and cheap device and also because some physical restrictions coming from the generators themselves (for example introduction at the bottom of the fluidized bed). Finally we selected and archimedean screw fed by a hopper containing a fixed bed (see figure 3). Pratically we use two models of screws : for free-flowing powder a home-made teflon screw and for the others (which are numerous) a spiral metallic brush of that sort use for cleaning plumbing. This last solution, in spite of his simplicity, gave good results.

Very often the output of the archimedean screw is at the bottom of the fluidized bed for reasons that will become clear later on. With free-flowing powder, we saw that the teflon screw is able to push the powder into the fluidized bed only until the pressure at the end is not too high. For the spiral brush this pression is zero. This last properties give us the possibility, if necessary, to build a constant level fluidized bed where the height is fixed by the position of the screw. If we want (with one or the other type of screw) introducing powder at the bottom of the fluidized bed, it is necessary to compensate the opposing pressure. A simple solution is to make tight the hopper and to connect its top to the bottom of the bed via a tube, the output of which is protected by a fine mesh (by this way the pressure is sent to the hopper).

If we use fine or sticky powder, only the metallic brush is able to hold a constant and regular flow of matter. It is further necessary to vibrate the hopper, using some electromagnetic vibrator, to ensure a constant flow of the fixed bed into the screw. On the other hand, we observed that the screw is able to push such cohesive powder against higher pressure than in the case of free-flowing powders. In any case, the flow of matter can be adjust using different rotation speed of the screw.

When it is necessary to maintain a constant level of the fluidized bed (for example because we want a constant pressure at a point inside the bed) we can solder an overflow where it is needed.

II. SYSTEMS USING LIQUID-LIKE PROPERTIES OF FLUIDIZED BEDS

The best known way to disperse powder into aerosol is to send it through an air ejector. Generally the desagglomeration is good but the emission is not constant because of the cohesive property of the powder to be dispersed. A lot of work was done to improve this system and it was natural to try a solution taking into account the liquid-like properties of fluidized beds. Nevertheless it is not enough to introduce the sucking tube of the air ejector into the fluidized bed and some device has to be conceived to solve the problem. We shall describe two of them.

II.1 - The "pulverotron" (see figure 1)

The fluidized state is made possible by a Cagniard-Latour Siren which turns at 1500 rpm [2]. The sucking tube crosses the fluidized bed near its bottom. The idea is to try to deliver the matter across a fine hole drilled in it. The formula (1) shows us that it is necessary to have a constant pressure at the level of this hole however the depth of the fluidized bed should be. For this reason the fluidized bed is put under pressure by a calibrated capillary which restricts the output of the airflow. When the level of the bed sinks, the airflow blown by an air compressor increases due to the decrease of the total pressure and the pressure over the bed increases tending to restore the initial value at the level of the hole. A special chamber went to control and to regulate the flow of matter surrounds the hole. This chamber, constituted by a tube 4 mm diameter, is soldered outside the sucking tube. Its axis is that of the hole, inclined about 15° below the horizontal because this is the convenient position to mininize the perturbations induced in the bed. This space is surrounded by a bigger tube and its walls are perforated by a lot/fine holes. This /of system allows to blow clean air inside the chamber. The space is closed (side of the bed) by a hole drilled in a thin wall. The diameter of this hole gives a means of controlling the flow of matter. The clean air which is blown into the chamber plays several roles. Firstly it avoïds blocking the hole by the powder, secondly it increases the effective pressure in the chamber and consequently reduces the pressure drop across the hole, thirdly it disperses roughly the powder avoïding big agglomerates which could interfere with the regularity of the emission. When the diameter of the hole is chosen, any flow of powder can be obtained by adjusting the clean air blown into the chamber. An example of this is given by the figure 2; the air flow in the sucking tube modifying the effective pressure drop across the chamber.

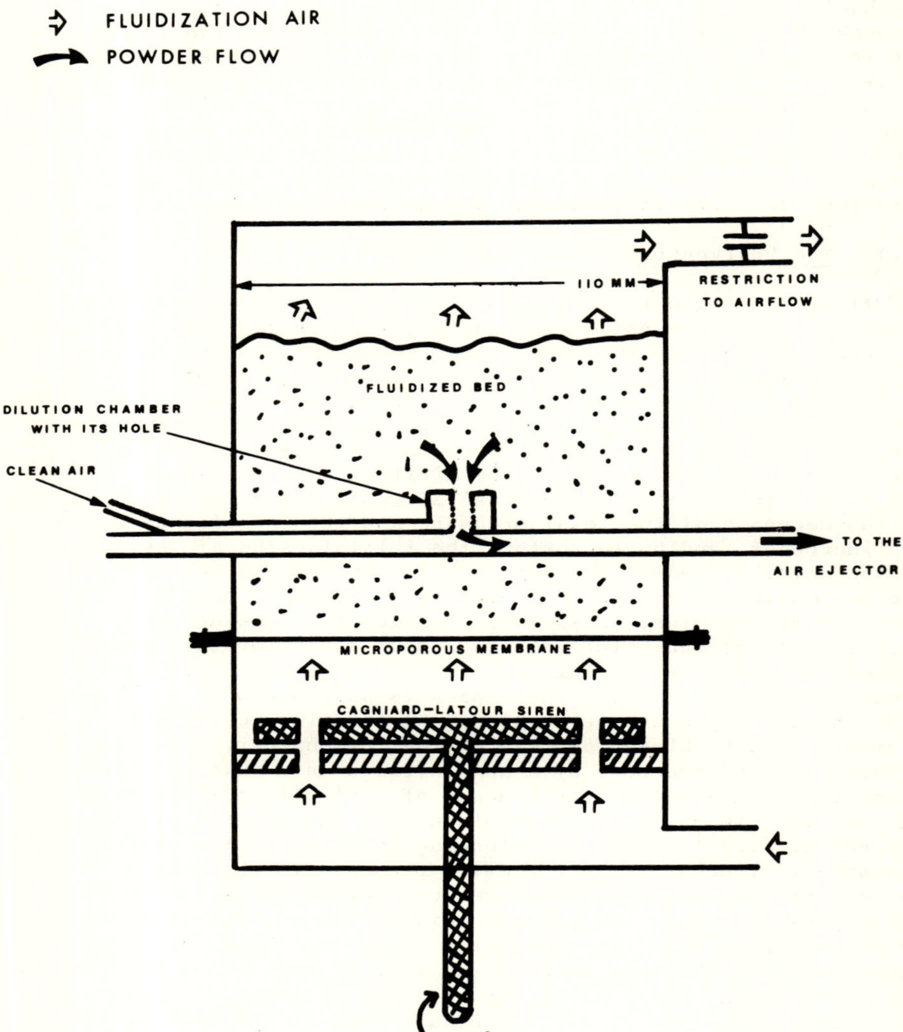

Figure 1. General Structure of a "Pulverotron".

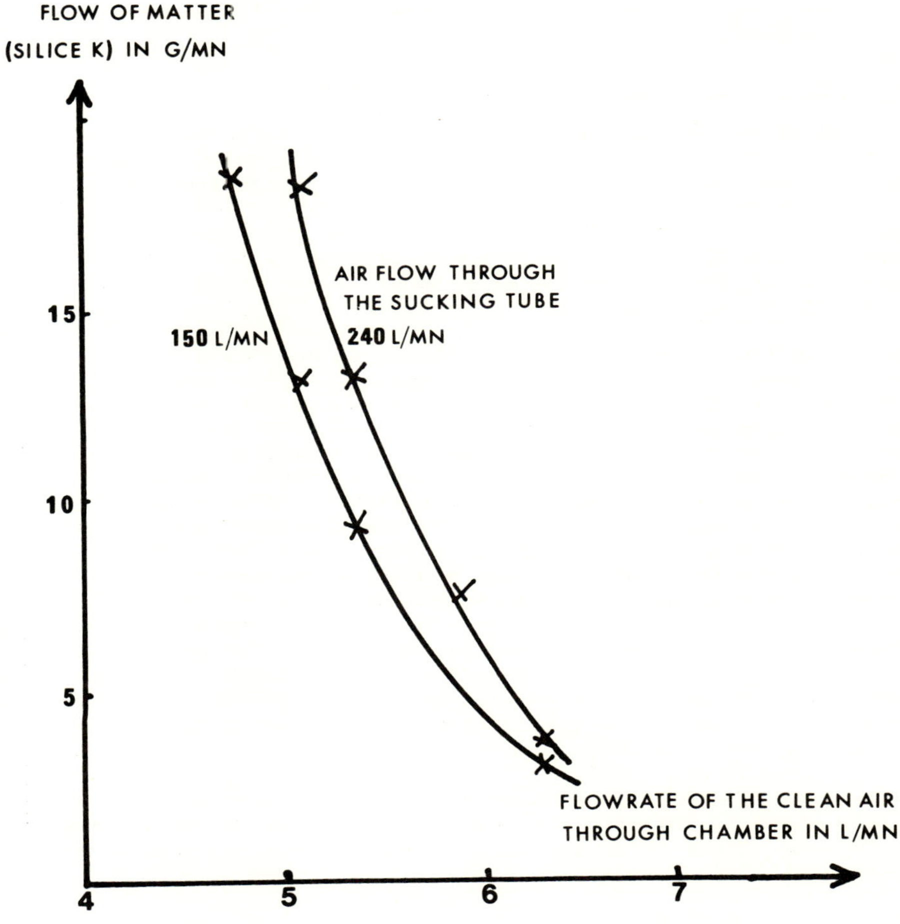

Figure 2. Flow of matter against flow rate of the clean air for a "Pulverotron" equipped with a hole 1.3 mm ϕ
Total pressure at the level of the hole 40 cmWG

This device called "pulverotrom" is well suited for dispersion of fine powders from some g/mn to 100 g/mn.

II.2- The fluidized bed pulverizer (see figure 3)

With the fluidized bed pulverizer, the powder is pushed out of the bed and the sucking tube of an air ejector collects it. The two stages are completely independant. The complementary air comes from the ambiance, when the air flow sucked by the air ejector is larger than those coming from the tube which is immersed in the fluidized bed. So modifications of the working conditions of the air ejector are without effect on the performances of the pulverizer.

The tube which is immersed in the fluidized bed has a 22 mm inside diameter and is perforated (about 30 mm of it end) by four holes of 10 mm diameter. Each hole is covered by a fine mesh. A larger tube surrounds it and allows clean air to be blown into the fluidized bed across the meshes. For the particular model which is presented on figure 3, the fluidization is obtained using an electromagnetic vibrator of 500g weight. The powder is introduced at the bottom of the fluidized bed by the metallic screw and the level is fixed by an overflow. The air which is blown into the fluidized bed by the perforated tube carries up the powder at a rate which increases with the air flow. The air ejector collects the totality and disperses it into airborne particles. Another way to control the powder flow is to change the depth where the clean air is blown but is appears that 7 centimeters under the surface of the bed is a suitable disposition. The figure 3 shows how the air allows to control the flow of aerosol.

The maximum flow of solids which can be obtained is a function of that which is introduced by the screw ; it is currently about 20 g/minute. There is no limitation towards weak concentrations except stability of emission.

III. SYSTEMS WHERE THE FLUIDIZED BED IS THE DEVICE WHICH DISPERSES THE FINE POWDER

The elutriation of fluidized beds is a well-known phenomena and was the subject of numerous publications[4]. Neverthless these works have not given, for the time being, a satisfactory knowledge of the mechanisms involved or a full experimental description valid for each of the different systems. In fact this is a complicated phenomenon and we think that the failure of many a

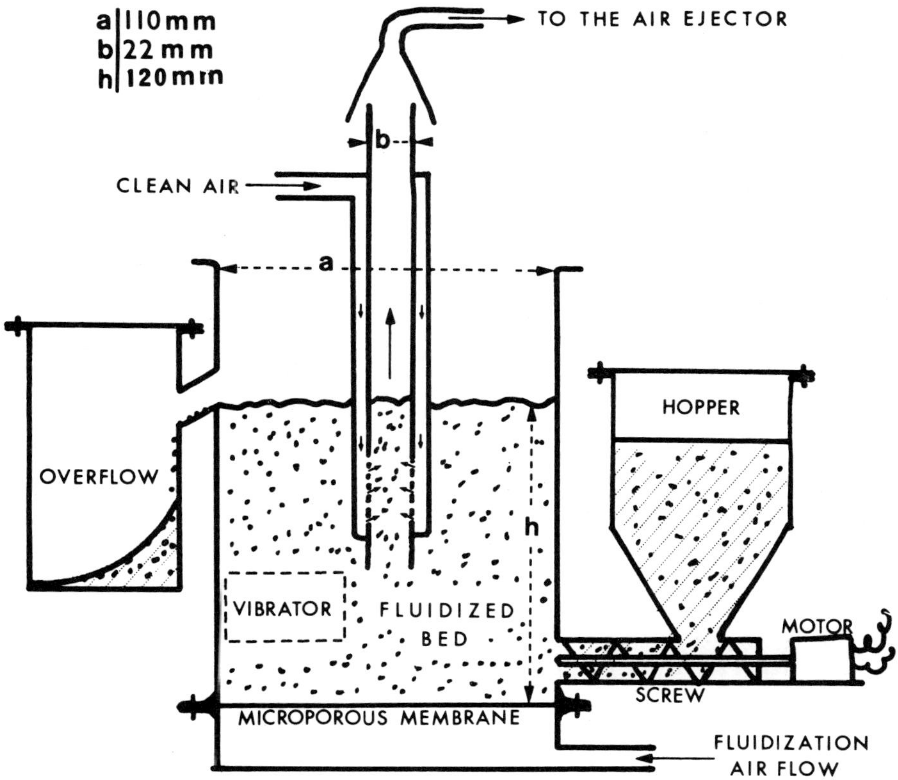

Figure 3. The fluidized bed pulverizer

research was the lack of simple and well defined system for which the analysis could be successful. We will present below two such systems which are useful to study this problem of fluidized bed elutriation but which are also new systems to disperse fine powders into airborne particles.

Notice - We did not make use of systems where a powder is fluidized at flowrate such that the surface of the bed is elutriated, except for monodisperse particles. The reason is the lack of laws discribing the phenomena and, consequently, difficulties to operate satisfactory systems of this type.

III.1 - Generators of monodisperse pollens

Pollens are equipped with special structures which facilitate wind transportation and desagglomeration. For this reason they give easily good fluidized beds even if they are fine. When elutriation is active for such fluidized beds, the pollens are carried away in form of monodisperse aerosols. One example is lycopodium perlatum (26µ) largely used by pharmaceutical industry and which is commercially available. We made a detailed study of the laws governing the elutriation of such a fluidized bed and we came to the formula [5]:

$$n = 7.96\ (q-150)e^{-\frac{20.7\ -\ 4.74\ \text{LnD}}{h}} \qquad (2)$$

where n is the numerical concentration in the aerosol phase measured at fifty centimers maximum above the bed
D is the column diameter
q is the flowrate of the fluidizing air for one cm^2 of free surface of the bed.
h is the height of the bed

In a more general form (2) can be writed

$$n = k\ (q - q_o)e^{-\frac{a\ -\ b\ \text{LnD}}{h}} \qquad (3)$$

where q_o is the airflow at the incipient fluidization
k, a and b are constants which characterize each powder

Starting from this law it is easy to build any generator with constant level fluidized bed. The concentration will be adjusted only by the airflow because n varie linearly with it (see also figure 7)

Notice - The generators with forced elutriation - A microscopic examination of a boiling fluidized bed shows that the bubble

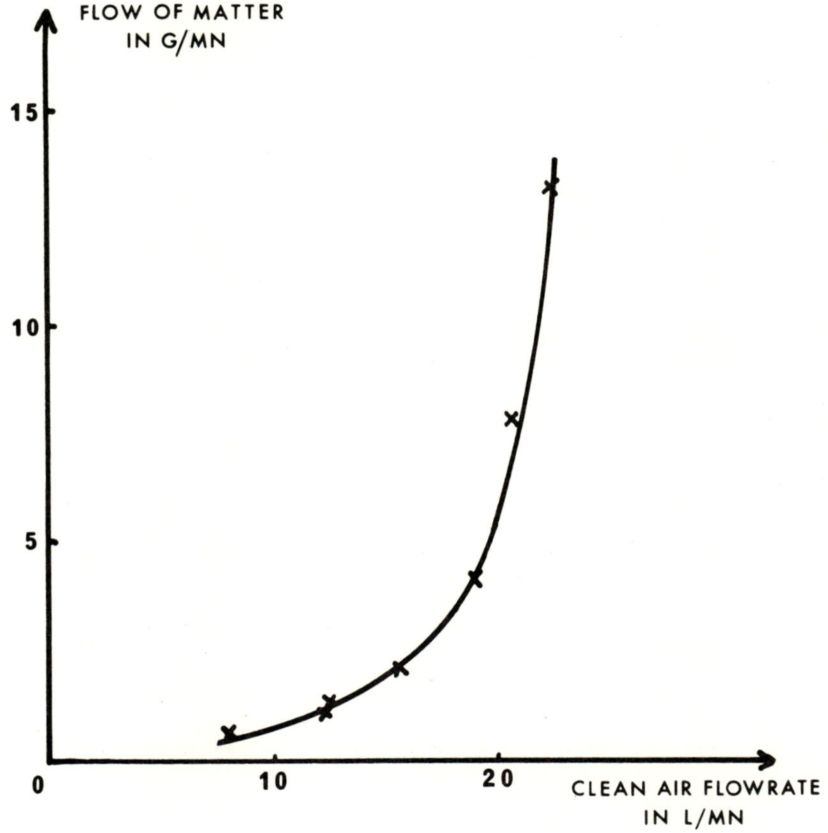

Figure 4. Powder flow of a fluidized bed pulverizer

plays an important role in the phenomenon of elutriation. Firstly the bubbles take an aerosol during their ascension and particles are liberated when they burst. Secondly when they burst at the surface, they throw up pieces of the fluidized bed which can be disintegrated and carried away if the mean air velocity is sufficient. It is also why the concentration increases with the air flow. An effective idea is to collect bubbles inside the bed and to force them to burst through a smaller surface. That is done by an inversed funnel (see figure 5). The concentration of the emerging aerosol increases when the funnel is sunk into the bed (that is equivalent to increase q in the formula 3).

This system allows weak flow of aerosol from a larger fluidized bed. It can work with any fluidized bed but the fine particles are desagglomerated only in the case of pollens.

III.2 - Systems using fluidized beds with two differents constituants - devices of the "puldoulit line".

Let us considerer a fluidized beds of glass beads diameters between 100 and 200 µ. A good fluidization is obtained for air flow giving a mean velocity above the bed around 5.3 cm/s. Let us suppose that fine particles are clung on to their surface. From an experimental point of view we observed that such particle are progressively loosened and evacuated in the airborne state if their elutriation velocity is less than 5.3 cm/s (corresponding to particle of 42 µ for a density of unity). This experiment gives the principle of the aerosol generation by a fluidized bed with two constituants. It is possible to make use of this principle by different devices. We shall give the description of three of them which are able of producing weak medium and high aerosol concentration. They have common features which are :

- good desagglomerating efficiency which was also noticed by Liu [6-7]
- good stability of the concentration without any instantaneous fluctuations
- particle size distribution of the aerosol identical with that of the fine powder

The "puldoulit" model A [5]- The general arrangement of the device is identical with that shown by figure 3. In place of fine powder, there is a mixture of glass beads and powder in the hopper. When the screw pushes this mixture in the fluidized bed, an equal quantity of glass beads falls down into the overflow and the

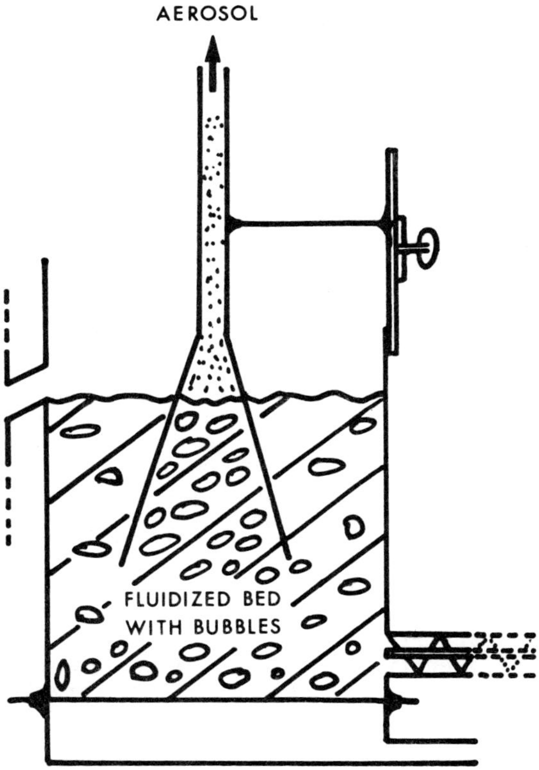

Figure 5. Principle of the generator with forced elutriation.

fine powder is dispersed as airborne particle. It is possible to get some idea concerning the emission mechanisms at work by considering the simple situation where the fine particles are totally adhering on the glass beads.

The inside of the fluidized bed is a very turbulent universe and the air velocity fluctuates : it can reach locally the values needed for loosen the fine particles. Further, the bubbles bursting at the surface throw up pieces of fluidized bed and liberates inner aerosols as explained before .

The glass beads are in quasi-brownian motion and collide. This mechanical effect can also loosen fine particles as it is currently observed at macroscopic scale. Besides, there is another phenomenon which plays a major role. The fluidized bed is a very efficient filter for aerosols and the particles liberated near the bottom have little chance to reach the surface. In fact they are captured by some bead and re-emitted later. This mechanism has probably a favorable effect on the desintegration of agglomerates and it shows us that different parts of the fluidized bed play different roles in the emission of aerosols. The main consequence is the build of a gradient concentration in a complex manner. A good approximation, useful in practice,is that of the aerosol emission kept under control by a top layer of fluidized beads (for example this layer would be 15 mm thick for a bed 120 mm height fluidized at 301/mn) We also understand that the best result are obtained when the introduction of mixing is made at the bottom of the fluidized bed.

The preceding description shows the importance of adhesive forces and consequently we are waiting efficient parameters correlated with this forces. In fact, for a stable and reproducible emission, it is necessary :

- to work with constant humidity of the air and to avoïd large temperature fluctuations. The best is often to operate with dry air.

- to use a mixture where the humidity is kept under control - the most simple being to dry the mixture before use.

- to ground the fluidization column in view of electrostatics phenomena existing in the fluidized bed[8].

- to use the same fine powder because adhesive forces depend on the physico-chemical properties of the particles but also on their previous history (grinding for example).

- on the contrary, for the conditions prevailing in a "puldoulit",although the fact that adhesives forces are size dependent, it seems that the probability of reemission for a particle is independent of its size and it is why the particle size distribution of the aerosol is just that of the powder (if the desagglomeration has been good)

Always starting from the preceding description, we understand how differents parameters control the emission. Let us consider a fluidized bed where the fine particle concentration is kept constant.

The aerosol concentration increases with the concentration of the powder in the bed (pratically we can increase the flow of mixture or increase the partial concentration in the mixture).

The aerosol concentration increases linearly with the air flow as shown below (the law is similar to 3)

But this static description does not apply exactly to the device working continuously because of an equilibrium between the feed and the output. Taking into account the hypothesis of a top layer controlling the aerosol emission. It is easy to write the fundamental equation :

$$M\,\theta = M\,(1-\theta)\,\alpha + CQ \qquad (4)$$

where
- Q is the volumetric air flow
- M the weight flow of the mixture
- θ the weight concentration of fine powder in the mixture
- α the weight concentration of fine powder in the fluidized bed falling into the overflow
- C the weight concentration of aerosol

This equation allows to calculate C if the other quantities are know or to calculate θ and α to obtain an emission of a definitive C. In this latter case, it is necessary to know the relationship $C = F(\alpha)$, which is fundamental for a good knowledge of the device. This law depends on the physical nature of the fine powder and is only experimentally known. An example is shown by the figure 6, for an aloxite powder with particles between 1 and 20μ and a median diameter, by weight, of 5.5μ.

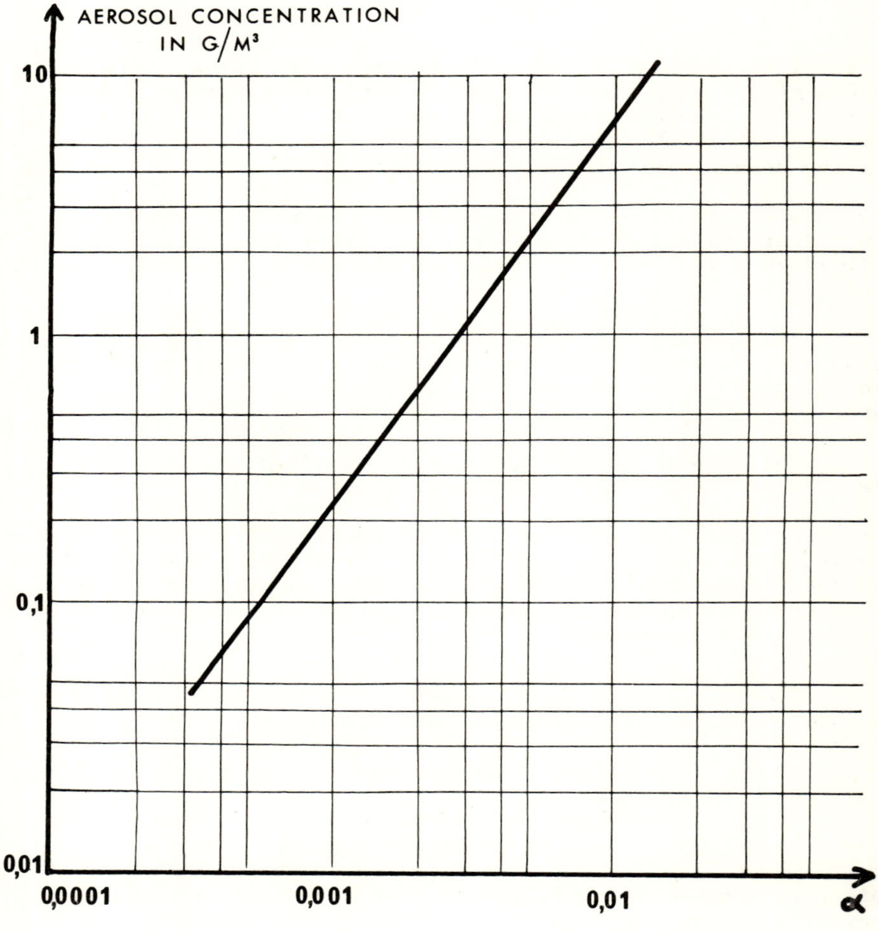

Figure 6. Relationship C=f(α) for a "puldoulit" working with glass beads (100 to 200μ) and aloxite powder (1 to 20μ with 5.5μ median diameter by weight)

For the model currently used in our laboratory (see figure 3) we have obtained good results with aerosol concentration from some mg/m3 to 1 g/m3 for an air flow between 20 and 60 l/mn. The emission is very stable but there is some details to observe if we went reproductible emission. The main problem is a good preparation of the mixture. It is sufficient to mix the two constituants in any bottle by shaking, if we want only a stable emission. But for any mixture a part of the fine powder is adhering onto the beads and the other is free in the insterstices of the bed. If the proportion between the two states of particles is not the same from one to another experiment it could result a different aerosol concentration. For this reason it is advisable to fluidise independantly the mixture during a few minutes before use. The shortcoming is the necessity to determine directly the real value of θ in the mixture.

Notice - If we want to obtain a known value of C when starting with a fine powder that has not been used previously, it appears the problem of choosing an initial partial concentration for the fluidized bed in view of a minimum delay before reaching equilibrium.

Let us consider a fluidized bed without fine particles and at t = o let us introduce the mixture. Let us suppose that the concentration α of fine particles is instantaneously the same inside the layer controlling the emission. If we borrow the simple hypothesis that the relationship between C and α is approximated by $C = k\alpha$, the law governing the aerosol concentration is

$$C = Cm \left[1 - e^{-\left(\frac{M + k\varphi}{p}\right) t} \right] \tag{5}$$

where
- Cm is the equilibrium concentration
- p is the weight of the layer
- t is the time

We see that the delay before equilibrium is independent from θ and is reduced if M increases. This delay is around ten minutes with the conditions specified by the figure 6, and a motor running at 10 rpm (M in the order of 0.9 g/s). But it is more lenghty with a motor 1 rpm and it is advantageous to start with a predusted fluidized bed.

If all the quantities of equation (4) are know, it is not possible to calculate the partial concentration of fines inside the fluidized bed because α is valid only for a top layer. From a practical point of view, taking into account that α is a maximum for the unknow concentration, it is convenient to try $\frac{\alpha}{3}$

<u>The "puldoulit" model B</u> - The device is identical with model A but without overflow and with an electromagnetic vibrator (Approximatively 100 g) fastened on the hopper. The induced vibrations help to have a good flow of the fixed bed here which is of fine powder alone. The archimedean screw is of metallic brush type and is currently able to send between 1 and 10 g/mn of the fine powder at the bottom of the bed. Another useful detail is the possibility to separate immediately the hopper and the screw from the device and consequently to know by weighting the quantity of powder used up during an emission.

With this model where the powder is not premixed with glass beads, the mechanism of capture and reemission plays a major role. But the lack of predispersion as for the model A explains that the desagglomeration efficiency is lower for sticky powder. Consequently it is necessary to introduce the powder near the bottom, to use a thicker fluidized beds/to work with dry / and powders. Nevertheless it can happen that somme agglomerates are not broken. In this case they rise and float on the surface giving a layer which has to be eliminated from time to time. Finally it appears that the desagglomeration capacity related to the flow of solids introduced is limited but this phenomenon was not studied quantitatively.

With the present model of "puldoulit B" it is possible to generate aerosols with concentrations from some 1 g/m3 to 100 g/m3, the air flow being between 15 and 60 l/mn.

<u>Generator giving weak concentrations</u> - Starting from the same idea it possible to have a generator giving weak aerosol concentrations which can be measured with an optical counter. That device is very useful for experiments in aerosol physics. Let us consider a fluidization column in which we poured a mixture of glass beads and fine powder prepared as explained earlier, let us follow the concentration function of time. At the beginning of fluidization we observe a high concentration that decreases rapidly and appears stable after some hours. At this point the consumption of powder is small compared to the quantity available inside the bed and which is totally sticked on the surface of the beads (the powder that was free has been eliminated). The stability of concentration and particle size distribution is maintained during a long time giving a calibrated generator. Nevertheless three points have to be noticed.

- At the beginning of each experiment, there is a peak concentration. A progressive increase in place of starting with full airflow, smoothens this phenomenon.

- Starting from two mixtures with the same value of θ , we do not generally end with the same numerical concentration of aerosol.

- It is necessary to operate with air flow of controlled humidity and to keep the fluidized bed at this level of humidity between two successives operations.

This generator gives us a useful way to study the fundamental problem of fluidized bed elutriation in the particular case of two constituants. Such studies have not been systematically undertaken. Nevertheless the figure 7 shown the relationship between numerical concentration and air flow for a fluidized bed where the initial value of θ was 1% and the equilibrium concentration in the top layer 0.014%. The concentration increases linearly with the total air flow and it is possible from the results of the figure 7 to show that the particle size distribution is independent of that air flow.

CONCLUSION

The details given in the text are sufficient to build with success the generator well adapted to the particular problem in view. The dispersion of fine powders using fluidized beds is the best solution for a lot of pratical problems.

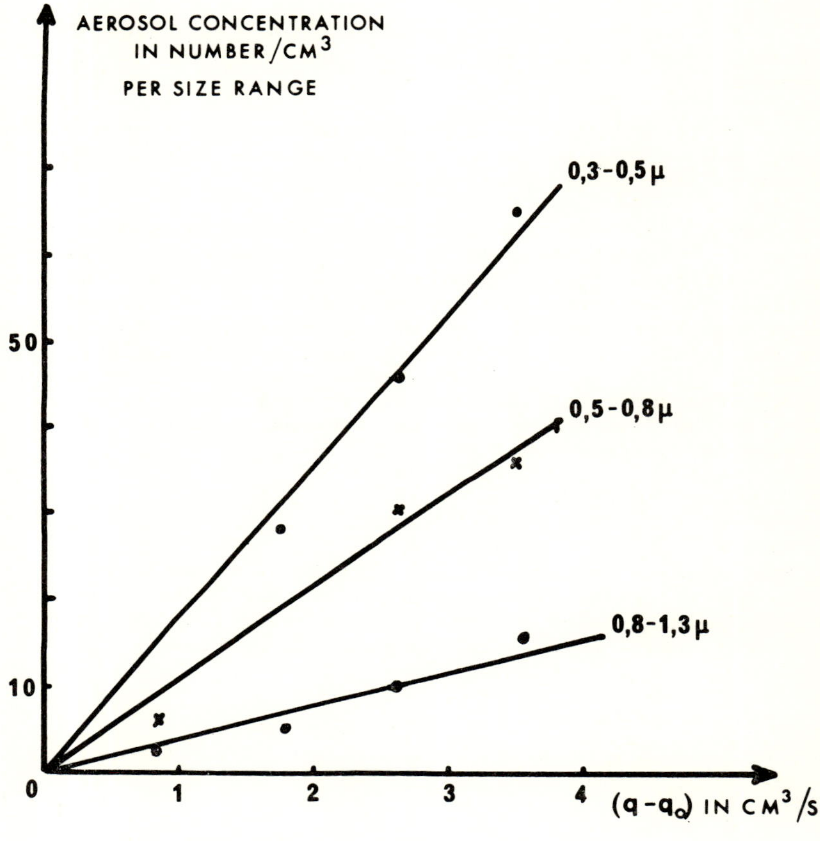

Figure 7. Aerosol concentration versus flowrate for a fluidized bed 110 mm diameter 40 mm height using a mixture of glass beads and aloxite powder

REFERENCES

1.- F.A ZENZ and D.F OTHMER - Fluidization and fluid-particle systems - Reinhold publishing Company - New York 1960

2.- J.C GUICHARD - Approvisionnement constant en poussières à l'aide d'un lit fluidisé (Presented at the 8ème Colloque International sur les poussières - IRCHA - STAUBFORSCHUNG-INSTITUT - Paris 17 - 18 Novembre 1966)

3.- L. MASSIMILA, V. BETTA and C. DELLA ROCCA - A study of streams of solids flowing from solid-gas fluidized beds A I Ch E Journal 502-508 September 1961

4.- J.F DAVIDSON and D. HARRISON - Fluidization Academic Press London and New York 1971

5.- J.C GUICHARD - Anwendung der wirbelbett-Ausschlämmung bi der Herstellung von Aerosolen Dechema Monographien 59 N°1045 : 247-269 - 1968

6.- B.Y.H. LIU, V.A MARPLE, K.T WHITBY and N.J BARSIC - Size distribution measurements of Airborne coal dust by Optical Particle Counters American Industrial Hygiene Association Journal 443-451 August 1974

7.- K. WILLEKE, C.S.K LO and K.T WHITBY - Dispersion characteristics of a fluidized bed. Aerosol Science 5 : 449-455 1974

8.- J.C GUICHARD - The triboelectrification of metallic particulate aerosols generated by a two-component fluidized bed Staub Reinhaltung der Luft (English) 33 : 174-179 April 1973

LARGE FLOW RATE REDISPERSION AEROSOL GENERATOR

Fred Moreno, Dale Blann
Aerotherm Division/Acurex Corporation
485 Clyde Avenue
Mountain View CA 94042

ABSTRACT

In Spring of 1974 Aerotherm began a program for the EPA Control Systems Laboratory to develop and construct a large flow rate (480 plus pounds per hour) aerosol generator for use on the Particulate Aerodynamic Test Facility. It was assumed that the designs employed in a small fluidized bed aerosol generator previously tested could be enlarged to meet the requirements of the larger system. However, an experimental program showed that a variety of scale-up phenomena prevented the simple fluidized bed concept from working effectively at large bed sizes and flow rates. A two stage concept was evolved utilizing a simple fluidized bed first stage feeding a second stage fluidized bed with a gas distributor capable of passing the particulate laden gases. The second stage completes the redispersion process initiated by the first stage. Preliminary testing with a two stage cyclone sampling system has shown that for some dusts as much as 40 percent by mass of the redispersed dust is less than 3.8 microns.

The paper will describe the nature of the problems encountered when the original concept was enlarged, and will describe the details of design of the new two stage concept currently under construction.

LARGE FLOW RATE REDISPERSION AEROSOL GENERATOR

Fred Moreno, Dale Blann
Aerotherm Division/Acurex Corporation
485 Clyde Avenue
Mountain View CA 94042

BACKGROUND

In Spring of 1974 Aerotherm began a program for the EPA Control Systems Laboratory to develop and construct a large flow rate redispersion aerosol generator for use with their Particulate Aerodynamic Test Facility. The test facility is a specially designed and constructed wind tunnel dedicated to aerosol research. The Facility specifications are shown in Table 1.

Table 1

Particulate Aerodynamic Test Facility Specifications

Test section size	2 ft diameter by 40 ft long
Test section velocity	5 to 90 ft/sec
Flow rate	Up to 17,000 cfm
Temperature	70°F to 450°F
Dew point	50°F to 130°F
Test gases	Air or combustion products
Configuration	Closed loop gas flow
	Open loop dust flow

Dust is injected upstream of the test section and removed from the gases downstream of the test section in a baghouse (see Figure 1).

When first constructed, the facility had a small aerosol injection system capable of generating very light loadings typical of stack outlet conditions in the tunnel test section. It was decided that it would be desirable to create dust loading conditions within the tunnel which are more typical of the conditions at the inlet of particulate control

Figure 1. EPA Particulate Aerodynamic Test Facility

devices so that instruments designed to operate in this environment could be evaluated and prototype-sized particulate control devices could be tested in a laboratory environment.

The general requirements of the aerosol system are as follows:

- Three grains per cubic foot dust loading in the test section at 17,000 cubic feet per minute flow rate implying a feed rate of 480+ pounds of dust per hour
- Capability for around-the-clock operation
- A large fine particulate fraction in the dust injected into the test section.

In addition, the following specific requirements peculiar to this installation had to be met:

- No dilution of the tunnel gases (which may consist of combustion products doped with SO_x and other gaseous constituents)
- Temperature matched flow streams such that the temperature of the gases injecting the particulate into the tunnel has the same temperature as the gases flowing through the mainstream of the tunnel
- Tight space and layout constraints dictated by the physical facility

APPROACHES CONSIDERED

A number of approaches were considered in the initial phases of the program and each was judged on its merit relative to this installation. The first considered was a single component fluidized bed utilizing particle dispersion by means of aspiration (see Figure 2). This system fluidizes dust directly and then extracts it in the fluidized form by means of an aspirator, as illustrated. The feed rate of dust into the aspirator is controlled by the flow of dilution air injected at the extraction point within the fluidized bed. Our experience with a prototype of this system indicated very good flow control capability and moderately good redispersion of the dust. However, this approach requires high pressure air which must be carefully dried to prevent icing within the aspirator. Reliability was questionable because erosion frequently takes place within the aspirator. There is an additional problem of dilution of the tunnel gases by means of the compressed air used to drive the aspirator and convey the dust into the tunnel test section. Finally we noted high levels of static charge generated at the system outlet.

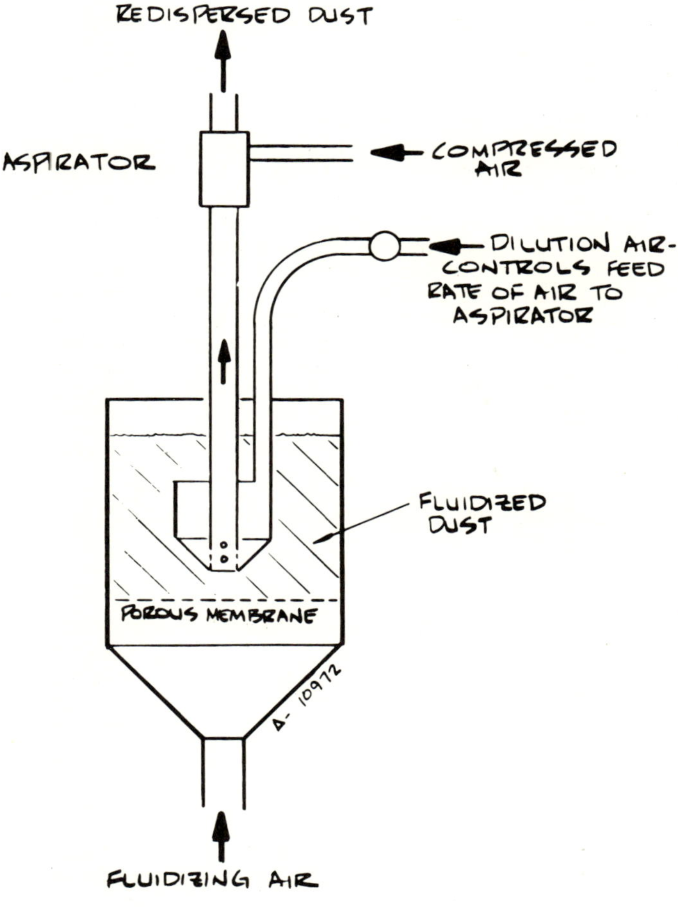

Figure 2. One Component Fluidized Bed with Aspirator

The second approach considered is the two component fluidized bed illustrated in Figure 3. This approach is based upon a design originated by Dr. J. C. Guichard of I.R.C.H.A., France[1]. The system utilizes a fluidized bed of glass beads of approximately 0.010 to 0.030 inch diameter fluidized by the passage of the appropriate gases. The dust to be dispersed is injected into the side of the fluidized bed and subsequently coats the glass beads. Dust particles are removed by the aerodynamic action of the fluidizing gases and by the mechanical action of the glass beads as they impinge upon one another. This system utilizes low pressure gases and therefore has low power consumption. Further, it has demonstrated good dispersion efficiency. The bed can be fluidized by means of the gas available within the tunnel, and the flow rate of dust is controllable by means of a variable speed screwfeeder as illustrated in Figure 3. By calibrating the screwfeeder for each dust of interest, the operator can inject a known and controllable rate of dust into the tunnel test section. We experimented with this approach previously in work done for the EPA Chemistry and Physics Laboratory wherein we developed a redispersion aerosol generator having a maximum feed rate in the range of 60 to 80 pounds per hour. Table 2 summarizes the advantages and the disadvantages of the two approaches considered as the most viable for this particular application.

Table 2

Advantages and Disadvantages of Various Approaches

Fluidized dust with aspiration	Compact, small feed lines, large turndown ratio, simple design.	Requires dust that can be fluidized; high pressure air. Air dilutes tunnel gases. Possible erosion of aspirator. Generates high static charge.
Two component fluidized bed	Can utilize tunnel gases, good dispersion, low power consumption, low static charge, flow rate controlled by screwfeeder.	Turndown limited by screwfeeder. More complex system, possible attrition of beads.

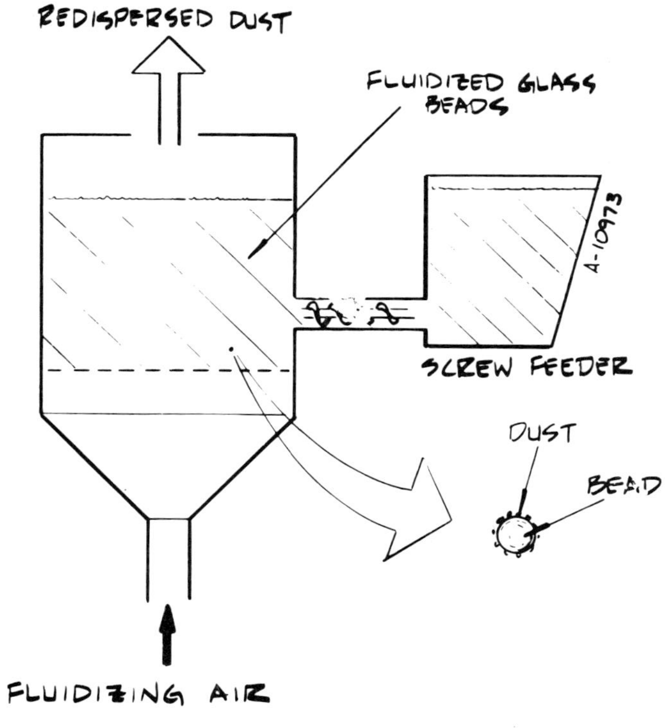

Figure 3. Two Component Fluidized Bed

The approach selected was to utilize a large version of the two component fluidized bed examined in the earlier program. The approach was simply to increase the bed diameter until an adequate area was obtained for conveying dust into the airstream at the desired flow rate. It was quickly observed, however, that there were problems with dust injection and mixing in the bed. The dust was fed into the side of the bed by means of the screwfeeder as before, but failed adequately to mix throughout the bed before being elutriated into the fluidizing gas stream. As a result there was a stratification of the exit gas stream with a higher dust loading in the vicinity of the injection location, and a lower dust loading on the opposite side of the fluidized bed. In addition, the excessive concentration of dust in the region of the screwfeeder inlet resulted in poor dust dispersion because of saturation of the glass bead media.

The fluidized bed aerosol generator requires a fairly good quality of fluidization with no violent bubbling, slugging or fissuring within the glass media of the bed. If any of these conditions occur, it permits the passage of dust out of the bed without being first acted upon by the glass beads. In an attempt to increase the mixing in the bed the bed depth was increased. This succeeded only in increasing the size of the bubbles which were generated in the bed, and also increased the tendency of the bed toward a slugging mode of operation. The result of the increase in bed depth was thus a poor quality of fluidization and thus poor redispersion of the dust into the gas stream.

A variety of other dust injection schemes were considered in order to circumvent the problems experienced with the large diameter fluidized bed. The first and most obvious is multiple injection ports at various locations around the circumference of the circular bed as illustrated in Figure 4. The problem with this type of approach is that the larger fluidized bed exhibits recirculation cells in the glass media, as illustrated in Figure 5. The circulation cells are fairly stable in their behavior and are quite effective in preventing dust injected at the perimeter of the bed from mixing into the center of the bed. Thus the problem of saturation of the glass beads media in the vicinity of the injection ports would remain.

Injection through the gas distributor plate at the bottom of the fluidized bed was examined but rejected because of the expectation that there would be poor quality fluidization in the region of the injection location due to the lack of fluidizing air in this area. In addition, this approach would require the operation of the screwfeeder in a vertical mode while in fact the screwfeeder operates satisfactorily only in a horizontal orientation.

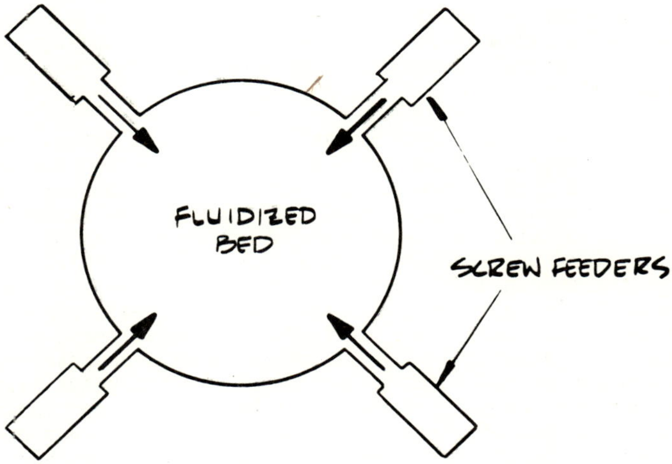

Figure 4. Circular Fluidized Bed with Multiple Injection Parts (Plan View)

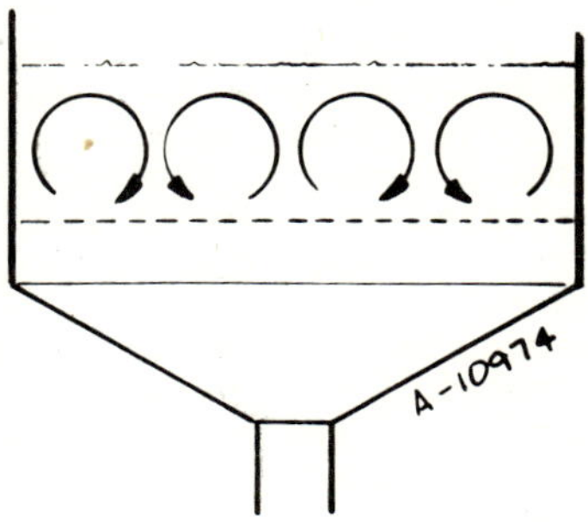

Figure 5. Recirculation Cells in Larger Circular Fluidized Beds

The final approach considered was a rectangular fluidized bed with multiple injection locations as illustrated in Figure 6. The rectangular fluidized bed removes the problem of circulation cells if the bed width is carefully selected. However this approach requires numerous screwfeeders in order to effect multiple-injection locations and this would be attended by a complex materials handling system to feed dust to the several screwfeeders as would be required for round the clock operation.

Because of the problems and complexities associated with large scale two component fluidized beds, a new approach, the three component fluidized bed (based upon an idea of Dr. J. C. Guichard) was considered. The three component system shown in Figure 7 utilizes a screen supporting a layer of steel or lead beads which in turn is covered by another layer of glass beads of comparable size. The fluidizing gas passes through the screen and is distributed by the layers of heavy beads. The gas velocity is selected such that the heavy beads remain stationary while the glass beads are fluidized. The heavy beads are, however, near incipient fluidization. The advantage of this approach is that dust laden gas can be fed into the bottom of the apparatus passing through the screen and the stationary bed of heavy beads and subsequently into the fluidized bed of glass beads where the normal redispersion action takes place. As the heavy beads become plugged with dust the pressure drop across the layer of heavy beads achieves the point where fluidization occurs. At this point the layer fluidizes thereby cleaning itself, and then quickly reestablishes stationary behavior when the pressure drop falls. The result is a fluidized bed capable of being operated with dust laden air without plugging of the gas distributor.

The primary problem in the initial design was determining the best technique to introduce dust into the bottom of the system. The selected approach was to utilize a smaller diameter two component fluidized bed using steel or lead beads as the fluidizing media. Dust is fed into the side of the two component bed by means of a screwfeeder as illustrated in Figure 8. The lower bed operated in a violent bubbling mode which serves to break up the agglomerates. The gas stream coming through the lower bed carries the dust to the bottom of the second fluidized bed. Redispersion is completed by passage through the second bed. The advantage of the two stage systems created by the vertical stacking of two fluidized beds is that (in theory) the system can be scaled to virtually any size because the dust stream conveyed into the bottom of the fluidized bed can assume virtually any diameter.

It was anticipated that a "puffing" type of behavior might occur when the stationary heavy bead layer fluidized during the cleaning cycle. It was decided that if this problem were severe it could be solved by

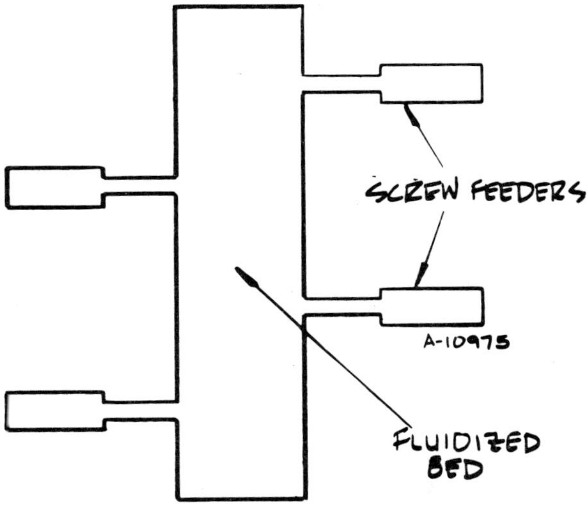

Figure 6. Rectangular Fluidized Bed with Multiple Injection Ports

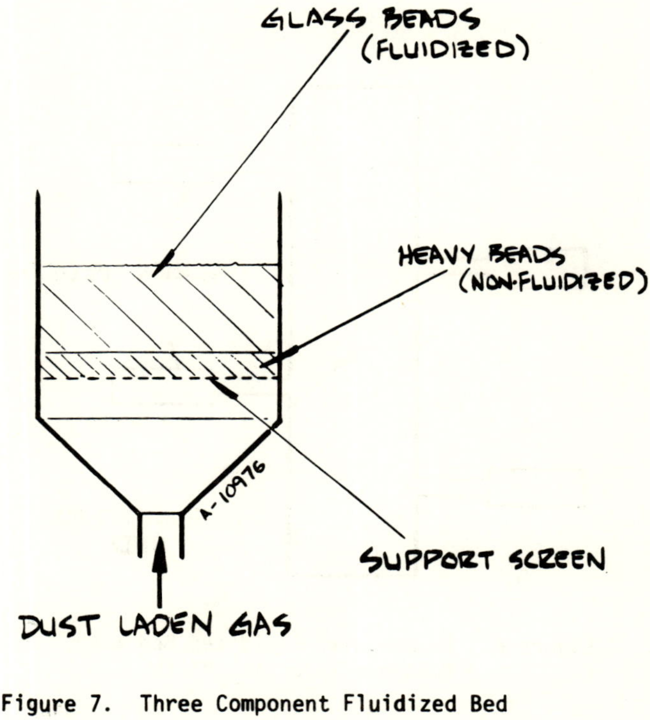

Figure 7. Three Component Fluidized Bed

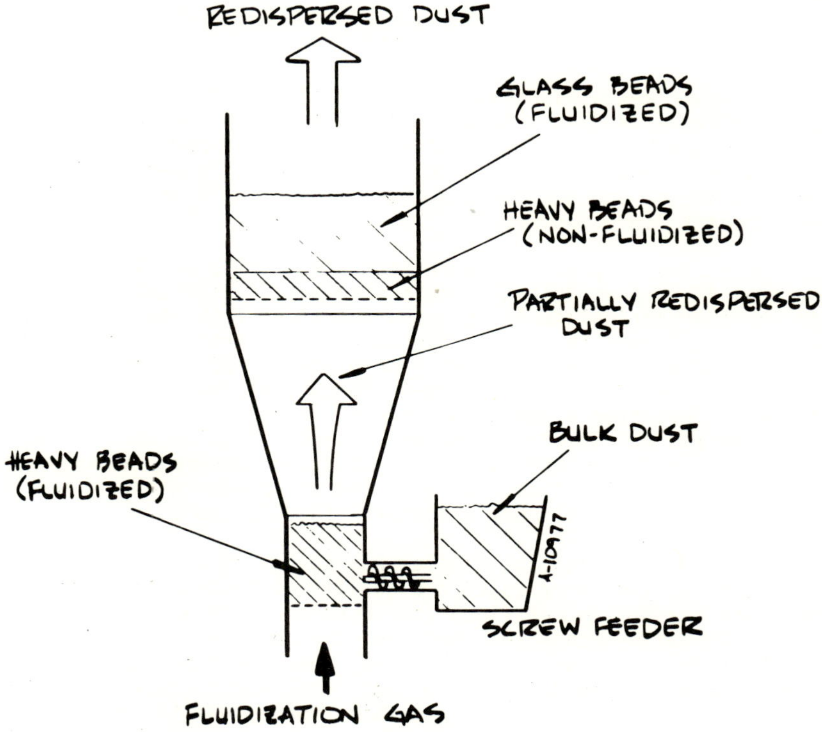

Figure 8. Two Stage Aerosol Generator

stirring the layer of heavy beads with moving fingers reaching through the fluidized bed. The stirring action would then keep the layer comparatively clean. An experimental program was designed to examine this and other potential problems.

EXPERIMENTS

The experimental apparatus utilized an Acrison Model 105Z variable speed screwfeeder feeding into an 8-inch diameter fluidized bed utilizing 0.020 to 0.030-inch diameter steel beads as the bed media. The gases passing through this fluidized bed subsequently passed through a low angle diffuser and thence into the bottom of a 16-inch diameter three component fluidized bed.

The initial experiments were conducted utilizing flyash derived from a coal fired power plant. The flyash was fed through the screwfeeder and the particulate laden gases emanating from the top of the system were collected in a small baghouse. The first problems noted when operating at the design point (approximately 500 pounds per hour feed rate) was that the bottom screen assembly of the top fluidized bed plugged with dust over a period of several minutes. The solution to this problem was to vibrate the screen assembly from below by means of a rod which passed from the bottom of the screen assembly through the center of the lower fluidized bed into the air plenum where it was connected to a variable amplitude vibrator. (See Figure 9.)

The immediate plugging problem was solved by vibration, but it was observed that as time progressed the quality of fluidization began to degrade with channeling and fissuring occurring within the upper fluidized bed. It was hypothesized that the poor quality fluidization was probably due to the plugging of the heavy bead layer by dust because flow rates were substantially below those required to permit fluidization of the heavy layer. The solution to this problem was to increase the depth of both the glass and steel bead layers, increase the flow rate of the air, and increase the vibration amplitude of the supporting screen assembly. It is postulated that the increased vibration gives the heavy beads sufficient relative motion with respect to one another that they can continually clean themselves without having to be fluidized.

The above modifications permitted operation for a greater period of time before air channeling and fissuring were again observed in localized regions of the upper bed. Unlike the previous case, the regions of poor quality fluidization appeared to be fixed in location rather than random in occurrence as was the case previously. There appeared to be two reasons for this. First, there was a selective flow of air through one side of the bed which was traced to a failure to adequately level of assembly thereby causing a greater depth of beads to

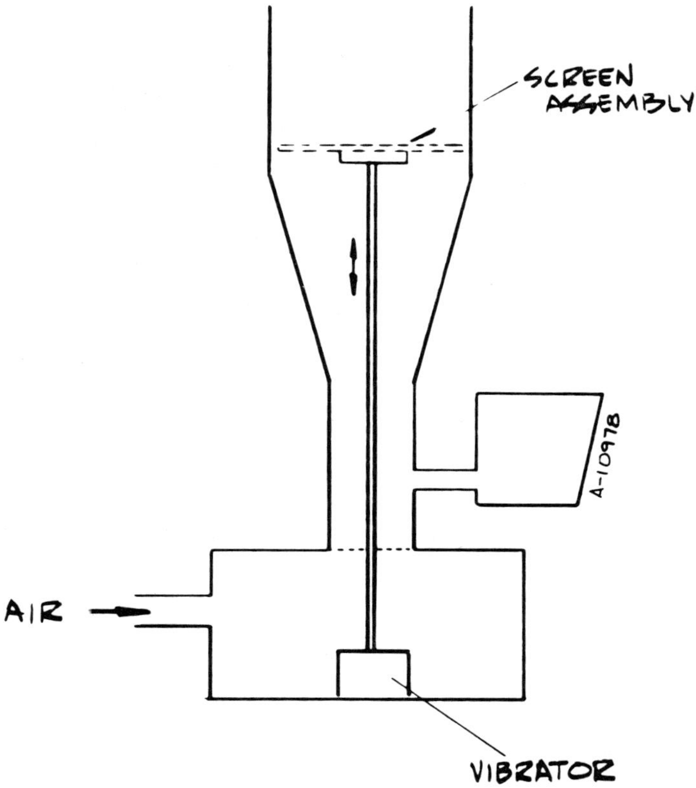

Figure 9. Vibrator Arrangement for Upper Bed Screen Assembly

migrate to one side of the bed than the other. Air then passed preferentially through the shallower portion of the bed. This gross maldistribution of air flow was fixed by careful leveling of the screen and bed assembly. Second, it was necessary to reconstruct the screen assembly in order to reduce the size and number of vibratory nodes. These regions were regions of inadequate agitation and thus plugging of heavy beads occurred within these locations.

Correction of these minor mechanical problems permitted stable operation for a still greater period of time. However it was observed that under some conditions an unusual instability would occur wherein the upper fluidized bed exhibited an unusual "hopping" or "slugging" behavior. The behavior of the bed was such that the entire mass of glass beads contained in the upper fluidized bed would jump up and down in an oscillatory behavior. This unusual phenomena was traced to a pneumatic coupling of the upper and lower bed through the air stream. Instabilities in the upper bed would change the pressure drop of the system. The reduced pressure drop in the upper bed would then cause an increased air flow in the lower bed which in turn caused another change in flow rate. The result was a feedback system with a resonance occurring at a frequency of one to two cycles per second. During the hopping behavior both beds would jump up and down in concert with one another. The problem was ultimately corrected by adjusting the air flow rate and the depths of the upper and lower beds in order to get an appropriate pressure drop regime such that instabilities did not occur.

The solution of the above problems led to satisfactory operation with flyash. The system operated successfully at feed rates of up to 600 pounds an hour in a consistent and reliable fashion.

Once operation with flyash was established the dust was switched to black iron oxide particulate in order to see if the system could handle a dust with a much finer size distribution. Initial tests again showed a plugging problem associated with the support screen of the upper fluidized bed. The plugging was avoided by increasing the screen size. This largely eliminated any problems with plugging except for occasional erratic behavior of the upper bed. Several times the system exhibited poor quality fluidization and ejection of glass beads from the bed indicating severe channeling and constriction of flow. The source of this final problem was not ultimately explained because it happened so irregularly. It is theorized that it was due to the moisture content in the dust or the air stream which causes this particular dust to assume a caking behavior which makes it difficult to handle.

SAMPLING RESULTS

A brief set of particle sampling experiments were conducted towards the end of the experimental program. The objectives of these tests was to give a quick look to evaluate the redispersion efficiency of the system. It was felt that an in-depth measurement program was not required because the fluidized bed technique has been well examined in the past.

The sampling apparatus employed utilized a series of staged cyclones calibrated and loaned by Joe McCain of Southern Research Institute. The apparatus is shown schematically in Figure 10. The cyclones have a fairly well defined cutoff point with the larger cyclone having a 50 percent cut point of 3.8 microns while the smaller cyclone had a 50 percent cut point of 0.63 microns. Normally the cyclones are followed by a filter, but a variety of problems dictated that only the approximate percentage of dust falling in the smaller cyclone cup be determined.

The results of the brief sampling experiments are shown in Figures 11 and 12 for flyash and iron oxide respectively. It should be noted that in some conditions with iron oxide in excess of 40 percent of the total dust emitted from the bed was smaller than 3.8 microns. There is an unexplained trend wherein the dispersion efficiency of the system appears to increase with higher feed rates of dust when one would expect a decrease. If one were to continue increasing the feed rate one would expect saturation to occur, but apparently the system was operating in a regime sufficiently far removed from saturation that dispersion efficiency was maintained over the range of the feed rates examined. We have no current explanation for this anomolous behavior.

OVERALL SYSTEM DESIGN

Based upon the excellent experimental results derived from the crude laboratory apparatus, an overall system has been designed. It is currently being fabricated for installation at the Particulate Aerodynamic Test Facility. The system block diagram is illustrated in Figure 13. The gas utilized to operate the fluidized bed is extracted from the wind tunnel and under conditions when high moisture content is being used in the tunnel gases, the gas is passed through a chiller to remove excessive amounts of water. The gases then pass through a blower, and then through an electric heater which heats the gas stream such that the mixture of gas plus dust has a temperature matching that in the wind tunnel test section. Gas is then fed into the lower fluidized bed where dust is injected by a variable speed screwfeeder with dust supplied from an integrated materials handling system drawing upon bin storage located in an adjoining room. The dust is dispersed in the first fluidized bed and then passes through the second fluidized bed for final redispersion. The dust exiting the top fluidized bed enters eight 2-inch diameter transport tubes which carry the aerosol mixture into the wind tunnel for injection upstream of the tunnel test section. The aerosol is injected through nozzles with tip diameters selected to optimize the mixing

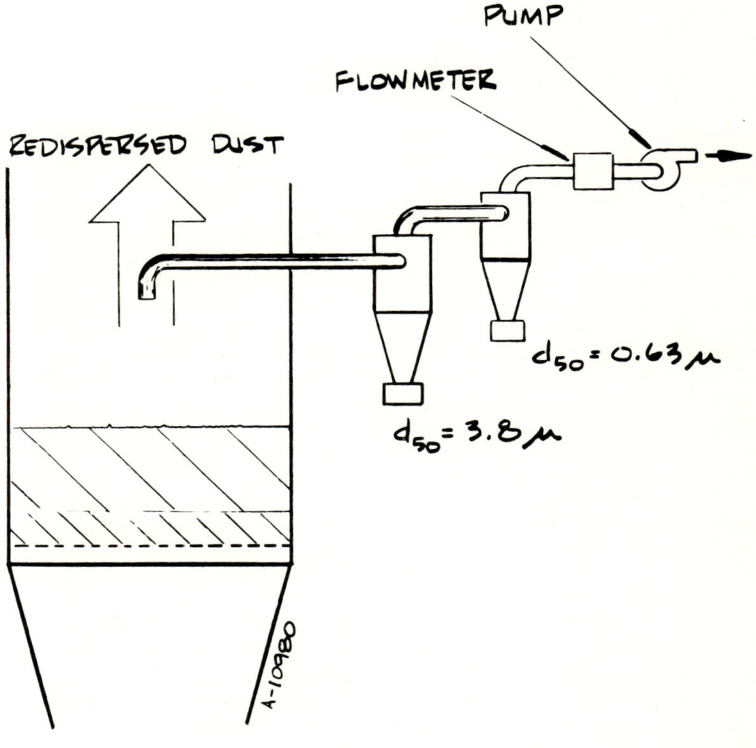

Figure 10. Sampling Apparatus

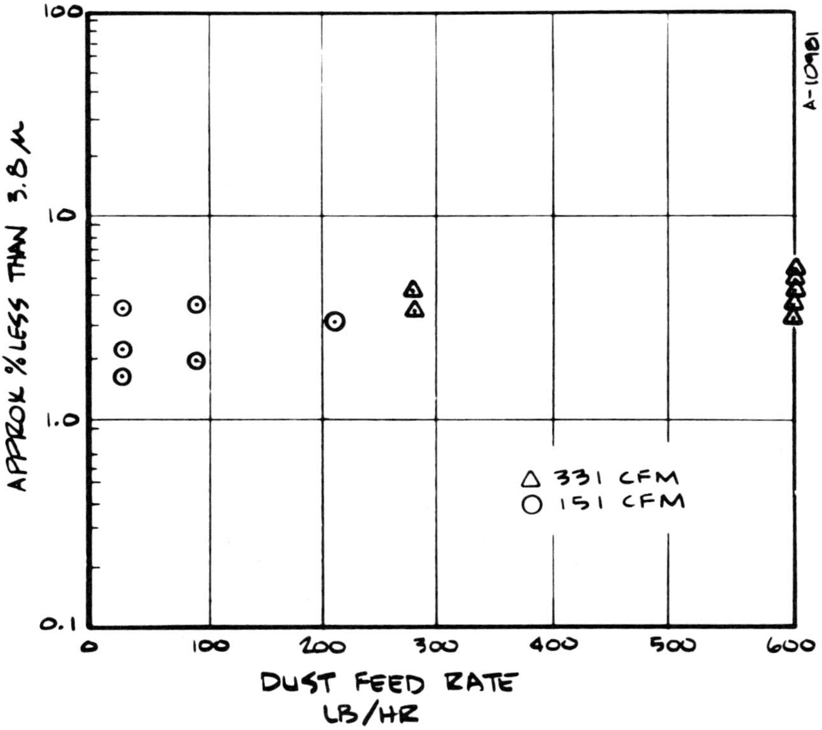

Figure 11. Sampling Results for Flyash

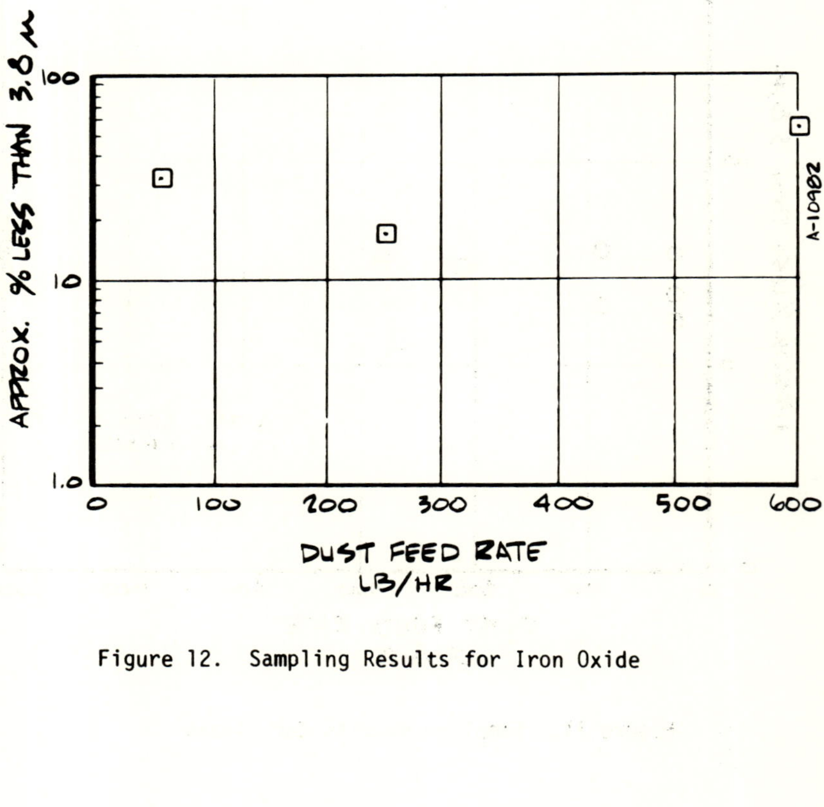

Figure 12. Sampling Results for Iron Oxide

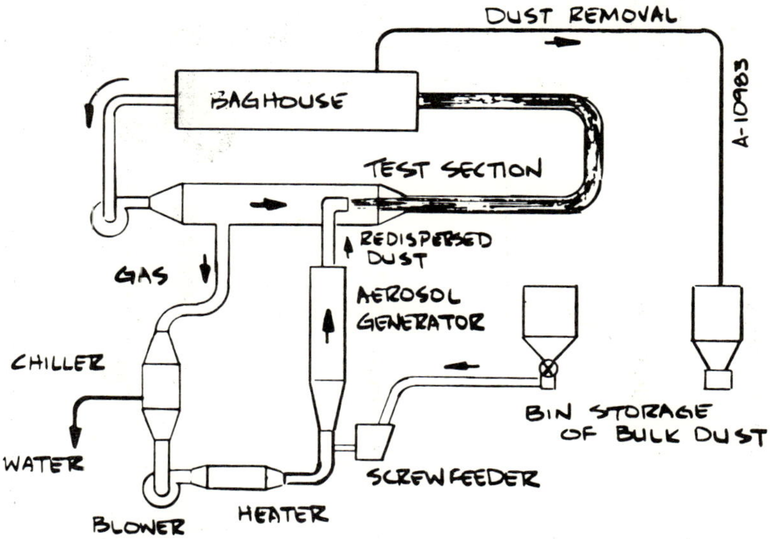

Figure 13. Block Diagram of Aerosol Generation System

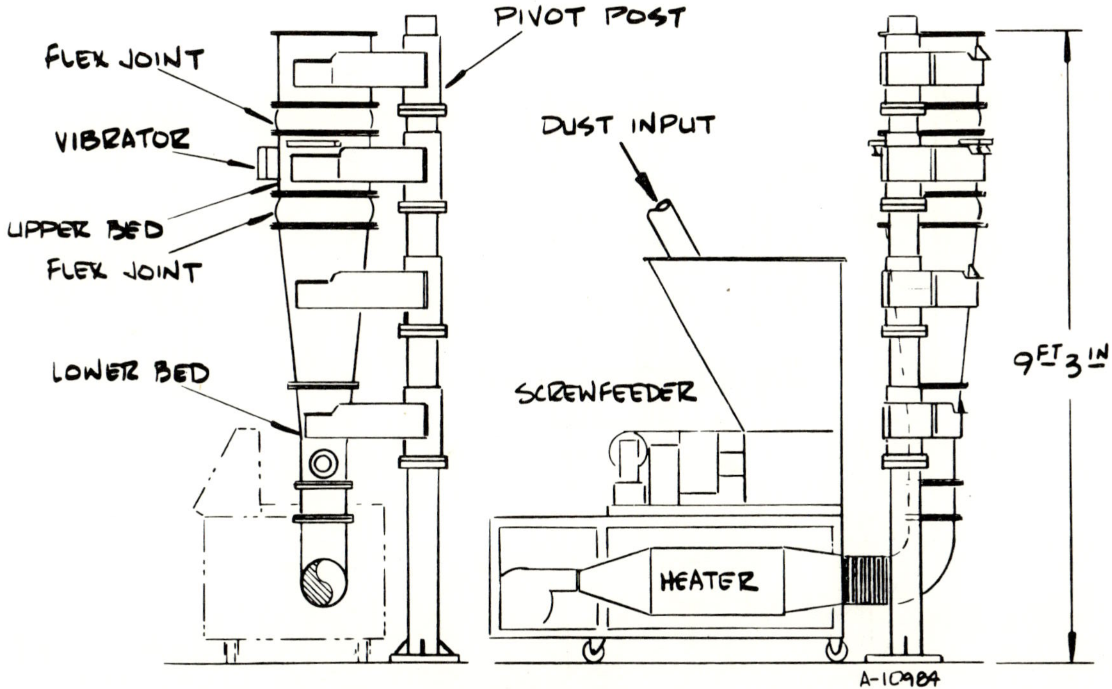

Figure 14. Detail of Aerosol Generator

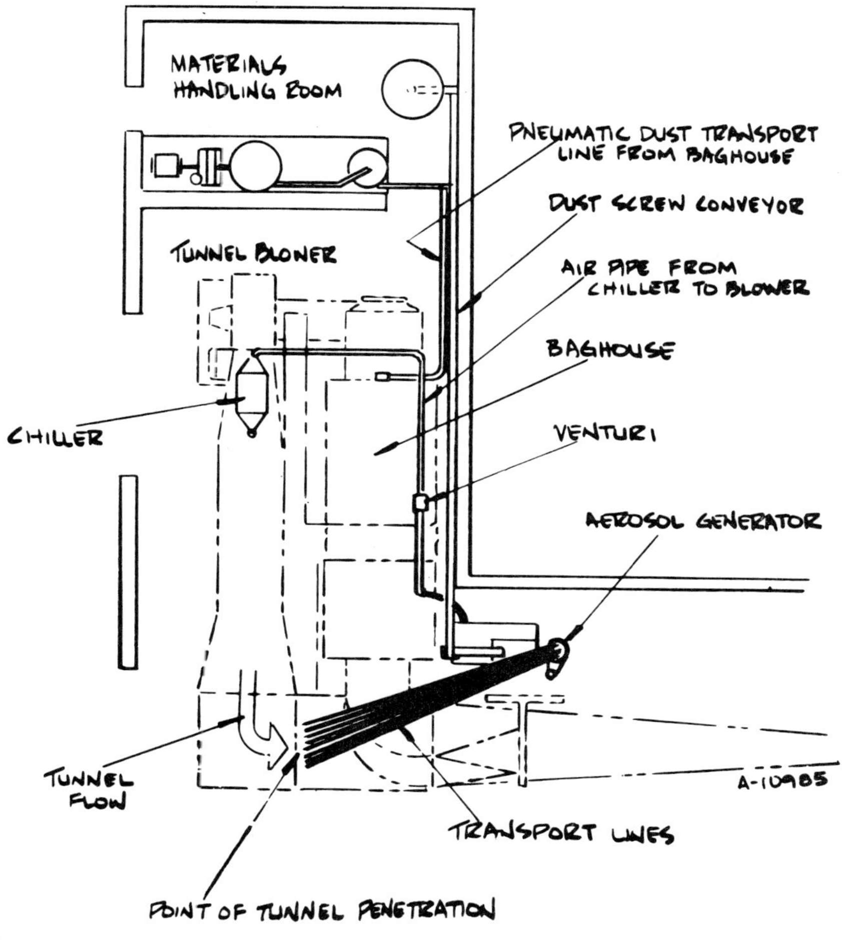

Figure 15. Facility Layout Showing Wind Tunnel, Aerosol Generator, and Dust Conveying System

diameter of the jet issuing from each of the eight tubes. Dust mixes with the tunnel gases within the stilling chamber upstream of the test section and then passes through the contraction into the test section where instruments can be tested or prototype particulate collection devices can be evaluated. Figure 14 shows a detailed side view of the aerosol generator itself and illustrates the sizes of the equipment that will be installed. The system is currently under construction and will be operational in the fall of 1975.

CONCLUSIONS

- The two component fluidized bed has been demonstrated to be incapable of operating at high flow rates of dust because of a variety of scale-up problems.

- The new three component fluidized bed combined with a two component fluidized bed in a two stage system appears to be scaleable to any reasonable size.

- The two stage system has demonstrated capability to redisperse collected dusts with substantial amounts of particulate in the fine particle range, and thus the system is appropriate for fine particulate research involving either instrumentation or particulate control device technology.

REFERENCES

Guichard, J.C., and Magne, J.L., Applications of Liquified Bed in Gaseous Atmosphere to the Production of Aerosols. Part IV: Study of Elutriation of Certain Simple Liquified Bed - Applications. July 28, 1967. Translated from the French for the National Air Pollution Control Administration. APTIC-Tr-0169.

GENERATION OF INORGANIC AEROSOLS
FOR WEATHER MODIFICATION EXPERIMENTATION

William G. Finnegan and John W. Carroz
Naval Weapons Center
Michelson Laboratory
China Lake, California

ABSTRACT

Since 1947 submicron-size aerosols of materials which can initiate ice nucleation activity have been used in weather modification experiments and in operational programs. Many aerosol generating methods have been tried; the two methods most used at this time are combustion of certain flammable solutions and the burning of specially formulated pyrotechnics. Other aerosols have also been generated for atmospheric tracer and warm cloud modification experiments. These aerosol generation techniques are quite flexible, in that many different materials can be generated in a wide range of particle size distributions. Aerosol technology has been and is continuing to be advanced by the weather modification effort.

GENERATION OF INORGANIC AEROSOLS

FOR WEATHER MODIFICATION EXPERIMENTATION

William G. Finnegan and John W. Carroz
Naval Weapons Center
Michelson Laboratory
China Lake, California

INTRODUCTION

This paper describes methods used to generate ice nucleating aerosols for weather modification experimentation. The earliest methods used were sparks, hot filaments, burning charcoal, and burning solutions. Some of the propellant and pyrotechnic formulations used are discussed in some detail. Other aerosols used to release tracer elements and hygroscopic salts into the atmosphere, particle sizes, and a computer program for calculating the chemical species in the hot gases of combustion are also briefly described.

After Schaefer (1946)[1] demonstrated that supercooled clouds could be greatly changed by seeding them with many small ice crystals produced by falling "dry ice" and after Vonnegut (1947)[2] observed that silver iodide (AgI) particles serve as nuclei for the formation of snow crystals, there was considerable interest in generating ice nucleating aerosols to test the possibility that the weather might be influenced over large areas by dispensing large numbers of ice nucleating particles into the atmosphere.

The efforts to generate aerosols for weather modification started in 1947 and have continued to the present. Many different methods have been tested. The two methods most used are burning solutions and burning pyrotechnics; however, several of the other methods may be as useful to environmental protection technology. This paper will attempt to cover most of the methods used.

EARLY EFFORTS

The production of submicron-size AgI smokes is easily accomplished by first evaporating AgI at a high temperature and then rapidly cooling the vapor so that it condenses into many small particles. Early methods used to produce AgI nuclei for laboratory studies were to pass a spark between silver electrodes in the presence of iodine vapor, to heat AgI on a hot filament or to disperse it in a flame (Vonnegut, 1947)[2]. Another method of doing this was to burn AgI-impregnated charcoal in a stream of air (Vonnegut, 1951)[3]. The heat of the burning charcoal evaporates the AgI on the surface of the charcoal and the airstream quenches the AgI vapor, as well as promoting the combustion of the charcoal.

Although AgI is highly insoluble in both water and organic solvents, it is, nevertheless, quite soluble in acetone solutions of a soluble iodide, such as NaI, KI, and NH_4I. Acetone solutions can be diluted with acetone to any concentration without precipitation of the AgI. Accordingly, acetone solutions were and still are widely used in weather modification.

Using solutions in the spray nozzle type of generator is one of the simplest and most efficient methods for producing large numbers of AgI nuclei. The early method used a two-phase conventional nozzle; the AgI solution was atomized by the action of a jet of compressed gas (Vonnegut, 1949)[4]. Because of the small size of the solution droplets, they were very rapidly vaporized in the flame. The AgI and NaI vapor in the flame rapidly condensed when they mixed with the cool air of the atmosphere to form a smoke of very small particles. By varying the concentration and the rate of flow of the solution to the spray nozzle, the size of the resulting particles could be varied over a wide range. In general, the greater the flow rate to the nozzle, the larger were the resulting particles, Vonnegut, 1949[4]. Solutions containing up to 10% of AgI can be used in spray nozzle type burners.

Additional methods were developed after 1951. They are burning of a celluloid film, toilet tissue, or confetti impregnated with AgI (Vonnegut, 1957)[5], and vaporization of a solution of AgI in anhydrous ammonia, or isopropylamine (Davis and Steele, 1968)[6].

PYROTECHNICS AND PROPELLANTS FOR WEATHER MODIFICATION

"Pyrotechnics, the Fire Art - from the Greek words pyr (fire) techne (an art) - is one of three closely related technologies, those of explosives, propellants, and pyrotechnics proper . . . . Explosives perform at the highest speed of reaction, leaving gaseous products; propellants are gas formers of brisk reactivity, but slower than explosives; and pyrotechnic mixtures react mostly at visibly observable rates with formation of solid residues. Numerous exceptions to these definitions may be cited" (Ellern, 1968)[7]. A fourth class of materials can be defined as gas-generating substances which are not intended to produce heat or light as such but are sources of gas used to disperse other materials (e.g., colored organic dyes).

Propellants usually are burned under pressure; pyrotechnics usually are burned at ambient pressure up to altitudes of 40,000 feet.

PRECAUTIONS

A word of warning may be unnecessary to most readers but may not be amiss for some who may wish to make their own pyrotechnics. The mixtures described are all dangerous if not properly handled. Mixing and firing pyrotechnics is safe only when done under proper conditions, and these should be rigorously observed. Advice should be sought from someone skilled in the field before undertaking such work.

PROPELLANTS FOR FREEZING NUCLEANTS

Vetter et al. (1970)[8] summarized the production of cloud nucleants from propellant formulations.

While working on the production of colored smokes, Drs. Burkardt and Finnegan of the Naval Weapons Center (NWC) generated pure AgI aerosols by burning mixtures of organic fuel binders and silver iodate ($AgIO_3$); the $AgIO_3$ is the oxidizer. These first studies (1957) at NWC were made using the exotic fuel poly(2-methyl-5-vinyl)tetrazole, $C_4H_6N_4$. The general reaction, balanced to produce both CO and CO_2, is

$$3AgIO_3 + C_4H_6N_4 \longrightarrow 3AgI + 2CO + 2CO_2 + 3H_2O + 2N_2$$

This composition produced copious smoke, the solid particles of which were identified by X-ray and chemical analysis as AgI and a trace of metallic silver. This composition, an ideal choice from a chemical point of view, is not widely used, however, because of the unavailability of the fuel-binder.

Polybutadiene is a less expensive binder. Several satisfactory polybutadiene compositions for the dissemination of AgI have been developed.

A large number of polybutadiene formulations were investigated using a computer to find the best possible mixture. The basic chemistry is shown in the unbalanced equation:

$$AgIO_3 + [-CH_2-]_n + Al \longrightarrow$$

$$Ag + AgI + H_2O + nCO_2 + nCO + HI + H_2 + Al_2O_3$$

Polybutadiene has a high oxygen requirement for complete combustion, and $AgIO_3$ is a poor oxidizer. It is impossible to use enough $AgIO_3$ to completely oxidize the polybutadiene and still have enough polybutadiene to mix and cast the propellant. The heat of combustion of polybutadiene is not great enough to vaporize all AgI, so aluminum is added to the formulations to raise the heat of combustion and the flame temperature. The presence of aluminum presents an additional oxygen demand which can only be met by adding another oxidizer or a high-energy explosive plasticizer.

Double base propellant systems are interesting because they produce an aerosol of only small particles. When burning they do not produce a visible smoke. A maximum of 15% by weight unrecrystallize $AgIO_3$ can be added to this propellant. A 5% formulation is shown below:

	Wt. %
Nitrocellulose	47.3
Nitroglycerin	38.0
Di-N-Propyl adipate	2.9
2-Nitrodiphenylamine	1.9
Lead beta resorcylate . . .	2.4
Cupric salicylate	2.4
Candelilla wax	.1
$AgIO_3$ powder	5.0

The products of combustion are AgI, a trace of copper oxide and lead oxyiodide, and large amounts of water vapor, N_2, CO_2, and CO. The high ratio of gas to solids minimizes agglomeration and the yield of nuclei per gram of AgI is similar to the high values from NH_4I-AgI-acetone solutions which when burned also have very high gas to solid particulate ratios.

PYROTECHNICS FOR FREEZING NUCLEANTS

Since several hundred different pyrotechnics have been formulated for atmospheric modification, this paper will cover only a small portion of what has been done. St.-Amand et al. (1970)[9] summarized the production of nucleants from pyrotechnic compounds.

The simplest composition is an explosive mixture of pyrotechnic-grade magnesium and $AgIO_3$ powder. The mixture is sensitive, quite dangerous, and should be kept dry.

$$AgIO_3 + 3Mg \longrightarrow AgI + 3MgO$$

A rapid burning composition suitable for a pressed flare or fusee for use at high altitudes is

	Wt. %
$AgIO_3$	84
Mg	8
Binder	8

Slower burning mixtures may be made with aluminum.

The reaction with aluminum is

$$AgIO_3 + 2Al \longrightarrow AgI + Al_2O_3$$

Mixtures containing only aluminum, $AgIO_3$ and binder are hard to ignite, so enough magnesium is usually added to ensure an easy ignition and a quiet burn. An example of a practical formula is

	Wt. %
$AgIO_3$	78
Al	12
Mg	4
Binder	6

When burned, the solid products are AgI, and a mixture of magnesium and aluminum oxides; the binder is oxidized to H_2O, CO_2, and CO.

It is frequently desirable to decrease the amount of $AgIO_3$ in a mixture in order to reduce the particle size of the resultant AgI or to produce a hygroscopic wetting agent to go along with the AgI. For this purpose another oxidizer, such as KNO_3 or KIO_3, may be added to the formulation. One such mix is

	Wt. %
$AgIO_3$	28
KNO_3	44
Mg	11
Al	11
Binder	6

The smoke from this formulation is yellowish white in color and consists of a mixture of AgI, K_2CO_3, MgO, and Al_2O_3.

Drew (1966)[10] has shown that aluminum oxides produced from the combustion of aluminum powder consist of spheres of a variety of sizes dependent primarily upon the size of the starting material and the nature of the environmental flame gases. The aluminum oxides produced occur in two particle size distributions. The larger material, consisting of spheres 1 to 100 μ in diameter, is believed to be produced from accumulations of oxide on the particle surface during combustion. The smaller material, comprising perhaps 30% of the oxide produced, is spherical powder usually less than 1 μ with the smallest considerably less than 50 Å in diameter. Under some conditions, thin-walled, hollow spheres are produced which reach a maximum of about three times the diameter of the original particle of aluminum. For example, with a 100 μ diameter starting material, 300 μ diameter bubbles have been measured.

If an alloy of aluminum and magnesium is used, spinel oxide ($MgO \cdot Al_2O_3$) is produced, its yield being dependent on the ratio of aluminum to magnesium in the combustion process.

In general, the hottest reactions occur when stoichiometric mixtures are used. With AgI, however, any temperature above 552°C can be used for dissemination. The temperature can be controlled while the reaction is still properly oxidized by loading the mixture with an excess of AgI. In this manner the temperature can be reduced to as low a value as desired, compatible with satisfactory burning.

Most of the $AgIO_3$ pyrotechnic formulations contain ingredients which react with or shift the final chemical equilibrium away from pure AgI particulates. Thus, each formulation has to be carefully and thoroughly studied before its cloud seeding potential is understood. Some materials when added to a formulation greatly lower or eliminate the nuclei output.

An important consideration for environmental protection technology as well as weather modification is the particulate mass concentration in the pyrotechnic flame, the size of the flame, and the flame temperature. They all affect the size of the resulting particulates of most materials.

SOLUTIONS AND SOLUTION BURNERS

Bernard Vonnegut (1949)[4] published a paper on an acetone solution burner for seeding clouds. Because AgI will not dissolve in acetone without excess iodide ions, he added NH_4I to the acetone solution. Then, in a later paper, he discussed use of NaI as a solution additive to replace the NH_4I (Vonnegut, 1950)[11]. Later, Vonnegut pointed out that the AgI nuclei produced with NaI were actually complexes. Although these complex nuclei have a much poorer nucleating ability than the AgI nuclei produced from NH_4I solutions, Vonnegut's

1950 paper had a great deal of influence, and for 20 years the scientists who were engaged in weather modification used the NaI to dissolve AgI in acetone. Consequently, from 1950 to 1970 the results of weather control efforts were poor and indeterminate.

Thompkins et al. (1963)[12] carefully analyzed the burner situation, with particular regard to the use of KI as a solubilizing agent. They (and Vonnegut) pointed out the importance of the chemistry involved. Finnegan et al. (1971)[13] again emphasized the importance of differences in the chemistry of the solutions, and most weather modifiers became convinced that NH_4I is a better solubilizing agent. Currently, solutions combining AgI with NH_4 are used almost universally in acetone burners. The NH_4I is destroyed in the flame and pure AgI is the only solid emitted from the burner.

Several burners were developed to burn AgI-acetone solutions. Some are ground based, others are operated from aircraft. Most of the ground-based burners were designed to burn propane (or a similar fuel) along with the acetone solution. The airborne burners often were designed to operate without an auxillary fuel. In Patent 2,527,231, Vonnegut (1950)[14] describes a ground-based solution burner with and without auxillary fuel. Vonnegut and Maynard (1952)[15] described the design and operation of an airborne burner which burned butane and acetone solution. Fuquay (1960)[16] described the development and calibration of an airborne solution burner without auxillary fuel. Carroz and Noles (1972)[17] described a ground-based solution burner which operates without an auxillary fuel (see Figure I). Carroz (1973)[18] described an airborne burner which generates more measured nuclei at -5°C per gram of AgI than any other system (see Figures II and III).

OTHER AEROSOLS

Aerosols of the tracer elements lithium and indium have been used to aid in the identification and understanding of atmospheric processes. Efforts to formulate both solutions and pyrotechnics for rapid release of kilogram quantities of a water soluble lithium tracer aerosol were unsucessful because of the low atomic weight of lithium. Large kilogram quantities of water soluble cesium tracer, however, can be produced by burning a pyrotechnic containing 70% by weight C_sNO_3 (Finnegan and Carroz, 1974)[19]. The weight % of elemental cesium in this pyrotechnic formulation is 47.7%. A water soluble rubidium tracer can also be formulated by using $RbNO_3$ in place of $CsNO_3$. A high purity aerosol of In_2O_3, a water insoluble tracer, was produced by burning a pyrotechnic made by mixing indium metal with nitroplasticized smokeless gun powder. The formulation contains 16.7% by weight elemental indium.

Figure I. Ground-based solution burner; 6-gallon capacity

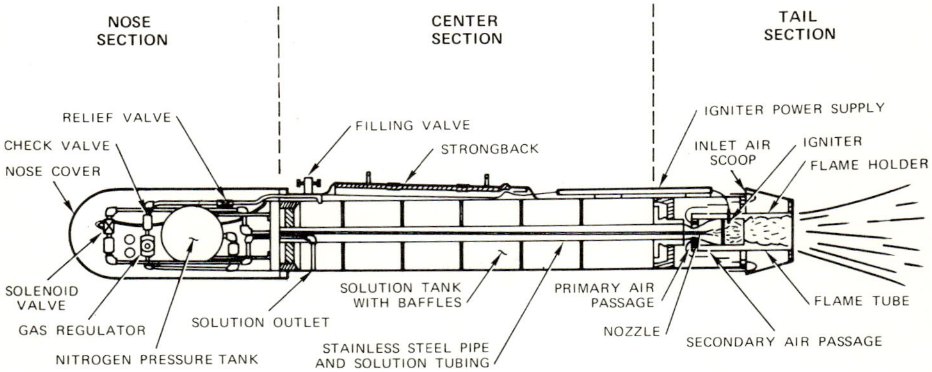

Figure II. Airborne burner with solution supply system

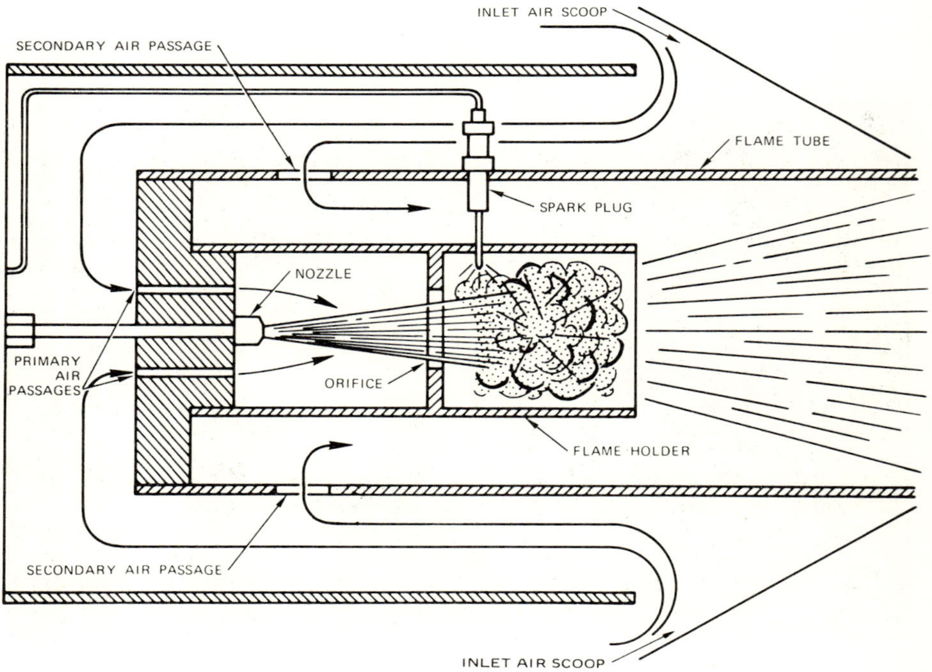

Figure III. Airborne burner

The smoke from this pyrotechnic is free of a contamination from other water soluble particulates. Except for the indium metal and trace amounts of lead and copper, all of the ingredients decompose to gaseous products.

Large 250-pound pyrotechnic units were manufactured to dispense NaCl-KCl-LiCl aerosols. These and smaller units were used in warm cloud modification experiments.

AEROSOL PARTICLE SIZES

The size range for AgI-NaI particles generated by the Skyfire generator was found by Kissinger and Mitchell (date unknown) to be 10^{-2} to 10^{-1} μ (Fuquay, 1960)[16]. Mossop and Tuck-Lee (1968)[20] reported the size distribution of AgI-NaI particles generated by the Warren-Nesbitt generator. The particle distribution followed a log probability law with a median diameter of 0.085 μ and a standard deviation factor of 1.47. Combustion of an atomized ethyl alcohol (95%) solution with 18.9% $LiClO_4$ yielded particles ranging between 0.27 to 0.63 μ with a mean of 0.45 μ.

In many cases the nuclei measurements of active nuclei exceed the measurements of particles. It is difficult to capture and to measure all of the smallest particles in an aerosol.

MEASUREMENTS OF NUCLEI

In an effort to compare one generator with another and with pyrotechnic formulations the National Science Foundation has for several years supported the Cloud Simulation and Aerosol Laboratory at Colorado State University. Here generators and pyrotechnics are tested for measured nuclei output. Even though the measured nuclei may differ from the actual nuclei, this facility does provide some comparison of generator output. The many complications of nuclei measurements will not be covered in this paper because of their limited application to pollution technology. A sample of the measured nuclei output from a solution burner is shown in Figure IV.

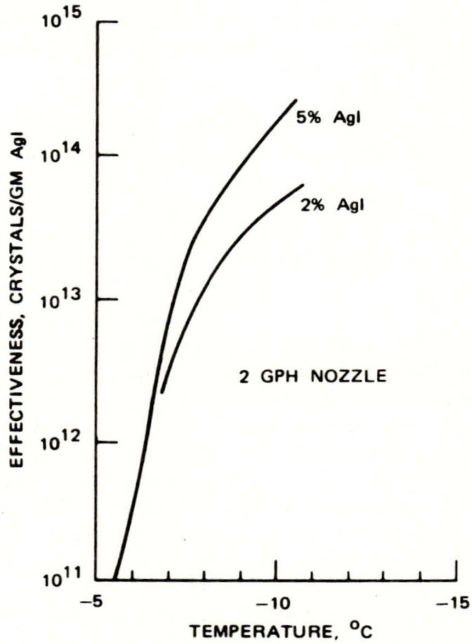

Figure IV. A sample of the measured nuclei output

COMPUTER CALCULATIONS

The NWC Propellant Evaluation Program (PEP) can be used to calculate flame temperatures. The calculations are more realistic for chemical reactions at equilibrium (e.g., rocket motors, power plants, etc.) than for small pyrotechnics burning in air.

The PEP is based on a straightforward thermodynamic model consisting of two processes: (1) constant pressure, or adiabatic combustion, and (2) isentropic pressure, or adiabatic expansion.

The assumptions behind the combustion process include:

1. Reaction kinetics are fast enough so that chemical equilibrium is attained before the products leave the combustion chamber.

2. No heat is exchanged between the system and the surroundings. (In ramjets, the stagnation energy of the incoming air becomes part of the system. This may simply be added to the heat of the formation of air.)

3. Gaseous species individually obey the perfect gas law and collectively obey Dalton's law of partial pressures.

When such assumptions are made, the system enthalpy and system pressure completely determine the final state and chemical composition of the system after combustion. The solution to this state and composition is found by a computer technique called "enthalpy balance."

An example of a computer calculation is shown below for an AgI pyrotechnic formulation burning at one atmosphere of pressure.

INGREDIENTS	WEIGHT
SILVER IODATE	78.300
ALUMINUM (PURE CRYSTALLINE)	10.800
MAGNESIUM (PURE CRYSTALLINE)	5.200
100DER321/43DEH14 (EPOXY BINDER)	5.700

GRAM ATOM AMOUNTS FOR PROPELLANT WEIGHT OF 100.000

(H)	(C)	(N)	(O)	(MG)	(AL)	(AG)	(I)
.461183	.339340	.012526	.892143	.213816	.400297	.276884	.276884

Reaction Pressure: 1 atmosphere
Flame Temperature: 2346° K

Chemical Species in Flame: Number of Moles

AL	.0002/ 2.18-04	ALH	.0000/ 1.08-05	ALHO2	.0000/ 2.59-11	ALO	.0000/ 8.12-07
AL2O	.0000/ 3.51-05	C	.0000/ 6.16-11	CH4	.0000/ 1.44-08	CO	.3391/ 3.39-01
CO2	.0002/ 2.27-04	C2H2	.0000/ 7.92-09	CNH	.0000/ 5.73-06	C3	.0000/ 5.14-16
H	.0065/ 6.49-03	MGH	.0015/ 1.49-03	NH	.0000/ 1.70-08	HO	.0000/ 4.37-06
H2	.2129/ 2.13-01	H2O	.0008/ 8.07-04	NH3	.0000/ 5.90-08	MG	.2123/ 2.13-01
MGO	.0000/ 1.41-05	N	.0000/ 5.81-09	NO	.0000/ 4.16-08	N2	.0063/ 6.26-03
O	.0000/ 8.72-08	O2	.0000/ 1.39-10	HI	.0249/ 2.49-02	I	.2192/ 2.19-01
I2	.0003/ 2.57-04	ALI	.0322/ 3.22-02	ALI3	.0000/ 7.14-06	NHO	.0000/ 7.09-12
AGH	.0008/ 8.14-04	AGO	.0001/ 5.45-05	AG	.2760/ 2.76-01	ALN$	.0000/ 1.00-25
AL203$	.0000/ 1.00-25	AL203*	.1839/ 1.84-01	C$	.0000/ 1.00-25	MGO$	.0000/ 1.00-25
AGI$	.0000/ 1.00-25	AG$	.0000/ 1.00-25				

CONCLUSION

The preceding brief description of the methods used to generate aerosols for atmospheric experimentation is necessarily limited. Weather modification experiments during the past 25 years have provided much experience in aerosol generation. Aerosol generation technology has been advanced by the weather modification effort and we look forward to applying our knowledge of these advances to meet the requirements of environmental protection technology.

REFERENCES

1. Schaefer, V. J., 1946: The production of ice crystals in a cloud of supercooled water droplets. Science, pp. 457-459.

2. Vonnegut, Bernard, 1947: The nucleation of ice formation by silver iodide. J. Appl. Phys., Vol. 18, pp. 593-595.

3. Vonnegut, Bernard, 1951: Final Report Project Cirrus. General Electric Research Laboratory, Report No. RL-566, pp. 27-56.

4. Vonnegut, Bernard, 1949: Nucleation of supercooled water clouds by silver iodide smokes. Chem. Rev., Vol. 44, pp. 277-289.

5. Vonnegut, Bernard, 1957: Early work on silver iodide smokes for cloud seeding. Final Report to the Advisary Committee on Weather Control, Vol. II, pp. 283-285.

6. Davis, C. L. and R. L. Steele, 1968: Performance characteristics of various artificial ice nuclei sources. J. Appl. Meteor., Vol. 7, No. 4, pp. 667-673.

7. Ellern, Herbert, 1968: Military and Civilian Pyrotechnics. Chem. Publishing Co. N. Y., p. 3.

8. Vetter, R. F., et al., 1970: Pyrotechnic production of nucleants for cloud modification - Part III: Propellant compositions for generation of silver iodide. J. Weather Modification, Vol. 2, pp. 53-64.

9. St.-Amand, P., 1970: Pyrotechnic production of nucleants for cloud modification - Part II: Pyrotechnic compounds and delivery systems for freezing nucleants. J. Weather Modification, Vol. 2, pp. 33-52.

10. Drew, C. M., 1966: Aluminum particle combustion in gas burner flames. Naval Ordnance Test Station, China Lake, TPR 415, NOTS 3916, pp. 8-56.

11. Vonnegut, Bernard, 1950: Techniques for generating silver iodide smoke. J. Colloid Sci., Vol. 5, No. 1, pp. 37-48.

12. Thompkins, L. M., et al., 1963: Water adsorption in the system AgI-KI-H_2O. J. Geophys. Res., Vol. 68, pp. 3537-39.

13. Finnegan, W. G., P. St.-Amand, and L. A. Burkardt, 1971: An evaluation of ice nuclei generator systems. Nature, Vol. 232, pp. 113-114.

14. Vonnegut, Bernard, 1950: Method of generating silver iodide smoke. U.S. Patent 2,527,231.

15. Vonnegut, Bernard and Kiah Maynard, 1952: Spray nozzle type silver-iodide smoke generator for airplane use. Bull. Am. Meteor. Soc., Vol. 33, No. 10, pp. 420-428.

16. Fuquay, D. M., 1960: Generator technology for cloud seeding. ASCE Proceedings, J. Irrigation Drainage Division, Vol. 86, pp. 79-91.

17. Carroz, J. W. and R. C. Noles, 1972: A new ground based solution burner for cloud seeding and production of atmospheric tracer materials. Proceedings Third Conf. Weather Modification, Am. Meteor. Soc. pp. 37-40.

18. Carroz, J. W., 1973: Airborne jet seeder solution burner. J. Weather Modification, Vol. 5, pp. 166-177.

19. Finnegan, W. G. and J. W. Carroz, 1974: Tracer element from pyrotechnics and solution burners. Proceedings Fourth Conf. Weather Modification, Am. Meteor. Soc., pp. 214-217.

20. Mossop, S. C. and C. Tuck-Lee, 1968: The composition and size distribution of aerosols produced by burning solutions of AgI and NaI in acetone. J. Appl. Meteor., Vol. 7, No. 2, pp. 234-240.

GENERATION OF AEROSOLS BY BURSTING OF SINGLE BUBBLES

Milos Tomaides*, and K. T. Whitby**

Interpoll Inc., St. Paul, Minnesota, and
University of Minnesota, Minneapolis

ABSTRACT

The mechanism of aerosol generation by bursting of single bubbles from aqueous solution of sodium chloride has been studied. A single bubble aerosol generator has been designed to evaluate the size and concentration of droplets formed by bursting of a bubble film and also by the liquid jet action subsequent to the bubble burst. A solution of 0.1 percent sodium chloride in distilled water was used in these experiments to generate bubbles of 1.4 and 5.5 mm in diameter. While bubbles of 1.4 mm size generated both the film and also jet droplets the 5.5 mm bubbles result in generation of the film droplets only. The burst of a 1.4 mm bubble resulted in a generation rate of twenty film droplets and four jet droplets on the average. The 5.5 mm bubbles formed two-hundred film droplets. Semi-empirical equations have been developed to describe the total mass of droplets generated by bursting of bubbles of various sizes. A superior monodispersity of jet droplets over film droplets has been found. The monodispersity of film droplets improves rapidly with increasing bubble size. Practical aspects of the experimental results on the design of a bubble aerosol generator are presented.

This paper is based on work performed at the University of Minnesota and in the laboratories of Interpoll Inc., St. Paul, Minnesota.

* Present Address - Interpoll Inc., 1996 W. Co. Rd. C., St. Paul, Minnesota 55113

** Present Address - University of Minnesota Particle Technology Laboratory, Minneapolis, Minnesota 55455

NOMENCLATURE

B	-	Constant
C	-	Constant
d_B	-	Bubble diameter as measured on the solution surface
d_{DJ}	-	Volume mean droplet diameter, μm
d_D	-	Droplet diameter, μm
d_P	-	Residue particle diameter, μm
E	-	$\frac{6B}{\pi C^3 \rho_D}$ = Constant
k	-	Solution concentration, g/g
K_1	-	Constant depending on the air flow conditions
K_2	-	Constant
M_J	-	Total mass of jet droplets generated from one bubble
n_D	-	Number of droplets per bubble
n_{DJ}	-	Number of jet droplets per bubble
n_{DF}	-	Number of film droplets per bubble
n_B	-	Number of air bubbles
ρ_D	-	Density of bubble liquid
ρ_L	-	Solvent density, g cm^{-3}
ρ_P	-	Solute density, g cm^{-3}
t	-	Time

GENERATION OF AEROSOLS BY BURSTING OF SINGLE BUBBLES

Milos Tomaides*, and K. T. Whitby**

Interpoll Inc., St. Paul, Minnesota, and
University of Minnesota, Minneapolis

INTRODUCTION

The behavior of single bubbles and also of a cloud of bubbles rising in liquids has been studied by many investigators both theoretically and experimentally. Studies reported by K. Koide, et al.[1], E. Kojima, et al.[2], S. Krishnamurthy, et al.[3], V. G. Glejm, et al.[4], V. G. Glejm and V. M. Vilenskij[5], and R. L. Datta and R. Kumar[6] are only some of many others. The majority of these studies have primarily been initiated by needs of the chemical industry. Variety of mass and energy transfer operations and processes in which a gas is dispersed in a liquid are usually accompanied by more or less vigorous entrainment of liquid in the form of droplets as for example studied by F. M. Garner, et al.[7]. An original work by D. C. Blanchard[8] has been devoted to the generation of droplets by bursting of bubbles on the sea as a part of a meteorological and oceanographical research.

The generation of small droplets by bursting of bubbles started to be considered a possible, simple source of artificial aerosols only recently. For example, an aerosol generator in which a flow of air bubbles through a layer of aqueous solution of methylene blue has been studied by J. C. Guichard and J. L. Magne[9]. These authors produced successfully an artificial aerosol of a monodispersity which is comparable to the quality of artificial aerosols generated by the Collison atomizer aerosol generator which has been accepted in a variety of aerosol studies. Because the "foaming" aerosol generator has an advantage of being easy to construct a question arises how versatile device it is and up to what extent the quality of the aerosol generated by this approach can be maintained, varied or improved comparing to other aerosol generators. Trying to analyze these questions a lack of technical information on generation of droplets by bursting of bubbles became obvious.

An attempt was made in this work to develop a technique which can be utilized to study the physical properties of aerosols generated during the burst of the individual bubbles floating on a liquid surface.

MECHANISM OF BUBBLE BURSTING

Two independent physical processes are involved in generation of aerosols by bursting of bubbles. The first one is the formation of gas bubbles and their motion through a liquid. This process is followed by a decay of bubbles after they penetrate the liquid surface.

The formation of bubbles and their behavior in liquid is an important factor which influences the final size and number of bubbles penetrating the liquid surface film. It has also direct impact on the mass of droplets generated during the bubble decay. The size of bubbles is also a basic parameter which can be directly related to the size of minute droplets generated during the bubble burst. The formation and dynamics of bubbles has been subjected to many investigations. The results of these studies are well described and summarized in the literature[1,2,3,6].

Of primary interest in this study was the process of the generation of minute droplets during the bubble decay and the evaluation of their physical properties. The minute droplets are the result of two independent physical mechanisms. A bubble rising through a liquid penetrates eventually the liquid surface forming a thin, dome-shaped film. This film is stretched above a more or less spherical, shallow depression in the liquid surface. Formation of the bubble film is accompanied by the liquid drainage from the dome apex causing the film thinning in this area. The drainage is very rapid with pure liquids. Any impurities, monomolecular films or dissolved materials tend to stabilize the bubble and the liquid drainage is slower. If these stabilizing agents are present in large quantities a very stable foam is usually formed. Thinning of the bubble film results eventually in its rupture forming an opening on the top of the dome. The gas trapped inside the bubble under positive pressure leaves the bubble rapidly. The high velocity gas jet helps to break up the bubble film surrounding the initial opening. A cloud of so called film droplets is formed. The rest of the bubble film is rapidly attracted to the liquid surface and disappears.

In the process of the bubble film rupture the decreasing pressure within the bubble is followed by a motion of the surrounding liquid into the bubble crater. For small bubbles the potential energy at the bottom of the bubble crater is high enough to eject a jet of liquid from that area. The jet rises above the liquid surface and disintegrates partially into droplets before it disappears back into the liquid. The droplets resulting from this liquid jet action are called jet droplets. They are generally of much larger size comparing to film droplets.

FILM DROPLETS

The process of the generation of film droplets is very complex and very sporadical experimental results have been published so far on this subject. Some of the recent studies suggest that the air jet escaping from the bubble has only limited influence on the generation of film droplets. Blanchard[8] came to the conclusion that hydrodynamical disturbances in the bubble film after its puncture are the primary mechanism of film droplets generation. This opinion seems to be realistic and can be supported by a simple calculation. It is known that depending on surface tension and viscosity of liquid the ruptured bubble film moves outwards with the speed of several hundred to several thousand cm per second. At this speed a 0.25 mm water bubble collapses in about 3×10^{-7} seconds. Even assuming that the gas leaving the ruptured bubble attains speed of the sound instantly, it travels only a distance of 0.1 mm during the above calculated rupture time. It is less than a half of the bubble diameter. This supports the opinion that the film droplets probably originate during the decay of a disturbed liquid toroid whose circumference is proportional to the bubble film diameter.

The published experimental data on number of film droplets generated by rupture of bubbles of various sizes are scarce. Only qualitative conclusions can be drawn from the published data. The number of film droplets can be considered approximately directly proportional to the diameter of the film cap. No film droplets can be expected for water bubbles smaller than 0.2 mm, but this limit depends strongly on even very small amount of organic impurities in the liquid. The impurities may prevent generation of film droplets even from 1.6 mm bubbles. To our knowledge no reliable systematic data on the size distribution of the film droplets are available.

JET DROPLETS

More attention has been paid in various research activities to the jet droplets. As described above the jet droplets originate from a breakup of a bubble jet. These droplets are ejected upward to the heights as great as 20 cm and with speeds that reach several thousand cm per second initially. The energy for this process results from the conversion of a part of the bubble surface free energy into the kinetic energy of jet droplets. Comparing the surface free energy of a bubble with the energy of jet droplets Glejm[4] has expressed total mass of droplets generated by a single bubble jet as

$$M_J = B \times d_B^2 \tag{1}$$

Based upon the experiments with sea water bubbles by Hayami and Tobal[10] the mean diameter of jet droplets generated from bubbles larger than 0.3 mm can be described by the following empirical equation

$$d_{DJ} = C \times d_B^{\,1.3} \tag{2}$$

Number of jet droplets liberated from one bubble can also be calculated from

$$M_J = \frac{1}{6}\pi\, d_{DJ}^3 \times n_{DJ} \times \rho_D \tag{3}$$

Plugging equation (2) into (3) and equating equation (3) and (1) results in

$$n_{DJ} = \frac{E}{d_B^{1.9}} \tag{4}$$

Because term E is a constant the equation (4) predicts a quite rapid decrease of the jet droplets number concentration with increasing bubble size. On the contrary a large number of jet droplets must be expected from the burst of small bubbles. Experimental results have shown the equation (4) be valid only up to the bubble diameter of approximately 5 mm for pure water. For 6 mm diameter bubble of pure water the jet ejection velocity is already so small that no jet droplets are formed any more.

EXPERIMENTAL PROCEDURES

As a part of a program to evaluate the feasibility of bubbling aerosol generators an attempt was made first to collect more information on the size and number of droplets generated by bursting of small bubbles. For this purpose a single bubble aerosol generator has been constructed to study the properties of both the jet and also the film droplets. This device proved to be a useful tool for this purpose.

SINGLE BUBBLE AEROSOL GENERATOR

The single bubble aerosol generator is shown in Figure 1. Its design allows only one bubble to burst at a time. It consists of the air humidifier, the aerosol generating section and of the aerosol sampling filter. The humidifier is made from a piece of 18 mm I.D. glass tube being plugged from both sides and nearly filled up with distilled water. Compressed air enters and rises through water in a form of small bubbles formed on the tip of a glass capillary. The bubble-water contact time is sufficient to nearly saturate bubbles with water vapor. This is necessary for trouble-free operation of the aerosol generator which follows the humidifier section. The jet droplets generated by bursting of bubbles in the humidifier are removed by an impaction disc which is mounted very close to the water surface. Humid air continues to flow into the aerosol generator through another capillary tube. The generator section is of a design similar to the humidifier. The diameter of a capillary tube used to form bubbles is selected such to form a bubble size desired. The generator is filled with an aqueous solution as for example of sodium chloride. This helps to determine the size of original droplets generated if the size of residue particles is measured. All droplets generated by bursting of bubbles on the solution surface are removed from the generator by a specially designed cone-shaped flushing air deflector. They are transported in the stream of clean, dry flushing air and dried completely before reaching the Millipore membrane and collected on its surface. This generator was utilized successfully to generate aerosol by bursting of single bubbles of 1.4 and 5.5 mm diameter. The aqueous solution used in these experiments was 0.1 per cent of sodium chloride.

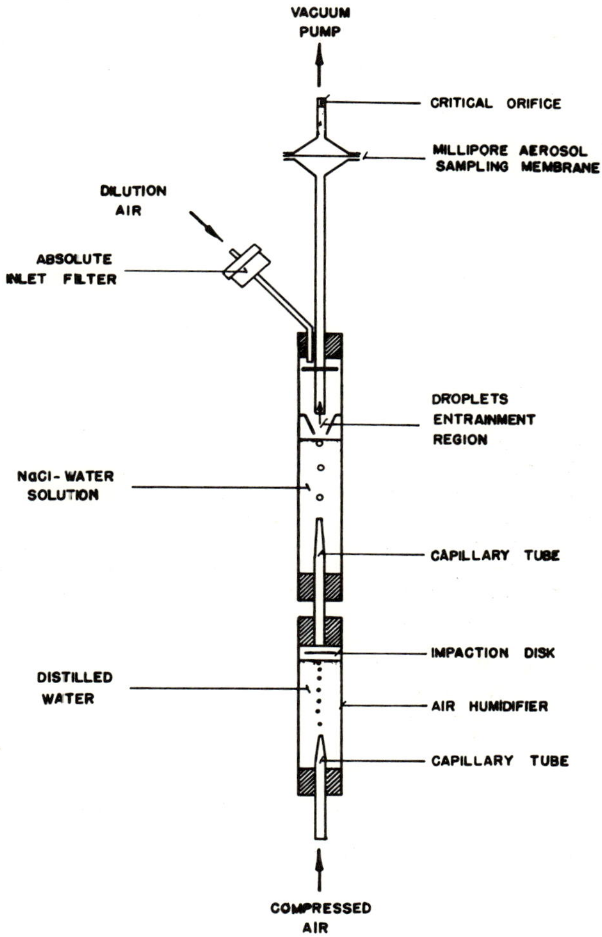

Figure 1. Single Bubble Capillary Tube Aerosol Generator

The bubble size was measured using a direct and also an indirect technique. In the direct sizing the bubble size was compared with a scale engraved on a thin glass slide which was immersed in the solution close to the capillary tube outlet. The indirect technique was based on the measurement of the bubble rising velocity which was compared with the experimental results of Harberman and Morton[11]. To determine the bubble rising velocity the total number, N, of bubbles seen stationary within the distance, H, when illuminated by the stroboscopic light of frequency, f, were counted and rising velocity calculated from the following equation

$$u = \frac{H\,f}{N} \tag{5}$$

The preference was given to the direct technique because the bubble rising velocity does not vary significantly within the bubble size range of 1 to 10 mm and the indirect technique is thus not too accurate.

AEROSOL SIZE DISTRIBUTION AND CONCENTRATION

The residue aerosol particles generated by the single bubble aerosol generator were collected on a 0.45 μm pore size Millipore membrane. The membrane was divided into five equal annular areas and a replica prepared of a small portion of each of these areas. The particles were sized and counted under the transmission electron microscope. The size of droplets generated by the bubble rupture was then calculated knowing the size of sodium chloride particles collected on the membrane with the use of the following equation

$$\frac{d_D}{d_P} = \sqrt[3]{\frac{k\,\rho_L + \rho_P}{k\,\rho_L}} \tag{6}$$

Total number of droplets generated during each generation experiment was calculated knowing the surface of the Millipore membrane and the area of the membrane replica in which particles were counted.

RESULTS

As an example of the performance of the single bubble aerosol generator the results of the size distribution of droplets generated by bursting of 1.4 and 5.5 mm bubbles of 0.1 per cent sodium chloride aqueous solution are presented and discussed. The size of the droplets generated by bursting of 1.4 mm bubbles is shown in Figure 2 and for 5.5 mm bubbles in Figure 3.

The size distribution of droplets shown in Figure 3 has a clear bimodal characteristic. This means that both the film and also the jet droplets were generated from 1.4 mm bubbles. Larger droplets resulted from jet action while smaller droplets are the result of the film rupture. Very few submicrone droplets were found in samples and the size distribution curve falls off rapidly for sizes below 1 μm for both bubble sizes tested. As it follows from Figure 3 the spread of the sizes of the film droplet is about 1:50 while the majority of jet droplets is within 1:4 limits. Superior monodispersity of jet droplets for 1.4 mm bubbles is obvious. On the average 20 film droplets and 4 jet droplets were liberated by each bubble.

Examining the data plotted in Figure 3 more carefully a solid line can be drawn through the data points having at least three maximas for film droplets and two for jet droplets. It is not difficult to believe that this curve is much closer to reality than the dashed curve is. The fact that the jet droplet created on the top of the jet is always larger in comparison to the rest of jet droplets generated closer to the liquid surface explains the presence of a little peak in the jet droplet size distribution curve. For very small bubbles even another peak can be probably expected located toward smaller droplet sizes that would correspond to generation of small satellite droplets. These satellites are created during the separation of larger droplets along the bubble jet. This mechanism is well known when aerosols are generated by spinning disk aerosol generators.

The 5.5 mm bubbles do not result in any jet droplets as seen in Figure 2. The ratio of the smallest and the largest film droplet size is about 1:20 because only insignificant number of droplets was generated in the submicrone range. Experiments with larger bubble sizes have shown that the monodispersity of film droplets is improving with increasing bubble size. Number of film droplets generated by 5.5 mm bubble is about 200.

Similarly to Figure 3 several maximas can be visualized on the size distribution curve in Figure 2. This fact may support the idea of a wave-form disturbance in the bubble film during the film rupture and decay. Such a disturbance would break the bubble surface into several concentric toroids. Each toroid breaking up into nearly monodisperse droplets of the size which corresponds to the toroid diameter.

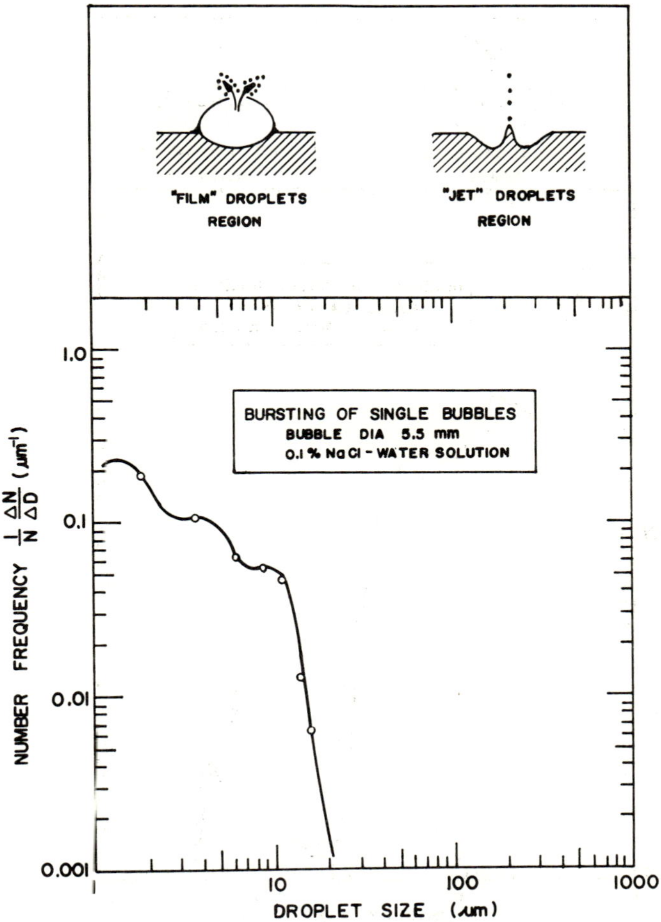

Figure 2. Size Distribution of Droplets Generated by Bursting of 5.5 mm Diameter Bubbles on a Surface of 0.1 Per Cent Sodium Chloride Solution

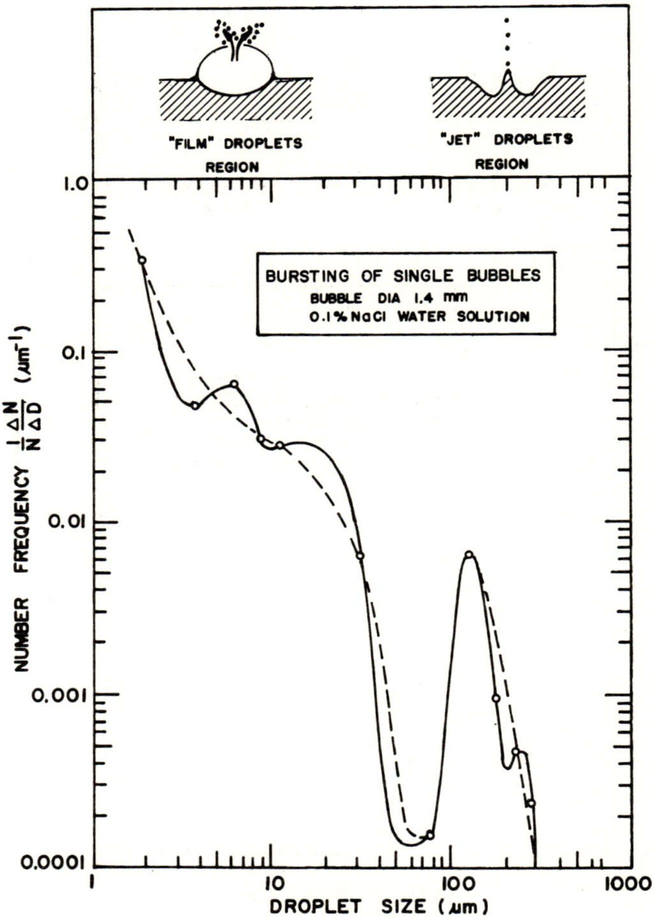

Figure 3. Size Distribution of Droplets Generated by Bursting of 1.4 mm Diameter Bubbles on a Surface of 0.1 Per Cent Sodium Chloride Solution

DISCUSSION

The design of an aerosol generator which would employ bursting of bubbles to generate artificial aerosols can be discussed based upon the single bubble experiments. It is obvious that the best aerosol monodispersity can be achieved by bursting of very small bubbles. The bubbles would have to be below the critical size at which no film droplets are already generated. Monodispersity of the aerosol of geometric standard deviation of about 1.2 would be quite feasible. Unfortunately, it is very difficult to generate larger number of very fine bubbles and prevent their coalescence on the solution surface. For this reason, the generation of artificial aerosols from very large bubbles is more promising. The monodispersity of such an aerosol would not be much better than 1.4. This aerosol quality is already very close to the quality of aerosols generated by the best atomization aerosol generators. The aerosol generation from bubbles between 0.5 mm to 6 mm diameter should be avoided because these bubble sizes result in generation of both the jet and also the film droplets.

An estimate of the aerosol output from a bubbling aerosol generator can be made based partially upon the results of this study. Let us suppose the case of generation of droplets from bursting of single monodispersed bubbles on a solution surface in the absence of any foam that would otherwise change drastically the aerosol generation conditions. If a porous plate with the air bubbling through it is used to generate bubbles and the flow rate of the air is constant and only the plate pore size changes, then the number of bubbles generated for these conditions can be expressed by

$$n_B = \frac{K_1}{d_B^3} \tag{7}$$

The rate of droplet generation is described by

$$\frac{dn_D}{dt} = n_D \times n_B \tag{8}$$

The number concentration of film droplets can be calculated from

$$n_{DF} = K_2 \times d_B, \tag{9}$$

as described earlier in the text.

Plugging (9) and (7) into (8) the rate of the film droplet generation obeys the equation

$$\frac{dn_{DF}}{dt} = \frac{K_1 \times K_2}{d_B^2}, \tag{10}$$

indicating a rapid decrease of number of particles generated with increasing bubble size for constant air flow rate.

Similar tendency can be expected for jet droplets as calculated from equation (8) with equations (4) and (7) plugged in, which results in

$$\frac{dn_{DJ}}{dt} = \frac{E \times K_1}{d_B^{4.9}} \tag{11}$$

This shows an extremely fast decrease in number of jet droplets with increasing bubble size for constant air flow rate.

The same procedure can be used to evaluate the mass rate of droplets generated by bursting of bubbles. This can be done without problems for jet droplets using equation (1) but lack of data on total mass of film droplets makes this calculation impossible for film droplets. But the experimental result published by Garner et al.[7] shows that the total mass of both the jet and also the film droplets decreases with increasing bubble size. The interesting phenomena found by these authors is levelling off of the total particle mass for bubbles larger than approximately 10 mm for water. This suggests limited validity of equations (11) and (10).

REFERENCES

1 Koide, K., S. Kato, Y. Tanaka, and H. Kubota. Bubbles Generated from Porous Plate. J. of Chem. Eng. of Japan, 1968, Vol. 1, p. 51-56.

2 Kojima, E., T. Akehata, and T. Shirai. Rising Velocity and Shape of Single Air Bubbles in Highly Viscous Liquids. J. of Chem. Eng. of Japan, 1968, Vol. 1, p. 45-50.

3 Krishnamurthi, S., R. Kumar, and N. R. Kuloor. Bubble Formation in Viscous Liquids Under Constant Flow Conditions. I and EC Fundamentals, 1968, Vol. 7, p. 549-554.

4 Glejm, V. G., I. K. Shelomov and B. R. Shidlovskij. O processach privodjascich k generacii kapel pri razryve puzyrej na poverchnosti radela zidkost-gaz. Zhurnal Prikladnoj Chimii, 1959, Vol. XXX11, p. 218-222.

5 Glejm, V. G., and V. M. Vilenskij. Vlijanie stepeni dispersnosti parovych (gazovych) puzyrej na kapelnyj vynos pri kipenii i barbotaze. Zhurnal Prikladnoj Chimii, Vol. XXX11, 1966, p. 82-84.

6 Datta, R. L., and R. Kumar. Gas Bubbles in Liquids (Part 1). Indian Chemical Engineer, 1961, Vol. 3, p. 71-75.

7 Garner, F. H., S. R. M. Ellis, and J. A. Lacey. The Size Distribution and Entrainment of Droplets. Trans. Inst. Chem. Engrs., 1954, Vol. 32, p. 222-235.

8 Blanchard, D. C. The Electrification of the Atmosphere by Particles from Bubbles in the Sea. Progress in Oceanography, The Macmillan Company, Vol. 1, 1963.

9 Guichard, J. C., and J. L. Magne. Etude D'Un Pulverisateur A Membrane. Institut National de Recherche Chimique Appliquee. VERT-le-PETIT. 1966.

10 Hayami, S., and Y. Toba. Drop Production by Bursting of Air Bubbles on the Sea Surface. J. Ocean. Soc. Japan, 1958, Vol. 14, p. 145-150.

11 Harberman, W. L., and R. K. Morton. An Experimental Study of Bubbles Moving in Liquids. Trans. Am. Soc. Civil Engrs., 1956, Vol. 121, p. 227-250.

AN INVESTIGATION OF AN EXPLODING WIRE AEROSOL

by

James E. Wegrzyn
Cloud Physics Research Center
University of Missouri-Rolla
Rolla, Missouri 65401

ABSTRACT

The concentration, shape, mass, and total charge of a gold aerosol (produced by the exploding wire technique) is monitored leading to discussion of the effect that a bipolar charge has on the stability of a submicron aerosol. Values of K (coagulation kernel) and β (wall loss coefficient) for the bipolar aerosol are compared against the values of K and β determined from an exploding wire aerosol that was neutralized to their bipolar charge equilibrium. This aerosol is neutralized by use of a K-85 neutralizer. The results of this investigation showed that the charges on the gold aerosol enhance the coagulation in the first 600 seconds, and that gas adsorption on the newly formed aerosol is considerable.

AN INVESTIGATION OF AN EXPLODING WIRE AEROSOL

by

James E. Wegrzyn
Cloud Physics Research Center
University of Missouri-Rolla
Rolla, Missouri 65401

INTRODUCTION

The field of aerosol mechanics can be divided into the following three groups: aerosol generation, aerosol sampling, and aerosol analysis. This report deals mainly with the generation of a gold aerosol by an exploding wire (E.W.) technique. Aerosol generation, however, has to be complimented by correct sampling procedures and analysis if it is to be understood. Therefore, this paper will also discuss the use of the Electrostatic Aerosol Sampler[1] (produced by Thermo-Systems Inc.) as a sampling instrument. Other instruments used in this experiment are: the Particle Mass Monitor[2] (produced by Thermo-Systems Inc.), a Gardner counter, a diffusion battery[3], a K-85 neutralizer[4] (produced by Thermo-Systems Inc.), and a second electrostatic precipitator which was designed in this lab. The analysis will consist of intercomparison of these instruments for measurements on the concentration, shape, size, total charge, and reproducibility of an aerosol generated by exploding a wire. Values of K (coagulation kernel) were determined from the solution of the equation,

$$-\frac{dn}{dt} = Kn^2 + \beta n$$

where β (wall loss coefficient) was evaluated experimentally. The K-85 neutralizers effect on the coagulation kernel was separately evaluated. The overriding purpose of this investigation is to determine the feasibility of using an exploding wire aerosol as a trace aerosol in a scavenging experiment.

PROPERTIES OF AN EXPLODING WIRE

The physics of the E.W. phenomenon is well documented in the literature[5]. Likewise, the aerosols generated from this explosion have been studied by several investigators[6-12]. The report by R. F. Phalen[2] in 1972 gives a thorough review on this subject, and, in general, the previous studies indicate that if the E.W. is completely vaporized it results in the following characteristics: Pictures 1, 2, 3.

1) The primary particles form a chain-like structure of smooth spherical particles whose size distribution can be characterized by a log normal or normal distribution.

2) The mean diameter of these primary particles are between .01 and .1 micron in size depending on the parameters of their generation

3) There exists a symmetric bipolar charge on the aerosols.[13,14]

4) The aerosols are reproducible if generated in the same manner.

Unless the joule heating is enough to completely vaporize the wire in a manner of 300 nanoseconds or less, the E.W. aerosol will not have all these characteristics. Karoisis and Fish[4] illustrated the decrease in the primary size of the particles with increasing electrical energy. The reason being that some of this electrical energy goes into creating new surface area. Thereby, a larger input of energy yields more surface area and results in smaller particles. The size of the primary particles has little dependence on the temperature of the surrounding gas since the temperature of the plasma is between 7000° and 8000°K[6]. Several explosions were done at the liquid nitrogen temperature of 77°K with the primary size of these particles generated, remaining the same. However, the pressure does influence the size of the particle because at reduced pressures the plasma spreads over a larger region. With a pressure near 1mm Hg absolute, an aerosol is formed that is too small to be detected by the aerosol sampler and the mass monitor; and yet, the Gardner counter gave readings in excess of 100,000 particles per cm^3 range. To insure that this aerosol was due to the gold wire and not an artifact of the high temperature and radiation of the arc, a similar experiment was repeated without the gold wire. An arc was intentionally produced between the electrode and ground at the pressure of 1mm Hg, and in all cases the condensation nuclei was less than 1,000 particles per cm^3.

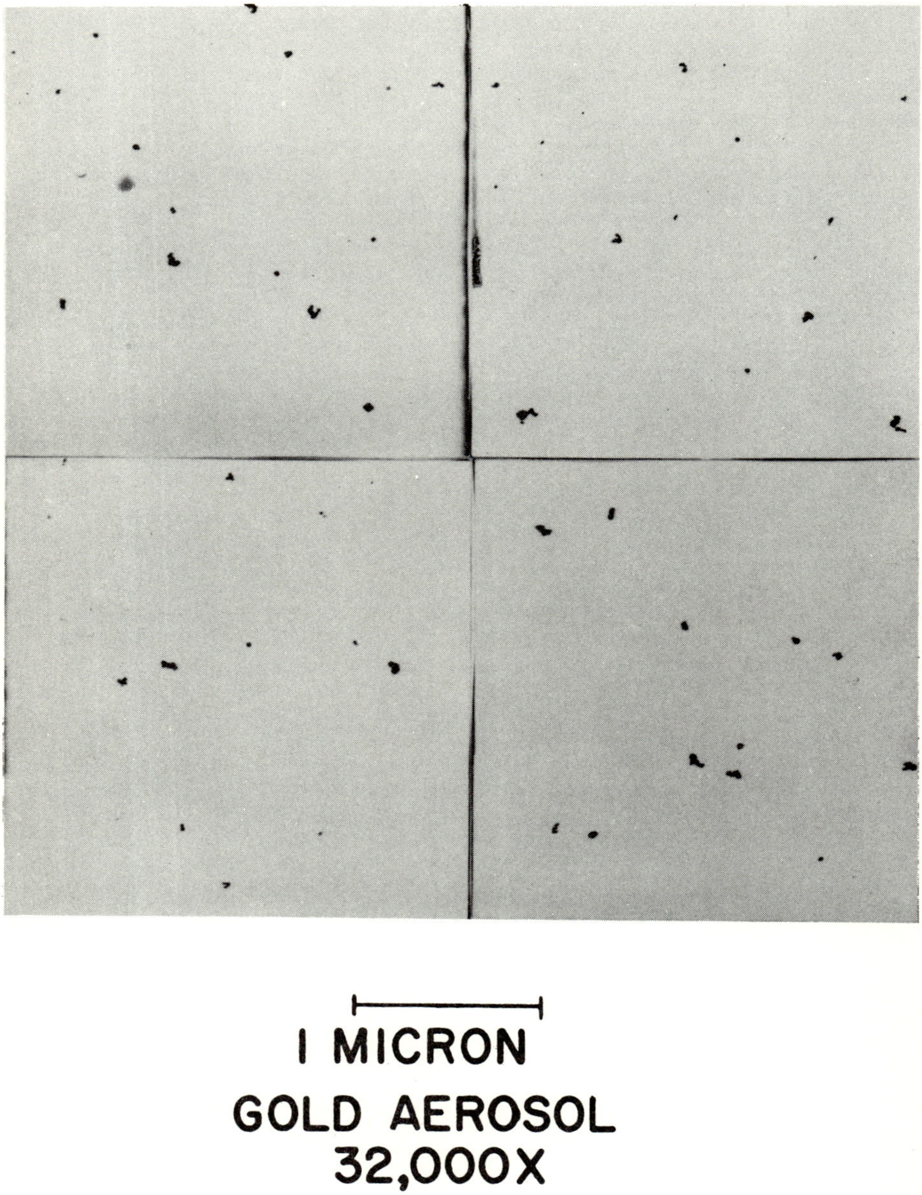

Picture 1 Gold Aerosol x 32,000 Magnifications.

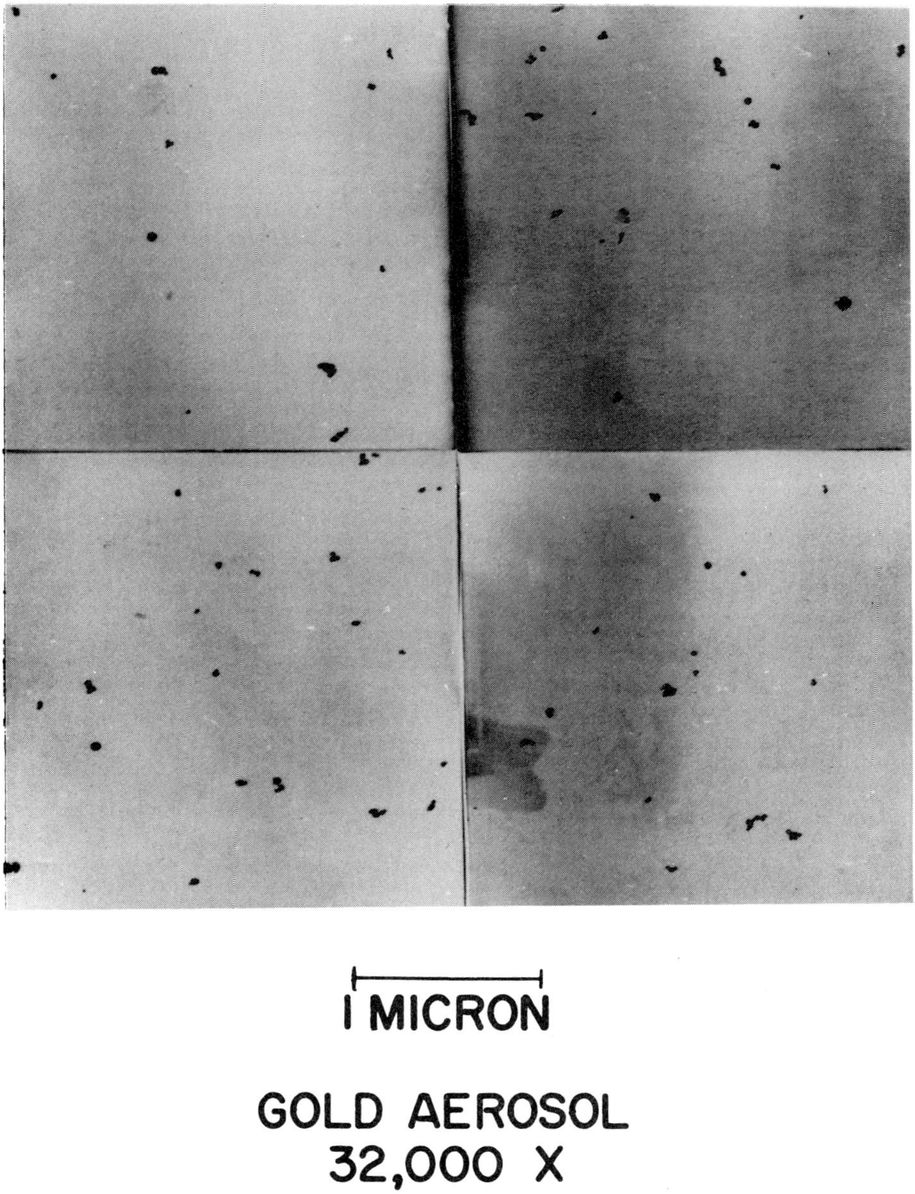

Picture 2 Gold Aerosol x 32,000 Magnifications Neutralized by Kr-85.

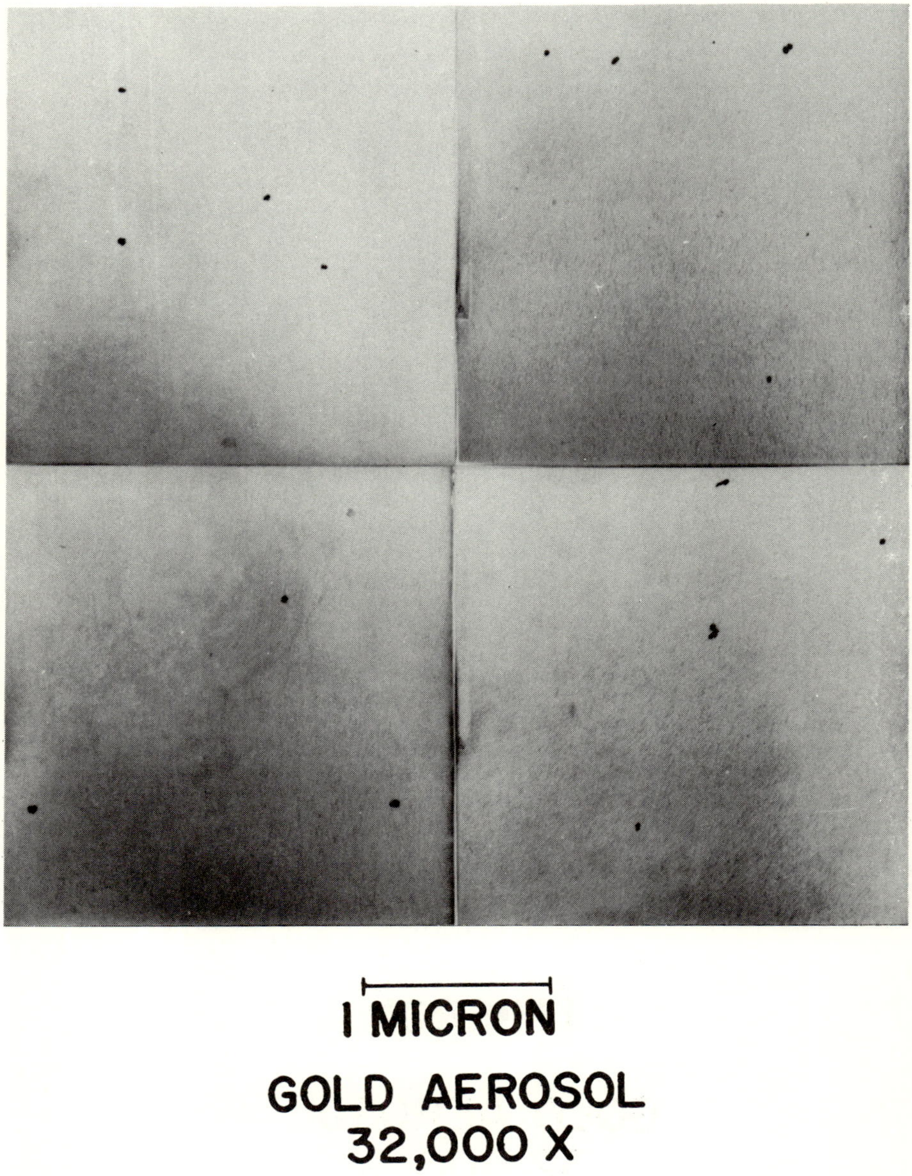

Picture 3 Gold Aerosol x 32,000 Magnifications Collected Without the Use of the Corona Field.

EXPLODING WIRE AEROSOL

In order to gauge the reproducibility of the E.W. aerosols, the parameters of generation were kept the same throughout the experiment. These parameters are: diameter of wire - .002 inch, length of wire - 2 cm, composition of wire - gold, pressure in chamber - 20mm Hg absolute, electrical energy - 2000 volts at 3 micro-farads, and carrier gas - argon. Argon was used to reduce any particulate matter of nitrogen that might form from an arc in a nitrogen, oxygen, and water environment. Another problem at reduced pressure in argon gas is that some of the electrical energy can be discharged around the wire rather than through it.[8] To overcome this problem, the wire was kept hot, and a hydrogen thyratron (EG&G 7322/1802) was used to short the wire to ground.

To reduce the agglomeration of the E.W. particles, it was necessary to use a two stage diluation system (Figure 1). This two stage dilution system can control the agglomeration size and concentration of the E.W. aerosol. The mean number of particles in the agglomerates can be varied between two and several hundred while the size distribution of the primary particles remain the same.

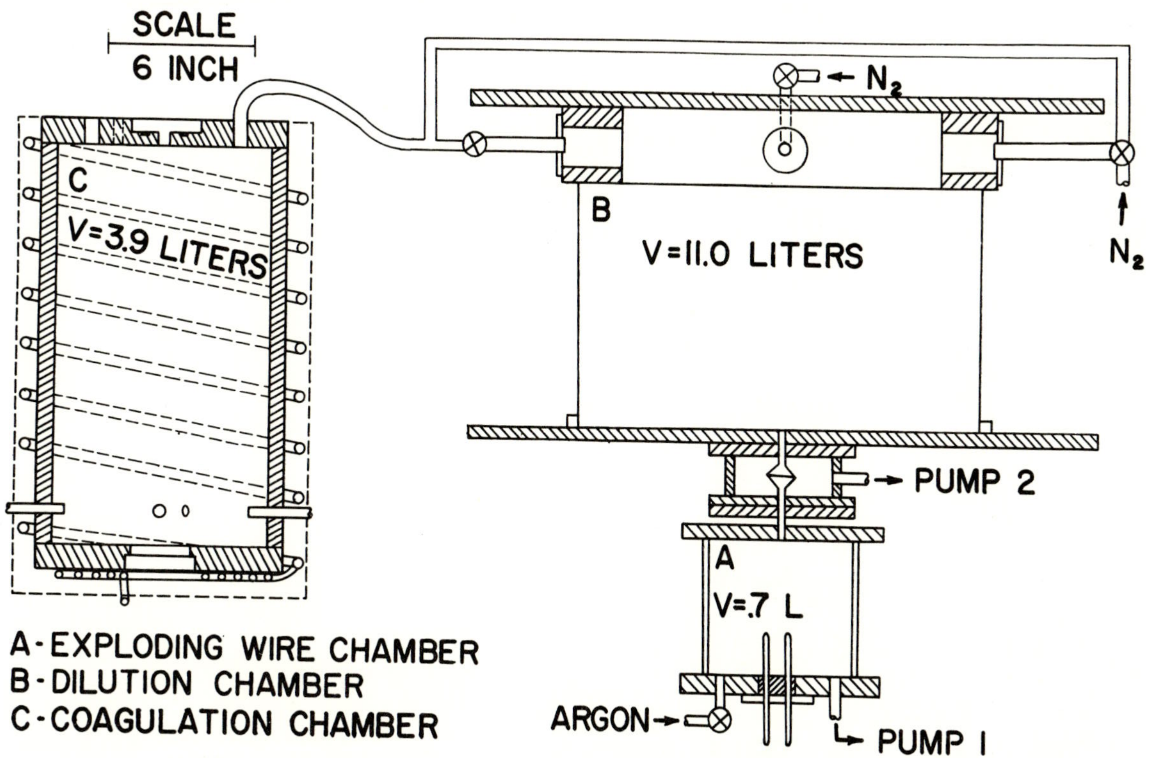

Figure 1 Two Stage Dilution System.

GENERATION PROCEDURE

The procedure used for aerosol generation was as follows: A gold wire was mounted between the electrodes, and a fine stainless steel mesh (175 per inch) was placed between the exploding wire chamber and the dilution chamber. To clean the system, both chambers were pumped down to less than 1mm Hg absolute and immediately refilled with filter argon gas. This procedure was repeated at least three times. The coagulation chamber was simultaneously purged with filter N_2 gas until the concentration of Aitken nuclei measured by the Gardner counter was less than 100 particles per cm^3. Also, the total mass detected by the T.S.I. mass monitor had to be less than $\pm 3.0 \mu\mu g/cm^3$, and the background current of the Keithley 410 electrometer had to be less than $\pm .25 \times 10^{-13}$ amps before the experiment was run. The pressure in the coagulation chamber was monitored by a Magnehelic pressure gauge (±2 inches of water), and was kept at a slight overpressure to prevent any contamination from the room.

The experimental setup is shown in Figure 2. This figure illustrates the overall goal of this investigation which is to measure the scavenging efficiency of the E.W. aerosol with a larger particle. The time of fall of these larger particles will be between 5 and 30 minutes. This paper deals with the first stage of the experiment which is to measure the characteristics of the E.W. aerosol over this time span. In Figure 2, blocks D, E, and F were not employed for this investigation. The electrostatic precipitator is shown in Figure 3. The charge on the top plate was ±2200 V.D.C., and the bottom plate was connected to the Keithley 410 electrometer. Responce frequency of this electrometer is 8.7 sec. for an eternal capacitance of 50μμ.f. The capacitance of this setup, which includes the connecting leads, was only 14μμf. Picture 4 shows the output from the electrometer over a five minute span.

A- EXPLODING WIRE CHAMBER
B- DILUTION CHAMBER
C- COAGULATION CHAMBER
D- SETTLING CHAMBER I
E- FLOW REACTOR
F- SETTLING CHAMBER II
G- T.S.I. K-85 NEUTRALIZER
H- LA MER GENERATOR
I- MAGNEHELIC PRESSURE GAUGE
J- T.S.I. MASS MONITOR
K- GARDNER COUNTER
L- DIFFUSION BATTERY
M- T.S.I. MASS MONITOR
N- ELECTROSTATIC PRECIPITATOR
O- KEITHLEY 410 ELECTROMETER
P- N_2 PRESSURE TANK I
Q- N_2 PRESSURE TANK II
R- ARGON PRESSURE TANK
S- EG & G 7322/1802 HYDROGEN THYRATON
T- 0-6 KV POWER SUPPLY

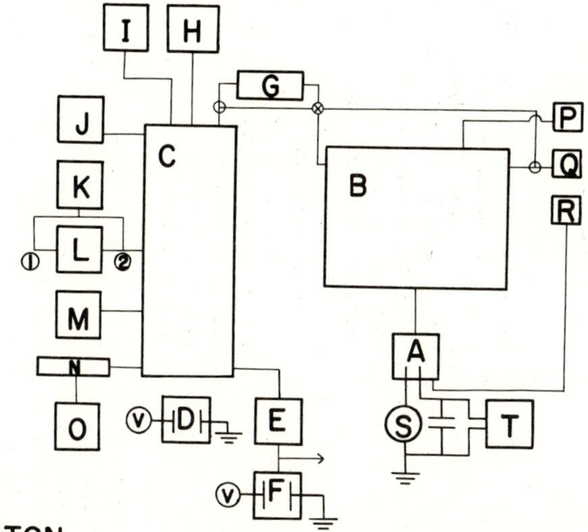

Figure 2 Block Diagram of the Experiment.

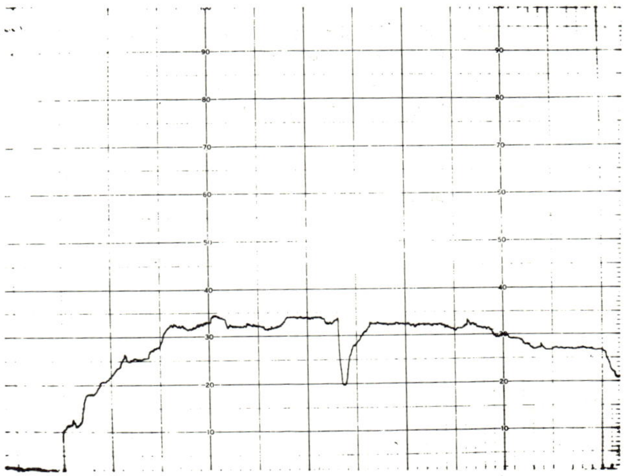

SCALE

2 IN.

TOTAL CHARGE MEASUREMENT
KEITHLEY 410 ELECTROMETER
FULL SCALE - 10 X 10^{-13} AMPS
FLOW RATE - 60 CM/SEC
CHART SPEED - 20 SEC/IN
CONCENTRATION - 125,000/CM

Picture 4 Output from Keithley 410 Electrometer.

RESULTS

The diffusion battery (L) was designed by T. A. Rich, and is used for the detection of particles too small to be precipitated out by the aerosol sampler (less than .01μ). Table 1 gives the results from the use of the diffusion battery.

time (sec)	conc/cm^3 at position 2	conc/cm^3 at position 1	conc/cm^3 at position 2	conc/cm^3 at position 1
	charger current-10 microamps		charges current=0.0	
t<0	500	500	200	200
60	8,000	--	55,000	--
90	--	4,800	--	45,000
120	6,000	--	66,000	--
150	--	4,400	--	47,000
180	5,000	--	59,000	--
210	--	3,500	--	42,000
240	4,800	--	50,000	--
270	--	3,000	--	36,000
300	3,500	--	45,000	--

Table 1 Diffusion Battery Data.

A diffusion battery along with the Gardner counter were placed between the aerosol sampler and its pump. The first column on the left hand side of Table 1 gives the concentration of particles that passed through the aerosol sampler when it was operating in the continuous mode with a charger current of 10 microamps. The second column gives the concentration of particles that passed through both the aerosol sampler and diffusion battery. With a background count of 120,000 particles per cm^3, it is seen that the aerosol sampler is 95% efficient in removing the E.W. aerosol, and the diffusion battery has a 75% transmission rate. Using the table published in Rich's paper, this 75% transmission rate corresponds to an effective diameter for this aerosol of .032 microns. If the particles were less than .01 microns in diameter, the transmission rate would be less than 15%. This experiment indicates that the number of particles too small to be precipitated out, yet able to act as condensation nuclei, are negligible. The right hand side of Table 1 was run with the charger current of the aerosol sampler set to zero so that only the previously charged E.W. aerosols would be collected. (See picture 3.)

Again with a background count of 120,000 particles/cm^3, it is seen that the percentage of charged particles is a little less than 50%. This agrees well with the other electrostatic precipitator. The transmission rate was also 75% which yielded the same effective diameter of .032 micron for the aerosol.

The data taken from the E.W. aerosol over a thirty min. time span is presented in Table 2 and Table 3. Table 2 is the data taken without the use of the K-85 neutralizer and Table 3 is the data taken with the use of the K-85 neutralizer. The 100,000 charges per cm^3 at a flow rate of 10 Lpm is well within the operating range of the 2 milli-Curie K-85 neutralizer[4,15]. B. Y. H. Liu and D. Y. H. Pui[16] verified that for a monodisperse aerosol in the size range of .02 to 1.17 microns, the magnitude of the charge on the aerosol is reduced to the Boltzmann's equilibrium value which is:

$$\frac{N_n}{N_o} = \exp(-n^2e^2/2akT),$$

where N_n is the number of particles carrying n elementary units of charge, N_o is the number of neutral particles, e ($=4.8\times10^{-10}$esu) is the elementary unit of charge, a(cm) is the particle radius, k is the Boltzmann constant, and T(^oK) is the absolute temperature. The validity of the Boltzmann's distribution of charge for the smaller sized particles has been questioned, however, by N. A. Fuchs[17]. The values of K was determined from the solution of the equation

$$-\frac{dn}{dt} = Kn^2 + \beta n$$

which is

$$K = \frac{n_o - n}{n\,n_o}\;\frac{\beta}{(e^{\beta t}-1)}$$

where β = .00009/sec.

Use of this equation has been criticized since it applies only for a monodisperse aerosol; but it is also found that this equation fits emperically the case of the coagulation of a polydisperse aerosol. From Table 2, the values of K range from 21 to 29 cm^3/sec, and there appears to be no dependence with time. These values compare well with the experimental value of 20×10^{-10} cm^3/sec found by Phalen for a silver E.W. aerosol; and also with Rosinski values of 24.3 to 28.2 x 10^{-10} cm^3/sec[18] for a gold E.W. aerosol. It also compares favorably with Zebel's[18] theoretical value (taking into account

5/7/75-6	150,000	76.6	55%	1,800	90,000	22	--
4/30/75-5	115,000	79.4	39%	1,800	78,000	21	103.2
4/30/75-6	110,000	59.3	41%	1,350	82,000	21	57.7
4/30/75-4	110,000	63.8	43%	1,200	85,000	21	61.6
4/30/75-7	105,000	69.4	39%	1,050	87,000	17	60.7
5/7/75-5	140,000	66.0	55%	900	98,000	32	91.0
4/30/75-3	105,000	68.3	43%	900	80,000	31	68.3
5/7/75-4	120,000	71.6	-	600	90,000	45	62.2
4/30/75-2	95,000	--	50%	600	75,000	45	71.6
5/7/75-3	140,000	63.8	-	300	112,000	58	89.3
4/30/75-1	145,000	61.6	-	300	118,000	51	65.2

Table 3 Aerosol Data without Neutralization.

DATE	INITIAL CONC/CM^3	MASS/CM^3 (μμg)	PER CENT CHARGED	COAGULATION TIME (SEC)	FINAL CONC/CM^3	COAGULATION KERNEL x10^{-10} cm^3/sec	MASS/CM^3 (μμg)
5/4/75-9	125,000	97.7	38%	1,800	74,000	28	143.7
5/1/75-3	130,000	127.6	-	1,800	74,000	29	--
5/2/75-1	110,000	110.4	27%	1,500	75,000	26	--
5/4/75-8	118,000	94.9	34%	1,500	84,000	21	104.9
5/2/75-5	130,000	84.3	-	1,350	85,000	28	127.6
5/2/75-2	90,000	88.8	31%	1,200	70,000	25	81.6
5/4/75-5	100,000	67.9	31%	1,200	74,000	27	81.5
5/4/75-7	88,000	93.2	31%	900	74,000	22	112.6
5/2/75-3	128,000	103.8	32%	900	100,000	23	135.4
5/2/75-4	118,000	97.6	-	600	100,000	24	123.2
5/4/75-6	128,000	98.8	33%	600	98,000	25	149.8
5/4/75-4	122,000	87.7	33%	300	110,000	29	90.5

Table 2 Aerosol Data with Neutralization.

the gas kinetic correction) of 21×10^{-10} cm^3/sec for a .02 micron monodisperse aerosol.

The data run, without the use of the K-85 neutralizer, gives another picture. The values of the coagulation kernel, in this case, is time dependent with higher values of K being recorded in the first 600 seconds than in the longer time runs. Since the aerosol had essentially an identical size distribution (Pictures 1 and 2), this effect has to be assumed due to the higher bipolar aerosol. This was also observed by Jerry Burford[19] of this lab for his Master's Thesis. This higher initial coagulation rate can be explained several ways. If the charge distribution was not completely symmetric, the excess charge would be forced to the walls due to their mutual electrostatic dispersion. This would, in effect, lower the concentration until a symmetric charge distribution could be established. Another reason could be that some of the E.W. aerosol had multiple charges and this would increase their coagulation rate. It is also interesting to note that after 900 seconds the aerosols in Table 2 appear to be more stable than those in Table 3. All this speculation leads to the conclusion that the charge distribution as a function of time must be known before this phenomenon can be understood.

The electrostatic precipitator shown in Figure 3 was used to determine the total charge per cm^3 for the aerosol. Making the approximation of only one charge per particle, the aerosols which were subject to the K-85 radiation were, on the average, 32% charged; and the aerosols coming directly from the dilution chamber were 47% charged. Theory does not predict that the difference of total charge of 15% would effectively double the coagulation rate. Again, the answer to this problem must lie in the charge distribution.

The concentration was measured by a Gardner counter, which was recently recalibrated by Gardner Associates, Inc. of Schenectady, New York. The calibration curve supplied by Gardner Associates was used to convert the output signal from the photomultiplier tube to a concentration reading. In light of the results of the Fort Collins Symposium[20] where the Gardner instrument counted somewhat less than the other condensation nuclei counters, a recalibration is in order. This calibration will be done by a comparison with the University of Missouri-Aitken Nuclei Counter designed by Dr. Kassner[21]. If this Gardner counter is found to count low, this will lower somewhat all the values of the coagulation kernels listed in Tables 2 and 3.

The size of the E.W. particles can also be determined by comparing the Gardner counter with the mass monitor. This comparison yielded an effective diameter of between .03 and .04 microns for

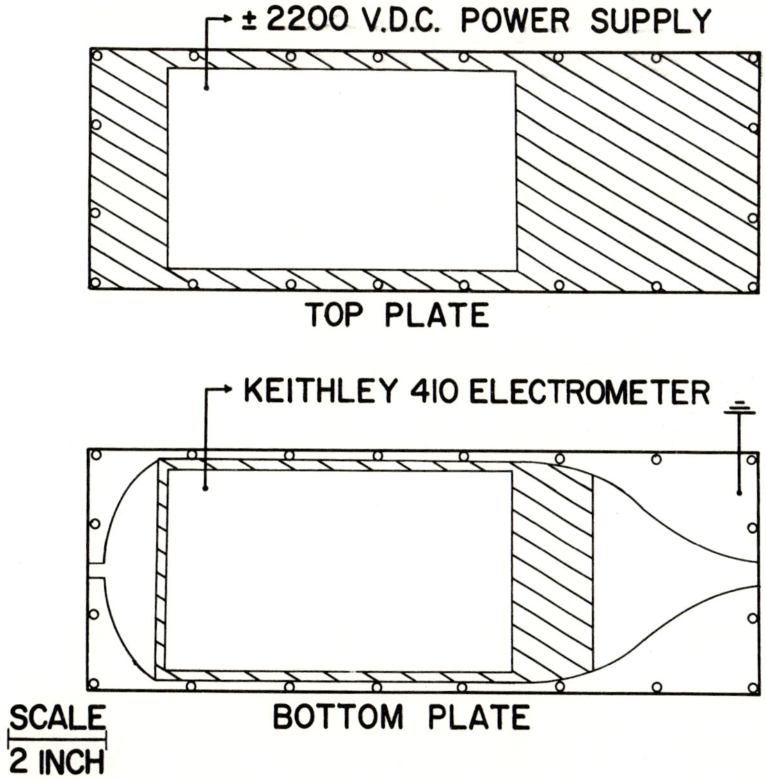

Figure 3 Electrostatic Precipitator.

the particles, and this agrees well with the electron microscope pictures. An interesting side light to this investigation, is the ability of the exploding wire aerosol to adsorb gas vapors. Looking at the total mass/cm^3 data listed in Table 2 and 3, it is seen that in nearly all cases an increase with total mass is recorded with time. Since the coagulation chamber is a closed system, one would expect a small decrease in total mass due to particle wall losses. A net increase in total mass can be explained by condensation of gas vapors in the gaps between the primary gold particles. On the average, the total mass listed in Table 3 is greater than the total mass in Table 2 yet the concentrations remain about the same. This indicates that the K-85 neutralizer or the connecting tubing had been previously contaminated with some form of condensable vapor.

CONCLUSION

This paper reinforces the previous findings that the exploding wire aerosol is reproducible, submicron in primary size, highly charged, and distributed normal or log normal in primary size. The initial aggregation can be controlled by a two stage dilution system, but a comparison of pictures 1 and 2 with picture 3 shows larger agglomerates in the first two pictures. Since picture 3 was collected with the corona field turnoff, it must be assumed that the field or ions from the corona wire enhances coagulation before the particles are precipitated. This has serious consequences when a representative sample is needed for analysis. It must also be pointed out that this phenomenon is peculiar to E.W. aerosols since a high concentration of .02 micron NACL aerosols showed no agglomeration when collected by the same aerosol sampler.

This question could be answered by a paper recently published by W. H. Marlow and J. R. Brock[22], who found that the charging rate for a conducting particle depend importantly and in a complex manner on the particle size and total particle charge.

The data indicates that the initial coagulations of the E.W. aerosols is more pronounced than the theory predicts, but, as stated before, the charge distribution rather than the total charge must be monitored as a function of time if this reaction is to be understood.

Perhaps the thing that is most important about the E.W. aerosol is its shape factor. Vomela[5] found the charge on the E.W. aerosol to be between 50% and 70% higher than on an equivalent volume aerosol. The influence of the shape factor on the diffusion coefficient of the particles still needs to be determined experimentally for this size particle. Finally, the large surface area to volume characteristic of the combustion aerosol provides many sites for the adsorption of gas vapor. Corn and Reitz[23] discovered that the specific surface of an urban aerosol taken in Pittsburgh would increase by a factor of about two after degassing at 200°C.

This large adsorption of gas by this type of Aitken nuclei could be an effective mechanism for cleansing the condensable vapors from the atmosphere.

ACKNOWLEDGEMENTS

I wish first to thank Austin W. Hogan and Dr. Volker Mohnen of the State University of New York at Albany for their generous loan of the diffusion battery. Further I express my regards to Dr. Kassner for his initial guidance and suggestion of the problem. I also must thank Dr. Podzimek for his steadfast advice throughout this investigation, and finally, I must thank Mrs. Sandy Shults for typing this manuscript. This work is being done as a partial fulfillment of my Ph.D. Dissertation.

REFERENCES

1. Liu, B. Y. H., K. T. Whitby, and H. H. S. Yu. Electrostatic Aerosol Sampler for Light and Electron Microscopy. Rev. Sci. Inst. 38: No. 1, 100-102, Jan., 1967.

2. Olin, J. G., and J. S. Gilmore. Piezoelectric Microbalance for Monitoring the Mass Concentration of Suspended Particles. Atmospheric Environment, Vol. 5: 653-668, 1971.

3. Rich, T. A. Apparatus and Method for Measuring the Size of Aerosols. J. Rech. Atmos. Vol. II, 2e Anneé: 79-86, 1966.

4. Liu, B. Y. H., and D. Y. H. Pui. Electrical Neutralization of Aerosols. Aerosol Science. Vol. 5: 465-472, 1974.

5. Chace, W. G. and H. R. Moore, eds. Exploding Wires, Vols. 1-4. Plenum Press, N.Y. 1958-1968.

6. Phalen, R. F., Evaluation of an Exploded Wire Aerosol Generator for use in Inhalation Studies. Aerosol Science. 3: 395-406, 1972.

7. Karoisis, F. G. and B. R. Fish. An Exploding Wire Aerosol Generator. Jour. Colloid Sci., 17: 155-161, 1962.

8. Karoisis, F. G., B. R. Fish, and G. W. Royster, Jr. Exploding Wires, Vol. II (eds) Chace, W. G. and H. K. Moore. Plenum Press, 1962, 299-311.

9. Vomela, R. A. and K. T. Whitby. The Charging and Mobility of Chain Aggregate Smoke Particles. Jour. Colloid Sci., 25: 568-576, 1967.

10. Rosinski, J., D. Werle and C. T. Nagamoto. Coagulation and Scavenging of Radioactive Aerosols. Jour. Colloid Sci. 17: 703-716, 1962.

11. Sherman, P. M. Generation of Submicron Metal Particles. Jour. Colloid Sci. 51: 87-93, April, 1975.

12. Harvey, J., H. I. Matthews and H. Wilson. Crystal Structure and Growth of Metallic or Metallic-Oxide Smoke Particles Produced by Electric Arcs. Discussion of the Faraday Society, 30: 113-124, 1960

13. Whytlaw-Gray, R. and H. Patterson. Smoke, E. Arnold, London (1932).

14. DallaValle, J. M., C. Orr, Jr., B. L. Hinkle. The Aggregation of Aerosols. Brit. Journ. App. Physics, 5: 5198-5208, 1954.

15. Cooper, D. W., and P. C. Reist. Neutralizing Charged Aerosols with Radioactive Sources. Jour. Colloid Sci. 45, Vol. 1: 17-26, Oct., 1973.

16. Liu, B. Y. H., and D. Y. H. Piu. Equilibrium Bipolar Charge Distribution of Aerosols. Jour. Colloid Sci., Vol. 49, No. 2: 305-312, Nov., 1974.

17. Fuchs, N. A. and A. G. Sutugin. Topics in Current Aerosol Research. G. M. Hidy and J. R. Brock (eds.). Pergamon Press Ltd., p. 42-47, 1971.

18. Zebel, G. Aerosol Science. (ed.) C. W. Davies. Academic Press Inc. Chapter II, 31-57, 1966.

19. Burford, J. N. The Effects of Washout in Polydisperse Metalic Aerosols. Master's Thesis, University of Missouri-Rolla, 1974.

20. The Second International Workshop on Condensation and Ice Nuclei (held at Colorado State University) National Science Foundation, Aug., 1970.

21. Kassner, J. L, J. C. Carstens, M. A. Vietti, A. H. Biermann, Paul C. P. Yue, L. B. Allen, M. R. Eastburn, D. D. Hoffman, H. A. Noble and D. L. Packwood. Expansion Cloud Chamber Technique for Absolute Aitken Nuclei Counting, Jour. Rech. Atmos. 1968.

22. Marlow, W. H. and J. R. Brock. Unipolar Charging of Small Particles. Jour. Colloid Sci., Vol. 50, No. 1, 32-37. Jan., 1975.

23. Corn, M. and R. Reitz. Atmospheric Particulates: Specific Surface Areas and Densities, Science 159: 1350 (1968).

AEROSOL PARTICLE FORMATION FROM PHOTO-OXIDATION OF SULFUR DIOXIDE VAPOR IN AIR

Kanji Takahashi and Mikio Kasahara

Institute of Atomic Energy, Kyoto University
Uji, Kyoto 611, Japan

ABSTRACT

Aerosol formation from photo-oxidation of SO_2 in air is studied both experimentally and theoretically. Experiments on the effects of environmental factors, such as SO_2 concentration, relative humidity, UV light irradiation intensity, irradiation time, and the presence of foreign aerosol particles, on the formation and the evolution of aerosols are conducted. A kinetic model for aerosol formation is also presented, and calculated examples are shown.

Aerosol formation and evolution are considerably influenced by SO_2 vapor concentration and relative humidity, and it is noticed that the effects of those environmental factors are markedly varied with the irradiation time.

AEROSOL PARTICLE FORMATION FROM PHOTO-OXIDATION OF SULFUR DIOXIDE VAPOR IN AIR

Kanji Takahashi and Mikio Kasahara

Institute of Atomic Energy, Kyoto University
Uji, Kyoto 611, Japan

INTRODUCTION

Sulfur dioxide vapor in air is photo-oxidized by ultraviolet(UV) irradiation in sunlight to form sulfuric acid aerosols, and this process plays an important role in air pollution.

The process of aerosol formation from photo-oxidation of SO_2 vapor is considered to consist of three stages,i.e.: 1)photochemical oxidation of SO_2 to SO_3, and followed production of H_2SO_4 vapor through the combination with H_2O molecule, 2)heteromolecular nucleation of H_2SO_4 vapor to a critical sized cluster or an embryo through the combination of several number of molecules of H_2SO_4 and H_2O, and growth of the nucleated embryo to larger aerosol particle through the condensation of H_2SO_4, H_2O and SO_3 molecules, and through the coagulation with other particles.

Number of experimental works have been reported on photochemical aerosol formation from SO_2 in air by Gerhard and Johnstone[1], Renzetti and Doyle[2], Cox and Penkett[3], Quon et al.[4], and Clark[5]. We have also shown some experimental results[6] and a kinetic model[7] on the photochemical aerosol formation. In this work, those results of our previous studies on the effects of various envirnmental factors, such as SO_2 vapor concentration, relative humidity, UV light intensity and irradiation time, on the formation and the evolution of aerosol particles are summarized, and an additional study on the influence of the presence of foreign aerosol particles on sulfuric acid aerosol formation are conducted.

EXPERIMENT ON FORMATION AND EVOLUTION OF SULFURIC ACID AEROSOLS FROM SO_2 - H_2O - AIR SYSTEM

Sulfur dioxide vapor was mixed with carefully purified and conditioned air, and passed through the reaction chamber surrounded by blacklight source. SO_2 vapor concentration ranged from 0.05 to 10 ppm. Relative humidity was kept at about 80 %, 60 % or less than 10 %. The maximum irradiation intensity of UV light was about 0.15 mW/cm^2 sterad, and this value is equivalent to 0.05 hr^{-1} for the specific absorption rate constant of SO_2 molecule(relative light intensity to the maximum is denoted by I_r, hereafter). The maximum irradiation time was 750 sec. Particle number concentration was measured with a condensation nuclei counter, and the smallest size of detectable particles with this type of counter is about 0.0025 μm. Size distribution of formed particles was determined by diffusion tube method.

Generally, as shown in Fig.1, the number concentration of formed aerosol particles increases rapidly after some period from the onset of irradiation, and decreases gradually after reaching a maximum value. The volumetric concentration of particles increases almost proportionally to the increase of irradiation time, and the surface area concentration approaches an equilibrium value. Aerosol particle number concentration is strongly dependent on SO_2 vapor concentration and the relative humidity as shown in Fig.2. The particle number formation rate is in proportion to the product of SO_2 concentration and the irradiation intensity, particularly in the case of lower formation rate. The maximum particle number concentration and the irradiation time required to reach the maximum value are expressed as a function of formation rate. Particle size just after formed is found to be smaller than 0.0025 μm and almost homogeneous. Fig.3 shows variations of average particle size which are increasing with the irradiation time.

As a whole, particle number concentration is strongly dependent on both SO_2 vapor concentration and relative humidity, and the particle size is mainly influenced by SO_2 concentration. Volumetric formation rate of particles is also dependent on both SO_2 vapor concentration and relative humidity, and the value ranges 0.15 to 18.7 $\mu m^3/cm^3/hr$ within this experimental condition. The oxidation rate and the over-all quantum yield of photochemical reaction of SO_2 are 0.04 %/hr and 8 x 10^{-3}, respectively. And the former value is equivalent to be 0.7 %/hr for noonday sunlight in summer of the middle Japan.

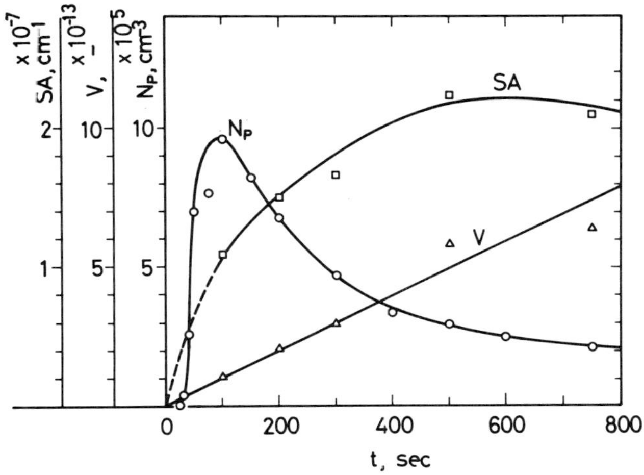

Fig.1 Variation of particle number concentration(N_P), volume concentration(V) and surface concentration(SA) with irradiation time: SO_2=1 ppm, r.h.=60%, I_r=1.0

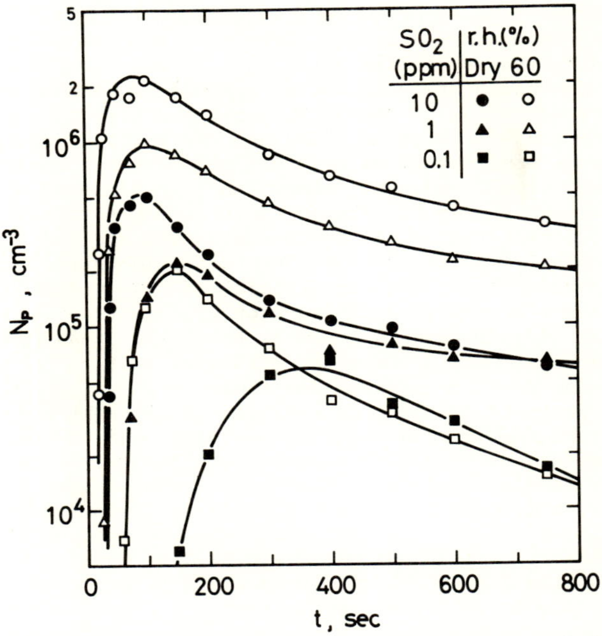

Fig.2 Variation of particle number concentration of formed aerosols with irradiation time: I_r=1.0

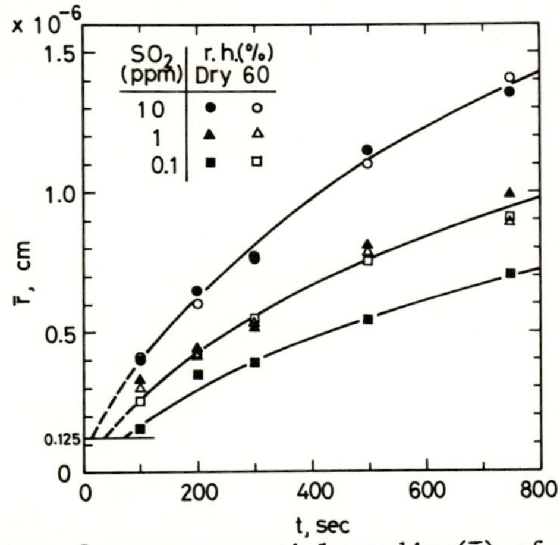

Fig.3 Variation of average particle radius($\bar{r}$) of formed aerosol particles with irradiation time: I_r=1.0

EXPERIMENT ON SULFURIC ACID AEROSOL FORMATION FROM SO_2 - H_2O - PARTICLE - AIR SYSTEM

Metallic fume particles were generated using an electrical furnace in N_2 atmosphere. A modified Rapaport type generater was also utilized for production of searic acid aerosols. Particle size distribution was measured by Electrical Aerosol Size Analyser(Thermo-System Inc., USA). Aerosol particles were neutralized electrically before mixed with an irradiation system of SO_2 - H_2O - Air by passing over a radioactive source(10 mCi of Kr-85). The irradiation and the measurement procedures were almost the same as those used for the experiments of foreign particle free.

Fig.4 shows the time change of number concentration of formed aerosol particles in both cases of foreign particle free and added. At the early stage of irradiation, aerosol formation is markedly hindered by the addition of foreign particles. However, as the irradiation proceeds this hinderance effect becomes smaller, and at a period of longer irradiation time, number concentration of formed particles appears sometimes larger than in the case of foreign particle free.

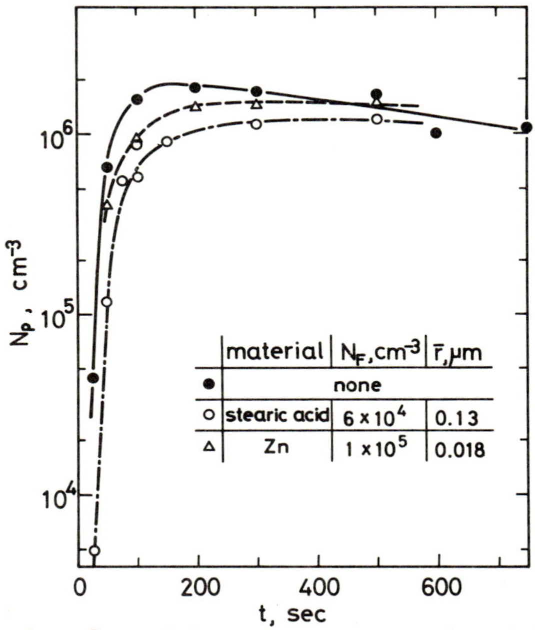

Fig.4 Variation of particle number concentration with irradiation time: SO_2= 1ppm, r.h.=60%, I_r=0.2: N_F, $\bar{r}$; number concentration, average radius of foreign particles, respectively.

In Fig.5, the number concentration of formed particles is shown as a function of surface concentration of added foreign particles at the fixed irradiation time of 50 sec. Even though the material and the size of added particles are varied, aerosol formation is markedly affected when the surface area concentration is ranged from 10^3 to 10^4 $\mu m^2/cm^3$ and almost disappears in the case of larger surface area concentration than 10^4 $\mu m^2/cm^3$.

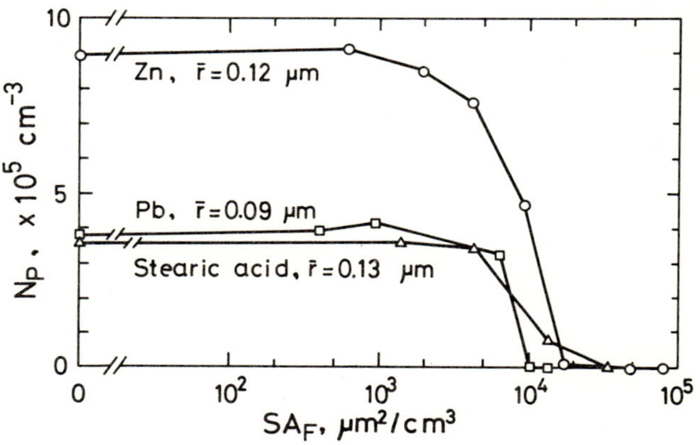

Fig.5 Particle number concentration of formed aerosol particles as a function of surface area concentration(SA_F) of added particles: SO_2= 1ppm, r.h.=60%, I_r=0.2, t=50sec.

KINETIC MODEL OF SULFURIC ACID AEROSOL FORMATION

The nucleation rate for a system of H_2SO_4 - H_2O vapor was calculated[7] at various relative humidities according to the heteromolecular nucleation theory. Nucleation rate is remarkably dependent on both SO_2 vapor concentration and relative humidity. The embryo usually consists of several molecules of H_2SO_4 and ten or twenty molecules of H_2O, and the size is about 10 to 15 Å.

Taking into account the concentration of H_2SO_4 vapor and the formed particles, kinetic model for each component can be expressed as,

$$\frac{dN_G}{dt} = R - (\alpha + \beta_G + \gamma N_P) N_G \qquad (1)$$

$$\frac{dN_P}{dt} = \frac{\alpha}{i_o} N_G - \beta_P N_P - K N_P^2 \qquad (2)$$

where, N_G is the number concentration of H_2SO_4 molecule, N_P is the number concentration of aerosol particles formed including embryo, t is the time of aerosol formation and irradiation continues throughout this time, R is the conversion rate of SO_2 to H_2SO_4 by photo-oxidation and is given as a function of SO_2 vapor concentration and irradiation light intensity, α is the conversion rate of H_2SO_4 vapor to embryo, i_o is the number of H_2SO_4 molecules contained in a nucleated embryo, β_G (or β_P) is the deposition rate of H_2SO_4 vapor (or particle) to boundary surfaces, γ is the deposition rate of H_2SO_4 vapor to already formed aerosol particles, and K is the coagulation constant.

Calculations of aerosol free conditions are performed for various values of relative humidity and of photo-oxidation rate of SO_2. Fig.6 shows one of the calculated examples of aerosol formation. by Eqs.(1) and (2). Generally, vapor concentration of H_2SO_4 molecule increases rapidly at the start of irradiation and decreases slowly after reaching a maximum value. The maximum particle number concentration appears a little later than for vapor concentration, but this time lag is not so large. Surface area concentration approaches an equilibrium value. This implies that at the early stage of the process nucleation is predominant, however, condensation of vapor on already formed particles becomes significant in later stage.

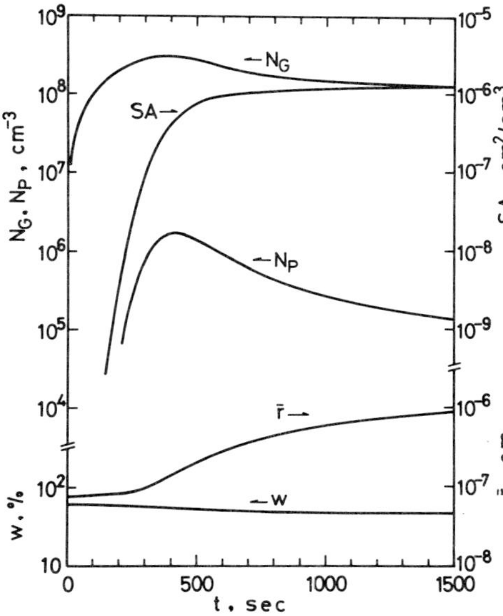

Fig.6 Time change of some characteristic values in aerosol formation: $R=10^6 cm^{-3} sec^{-1}$, r.h.=50%, w is the weight fraction of H_2SO_4 in a particle, other notations are the same as in Figs. 1 through 3.

When foreign particles exist, Eqs.(1) and (2) are also applicable by assuming that foreign particles act as adsorbing matter for vapor and that all other effects such as catalytic reaction and coagulation are negligible. Adsorption effect of foreign particle can be taken into account by letting $\beta_G = \gamma_F N_F$, where N_F is the number concentration of foreign particles, and γ_F is the adsorption rate of vapor molecule on a foreign particle surface and is given by,

$$\gamma_F = \frac{\pi r^2 v}{\dfrac{1}{\delta} + \dfrac{r^2 v}{4(r + l)D}} \left(1 - \frac{S}{4 \pi r^2} \right) \tag{3}$$

where, r is the radius of foreign particles, v is the mean thermal velocity of vapor molecule, δ is the condensation probability of vapor molecule onto a particle, l is a length nearly equal to mean free path, D is the diffusivity of vapor molecule in air, and S is the covered area of particle surface by adsorbed vapor molecule and is regarded zero for perfect sink surface. Eq.(3) is reduced to $\gamma_F = \delta \pi r^2 v$ for very small particle with perfect sink surface.

Judeikis and Siegel[8] estimated that the value of δ is 1.0 to 20 x 10^{-6} for the mixture of SO_2 and atmospheric dust and it is raised more than 0.5 to 10 x 10^{-3} in the presence of trace quantities of metal salts. According to our previous experiments[9], adsoption of SO_2 vapor onto lead fume particles in humid air is much greater than monolayer, and the adsorption rate is so slow that the order of magnitude of δ is 10^{-4}. Adsorption of SO_2 vapor by other particles such as iron fume is found to be very slight. The values of δ of H_2SO_4 vapor for various kinds of aerosol particles are unknown, however, they are considered much larger than the value of SO_2 vapor. Hence, in this calculation, adsorption of SO_2 vapor is neglected, and it is assumed that foreign particle has perfect sink surface for H_2SO_4 vapor. Various values of δ are tried to use for calculations, and it is found that the magnitude in the order of 10^{-2} is most suitable to the experimental results.

Fig.7 shows a calculated example of the time change of the number concentration of formed aerosol particles, where the value of R is consistent with that obtained from the experiments. In Fig.8, the number concentration of formed aerosol particles at various irradiation time is given as a function of surface area concentration of foreign particles. Within the calculations for such a short irradiation time, the value of $S/4\pi r^2$ is so small that the assumption of perfect sink surface has insignificant meaning.

DISCUSSION AND CONCLUSION

Due to the different conditions between the experimental and the theoretical studies, such as the discrepancies in the smallest size of

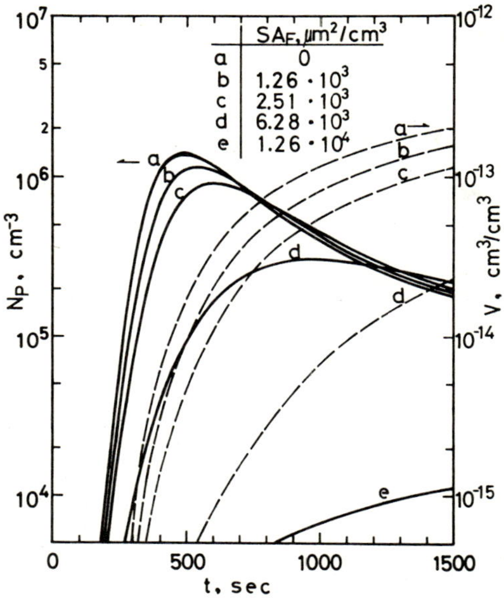

Fig.7 Variation of number(—) and volume(– –) concentration of formed aerosol particles with irradiation time: $R=5 \times 10^5$ $cm^{-3}sec^{-1}$, r.h.=50%, δ =0.01

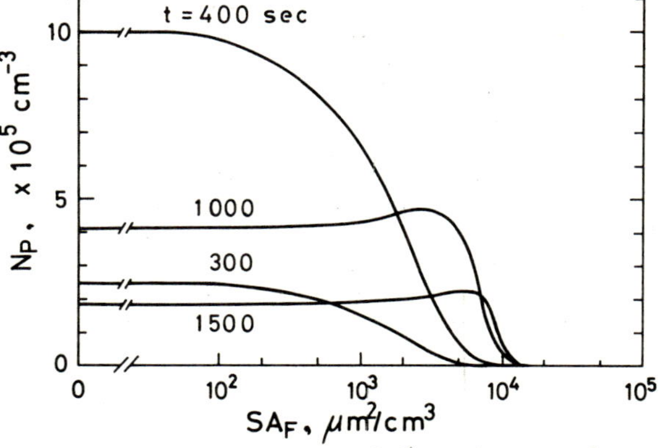

Fig.8 Particle number concentration of formed aerosols as a function of surface area concentration of foreign particles at various irradiation times: conditions used for calculations are same as in Fig.7

particles taken into account and in the boundary conditions of wall, the agreement and the validity of the kinetic model are unable to be discussed in detail. However, general tendencies and the order of magnitude of various factors obtained from the experiments are well elucidated by the present kinetic model in all cases.

As to the influence of foreign particles on sulfuric acid aerosol formation, some other workers[1,10] have shown enhancement or neutral effect. However, it is noticed that, even in the presence of adsorbing foreign particles, number concentration of formed aerosol particles can be enhanced sometimes in the course of long irradiation as shown both experimentally and theoretically. In such a case volumetric concentration of formed aerosol particles is always lowered by the presence of adsorbing foreign particles.

REFERENCES

1. Gerhard, E. R., and H. F. Johnstone. Photochemical Oxidation of Sulfur Dioxide in Air. Ind. Eng. Chem. 47:972-976, 1955
2. Renzetti, N. A., and G. J. Doyle. Photochemical Aerosol Formation in Sulfur Dioxide-Hydrocarbon System. Int. J. Air Poll. 2:327-345, 1960
3. Cox, R. A., and S. A. Penkett. The Photo-oxidation of Sulfur Dioxide in Sunlight. Atmos. Environ. 4:425-433, 1970
4. Quon, J. E., R. P. Siegel, and H. M. Hulburt. Particle Formation from Photolysis of Sulfur Dioxide in Air. 2nd Int. Clean Air Congress, Wash. D. C., CPID. p.330-335, 1970
5. Clark, W. E. Measurements of Aerosol Produced by the Photochemical Oxidation of SO_2 in air. Thesis, Univ. of Minnesota, 1972
6. Kasahara, M., and K. Takahashi. Experimental Studies on Aerosol Particle Formation from Photochemical Oxidation of Sulfur Dioxide. submitted to Atmos. Environ. 1974
7. Takahashi, K., M. Kasahara, and M. Itoh. A Kinetic Model of Sulfuric Acid Aerosol Formation from Photochemical Oxidation of Sulfur Dioxide Vapor. J. Aerosol Sci. 6:45-55, 1975
8. Judeikis, H. S., and S. Siegel. Particle-Catalyzed Oxidation of Atmospheric Pollutants. Atmos. Environ. 7:619-631, 1973
9. Takahashi, K., and T. Tamachi. Some Experiments on Adsorption of SO_2 Gas on Metallic Fume Particles. Taikiosen-Kenkyu(J. Japan Soc. Air Poll.). 7:1-6, 1972, in Japanese
10. Urone, P., H. Lutsep, C. M. Noyes, and J. F. Parcher. Static Studies of Sulfur Dioxide Reactions in Air. Environ. Sci. Techn. 2:611-618, 1968

PART III

AEROSOL SAMPLING

SIZE-SELECTIVE SAMPLING FOR INHALATION HAZARD EVALUATIONS

Morton Lippmann, Ph.D.

Institute of Environmental Medicine

New York University Medical Center

New York, New York 10016

The basis for the increasing application of size-selective aerosol samplers is reviewed in terms of the factors affecting and parameters describing particle deposition and retention in the human respiratory tract. The available experimental data describing total and regional particle deposition in man are discussed, as are the sampler acceptance criteria adopted by BMRC, AEC, and ACGIH, and the ICRP Task Group deposition model. Techniques and equipment used for size-selective aerosol sampling are evaluated in terms of the design principles applied, the correspondence between design and performance of specific samplers, their applicability to field conditions, and their ability to satisfy sampler acceptance criteria.

INTRODUCTION

A given aerosol can be described by at least as many characteristic parameters as there are measurement techniques. Several papers in this session describe techniques for determining aerodynamic diameter distributions using cascade impactors and aerosol centrifuges. Papers in the following sessions describe techniques for characterizing aerosols by their optical and electrical properties, by total particle counts, and by total mass concentrations. However, none of these parameters provide any direct measure of inhalation hazard.

The human respiratory tract is a size selective particle trap with a continuously variable collection efficiency at all particle sizes below its effective upper cut-size. Furthermore, inhalation hazards are usually more dependent on the concentration of deposits at specific sites or regions within the respiratory tract, than on the total amount deposited. The deposition pattern is particle size dependent, and the problem is further compounded by a very large intersubject variability in deposition pattern even under identical exposure conditions. The deposition characteristics of the human airways must be known in order to realistically evaluate the inhalation hazard potential associated with a given aerosol.

INFLUENCE OF DEPOSITION PATTERN ON INHALATION HAZARD

The toxic dose delivered to respiratory epithelium by inhaled particles depends on their surface density and residence time. In the conducting airways, where the air velocity is relatively high and there are sudden changes in air path directions, the particles which deposit tend to be concentrated on a small fraction of the interior surface. Such areas, *i.e.*, the dividing spurs of large bronchial airway bifurcations, the larynx, and the nasal septum, also happen to be the locations most frequently identified as sites of cancer origin among populations exposed to occupational and/or environmental carcinogens. Although the residence times on the surfaces of these airways are relatively short, *i.e.*, generally less than about 2 hours, the exposures may be frequent or continuous, and a limited number of cells may receive very high dosages.

Another region where the dosage may be high is much deeper in the lungs, *i.e.*, in the alveolar region where the gas exchange takes place. Here, most of the deposition takes place by sedimentation and/or diffusion, and therefore the particles deposit relatively uniformly over a huge surface area, estimated to be 70-80 m^2 1/. Insoluble particles which deposit in this region have very long retention times, which vary with the physical and chemical properties of the particles.

In some cases, these can be characterized by half-times measured in months or years. Furthermore, such particles tend to be gradually concentrated in semi-permanent storage depots as peribronchiolar dust foci. Such accumulations are associated with chronic lung diseases, e.g., silicosis, coal-workers pneumoconiosis (black lung) and emphysema.

EFFECT OF PARTICLE SIZE ON REGIONAL DEPOSITION

It follows from the preceding discussion that the hazards associated with airborne particles are dependent on where they deposit within the respiratory tract, or if in fact, they deposit at all. A summary of the best available human in vivo regional deposition data was recently completed 2, 3/. For very large particles, i.e. > 50 μm, the suction created by the nose or mouth is insufficient for them to be inhaled. Particles which are drawn into the nose will all deposit there unless they are smaller than ~ 9 μm. For 3 μm particles, nasal deposition falls of to ~ 50%, and tracheobronchial deposition will remove about 20% of the remainder, i.e., ~ 10% of the amount inhaled. Of the ~ 40% which reaches the alveolar zone, about ½ will deposit and the balance will remain airborne and be exhaled. Head and bronchial deposition of ½ μm particles are very low, and almost all of these particles penetrate to the alveolar region with the inspired air. However, only about 20% will deposit there and the balance will be exhaled. For still smaller particles, deposition by diffusion increases, and the exhaled fraction decreases. For molecular sized particles, i.e., unattached radon daughters, the rms diffusional displacement becomes so large that these particles deposit efficiently in the upper airways.

The particle sizes of interest in inhalation hazard evaluations obviously depend on the toxic material involved. For materials associated with an elevated incidence of bronchial cancer, the particles depositing on the bronchial tree are of primary concern. On the other hand, for silicosis hazard evaluations, the only particles of interest are those which deposit in the alveolar region.

AEROSOL SAMPLING FOR INHALATION HAZARD EVALUATIONS

Two different sampling approaches can be followed in evaluating inhalation hazards. One approach, which has the virtue of general applicability, and the possibility of retrospective reinterpretation, is to determine the aerodynamic size distribution of the whole aerosol. With this information, the particle fraction in every size range of interest is known. The disadvantage of this approach is in the complexity and cost of the sampling and/or analytical effort. The alternate approach is selective sampling, i.e., to collect only that fraction of the aerosol which is capable of reaching the target sites. This approach has been widely used for sampling pneumoconiosis producing dusts, where the target sites are in the alveolar spaces. Selective sampling greatly simplifies the analytic task. The major problem and limitation of this approach is the requirement for a definition of the aerosol fraction which is potentially hazardous. Two similar, but somewhat different definitions of pneumoconiosis producing dusts are in widespread use, as will be discussed in the next section. For some of the carcinogenic aerosols, where the large airways may represent the critical deposition region, there have been no recommendations on sampler acceptance criteria proposed to date.

Since the particle size parameter which has the greatest influence on respirable mass deposition is the aerodynamic diameter, the most reliable measurements can be made with instruments which classify airborne particles by this parameter during the sampling process. It is seldom possible to perform an accurate size distribution analysis on collected samples, since the particles can no longer be restored to their original state of dispersion. Thus, particles which were unitary in the air may be analyzed as aggregates and vice versa. Furthermore, particles analyzed by microscopy will be graded by a linear dimension or by projected area diameter, and these are normally larger than the true average diameter.

HISTORICAL BACKGROUND FOR RESPIRABLE MASS SAMPLING

The first standard sampler for pneumoconiosis producing dusts, the Greenburg-Smith impinger was developed in 1922-25 through the cooperative efforts of the U.S. Bureau of Mines, U.S. Public Health Service, and the American Society of Heating and Ventilating Engineers 4/. This impinger and the midget impinger, developed in 1928 by the Bureau of Mines 5/, collect most particles larger than about 0.75 micron in a liquid medium. Samples collected in impingers are analyzed by counting the particles which settle to the bottom of a wet counting cell and are visible when viewed through a 10 x objective lens. Particles larger than 10 μm observed during the count are rejected by many industrial hygienists as "non-respirable." The only alternative approach available prior to 1952 was gravimetric analysis of the total airborne particulate sample. With total mass sampling there is no practical way to discriminate against the oversized particles which tend to dominate the sample mass.

STANDARDS AND CRITERIA FOR RESPIRABLE DUST SAMPLERS

British Medical Research Council (BMRC)

In 1952, The British Medical Research Council adopted a definition of "respirable dust" applicable to pneumoconiosis producing dusts. It defined respirable dust as that reaching the alveoli. The BMRC selected the horizontal elutriator as a practical size selector, defined respirable dust as that passing an ideal horizontal elutriator, and selected the elutriator cut-off to provide the best match to experimental lung deposition data. The same standard was adopted by the Johannesburgh International Conference on Pneumoconiosis in 1959 6/.

In order to implement these recommendations, it was specified that:

1. "For purposes of estimating airborne dust in its relation to pneumoconiosis, samples for compositional analysis, or for assessment of concentration by a bulk measurement such as that of mass or surface area, should represent only the 'respirable' fraction of the cloud.

2. "The 'respirable' sample should be separated from the cloud while the particles are airborne and in their original state of dispersion.

3. "The 'respirable fraction' is to be defined in terms of the free falling speed of the particles, by the equation $C/C_o = 1-f/f_c$, where C and C_o are the concentrations of particles of falling speed f in the 'respirable' fraction and in the whole cloud, respectively, and f_c is a constant equal to twice the falling speed in air of a sphere of unit density 5 μm in diameter."

<u>U.S. Atomic Energy Commission (AEC)</u>

A second standard, established in January 1961 at a meeting sponsored by the AEC Office of Health and Safety, defined "respirable dust" as that portion of the inhaled dust which penetrates to the non-ciliated portions of the lung 7/. This application of the concepts of respirable dust and concomitant selective sampling was intended only for "insoluble" particles which exhibit prolonged retention in the lung. It was not intended to include dusts which have an appreciable solubility in body fluids and those which are primarily chemical intoxicants. Within these restrictions, "respirable dust" was defined as follows:

Size*	10.	5.	3.5	2.5	2
Respirable (%)	0	25	50	75	100

*Sizes referred to are equivalent to an aerodynamic diameter having the properties of a unit density sphere.

For these "insoluble" internal emitters, the concentration of the respirable fraction can be compared to a modified maximum permissible concentration (MPC_a) on a simple basis. In the procedure for calculating MPC_a values as established by the ICRP, the equation is:

$$MPC_a = \frac{10^{-7} q f_2 \ \mu Ci}{T f_a (1 - 2^{-0.693\frac{t}{T}}) cm^3}$$

where

q = maximum permissible body burden, μCi

f_2 = fraction of isotope in critical organ, relative to whole body burden

f_a = fraction of inhaled aerosol that reaches critical organ

t = period of exposure (set at 50 years)

T = effective half-life, days.

The published MPC_a standards for insoluble particles were calculated from this equation using the assumption that f_a = 0.12 in all classes where the lung is the critical organ, i.e., that 25% of the airborne is "respirable" and one-half of the "respirable" dust is retained. Thus, the MPC_a for insoluble "respirable" dust should be 25% of the MPC_a for total airborne dust.

American Conference of Governmental Industrial Hygienists (ACGIH)

The application of respirable dust sampling concepts to other toxic dusts and the relations between respirable dust concentrations and accepted standards such as the ACGIH Threshold Limit Values (TLV's) are more complicated. Unlike the MPC_a's for radioisotopes, which are based on calculation, most TLV's are based on animal and human exposure experience. Thus, even if the data on which these standards were based could be related on the particle size of the dust involved, which unfortunately is unlikely, there would probably be a different correction factor for each TLV, rather than a uniform factor such as 0.25.

The task of establishing alternative or revised TLV's based on "respirable" dust concentrations, was begun by ACGIH at its annual meeting in 1968, when ACGIH announced 8/ in their "Notice of Intended Changes" alternate mass concentration TLV's for three forms of crystalline free silica to supplement the TLV's based on particle count concentrations.

The U.S. Department of Labor 9/ adopted the ACGIH size-selector criteria for respirable dust and extended its application to coal dust and inert or nuisance dust.

Discussion of Standards for Respirability

Basically, the two sampler acceptance curves described in the preceding discussion have similar, but not identical, characteristics. This is illustrated in Figure 1. The shapes of the curves differ because they are based on different types of collectors. The BMRC curve was chosen to give the best fit between the calculated characteristics of an ideal horizontal elutriator and lung deposition data, while the AEC curve was patterned more directly after the Brown, et al. 10/ upper respiratory tract deposition data and is simulated by the separation characteristics of cyclone type collectors. Some of the practical considerations and problems in using elutriator and cyclone samplers for these applications will be discussed in a subsequent section. In most field situations, where the geometric standard deviation (σ_g) of the particle size distribution is greater than two, samples collected with instruments meeting either criterion will be comparable. For example, Mercer calculated the predicted pulmonary (alveolar) deposition according to the ICRP Task Group deposition model 11/ for a tidal volume of 1450 cm^3 and aerosol with $1.5 < \sigma_g < 4$. He found that a sampler meeting the BMRC acceptance curve would have about 10% more penetration than a sampler meeting the AEC curve 12/. Calculations by Lynch 13/, Moss and Ettinger 14/, and Coenen 15/ indicated cyclone/elutriator penetration ratios of 0.81 to 0.83.

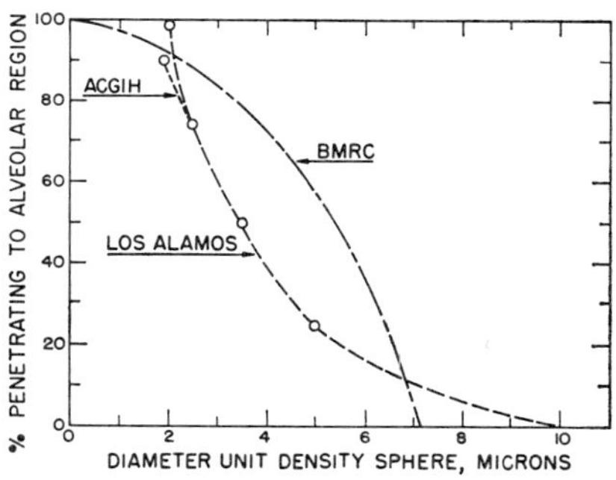

Figure 1. BMRC, Los Alamos (AEC), and ACGIH sampler acceptance curves. (Reprinted courtesy American Industrial Hygiene Association).

HISTORICAL BACKGROUND FOR SIZE-MASS DISTRIBUTION DETERMINATIONS

For practical purposes, aerodynamic size classification of "respirable" aerosols began with the development of the cascade impactor by K.R. May 16/ during World War II. Up until this time, essentially all particle size analyses were done by optical microscopy, which was inherently tedious. Furthermore, optical measurements cannot provide useful data on airborne density, or the aerodynamic drag of nonspherical particles. Cascade impactor samples could be analyzed chemically, and therefore could not only avoid these limitations, but could also be used to determine the size-mass distributions of multiple aerosol constituents. The real advantages of this approach led to a proliferation of instrument designs 17-22/. Unfortunately, the cascade impactor has some inherent limitations which are too often not recognized by the users. There is a tendency to overload the collection surface in the small particle stages in misguided attempts to collect measurable sample masses. Attempts to avoid this problem by the use of adhesive layers or fibrous filters on the impaction plates have often created other unanticipated problems, which will not be discussed here since another paper in this morning's session will be discussing inertial impactors.

The size-mass distributions of industrial and ambient aerosols can also be measured using aerosol centrifuges, cascade centripeters and multicyclone samplers 23-27/. The aerosol centrifuges, which will be discussed in two papers later in this session, provide better resolution and lower size limits than the other instruments. The cascade centripeter and multicyclone have poorer resolving power, but are least subject to problems associated with overloading, *i.e.*, they can be run for relatively long intervals and collect relatively large sample masses. The performance characteristics of small cyclones used in air sampling will be discussed in a later section of this paper.

The criterion most often followed in applying size-mass distribution data to hazard evaluations has been the 1966 Task Group Deposition Model 11/. This model received the approval of ICRP in 1973, and may be promulgated by ICRP during 1975.

The Task Group defined a model respiratory tract described as follows:

"(1) The nasopharynx, (N-P)-This begins with the anterior nares and extends through the anterior pharynx, back and down through the posterior pharynx (oral) to the level of the larynx or epiglottis...

"(2) Continuing caudally, the next component, (T-B), consists of the trachea and the bronchial tree down to and including the terminal bronchioles...

"(3) We recommend the third compartment be entitled pulmonary (P). This region consists of several structures, viz. respiratory bronchioles, alveolar ducts, atria, alveoli and alveolar sacs... The region can be regarded as the functional area (exchange space) of the lungs. Its surface consists of non-ciliated, moist epithelium with none of the secretory elements found in the tracheobronchial tree..."

Assuming all dusts to occur as log-normal distributions, deposition calculations were made for three different ventilation states typified by tidal volumes of 750, 1450 and 2150 cm^3, respectively, at a respiratory frequency of 15 cycles/min.

One of the more significant conclusions of the Task Group study was that the regional deposition within the respiratory tract can be estimated using a single aerosol parameter, i.e., the mass median diameter. For a tidal volume of 1450 cm^3, there were relatively small differences in estimated deposition over a very wide range of geometric standard deviations ($1.2 < \sigma_g < 4.5$). Also, there were relatively small variations associated with the substitution of 750 and 2150 cm^3 tidal volumes.

Using the mass median diameter, as determined using any of the instruments discussed in this section, and the Task Groups estimates of regional deposition for that size, one can produce predictions of the fractional deposition in each component region. These can be converted to predictions of inhaled dose, using the measured value of the overall mass concentration of the aerosol.

RELIABILITY OF BMRC, ACGIH, AND ICRP TASK GROUP DEPOSITION CRITERIA

The human in vivo deposition data base used by BMRC, ACGIH, and the Task Group was not very strong, as was freely acknowledged by all concerned. Many previous reviews on deposition have called attention to the very large differences in the reported results 11, 28-31/.

COMPARISON OF DATA USED IN CRITERIA DEVELOPMENT WITH RECENT MEASUREMENTS

Much of the discrepancy can be attributed to uncontrolled experimental variables and poor experimental technique. The major sources of error have been described by Davies 32/. Figure 2 shows data from studies of total respiratory tract deposition which have been performed with good techniques and precision. None of these data were available to BMRC. The data of Altshuler et al. 33/ and Landahl 34, 35/ were not used by ACGIH or the Task Group, and the others were not available. All were done with mouth breathing at respiration frequencies of from 12 to 16. Tidal volumes varied from ½ to 1.5 liter. All appear to show the same trend with a minimum of deposition at ∿ ½ μm diameter. The four studies using di-2-ethyl hexyl sebacate (DES) 36-40/ all appear to have somewhat lower absolute values.

It is also apparent that in most studies involving more than one subject, there was considerable individual variation among the subjects. Davies et al. 37/ showed that some of this variation could be eliminated by standardizing the expiratory reserve volume (ERV) and thereby the size of the air spaces. They found that deposition decreases as ERV increases. This was confirmed by Heyder et al. 38, 39/ who reported that there was little intrasubject variation among six subjects when their deposition tests were performed at their normal ERV's.

The DES data of Heyder et al. appear to represent deposition minima for normal men. Their test protocols were precisely controlled. There were no electrical charge on particles. With more natural aerosol and respiratory parameters, higher deposition efficiencies would be expected. The Landahl, Altshuler, Giacomelli-Maltoni, and Martens data provide the best available estimate of total deposition in normal humans breathing through the mouth. For nose breathing the rate of increase in deposition with particle size above 1 μm would be greater.

DEPOSITION IN THE ALVEOLAR ZONE

Brown et al. 10/ studied regional deposition during nose breathing using china clay aerosols in narrow size ranges ($\sigma_g = 1.25$), with the count median diameters between 0.9 μm and 6.5 μm, and these data formed the major basis for the BMRC and ACGIH criteria. They collected the exhaled air in seven sequential components, and used the CO_2 content of each fraction as a tracer to identify the region from which the exhaled air originated. The validity of these data depends upon the accuracy of the association between the various exhaled air fractions and their presumed sources. The Brown data have been criticized 41/ on the basis that their simple two filter model was inadequate, and that the rapid diffusion of CO_2 caused the measured CO_2 values to differ significantly from the corresponding CO_2 concentration in the alveolar spaces. The resulting error in the volume partitioning caused an underestimation of the alveolar deposition.

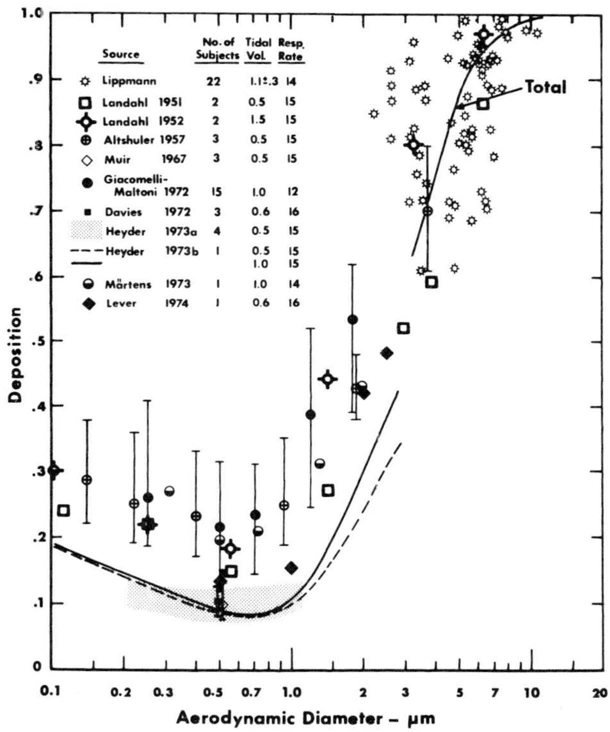

Figure 2. Total respiratory tract deposition during mouthpiece inhalations as a function of aerodynamic diameter, except below ½ μm, where deposition is plotted vs. linear diameter. Data of Lippmann 2, 3/ are plotted as individual tests, with eye-fit average line. Other data on multiple subjects are shown with average and range of individual tests. Monodisperse test aerosols used were Fe_2O_3 (Lippmann), triphenyl phosphate (Landahl 34, 35/ and Altshuler 33/), carnuba wax (Giacomelli-Maltoni 55/), polystyrene latex (Märtens 56/), and di-2-ethyl-hexyl-sebacate (Muir 36/, Davies 37/, Heyder 38, 39/, and Lever 40/). (Reprinted courtesy The American Physiology Society).

Altshuler, Palmes and Nelson 41/ estimated regional deposition from mouth breathing experiments on three subjects in which measurements of both the concentration of a monodisperse triphenyl phosphate aerosol and the respiratory flow were made continuously during individual breaths 33/. Using a tubular continuous filter bed model as a theoretical analog for the respiratory tract, regional deposition in the upper and lower tract components was calculated for various values of anatomic dead space. The upper tract penetration during inspiration, pause and expiration was derived from the expired aerosol concentration corrected for aerosol mixing. The alveolar deposition estimates varied from subject to subject, and for each subject varied with the volume of anatomic dead space. The particle size for maximum alveolar deposition was estimated to be greater than 2 μm, but its value could not be estimated since only one particle size > 2 μm was used in the experiments. The alveolar deposition estimates of Altshuler et al. in the 0.1 to 3.6 μm range and the alveolar deposition values for 2.5 to 12 μm particles measured by Lippmann 2, 3/ using γ-tagged aerosols and external *in-vivo* retention measurements are in good agreement in the region of particle size overlap. The alveolar deposition curve in Figure 3 labelled "*in-vivo* via mouth" is based on the Altshuler and Lippmann data.

Figure 3 also shows an estimate of the alveolar deposition which could be expected when the aerosol is inhaled via the nose. It is based on the difference in experimental *in vivo* head retention data during both nose breathing and mouth breathing tests 2, 3/. It can be seen that for mouth breathing the size for maximum deposition is ∿ 3 μm, and that ∿ ½ of the inhaled aerosol at this size deposits in this region. For nose breathing, there is a much less pronounced maximum of ∿ 25% at 2.5 μm, with a nearly constant alveolar deposition averaging about 20% for all sizes between 0.1 and 4 μm.

The sampler acceptance criteris of ACGIH, BMRC, and the alveolar deposition according to the *in vivo* data of Figure 3 and the ICRP Task Group Model are illustrated on Figure 4. It can be seen that the Task Group's model, which is based on nose breathing, overestimates alveolar deposition at all particle sizes. It should be noted that some of the difference is due to the fact that the Task Group's model is based on the MMD of a polydisperse aerosol rather than a monodisperse aerosol. The overestimates would tend to be somewhat lower if the Task Group calculations were repeated for monodisperse aerosols.

Samplers meeting the BMRC and ACGIH criteria would collect the larger airborne particles with very similar efficiencies as the conductive airways of normal mouth breathing humans. However, they would remove less of the large particles than most nose breathing humans, and "respirable" dust samples as defined by these criteria

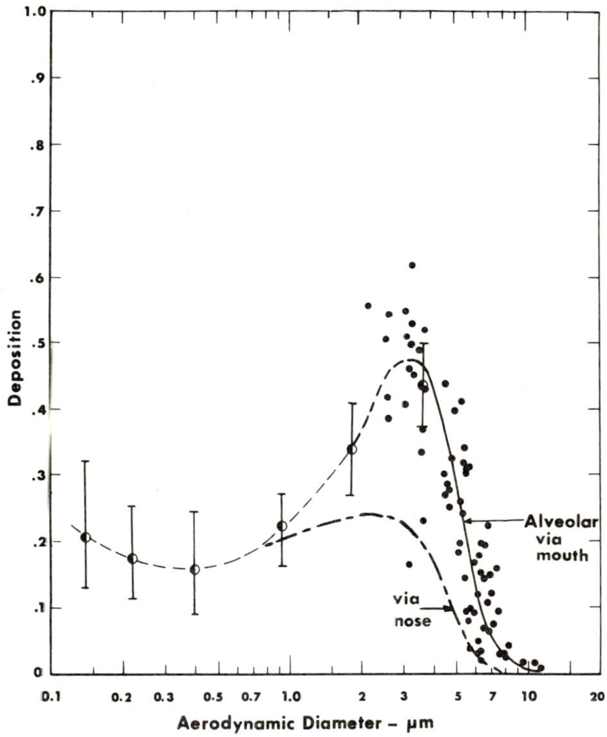

Figure 3. Deposition in the nonciliated alveolar region, in % of aerosol entering the mouthpiece as a function of aerodynamic diameter, except below ½ μm, where linear diameter was used. Individual data points and eye-fit solid line are for the same Fe_2O_3 aerosol tests plotted in Figure 2. The dashed line is an eye-fit through the median best estimates of Altshuler 41/ on 3 subjects who varied as shown by the vertical lines. The lower curve is based on the additional head correction resulting from inhalation via the nose. (Reprinted courtesy The American Physiology Society).

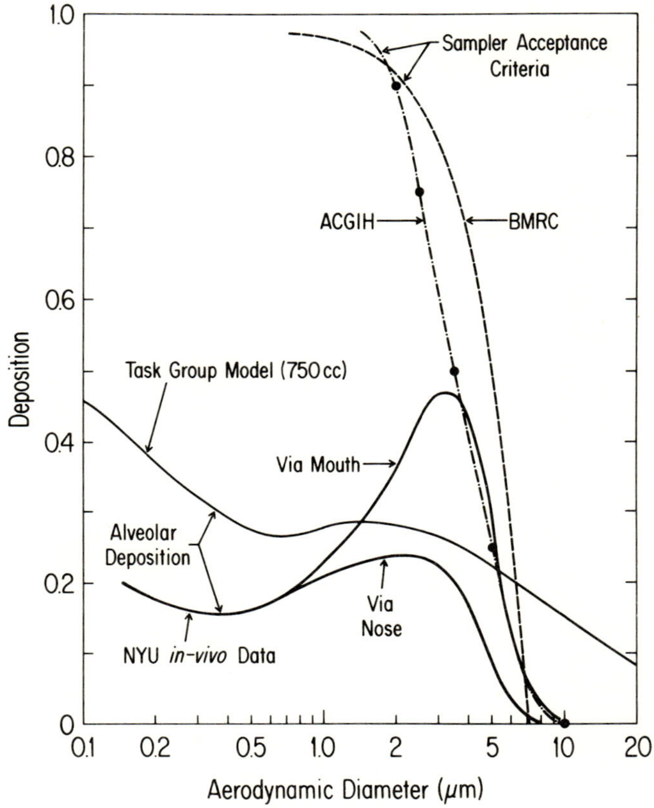

Figure 4. Comparison of sampler acceptance curves of BMRC and ACGIH with alveolar deposition according to ICRP Task Group Model and median human *in vivo* data from Figure 3.

would provide conservative estimates of the pneumoconiosis hazard. It should also be noted that for nose breathing, the deposited dust will most likely be about 20% of the "respirable" dust.

PERFORMANCE CHARACTERISTICS OF "RESPIRABLE" DUST SAMPLERS

This discussion will emphasize data developed since 1969, when an earlier review of the literature was prepared as a background document 30/ for the AIHA's Guide for Respirable Mass Sampling 42/. At that time, the field performance was not fully satisfactory for either the elutriator samplers designed to meet the BMRC criteria, or for the cyclone samplers selected on the basis that they followed the AEC-ACGIH criteria.

The chief advantage of the horizontal elutriator over the cyclone as a pre-collector is that its performance can be predicted on the basis of gravitational sedimentation theory and the physical dimensions of the device. Thus, it was sometimes claimed that laboratory calibrations with carefully characterized test aerosols were not needed. This generalization was true, at least in a relative sense, in comparison to cyclone collectors where no adequate predictive relations exist for collection efficiency. However, in a absolute sense, the prediction of performance of actual elutriator samplers is not completely reliable, as discussed in the earlier review 30/. Some of the discrepancies observed in elutriator performance have been attributed 43/ to technical shortcomings in the manufacture of the elutriators, especially in the non-uniformity of the plate spacings.

One of the major limitations of all elutriator pre-collectors is that it is difficult if not impossible to recover the collected material for analysis. In many cases, it is even difficult to periodically clean out the collected dust to minimize contamination of the second stage collection by reentrained dust. For research studies and other situations where the concentrations of both fractions are to be determined, cyclone pre-collectors are generally used.

Another limitation of the elutriator type of sampler is that it must be operated in a fixed horizontal position. Cyclones, on the other hand, can be operated in any orientation without significant change in their collection characteristics 26, 44/. This independence of orientation, combined with their smaller physical size at comparable flowrates, are among the reasons that most of the recent two-stage personal sampler designs have been built around miniature cyclones as the pre-collectors. These samplers, combined with filter collectors as the second stage and miniature battery powered air pumps are small and light enough to be worn throughout a work shift.

Most of the miniature battery powered pumps are diaphragm or piston type air movers and therefore produce a pulsating flow, which is basically incompatible with their usage with pre-collectors whose collection characteristics are flowrate dependent. Reports by Anderson, Seta, and Vining 45/, Lamonica and Treaftis 46/, Caplan et al. 47/ and Blachman and Lippmann 26/ have demonstrated that the instantaneous flow can be as much as four times the average. Since there is a greater increase in collection efficiency at flows above the average than there is a decrease for flows below it, the net effect is to produce reduced cyclone penetration in pulsating flow as compared to a constant flow at the average rate. Earlier field samplers with pulsating flows underestimated the respirable mass, and more recent models have been equipped with pulsation dampers in order to overcome this problem.

Small variations in steady-state flowrate in cyclone collectors may not be a severe problem, at least for those applications when the parameter of interest is the "respirable" mass measured on the second stage. Knight and Lichti 48/ demonstrated that variations in airflow are corrected to some extent by changes in cyclone collection efficiency. For nonfibrous test aerosols, including mica and silica, there was essentially no change in the mass collected on the filter for flowrates between 1.3 and 2.65 lpm. As the flowrate increases, the aerosol mass entering the cyclone increases proportionally, but so apparently does the collection efficiency. In an elutriator, the effect would be the opposite; an increase in sampling rate would result in an increase in penetration to the filter.

Despite all of their practical advantages as pre-collectors, cyclones have not met with universal favor. Their major limitation is the continuing uncertainty about their efficiency calibrations. There are no predictive equations capable of providing useful calibration estimates for the small cyclones used as pre-collectors 49/. Furthermore, for most cyclones, there is insufficient and/or conflicting empiric data. The greatest amount of attention has been directed twoard the Dorr-Oliver 10 mm nylon cyclone used extensively in underground coal mines, and most recent studies 26, 47, 50/ confirm the AIHA Guide 42/ which indicated that it comes closest to matching the ACGIH criteria when operated at a 1.7 lpm flowrate. However, the developing consensus does not include MESA, and they continue to require the 2.0 lpm flowrate indicated by their own laboratory calibration 51/ for samplers used in the mines.

The recent calibrations of the 10 mm cyclone have also clearly established that this cyclone cannot really match the ACGIH criteria at any flowrate, because it has a cut-off characteristic which is much sharper than that of the criteria. In other words, when the cyclone is operated at a flowrate which produces a 50% cut at 3.5 μm, it collects more than 75% of 5 µm particles, and less than 25% of 2.5 µm particles. This relatively sharp cut-off characteristic makes this cyclone useful in a multicyclone sampler which is used to determine aerodynamic size distributions 26/.

One other problem with the nylon cyclone is that, being an insulator, it can accumulate a static charge. When sampling aerosols with very high charge levels, this can significantly affect collection efficiency. Blachman and Lippmann 26/ showed that highly charged aerosols with aerodynamic diameters below ~ 4 µm were collected with higher efficiencies than charge neutralized aerosols of the same aerodynamic diameter. Almich and Carson 52/ reported that the average collection efficiency for 4 to 5 µm charged particles was not significantly increased, but that the variability in collection efficiency in replicate runs was increased. This variability was absent when using 10 mm cyclones of the same design which were constructed of stainless steel.

While the 10 mm cyclones are injection molded and virtually identical to each other, all of the larger cyclones used as pre-collectors are hand assembled, and the quality control exercised by their manufacturers has frequently been quite poor 53/. On the other hand, the larger cyclones have the advantage, at least for their application as respirable dust pre-collectors, of much less sharp cut-off characteristics. Their cut-offs turn out, probably fortuitously, to closely match the ACGIH criteria, as illustrated in Figure 5. The flow-rates at which these cyclones most closely match the ACGIH sampler acceptance criteria are 9, 25, 75 and 430 lpm for the ½-inch HASL, Aerotec 3/4, 1-inch HASL, and Aerotec 2 respectively. For still higher flowrates, Battelle 54/has developed what they are calling a massive volume air sampler. It includes an impactor collector which was designed to match the ACGIH criteria at 1.3 x 10^6 ft^3/day (21.2 m^3/min). This multistage series sampler also includes an entry section with a cut-off at 20 µm, an impactor with a 1.7 µm cut-off, and an electrostatic collector to remove the particles penetrating the 1.7 µm impactor. This sampler is currently being evaluated by EPA, and its ability to avoid the characteristic limitation of impactors, i.e., particle bounce and reentrainment, remains to be demonstrated.

SUMMARY AND CONCLUSIONS

Inhalation hazards are dependent upon the aerodynamic size distributions of the airborne particles, and the size-selective deposition characteristics of the human airways. Realistic hazard evaluations must therefore be based either on the determination of the aerosol's aerodynamic size-mass distribution, or on concentrations determined from samples collected with devices which simulate the human airway's deposition characteristics. The latter approach is more direct and

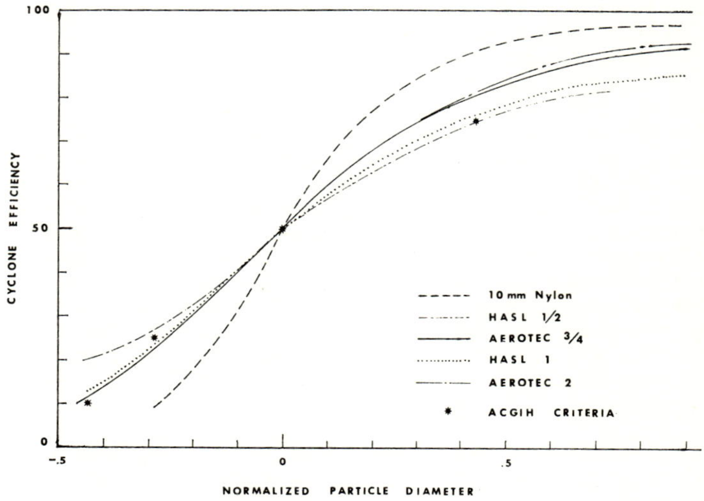

Figure 5. Collection efficiency vs normalized particle size for four dual-inlet stainless steel cyclones and the 10 mm Dorr-Oliver nylon cyclone. (Reprinted courtesy American Industrial Hygiene Association).

economical, but requires an accurate quantitative knowledge of human regional particle deposition.

Recent human in vivo deposition data have helped to resolve some of the long-standing confusion concerning regional deposition in man, and provide an improved basis for the utilization of size-selective sampling for inhalation hazard evaluations. Samplers which satisfy the BMRC or ACGIH sampler acceptance criteria simulate the collection characteristics of the conductive airways of normal mouth breathing humans. For particles between ~ 1½ μm and the samplers' upper cut-off, the alveolar zone deposition is about twice as high for mouth breathers as compared to nose breathers. For particles between 0.1 and 1.5 μm, the route of entry has little effect on alveolar deposition. For nose breathing, the fraction deposited in the alveolar zone remains nearly constant at ~ 20% for all sizes between 0.1 and 3.5 μm. The ICRP Task Group Model overestimates alveolar zone deposition at all particle sizes, with the greatest discrepancies for particles > 5 μm and < 0.5 μm.

A large number of elutriators which can satisfy the BMRC criteria are available, as are a large number of cyclones which can satisfy the ACGIH criteria. In most cases, elutriator performance can be predicted from physical theory, but the samplers are large in comparison to their sampling rate, and must operate in a horizontal position. Cyclones are much more convenient and compact field instruments, but must be calibrated empirically. Unfortunately, the quality control in small cyclone manufacture has frequently been poor, and the calibration data needed to verify adequate performance have frequently been lacking or of questionable reliability.

Despite the somewhat unhappy past history of cyclone usage in inhalation hazard evaluations, their practical advantages are considerable and there are no fundamental limitations to their proper usage. With better quality control and the use of reliable calibration procedures, they should continue to be widely used.

ACKNOWLEDGEMENTS

This investigation was supported by Grant #ES-00881 and is part of a center program supported by Grant ES-00260, both from the National Institute of Environmental Health Sciences. It is also part of the center program supported by Grant #CA-13343 from the National Cancer Institute.

REFERENCES

1. Liebow, A. A. Recent Advances in Pulmonary Anatomy. In: Ciba Foundation Symposium on Pulmonary Structure and Function, De Reuck, A. V. S. and M. O'Connor (eds.). Boston, Little, Brown and Co., 1962. p. 2-25.

2. Lippmann, M. Regional Deposition of Particles in the Human Respiratory Tract. In: Handbook of Physiology (Environmental Physiology), Lee, D. H. K. and S. D. Murphy (eds.). The American Physiological Society, Bethesda, Md., 1975.

3. Lippmann, M., and B. Altshuler. Regional Deposition of Aerosols. In: Proceedings of 20th OHOLO Biological Conference at Ma'alot, Israel, 16-19 March, 1975. (In Press. John Wiley and Sons, Ltd.).

4. Katz, S. H., G. W. Smith, W. M. Myers, L. J. Trostal, M. Ingels, and L. Greenburg. Comparative Tests of Instruments for Determining Atmospheric Dust. Public Health Bulletin No. 144, Public Health Service, DHEW, Washington, D. C., 1925.

5. Littlefield, J. B., and H. H. Schrenk. Bureau of Mines Midget Impinger for Dust Sampling. Bureau of Mines RI 3360, U. S. Department of Interior, Washington, D. C., Dec. 1937.

6. Orenstein, A. J. (ed.). Proceedings of the Pneumoconiosis Conference, Johannesburg, 1959. London, J. and A. Churchill, Ltd., 1960.

7. Lippmann, M., and W.B.Harris. Size-Selective Samplers for Estimating "Respirable" Dust Concentrations, Health Physics 8:155-163, 1962.

8. Threshold Limits Committee. Threshold Limit Values of Air Borne Contaminants for 1968. American Conference of Governmental Industrial Hygienists, 1014 Broadway, Cincinnati, Ohio, 1968.

9. U. S. Dept. of Labor. Public Contracts and Property Management. Federal Register 34 (No. 96):7946, May 20, 1969.

10. Brown, J. H., K. M. Cook, F. G. Ney, and T. Hatch. Influence of Particle Size Upon the Retention of Particulate Matter in the Human Lung. Amer. J. Public Health 40:450-457, 480, 1950.

11. Task Group on Lung Dynamics Committee II - ICRP. Deposition and Retention Models for Internal Dosimetry of the Human Respiratory Tract. Health Physics 12:173-208, 1966.

12. Mercer, T. T. Air Sampling Problems Associated with the Proposed Lung Model. Presented at the 12th Annual Bioassay and Analytical Chemistry Meeting, Gatlinburg, Tenn., Oct. 13, 1966.

13. Lynch, J. R. Evaluation of Size-Selective Presamplers. I. Theoretical Cyclone and Elutriator Relationships. Amer. Ind. Hyg. Assoc. J. 31:548-551, 1970.

14. Moss, O. R., and H. J. Ettinger. Respirable Dust Characteristics of Polydisperse Aerosols. Amer. Ind. Hyg. Assoc. J. 31:546-547, 1970.

15. Coenen, W. Berechung von Umrechnungsfaktorem fur Verscheidene Feinstaubmessverfahren. In: Inhaled Particles III, Walton, W. H. (ed.). London, Unwin Bros., 1971.

16. May, K. R. The Cascade Impactor, An Instrument for Sampling Coarse Aerosols. J. Sci. Inst. 22:187-195, 1945.

17. Mitchell, R. I., and J. M. Pilcher. Improved Cascade Impactor for Measuring Aerosol Particle Sizes. Ind. and Eng. Chem. 51:1039, 1959.

18. Brink, J. A. Cascade Impactor for Adiabatic Measurements. Ind. and Eng. Chem. 50:645, 1958.

19. Andersen, A. A. A New Sampler for Collecting, Sizing and Enumeration of Viable Airborne Bacteria. J. Bact. 76:471, 1959.

20. Lippmann, M. Compact Cascade Impactor for Field Survey Sampling. Amer. Ind. Hyg. Assoc. 22:348-353, 1961.

21. Lundgren, D. A. An Aerosol Sampler for Determination of Particle Concentrations as a Function of Size and Time. J. Air Poll. Cont. Assoc. 17:4, 1967.

22. Mercer, T. T., M. I. Tillery and G. J. Newton. A Multi-Stage Low Flow Rate Cascade Impactor. J. Aerosol Sci. 1:9-15, 1970.

23. Stober, W., and H. Flachsbart. Size-Separating Precipitation in a Spinning Spiral Duct. Env. Sci. & Tech. 3:1280-1296, 1969.

24. Kotrappa, P., and M. E. Light. Design and Performance of the Lovelace Aerosol Particle Separator. Rev. Sci. Inst. 43:1106-1112, 1972.

25. Hounam, R. F., and R. J. Sherwood. The Cascade Centripeter: A Device for Determining the Concentration and Size Distribution of Aerosols. Amer. Ind. Hyg. Assoc. J. 26:122-131, 1965.

26. Blachman, M. W., and M. Lippmann. Performance Characteristics of the Multicyclone Aerosol Sampler. Amer. Ind. Hyg. Assoc. J. 35:311-316, 1974.

27. Bernstein, D., M. T. Kleinman, T. J. Kneip, T. L. Chan, and M. Lippmann. Development of a High-Volume Sampler for the Determination of Particle Size Distributions in Ambient Air. J. Air Pollut. Control. Assoc. (In Press)

28. Davies, C. N. Deposition and Retention of Dust in the Human Respiratory Tract. Ann. Occup. Hyg. 7:169-183, 1964.

29. Hatch, T. F., and P. Gross. Pulmonary Deposition and Retention of Inhaled Aerosols. New York, Academic Press, 1964.

30. Lippmann, M. 'Respirable' Dust Sampling. Amer. Ind. Hyg. Assoc. J. 31:138-159, 1970.

31. Stuart, B. O. Deposition of Inhaled Aerosols. Arch. Intern. Med. 131:60-73, 1973.

32. Davies, C. N. The Deposition of Aerosol in the Human Lung. In: Aerosole in Physik, Medizin und Tecknik. Bad Soden, W. Germany, Gesellschaft fur Aerosolforschung, 1973. p. 90-99.

33. Altshuler, B., L. Yarmus, E. D. Palmes, and N. Nelson. Aerosol Deposition in the Human Respiratory Tract. AMA Arch. Ind. Health 15:293-303, 1957.

34. Landahl, H. D., T. N. Tracewell, and W. H. Lassen. On the Retention of Airborne Particulates in the Human Lung II. AMA Arch. Ind. Hyg. Occup. Med. 3:359-366, 1951.

35. Landahl, H. D., T. N. Tracewell, and W. H. Lassen. Retention of Airborne Particulates in the Human Lung III. AMA Arch Ind. Hyg. Occup. Med. 6:508-511, 1952.

36. Muir, D. C. F., and C. N. Davies. The Deposition of 0.5 μm Diameter Aerosols in the Lungs of Man. Ann. Occup. Hyg. 10:161-174, 1967.

37. Davies, C. N., J. Heyder, and M. C. Subba Ramu. Breathing of Half-Micron Aerosols I. Experimental. J. Appl. Physiol. 32:591-600, 1972.

38. Heyder, J., J. Beghart, G. Heigwer, C. Roth, and W. Stahlhofen. Experimental Studies of the Total Deposition of Aerosol Particles in the Human Respiratory Tract. Aerosol Sci. 4:191-208, 1973.

39. Heyder, J., L. Armbruster, J. Gebhart, and W. Stahlhofen. Deposition of Aerosol Particles in the Human Respiratory Tract. In Aerosole in Physik, Medizin und Technik. Bad Soden, W. Germany, Gesellschaft fur Aerosolforschung, 1973. p. 122-125.

40. Lever, J. In Press, Aerosol Sci. 1974, quoted by C. N. Davies in Reference #32.

41. Altshuler, B., E. D. Palmes, and N. Nelson. In: Inhaled Particles and Vapors II, Davies, C. N. (ed.). Oxford, Pergamon Press, 1966. p. 323-335.

42. AIHA Aerosol Technology Committee. Guide for Respirable Mass Sampling. Amer. Ind. Hyg. Assoc. J. 31:133-137, 1970.

43. Thurmer, H. Investigations with the Horizontal Plate Precipitator. Staub (English Translation) 29:35-42, 1969.

44. Treaftis, H. N., and T. F. Tomb. Effect of Orientation on Cyclone Penetration Characteristics. Amer. Ind. Hyg. Assoc. J. 35:598-602, 1974.

45. Anderson, D. P., J. A. Seta, and J. F. Vining III. The Effect of Pulsation Dampening on the Collection Efficiency of Personal Sampler Pumps. Report #TR-70. Cincinnati, Ohio, Nat'l Inst. for Occ. Safety and Health, Sept. 1971.

46. Lamonica, J. A., and H. N. Treaftis. The Effect of Pulsation Dampening on Respirable Dust Collected by Coal Mine Personal Samplers. (Presented at 1972 Amer. Ind. Hyg. Conference. San Francisco, California. May 1972).

47. Caplan, K., S. Sorenson, and L. Doemeny. Evaluation of a Coal Mine Personal Dust Sampler. (Presented at 1972 Amer. Ind. Hyg. Conference. San Francisco, Calif. May 1972).

48. Knight, G., and K. Lichti. Comparison of Cyclone and Horizontal Elutriator Size Selectors. Amer. Ind. Hyg. Assoc. J. 31:437-441, 1970.

49. Chan, T. L., and M. Lippmann. Prediction of Cyclone Collection Efficiency. (Presented at Amer. Ind. Hyg. Conference. Miami Beach, Fla. May 1974).

50. Seltzer, D. F., W. J. Bernaski, and J. R. Lynch. Evaluation of Size-Selective Presamplers II. Efficiency of the 10 mm Nylon Cyclone. Amer. Ind. Hyg. Assoc. J. 32:441-446, 1971.

51. Tomb, T. F., and L. D. Raymond. Evaluation of Collecting Characteristic of Horizontal Elutriator and 10 mm Nylon Cyclone Gravimetric Dust Samplers. (Presented at the 1969 Annual Meeting of the American Industrial Hygiene Assoc. Denver, Col. May 1969).

52. Almich, B. P., and G. A. Carson. Some Effects of Changing on 10 mm Nylon Cyclone Performance. Amer. Ind. Hyg. Assoc. J. 35:603-612, 1974.

53. Lippmann, M., and T. L. Chan. Calibration of Dual-Inlet Cyclones for "respirable" Mass Sampling. Amer. Ind. Hyg. Assoc. J. 35:189-200, 1974.

54. Mitchell, R. I. Massive Volume Air Sampler for Large Quantity Particulate Size Fraction Collector. Internal Report. Battelle Columbus Laboratories, 505 King AVe., Columbus, Ohio 43201.

55. Giacomelli-Maltoni, G., C. Melandri, V. Prodi, and G. Tarroni. Deposition Efficiency of Monodisperse Particles in Human Respiratory Tract. Amer. Ind. Hyg. Assoc. 33:603-610, 1972.

56. Märtens, A., W. Jacobi. Die in vivo Bestimung der Aerosolteilchendeposition im Atemtrakt bei Mund-bzw. Nasenatmung. In Aerosole in Physik, Medizin und Technik. Bad Soden, W-Germany, Gesellschaft für Aerosolforschung, 1973.

DICHOTOMOUS VIRTUAL IMPACTORS FOR LARGE SCALE MONITORING OF AIRBORNE PARTICULATE MATTER *

Billy W. Loo, Joseph M. Jaklevic, and Fred S. Goulding

Lawrence Berkeley Laboratory
University of California
Berkeley, California 94720

ABSTRACT

The bimodal particle size distribution in the urban aerosol suggests the use of a dichotomous sampling method with a cut point near two microns. The collection of the two size fractions separately should aid the identification of pollutant sources and the evaluation of fine particle-related health hazards. We have constructed such a sampler in which size segregation is effected by inertial impaction across a virtual surface into a volume of relatively stagnant air. Following a two stage separation, the particle size fractions are deposited on membrane filters. One filter collects 95% of uncontaminated fine particles while the other collects all of the coarse particles along with 5% of the fine fraction.

The sampling scheme is tailored to match the requirements of elemental analysis by X-ray fluorescence and total mass measurements by beta gauging. The design parameters of the virtual impactor have been optimized for minimum loss and sharp cut characteristics. A servo flow controller is utilized to maintain a constant sampling rate of 50 ℓ/min with a particle size cut at 2.4 μm Stokes' diameter. Fully automated units have been developed for a computer controlled monitoring network.

The compatibility of the virtual impactor with other analytical instrument and its adaptibility to automation suggest that it may be suited for wider use.

* This work was supported by the Environmental Protection Agency under Interagency Agreement with the U. S. Energy Research and Development Administration.

DICHOTOMOUS VIRTUAL IMPACTORS FOR LARGE SCALE MONITORING OF AIRBORNE PARTICULATE MATTER

Billy W. Loo, Joseph M. Jaklevic, and Fred S. Goulding

Lawrence Berkeley Laboratory
University of California
Berkeley, California 94720

INTRODUCTION

Airborne particulate matter consists primarily of natural aerosols such as soil dust, pollens and sea spray as well as a multitude of man made pollutants in the form of hydrocarbons, sulphates, nitrates and a broad spectrum of trace elements, many of which are considered potentially toxic. Any effective air quality control strategy must be based on detailed knowledge of the generation, transformation, dispersion and depletion of those undesirable substances. Since gas-particle interaction is recognized to play a key role in forming secondary aerosol from primary pollutants, the study of particulate pollutant is an important facet of the total problem.

Unfortunately the problem is made unwieldy by the complex meteorological, geographical conditions and anthropogenic processes. The system under study has no steady state. Any comprehensive air pollution study will require the large scale deployment of instruments to provide both time and position data on the concentrations of the polluting species so that practical models of the systems might be developed.

Energy dispersive X-ray fluorescence analysis has now emerged as a potent and versatile technique for trace element analysis due to its capability for simultaneous rapid multi-element analysis. Quantitative non-destructive analysis on a large and economical scale is possible for most elements of interest[1]. A fully automated analysis system can measure 35 elements ranging from Al to Pb collected on two-hour air filter samples through a sequence of three five-minute runs with a detection limit on the order of 10 ng/m^3 [2]. Such a system can comfortably analyze some 20,000 samples a year.

In understanding the evolution of pollutants and in assessing potential hazards, it is desirable to obtain particle size information in addition to elemental concentrations. It has been observed that urban aerosols tend to have a bimodal size distribution with a minimum at about 2 μm, (Whitby, et al[3]). The fine particle fraction peaks at about 0.3 μm and is considered primarily to be the accumulation of combustion products by condensation and coagulation. The coarse particle fraction consists mainly of mechanically produced aerosol with particle size peaking around 10 μm, although this is subject to large fluctuations caused by gravitational settling and impaction losses. There appears to be little mass transfer between the two modes[4]. The occurance of a fairly distinct minimum may correspond to a natural physical effect since the creation of larger surface areas by mechanical means becomes energetically unfavorable at sub-micron sizes. Further interest in separating particles whose sizes lie on either side of the 2 μm point arises because the deposition and retention of aerosols in the human respiratory system is such that particles smaller than 2 or 3 microns are not efficiently removed in the nasal-pharyngeal region and penetrate to the tracheo-bronchial and pulmonary regions[5]. The large surface area (some 60 m^2) in the lungs renders the human body vulnerable to sub-micron pollutants. The coincidence of the size transition at 2 μm in both the particle production and the health effects is a prime motivation for developing a dichotomous sampler to collect size-segregated fractions from each size range for subsequent X-ray fluorescence analysis.

OBJECTIVE

As part of the St. Louis Regional Air Pollution Study (RAPS), a network of 25 monitoring stations (RAMS) has been set up to collect meteorological, gaseous and particulate pollutant data. The initial plan called for the installation of automated dichotomous air samplers (ADAS) in ten of these stations to be operated under the control of a central computer. Figure 1 illustrates an integrated program of sampling and analyzing some 20,000 samples per year. To facilitate data handling, each filter carries a computer readable digital label. Total mass loading on filters will be determined by beta-gauge measurements before and after exposure and elemental analysis will be performed by X-ray fluorescence analysis. These results will then be merged with meteorological and other relevent data for inclusion in the main data bank.

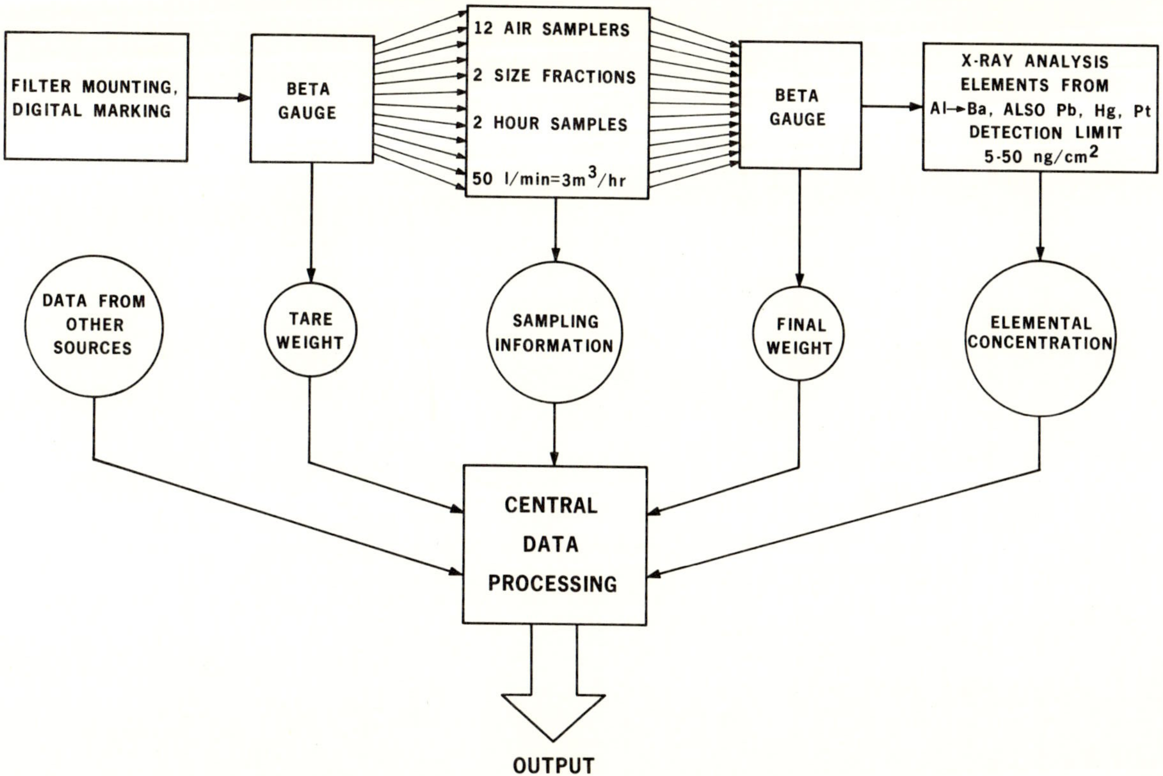

XBL 743-538

Fig. 1 Block diagram of a sampling and analysis system.

Our objective was to develop a dichotomous sampler that matches the requirements of analysis systems. It should collect uniform depositions on a thin substrate of low atomic number and should exhibit a sharp particle size cut with very small or zero particle losses in the apparatus. It should provide a high sampling rate to permit adequate sample collection in a reasonably short time. Ease of sample handling and reliable operation are also very important considerations for large scale monitoring. The desired sampling rate and particle size cut point have been chosen to be 50 ℓ/min and 2 μm respectively.

APPROACH

One of the prime considerations in particle sizing is the selection of a physical mechanism by which size separation can be accomplished. Since the shape, density and dielectric properties of the atmospheric aerosols cannot be predicted, the interpretation of the term "particle size" depends largely on the method of measurement. Since effects of particulates depend on their transport processes in the atmosphere and on their uptake and retention through respiration, inertial methods of separation should provide more relevant data than electrical, thermal or optical methods. We shall therefore regard the size of a particle as being the diameter of an aerodynamically equivalent unit density sphere (i.e., the Stokes' diameter).

Conventional methods of inertial separation in the 2 μm range often uses an impactor which is an air jet impinging on a collection plate as shown in Fig. 2. The jet size, velocity and geometry are controlled so that the inertia of particles above a certain size causes them to overcome the drag force as they leave their deflected stream lines and to impact onto the plate. A useful concept in impactor studies is the Stokes' number Stk defined as

$$\mathrm{Stk} = \frac{\rho_p\, C\, V_o\, D_p^{\,2}}{9\, \mu\, D_1} \tag{1}$$

where ρ_p = particle density

C = Cunningham slip correction for the discontinuous nature of fluid interaction when the pressure is low or when the particle size is small compared with the molecular mean free path in air

V_o = mean fluid velocity of the jet

D_p = particle diameter

μ = viscosity of air

D_1 = a characteristic size usually taken as the jet diameter.

At low Reynolds' numbers, the drag force on a particle is $3\,\pi\mu\,VD_p$ as given by the Stokes' law, where V is the relative velocity between the fluid and the particle. The Stokes' number is a measure of the ratio of inertial force (in this case centrifugal force) to the drag force. It is also equal to the stopping distance of a particle in stagnant air divided by the jet diameter. Under a given set of flow conditions, particles with a Stokes' number higher than a critical value are impacted onto the collection plate. It can therefore be regarded as useful for scaling purposes.

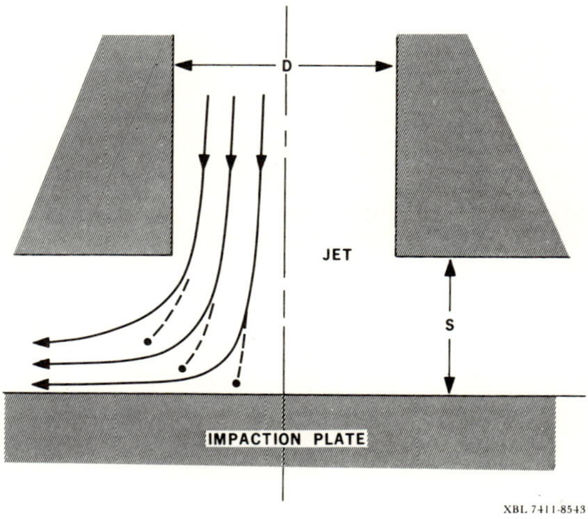

Fig. 2 Schematic of a conventional impactor

Impactors with one or more size cuts have been widely used for collecting size-segregated particulate samples. However, the inherent difficulties with particle bounce, reentrainment, non-uniform deposition and cumbersome sample handling have limited their potential for large scale applications. A virtual impactor uses the principle of inertial separation, but the impaction plate is replaced by a region of relatively stagnant air (receiving tube in Fig. 3 and 4). The virtual surface formed by the deflected streamlines realizes similar boundary conditions to those in real impactors. Large particles will pass into the forward low-flow region while small particles will remain mostly in the high-flow air stream deflected radially around the receiving tube. Both size fractions can subsequently be deposited onto separate filters.

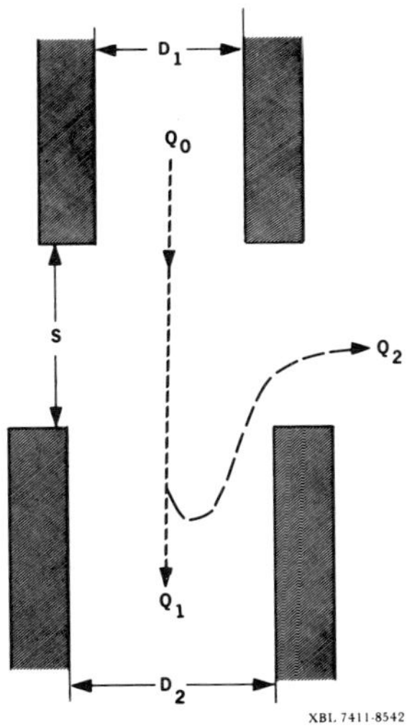

Fig. 3 Critical parameters of a virtual impactor

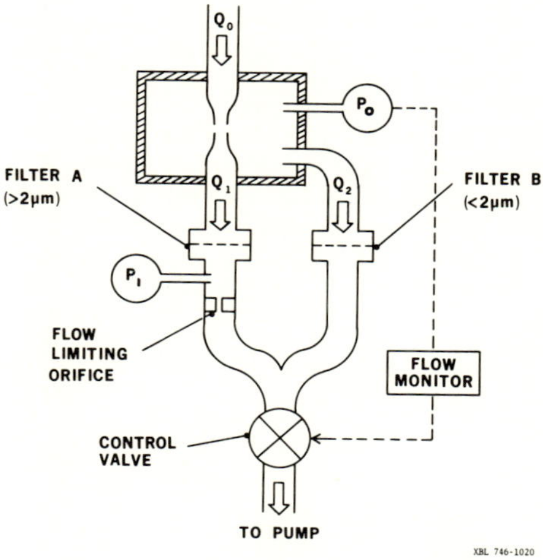

Fig. 4 Schematic of a single-stage dichotomous virtual impactor

A virtual impactor possesses several distinct advantages:

1) Particle depositions on the filters can be made quite uniform, which is ideal for photon excited X-ray fluorescence analysis and total mass measurement using a beta gauge.

2) The size and pattern of deposition can be controlled to match the optimum requirement of the analysis system.

3) Since the object is to avoid collection within the apparatus, particle bounce is a favorable phenomenon. In fact it is desirable to shape streamlines as nearly tangential to the surfaces as possible to encourage bounce and thereby reducing losses.

4) Reentrainment of particles in the air stream is a second order effect since only particles lost in the apparatus are subject to blow off.

5) Collecting samples external to the apparatus not only facilitates automatic sample handling but also permits the design of the impaction geometry to concentrate on maintaining precise and consistant separation characteristics.

6) The virtual impactor can serve as an input stage for any downstream instrument.

VIRTUAL IMPACTOR DESIGN CONSIDERATIONS

An early study of a virtual impactor was done by William Conner[6], who utilized ideas from the cascade centripeter[7]. Although the measurements and analysis were inaccurate, it did demonstrate the size separating power of the virtual impaction principle. A more sophiscated version using two cascade separation stages was developed by Carl Peterson of the Environmental Research Corporation. We have made a detailed evaluation on the ERC unit to assess the feasibility of the virtual impaction scheme for large scale sampling[8]. Emphasis in our work has been placed on improving the sharpness of the particle size cut characteristics and reducing losses. The result of this and other related studies revealed the importance of several factors which govern the performance:

1) Jet Reynolds' number - The Reynolds' number is defined as $\rho VD/\mu$, where ρ and μ are the density and viscosity of air and V is the mean air velocity and D is the diameter of the jet. Excessive turbulence which tends to occur at high Reynolds' numbers sets an upper limit to the maximum flow in a jet for a given size cut.

2) Flow symmetry and alignment - The sharpness of the cut characteristic is degraded by the azimuthal flow asymmetry about the axis of each jet. In our design this effect is minimized. It is clear that rectangular jets have inferior performance in this regard and stable symmetrical flow configurations are difficult to maintain. The coaxial alignment of jet and receiving tube is obviously essential.

3) Flow control - A feedback system is needed to maintain a constant operating point independent of variations in the filter impedance (including changes due to particle loading).

4) Particle bounce - Taking advantage of the non-sticking probability of solid particles, streamlines are designed to be as nearly tangential to the physical surfaces of the impactor as

is possible. Contours of parts are shaped to reduce impaction losses.

5) Critical parameters - Referring to the notations in Fig. 3 on the relevant parameters of a simple jet and impaction tube, it is necessary to determine an optimum set of parameters Q_0, D_1, Q_1/Q_0, D_2/D_1 and S/D_1 for minimum loss at a given size cut. A fundamental study of conventional impactors has been done by Marple[9]. However, the theoretical analysis is not readily applicable to the case of the virtual impactor due to the much more complex boundary conditions. We have adopted an empirical approach, first measuring the range and sensitivity of each parameter then converging onto a region where the cut point is relatively definitive and stable and where remaining parameters are optimized.

METHODS OF MEASUREMENT

Due to the close approximation of their geometric diameters to unit density Stokes' diameters, dioctyl phthalate (DOP) droplets are well suited as test particles. Particles between 1 to 10 μm in size can very readily be produced in a Berglund-Liu monodisperse aerosol generator with uranine (fluorescein sodium) used as a tracer for quantitative measurements. The generator utilizes the uniform breakup of a liquid jet into droplets as it passes through a vibrating orifice. The final size of a liquid or solid particle is calculated from the initial solvent concentration, jet flow rate and vibration frequency. The monodisperse aerosol output is then used to test a particular impactor configuration. Particle depositions can be dissolved in water and the uranine content of the solution determined by UV fluorescence techniques. Such procedures to determine deposits on parts have been described in detail in an earlier report[8].

The Stokes' diameters of solid particles are less well defined because of uncertainties in the final density of the particles. Figure 5 is a picture of an "8.8" μm NaCl particle as viewed by a scanning electron microscope at 7000 magnification. The void fraction after the solvent is evaporated is estimated at 50%. The sizes of solid particles (uranine) used in our final evaluation were assigned using the size cut characteristics of the virtual impactor as measured with liquid DOP particles in order to compensate for such uncertainties.

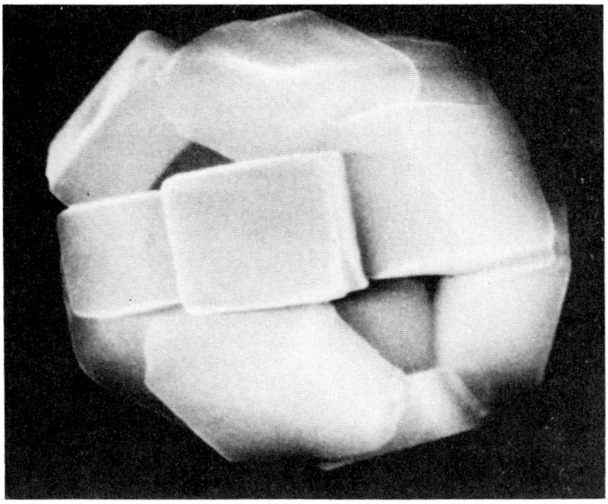

Fig. 5 A "8.8" μm NaCl particle as viewed by a scanning electron microscope.

Flow measurements were made with pressure-corrected rotameters and cross checked with a temperature-compensated integrating flow meter. All calibrations were performed at 20°C and 735 mm Hg pressure.

We shall now digress to a problem of special interest which is the measurement of the flow division within a two-stage virtual impactor where the internal flows are not directly assessable. Referring to the schematics of a two-stage virtual impactor (Fig. 15), the flow conditions may be closely approximated by the equivalent circuit shown in Fig. 6. R_0, R_1 and R_2 represent the non-linear flow impedances associated with the inlet orifice, and the orifices in the coarse and fine particle streams respectively. The flow division Q_1/Q_0 may be derived from the measurement of the total external flow as a function of a single static differential pressure P_0 between the inlet and the second stage of the impactor as shown in Fig. 7. Curve A is the characteristic when R_1 and R_2 are made zero by removing the inner section of the virtual impactor. Curve B is the characteristic with all the parts in place and Curve C is similar to B except that R_1 is made infinite by temporarily sealing the orifice such that R_0 and R_2 are in series. Under normal operation, the total flow is Q_0 and XY is the pressure differential across R_2. Hence, Q_2 is determined graphically by locating UV = XY. In this case Q_1/Q_0 is measured to be 24.9%.

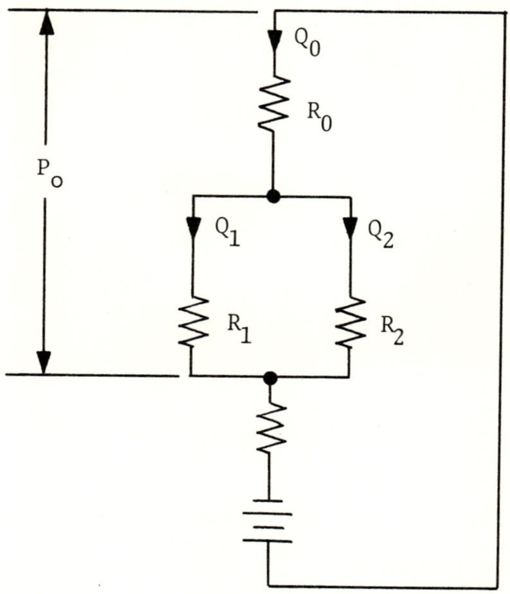

Fig. 6 An equivalent circuit of fluid flow in a virtual impactor.

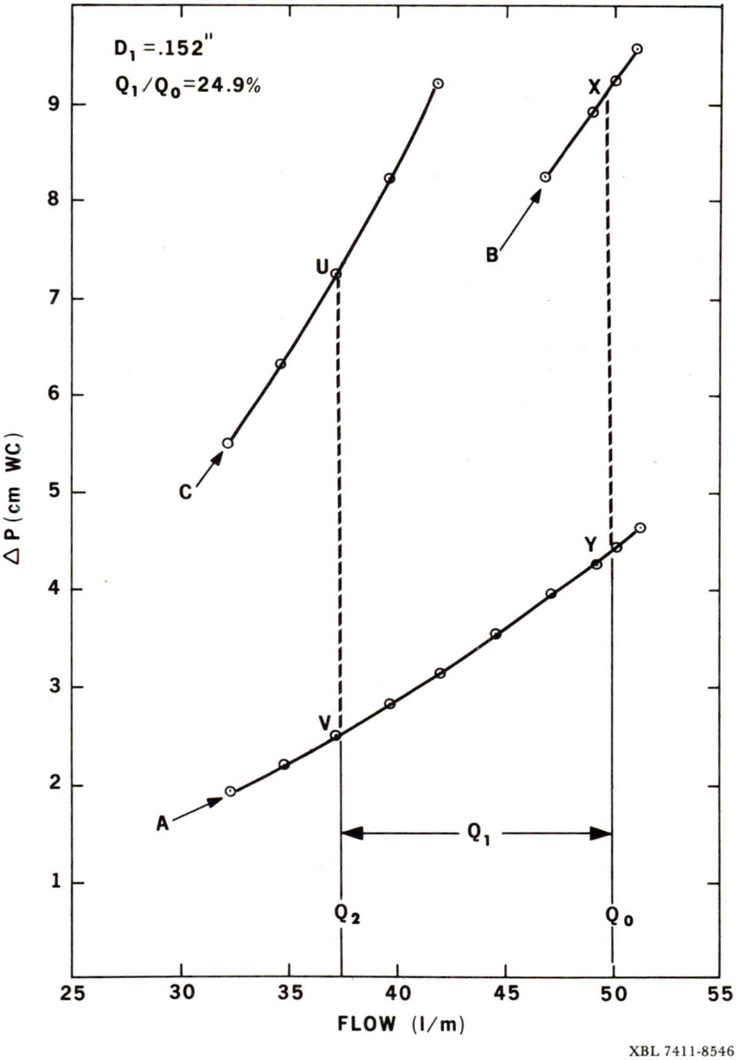

Fig. 7 A graphical solution to the internal flow distribution problem in a two-stage virtual impactor.

RESULTS AND DISCUSSION ON DESIGN STUDIES

A series of measurements have been made to probe the behavior of a single-jet virtual impactor. The coarse and fine particles were collected on filters A and B corresponding to greater and less than 2 μm respectively. For each set of measurement, the collection efficiency, defined as E = A/(A+B), and/or particle losses were observed as a function of the parameter being varied. Starting with a set of parameters near the 2 μm cut point, the objective was to develop an optimum set of parameters with a minimum number of excursions in the multiparameter space. The selection of 2 and 10 μm test particles were used as fiducial sizes in this regard.

The results of such measurements are shown in Figs. 8 through 14. Table 1 is a summary of the conditions under which each profile is observed, using the symbols of Fig. 3.

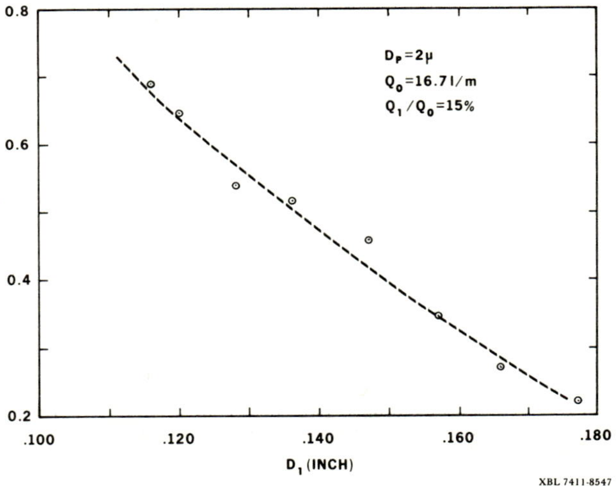

Fig. 8 Behavior of E vs. D_1

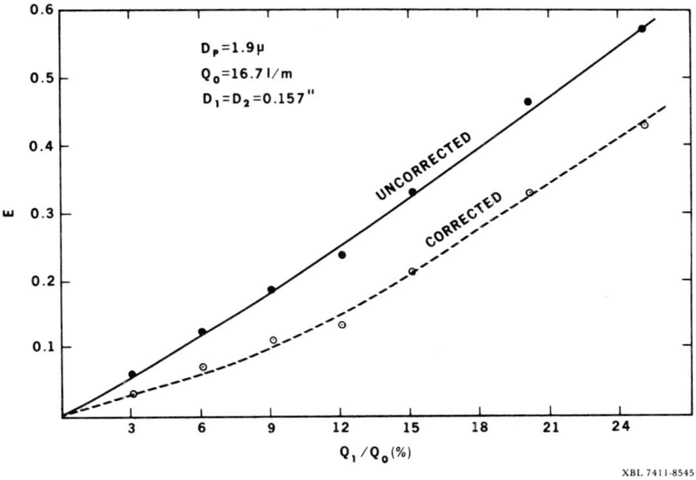

Fig. 9 Behavior of E vs. Q_1/Q_0

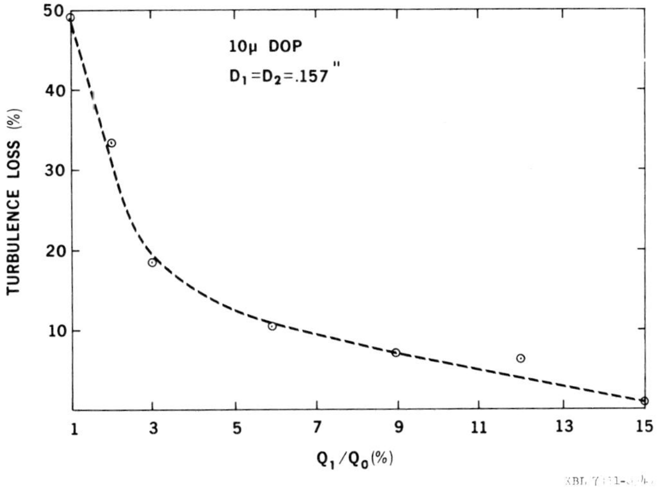

Fig. 10 Turbulent mixing as a function of Q_1/Q_0

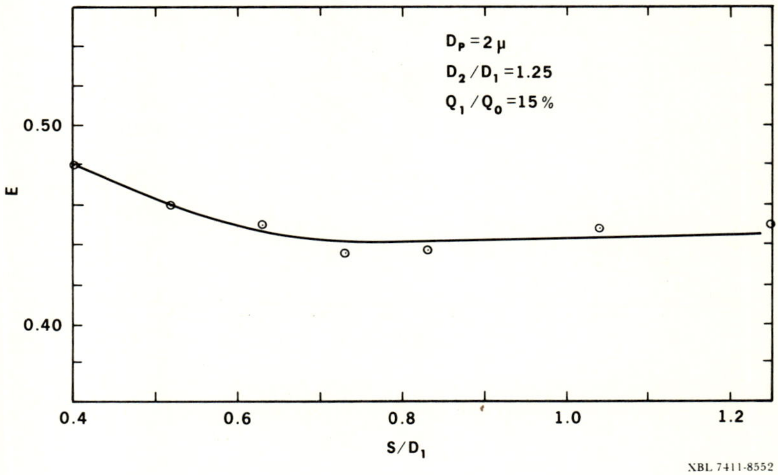

Fig. 11 Behavior of E vs. S/D_1

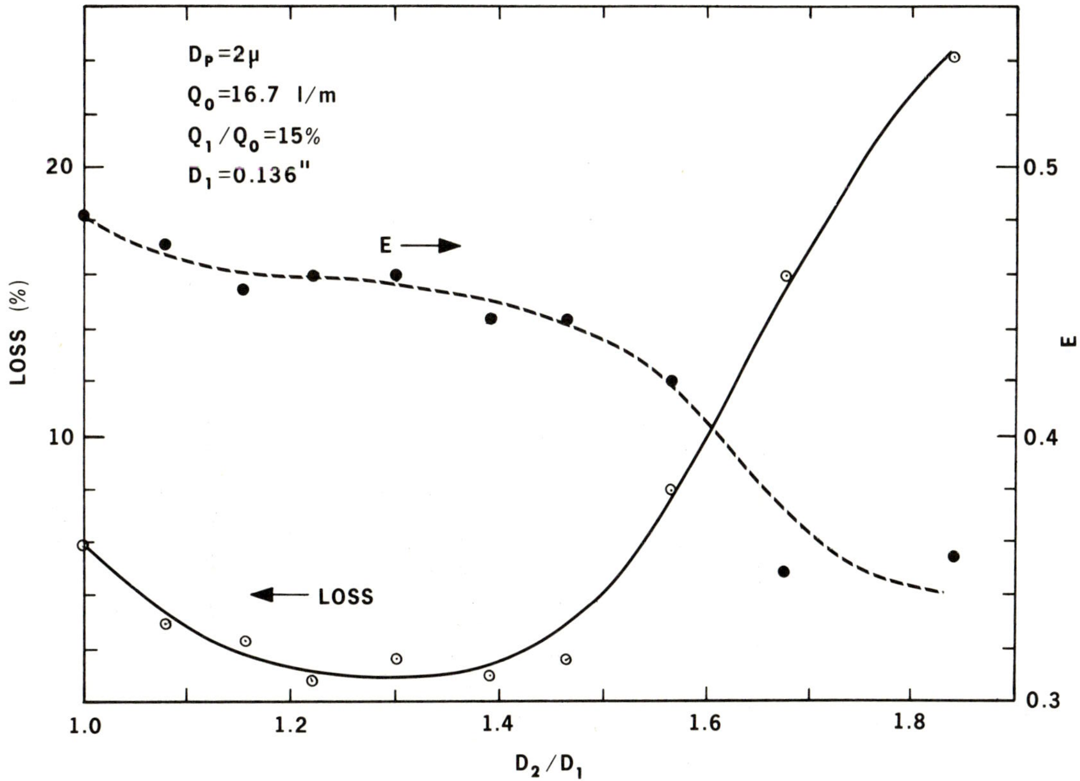

XBL 7411-8550

Fig. 12 Behavior of E and losses vs. D_2/D_1

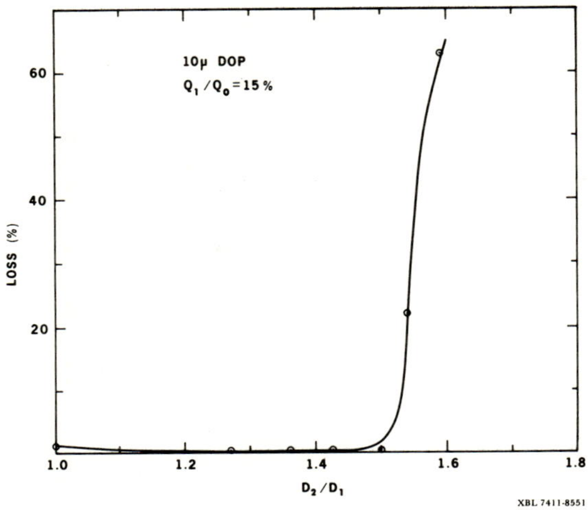

Fig. 13 Losses of 10 µm DOP particles as a function of D_2/D_1

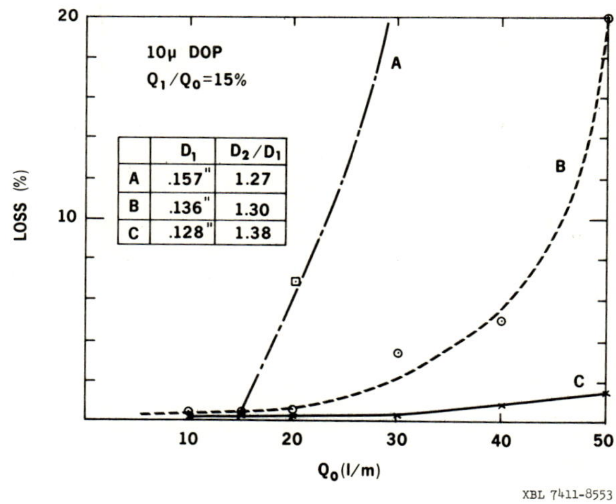

Fig. 14 Losses of 10 µm DOP particles as a function of Q_0

FIGURE NO.	DEPENDENT VARIABLE	Q_0 (ℓ/m)	Q_1/Q_0 (%)	D_1 (mm)	D_2/D_1	S (mm)	D_p (μm)	PARTICLE
8	E	16.7	15	VARIABLE	1.0	3.2	2	URANINE
9	E	16.7	VARIABLE	4.0	1.0	4.0	2	URANINE
10	LOSS	16.7	VARIABLE	4.0	1.2	3.5	10	DOP
11	E	16.7	15	4.0	1.0	VARIABLE	2	URANINE
12	E	16.7	15	3.5	VARIABLE	3.2	2	URANINE
12	LOSS	16.7	15	3.5	VARIABLE	3.2	2	URANINE
13	LOSS	16.7	15	4.0	VARIABLE	3.2	10	DOP
14	LOSS	VARIABLE	15	4.0 3.5 3.3	1.27 1.30 1.38	3.2	10	DOP

TABLE 1 Summary of conditions underwhich initial virtual impactor testings were conducted

Figure 8 shows the variation of E with D_1. This strong dependence on jet velocity is expected to be the dominant factor governing the cut point. Figure 9 shows the variation of E with the flow division Q_1/Q_0. The corrected values of E, after contamination of fine particles in the coarse particle stream has been subtracted, indicate the need for a significant amount of flow in the forward direction, particularly for the particles near the cut point. At a given Q_1/Q_0 there will always be the same fractional contamination of fine particles in the large particle stream but more than one stage of separation can be used to reduce this below the amount desired. The results in Fig. 10 were taken with a single-stage virtual impactor with three jets in parallel. The cross-contamination of 10 μm particles in the fine particle stream is taken as a measure of turbulent mixing. It is obviously desirable to use a value of Q_1/Q_2 greater than 15% for each stage of separation.

The dependence of E on the spacing S between the inlet jet and the receiving tube is shown in Fig. 11. It is seen that E is about constant for S/D_1 greater than 0.5.

Figure 12 shows both E and the losses as a function of D_2/D_1. Note that E has a flat region for $1.1 < (D_2/D_1) < 1.5$ and the impaction loss on the lip of the receiving tube has a pronounced minimum at $D_2/D_1 = 1.3$ and increased rapidly beyond 1.5. The same sharp increase in losses at $D_2/D_1 \simeq 1.5$ is also observed (Fig. 13) for 10 μm particles, suggesting that this extreme sensitivity is related to flow geometry rather than to particle size. This may explain the poor performance of virtual impactors with a rectangular slit geometry.

All these measurements were taken with the assumption that a total flow rate of 50 ℓ/min will be drawn through three inlet jets in parallel to achieve a cut point at 2 μm and the jet Reynolds' number will typically be in excess of 5000. For best performance Marple has shown that the optimum jet Reynolds' number is about 3000 for a conventional impactor. Figure 14 illustrates the effect of particle losses as Q_0 is increased. It appears that the wall losses are the result of the increase in stopping distance of the very large particles as they leave the diverging streamlines at the region of separation, although a small component due to turbulence losses has not been ruled out. It might be possible to overcome this impaction loss by using a more complex design of the receiving tube.

The considerations of jet Reynolds' number, flow symmetries, turbulent mixing, stability of the cut characteristic and flow geometry for minimum loss have led to the conclusion that sampling 50 ℓ/min with a 2 μm cut point with a tolerable fine particle contamination in the coarse particle stream requires two series stages of separation with three parallel jets for the inlet stage and a single jet for the second stage. The appropriate parameters will be $Q_1/Q_0 \geq 15\%$, $D_2/D_1 = 1.3$ and $S \approx 1$.

DESCRIPTION OF VIRTUAL IMPACTOR

The schematic of the complete virtual impactor design is shown in Fig. 15. It consists of the following main components:

PARTICLE SIZE FRACTIONATOR

Details of this component are illustrated in Fig. 16 with the key parts numbered. Air is drawn through three inlet jets (part 1) in parallel. Their protrusion into the first stage cavity is necessary to eliminate the "backwall" losses on part 2 due to the spatial oscillation of streamlines as found in the ERC design, as well as in some conventional impactors. Part 3 forms the first stage cavity. The three small holes in this part are symmetrically located about the central axis but are offset 60° azimuthally with respect to the coarse particle receiving tubes (parts 4) to minimize flow interference. These holes, in combination to the one in part 7, also govern the internal flow distribution. The Q_1/Q_0 for the first stage was adjusted to be 25% to minimize wall losses in the cavity. The tapered lips on the tubes have no significant effect on the cut point although they do tend to defocus the streamlines and reduce cavity losses.

The three coarse particle jets are then converged by a 15° cone (part 6) onto the second stage of separation after passing through the drift tube (part 5). Parts 8 are three positioning rods which forms an open cavity for the second stage jets. The ratio Q_1/Q_0 here is chosen to be 20%. Thus 2.5 ℓ/min of air will pass through filter A carrying all the coarse particles along with 5% of the fine particles. The fine particle stream of the second stage will merge with that from the first stage and be deposited on filter B. In analysis, a correction for the 5% contamination on filter A can be made based on the amount of the uncontaminated 95% of the fine particles on filter B. Table 2 summarizes the actual parameters used.

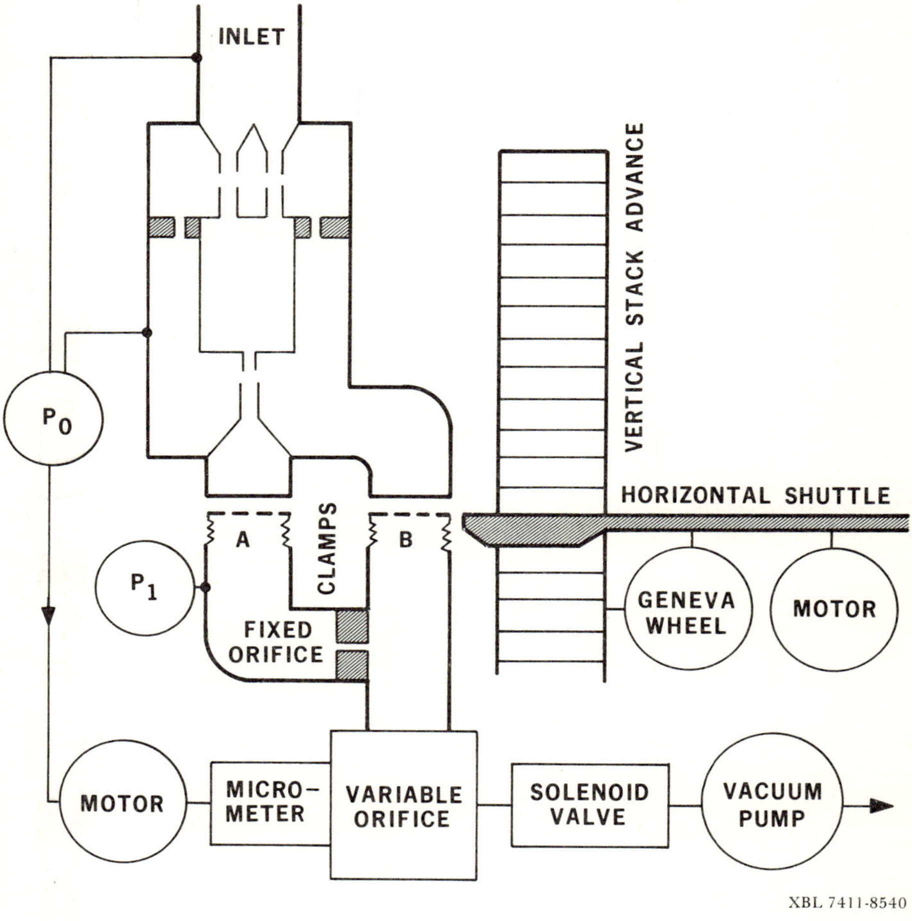

Fig. 15 Schematic of the automated dichotomous air sampler.

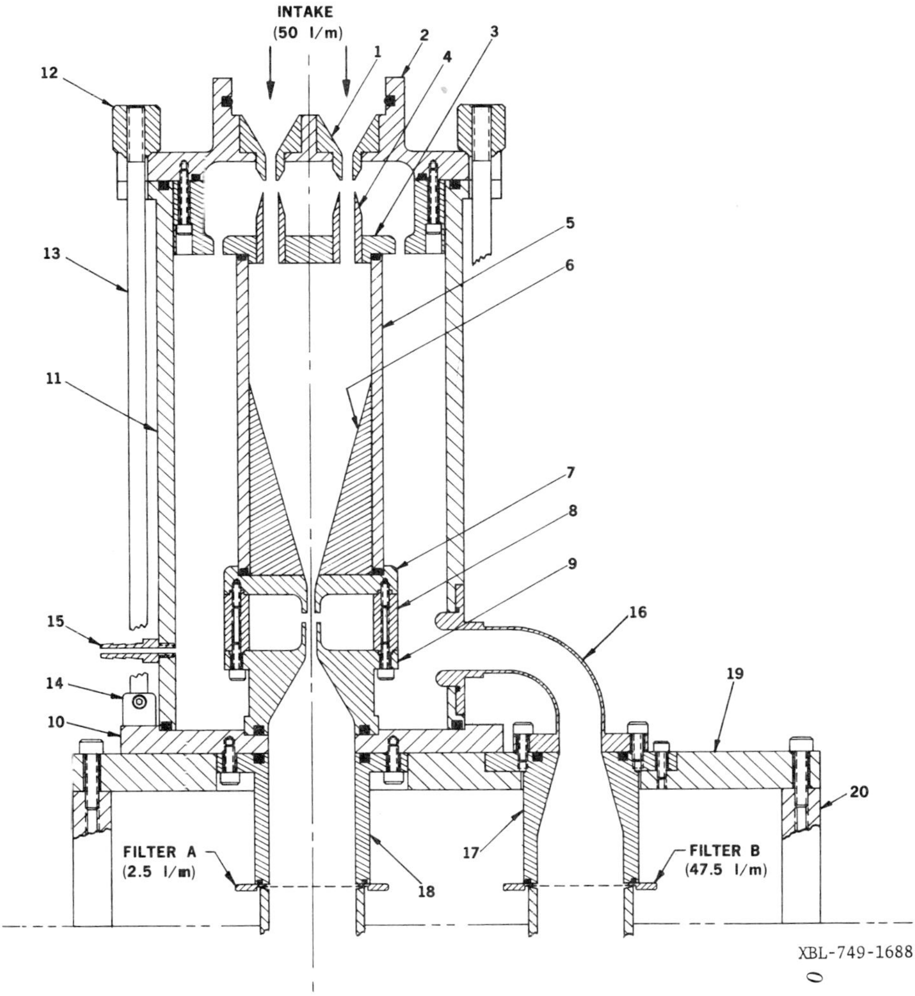

Fig. 16 Details of the virtual impactor.

PARAMETERS	Q_0 (ℓ/min)	Q_1/Q_0 (%)	D_1 (mm)	D_2 (mm)	S (mm)
First Stage Each of 3 Jets	16.7	25	3.86	5.05	3.81
Second Stage Single Jet	12.5	20	2.87	3.86	3.18

Table 2 Final choice of virtual impactor parameters

The overall construction utilizes all stainless steel parts (excepting part 6) for mechanical integrity and corrosion resistance, compression O-ring seals and tie rods (parts 13) with thumb nuts (part 12) for easy disassembly. Strategic corners are shaped to minimize losses. Figure 17 shows the actual components of a virtual impactor.

FLOW CONTROLLER

Flow regulation is essential for precise measurement of the air volume sampled and the maintenance of a fixed particle-size cut point. This is accomplished as shown in Fig. 15 by sensing (through part 15) the pressure differential P_0 between the inlet and the second stage of the impactor with a diaphram operated null switch (Dwyer Model 1640-5) which, in turn, causes the opening in a motor driven valve to be increased or decreased to maintain the preset null condition. The valve is simply a 5.1 mm diameter orifice pierced by a travelling micrometer shaft with a 2° taper. A fixed orifice limits the flow through filter A to 2.5 ℓ/min. The variable orifice and the null switch thus form a feedback loop to compensate any impedance change in filter B. The carbon-vane vacuum pump used (Gast Model 0522-103-G18D) has adequate pumping power to overcome an increase of about 70% in impedance from a typical initial value of 26.2 torr-cm^2/ℓ/min (1.2 μm cellulose membrane filter manufactured by Nuclepore Corporation).

Fig. 17 Disassembled components of the cirtual impactor.

AUTOMATED SAMPLING SYSTEM

Our objective has been to develop a fully automated sampling system for extended continuous operation. It is designed to cycle samples with the minimum amount of handling and preparation. A network of such samplers is to be controlled by a remote computer which also monitors the system status and possible failure modes. Key components include:

FILTERS

The filters used are 37 mm discs of cellulose membrane filters supplied and mounted in 5.1 cm x 5.1 cm plastic frames by the Nuclepore Corporation. Up to 36 of these slides are carried in a linear array in standard 35 mm projector cartridges (Argus Camera). Figure 18 shows such a pair of cartridges containing the digitally labeled filter holders.

SAMPLE CHANGER

The function of the slide changer is to extract a matched pair of filters from side by side slide trays corresponding to the A and B filter stacks, (only one stack is illustrated in Fig. 15). A horizontal shuttle manipulates the slides into their sampling positions where they are clamped in the output tubes of the virtual impactor. Upon the completion of the sampling interval, they are unclamped and withdrawn back into the slide trays. The over-travel of the shuttle actuates a "Geneva wheel" which advances the stack by one vertical increment to be ready for the next insertion. A single motor drives the shuttle which performs the function of transporting, clamping and unclamping slides together with advancing the trays with a single forward and return stroke.

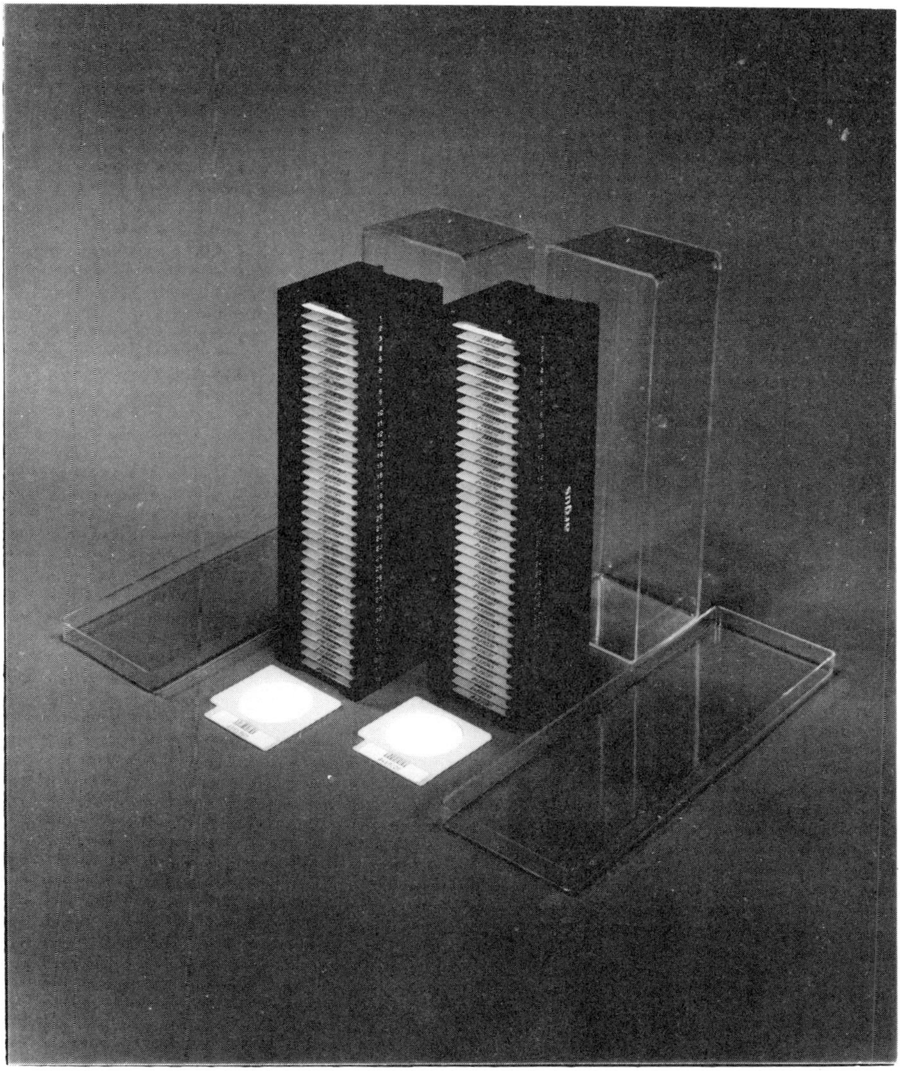

Fig. 18 Typical slides and slide trays used in automated samplers.

FLOW MONITOR

Several types of out-of-range conditions in the flow circuit are detected and indicated by the system. Excessive travel of the micrometer valve due to the presence of leaks or broken filters causes out-of-range switches to be activated. An auxiliary pressure sensing snap switch P_1 (Fig. 15) is used to detect improper clamping or a broken filter. Since the vacuum needed for the fixed limiting orifice results from the proper flow condition in the fine particle stream, P_1 actually monitors the conditions at filter B even before the micrometer valve reaches its limits.

ELECTRONIC CONTROLLER

The selection of sampling intervals, execution of the sequential steps, regulation of flow, detection of errors, monitoring and display of the system status and communication to an optional remote computer are performed via the control module shown in Fig. 19.

In order to maintain the synchronism of the samplers with the clock, ten seconds are allowed for a sample insertion or withdrawal cycle, which normally requires about seven seconds, to complete. Figure 20 illustrates the time sequence of a typical sampling period. While the vacuum pump is turned on continuously, actual sampling starts at the twenty second mark when a solenoid valve is opened. Another ten seconds are allowed for steady flow conditions to be established before the flow controller is enabled. The right hand column of the figure indicates the sequence in which error conditions are checked. The maximum times allowed to complete a sample transport and flow adjustment are 10 seconds and 12 minutes respectively.

To ensure synchronization in the event of short ac power failures (< 10 min) the elapsed time clock and logic control circuits are automatically switched to a rechargeable battery.

Figure 21 shows a complete sampler. It is contained in a portable soundproofed electronic rack with a plastic dust cover and a special inlet pipe designed to draw an isokinetic sample from a larger rooftop sampling port. Figure 22 is a closer view of a sampler with the dust cover removed. It shows the final modification of adding a 7.6 cm section to the height of the virtual impactor to reduce losses in the drift tube.

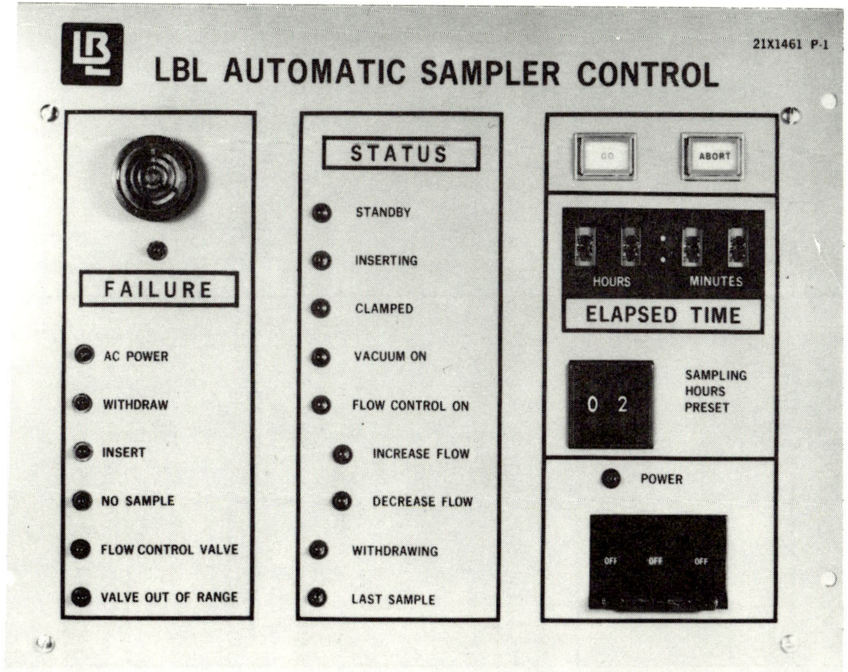

Fig. 19 Front panel of the electronic controller.

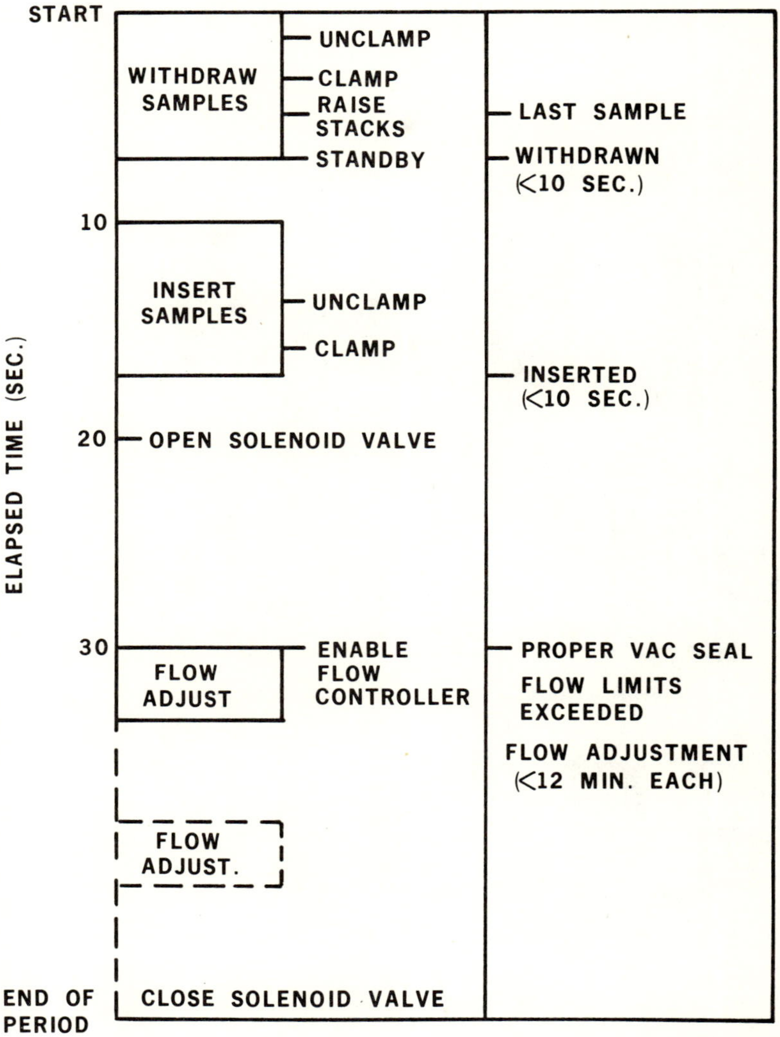

XBL 7411-8538

Fig. 20 The functional time sequence of a sampling period.

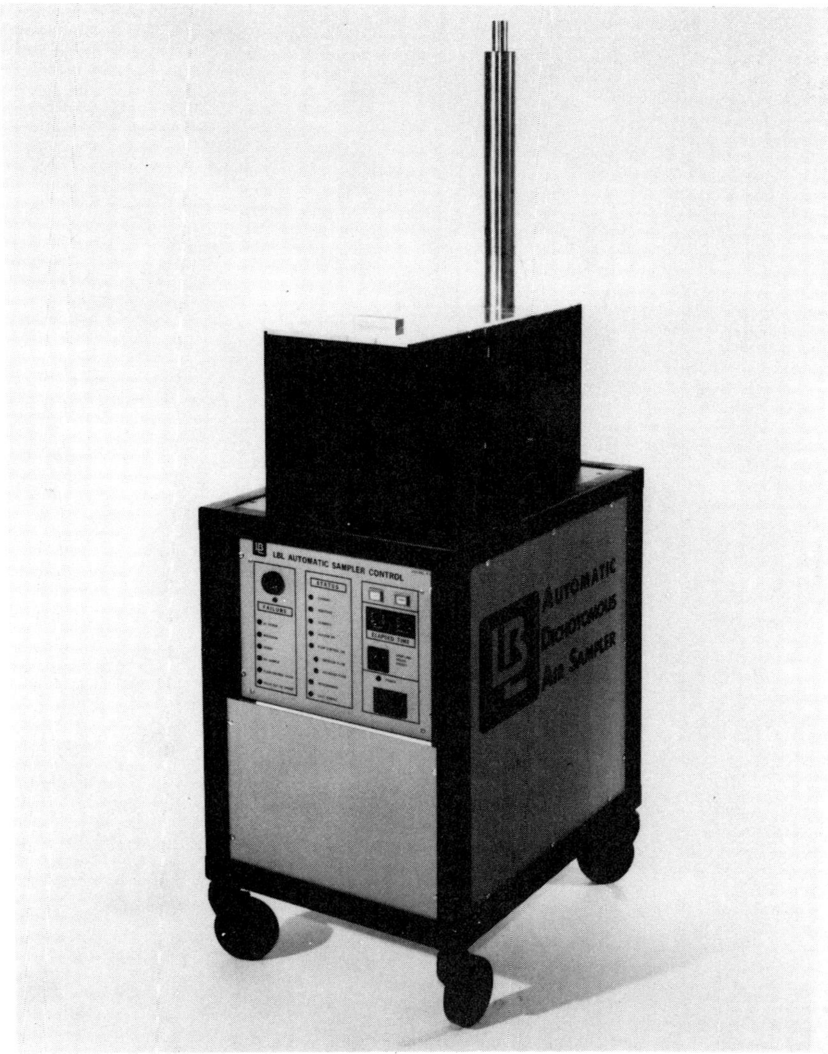

Fig. 21 Oblique view of a completed ADAS.

Fig. 22 A close view of the ADAS with dust cover removed.

RESULTS AND DISCUSSION

Thirteen sampling units of the type described have been built. A high degree of uniformity has been achieved particularly on the critical virtual impactor parameters. For example, the spread of the pressure differentials P_1 required to draw 50.1 ℓ/min of air is less than 1% among all the samplers. The difference in cut point between two different units was measured to be less than 0.1 μm.

SIZE SEPARATION AND LOSSES

The results of an evaluation of a typical sampler are summarized in Table 3 and 4 and the overall results are plotted in Fig. 23. Loss measurements are tabulated for specific regions. Region 1 losses include those washed from parts 1 and 2 of Fig. 16, region 2 from parts 3 and 4, region 3 from parts 5 and 6, and regions 4 and 5 are from parts 7 and 9 respectively. There are two obvious components of losses. A loss peak near the size cut point reflects the intrinsic tendency for particles of the cut point size to be intercepted by the physical surface which deflects the streamlines (in this case the inner rim of the receiving tubes). The observed fact that there is no deposition on the top of the tubes suggests that the virtual impaction surface is somewhat below the opening to the tubes. The coarse particles loss component is mostly gravitational settling on the cone (part 6) in the drift tube. In the specific design shown in Fig. 16, impaction losses were found for large particles on part 6. A later modification added 7.6 cm to the length of the drift tube to eliminate this component. The tabulated measurements reflect this modification.

The high sticking probability of the liquid DOP particles to the wall and to each other leads to the exaggerated loss condition. A pronounced reduction in losses is observed for solid uranine particles. The E vs. D_p plot shows a very sharp cut characteristic. Its long term stability (months) enables us to define the aerodynamic size of solid particles whose final density is less well known. The slight down shift of the solid particle loss peak compared with that of the liquid is consistent with the expectation that large solid particles are more likely to bounce.

PARTICLE SIZE (μm)	1.34	1.81	2.21	2.64	3.10	4.11	6.05	8.06	10.04
REGION 1	0	0	0.5	0.8	0	0	0	0.1	0.3
REGION 2	0	0.7	4.1	11.9	11.0	0	0	0	0
REGION 3	0	0	0	0	0.1	0.2	0.5	0.8	1.0
REGION 4	0	0	0.7	0	0	0	0	0	0
REGION 5	0.1	1.9	4.9	10.7	8.0	0.6	0.6	0	0
TOTAL LOSS	0.1	2.6	10.2	23.4	19.1	0.8	0.5	0.9	1.3
FILTER A	6.5	11.9	23.6	6 0.1	79.2	99.1	99.4	99.1	98.7
FILTER B	93.4	85.5	66.4	16.7	1.7	0	0	0	0
E = A/(A + B)	6.5	12.2	26.2	78.3	97.9	100	100	100	100

TABLE 3 Summary of evaluation with liquid DOP particles (Percent deposition of particles on sampler and filters)

PARTICLE SIZE (μm)	1.00	1.61	2.18	2.94	3.50
REGION 1	0	0	0	0	0
REGION 2	0	0	0.1	0.6	0
REGION 3	0	0	0	0	0.1
REGION 4	0	0	0	0	0
REGION 5	0	0	1.3	1.6	0.2
TOTAL LOSS	0	0	2.4	2.2	0.3
FILTER A	4.7	8.7	24.4	93.0	99.6
FILTER B	95.3	91.3	74.2	4.8	0.1

TABLE 4 Summary of evaluation with solid uranine particles (percent deposition of particles on sampler and filters)

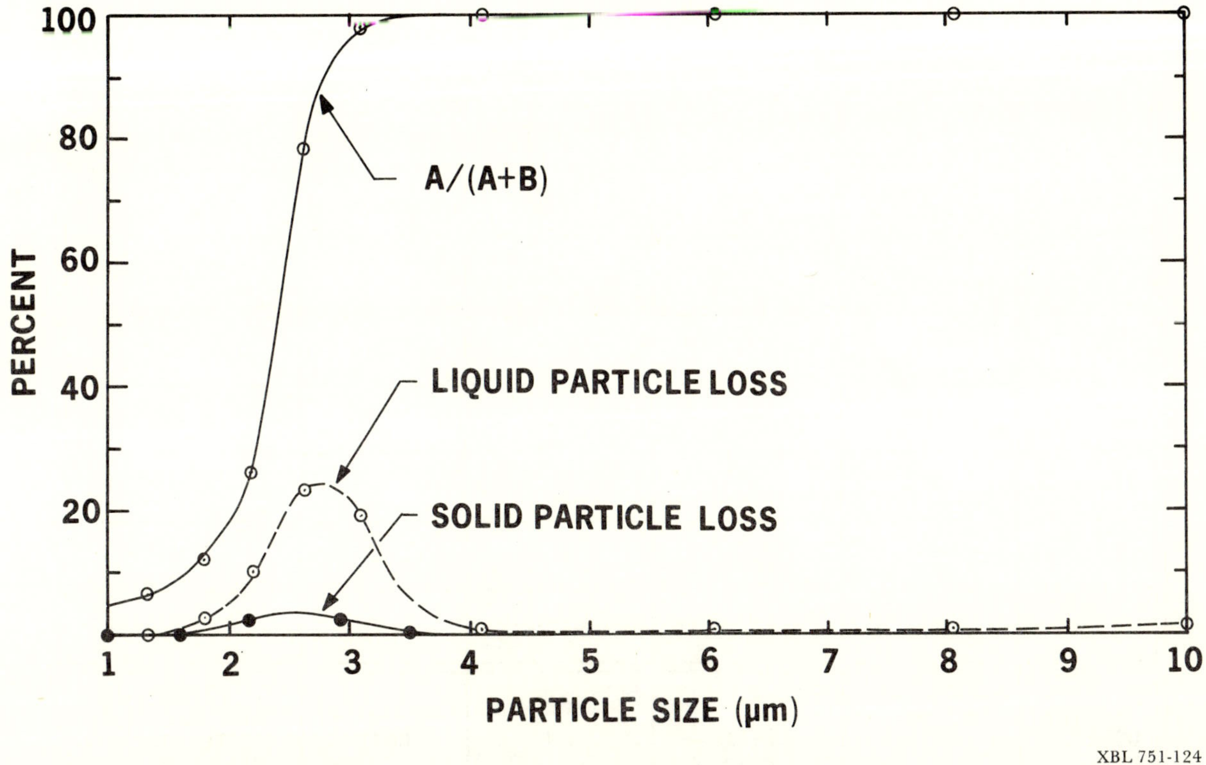

Fig. 23 Particle size segregation characteristics and losses as a function of particle size.

FLOW CONTROL

The null switch for the pressure sensor is set to reduce hysteresis errors to less than 0.5%. The repeatability in flow rate is typically better than 0.2%. If constant pressure drop is maintained across a set of fixed orifices, the mass flow, to a first approximation, is inversely proportional to the square root of the absolute temperature. Aerosol concentrations are conventionally expressed in units of mass per unit volume. Thus, the volume flow needed is directly proportional to the square root of the absolute temperature.

As the in-take air is cooled from 20°C to -35°C, the automatic micrometer valve shows little adjustment as long as clean filters are used in the device. When heavily-loaded filters are in place, a slight closing of the micrometer vavle is observed as the temperature is lowered, suggesting the larger viscosity contribution to the pressure drop across the loaded filter.

STABILITY AND RELIABILITY

The flow calibration drift over a three month period was measured to be under 0.5%. The almost continuous sampling over the same period produced no detectable change in the 2.5 ℓ/min limiting orifice behind filter A.

The samplers have been extensively tested under laboratory conditions by continuously recycling filters over an extended period. An equivalent of 15,000 samples have been run as part of this study. After eliminating obvious problems in the initial debugging period, the average failure probability has been reduced to less than 0.1% per sample.

LIMITATIONS

A real instrument often falls short of ideal performance due to compromises made to satisfy practical boundary conditions. The physical size of the apparatus sets an upper limit on the largest particle that may be efficiently sampled. For example, the 10 μm particle loss is less than 1.5% for liquid particles, but there is a sharp rise to 70% at 20 μm caused by impaction on sidewalls and by gravitational settling.

Some of the limitations arise from the properties of the filter medium selected. An ideal filter should exhibit high filtration efficiency, homogeneity, mass loading capability, and mechanical strength, and should exhibit low trace impurity content, flow impedance, mass thickness and moisture uptake, as required by beta gauge and X-ray fluorescence measurements. Some of these requirements are obviously mutually exclusive. The 1.2 μm cellulose membrane filter used is considered a good compromise. The power and weight considerations on the vacuum pump lead to the choice of a pump that will maintain the desired sampling rate of 50 ℓ/min for up to a 70% increase in filter impedance over its nominal value of 26.2 torr-cm^2/ℓ/min. The corresponding loading on the fine particle filter, which bears the main flow, is about 200 μg/cm^2. This leads to the limitation that in heavily polluted air with 100 μg/m^3 of fine particles, the maximum sampling time will be limited to approximately 4.5 hours.

As presently packaged in the sound insulated chamber, the pump requires cooling air at a temperature below 35°C to avoid accelerated wear.

CONCLUSION

Of the thirteen samplers, ten are installed and operating in the St. Louis RAMS network. It has been demonstrated that the virtual impactor, with its distinct advantages over its conventional counterpart, has fulfilled the need for an instrument to collect aerosols in two strategic size ranges. The adaptability of the virtual impactor to automation and its compatibility to other analytical instruments suggests that it may be suited for wider use. Refinements on the range of acceptable particle size, filter impedance and filter mass loading should further improve its potential for large scale deployment.

ACKNOWLEDGEMENTS

The authors wish to express their gratitudes to T. Dzubay of EPA and C. Peterson of ERC for making the ERC virtual impactor available for evaluation. We acknowledge the contributions of the following LBL staff: R. Adachi, O. Arrhenius, C. Cyder, B. Jarrett, N. Madden, J. Meng, H. Riebe, A. Roberts, W. Searles, D. Vanacek and S. Wright. The fabrication of thirteen samplers would not have been possible without the great support from the personnels of the Electronics and Mechanical Shops.

We also appreciate the cooperation of H. Schneider and R. Leiser of the Nuclepore Corporation in developing cellulose membrane filters of acceptable quality and mounted appropriately. We are indebted to B. Liu of the University of Minnesota for evaluating filter efficiency, and R. Giauque of LBL for analyzing filter impurity content.

REFERENCES

1. Goulding, F. S. and J. M. Jaklevic, Photon-Excited Energy Dispersive X-ray Fluorescence Analysis for Trace Elements. Annual Review of Nuclear Science, 23:45-74, 1973.

2. Goulding, F. S. and J. M. Jaklevic, X-ray Fluorescence Spectrometer for Airborne Particulate Monitoring. Environmental Protection Agency Publication No. EPA-R2-73-182, April 1973.

3. Whitby, K. T., R. B. Husar and B. Y. H. Liu, The Aerosol Size Distribution of Los Agneles Smog. In: Aerosols and Atmospheric Chemistry, Hidy, G. M. (ed.). Academic Press, 1972. p. 237-264.

4. Whitby, K. T., On The Multimodal Nature of Atmospheric Aerosol Size Distribution. Particle Technology Lab Publication No. 218. University of Minnesota, 1973.

5. Air Quality Criteria for Particulate Matter, Chapter 9. National Air Pollution Control Administration Publication No. AP-49, 1969.

6. Conner, W. D., An Inertial-Type Particle Separator for Collecting Large Samples. Journal of the Air Pollution Control Association. Vol. 16, No. 1:35-38, January 1966.

7. Hounam, R. F. and R. J. Sherwood, The Cascade Centripeter: A Device for Determining the Concentration and Size Distribution of Aerosols. American Industrial Hygiene Association Journal. Vol. 26, Nc. 2:122-131, March-April 1965.

8. Loo, B. W. and J. M. Jaklevic, An Evaluation of the ERC Virtual Impactor. Lawrence Berkeley Laboratory Report No. LBL-2468. January 1974.

9. Marple, V. A., A Fundamental Study of Inertial Impactors. Doctoral Thesis, Department of Mechanical Engineering, University of Minnesota. December 1970.

DESIGN, PERFORMANCE AND APPLICATIONS OF SPIRAL DUCT AEROSOL CENTRIFUGES

Werner Stöber

Fraunhofer-Gesellschaft zur Förderung der angewandten Forschung
Institut für Aerobiologie
5948 Schmallenberg, Germany

ABSTRACT

This paper gives a review of the design, the performance and the applications of spiral duct aerosol centrifuges as they are increasingly used in aerosol particle size spectrometry and related aerosol research. Background, advantages and limitations of this type of instrument and its modifications will be discussed. Precision measurements of dynamic shape factors on certain types of nonspherical particles as well as applications to high-resolution size distribution analysis will be reported. Comparative studies of different authors on the possibility of using the spiral duct aerosol centrifuge as an absolute instrument for aerodynamic size distribution measurements are compiled from literature. Data on aerosol particle densities obtained with spiral duct centrifuges are reported and problems of measuring cigarette smoke and dense aerosols are discussed. An adaptations of a short spiral duct centrifuge for applications to ambient particulate air pollution is described and a recent feasibility study of developing the instrument into an aerosol mass distribution monitor by way of using piezo-electric quartz crystals as size-selective mass sensors is reported.

DESIGN, PERFORMANCE AND APPLICATIONS OF SPIRAL DUCT AEROSOL CENTRIFUGES

Werner Stöber

Fraunhofer-Gesellschaft zur Förderung der angewandten Forschung
Institut für Aerobiologie
5948 Schmallenberg, Germany

INTRODUCTION

Experimental research projects on airborne particulate matter generally necessitate the analysis of aerosol particle size distributions at some point along the course of the investigations. To serve this purpose, various analytical methods, procedures and instruments have been devised and used in the past. Most of these techniques involve a straightforward precipitation of samples of the aerosol and a subsequent statistical analysis by microscopic evaluation. During the last 30 years, however, beginning with the development of cascade impaction by May [33] in 1945, a size-related dynamic separation of particles prior to their precipitation became of increasing interest to aerosol scientists. An obvious reason for this was, of course, the simplification of the distribution analysis by the preceding experimental size fractionation, but there was also the additional advantage that the forces effecting the dynamic size separation were directly related to the stopping distance and the aerodynamic diameter of the particles, both of which parameters are of fundamental importance in aerosol dynamics and in health hazard evaluations of inhaled industrial aerosols. Thus, dynamic size separation for distribution analyses was very desirable.

The first real aerosol particle size spectrometer actually providing a continuous size spectrum in terms of aerodynamic diameters was built in 1950 by Sawyer and Walton[43]. Their centrifugal device, called a conifuge, deposited the particles according to their aerodynamic diameter in a size range between 0.5 and 30 µm on the outer wall of a rotating conical annular duct. Figure 1 shows a schematic diagram. The size separation was achieved by a laminar stream of clean air enveloping the aerosol entrance at the apex of the cone and flowing down toward the base. The aerosol particles, when entrained and leaving the apex, were subject to the centrifugal forces and, thus, traversed the clean air layer in a radial direction at a velocity determined by their aerodynamic size. Due to this effect, a continuous size separation was obtained and particles of equal aerodynamic diameter were collected in concentric rings on the outer wall of the duct. At 3000 rpm, the instrument would permit an aerosol sampling rate of 25 cm^3 min^{-1}.

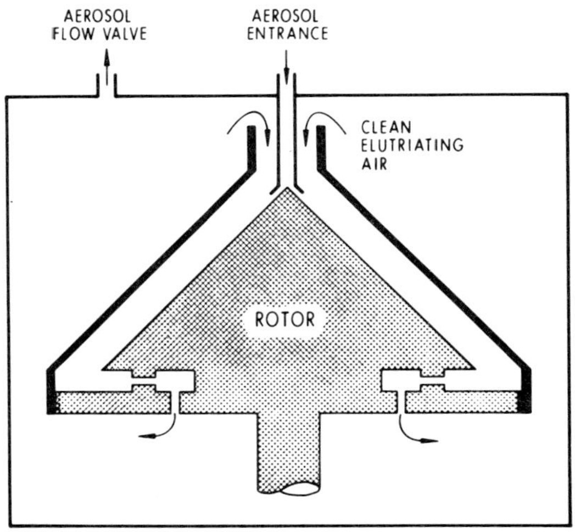

Figure 1: Schematic Diagram of the Conifuge Rotor and Housing

In spite of the promising aspects of this design concept, the conifuge did not receive much recognition for more than 15 years, although Keith and Derrick[17] extended the lower size limit for deposited particles to 0.05 μm and increased the sampling rate to 300 cm^3 min^{-1}. The lack of specific interest in the conifuge is surprising in view of the fact that gravitational versions of aerosol size spectrometers as introduced later by Timbrell[62] and Boose[4] could not match the performance data of the conifuge. The smallest sizes deposited in the gravitational devices were not below 1.5 μm and the sampling rate was only 1 cm^3 min^{-1}.

In reviewing the situation of centrifugal aerosol size spectrometry and after assessing the limitations of a semi-dispersive cone-shaped helical-duct aerosol centrifuge (Goetz, Stevenson and Preining[10]), it was suggested in 1965 (Stöber and Zessack[53]) that the performance of the conifuge type size spectrometers could be improved by employing ring slit aerosol entrances in modified designs featuring slender cones or cylindrical annular ducts. It was anticipated that ring slit aerosol inlets would permit increased sampling rates as desired for many practical purposes. The cylindrical design would have the additional benefit of facilitating an exact theoretical performance evaluation. An actual small-size instrument of the latter kind was subsequently built by Berner and Reichelt[1] and showed that the experimental deposit patterns did, in fact, follow the theoretical predictions (Stöber, Berner and Blaschke[54]).

In addition to the suggested conifuge modifications, the review also conceived an entirely different centrifuge design by combining elements of the semi-dispersive Kast centrifuge[16] with the winnowing air flow arrangements of gravitational size spectrometers: On a disk-shaped rotor, a duct of rectangular cross section was to be wound into a spiral in such a way that the aerosol inlet could be located at the axis of rotation.

During the following years, a variety of ring slit centrifuges of the conifuge concept as well as the first spiral duct centrifuge were built and tested (Stöber[45]; Berner and Reichelt[2]; Hochrainer and Brown[13]; Stöber and Flachsbart[48,49]). A comparison of the performance tests of these devices indicated that from almost all practical points of view the concept of the spinning spiral duct was superior to the other designs.

The most important drawback of the ring slit instruments was the difficulty of introducing the aerosol into the rotating duct without risking substantial particle losses or significant flow disturbances. In addition, a desirable increase of the aerosol sampling rate of a conifuge required a design of big instrument dimensions. Maximum rates reported in such a case (Stöber and Flachsbart[48]) were 1.2 liters min^{-1} with a range of deposited sizes about one order of magnitude (0.3 to 3.0 μm). Miniature designs suffered from greatly reduced sampling rates down to 12 cm^3 min^{-1} (Hochrainer and Brown[13]) and the gain in size resolution was practically offset by increases in particle losses. In contrast, the original spiral duct centrifuge (Stöber and Flachsbart[49]) permitted sampling rates of several liters per minute. Furthermore, the particles were deposited over a size range of almost two orders of magnitude, while losses could be kept very low with suitable aerosol inlets. In reducing the sampling rate to some 400 cm^3 min^{-1}, an excellent size resolution not surpassed by any other dynamic spectrometer design could be obtained.

LONG-SPIRAL-DUCT DESIGNS

The first spiral duct centrifuge was built at the University of Rochester (Stöber and Flachsbart[49]) with a total duct length of about 180 cm. Although it was not a perfect instrument from the engineering point of view and comprised certain elements of overdesign, the new device could immediately be applied to the research problems it was intended for.

Figure 2 shows a photograph of the instrument with the rotor lid removed. The disk-shaped rotor has a diameter of 26.2 cm. The essential part of it, the spiral duct, is of rectangular cross section and begins off-center. Then, with the inner wall touching the axis of rotation, the duct leads in a narrow semicircle from the center toward and parallel to

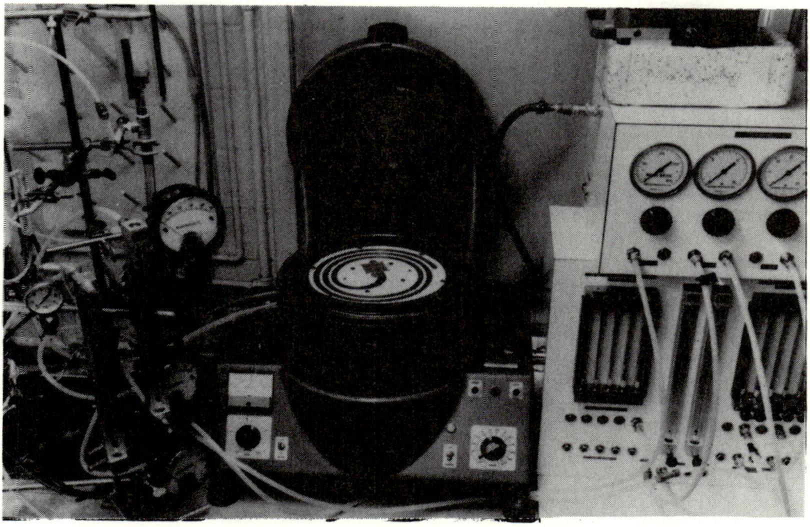

Figure 2: Original Spiral Duct Centrifuge (Rotor Housing Opened and Rotor Lid Removed)

the periphery of the rotor, from where it continues in wide curvatures for two and a half more turns. Figure 3 shows the original pattern which is composed of six semicircles. A later design (Moss, Ettinger and Coulter[35]) utilized an Archimedean spiral instead of the last five semicircles. This modification may have advantages in mathematical evaluations and descriptions but is immaterial with regard to the experimental performance of the instrument.

The depth of the spiral duct from the upper edge of the rotor was 3.30 cm in the original design, but some subsequent models were built with ducts cut 4.30 cm, 4.50 cm and 5.08 cm deep. The significance of these changes will be discussed later. The width of the spiral duct narrows down from 1.73 cm at the center of the rotor to 1.00 cm at the end of the first semicircle. The rest of the duct is of constant width. Along the outer wall, the spiral section of the duct has a small groove cut into the outer bottom edge, where a chromium-plated brass foil of 0.015 cm thickness can be inserted to cover the entire outer wall of the duct. The total foil length is 179.5 cm.

In operation, the closed rotor spins at a standard speed of 3000 rpm in a clockwise direction when seen from above. Clean air is blown into the

Figure 3: Original Spiral Duct Rotor (View of the Top of the Rotor Disk: Rotor Lid Removed, Laminator Block and Aerosol Inlet Section Inserted)

duct at the off-center inlet and passes first through an inserted laminator which subdivides the flow by five thin parallel foils extending in the coaxial direction. In this way, the clean air is quickly stabilized and emerges as a laminar flow from the down stream end of the inserted section. A new simplified laminator will be described later in this paper. The air then approaches an exchangable aerosol inlet at the center of the rotor, where the aerosol enters coaxially through a non-rotating center bore of 0.48 cm. The aerosol is released into the duct as a thin layer parallel to the inner wall. Subsequently, it is entrained into the laminar air flow.

Different aerosol inlets may be used for different purposes. For high size resolution, where reduced flow rates must be employed, a narrow slit design adjacent to the inner wall is very useful. For minimum particle losses, an open cut-away design of the rotating section of the

Figure 4: "Cut-away" Type of Aerosol Inlet Section for Minimum Aerosol Losses at Entrance

center bore extending into the duct is the best solution. Such an inlet section is shown in Figure 4. A systematic investigation of the influence of different inlet arrangements was made by Moss, Ettinger and Coulter[35].

The size separation in the spiral duct is a simple process. On leaving the center of rotation, the aerosol particles are subjected to centrifugal forces and start moving in a radial direction across the air stream. Their trajectories depend upon the operating conditions of the centrifuge and the aerodynamic size of the particles. Thus, while the air is drawn down the spiral duct toward the outlet, the particles are deposited, according to their size, in different locations on the chromium-plated foil or some other inserted collecting surface along the outer wall. The deposit represents a continuous size spectrum in terms of decreasing aerodynamic diameters beginning near the aerosol inlet and extending to the end of the spiral. To facilitate the analysis, the collecting foil can easily be removed from the instrument.

The original arrangement for the flow controls of the spiral centrifuge was essentially an open system relying on a source of compressed clean air and a suction line. In principle, the suction line valve controlled the total flow through the spiral duct and the clean air pressure had to be adjusted in such a way that only the desired small fraction of the total flow was drawn in through the aerosol inlet. This arrangement requires extremely stable pressures and subpressures but it is not very sensitive to small leakages which occur particularly when the sealed bearings of the shaft housing are worn.

In more recent spiral centrifuge designs, modern V-rings replaced or tightened the sealed bearings (e.g. Oeseburg and Roos[37]; Stöber, Franzes and Steinhanses[57]) so that a simple closed air circulation system could be devised where the total flow is set by a valve while the controlled bleeding from the pressure line determines the aerosol sampling rate.

Due to the complexity of the geometry and the physical conditions in the spinning spiral duct, no quantitative theoretical model for the deposit pattern was devised. Instead, empirical calibrations were made for different operating conditions. Typical results for the original design

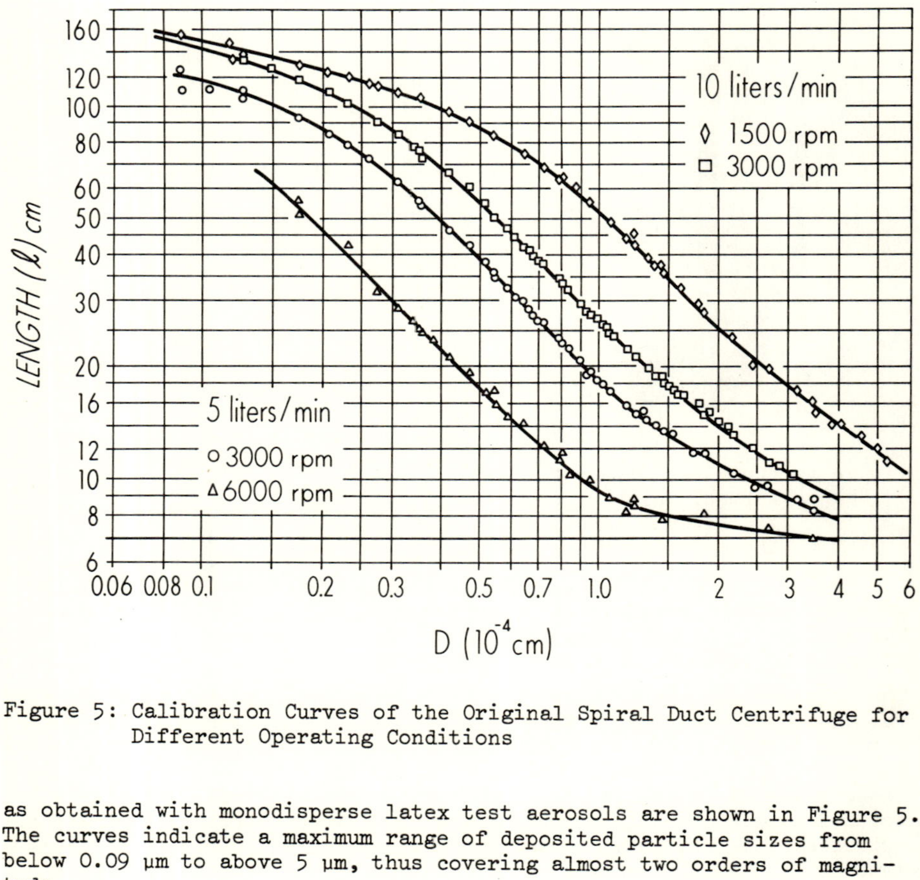

Figure 5: Calibration Curves of the Original Spiral Duct Centrifuge for Different Operating Conditions

as obtained with monodisperse latex test aerosols are shown in Figure 5. The curves indicate a maximum range of deposited particle sizes from below 0.09 µm to above 5 µm, thus covering almost two orders of magnitude.

The graph also reveals another significant feature of the long-spiral-duct centrifuge: For small particle sizes around 0.1 µm, the deposit location of the particles is no longer strongly dependent upon the centrifugal forces acting on the particles. The upper two curves of Figure 5 show this quite clearly. They indicate that near the end of the spiral duct the motion of the particles toward the collecting surface is predominantly controlled by factors other than the centrifugal forces. There is, indeed, experimental evidence (Stöber and Flachsbart[49]) that the deposition of particles at long distances down the duct is primarily due to an entrainment and transport of the particles by a slow secondary

Figure 6: Secondary Double Vortex Flow as Imaged by an Aerosol Deposit on a Filter in a Curved Duct Centrifuge (Tillery[61])

double vortex flow in the cross sectional plane of the duct. In a spinning concentrical duct, Tillery[61] succeeded to obtain a visible track of the double vortex flow by collecting a fluorescein aerosol on a filter placed into the cross section near the end of the duct. Figure 6 shows a photograph of the filter deposit. This double vortex flow can be expected to exist because of the curvature of the duct (Dean[7]) and, on a spinning rotor, even more so because of the Coriolis forces. These forces are created by the motion of the air relative to the spinning rotor. Under the standard direction of rotation of the spiral duct centrifuge they act perpendicularly toward the outer wall of the duct and, except for a slight reduction at the center of the rotor where the duct is wider and the flow velocity reduced, they remain at constant strength throughout the entire duct. A different assessment made earlier (Stöber and Flachsbart[49]) was incorrect because it accounted only for the circular component of the Coriolis force.

Theoretical attempts of computing the secondary flow by assuming steady-state conditions (Stöber, Hederer and Horvath[58]) were made according to a solution from boundary layer theory (Ludwieg[30]) and by deriving a numerical solution of the simplified Navier-Stokes equations for creeping flow. In both cases, however, the results were quantitatively not in keeping with experimental evidence. Instead, the Ludwieg approach confined the secondary flow to the boundary layer while the creeping flow solution gave flow velocities for the double vortex which were higher than compatible with the undisputedly proper function of the spiral duct centrifuge as a size spectrometer. This suggests strongly that the actual secondary flow situation in the duct is a case of developing double vortex flow. Figure 7 shows the streamline pattern of a fully developed double vortex flow in the cross sectional plane of the duct as calculated under creeping flow conditions.

Apparently, a full development of the secondary double vortex flow will be detrimental to the instrument performance and has to be avoided. As known from experimental and theoretical evidence, the vortices can be confined to areas adjacent to the small sides of the cross section if the duct approaches the shape of a slot. In other words, a high aspect ratio of height to width of the duct is desirable. Thus, the development of the double vortex will become less influential by either increasing the depth or reducing the width of the duct. The former was done in the modified designs mentioned earlier in this paper. Figure 8 shows a comparison of two calibration curves obtained under comparable operating

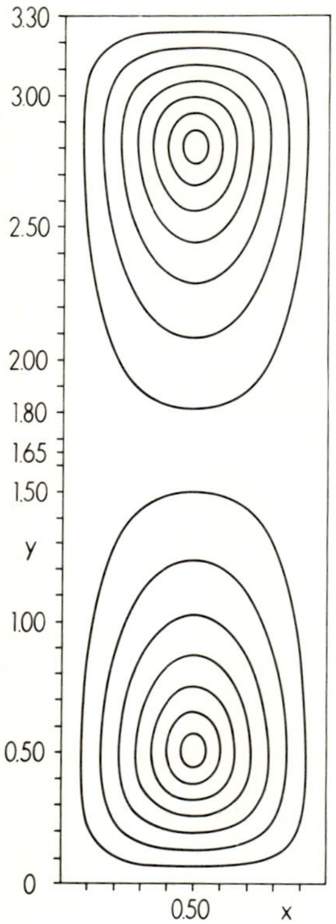

Figure 7 :
Theoretical Pattern of a Fully Developed Secondary Double Vortex Flow under Creeping Flow Conditions

conditions for the original design and for the 5.08 cm deep duct modification (Moss, Ettinger and Coulter[35]), respectively. Evidently, the development of the double vortex in the latter design (aspect ratio 5.1 : 1) is less pronounced so that the deposition of small particles around 0.1 μm diameter requires greater distances down the duct than in the original instrument. The improvement is gradual only. Since technical reasons practically inhibit substantial further deepening of the duct, a reduction of the width would be the method of choice for more effective vortex suppression.

In all performance tests and applications of the original spiral duct centrifuge at ambient temperature, it became apparent that the thermostat arrangement for the rotor housing was actually not needed and the housing itself was a source of additional air friction and heat generation (Moss, Ettinger and Coulter[35]). Thus, in subsequent models intended for work at ambient temperature, the rotor housing was omitted and cooling arrangements were confined to the shaft bearings (Stöber, Franzes and Steinhanses[57]). Without the rotor housing, a modified design of the non-rotating aerosol inlet tube as introduced by Moss, Ettinger and Coulter[35] became necessary. Figure 9 is a schematic drawing of a more recent design.

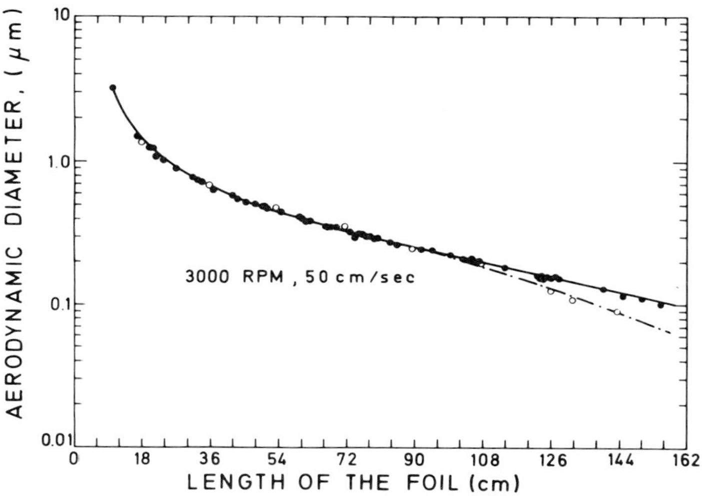

Figure 8: Influence of the Different Depth of the Spiral Duct on the Calibration Curves under Comparable Operating Conditions (Open Circles: Duct Depth 3.3 cm; Closed Circles: Duct Depth 5.08 cm)

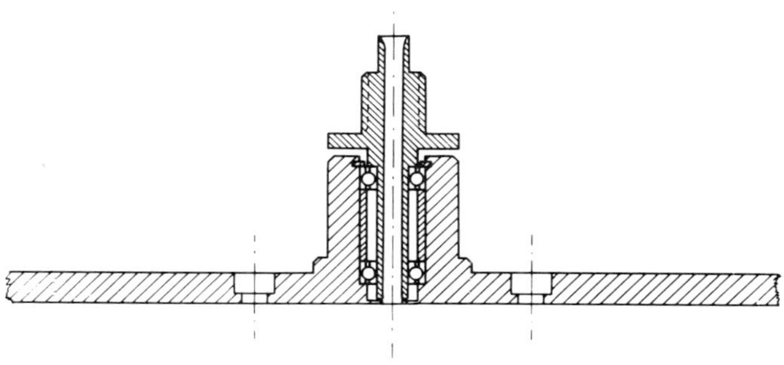

Figure 9: Design of a Non-Rotating Aerosol Intake Tube on the Rotor Lid of a Spiral Duct Centrifuge without Rotor Housing

PERFORMANCE AND APPLICATIONS OF LONG-SPIRAL-DUCT CENTRIFUGES

In the paper describing the first spiral duct centrifuge (Stöber and Flachsbart[49]), the instrument was applied to the determination of relative aerodynamic diameters of clusters and chains of latex spheres of uniform size. By exploiting the unprecedentedly high size resolution of the instrument at low sampling rates around 1% of the total flow, it was possible to discriminate clusters of up to 23 primary latex spheres. Thus, their aerodynamic size could be measured directly by the location of their discrete deposits on the collection foil as shown in Figure 10. Electron microscopic work was merely required to confirm the identity of the clusters making up the particular deposits. Other electron micrographs further revealed that chain aggregates of more than three spheres, which were less frequent and not found as discrete deposits, were interspersed among the clusters and could be identified in specific locations by electron microscopic screening. Figure 11 shows a deposit of quintuplet chains interspersed with other aggregates of various forms.

Figure 10: Photograph of Deposits of an Aerosol of Uniform Latex Spheres of 0.71 μm Diameter and their Aggregates on the Foil Strip of the Original Spiral Duct Centrifuge as Obtained under high Size Resolution Conditions

Tables 1 and 2 give a summary of the data obtained (Stöber[46]). The second columns show the relative aerodynamic diameter f_n, which is the ratio between the aerodynamic diameters of an aggregate and its n uniform primary spheres. The third columns give the parameter $\overline{\kappa}$, defined as the square of the ratio between the equivalent volume diameter and the Stokes diameter of the aggregate, thus being equivalent to the dynamic shape factor κ except for the influence of the slip correction.

The most interesting experimental result for the aggregate particles was the fact that f_n and $\overline{\kappa}$ were practically independent of the absolute size of the primary spheres. The mean error of the mean values of the measurements with primary latex spheres of sizes between 3.5 and 0.13 μm ranged from 0.2 to 0.8%. Systematic changes due to the decreasing absolute sizes must have remained within these limits.

Table 1

Dynamic Data of Cluster Aggregates

($0.13\ \mu m \leq D_1 \leq 3.5\ \mu m$)

n	f_n	$\bar{\kappa}$	κ, D_1 1.0 μm	κ, D_1 0.1 μm
2	1.189	1.123	1.115	1.080
3	1.343	1.153	1.144	1.102
4	1.471	1.165	1.156	1.113
5	1.568	1.189	1.180	1.130
6	1.676	1.175	1.167	1.123
7	1.748	1.198	1.189	1.140
8	1.812	1.218	1.209	1.155
9	1.887	1.215	1.206	1.155
10	1.936	1.238	1.228	1.172
11	1.996	1.241	1.231	1.175
12	2.043	1.256	1.246	1.187
13	2.107	1.246	1.237	1.181
14	2.159	1.246	1.237	1.182
15	2.211	1.244	1.235	1.182
16	2.254	1.250	1.235	1.187
17	2.297	1.253	1.244	1.190
18	2.328	1.267	1.258	1.201
19	2.370	1.267	1.258	1.201
20	2.428	1.250	1.242	1.190
21	2.487	1.230	1.223	1.176
22	2.537	1.220	1.213	1.169
23	2.577	1.217	1.210	1.167

Table 2

Dynamic Data of Chain Aggregates

($D_1 > 0.3\ \mu m$)

n	f_n	$\bar{\kappa}$	κ, D_1 1.0 μm	κ, D_1 0.1 μm
2	1.189	1.123	1.115	1.080
3	1.280	1.270	1.254	1.177
4	1.380	1.323	1.305	1.214
5	1.420	1.450	1.425	1.297
6	1.450	1.570	1.538	1.376
7	1.480	1.671	1.634	1.442
8	1.520	1.731	1.691	1.483

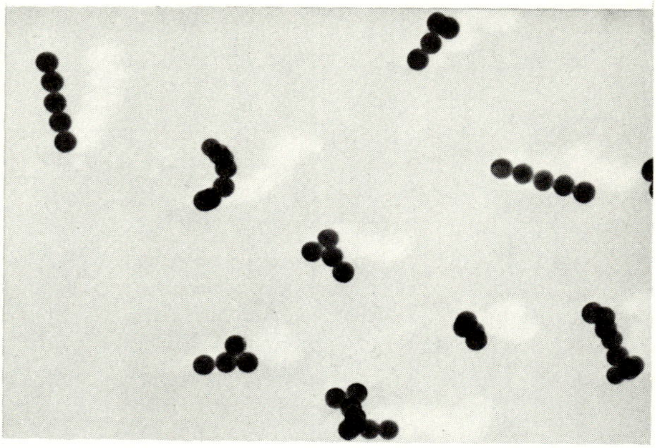

Figure 11: Electron Micrograph of a Deposit of Quintuplet Chain Aggregates Interspersed Among Other Aggregates of Uniform Latex Spheres of 0.71 µm Diameter as Obtained under High Size Resolution Conditions on the Foil Strip of the Original Spiral Duct Centrifuge

This finding has an important consequence. Since $\bar{\kappa}$ is practically constant for an aggregate of given shape regardless of the absolute size within the slip regime, it is acceptable to calculate the actual dynamic shape factor by

$$\kappa = \bar{\kappa} \frac{C(n^{1/3} D_1)}{C(f_n D_1)}$$

where

$$C(D) = 1 + \frac{2\lambda}{D} (A + Q\, e^{-\frac{bD}{2\lambda}})$$

is the Knudsen-Weber slip correction with empirical constants ($A = 1.246$, $Q = 0.42$, $b = 0.87$) and the mean free molecular path length of air molecules ($\lambda = 6.53 \times 10^{-6}$ cm). Values of κ at primary sphere sizes D_1 of 1.0 and 0.1 µm are included in Tables 1 and 2.

Investigations on several shape factors of irregular particles and aggregates were made with the original long-spiral-duct centrifuge by Kotrappa[20,21]. Particles of a low grade coal, uranium dioxide, thorium dioxide and quartz in a range of respirable sizes between 0.2 and 5 μm were deposited and analyzed. Table 3 summarizes the results, which were discussed by Davies[6]. It was argued that the shape factors $\bar{\kappa}$ and α, which both depend upon the particle mass, were too high for coal and quartz to be consistent with other data. This suggests that the mass determinations by beta-activity measurements of the activated ash content of the coal and by colorimetric chemical analysis of the quartz may have been in error. In contrast, the mass-independent ratio between the projected diameter as observed in the electron microscope and the Stokes diameter as obtained with the spiral duct centrifuge showed regular values.

Table 3

Shape Factors of Respirable Particles of Low Grade Coal, Uranium Dioxide, Thorium Dioxide and Quartz

	Size range D_p (μm)	κ	α	D_p/D_{st}
Low Grade Coal	0.56 ÷ 4.27	1.80	0.38	1.42
	1.0	1.95	±0.02	1.53
Uranium Dioxide	0.21 ÷ 1.68	1.22	0.33	1.39
	1.0	1.28	±0.03	1.42
Thorium Dioxide	0.23 ÷ 3.38	1.06	0.22	1.67
	1.0	0.99	±0.02	1.57
Quartz	0.7 ÷ 2.0	1.90	0.35	1.55
	0.2 ÷ 2.0	-	-	1.53
	1.0	1.82	0.34	1.53

D_p : Projected diameter in the electron microscope
D_{st}: Stokes' diameter
κ : Dynamic shape factor (average of
α : Volume shape factor (size range stated

Another study on dynamic shape factors focussed on elongated particles like asbestos fibers (Stöber, Flachsbart and Hochrainer[56]). The results, when compared to theoretically available dynamic shape factors of prolate spheroids, favored the assumption that, in laminar duct flow, the fibers are predominantly orienting their polar axis parallel to the streamlines. The study also showed that the aerodynamic size of a fiber is closely related to the actual diameter of the fiber while the length is almost immaterial. Figure 12 is an electron micrograph of a spiral duct deposit of amosite asbestos fibers, all of which have an aerodynamic diameter of 1.675 μm as indicated by the black sphere in the graph. By extrapolating the empirical findings with chain aggregates of uniform latex spheres to long fibers, a semi-theoretical model permitted the derivation of a relationship

$$D_{ae} = \left(\frac{9}{4}\right)^{1/3} \left(\frac{\rho}{\rho_0 k}\right)^{1/2} (\ell/D)^{1/6}\, D$$

between the aerodynamic diameter D_{ae}, the actual diameter D, the length ℓ and the density ρ of the fiber. With ρ_0 representing unit density, the empirical factor κ can be calculated from experimental data obtained with the spiral duct centrifuge. For amosite and crocidolite, a value of $\kappa = 1.08$ was found assuming $\rho \simeq 3$ grams cm^{-3}. This compares reason-

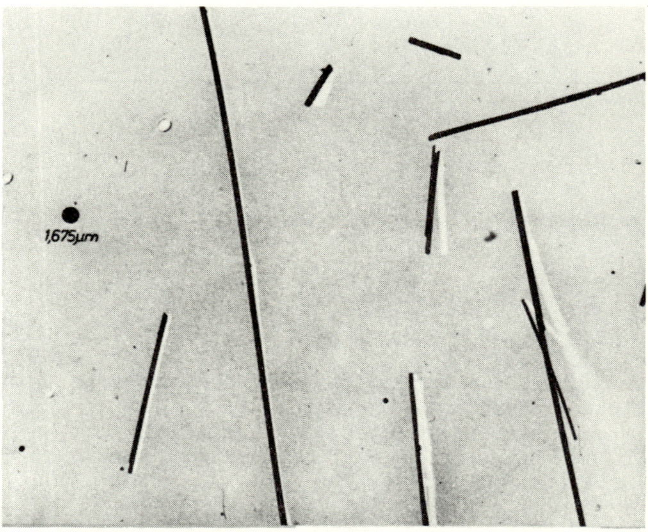

Figure 12: Electron Micrograph of a Deposit of Asbestos Fibers (Amosite) of 1.675 μm Aerodynamic Diameter as Obtained under High Size Resolution Conditions on the Foil Strip of the Original Spiral Duct Centrifuge

ably with k = 0.36 for chain aggregates, although the asbestos data showed considerable scatter around the regression line as indicated by Figure 13.

Recently, a study by Kops, Dibbets et al.[18] indicated that the empirical relations obtained for the aerodynamic diameters of chain and cluster aggregates of uniform spheres can be extended to aggregates of a large number of small spheres as produced by the exploding wire technique. The authors measured the aerodynamic diameters of aggregates of six labeled iron oxides (^{59}Fe) and two gold aerosols (^{198}Au) of different primary particle size distribution in a long-spiral-duct centrifuge. Subsequently, they evaluated the electron micrographs taken of the deposits. For the primary particle sizes, log-normal distributions were obtained from the micrographs while the pattern of the logarithms of the equivalent volume diameters, as represented by the logarithms of the primary particle numbers n of the aggregates, versus the logarithms of the aerodynamic diameters had two apparent regimes divided at a value of $n \leq 10^4$. Within these regimes, the log-log size pattern followed distinct relationships correspondig to those which were respectively derived for chain and cluster aggregates of a few uniform spheres (Stöber, Flachsbart and Hochrainer[56]). Figure 14 gives an example for iron

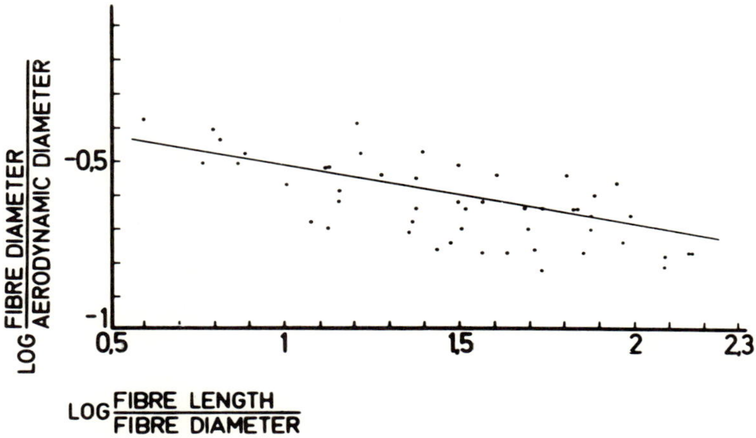

Figure 13: Regression Line of Experimental Data of Asbestos Fibers (Crocidolite) Arranged According to a Semi-Theoretical Aerodynamic Diameter Model

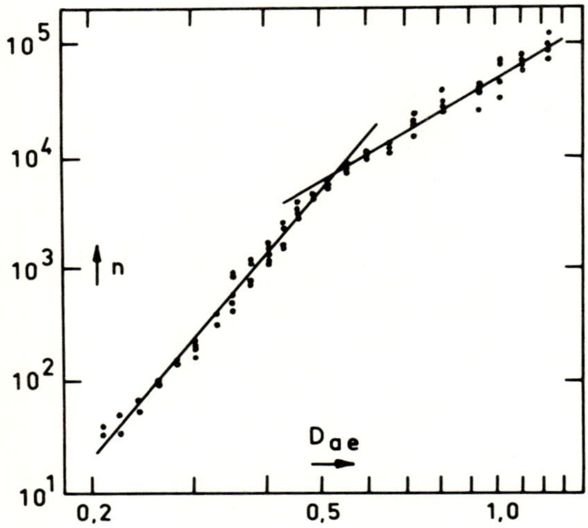

Figure 14: Relationship between die Number n of Primary Particles and the Aerodynamic Diameter D_{ae} (in µm) of Iron Oxide Aggregates Generated by an Exploding-Wire Technique (Kops et al.[18])

oxide aggregates of primary particles of a median size of $\bar{D}_1$ = 0.027 µm and a geometric standard deviation of σ_g = 1.8. In the lower regime, the aggregates, shown in the left half of Figure 15, resemble chain aggregates. The regression line corresponds here to

$$D_{ae} = \left(\frac{\rho}{\rho_0 k}\right)^{1/2} n^{1/6} \bar{D}_1 e^{2(\ln \sigma_g)^2}$$

In contrast, the aggregates in the upper regime as shown in the right half of Figure 15 are closer to a cluster shape and follow a relationship

$$D_{ae} = \left(\frac{\rho}{\rho_0 k}\right)^{1/2} n^{1/3} \bar{D}_1 e^{\frac{3(\ln \sigma_g)^2}{2}}$$

With these designations, the values of k and $\bar{\kappa}$, the latter representing the dynamic shape factors of the aggregates except for the influence of particle slip, can be calculated in both regimes and, by comparison, a value of n for the transitional range can be found. Table 4 gives the data.

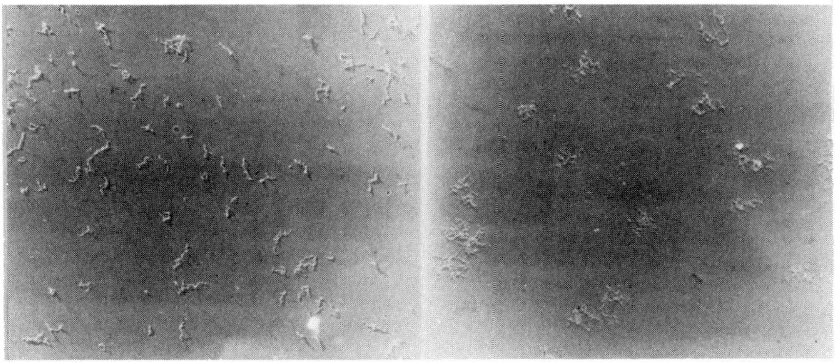

Figure 15: Electron Micrographs of Iron Oxide Aggregates Generated by an Exploding-Wire Technique and Deposited in the Spiral Duct Centrifuge; left side: Strand Shapes (Low Number of Primary Particles, D_{ae} = 0.259 μm); right side: Cluster Shapes (High Number of Primary Particles, D_{ae} = 0.925 μm); (Kops et al.[18])

Table 4

Dynamic Shape Factors $\bar{\kappa}$ for Aggregates of Iron Oxide spheres from an Exploded Wire

primary size		chain shape (lower regime)		cluster shape (upper regime)	transition range
D_1 μm	σ_g	k	$\bar{\kappa}$	$\bar{\kappa}$	n
0.020	1.8	0.842	$0.596\ n^{1/3}$	7.63	2100
0.024	1.8	0.939	$0.665\ n^{1/3}$	13.13	7700
0.027	1.8	1.050	$0.743\ n^{1/3}$	14.46	7400
0.041	1.8	1.294	$0.916\ n^{1/3}$	15.74	5100
0.043	1.8	1.349	$0.955\ n^{1/3}$	15.50	4300
0.047	1.8	1.459	$1.033\ n^{1/3}$	15.02	3100

The high size resolution capability of the long-spiral-duct centrifuge, which was successfully exploited for dynamic shape factor studies and similar investigations, can also be utilized for dynamical measurements of the particle size distribution of nearly monodisperse aerosols. This was done in a study of the parameters which influence and limit the size resolution of the long-spiral-duct centrifuge (Stöber and Flachsbart[50]). The study showed that the size resolution is directly proportional to the absolute value of the slope of the calibration curves as plotted in Figure 5. Thus, for the favorable range between 18 and 70 cm along the sampling foil, it could be shown that the size resolution primarily depends upon the ratio of the aerosol sampling rate F_{ae} to the total flow F, and a relationship

$$\left| \frac{\Delta D}{D} \right| \lesssim \frac{F_{ae}}{F}$$

was derived. This was confirmed experimentally by the fact that the area covered by the deposits of uniform latex spheres could be reduced with decreasing aerosol sampling rates. However, no further reduction was obtained for aerosol sampling rates below 2 %. At this rate, the size resolution seemed to be better than 0.5 %, which can be concluded from the fact that very small relative standard deviations of 1.4 to 2.4 % as obtained for latex spheres in the electron microscope (Heard, Wells and Wiffen[11]) were redetermined with the long-spiral-duct centrifuge within 0.4 % or better. Figure 16 gives the number distribution of quasi-monodisperse latex spheres of 0.357 µm diameter as obtained by electron micrographic count evaluation of a spiral duct deposit.

Similarly, in an investigation by Oeseburg, Benschop and Roos[38], the narrow size distributions of dioctyl phthalate (DOP) aerosols from a condensation generator (Lassen[29]) were studied with a long-spiral-duct centrifuge. The authors found relative standard deviations of about 10 % in the micron size range and felt that small experimental deviations from a theoretically expected log-normal distribution could be attributed to coagulation processes in the generating system. Figure 17 shows some of their normalized data and the curves of the log-normal distributions they approximate.

When utilizing the long-spiral-duct centrifuge as an absolute instrument for sampling and analyzing polydisperse aerosols, two adverse influences have to be considered: With increasing size, the particle losses in the aerosol inlet will become increasingly significant, and, for small sizes precipitated way down the duct, the deposit concentrations may be distorted by the secondary double vortex flow discussed earlier. To assess these effects, a comparative study was made (Stöber, Flachsbart and Boose[55]) to detect systematic deviations between the spiral duct centrifuge, an electrostatic and two thermal precipitators as well as a cascade impac-

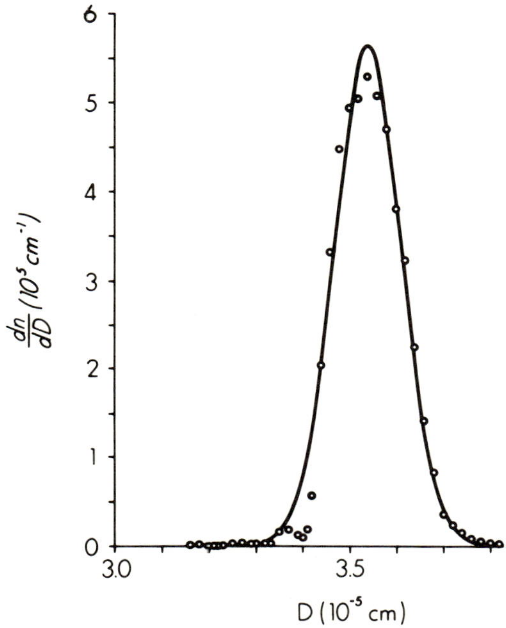

Figure 16: Size Distribution of Quasi-Monodisperse Latex Spheres of 0.357 µm Nominal Diameter as Obtained by Evaluating the Deposit Concentrations of an Aerosol Sampled under High Size Resolution Conditions in the Spiral Duct Centrifuge (Data Points and Approximated Normal Size Distribution Curve)

tor by analyzing model aerosols of fluorescein deposited simultaneously in these instruments. For three different test aerosols, it appeared that practically no corrections were necessary because of the random nature of the deviations between the instruments. The data of the long-spiral-duct centrifuge were quite consistent but, possibly, somewhat insensitive for sizes near 0.1 µm. Figure 18 gives a typical result for the size distribution analyses to be compared. Furthermore, a comparison of the cumulative data of the centrifuge and the cascade impactor revealed that, with the particular aerosol inlet system employed in the centrifuge for this study, there was an upper size limit around 2 µm for particle sizes actually sampled on the centrifuge foil. This cut-off size was lower than theoretically expected, a result which was also observed by Ferron and Bierhuizen[8]. Figure 19 gives the probability plots of the cumulative mass distributions to be compared.

In a very careful absolute calibration study which was made by Kops, Hermans and van de Vate[19] with regard to the deposition along the center line of the sampling foil of a long-spiral-duct centrifuge with a duct of 4.5 cm depth, the authors did find systematic deviations be-

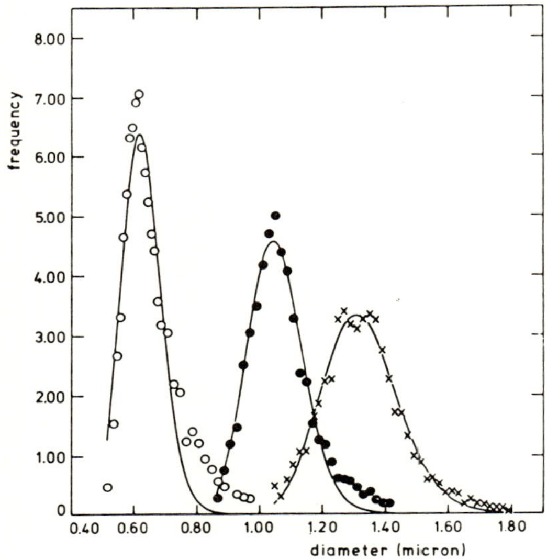

Figure 17: Size Distributions of Dioctyl Phthalate Aerosols and Their Approximations by Logarithmic Normal Distribution Curves (Median Diameters of 0.64 μm, 1.07 μm and 1.33 μm, respectively; relative standard deviations 10 to 12 %); Oeseburg et al.[38]

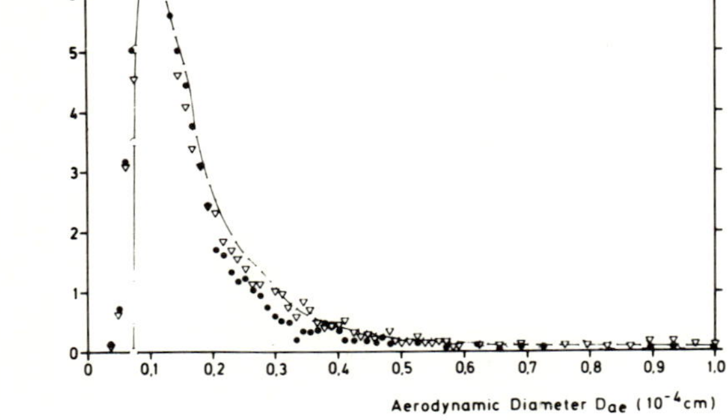

Figure 18: Particle Size Distribution of a Fluorescein Aerosol Measured with the Spiral Duct Centrifuge (Open Circles and Curve), a Thermal Precipitator (Closed Circles) and an Electrostatic Precipitator (Triangles)

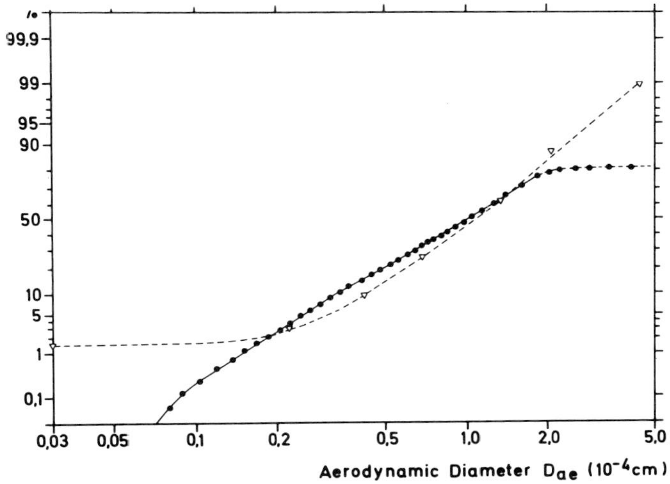

Figure 19: Logarithmic Probability Plots of the Cumulative Mass Distribution Data of a Fluorescein Aerosol Measured with the Spiral Duct Centrifuge (Circles) and a Cascade Impactor (Triangles)

tween the size distribution analyses of the centrifuge and an electrostatic point-to-plane precipitator. Instead of using the derivative of the calibration curve $\ell = f(D_{ae})$ for the conversion of deposit concentrations into size frequencies as required by a simple mathematical model (Stöber and Flachsbart[50]), they determined a special experimental correction function for the actual operating conditions employed in the study. Figure 20 shows the corresponding graphs. The experimental data show good reproducibility and although the validity of the correction function hinges somewhat on the assumption that sampling with the electrostatic sampler was not size-selective (van de Vate[64]), the relative pattern of the correction factor P applicable to the derivative of the calibration curve is physically meaningful. Figure 21 is rescaled from the data by Kops and coworkers to bring P close to unity in the almost size-independent range between 0.2 and 0.4 μm. Then, for smaller sizes the increase of $P > 1$ indicates the compounding of particles along the center line of the collecting foil due to secondary double vortex flow, while the decrease $P < 1$ for larger sizes may be due to particle losses in the aerosol inlet system used in the study.

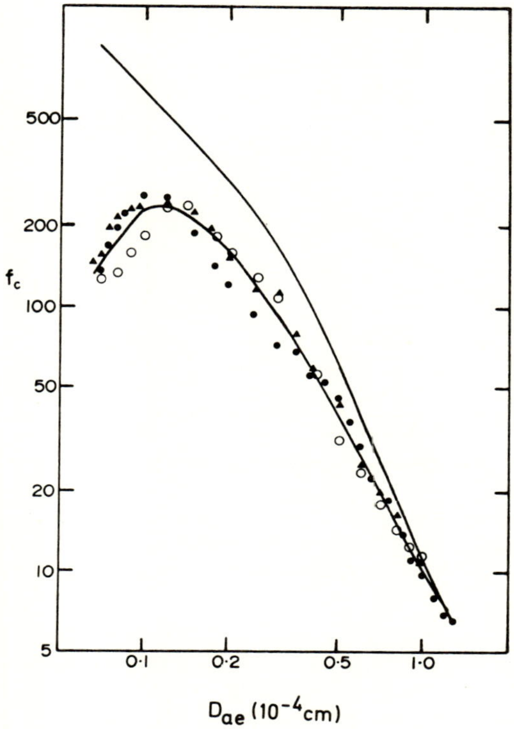

Figure 20: An Experimental Correction Function $f_c(D_{ae})$ by Kops et al.[19] Replacing the Derivative of the Size Calibration Curve ($d\ell(D_{ae})/dD_{ae}$, upper curve) in Computing Size Distribution Data from Deposit Concentrations in the Spiral Duct Centrifuge

Kops and coworkers[19] also investigated the influence of temperature imbalances of the rotor housing on the deposition patterns of aerosols of uniform latex spheres and their aggregates. They found that the deposit location remained unchanged but that the regular shape and the symmetry of the deposits could be distorted by thermal forces. Figure 22 shows typical results.

The influence of the curved contours of the deposits obtained under regular operating conditions and the changes of the deposit concentration across the foil at given foil lengths have been investigated by Ferron and Bierhuizen[8]. These authors used radioactive NaCl aerosols in their study and found that, for given strip widths of 2.4 and 0.8 cm around the center line of the foil, the deposit concentration was representative for mass distribution measurements within the first half of the

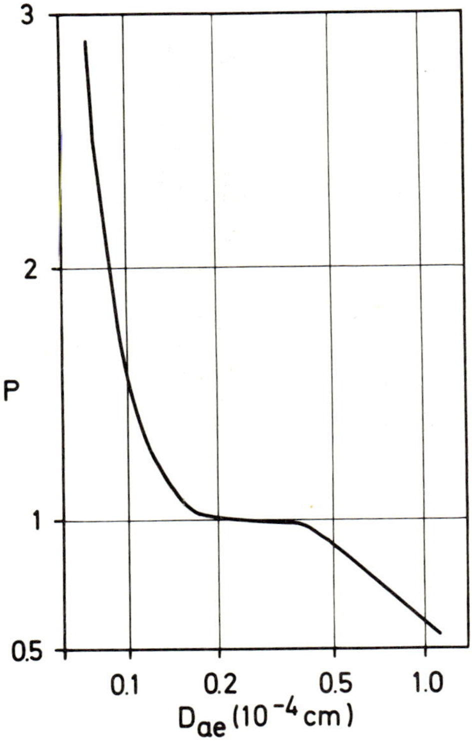

Figure 21: Relative Correction Factor P for Transforming the Experimental Correction Function $f_c(D_{ae})$ into the Derivative of the Size Calibration Curve (rescaled after Kops et al.[19])

length of the foil. Further down the duct, there occurred erratic deposits along the edges of the downstream half of the foil. This effect is apparently due to the double vortex secondary flow distortion and has been found for other polydisperse aerosols too (Figure 23). A typical result obtained by Ferron and Bierhuizen[8] at 1500 rpm and a total flow of 10 liters min^{-1} is shown in Figure 24 for seven locations along the foil. The authors gave no correction function, although it would be worthwhile to establish such function by disregarding the parasitic deposits within about 5 mm of the edges of the foil or utilizing only the center line area as done for number distributions by Kops and coworkers[19] (Fig. 21).

An interesting study of the size distribution of diluted cigarette smoke,

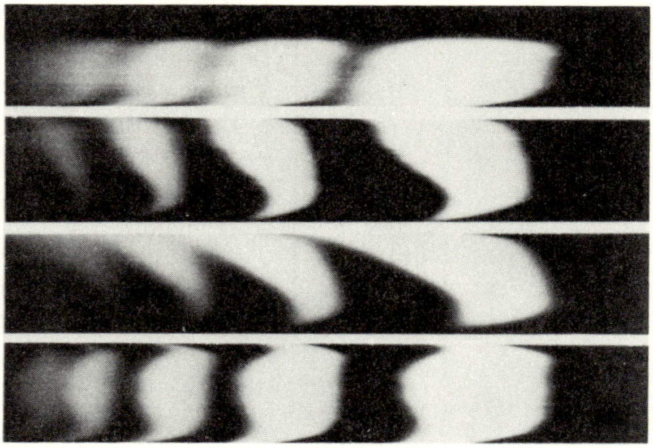

Figure 22: Deposit Patterns of a Latex Aerosol (Primary Particle Size 0.357 μm) at Different Thermal Gradients Maintained at the Centrifuge Rotor; (Temperatures T_1 at the Rotor Top, T_2 at the Rotor Housing, T_3 at the Bearings) from Top to Bottom:
a) $T_1 = 35°C$, $T_2 = 25°C$, $T_3 = 15°C$;
b) $T_1 = 15°C$, $T_2 = T_3 = 25°C$;
c) $T_1 = 15°C$, $T_2 = 25°C$, $T_3 = 35°C$;
d) $T_1 = T_2 = T_3 = 25°C$
(Kops et al.[19])

undiluted particulate car exhaust and atmospheric aerosols by means of the long-spiral-duct centrifuge was made by Porstendörfer[40]. The author labeled the particles by attaching ^{220}Rn decay products to the aerosol particles. Then he used a multi-channel analyzer for the number distribution analysis of the deposit. Depending upon the age and the dilution of the cigarette smoke, he found mean particle sizes from 0.21 to 0.44μm diameter. For the particulate matter in Diesel and gas engine exhaust, mean diameters of 0.14 and 0.18 μm, respectively, were determined. The atmospheric aerosol had a distribution similar to the D_{ae}^{-4}-distribution postulated by Junge[15].

In his measurements with cigarette smoke, Porstendörfer used a maximum aerosol flow of about 2 % of the total flow through the centrifuge. The aerosol sample consisted of at least 10-fold diluted cigarette smoke so that the smoke flow in the centrifuge was only 0.2 % or less. This arrangement gave apparently reasonable results for the mean diameters of the smoke particles. In contrast, long-spiral-duct centrifuge experiments with undiluted cigarette smoke at a high relative aerosol flow ra-

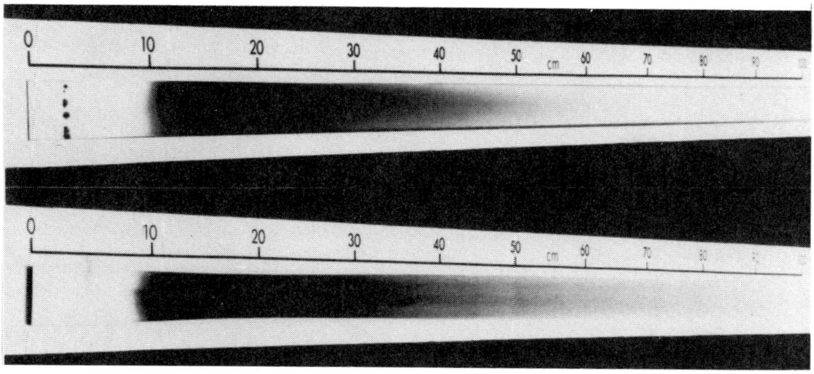

Figure 23: "Parasitic" Deposits of Polydisperse Aerosols along the Edges of the Foil Strip of the Spiral Duct Centrifuge;
Top: Cigarette Smoke with Heavy Edge Deposits beyond 30 to 40 cm Foil Length;
Bottom: Fluorescein Aerosol with Significant Edge Deposits beyond 80 cm Foil Length;
(unpublished University of Rochester Photographs, L. Schwartz and H. Flachsbart, 1970)

te of 10 % (Stöber and Osborne[52]) gave rather irregular results. A routine evaluation of the surprisingly short deposit obtained under this condition indicated an improbably high minimum size of 0.88 μm for the diameter of the smoke particles. On the other hand, running the instrument as a semi-dispersive device by using no clean air and, at a comparable total flow rate, filling the whole cross section of the duct with cigarette smoke through the aerosol inlet, a minimum diameter of 0.26 μm or less could be observed. This suggested that either the deposition pattern in the spectrometric test involved a cloud settling effect of the dense cigarette smoke or a different density of the gas phase caused a distortion, if not both effects contributed.

The possibility of density influences was systematically investigated (Martonen and Stöber[32]) by testing a variety of clean winnowing gases against latex aerosols with gas phases other than air. The lower photograph in Figure 25 shows an impressive case of deposit distortion caused by a relatively heavy aerosol gas phase consisting of carbon dioxide entrained by air as the clean winnowing gas. This result indicates that precise measurements with the spiral duct centrifuge should not involve

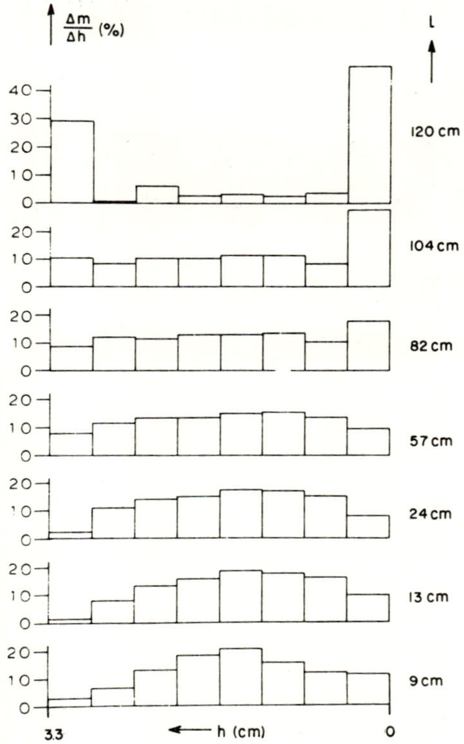

Figure 24: Mass Distribution across the Foil Strip of a Deposit of a Polydisperse NaCl Aerosol at Different Foil Lengths of the Spiral Duct Centrifuge (Ferron and Bierhuizen[8])

the confluence of gas phases of significantly different density. The gas phase of cigarette smoke is about 6 % denser than air (Martonen[31]) which certainly accounts for some of the distortions of an undiluted relative smoke flow rate of 10 % in the spiral duct centrifuge. The magnitude of the cloud settling effect of undiluted cigarette smoke is still under investigation.

For all aerosol particles of known dynamic shape factor, the spiral duct centrifuge can be used to determine the density of the particles. To achieve this, a suitable electron microscopic inspection of the deposited particles permitting the determination of the particle volumes must be performed. Such measurements can actually be made on single, selected particles, but there is the disadvantage that only the square

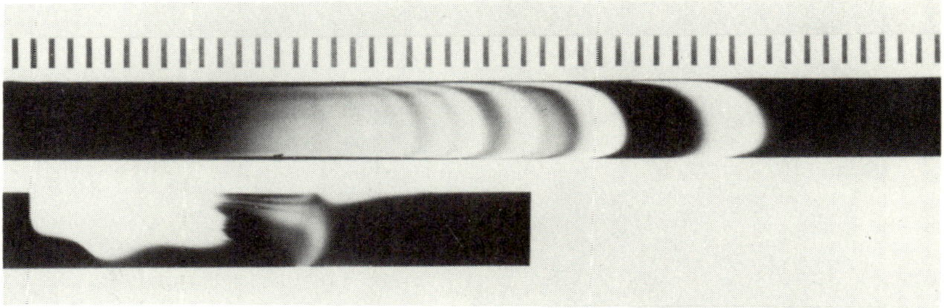

Figure 25: Foil Deposits of a Latex Aerosol; Top and Bottom at Same Scale in cm (Primary Particle Size 0.48 μm)
Top: Regular Deposit (Aerosol Gas Phase: Air; Winnowing Gas: Air)
Bottom: Irregular Deposit (Aerosol Gas Phase: Carbon Dioxide; Winnowing Gas: Air)

root of the density is obtained experimentally and the precision is limited to the accuracy of the determination of the aerodynamic diameter of the particle. For this diameter, the absolute accuracy of the spiral duct aerosol centrifuge is estimated as 5 % so that an error of some 10 % for the density value of a single particle is not unusual. However, when a large number of chemically identical particles of different size is investigated, the data may become quite consistent. In the simple case of spheres, such investigations are reported. Moss, Ettinger and Coulter[35] used a long-spiral-duct centrifuge for measuring the density of submicron iron oxide spheres (0.12 to 0.8 μm in diameter) and obtained a mean value of 2.53 grams cm^{-3}, which is within one per cent of the value found for larger iron oxide spheres with independent methods by Spertell and Lippmann[44]. This indicated that there was no significant change in the density of iron oxide particles over a wide range of sizes. Similar conclusions could be drawn from other measurements by Moss and coworkers concerning spherical fly ash particles in the 0.4 to 2.9 μm diameter range (ρ = 3.83 grams cm^{-3}) and a laboratory aerosol (4 : 1 methylene blue: uranine) between diameters of 0.6 and 2.2 μm (ρ = 1.37 grams cm^{-3}).

Aerodynamic particle density determinations of this kind can reveal systematic errors of the calibration of the spiral duct centrifuge if it is safe to exclude density variations. This is shown in a study on the density of laboratory aerosols nebulized from diluted ammonium fluorescein solutions (Stöber and Flachsbart[51]). When these polydisperse aerosols of spherical particles were sampled under high size resolution conditions, a correction of the calibration curve became necessary to obtain consistent density data for the different particle sizes.

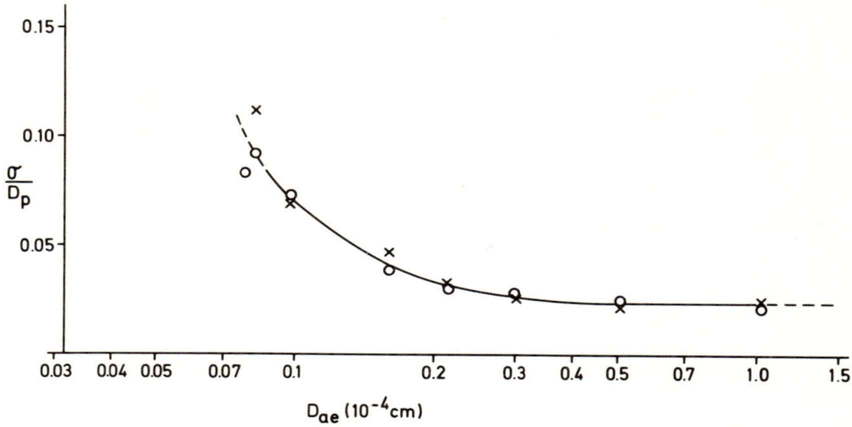

Figure 26 : Relative Standard Deviations of Local Deposits on Electron Microscopic Grids along the Foil Strip of a Spiral Duct Centrifuge after Drying and Sampling Polydisperse Fluorescein Aerosols Generated from Ammonium Fluorescein Solutions of 1 % (open circles) and 2.5 % (crossmarks)

In the same study, the evaluation of the electron micrographs further revealed that in all locations along the foil strip of the centrifuge, the local deposits had narrow Gaussian normal distributions whose standard deviations reflected the size resolution in that location. Figure 26 presents the results in terms of relative standard deviations for the lower end of the size range of a long-spiral-duct centrifuge where the size resolution deteriorates for reasons discussed earlier. Nevertheless, even at the small aerodynamic diameter of 0.08 μm, the standard deviation of the sizes found in that location is still less than 12 %. This is a value which, by several definitions, still permits the term "monodisperse" to be used for the size dispersity of such local deposits (Fuchs and Sutugin[9], VDI Guidelines[65]).

The occurrence of highly uniform sizes at any given location along the foil strip of the long-spiral-duct centrifuge was exploited by Kotrappa and Moss[25] for preparing monodisperse samples from polydisperse aerosols of insoluble, almost spherical fused clay particles. After deposition, the size-selected samples were resuspended in air, thus forming the desired monodisperse aerosols of the clay material as needed for inhalation studies.

In many other applications of the long-spiral-duct centrifuge, the instrument has been used as a convenient laboratory tool to characterize the aerodynamic size distribution of test aerosols. An example of this kind is the study by Blachman and Lippmann[3], who investigated the performance of multicyclone aerosol samplers. Similarly, van Buitenen and Oeseburg[63] used a long-spiral-duct centrifuge for comparing "light scattering diameters" of aerosol particles with aerodynamic data. Heyder and Porstendörfer[12] also reported optical and aerodynamic aerosol measurements involving a long-spiral-duct centrifuge. The literature on applications of the long-spiral-duct centrifuge in the laboratory seems to be expanding. However, no field applications have been reported so far, although there is a potential for this kind of work in view of the feasibility of sampling rates of some 3 liters min^{-1} at size resolutions around 15 % .

SHORT SPIRAL DUCT DESIGNS

For many applied research problems, particularly in dust control or health hazard evaluations of aerosols in industrial hygiene, it would be sufficient to obtain simply a gross value of the total mass concentration of all particles of less than about 0.3 μm aerodynamic diameter rather than the actual mass distribution below this size. Thus, it would be worthwhile to replace the corresponding far-end section of the long spiral duct by a suitable aerosol filter and to confine the aerodynamic distribution analyses to the larger sizes. For gravitational spectrometers, Walkenhorst[66] introduced such filter arrangement in 1965 at the end of his horizontal duct and actually obtained filter deposits with a continuous size separation.

The first short-spiral-duct centrifuge with an exit filter was built by Kotrappa and Light[23]. By leaving the range of smaller sizes to a back-up filter, they reduced the length of the collection foil to 46.2 cm and coiled it into a rotor of less than 18 cm diameter. Unfortunately, the authors misconceived their design called the Lovelace Aerosol Particle Separator (LAPS) by employing a duct of expanding radial width which cannot possibly bring an improvement of the instrument characteristics (Stöber[47]) because of reasons of secondary flow discussed earlier in this paper. The expanding spiral was supposed to deposit a given particle size at a shorter distance from the aerosol inlet than in a duct of constant width. According to Kotrappa and Light[24], an increase of the ratio of expansion of the duct would increase the range of deposited sizes. It appears obvious and could be shown though[47], that these expectations are clearly unfounded.

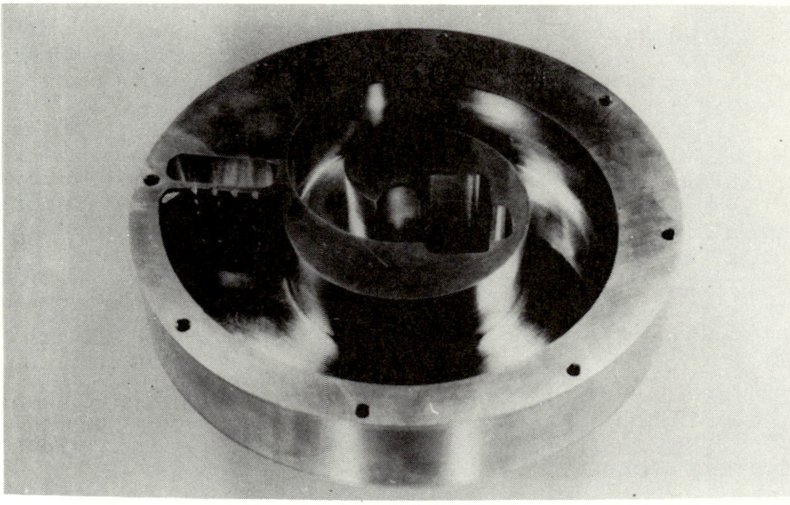

Figure 27 : Expanding Spiral Duct of the Lovelace Aerosol Particle Separator without Aerosol Inlet System and Laminator (Photograph by Lovelace Foundation, Albuquerque, N.M.)

Figure 27 shows a photograph of the expanding spiral duct of the LAPS. The increasingly unfavorable aspect ratio of the cross section of the duct (2.1 : 1 to 0.8 : 1) must necessarily increase the detrimental influence of the secondary double vortex flow so that a successful operation may be expected to be more stringently limited for the LAPS than for ducts with more favorable aspect ratios (3.3 : 1 to 5.1 : 1 for long-spiral duct centrifuges). From all this, it would appear that the LAPS is a lapse. However, Kotrappa and Light[23] claimed an excellent size resolution over the entire length of deposition at rotor speeds of 6 000 rpm for the LAPS and they mentioned operating characteristics at 4 500 rpm and total flow rates up to 15 liters min^{-1}.

These surprisingly favorable performance data triggered the construction of two experimental rotors with short spiral ducts of constant width (Stöber, Hochrainer and Flachsbart[59]). The rotor diameters (17.5 cm) were almost the same as for the LAPS rotor (17.78 cm) and the dimensions of the ducts were chosen so that one rotor would have a duct of the same cross section and aspect ratio as the upstream end of the expanding spiral of the LAPS, while the other rotor would have the same constant duct width but a better aspect ratio and a cross section close to that at the down-

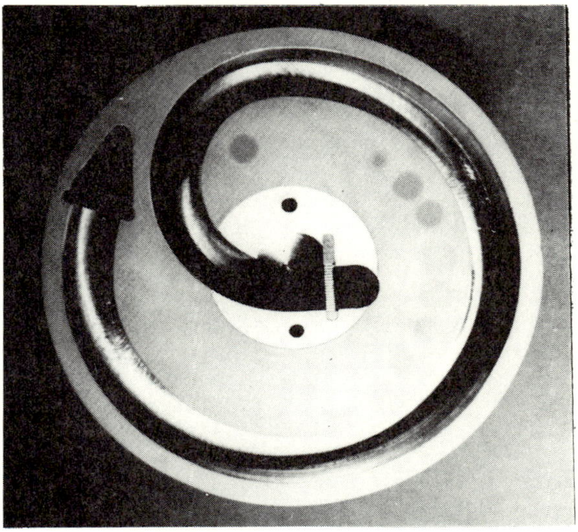

Figure 28 : Short-Spiral-Duct Design with a Wide Duct of Constant Width for an Experimental Aerosol Centrifuge (Aerosol Inlet System Removed, Laminator in Place)

stream end of the duct of the LAPS. Thus, with a width of 1.67 cm for the spiral duct, the rotor top geometry was the same for both versions (Figure 28) but the duct depths of 3.3 and 6.0 cm provided different aspect ratios of 2.0 : 1 and 3.6 : 1 , respectively. It was anticipated that these rotors would outperform the LAPS.

A comparison between the calibration curves of the LAPS and the two experimental short-spiral-duct centrifuges at 3 000 rpm and a total flow rate of 5 liters min^{-1} is presented in Figure 29. As shown for the long-spiral version[47], this comparison reveals again that the constant-width spirals do deposit smaller sizes at shorter distances from the aerosol inlet than the LAPS. This result is contrary to the stated purpose of the expanding spiral duct design[24], notwithstanding an author's incorrect denial[22] that such erroneous assumption was made.

The calibration tests of the constant-width spiral duct rotors at increased total flow rates were disappointing. Apparently, the relatively wide radial dimensions of these ducts, although generally smaller than in the LAPS design (1.5 to 3.9 cm) impaired a proper size spectrometer operation of the centrifuges at total flow rates as low as 10 liters min^{-1}. In view of this, it is very difficult to accept the statement by Kotrappa and Light[23] that they were able to establish operating characteristics of the LAPS at 4 000 rpm and 10 or 15 liters min^{-1}.

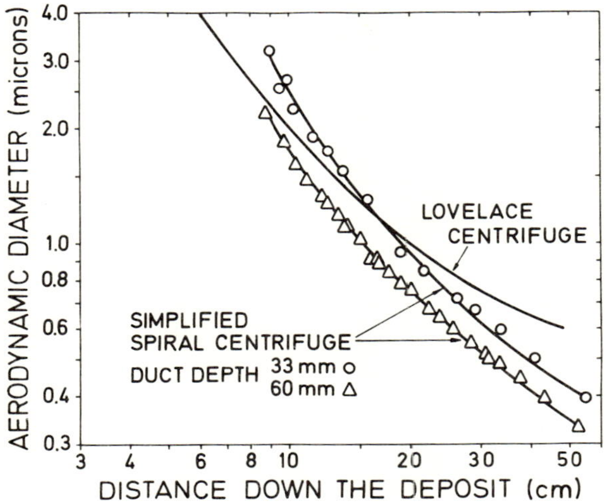

Figure 29 :

Comparison of the Calibration Curves of the Lovelace Aerosol Particle Separator and two Experimental Short-Spiral-Duct Centrifuges of Different Duct Depth under Equal Operating Conditions (3 000 rpm, total flow rate of 5 liters min^{-1})

There was also a limit to the rotor speeds applicable to the experimental constant width short-spiral-duct centrifuges. Only the version with the favorable aspect ratio of 3.6 : 1 could be operated at 6 000 rpm without showing excessive turbulent deposition (Figure 30). Thus, for a desirable increase of the total flow rate as well as for an extension of the range of deposited small sizes, a widening of the spiral duct beyond the width of the original long-duct design does not seem advisable. The contradicting results with the LAPS appear to be questionable.

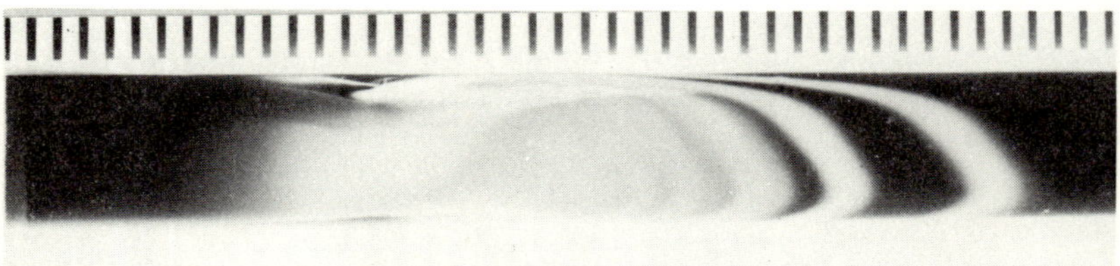

Figure 30 : Slightly Distorted Deposit of a Latex Aggregate Aerosol of Uniform Primary Particles of 0.234 μm Diameter in an Experimental Short-Spiral Deep-Duct Rotor Operated at 6 000 rpm and a Total Flow Rate of 5 liters min^{-1}

Figure 31 :
Short-Spiral-Duct-Centrifuge Rotor with a Narrow Duct (Aerosol Inlet System, Laminator and Back-up Filter Removed)

As concluded earlier by considering the secondary double vortex flow in the spiral duct, a practical improvement of the characteristics of a short-spiral-duct rotor can only be expected from a reduction rather than an expansion of the width of the spiral duct. Thus, an instrument of corresponding design was built recently (Stöber et al.[60]). While the rotor diameter of this device remained at 17.5 cm, its spiral duct narrows down from 1.6 cm at the center of rotation to 0.8 cm along the concentric part of the spiral. The duct is 3.4 cm deep and, thus, the aspect ratios range from 2.1 : 1 to 4.3 : 1. The sampling foil along the outer wall of the duct has a length of 52.3 cm. At the end of the duct, a new cylindrical back-up filter may be inserted. Figure 31 shows the geometry of the rotor. Like the preceding experimental models with wide ducts of constant width, the rotor has a new and simple laminator consisting of a brass plate of 0.27 cm thickness with a narrow array of 0.06 cm holes as shown in Figure 32. This arrangement can also replace the spacious laminator block of the original long-spiral-duct design.

Figure 32 :
New Aerosol Inlet Section and Laminator for Spiral Duct Centrifuges

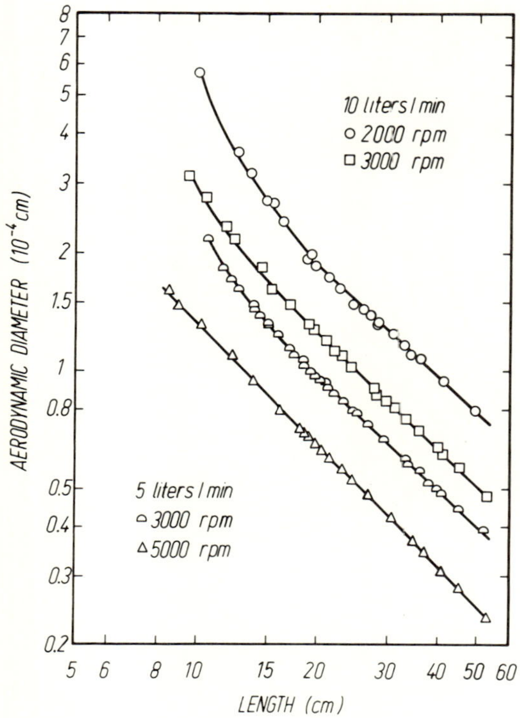

Figure 33 :

Calibration Curves for Various Operating Conditions of a Narrow-Duct Short-Spiral Centrifuge (Established by the Leading Edges of Latex Test Aerosol Deposits)

Calibration curves for a few selected operating conditions of the new short-spiral-duct centrifuge are given in Figure 33. The graph shows an obvious improvement over the experimental rotor designs. The narrow-width spiral duct permits now total flow rates of 10 liters min^{-1} at 3000 rpm. Under favorable conditions (5000 rpm and 5 liters min^{-1}) the size separation on the sampling foil can be extended to 0.23 μm .

PERFORMANCE AND APPLICATIONS OF SHORT-SPIRAL-DUCT CENTRIFUGES

Reports on the utilization of short-spiral-duct centrifuges are not abundant in the literature. Since the narrow duct design is a rather recent development which may still be considered as being in the experimental stage with regard to applications, all published studies applying short-spiral-duct designs to practical problems refer to the Lovelace Aerosol Particle Separator and all of these reports come from the same laboratory.

It appears from published data that the actual performance of the LAPS has not been investigated very systematically so far. In their evaluation of the instrument, Kotrappa and Light[23] made some performance statements which to date were not yet followed up by experimental data. For instance, no quantitative evidence is available for the allegedly excellent size resolution over the entire deposit length at 6 000 rpm. The same lack of data exists for the capability of the LAPS to separate and deposit sizes down to 0.2 μm diameters. No calibration curves are offered for quoted operating conditions at relatively high total flow rates. One exception will be discussed below. Similarly, Raabe[41] gave no evidence for his statement that the back-up filter of the LAPS collects the smaller particles so that they are also separated with respect to aerodynamic properties. Kotrappa alludes repeatedly[22,23] to the same effect, but if it is real, no quantitative investigation or analytic use of it has been published to date. There is little doubt that some of the performance claims for the LAPS need to be better documented or else, in view of the recent experience with short-spiral-duct designs not suffering from the misconception of an expanding duct, may be considered inflated.

In an applied study, similar to the work with the long-spiral-duct centrifuge on fused clay aerosols[25], Kotrappa, Wilkinson and Boyd[28] prepared monodisperse aerosols of plutonium oxide for inhalation studies by precipitating a polydisperse aerosol on the sampling foil of the LAPS and resuspending the local deposits of relatively uniform size. This study reports operating conditions of the LAPS of 4500 rpm and a total flow rate of 10 liters min^{-1}. In view of the failure to obtain regular deposits under comparable operating conditions with the experimental short-spiral-duct centrifuges described earlier, the question arises whether the quoted operating data were not in error. This conjecture is supported by the fact that a recent paper by Raabe et al.[42] dealing with a routine system for the production of monodisperse plutonium oxide aerosols calls the questioned paper a pilot procedure and uses four Lovelace Aerosol Particle Separators at 3600 rpm and 4.8 liters min^{-1}, although, in view of the use of four separators, higher flow rates would have been very desirable. However, mention is made of flow rates as being measured erroneously by Kotrappa et al.[28]. Thus, it may be concluded that the same is probably true for the earlier work with the LAPS.

Other potential applications of short-spiral-centrifuges were enumerated by Kotrappa and Light[23]. Their listing does not include any practical aspects which the long-spiral-duct centrifuges have not already been applie to. However, the authors stress a number of practical advantages, of which the more convincing ones are the smaller rotor and the shorter sampling foil. Both features facilitate an easier handling in routine operations. Surprisingly, the literature on LAPS apllications does not indicate any significant role or specific use of the 3 × 3 cm exit filter arrangement.

Kotrappa and Wilkinson[26] used the LAPS to investigate the mass-specific activity of polydisperse fused clay aerosols labeled with a variety of radioactive isotopes (^{91}Y, ^{90}Sr, ^{144}Ce, ^{137}Cs and ^{106}Ru). By comparing the calculated mass of the size-specific local deposits on the sampling foil with the radioactivity measured for such samples, they were able to prove that the mass-specific activity of the labeled aerosols did not change with particle size, except in one case involving a more volatile isotope (^{106}Ru).

In another study[27], the same authors applied the LAPS to particle density measurements of the kind described earlier in this paper. The investigated size range extended from 0.5 to 3.5 μm of aerodynamic diameter. In this range, the density of fused clay aerosol particles was found relatively constant and comparable to independent measurements with a Millikan chamber (2.1 grams cm^{-3}). Other aerosols of constant density were obtained by nebulizing and drying uranine (1.5 grams cm^{-3}) and iron oxide (2.2 grams cm^{-3}). However, oxide aerosols of uranium, cobalt and zirconium which were prepared by heat treatment of degradable salt aerosols showed density variations depending upon the parent material. For a plutonium oxide aerosol prepared by a two-stage heat degradation, a density of 10.5 grams cm^{-3} was measured which is within experimental error of the actual bulk density (11.5 grams cm^{-3}). However, the same value found by Kotrappa, wilkinson and Boyd[28] was questioned by Raabe et al.[42] because of low humidity conditions and erroneously measured flow rates. The latter authors used the LAPS at safe operating conditions and obtained densities of the plutonium oxide particles between 7 and 8 grams cm^{-3} for aerodynamic sizes of 0.65 to 1.33 μm.

Besides a new presentation by Newton et al.[36], there is a total of fifteen other applied studies contained in two annual reports[14] which make use of the LAPS. These contributions deal primarily with the production of monodisperse resuspensions of deposits of polydisperse aerosols for inhalation toxicology research as described before. It appears that this is now the main field of application of the LAPS.

No efforts have been published so far to employ spiral duct centrifuges for air pollution field studies or atmospheric aerosol research. This is obviously due to the relatively low aerosol sampling rates which, in most laboratory studies, seldom exceed values of 1 liter min^{-1} because of the associated loss of size resolution with increasing sampling rate. However, field studies on particulate air pollutants do not require size resolu-

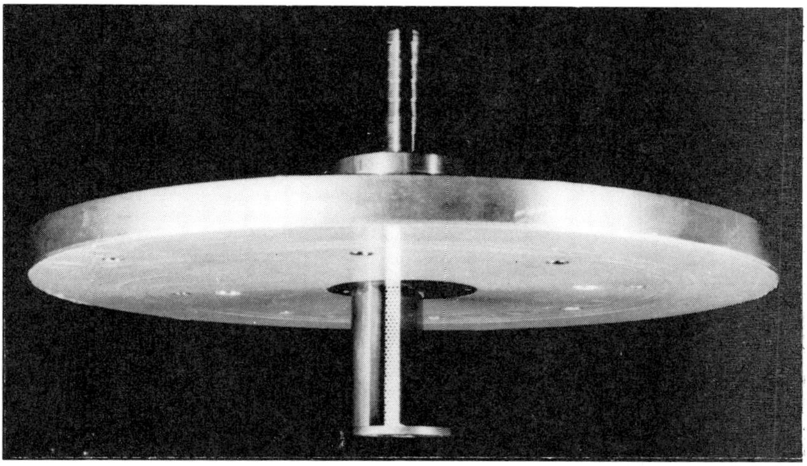

Figure 34 : Special Aerosol Inlet Section for Increased Aerosol Sampling Rates with a Narrow-Duct Short-Spiral Centrifuge (The Inlet Section is Mounted to the Rotor Lid)

tions to a few percent and a size resolution of some 50 % would still be considered an improvement over cascade impactor data, where the deposits on the various stages are of cumulative nature with size variations of more than 100 % . Thus, by accepting reduced size resolutions, the sampling rate of spiral duct centrifuges could be increased to values feasible to suit air pollution measuring requirements for particulate matter.

Such an attempt was made with a narrow-duct short-spiral centrifuge (Stöber et al.[60]) although the long-spiral-duct centrifuge which permits total flow rates up to 20 liters min^{-1} would have been preferable because of a better size resolution at high sampling rates. By way of a special aerosol inlet section of wide coaxial bore with a laminator insert as shown in Figure 34, the aerosol sampling rate of the short-spiral-duct centrifuge could be increased to 4 liters min^{-1}. A subsequent feasibility study at operating conditions of 2000 rpm and a total flow rate of 10 liters min^{-1} produced the upper calibration curve in Figure 33 and confirmed that the size resolution did not deteriorate beyond 40 % for sizes below 4 µm. For aerodynamic diameters of 8 µm, the size resolution was still at a reasonable value of 60 % . The sampling rate of 4 liters min^{-1} proved to be sufficient for the analysis of a 24-hour sample of a polydisperse fluorescein test aerosol at an airborne concentration of 50 micrograms m^{-3} which corresponds to moderate particulate air pollution levels. The gravimetrical mass distribution analysis required the use of a microbalance and special thin plastic collection films which were weighed before being placed on the sampling foil strip. With a detection limit of 1.8 micrograms cm^{-2} for the foil deposit at a 99 % confidence level, particulate air pollution mass distributions can be analyzed if the average airborne concentration within the sampling period was 40 micrograms m^{-3} or more between 0.8 and 8.7 µm of aerodynamic diameter. Modest as these results may be for air pollution surveillances, they certainly indicate a good potential for air pollution source emission measurements.

FUTURE MODIFICATIONS AND DEVELOPMENTS

From the past experience with various designs of spiral duct centrifuges, it appears that the long-spiral-duct versions permit more favorable operating conditions, i.e., in particular, higher total flow rates and, thus, higher sampling rates, than the short-spiral-duct designs. While a flow rate of some 19 liters min^{-1} is one of the standard operating conditions at 3000 rpm for the original spiral duct centrifuge[49], none of the small rotor designs can be operated at this flow rate. This came as a surprise for the narrow-duct short-spiral centrifuge design whose duct had a more favorable aspect ratio than the original instrument and, in proportion, should permit a total flow rate of 15 liters min^{-1}. The only explanation for the failure is that, contrary to theoretical estimates[58], the influence of the imbalanced centrifugal forces due to the duct curvature is competitive to the Coriolis forces and, thus, narrower curvatures on the small rotors increase the development of the secondary flow in the cross sectional plane more rapidly than the relatively wide curvatures of the original design. Therefore, future modifications of the spiral duct may have to go back to bigger rotors and ducts with larger radii of curvature and high aspect ratios.

In connection with modern elemental microprobe analysis or similar methods, it seems also possible to reduce the dimensions of the spiral duct. Although a decrease of the depth of the spiral duct requires a proportional decrease of the flow rate inorder to maintain the same flow velocities and deposition characteristics in the duct, such modification does not decrease the deposit concentration but only the area of deposition and, thus, the total amount of the sample due to a narrower deposition foil. Furthermore, a reduction of thw width of the duct will cause an inversely proportional increase of the flow velocities in the duct if the flow rates are kept constant. However, this may be permissible since the Reynolds number of the duct flow does not change. Hence, the deposition characteristics will not change significantly because the increase in flow velocities is compensated by a reduction of the distance the

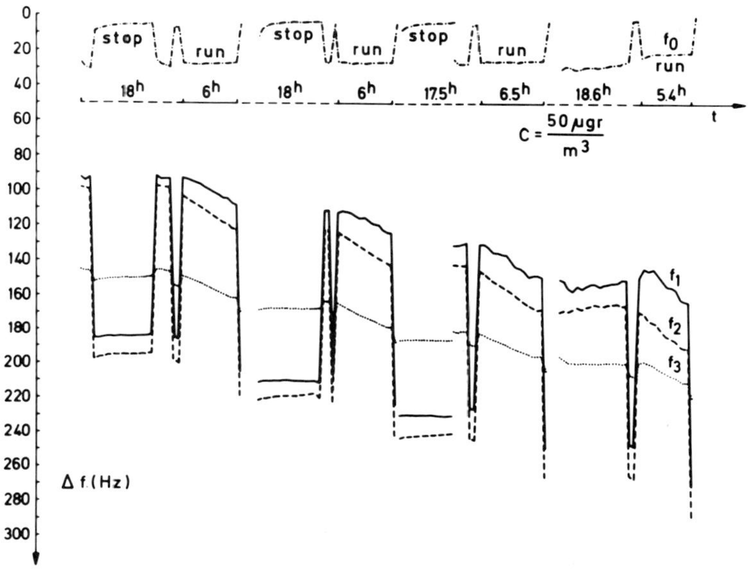

Figure 35 : Recorded Frequency Shifts during Sampling and Rotor Stops of 4 Quartz Oscillators Mounted along the Outer Wall of the Spiral Duct Centrifuge and Used for Monitoring the Deposition of a Fluorescein Test Aerosol of 50 micrograms m^{-3} Airborne Concentration ;
From Top to Bottom : Reference Crystal without Deposit, f_0 ; Crystals at Aerodynamic Sizes of 1.1 µm (f_1), 0.8 µm (f_2), and 0.5 µm (f_3)

aerosol particles have to travel across the duct. Thus, if deposit concentrations as presently achievable within reasonable sampling time are acceptable, they could probably easily be maintained in new designs with reduced cross sectional dimensions of the spiral duct. In addition, there is the chance that the smaller duct cross section may permit higher rotor speeds and, thus, an increase of the deposit concentrations.

It has been shown in a feasibility study (Mönig, Schwarzer and Stöber[34]) that the build-up of the foil deposit in the original spiral duct centrifuge is sufficiently fast to permit the use of piezo-electric quartz crystals as gravimetric sensors for measuring the deposit concentrations of moderate particulate air pollution without going to impractically long sampling times. This sensing technique, first utilized by Olin and Sem[39] in an aerosol mass monitor instrument, does not suffer seriously from the drawbacks described by Daley and Lundgren[5] when applied in a spiral centrifuge. The relative humidity identified as being the most influential disturbing factor can always be kept low or almost constant at the sensor locations when the winnowing gas is dry or kept at a constant humidity level. A rotor insert with four quartz crystals of a nominal resonance frequency of 10 megacycles was used in the feasibility study. A telemetric circuit permitted to measure the frequency while the rotor was spinning and sampling. With random frequency shifts of about ± 3 cycles due to variations of temperature and other surrounding physical conditions, the detection limit of the sensors was below a level of 20 nanograms cm^{-2}. Figure 35 shows a sequence of sampling runs and rotor stops of different duration in the range of hours. The corresponding frequency shifts are given on an arbitrary ordinate scale for the reference crystal receiving no deposit (f_o) and three sampling crystals at locations of 1.1 µm, 0.8 µm and 0.5 µm of aerodynamic size (f_1, f_2, f_3), respectively. The test aerosol was generated from ammonium fluorescein and had an airborne concentration of 50 micrograms m^{-3}, thus simulating a moderate particulate air pollution level. The graph shows that measurements during sampling as well as when the rotor was stopped were consistent, although a frequency shift between the two situations was involved. This shift was reproducible but individual for each crystal. The graph clearly indicates that significant readings were obtained in little more than an hour or two. This would make the spiral duct centrifuge an aerosol mass distribution monitot with a delay time of the order of an hour.

REFERENCES

1. Berner, A., and H. Reichelt, The Rotating-Slit Aerosol Spectrometer (ROSL-Spectrometer): Prototype, in : Aerosol Research at the First Physics Institute, Univ. of Vienna, Status Report, 1 - 18, January 1968

2. Berner, A., and H. Reichelt, Über Einlaß-Spaltsysteme in Konifugen, Teil I : Das ROSL-System, Staub 29 : 92 - 95, 1969

3. Blachman, M.W., and M. Lippmann, Performance Characteristics of the Multicyclone Aerosol Sampler, American Industr. Hyg. Assoc. Journ. 35 : 311 - 326, 1974

4. Boose, C., Ein Gerät zur fraktionierten Abscheidung von Staub für Korngrößenanalysen, Staub 22 : 109 - 112, 1962

5. Daley, P.S., and D.A. Lundgren, The Performance of Piezoelectric Crystal Sensors Used to Determine Aerosol Mass Concentrations, American Industr. Hygiene Conference, Paper No. 3, Miami Beach, Fla., May 12 - 17, 1974

6. Davies, C.N., Particle-Fluid Interaction, Proc. of Harold Heywood Memorial Symposium, Loughborough, England, September 17 - 18, 1973

7. Dean, W.R., The Stream-line Motion of Fluid in a Curved Pipe, Phil. Mag., Ser. 7, 5 : 673 - 695, 1928

8. Ferron, G.A., and H.W.J. Bierhuizen, The Measurement of Polydisperse Aerosols with the Spiral Centrifuge, J. Aerosol Sci. (in press)

9. Fuchs, N.A., and A.G.Sutugin, Generation and Use of Monodisperse Aerosols, in C.N. Davies (ed.): Aerosol Science, Academic Press, London and New York, 1 - 30, 1966

10. Goetz, A., H.J.R. Stevenson and O. Preining, The Design and Performance of the Aerosol Spectrometer, J. Air Poll. Contr. Assoc. 10 : 378 - 383, 1960

11. Heard, M.J., A.C. Wells and R.D. Wiffen, A Re-determination of the Diameters of Dow Polystyrene Latex Spheres, Atmosph. Environm. 4 : 149 - 156, 1970

12. Heyder, J., and J. Porstendörfer, Comparison of Optical and Centrifugal Aerosol Spectrometry: Liquid and Non-spherical Particles, J. Aerosol Sci. 5 : 387 - 400, 1974

13. Hochrainer, D., and P.M. Brown, Sizing of Aerosol Particles by Centrifugation, Environ. Sci. Technol. 3 : 830 - 835, 1969

14. Inhalation Toxicology Research Institute Annual Reports, Lovelace Foundation, National Technical Information Service, Springfield, Va., 22151, LF-46, 1972 - 1973 and LF-49, 1973 - 1974

15. Junge, C., Die Rolle der Aerosole und der gasförmigen Beimengungen der Luft im Spurenstoffhaushalt der Troposphäre, Tellus 5 : 1 - 26, 1953

16. Kast, W., Neues Staubmessgerät zur Schnellbestimmung der Staubkonzentration und der Kornverteilung, Staub 21 : 215 - 223, 1961

17. Keith, C.H., and J.C. Derrick, Measurement of the Particle Size Distribution and Concentration of Cigarette Smoke by the "Conifuge", J. Colloid Sci. 15 : 340 - 356, 1960

18. Kops, J., G. Dibbets, L. Hermans and J.F. van de Vate, The Aerodynamic Diameter of Branched Chain-like Aggregates, J. Aerosol Sci. 6: XX - XX, 1975 (in press)

19. Kops, J., L. Hermans and J.F. van de Vate, Calibration of a Stöber Centrifugal Aerosol Spectrometer, J. Aerosol Sci. 5 : 379 - 386, 1974

20. Kotrappa, P., Shape Factors for Quartz Aerosol in Respirable Size Range, J. Aerosol Sci. 2 : 353 - 359, 1971

21. Kotrappa, P., Shape Factors for Aerosols of Coal, UO_2 and ThO_2 in Respirable Size Range, in: T.T. Mercer, P.E. Morrow and W. Stöber (ed's.), Assessment of Airborne Particles, C.C. Thomas, Springfield, Ill., 331 - 356, 1972

22. Kotrappa, P., Reply to "Comments on 'Design and Performance of the Lovelace Aerosol Particle Separator'", Rev. Sci. Instrum. 44 : 1438, 1973

23. Kotrappa, P., and M.E. Light, Design and Performance of the Lovelace Aerosol Particle Separator, Rev. Sci. Instrum. 43 : 1106 - 1112, 1972

24. Kotrappa, P., and M.E. Light, Centrifuge Separator, U.S. Patent No. 3, 698, 626; October 17, 1972

25. Kotrappa, P., and O.R. Moss, Production of Relatively Monodisperse Aerosols for Inhalation Experiments by Aerosol Centrifugation, Health Physics 21 : 531 - 535, 1971

26. Kotrappa, P., and C.J. Wilkinson, Measurement of the Specific Radioactivity with Respect to Particle Size for Labeled Aerosols, J. Aerosol Sci. 3 : 167 - 171, 1972

27. Kotrappa, P., and C.J. Wilkinson, Densities in Relation to Size of Spherical Aerosols Produced by Nebulization and Heat Degradation, American Industr. Hyg. Assoc. Journ. 33 : 449 - 453, 1972

28. Kotrappa, P., C.J. Wilkinson and H.A. Boyd, Technology for the Production of Monodisperse Aerosols of Oxides of Transuranic Elements for Inhalation Experiments, Health Physics 22 : 837 - 843, 1972

29. Lassen, L., Ein einfacher Generator zur Erzeugung monodisperser Aerosole im Größenbereich 0.15 bis 0.70 μ (Teilchenradius), Z. angew. Physik 12 : 157 - 159, 1960

30. Ludwieg, H., Die ausgebildete Kanalströmung in einem rotierenden System, Ingenieur-Archiv 19 : 296 - 308, 1951

31. Martonen, T.B., to be published at the University of Rochester

32. Martonen, T.B., and W. Stöber, The Influence of the Gas Phase Density of an Aerosol on the Aerodynamic Size Separation of the Aerosol Particles in Spiral Duct Centrifuges, to be published

33. May, K.R., The Cascade Impactor: An Instrument for Sampling Coarse Aerosols, J. Sci. Instr. 22 : 187 - 195, 1945

34. Mönig, F.J., N. Schwarzer and W. Stöber, Bestimmung der Aerosol-Massenverteilung in einer Aerosol-Zentrifuge mit Hilfe von Schwingquarzen, in: V. Böhlau and H. Straubel (eds); Aerosole in Physik,

Medizin und Technik, Jahreskongress 1973 der Gesellschaft für Aerosolforschung, Bad Soden/Ts., pp. 58 - 61, 1973

35. Moss, O.R., H.J. Ettinger and J.R. Coulter, Aerosol Density Measurements Using a Modified Spiral Centrifuge Aerosol Spectrometer, Environ. Sci. Technol. 6 : 614 - 617, 1972

36. Newton, G.J., O.G. Raabe, R.L. Yarwood and G.M. Kanapilly, Generation of Monodisperse Aerosols of ^{67}Ga-Labeled Aluminosilicate an ^{198}Au-Labeled Gold Spheres, present at the Symposium on Fine Particles, Minneapolis, Minnesota, May 28 - 30, 1975

37. Oeseburg, F., and R. Roos, Technical Note: Improvement of the Bearing System of the Stöber Aerosol Spectrometer, Atmosph. Environm. 9 : XX - XX, 1975

38. Oeseburg, F., F.M. Benschop and R. Roos, Particle Size Analysis of Dioctyl Phthalate Aerosols Using the Stöber Aerosol Spectrometer, J. Aerosol Sci. 6 : XX - XX, 1975 (in press)

39. Olin, J.G., and G.J. Sem, Piezoelectric Microbalance for Monitoring the Mass Concentration of Suspended Particles, Atmosph. Environm. 5 : 653 - 668, 1971

40. Porstendörfer, J., Die Bestimmung der Größenverteilung von Aerosolen mit Hilfe der radioaktiven Markierung und der Spiralzentrifuge, J. Aerosol Sci. 4 : 345 - 354, 1973

41. Raabe, O.G., Instruments and Methods for Characterizing Radioactive Aerosols, IEEE Transact. Nucl. Sci. NS 19 : 64 - 75, 1972

42. Raabe, O.G., H.A. Boyd, G.M. Kanapilly, C.J. Wilkinson, and G.J. Newton, Development and Use of a System for Routine Production of Monodisperse Particles of $^{238}PuO_2$ and Evaluation of Gamma Emitting Labels, Health Physics 28 : 655 - 667, 1975

43. Sawyer, K.F., and W.H. Walton, The "Conifuge" - A Size-separating Sampling Device for Airborne Particles, J. Sci. Instr. 27 : 272 - 276, 1950

44. Spertell, R.B., and M. Lippmann, Airborne Density of Ferric Oxide Aggregate Microspheres, American Industr. Hyg. Assoc. Journ. 32 : 734 - 740, 1971

45. Stöber, W., Design and Performance of a Size-Separating Aerosol Centrifuge Facilitating Particle Size Spectrometry in the Submicron Range, in: Assessment of Airborne Radioactivity, Internat. Atom. Energy Agency, Vienna, 393 - 404, 1967

46. Stöber, W., Dynamic Shape Factors of Nonspherical Aerosol Particles, in: T.T. Mercer, P.E. Morrow and W. Stöber (ed's.), Assessment of Airborne Particles, C.C. Thomas, Springfield, Ill., 249 - 288, 1972

47. Stöber, W., Comment on Design and Performance of the Lovelace Aerosol Particle Separator, Rev. Sci. Instrum. 44 : 1149 - 1150, 1973

48. Stöber, W., and H. Flachsbart, Aerosol Size Spectrometry with a Ring Slit Conifuge, Environ. Sci. Technol. 3 : 641 - 651, 1969

49. Stöber, W., and H. Flachsbart, Size-Separating Precipitation in a Spinning Spiral Duct, Environ. Sci. Technol. 3 : 1280 - 1296, 1969

50. Stöber, W., and H. Flachsbart, High Resolution Aerodynamic Size Spectrometry of Quasi-Monodisperse Latex Spheres with a Spiral Centrifuge, J. Aerosol Sci. 2 : 103 - 116, 1971

51. Stöber, W., and H. Flachsbart, An Evaluation of Nebulized Ammonium Fluorescein as a Laboratory Aerosol, Atmosph. Environm. 7: 737 - 748, 1973

52. Stöber, W., and J.S. Osborne, On the Limitation of Aerodynamic Size Spectrometry of Dense Aerosols, to be published

53. Stöber, W., and U. Zessack, Zur Messung von Aerosol-Teilchengrößenspektren mit Hilfe von Zentrifugalabscheidern, Zentralbl. biol. Aerosolforsch. 13 : 263 - 281, 1966

54. Stöber, W., A. Berner and R. Blaschke, The Aerodynamic Diameter of Aggregates of Uniform Spheres, J. Colloid Interf. Sci. 29 : 710 - 719, 1968

55. Stöber, W., H. Flachsbart and C. Boose, Distribution Analyses of the Aerodynamic Size and the Mass of Aerosol Particles by Means of the Spiral Centrifuge in Comparison to Other Aerosol Precipitators, J. Colloid Interf. Sci. 39 : 109 - 120, 1972

56. Stöber, W., H. Flachsbart and D. Hochrainer, Der aerodynamische Durchmesser von Latexaggregaten und Asbestfasern, Staub 30 : 277 - 285, 1970

57. Stöber, W., N. Franzes and W. Steinhanses, Discussion: Improvement of the Bearing System of the Stöber Aerosol Spectrometer, Atmosph. Environm., in press

58. Stöber, W., E. Hederer and H. Horvath, unpublished data, Rochester, N.Y. and Vienna, Austria, 1971

59. Stöber, W., D. Hochrainer and H. Flachsbart, A Simplified Spiral Centrifuge for Aerosol Mass Distribution Analysis, American Industr. Hyg. Conference, Paper No. 96, Miami Beach, Fla., May 12 - 17, 1974

60. Stöber, W., D. Hochrainer, H. Flachsbart, N. Franzes and W. Steinhanses, Fraktionierung und gravimetrische Bestimmung atmosphärischer Partikel mit einer kleinen Spiralkanal-Zentrifuge, to be published

61. Tillery, M.I., A Concentric Aerosol Spectrometer, American Industr. Hyg. Assoc. Journ. 35 : 62 - 74, 1974

62. Timbrell, V., The Terminal Velocity and Size of Airborne Dust Particles, Brit. J. Appl. Phys. 5, Suppl. 3, 86 - 90, 1954

63. van Buitenen, C.J.P., and F. Oeseburg, Comparison of "Light Scattering Diameter" Based on Forward Scattering Measurements and Aerodynamic Diameter of Aerosol Particles, Atmosph. Environm. 8 : 885 - 896, 1974

64. van de Vate, J.F., The Safety of SNR-300 and the Aerosol Model, A Summary Report of the RCN Aerosol Research 1967 - 1971, Reactor Center Nederland Report No. 174, 23 - 24, 1972

65. VDI-Richtlinie 3491, Entwurf, Blatt 1: Messen von Partikeln, Prüfkriterien und Prüfmethoden für Verfahren und Geräte zum Bestimmen partikelförmiger Beimengungen in Gasen, Begriffe und Definitionen; Verein Deutscher Ingenieure, Düsseldorf, pp. 1 - 8, July 1975

66. Walkenhorst, W., Untersuchungen an einem nach Teilchengrößen geordneten Mischstaub im atembaren Korngrößenbereich, in: C.N. Davies (ed.), Inhaled Particles and Vapours II, Pergamon Press, Oxford, 563 - 571, 1967.

PROBLEMS IN STACK SAMPLING AND MEASUREMENT

D. B. Harris and W. B. Kuykendal
Process Measurements Section
Control Systems Laboratory
Environmental Protection Agency

ABSTRACT

Though many techniques for monitoring particulates have been studied in great detail under laboratory conditions, the application to field situations of these techniques has often been difficult. This paper describes the attempts to use cascade impactors for manual fractional efficiency and optical and beta gauging for instrumental particle sizing. Though each technique seemed to be ideally suited to particle measurement, the many different problems presented by industrial gas streams and the solutions used to overcome them are described. Particle bounce and reentrainment was the first problem encountered with the impactors and was quickly followed by substrate material selection due to reaction with gaseous components. Instrumental problems have centered on getting sensors to perform in the hostile stack environment and on making reasonable compromises necessary to apply these techniques to real stack environments.

PROBLEMS IN STACK SAMPLING AND MEASUREMENT

D. B. Harris and W. B. Kuykendal
Process Measurements Section
Control Systems Laboratory
Environmental Protection Agency

As the concern about the effects of air pollution became focused on meeting more stringent emission standards, it became apparent that some information on the degree of control of particulate emissions as a function of size was necessary. The Process Measurements Section of the Control Systems Laboratory quickly became convinced that external size measurements of collected masses of material did not yield useful data and that some means of sizing the particles as they exist in the ducting around the control device was needed. Thus, development efforts were initiated to provide the needed techniques.

MANUAL METHODS

The first effort was to examine those devices already available to determine if one could be used with little modification so that useful information could be gathered without embarking on a long and costly development program. Laboratory tests indicated that some problems existed from particle bounce off the hard collection surfaces of the commercial impactor units identified as the best existing method. Problems were also noted during weighing caused by the enormous ratio of the mass of the collector plates to the mass of material caught. Experimental low tare weight substrates were made and preliminarily tested in the lab with encouraging results. Because of the need for a field method, we moved to an extensive field test of the available commercial units and a few experimental devices.

A three-week test, including 142 separate sizing device tests, was conducted at the Marquette Station of the Upper Pennisula Generating Company. An example of the results obtained is given in Figure 1 and shows good agreement among the sizing devices tested. A full description of this program is presented in Reference 2. From this study, the first operational guides were set forth. To ensure that inlet and outlet measurements covered the same time span, low flow rates were suggested for the inlet measurements and high flow rates for the outlet measurements. Substrates were found not only to enhance the weighing, but also to reduce the bounce problem if greased metallic foils or fiber glass was used. Thus, we felt that we could recommend a method for determining fractional efficiency of particulate control devices based on commercially available impactors.

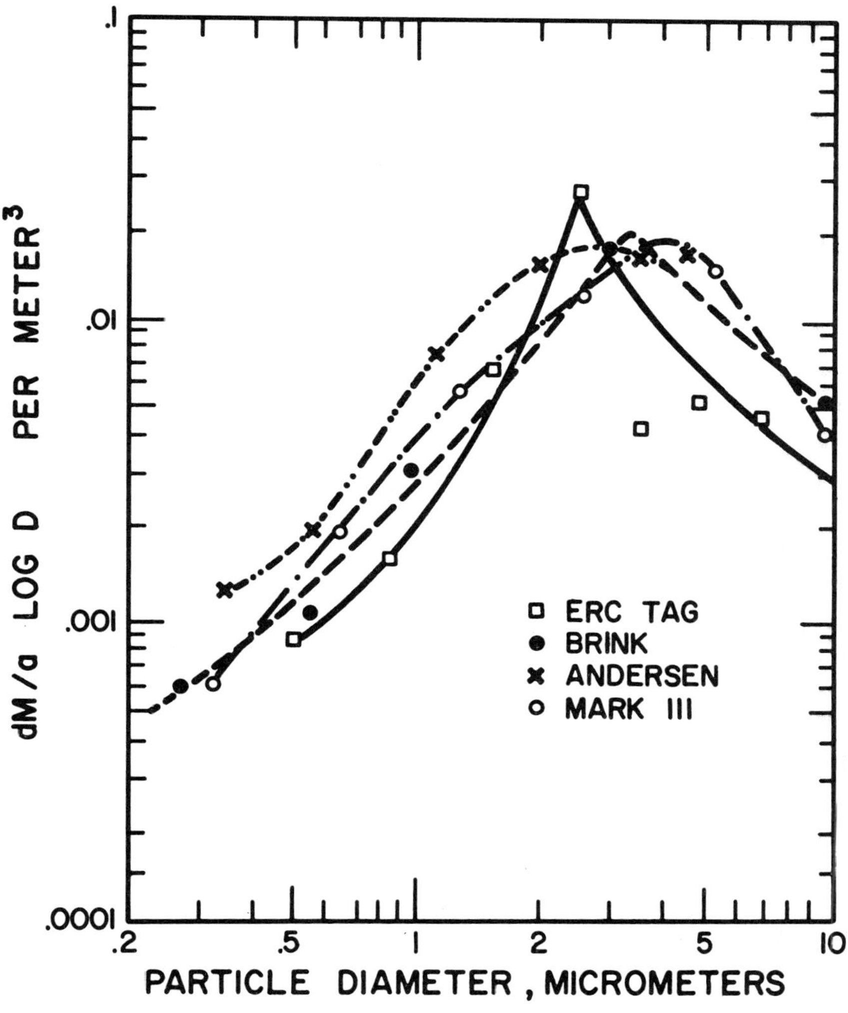

Figure 1 Impactor Performance Comparison.

Our testing proceeded following the new impactor method. Within a few months, reports of net weight losses were received from a number of our testing contractors when using fiber glass substrates. A number of quick tests were performed by several groups including EPA and it was concluded that the "type A" glass was susceptible to handling losses if anything other than extreme care was used.[3] The filter manufacturers suggested using a new "type E" glass with better mechanical properties. Laboratory tests indicated that it did improve the weighing accuracy, so one manufacturer began to supply its substrates in "type E" fiber glass.

A weight loss problem was also reported when using greased metallic foils. Most of the people reporting the problem were using silicone stopcock grease which had been used earlier with no abnormal weight loss. Tests run by EPA confirmed the degradation of some grease samples as a function of increasing temperature in the range found in process flow streams. We concluded that the problem was one of quality control of a mass produced item. As a result of our tests with a number of greases, we recommended that gas chromatographic support greases be used because of the high level of quality control applied to the product. The temperature stability could be quickly ascertained from the specification sheet.

The new grade of fiber glass performed adequately until our tests were conducted on high temperature or high SO_x gas streams. Excessive weight gains were noticed by many investigators working in these gas flows. It became apparent that the problem was serious enough to cast great doubt on the total results of a number of tests. The search for an answer to this gas reaction problem has been one of the longest and most intensive of the particle sizing program. A detailed description of the investigation will soon be available. Briefly, a number of filter materials were tested within these problem environments by using the impactor with a prefilter to remove the particulate. Any weight gained by a substrate should then be only by gas/solid reaction. The results are summarized in Tables 1 and 2. The tables show that Teflon is unreactive and a probable choice for temperature regimes under 260° C but we have not investigated this material for reentrainment caused by the slippery surface. Some fiber glasses were found to be significantly less reactive than those we had been using. We believe that the pH of the glass is the driving force behind the reactivity or lack thereof. Figure 2 shows the effect of stack temperature on the reactivity of these materials. As a result of this investigation we believe that a substrate material can be found in the group which shows low level reactivity; however, further testing is needed to find the right one. Thus, an approach to the development of an acceptable method has been the testing of a successful laboratory technique under various field applications. Other particulate measurement techniques are also undergoing a similar development process.

Table 1. ANOMALOUS FILTER WEIGHT GAINS[4]
Cement Plant

		CC-1	CC-2	CC-3	CC-4
	Date	2-12-75	2-18-75	2-19-75	2-21-75
mm-Hg	Amb. Pres.	754.4	749.3	754.4	759.5
°C	Amb. Temp.	16.6	20.0	8.88	14.4
°C	Stack Temp.	266	249	249	262
Std. liters	Gas Vol.	348	674	680	936
°C	Avg. Gas Meter Temp.	14.4	23.3	11.6	26.1
Min.	Run Time	60	120	120	240
mm	Ori. ID	1.42	1.42	1.42	1.42
ml	Cond. H_2O	74	121	131	97
	% H_2O	(29.6) 22% used for flowrate	(26.6) 22.0	(27.2) 22.0	(28.0) 22.0
Actual liter/min.	Flow Rate:				
	Ori.	11.72 } 12.06	10.84 } 10.95	11.07 } 11.32	4.16 } 4.24
	Gas Meter	12.40	11.04	11.61	4.34
mm-Hg	Avg. Probe ΔP	60.96	76.2	76.2	381
mm-"g	Ori. ΔP	20.7	19.4	19.6	4.9

Filter No.	CC-1			CC-2			CC-3			CC-4	
1a	GA*	3.22 mg		GA	10.58		GA	9.48		GA	4.38
1b	-	-	Silicone o-rings stuck to filters	GA	0.44		GA	0.84		MSA 1106 BH	2.14
2	SA*	0.28		SA	0.92	Slight brown ring	MSA 1106 BH	1.42		Teflon	-0.02
3	GA	-0.48		RA* 900 AF	1.16		RA 900 AF	1.70		GA	0.90
4	-	-		MSA 1106 BH	1.02	Slight brown ring	SA	1.66	Stuck to metal support	RA 900 AF	2.40
5	-	-		-	-					MSA 1106 BH	1.92
6	-	-		-	-					Teflon	0.00
				Teflon o-rings			Teflon o-rings			Teflon o-rings	

*GA - Gelman type A
RA - Reeves Angel
SA - Gelman Spectrograde

Table 2. ANOMALOUS FILTER WEIGHT GAINS[4]
Coal-Fired Boiler

		BRSP-1	BRSP-2	BRSP-3	BRSP-4	BRSP-5	BRSP-6	BRSP-7
	Date	2-25-75	2-25-75	2-26-75	2-26-75	2-27-75	2-28-75	2-28-75
mm-Hg	Amb. Pres.	738.6	738.6	747.7	747.2	745.7	739.1	739.1
°C	Amb. Temp.	12.0	16.0	11.6	20.5	17.7	5.0	10.0
°C	Stack Temp.	135	129	135	135	149	179	179
Std. liters	Gas Vol.	313	530	2,290	970	3,970	240	-
°C	Avg. Gas Meter Temp.	27.2	22.7	27.2	27.7	26.0	23.3	-
Min.	Run Time	60	60	240	120	480	30	0
mm	Ori. ID	3348-.059	3348-.059	3348-.059	3348-.059	3348-.059	3348-.059	-
ml	Cond. H_2O	6.0	11.4	57.5	23.8	90.8	4.8	-
	% H_2O	(3.8) 7.5 used in flow-rate calculations	(4.9) 7.5	(5.6) 7.5	(5.7) 7.5	(5.7) 7.5	(4.6) 7.5	-
Actual liters/min.	Flow Rate: Ori.	5.18	7.05	8.09	6.54	6.57	7.53	-
	Gas Meter	5.35	7.3	8.07	6.76	6.65	7.61	-
		5.26	7.19	8.09	6.65	6.57	7.58	
mm-Hg	Avg. Probe ΔP (+ΔP across orifice)	223	321	320	322	354	317.5	-
mm-Hg	Ori. ΔP	12.1	27.8	13.6	23.9	23.7	24.0	-
	1a	GA* 12.29 mg	GA 26.50	GA -	GA 37.99	GA -	GA 23.41	GA 0.34
	1b	MSA 1106 BH 0.43	GA 0.44	RA 900 AF 0.77	MSA 1106 BH 0.88	GA 3.27 Moderately brown on edge	MSA 1106 BH 0.69	GA 0.38
	2	GA 0.16	RA 900 AF 0.69	MSA 0.16	RA 900 AF 0.83	MSA 1106 BH 3.65 Moderately brown on edge	SA -0.12	MSA 1106 BH 0.77
	3	RA* 900 AF 0.37	MSA 1106 BH -0.14 Stuck to o-ring & support severely	Teflon 0.01	GA 0.24			
	4	Teflon 0.01	Teflon 0.00	SA* 0.12	SA 0.18	GA 0.53 No discoloration	SA 0.27	GA -0.10 Torn
	5	MSA 1106 BH 0.27	RA 900 AF 0.56	GA -0.04 Severely cut	GA 0.41	GA 0.53 No discoloration	SA 0.66	GA 0.38
	6	GA 0.19	GA -0.74 Stuck to o-ring & cut by it	SA 0.31	Teflon 0.07	Teflon 0.00	Teflon -0.01	MSA 1106 BH 0.06
						MSA 1106 BH 1.00 Slight browning	MSA 1106 BH 0.71	Teflon 0.06
			All filters stuck slightly to support. As much as possible recovered.	This (and all that follow) are Spectro Grade A from Batch 8192.				This set not run; placed in stack and taken out immediately.

* GA - Gelman type A
RA - Reeves Angel
SA - Gelman Spectrograde

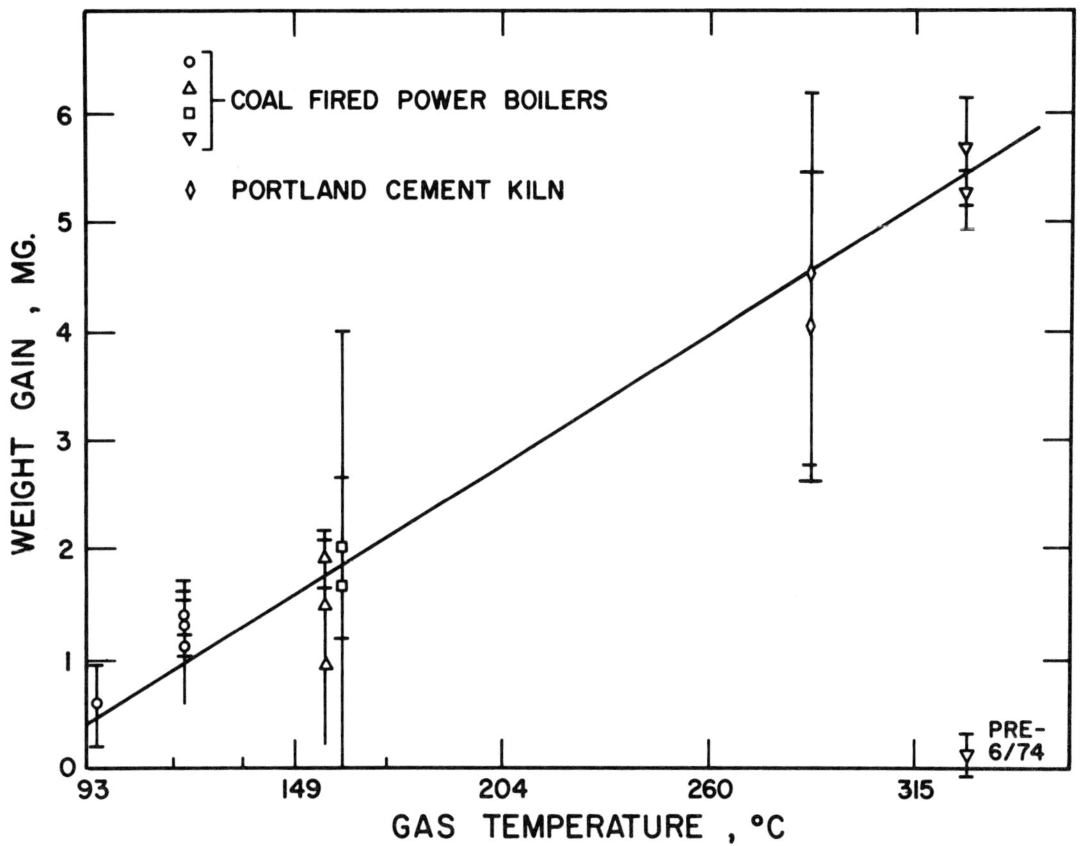

Figure 2 Anomalous weight increases of Andersen glass fiber impaction substrates at different flue gas temperatures.

INSTRUMENTAL TECHNIQUES

Although manual techniques have proven their usefulness and have generated the preponderance of particulate data, most will agree that they are cumbersome to implement. A more desirable approach would utilize an instrumental technique to overcome the drawbacks of the manual methods.

Before proceeding further it will be useful to define an ideal particle sizing instrument. To do this the purpose and application of the instrument must first be defined. For the purposes of this discussion the intended use of the data will be in the development and evaluation of control devices for stationary sources; more specifically for coal-fired utility boilers.

The ideal instrument must first measure the particle size in-situ without altering the flow field or size distribution. Although the instrument should be capable of sizing the particles in a nondisruptive fashion, it would also be desirable to collect a sufficient size fractionated sample for chemical and toxicological analysis. The data output from this instrument should be in real time so that control devices could be optimized on-line. In addition, the instrument should have two or more sensors so that both inlet and outlet measurements could be made simultaneously. The requirement for sampling before and after control devices is also an implicit requirement for a large dynamic range in particle mass and number concentration. Values for this concentration would typically be from 0.01 to 25 g/m^3 or 10 to 10^{10} particles/cc.

The question of the size range of the ideal instrument could well be the subject of a separate discussion in itself. First, the type of particle size to be measured must be determined. The linear dimension as measured by a microscope could be useful in determining filtration characteristics of particles. The optical equivalent diameter, measured by single particle light scattering instruments, is useful in determining the effect of particles on visibility. The aerodynamic diameter, measured in a variety of ways, is an indication of how particles behave in gas flow fields and appears to be the size parameter of most importance in air pollution research.[6] Other size parameters could also be selected and doubtless have specific applications, but these will not be considered here. Aerosols have been defined by Fuchs as gasborne solids or droplets whose size range is approximately 0.001 to 1000 micrometers.[7] Although this may serve as a classical definition, a more practical range would be 0.01 to 10 micrometers.

In addition to the above ideal characteristics, the instrument must be capable of performing the particle size measurements in the hostile stack environment. Temperatures can range as high as 450° C before a hot electrostatic precipitator, but a more typical value for industrial gas streams is 200° C. The components of the instrument should be portable and should require minimum support facilities.

Clearly, no existing or foreseeable instrument can meet all of these criteria. However, several instruments are in various stages of development which meet some of these requirements and at the same time illustrate some of the problems inherent in the development of particle sizing instruments.

One of the most ambitious efforts to date has been the development of an in-stack cascade impactor with real time mass detection for each individual stage via beta attenuation. The instrument, though in-stack, is still not truly in-situ. Because it uses an inertial impactor for aerodynamic size discrimination in the range of 0.2 to 5.0 micrometers, it shares some of the problems discussed above for manual impactors. A major problem in the development of the instrument has been the difficulty in obtaining beta detectors that will operate at 200° C. By the very nature of the detection principle employed, the instrument is complex and expensive.[8]

Another instrument is under development which uses the beta attenuation principle to detect the mass of particulate for each stage of an inertial impactor. In this instrument a virtual impaction scheme is used outside of the stack and the particulate sample is transported to the instrument in a probe. The aerodynamic size range is 0.07 to 10 micrometers. Although this approach sacrifices in-situ sizing, the operational problems of the sizing/detection system are significantly reduced by removing the instrument from the stack environment. Nevertheless, this instrument is still quite complex and expensive.[9]

Work is also underway to develop an in-stack cascade impactor with real time readout based on a detection system sensitive to the amount of particulate removed by each stage. A virtual impactor is used for the size discrimination over the range of 0.2 to 5.0 micrometers and a simplified sensor is used to measure the size fraction for each stage. Three candidate sensors are being evaluated: (1) light attenuation, (2) electric charge transfer, and (3) pressure drop across a filter. As with the in-stack beta sensing instrument, this instrument is still not truly in-situ. Because the sensor does not actually measure the mass of particulate collected in each size interval the instrument does not truly measure aerodynamic particle size. However, the detection principle is simpler and the cost of the instrument would be less than that of either of the beta sensing instruments.[10]

An instrument has also been developed which uses a laser light scattering technique to size particulate in-stack. Because the instrument optically defines the sensing volume it can be considered in-situ. Since it is an optical instrument it senses the optical equivalent diameter instead of the preferred aerodynamic diameter. This instrument, like all present optical sizing instruments, is a single particle counter and its maximum number concentration is limited by its sensing volume to 10^6 particles per cm^3. The size range for the instrument is 0.2 to 3.0 micrometers. Since a pulsed laser system is used and only one particle is sized per pulse, a finite time period (approximately five minutes) is required to measure the size distribution and the instrument is not actually real time. Unlike the impactor based systems discussed above, it is not possible to collect a particulate sample for subsequent analysis. The instrument is relatively simple and cost is less than for the other techniques.[11]

Having now defined an ideal instrument and discussed several instruments currently under development it would appear to be desirable to define a practical instrument in light of the problems discussed above. Unfortunately, this cannot be done au priori. Many factors must be taken into consideration. In previously defining an ideal instrument, cost was not considered whereas in any real system it will have a significant impact. The trade-off of whether or not to place the sensor in the stack must be made. In-situ measurements are always desirable, but if the stack environment becomes too hostile, a remote location would be the choice. The major decision point in the design process is the selection of the sizing/sensing mechanism. Each user must carefully trade off the available techniques and his requirements against the problems to be encountered in applying these techniques.

In short, the selection of any particulate sizing technique, be it manual or instrumental, must consider the practical limitations of that technique.

REFERENCES

1. McCain, J. D., K. M. Cushing, and A. N. Bird, Jr.. Field Measurements of Particle Size Distribution with Inertial Sizing Devices. U.S. Environmental Protection Agency, Washington, D. C. EPA-650/2-73-035, October 1973, p. 45.

2. Ibid.

3. Smith, W. B., K. M. Cushing, and G. E. Lacey. Andersen Filter Substrate Weight Loss. U.S. Environmental Protection Agency, Washington, D. C. EPA-650/2-75-022, February 1975.

4. Smith, W. B., K. M. Cushing, and G. E. Lacey. Particulate Sizing Techniques for Control Device Evaluation. Southern Research Institute, Birmingham, Alabama. EPA Contract 68-02-0273, Report No. 29, March 15, 1975, pp. 4-5.

5. Smith, W. B., K. M. Cushing, and G. E. Lacey. Particulate Sizing Techniques for Control Device Evaluation. Southern Research Institute, Birmingham, Alabama. EPA Contract 68-02-0273, Report No. 30, April 15, 1975, p. 14.

6. Lilienfield, P. and D. Cooper. Literature Review and Selection of Design Principle for the Development of a Fine Particulate Source Testing Instrument. GCA Technology Division, Bedford, Mass. Publication Number GCA-TR-73-22-G. October 1973. p. 30-31.

7. Fuchs, N. A. The Mechanics of Aerosols. New York, The MacMillan Co., 1964.

8. Lilienfield, P. Personal communication. May 1975.

9. Wagman, J. and C. M. Peterson. In: Proceedings of the 3rd International Clean Air Congress. Dusseldorf, Ger., October 8-12, 1973. p. C6-8.

10. Burckle, J. Personal communication. May 1975.

11. Shofner, F. M., G. Kreikebaum, H. W. Schmitt, and B. E. Barnhart. Environmental Systems Corp. In-Situ, Continuous Measurement of Particulate Size Distribution and Mass Concentration Using Electro-Optical Instrumentation. (Presented at the Fifth Annual Industrial Air Pollution Control Conference. Knoxville, Tenn. April 3-4, 1975).

INERTIAL IMPACTORS: THEORY, DESIGN AND USE

by

Virgil A. Marple and Klaus Willeke

Particle Technology Laboratory
Mechanical Engineering Department
University of Minnesota
Minneapolis, Minnesota 55455

ABSTRACT

Inertial impactors are devices used to classify particles with respect to their aerodynamic size. The theoretical analyzing techniques currently available are sufficiently accurate to predict the impaction characteristics of impactors, provided the impactor design conforms reasonably well with the boundary condition used in the theoretical analysis. Using results of a theoretical study employing these techniques, design criteria are given for such parameters as jet-to-plate distance, jet Reynolds number, and jet throat length to obtain sharp cut-off characteristics for both round and rectangular impactors. Precautions which must be taken in the use of impactors are discussed, and limitations of impactors due to interstage losses and particle bounce from various surfaces are presented. Various designs of impactor jets and impaction surfaces are examined with special attention given to the use of multiple jet impactors for the purpose of controlling the jet Reynolds number. Also, the principle operating characteristics of commercial cascade impactors are presented.

INERTIAL IMPACTORS: THEORY, DESIGN AND USE

by

Virgil A. Marple and Klaus Willeke

Particle Technology Laboratory
Mechanical Engineering Department
University of Minnesota

INTRODUCTION

Inertial impactors have been used extensively for many years to collect airborne particles for size and frequently also for composition analysis. Initially, they came into widespread use because of their simplicity of construction and operation. As modern instruments, they are becoming increasingly popular because recent studies[1,2] have shown that they will classify aerosol particles into very distinct size ranges, if they are properly designed and operated.

All impactors operate under the principle that if a stream of particle-laden air is directed at a surface, particles of sufficient inertia will impact upon the surface and smaller particles will follow the air streamlines and not be collected, as shown in Figure 1(a). Thus, an impactor consists simply of a nozzle, either round or rectangular in shape, and an impaction plate.

The configuration shown in Figure 1(a) is a single stage impactor which separates the particles into two size groups. The large particles are collected on the plate and the small ones remain airborne. One may define a cut-off size of a stage to be the particle size for which 50% of the particles are removed from the air stream and collected on the plate.

By operating several impactor stages at different flow conditions, the aerosol particles are classified into several size ranges from which the size distribution is determined. These single stages can be operated in a parallel or in a series (cascade) arrangement. In the parallel arrangement, each of the stages classifies the airborne particles at different cut-off sizes, so that the difference in the amount of the deposit of any two stages gives the quantity of particles in the particular size interval defined by the respective cut-off sizes of the two stages. In the series arrangement, also known as the cascade impactor, the aerosol stream is passed from stage to stage with continually increasing velocities and decreasing particle cut-off sizes. Thus, each stage acts as a differential size classifier. Of the two flow systems, the cascade arrangement is by far the most popular, as is evident by the many commercial cascade impactors currently available.

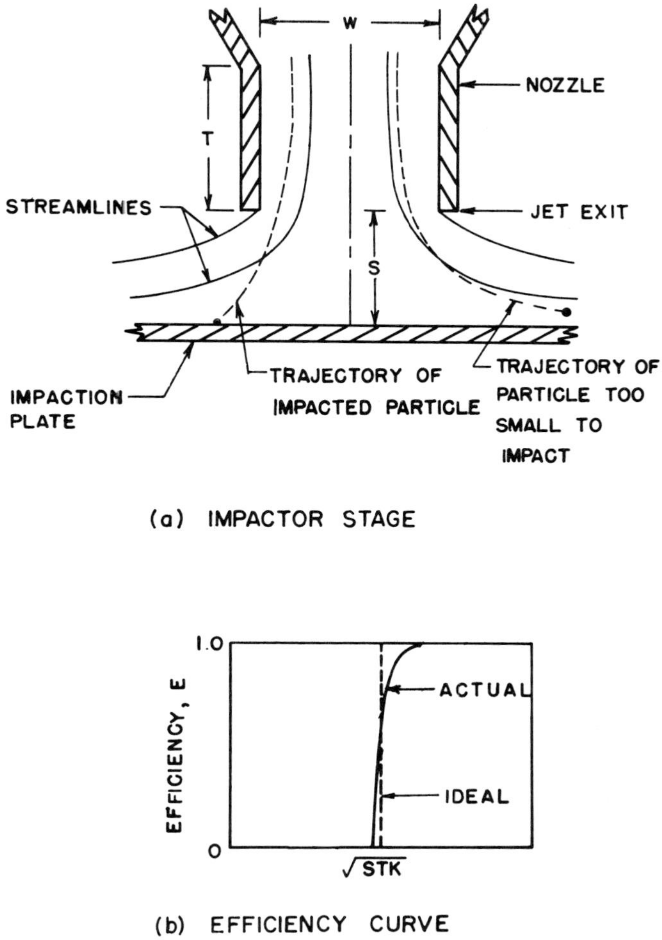

Figure 1 Streamlines, particle trajectories and efficiency curve for a typical impactor

In the conventicnal impactor, the jet is formed in a nozzle (internal flow) and then impacts onto a plate. It is also possible to pass the impaction plate through the particle laden air (external flow). The effectiveness of particle collection in the latter arrangement is comparable to that of conventional impactors[3]. In operation, these impactors normally consist of impaction plates (or cylinders) mounted at the ends of rotating arms.[4,5,6] As the arms are rotated through the air, particles are impacted onto the impaction surface. The size of the particles collected depends upon the speed and width (or diameter) of the impaction surface as well as the size and density of the particles.[3] These devices may be used to collect particles larger than 10 to 20 μm in diameter. Thus, for the collection of large particles, which may be difficult to sample efficiently in a conventional impactor, this type of impactor is a suitable alternative.

The single most important characteristic of an impactor stage is the collection efficiency curve, which gives the fraction of particles of a given size collected from the incident stream as a function of particle size. Ideally, an impactor should collect all particles larger than a certain size and none of the smaller ones, which would correspond to the sharp ideal cut-off illustrated in Figure 1(b). The efficiency curve of a typical real impactor stage, also shown in Figure 1(b), spans over a range of particle sizes, but still has a good "sharpness of cut".

It has been shown[2] that ideal impactor behavior can be obtained if two requirements are met: 1) in the region between the jet exit plane and the impaction plate, the y component of fluid velocity is a function of y only (y is parallel to the centerline of the nozzle); 2) the y component of the velocity of the particles at the jet exit plane is uniform across the jet. These two requirements are reduced to only the first requirement, if the velocity of the particles is equal to the fluid velocity at the exit plane of the jet. Such requirements are nearly met[2] in a real impactor as can be seen by the velocity profiles in Figure 2(a) and (b) for rectangular and round impactors, respectively. In both cases, V_y(or V_z) is quite independent of X(or r) over a major portion of the jet. However, this criteria is not met in the boundary layer near the wall. Particles passing through this region are the cause for the non-ideal cut-off characteristics at the larger efficiencies of the real impactor shown in Figure 1(b).

The determination of the impaction efficiency curves for real impactors has been the object of many studies reported in the literature. However, only recently, with the aid of modern computers, has the aerosol flow in impactors become thoroughly understood. This paper reviews these recent developments, describes how these developments can be used to design impactors with predictable operating characteristics, presents practical limitations of impactor use, and examines the operating characteristics of several commercial impactors which are currently available.

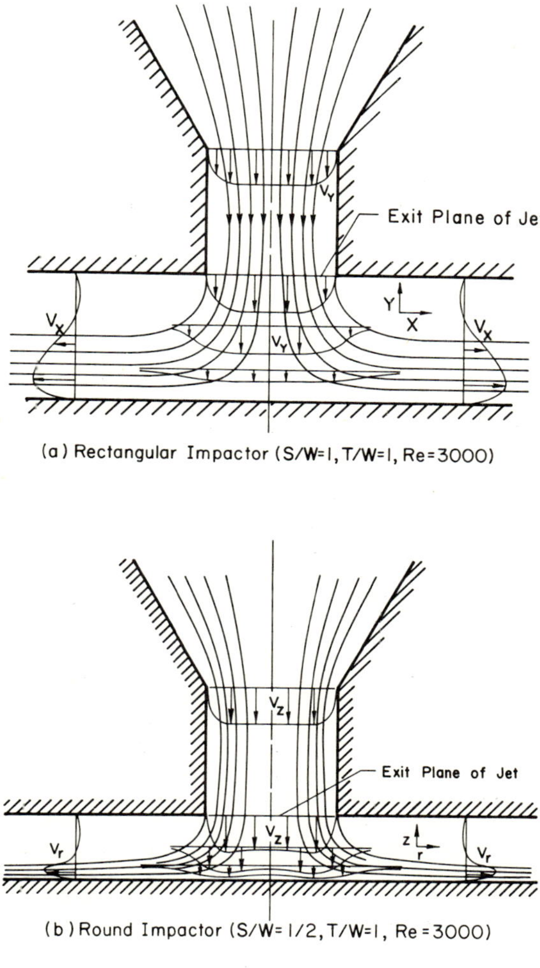

Figure 2 Flow fields in rectangular and round impactors.

THEORETICAL ANALYSIS

In performing a theoretical analysis of inertial impactors it is necessary to first calculate the flow field within the impactor and then determine the particle behavior in this flow field. Thus, an accurate theoretical analysis of particle impaction can only be made if the flow field is precisely known.

A review of the theoretical studies[7] prior to 1970 shows that most of the studies[8,9,10] assumed simple approximations to the flow field. One study[11] made a rigorous analysis based on the method of conformal mapping and on the solution of Euler's equation for the flow of a frictionless fluid through an impactor. However, to fully describe the flow field, the analysis must include the viscous effects of the fluid.[12]

An extensive theoretical study has been made[12] of the flow fields and of the impaction characteristics of rectangular and round impactors. First the flow fields were determined by using numerical methods to solve the Navier-Stokes Equations for viscous flow[13] following closely the technique described by Gossman, et al.[14] Then the particle trajectories were traced through these flow fields by numerically integrating the particles equation of motion[1] from which the impaction characteristics were subsequently found. In the study, the impactors were characterized by the specification of three dimensionless parameters: S/W, T/W and Re, where S = jet-to-plate distance, W = jet width or diameter, T = jet throat length, and Re = jet Reynolds Number based on the hydraulic diameter of the jet, D_h (D_h = W for round jets, D_h = 2W for rectangular jets). The resulting characteristic impaction curves are presented in Figure 3 showing the effects of the parameters S/W, Re, and T/W. Note that the abscissa of these curves, expressed in units of the square root of the Stokes number, $\sqrt{Stk}$, is a dimensionless particle size. The Stokes number is defined as the ratio of the particle stopping distance to the halfwidth or the radius of the impactor throat[15]

$$Stk = \frac{\rho_p C V_o \; D_p^2/18\mu}{W/2} \qquad [1]$$

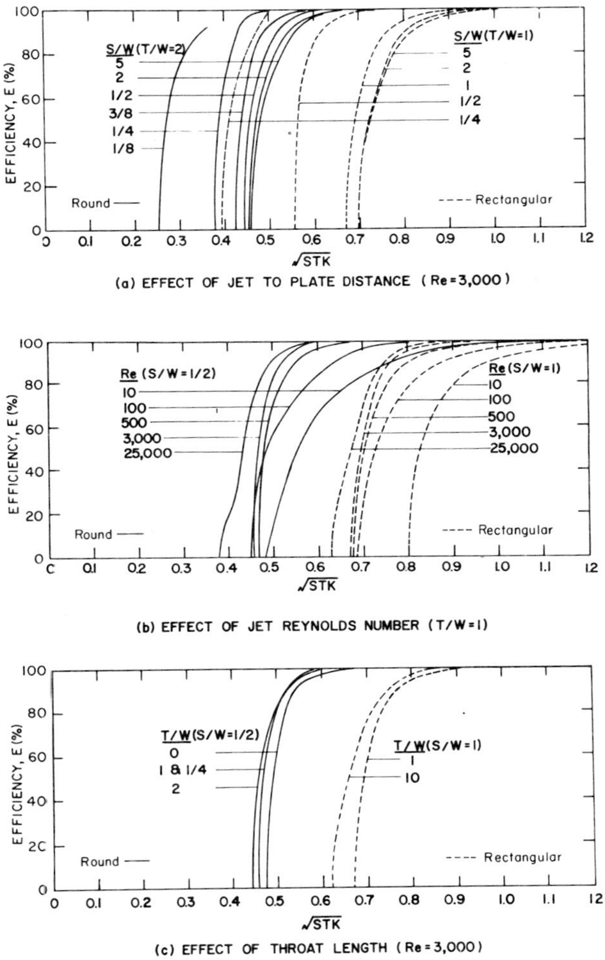

Figure 3. Impactor efficiency curves for rectangular and round impactors, showing the effect of jet-to-plate distance, Reynolds number and throat length. (From Marple, V.A. and Liu, B.Y.H. Environmental Science and Technology 8(7):648-654, 1974. Copyright 1974 by the American Chemical Society.)

where ρ_p is the particle density, C is the Cunningham slip correction factor, V_o is the mean velocity at the throat, D_p is the particle diameter, and μ is the fluid viscosity.

Comparisons of the theoretical efficiency curves of Figure 3 with those of experimental investigations[1,16,17,18,19] have shown that the agreement is very good for both round and rectangular impactors if the impactor inlet conditions, shape, and Reynolds number are similar. It has been shown,[16] that as experimental techniques have been improved, the experimental efficiency curves have approached the predicted theoretical curves.

Several interesting conclusions can be found by examining the curves of Figure 3. First, all of the efficiency curves are similar in shape except for the cases of low Reynolds numbers (Re < 500) or extremely high Reynolds numbers (Re = 25,000). For low Reynolds numbers the poor cut-off characteristics are caused by the thick viscous boundary layer in the jet of the impactor. For high Reynolds numbers, the knee in the efficiency curve at the low values of efficiency appears to be caused by a very thin boundary layer over portions of the impaction plate adjacent to the stagnation point[12]. This thin boundary layer, having a thickness about equal to the particle diameter, allows smaller particles to impact than in areas where the boundary layer is thicker.

Second, the efficiency curve's position on the $\sqrt{Stk}$ axis is relatively independent of these parameters, except for small numbers of S/W or Reynolds numbers. This can be seen more clearly in Figures 4 and 5, where the value of $\sqrt{Stk_{50}}$ is cross-plotted as a function of S/W and Re respectively. In Figure 4, it is important to note that $\sqrt{Stk}$ drops sharply with decreasing jet-to-plate distance for S/W < 1.0 for rectangular impactors and S/W < 0.5 for round impactors.

Third, the round impactor will collect smaller particles than the rectangular impactor for similar values of S/W and Re (i.e., the efficiency curves for round impactors are at smaller values of $\sqrt{Stk}$ than for rectangular impactors). However, it has been shown[17] that if a modified Stokes Number, Stk', is defined as the ratio of the particle stopping distance to half of the hydraulic diameter, D_h,

$$Stk' = \frac{\rho_p \; C V_o D_p^2/18\mu}{D_h/2} \qquad [2]$$

the efficiency curves for both round (D_h = W) and rectangular (D_h = 2W) impactors nearly coincide on the $\sqrt{Stk'}$ axis for the same values of Re, S/D_h and T/D_h .

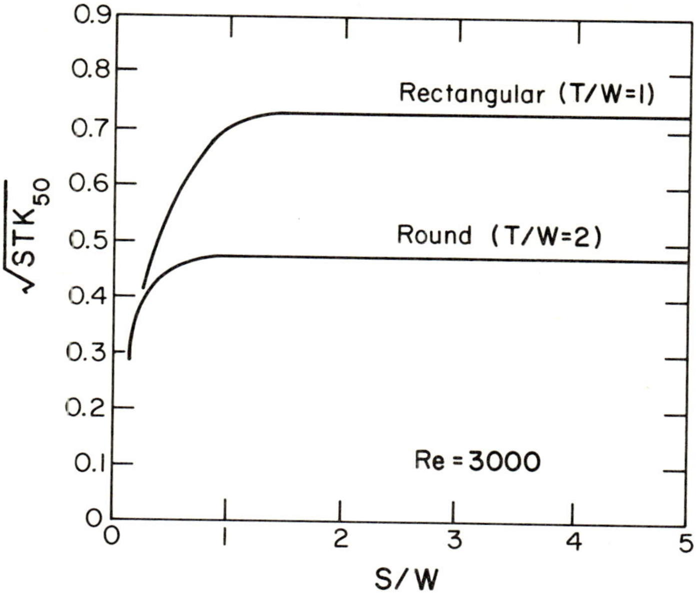

Figure 4 Impactor 50% cut-off size as a function of the jet-to-plate distance.

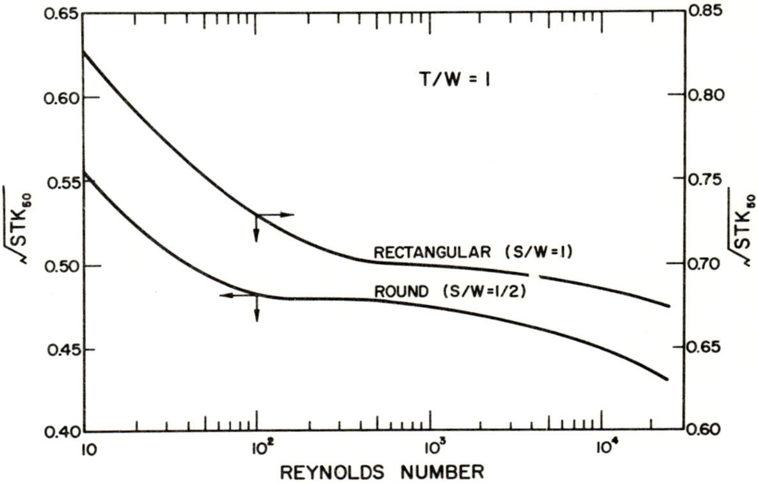

Figure 5 Impactor 50% cut-off size as a function of the Reynolds number.

DESIGN ARRANGEMENTS

Nearly all impactors can be classified into either one of two basic types: round or rectangular. The main variations within these basic types are given by the number of nozzles, by the arrangement of the nozzles, and the shape of the impaction plate.

Nozzle Types and Design Charts

Figure 6 shows the basic nozzle arrangements used in impactor designs. Nearly all impactors can be classified as having a single round nozzle, a single rectangular nozzle, multiple round nozzles, or multiple rectangular nozzles.

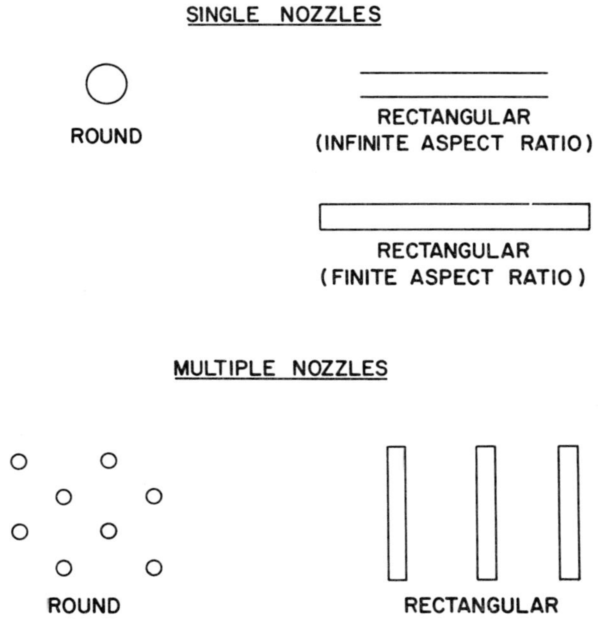

Figure 6 Basic impactor nozzle designs

Theory[12] can be used to predict the impaction characteristics for impactors with round nozzles and rectangular nozzles of infinite aspect ratio (Figure 6). This theory can also be applied to the flow in the central portion of an impactor with a rectangular nozzle of finite aspect ratio. However, because the flow at the ends of a rectangular nozzle is radial, as in the round impactor, smaller particles are impacted out in this region. In order to minimize this end affect, the aspect ratio should be as large as possible and the impactor should be designed so that the air at the ends of the jet is restrained from flowing radially outward. The impaction efficiency curve of a rectangular impactor with small aspect ratio matches that of a round impactor at efficiencies near zero, and that of a rectangular impactor with an infinite aspect ratio at higher efficiencies.

The theoretical predictions may also be applied to multiple nozzle arrangements, if the aerosol flow through each nozzle of a stage is identical. Multiple nozzle impactors of both round and rectangular design are commercially available. One reason for the use of multiple nozzles is the compactness of the design. This is especially true of the rectangular impactor where a long nozzle is divided into a number of small segments. In round impactors, the use of a large number of small diameter nozzles will allow a smaller jet-to-plate distance than for a large single nozzle.

Also, for a given flow rate through a multiple round nozzle impactor, variation in the number of nozzles allows the setting of the Reynolds number to a desired value (Figure 3). The relationship between the number of round jets, n, and Reynolds number, Re, is found by expressing the average velocity within the round jets, V_o, as

$$V_o = \frac{4Q}{\pi n W^2} \qquad [3]$$

where Q = total volumetric flow rate through the stage. This velocity expression is now substituted into the expressions for Reynolds number and Stokes number:

$$Re = \frac{\rho V_o W}{\mu} = \frac{4\rho Q}{\pi n \mu W} \qquad [4]$$

$$Stk_{50} = \frac{4\rho_p Q C D_{50}^{\,2}}{9\pi n \mu W^3} \qquad [5]$$

where D_{50} and Stk_{50} are, respectively, particle diameter and Stokes number at 50% efficiency, and ρ is the fluid density. By elimination of W from Equations 4 and 5, we obtain the expression

$$Q = \frac{\pi}{12}\left(\frac{\rho_p}{Stk_{50}}\right)^{1/2}\left(\frac{Re}{\rho}\right)^{3/2}\mu\sqrt{C}\;D_{50}n \qquad [6]$$

By assuming unit density particles (ρ_p = 1 gm/cm^3) and air flow at normal temperatures and pressures (ρ = 1.205 x 10^{-3} gm/cm^3, μ = 1.81 x 10^{-4} poise) and noting that Stk_{50} is a function of Re and a specific design, equation 6 can be expressed graphically as shown in Figure 7. This figure is for S/W = 1, T/W = 1, and Re = 500, 3000 and 10,000. With these values of S and T, small variations in S and T will have a negligible effect on the impaction efficiency (Figures 3 and 4).

Also note in Figure 7 that the parameter $\sqrt{C}\,D_{50}$ corresponds to a specific value of W. This relationship is found by eliminating Q/n from equations 4 and 5.

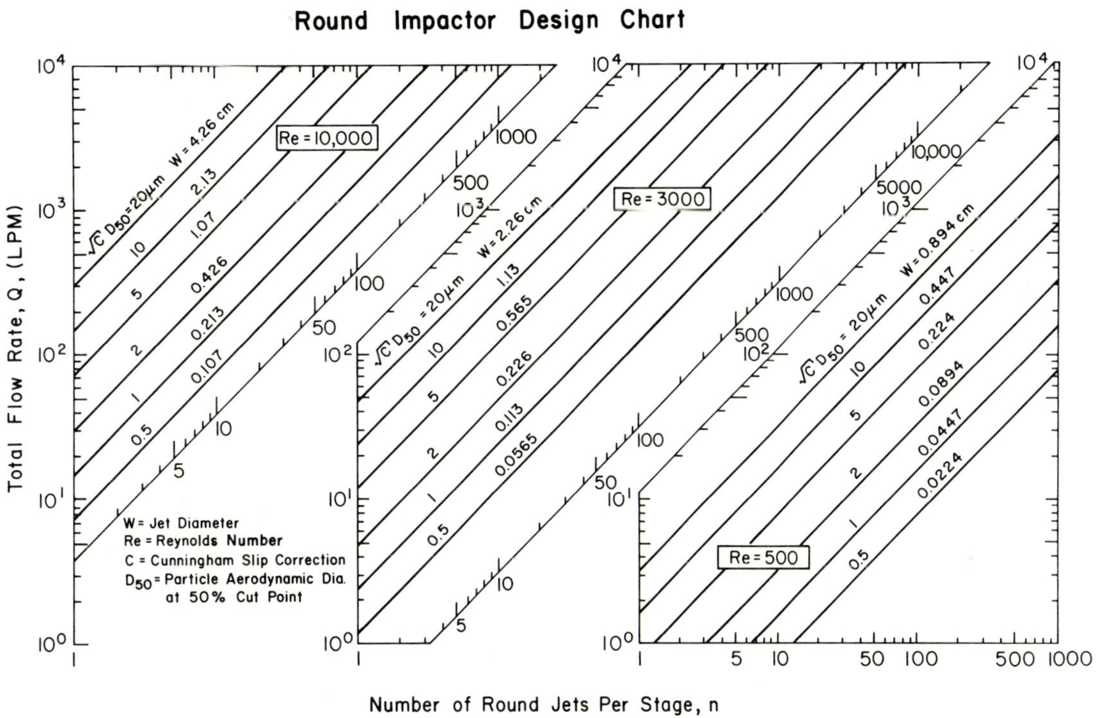

Figure 7 Design chart for round impactors. (D_{50} = aerodynamic diameter at 50% cut point).

$$W = \sqrt{\frac{\rho_p Re}{9 \rho Stk_{50}}} \sqrt{C}\, D_{50} \qquad [7]$$

As an example of the use of these curves, assume it is desired to have a cutoff size of $\sqrt{C}\, D_{50} = 2$ µm, a total flow rate of 40 lpm, and Re = 500. From Figure 7 it is found that this stage must have 120 holes of 0.0894 cm diameter. The number of holes can be reduced to 8 when each hole is 0.226 cm in diameter which results in Re = 3,000. The number may be further reduced to approximately one hole with a diameter of 0.426 cm and Re = 10,000.

A similar analysis can be made of rectangular impactors. However, instead of defining the number of jets, it is more conventient to define the total length, L, of the jets, irrespective of how this total length is divided into individual jets. Now the mean fluid velocity at the throat, V_o, is defined as

$$V_o = \frac{Q}{LW} \qquad [8]$$

and the Reynolds number (defined in terms of the hydraulic diameter) and Stokes number become

$$Re = \frac{2 \rho V_o W}{\mu} = \frac{2 \rho Q}{\mu L} \qquad [9]$$

$$Stk_{50} = \frac{\rho_p Q C D_{50}^2}{9 \mu L W^2} \qquad [10]$$

Since W does not appear in the expression for Re, we cannot eliminate W from equations [9] and [10] as we did for the round impactor. However, if the volumetric flow rate in a rectangular impactor of width W is expressed per unit length of the jet, the parameter Q/L versus W can be plotted from equation [10] by allowing $\sqrt{C}\, D_{50}$ to be a parameter as shown in Figure 8. The value of Stk_{50} in this equation is a function of Re (Figure 3) and, thus, a function of Q/L as defined by equation [9]. The corresponding value of Re is also shown in Figure 8. In this case S/W = 1.5 and T/W = 1.

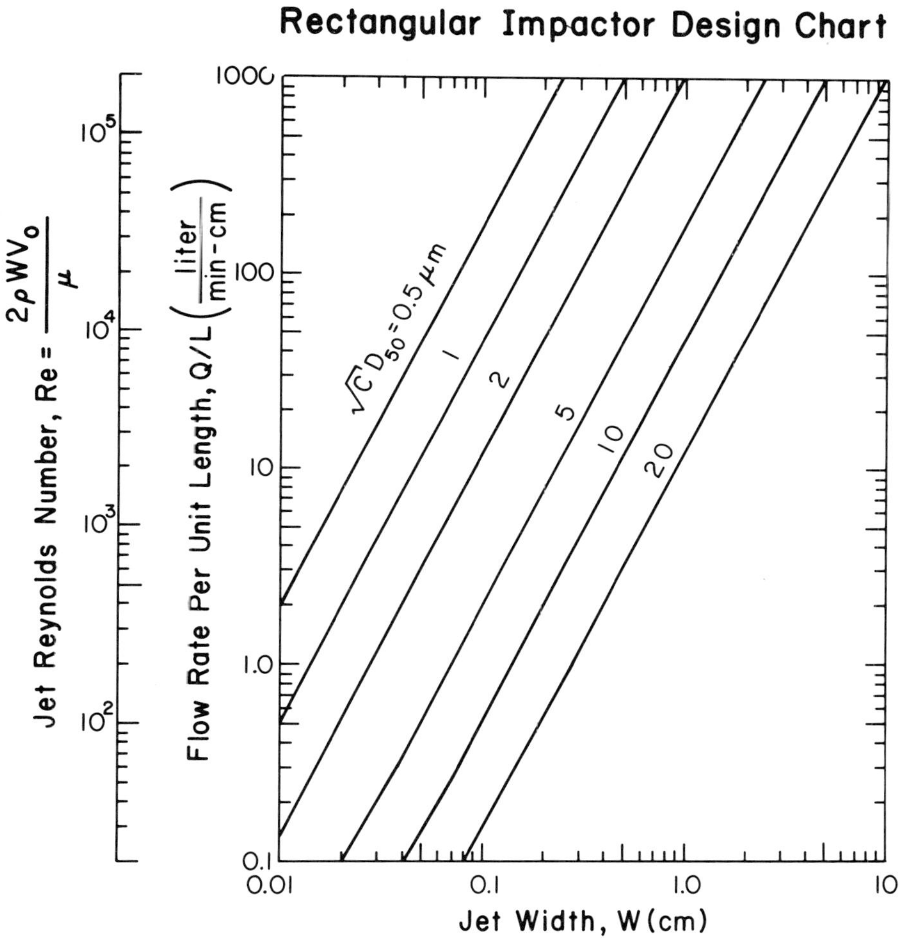

Figure 8 Design chart for rectangular impactors (D_{50} = aerodynamic diameter at 50% cut point).

Impaction Surfaces

Besides variations in jet design, there can also be variations in the design of the impaction plate, as shown in Figure 9. Although the most common design is a flat surface, the variations shown in Figure 9 have been made for good reasons and serve to expand the versatility of impactors.

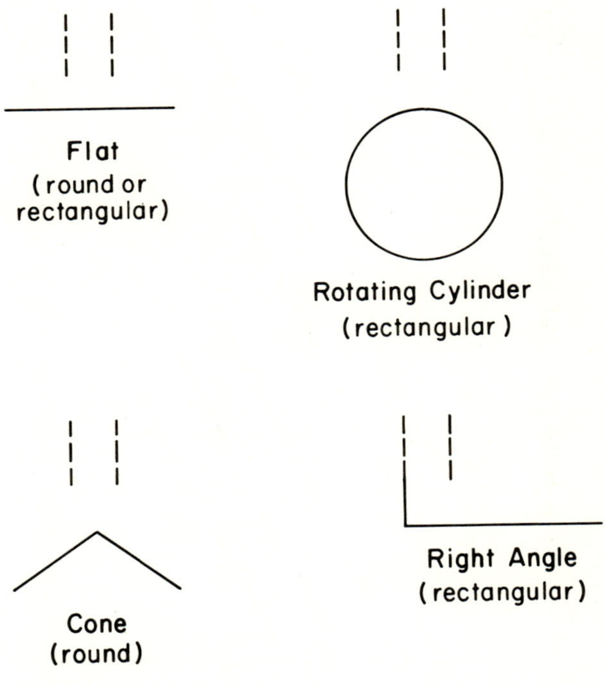

Figure 9 Types of impaction plates

Impaction from a rectangular jet onto a rotating drum is employed in the Lundgren impactor[20] for time-resolved studies of the deposit. This is accomplished by rotating the cylinder at rates from one revolution per 24 hours to one revolution per minute.

The cone was used in a round impactor by Schott[18]. The purpose was to develop an impactor which would allow particles which were blown or bounced off the cone, to be captured in a volume at the base of the cone, thus having the ability to collect large deposits.

One might expect that impaction onto a surface which is not perpendicular to the unimpeded jet flow, could be quite detrimental to the sharpness of cut of the efficiency curve. However, the experiments[18,20] have shown that the efficiency curves are very similar to those for a flat plate. In some of these experiments[18] the theoretical efficiency curves were also found for impaction onto a cone by employing the same technique used to obtain the efficiency curves for impaction onto a flat plate (Figure 3). The theoretical calculations agreed very well with the experimental data.

The final variation in impaction surfaces considered here is the "right-angle" arrangement of the rectangular impactor shown in Figure 9. This type of impaction surface must be used if the total flow is to be directed to one side or the other. For analysis purposes, it can be thought of as half of a rectangular impactor with a solid surface at the center plane. Since the boundary layer along the center plane hinders the development of a flat velocity profile at the exit plane of the jet, the efficiency curve will not have as sharp a cut-off as the conventional rectangular impactor.

OPERATIONAL PRECAUTIONS

As has been shown, an impactor has the ability to sharply classify particles into distinct ranges of aerodynamic size. However, to obtain reliable data from an impactor, one must be aware of their limitations. For example, precautions must be taken in the use of the design charts and of the calculated efficiency curves for conditions other than those specified. This is especially true if the inlet to an impactor stage is obstructed or if the collection of very small particles is desired. Also, an impactor cannot size particles which are not sampled. Thus, the inlet collection efficiency must be known. Once the particles are sampled, the particles must be deposited only upon the impaction plates and, once deposited, must remain on the plates. Particle loss and particle reentrainment may result in indicated size distributions which are significantly different from the actual ones.

Restricted Entrance Effects

Experiments[16,17,18,19,21] have shown that the theoretical efficiency curves presented in Figure 3 agree well with experiments, if the experimental design corresponds to the condition of the numerical model. However, if the flow conditions are different from those assumed in the theoretical treatment, the characteristic efficiency curve can be expected to be somewhat different. For example, some cascade impactors are designed for compactness, and the flow approaching the inlet will be horizontal with no tapered inlet, which is substantially different from the inlet shown in Figure 1 and assumed in the theory.

The influence of horizontal flow at the entrance on the efficiency curves has been shown [17,19] in an experimental evaluation of two commercial cascade impactors: The Sierra Instruments High-Volume Sampler, which has several parallel slots per stage, and the Andersen 2000 High-Volume Sampler, which has a multitude of circular holes per stage. The second stages of these impactors, which have comparable cut-off sizes, were tested with and without the presence of the first stage. The results, shown in Figure 10, reveal that, when the inlet flow is not obstructed (stage 2 alone), there is good agreement between the experimental efficiency curve and the theory. When the first stage is present, the agreement is still fairly good for the Andersen impactor, but a large shift and a decrease in sharpness of cut is seen with the Sierra impactor.

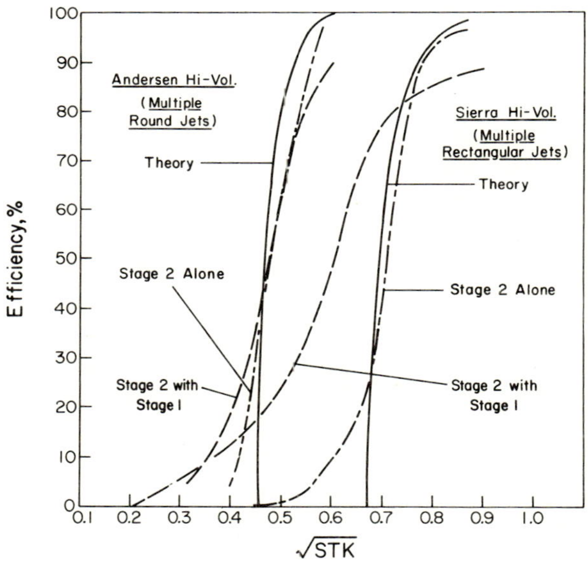

Figure 10 Second stage collection efficiency curves for the Andersen and Sierra high-volume impactors with and without stage 1 present (the curves are for liquid particles collected on a smooth plate).

The principle reason for this shift in the second-stage impaction efficiency curve of the Sierra impactor is probably due to a pronounced vena contracta formation in the throat (T/W = 0.8) of the rectangular jet when stage 1 was present. The velocity in the jet would thus be larger than for the case where the entrance is not obstructed, and smaller particles would be collected, with a corresponding shift of the impaction efficiency curve to a smaller Stokes number. This also means that the pressure drop through stage 2 will be larger, when stage 1 is present, than when it is not.

Lateral flow into the nozzles of an impactor stage and the resulting shift of the impaction efficiency curve does not mean it is detrimental to the impactor design. In simply indicates that the theoretical predictions for this type of inlet condition have not yet been made, and therefore, we have to rely on experimental calibrations.

High Velocity Effects

In using impactors, it is often desirable to be able to collect as small a particle as possible. As equation 1 implies, small particles can be collected by increasing the nozzle velocity, V_o. However, the theory previously presented for impactors assumes that the fluid is incompressible. The assumption of incompressibility[22] is normally assumed valid for air flows with Mach numbers, M, less than 0.2, for which the density is in error by only 2%. However, M = 1/3, for which the density is in error by about 5%, may be considered a practical upper limit[23].

High velocity through the nozzle also may bring about more particle losses between stages and from the impaction plate due to reentrainment (discussed later). These losses, however, can be minimized by the use of a proper impaction plate adhesive.

Inlet Losses

As is the case with any size analyzing instrument, the efficiency with which the impactor inlet is sampling the aerosol particles must be known before accurate size distribution and particle concentration data can be obtained.[24]

Recent work[25] has shown that large particles (up to about 100 μm) can be drawn into an inlet from still air with efficiencies of nearly 100%. However, there may still be considerable losses on the internal surfaces of the inlet due to impaction and settling. If sampling from a moving air stream is required, isokinetic conditions at the inlet will aid in efficient sampling of large particles.

If the sampling efficiency of large particles becomes too low for good data to be obtained, it may be necessary to use an impaction surface on a rotating arm, since this device has no inlet.

Interstage losses

Particle losses in an impactor, generally referred to as wall losses or interstage losses , is the deposition of particles on surfaces other than the impaction plate. Currently, no theory exists to predict these losses, and thus, they must be determined experimentally.

Recently, three commercial cascade impactors have been investigated experimentally to determine the interstage losses. In a detailed evaluation[21] of the Lundgrem impactor (rotating drum impaction plate) and of the Andersen viable aerosol sampler (28.3 lpm. 1 cfm, multiple round jets) and also in a similar evaluation of the Sierra high-volume cascade impactor[19] (1133 lpm, 40 cfm, multiple rectangular jets), the interstage losses were determined as a function of particle aerodynamic size, (i.e. equivalent diameter of unit density spheres), and were expressed as the fraction of the total number of particles entering the impactor. (Figure 11)

A study of Figure 11 shows that for the Lundgren and for the Andersen Viable impactors, interstage losses are not a serious problem for particles smaller than about 5 μm, but increase rapidly for larger particles. The Sierra impactor, however, has appreciable losses for particles smaller than 5 μm. It was found[19] that the interstage losses occurred primarily on both inner walls of the rectangular nozzle, apparently due to lateral impaction of the entering flow. Tapering of the nozzle inlets reduced these losses.

Particle Reentrainment (Bounce-off and Blow-off)

The limitations of impactors due to particles bouncing off the impaction plate or being blown off the impaction plate after collection are essentially the same: in both cases, particles which should have been collected are reentrained into the airstream. Thus, we shall use the term reentrainment for both blow-off and bounce-off. For single stage impactors, this means that the concentration of particles collected on the impaction plate will be too small. For cascade impactors, the resulting size distribution will be shifted to smaller sizes, since reentrained particles will be collected on stages intended to collect smaller particle sizes.

The degree of particle reentrainment is a function of the type of particle and the nature of the impaction surface. Liquid particles will impact on any type of surface with very little reentrainment. The reentrainment of solid particles, however, is a strong function of the type of

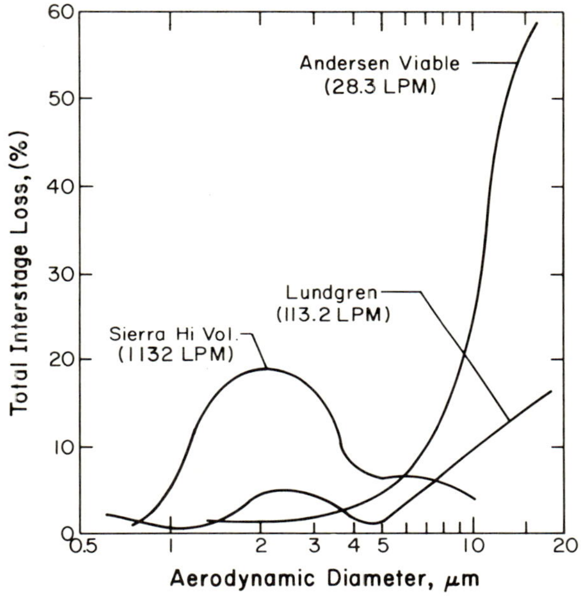

Figure 11 Total interstage losses for the Andersen (multiple round holes), Lundgren (single slots) and Sierra (multiple slots) impactors.

impaction surface used. The maximum amount of reentrainment is experienced with a dry, smooth surface and the least with a very sticky surface. Table I lists some of the sticky surfaces that have been used.

A detailed experimental study[21] to determine the degree of reentrainment from several types of impaction surfaces was made using single-stage impactors (single round jets) designed according to the theoretical criteria described above. The results for one of the impactors are presented in Figure 12. It should be noted that the test aerosols were polystyrene latex spheres, which are quite resilient. Thus, the curves of Figure 12 can be considered to be approximately the worst obtainable.

Figure 12 shows that with an oil coated glass plate there was essentially no reentrainment, since the efficiency curve agreed well with the theoretical curve and attained nearly 100% efficiency. At the other

Table I

Impaction Surface Adhesives Used in Impactor Studies

Reference	Adhesive
May[30] (1945)	- Fly paper mixture -- one part of rosin to three parts of castor oil (benzene solution) - Polyisobutane (chloroform solution) - One part of methylated starch to three parts of tricresyl phosphate (alcohol or amyl acetate solution) - Oil or vaseline
Davies et al.[31] (1951)	- Glycerin jelly
Stern et al.[32] (1962)	- Dow Corning Hi-Vac silicone grease
McFarland and Zeller[33] (1963)	- Dow Corning silicone oil (DC-200 fluid, 2% solution in hexane)
Zeller[34] (1965)	- One part dibutyl phthalate, 10 parts cellulose acetate butyrate and 90 parts cylohexanone - Scotch tape and other films
Lundgren[20] (1967)	- Viscous oils and sticky tapes - Dow Corning Hi-Vac Silicone grease - Minnesota Mining Kel-F Polymer wax
McFarland and Husar[35] (1967)	- 1000,000 cs. Dow DC-200 silicone oil (1% solution in hexane)
Mercer and Chow[9] (1968)	- Dow Corning antifoam A
Hogan[37] (1970)	- G.E. silicone resin SR-516
Berner[38] (1972)	- Vaseline dissolved in carbon tetrachloride or toluene
Wesolowski et al[39] (1973)	- Sticky film - Dow Corning Hi-Vac grease
McCain et al.[40] (1974)	- 10 to 15% solution in benzene of high vacuum silicone grease
Walkenhorst[41] (1974)	- Several oils, $CaCl_2$ solution

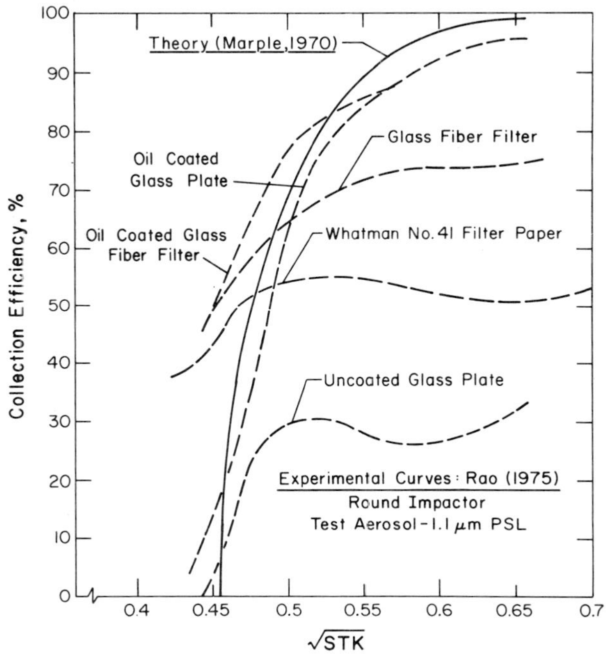

Figure 12 Collection efficiency of a typical single stage impactor for various collection surface media

extreme, the uncoated glass plate had severe reentrainment for particles larger than about $\sqrt{Stk} = 0.47$, and had a maximum collection efficiency of about 30%. When a filter paper was used as a collection surface, the efficiency curves were found to be between the two extreme cases of the oil-coated and the dry glass plates, and tended to have poor sharpness of cut characteristics.

Reentrainment in an impactor with several stages (Andersen viable) was found[21] to be similar to that in the single-stage impactor. As seen in Figure 13, the best collection efficiency was with liquid particles and the worst with solid particles impacted onto an uncoated plate. The curve for collection on a glass fiber filter paper reached an efficiency of about 90%. It appears that most of the particles get collected in the void spaces between the fibers, while some always bounce off the top

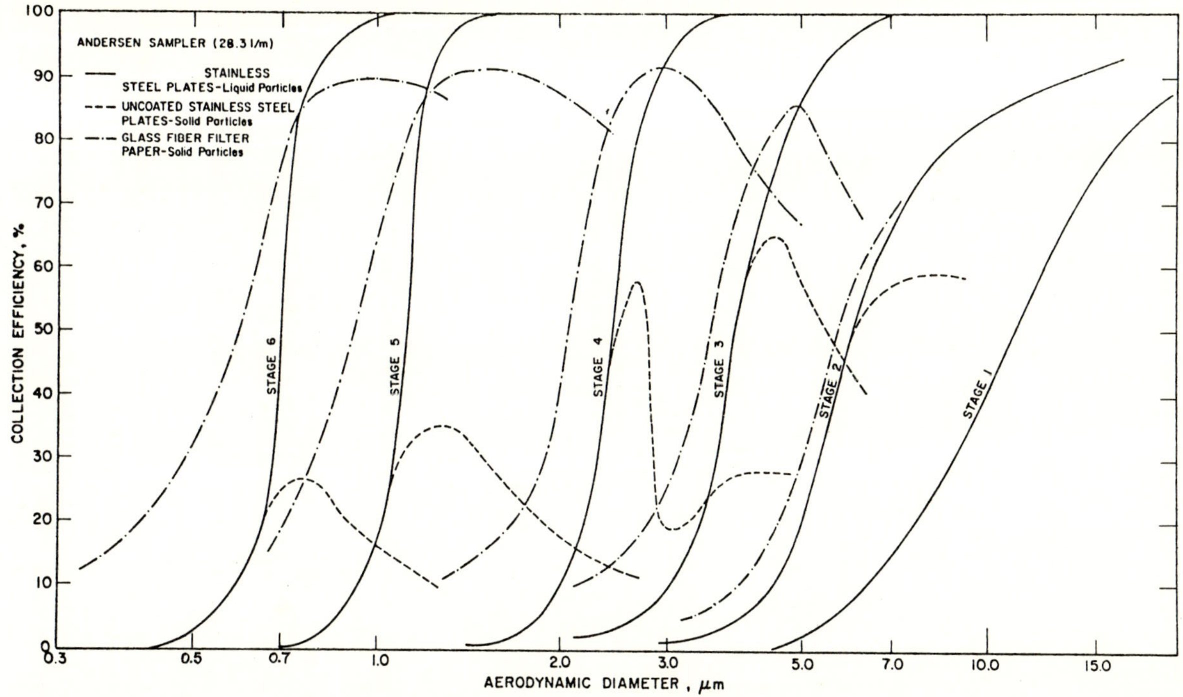

Figure 13 Collection characteristics of the Andersen viable sampler operated at 28.3 lpm.

fibers so that 100% collection efficiency cannot be attained with glass filter paper. Also, the cut-off characteristics are generally poorer for the glass fiber paper than for a coated plate.[17,19,21]

DATA ANALYSIS

The particles collected on the impaction plates are analyzed after the sampling is terminated. In the most common type of analysis, the concentration and size distribution (by number or mass) of the aerosol particles is determined from the deposits by one of the several methods of analysis of which only the more widely used ones will be discussed here.

The size distribution of the aerosol can be determined either from a single stage impactor or from a cascade impactor. For a single stage impactor the deposit must be analyzed with a microscope to get a number distribution. With a cascade impactor, the particle deposits of the various stages can also be analyzed with a microscope, but are more commonly analyzed gravimetrically to obtain a mass distribution.

The number or mass concentration of the aerosols can be determined from the size distribution data. If the size distribution is found from a portion of the deposit by microscope analysis, the concentration is calculated from the quantity of particle laden air sampled and from the fraction of the total deposit analyzed. If the gravimetric analysis is used, only the quantity of air flowing through the impactor needs to be known.

Methods other than the gravimetric analysis are also used to determine the mass of particles collected on an impaction plate. In one instrument[26] the particles are impacted onto a piezo-electric crystal which determines the mass by monitoring the change in the resonant frequency of the crystal. Another instrument[27] uses beta attenuation to detect the mass of particles collected.

Impactors also have been used in conjunction with an optical particle counter to obtain the number size distribution and particle concentration[28]. The aerosols were passed through an impactor with a variable jet width. The particles not collected on the impaction plate were counted with an optical particle counter. By varying the jet width in a stepwise manner, a size distribution was obtained.

The elemental composition of the deposits can be determined by techniques such as atomic absorption, neutron activation and x-ray fluorescence. The collection of the particles should be on a medium having a very low level of impurities. For instance, it has been found[29] that glass fiber filter paper is unacceptable for x-ray fluorescence analysis, but membrane filters, made of mixed esters of cellulose and having 0.8 μm pore size, is satisfactory.

In order to avoid the problems of particle reentrainment, a dichotomous (or virtual) impactor may be used.[29] In such a device the particles are impacted into a slowly pumped void and are then collected onto a membrane filter. The virtual impactor, however, is considerably more complicated than the normal type of impactor, and consists of only one classification stage.

DESIGN PROCEDURE

It is possible to design an impactor for a specific flow rate and be fairly certain of the particle cutoff size by making use of the results from the theoretical analysis described in a preceding section. The procedure for designing an impactor is essentially the same for either the round or the rectangular configuration. For a cascade impactor operating at a specific flow rate, a suggested procedure is as follows:

1. Choose a desired cutoff diameter, D_p. If the density, ρ_p is different from unity, calculate the value $\sqrt{C}\,D_{50}$ (where D_{50} is the equivalent aerodynamic diameter of a unit density sphere) from:

$$\sqrt{C}\,D_{50} = \sqrt{\rho_p}\,\sqrt{C}\;D_p \qquad [11]$$

   The variation in the Cunningham slip correction is usually negligible for small diameter differences. Thus, the values of C on both sides of equation 11 are approximately equal to each other and can be eliminated from the equation in most cases.

2. Round - Use Figure 7 and determine the number and size of jets required for the desired operating Reynolds number, Re, and aerodynamic cutoff size, $\sqrt{C}\,D_{50}$. Since the cutoff characteristics are fairly constant and sharp over a Reynolds number range from

500 to several thousand, a Reynolds number of 3,000 may be assumed for the calculations.

Rectangular - Use Figure 8 and choose a desirable Reunolds number (again Re = 3,000 is satisfactory) and determine the Q/L value and jet width, W, for the desired aerodynamic cutoff size $\sqrt{C}\ D_{50}$. Special note must be made of the resulting value of the jet aspect ratio, L/W, since end effects can become deterimental to the sharpness of cut for nozzles with small L/W values.

3. Select a convenient size of jet diameter or jet width which is close to the value found in step 1. For a round impactor this would be a standard reamer size.

4. For this value of jet diameter or jet width, check the Reynolds number using equation 4 or 9.

5. Determine the value of $\sqrt{Stk_{50}}$ from Figure 3 and calculate the cutoff size, $\sqrt{C}\ D_p$, by using equation 5 or 10.

6. Determine the pressure in the impaction region, P_2, by assuming that the pressure drop in the stage is equal to the dynamic pressure of the jet. Thus

$$P_2 = P_1 - 1/2\ \rho V_o^2 \qquad [12]$$

where P_1 = static pressure at the impactor stage inlet

P_2 = static pressure at the impaction plate

ρ = fluid density

V_o = average fluid velocity in the jet

7. Determine the cutoff particle size, by calculating the Cunningham slip correction[36] from the equation:

$$C = 1 + \frac{0.163}{D_pP_2} + \frac{0.0549}{D_pP_2} \exp\ (-6.66\ D_pP_2) \qquad [13]$$

where P_2 = static pressure (atmospheres)

D_p = particle diameter (μm)

8. Repeat steps 1 through 6, for the next stage of the impactor.

Besides determining the impactor jet diameter at each stage, it is also important to design the impactor so that the jet-to-plate distance, S/W, is greater than 0.5 and 1.0 for round and rectangular impactors, resepctively. In order to provide for a margin of safety, the criteria of

$$S/W = 1.0 \text{ (round impactor)}$$

$$\text{and} \quad S/W = 1.5 \text{ (rectangular impactor)} \qquad [14]$$

should be the minimum jet-to-plate distance used. Under this condition, small variations of jet-to-plate distance will not effect the value of $\sqrt{Stk_{50}}$.

As shown in Figure 3, the throat length, T/W, does not greatly influence the cut-off characteristics of the impactor. However, these curves were calculated for an impactor with a tapered or conical inlet section as shown in Figure 1. In the absence of the tapered or conical entrance, a short throat may not allow sufficient time for the particle to accelerate to the fluid velocity in the throat so that the efficiency curves may be different from those shown in Figure 3. Also, particle losses may be found at a sharp entrance. Thus, if possible, the entrance of an impactor jet should be tapered and the T/W value should be at least 1.

Design of the impactor between stages is also important since this is where interstage losses of the aerosol occur. In general, the fluid velocity must be kept large enough so that particles are not lost from settling and yet not so large as to lose particles from impaction in the interstage space. Also, obstructions and sharp corners should be kept at a minimum, since particles are deposited in the resulting turbulent areas behind these obstructions.

COMMERCIALLY AVAILABLE CASCADE IMPACTORS

Some of the cascade impactors which are commercially avaliable are listed in Figure 14. For each impactor the aerodynamic diameter at the 50% cut point is given for each stage at a typical flow rate. Most impactors, however, are not confined to the indicated flow rate and may be operated over a range of flow rates which would shift the cut points to larger or smaller diameters. Particles with a density different from unity would also shift the cut points. This shift may be calculated from equation 11. The variation in Cunningham slip correction is usually negligible for small diameter changes so that, equation 11 is reduced to

$$D_p = D_{50}/\sqrt{\rho_p} \qquad [15]$$

for the 50% cut point. When not sampling at normal temperatures and pressures, however, the slip correction factor in equation 11 may have to be calculated.

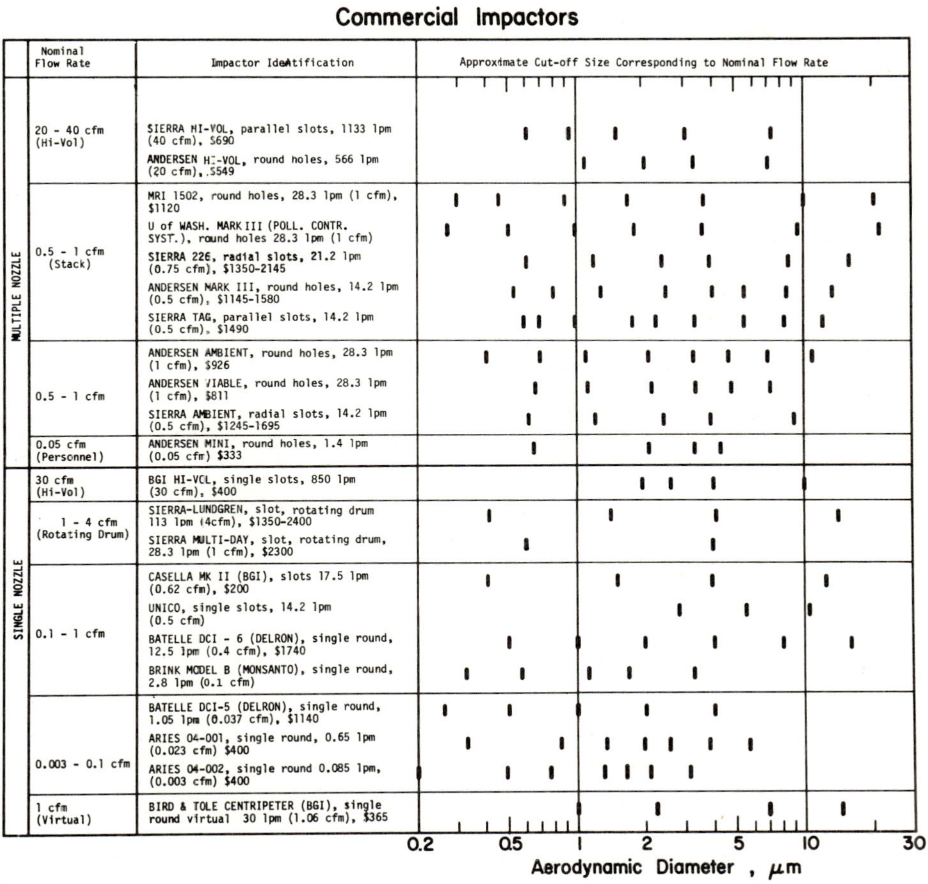

Figure 14 Some commercially available cascade impactors

The cut points indicated in Figure 14 should only be used as general guidelines, since several of the stage constants were only theoretically calculated but not experimentally calibrated. No measure of spread has been listed for the collection efficiency curves. Also, some of the impactors may also be purchased with fewer stages than indicated in the Figure.

Impactors for similar purposes are quite comparable in price. Some of the costs indicated also include auxiliary equipment, e.g. flow controllers and pumps.

The high-volume samplers are primarily used for ambient 24-hour sampling or in-plant short-term sampling. The samplers listed have 4 to 5 stages and differ from each other primarily through the shape and arrangement of the openings in each stage. The BGI Hi-Vol (BGI Inc., 58 Guinan Street, Waltham, Mass. 02154) has a single slot, the Sierra Instruments Inc., P.O. Box 909, Village Square, Carmel Valley, Calif. 93924) has 9 to 10 parallel slots, and the Andersen Hi-Vol (Andersen 2000, P.O. Box 20769, Atlanta, Ga. 30320) has 296 to 300 round holes arranged in several concentric circles. Their sampling efficiency of large particles depends on the sampling rate per overall inlet area, and if housed in a shelter, on the sampling efficiency of the shelter inlet. The latter may be affected by ambient air velocity and wind orientation with respect to the sampler inlet.

Cascade impactors used by industrial hygienists usually operate at about 30 lpm. (ca.1 cfm), if they are stationary, or at about 1 lpm (ca. 0.05 cfm) if they are carried by the worker.

In the Sierra and Andersen cascade impactors the plates containing the jet orifices also support the collection medium for impaction from the previous stage. The collection medium, usually glass fiber paper, has openings cut into it so that the flow may enter the orifices of the next stage. An exception to this is the Andersen Viable Sampler in which airborne micro-organisms are classified by six stages. Each stage collects viable and non-viable aerosols into an agar containing Petri dish, while the unimpacted aerosols flow around the Petri dish into the next stage.

When sampling effluents from industrial sources, the cascade impactor may be inserted directly into the effluent stream, usually the vertical stack, or a sample may be extracted for analysis outside the pollutant flow. Included in Figure 14 are several impactors used in stacks. The stages are generally contained inside a long cylindrical housing which is usually less than about 8 cm. in diameter and is topped by a conical nozzle inlet. The inlet is usually aerodynamically shaped and decelerates the effluent flow for first stage impaction inside the housing. Several nozzles are provided for isokinetic sampling from the effluent stream. The flow rates generally range from about 14 to 30 lpm (ca. 0.5 to 1 cfm). Most sampling is done in environments not exceeding 200°C (ca. 400°F). Exposure to higher temperatures necessitates the use of stainless steel in the construction. Particle bounce-off, reentrainment, and weight changes caused by gas-to-solid conversions or chemical reactions with the collection medium may make high temperature measurements more ambiguous than ambient air measurements. Further studies are needed to find suitable collection media for specific ranges of stack environments.

Stack samplers are presently produced by Sierra Instruments, Inc.; Andersen Inc.; Meteorology Research Inc. (464 W. Woodbury Rd., Altadena, Ca. 91001) and Pollution Control Systems Corporation (321 Evergreen Bldg., Renton, Wash. 98055). When effluent samples are extracted from the stack into an impactor outside the stack, a survey of the entire stack cross-section may be made more readily than with an instack device. However, particle losses in the sampling line may truncate the size distribution at the larger particle sizes.

Several cascade impactors have only one orifice per stage. Examples are the BGI Hi-Vol, the May/Casella MKII and MKIIa (C.F. Casella & Co. Ltd., Regent House, Britannia Walk, London, No. 1, England; U.S. distributor: BGI Inc.), the Battelle Impactors DCI-5 and DCI-6 (Delron Research Products Co., P.O. Box 95, Powell, Ohio 43065), the Unico Sampler (National Environmental Instruments, P.O. Box 590, Fall River, Mass. 02722), the Brink Model B Impactor (Monsanto Enviro-Chem Systems, Inc., 800 N. Lindbergh Blvd., St. Louis, Mo. 63166) and the Aris impactors (Aerosol Research Instrumentation Equipment and Services, Inc., P.O. Box 25541, Albuquerque, N.M. 87125).

In the Lundgren and in the Multi-Day Sampler (Sierra Instruments) the aerosols impact onto rotating drums for time resolved analysis of the deposits. The Centripeter (Bird & Tole, Ltd., Bledlow Ridge, High Wycombe, Buckinghamshire, England; U.S. distributor: BGI Inc.) is a dichotomous virtual impactor which splits the jet into a center flow with large particles and a laterally deflected flow of small particles. An impactor with a similar basic design concept has been developed[29] by the Environmental Research Corporation (3725 N. Dunlap St., St. Paul, Mn. 55112).

SUMMARY AND CONCLUSIONS

Inertial impactors have become a popular tool for the analysis of the size distribution and concentration of aerosols.

The performance of impactors can be accurately predicted by the use of modern numerical methods which solve the equations governing the fluid flow and particle motion in this flow. This theory shows that impactor stages which are properly designed and operated will provide sharp classification between the particles collected and those which are not. Also, the theory can be successfully used to aid in the design of multiple hole impactors. These results, presented in the form of design charts, indicate the number and size of holes needed, for a specific flow rate, to keep the flow conditions (Reynolds number) near optimum.

Although impactors are capable of making a sharply defined cut between particle sizes, special precautions must be taken to insure reliable and accurate data. Such problems as particle losses at the inlet, particle reentrainment from the impaction surface, and particle losses between stages of a cascade impactor can result in the indicated particle size distribution being considerably different from the one sampled. However, by the use of proper design and impaction surface coatings, these problems can be minimized.

The current popularity of cascade impactors is evident by the large number commercially available. The listing of some of these impactors indicates a variety of intended uses, flow rates and cut-off sizes.

References

1 Marple, V.A., B.Y.H. Liu. Characteristics of Laminar Jet Impactors. Journal of Environmental Science and Technology 8:648-654, July 1974.

2 Marple, V.A., and B.Y.H. Liu. On Fluid Flow and Aerosol Impaction in Inertial Impactors. Journal of Colloid and Interface Sci. (1975).

3 Ranz, W.E. Principles of Inertial Impaction. Bulletin No. 66, Dept. on Engineering Research, Pennsylvania State University. (1956).

4 Jaenicke, R. and C. Junge. Studien zur oberen Grenzgrösse des natürlichen Aerosoles (Studies of the Upper Size Limit of Natural Aerosols.) Beiträge zur Physik der Atmosphäre. 40:129-143, 1967.

5 Noll, K.E. and M. J. Pilat. Size Distribution of Atmospheric Giant Particles. Atmospheric Environment. 5:527-540, 1971.

6 Gruber, C.W. and G. Harms. Fugitive Dust Measurement by the Rotary Ambient Adhesive Impactor. Presented at 67th Annual APCA Meeting, Denver, Colorado. June, 1974.

7 Marple, V.A., B.Y.H. Liu and K.T. Whitby. Fluid Mechanics of the Laminar Flow Aerosol Impactor. Journal of Aerosol Science 5:1-16,1974.

8 Ranz, W.E., and J.B. Wong. Impaction of Dust and Smoke Particles. I & E Chem.44:1371, 1952.

9 Mercer, T.T. and H.Y Chow. Impaction From Rectangular Jets. J. Coll. and Interface Sci. 27:75, 1968.

10 Mercer, T.T. and R.G. Stafford. Impaction From Round Jets. Ann. Occup. Hyg. 12:41, 1969.

11 Davies C.N., M. Aylward and D.Leacey. Impingement of Dust From Air Jets. A.M.A. Arch. Inc. Hyg. Occup. Med. 4:354,1951.

12 Marple, V.A., A Fundamental Study of Inertial Impactors, Ph.D. Thesis, University of Minnesota, Particle Technology Laboratory, Publ. No. 144 1970.

13 Marple, V.A., B.Y.H. Liu and K.T. Whitby. On the Flow Fields of Inertial Impactors. ASME Journal of Fluid Engineering. 96:394-403, December, 1974.

14 Gosman, A.D., W.M. Pun, A.K. Ruchal, D.B. Spalding and M.Wolfshtein. Heat and Mass Transfer in Recirculating Flows, Academic Press. N.Y.

15 Fuchs, N.A., The Mechanics of Aerosols, Pergamon Press, New York, p. 154, 1964.

16 Jaenicke, R. and I.H. Blifford. The Influence of Aerosol Characteristics on the Calibration of Impactors. J. Aerosol Sci. 5:457, 1974.

17 Willeke, K. and J. J. McFeters "The Influence of Flow Entry and Collecting Surface on the Impaction Efficiency of Inertial Impactors. J. Colloid and Interface Science, in print ,1975.

18 Schott, J.H. Jet-Cone Impactors as Aerosol Particle Separators. M.S. Thesis, University of Minnesota (1973).

19 Willeke, K. Performance of the Slotted Impactor. Am. Ind. Hyg.Assoc. J., Sept. 1975.

20 Lundgren, D.A. An Aerosol Sampler for Determination of Particle Concentration of Size and Time. J. Air Poll. Cont. Assoc. 17:225, 1967.

21 Rao, A.K., An Experimental Study of Inertial Impactors. Ph.D. Thesis, University of Minnesota, Particle Technology Laboratory, Publ. No. 269 1975.

22 Olson, R.M., Essentials of Engineering Fluid Mechanics, International Textbook Co. Scranton, Pa. p. 165, 1961.

23 Cohen,J.J. and D. N. Montan. Theoretical Considerations, Design, and Evaluation of a Cascade Impactor. AIHA Journal. 28:95-104, March-April, 1967.

24 Fuchs, N.A. Sampling of Aerosols. Atmos. Env. 9:697-707, 1975.

25 Agarwal, J.K. Aerosol Sampling and Transport. Ph.D. Thesis, University of Minnesota, Particle Technology Laboratory, Publ. No. 265, 1975.

26 Chuan, R.L. An Instrument for the Direct Measurement of Particulate Mass. Journal of Aerosol Science 1:111-114, 1970.

27 Lilienfeld, P. and J. Dulchinos. Portable Instantaneous Mass Monitor for Coal Mines Dusts. Am. Ind. Hyg. Assoc. J., 33:136-145, 1972.

28 Cooper D.W. and L.A. Spielman. A New Particle Size Classifier: Variable-Slit Impactor with Photo-Counting, Atm. Env. 8:221,1974.

29 Dzubay, T.G. and R.K. Stevens. Ambient Air Analysis with Dichotomous Sampler and X-ray Flourescence Spectrometer, Env. Sci. and Tech. 9:663,1975.

30 May K.R.. The Cascade Impactor: An Instrument for Sampling Coarse Aerosols.J. Sci. Instr. 22:187, 1945.

31 Davies, C.N., M. Aylward, and D. Leacey. Impingement of Dust from Air Jets. A.M.A. Arch. Ind. Hyg. Occup. Med. 4:354,1951.

32 Stern, S.C., H.W. Zeller, and A.I. Schekman. Collection Efficiency of Jet Impactors at Reduced Pressures, I & EC Fundamentals 1:273, 1962.

33 McFarland, A.R., and H.W. Zeller. Study of a Large-Volume Impactor for High-Altitude Aerosol Collection, General Mills Inc., Electronics Division, Minneapolis, Minn. Rept.No. 2391, Contract AT (11-1), p. 401, 1963.

34 Zeller, H. Large Volume Impactor Collector, Litton Systems Inc., Applied Science Division, 2295 Walnut St., St. Paul, Mn. 55113, Rpt. No. 2893, Project No. 89125, 1965.

35 McFarland, A.R. and R.B. Husar. Development of a Multistage Inertial Impactor, Particle Tech. Lab. Publ. No. 120, Dept. Mech. Eng., U. of Mn. Mpls., Mn. 55455, 1967.

36 Wahi, B.N. and B.Y.H. Liu. The Mobility of Polystyrene Latex Particles in the Transition and the Free Molecular Regimes. J. of Colloid and Interface Science 37:374-381, Oct. 1971.

37 Hogan, A.W.. Evaluation of a Silicone Adhesive as an Aerosol Collecting Medium, J. Applied Meteorology, 10:592,1971.

38 Berner, A. Practical Experience with 20-Stage Impactor, Staub-Reinhaltung Luft, Engl. Ed. 32,8:1, 1972.

39 Wesolowski, J.J. Ambient Air Aerosol Sampling, Proceedings of "Second Joint Conf. in Sensing of Environmental Pollutants", Instrument Society of America, 400 Stanwix St. Pittsburgh, Pa. (1973).

40 McCain, J.D., K.M. Cushing and W.B. Smith. Methods for Determining Particulate Mass and Size Properties: Laboratory and Field Measurements. J. Air Poll. Control Assoc. 24:1173, 1974.

41 Walkenhorst, W. Untersuchungen über den Haftgrad von Staubteilchen. (Studies of the Degree of Adhesion of Dust Particles). Staub-Reinhalt. Luft (German ed.) 34:182, 1974.

THE CYLINDRICAL AEROSOL CENTRIFUGE+)

Mohammad Abed-Navandi[1)], Axel Berner und Othmar Preining[2)]

Ist Physics Institute, University of Vienna, A 1o9o, Strudelhofgasse 4

1) since July 1974 Bundesstaatliche bakteriologisch seriologische Untersuchungsanstalt, Abt. Lufthygiene, Wien
2) contact for reprints

ABSTRACT

A very important size parameter of aerosol particulates P is the aerodynamic diameter D_p. The distribution of the P's of an aerosol over D_p permits one to predict the mechanical behavior and is therefore of special interest. Instruments sep rating the P's according to D_p permits one to derive this size distribution experimentally; aerosol centrifuges are much used instruments for this purpose. Instruments sampling cummulatively have been developed as well as such with discrete sampling modes;only the latter permit one to derive the distributions directly. The principle is: A laminar clean air flow is superimposed by a thin sheet of aerosol flow in a strong centrifugal field. P's are deposited at different locations according to their D_p on a removable foil for further analysis. The forms of the flow chambers differ widely, e.g. spiral channels and hollow cones have been used successfully. The instrument described in the following uses a hollow cylinder, the aerosol is superimposed after the flow profile has developed, the P's are sampled on the wall of the outer cylinder. The aerosol inlet slit is perpendicular to the axis of rotation which is also the axis of the hollow cylinder. If the flow rate of the aerosol is kept small, size resolution can be increased to 0,5 %, i.e. two particles with D_p's of 1,000 µm and 1,005 µm are deposited at different locations. The instrument works best in the size range between 1 µm and 0,1 µm. The designe and the performance characteristics are discussed as well as the possibility of absolute measurements without calibration.

+) The paper covers partly the Ph.D. Thesis of Mohammad Abed-Navandi, December 1973.

THE CYLINDRICAL AEROSOL CENTRIFUGE

Mohammad Abed-Navandi[1)], *Axel Berner und Othmar Preining*

Ist Physics Institute, University of Vienna, A 1o9o, Strudelhofgasse 4

1) since July 1974 Bundesstaatliche bakteriologisch seriologische Untersuchungsanstalt, Abt. Lufthygiene, Wien

Aerosols are complex systems consisting of particulates, P, i.e. liquid and/or solid particles without reference to state, and a gas. To predict the mechanical behavior of the aerosols the sizes of all the P's have to be known. This is accomplished by acquiring the size distribution $f(D_p)$ experimentally. The size distribution is mathematically a probability density. The "sizes" of the P's have to be properly defined. A very useful and realistic measure is the aerodynamic diameter D_p, i.e. the diameter of a sphere of density one which assumes the same velocity in the gravity or in a centrifugal field as the P in question.

The reliable conversion of other size parameters, particularly from optical measurements, to D_p is very difficult and subject to great errors. Therefore the direct measurement of the f vs D_p relation becomes necessary, at least for the purpose of calibrating aerosol measuring devices. In the D_p-range below 1 µm the gravity field is insuffcent to separate the P's from the gas, hence centrifuges were developed at different places following different principles.

The main idea was, and is, to establish a laminar aerosol flow in a channel subject to a strong centrifugal field which separates the P's from the gas on a surface which can be further investigated. The location of deposition being a direct or indirect measure of the D_p of the P. The distribution of the deposited P's on the surface yields the size distribution directly, if only a thin sheet of aerosol flow is superimposed over a clean gas flow. The downstream location L of the deposited P is a measure of its D_p; L being the distance from the aerosol inlet to the location of deposition, called deposition length or sedimentation distance. So far quite a few different instruments have been designed. The idea arose in England, Sawyer and Walton[1]. The first instrument commercially available was the Goetz centrifuge[2,3,4], which is still used, e.g. Gerber[5]; in spite of the draw-back of being a cumulative sampling device. Keith and Derrick[6], Kast[7], Stöber and Zessack[8,9], Stöber and Flachsbart[10,11], Stöber et al[12], Stöber[13], Hauck and Schedling[14], Hochrainer and Brown[15], Hochrainer[16], Berner and Reichelt[17], Burson et al.[18] and Matteson et al[19] developed different solutions to the various technical problems of aerosol centrifuges.

THE CYLINDRICAL AEROSOL CENTRIFUGE

One particular solution is discussed in Stöber's contribution. Our centrifuge, designed primarily by A. Berner, used the following principle: The aerosol flow is superimposed on a laminar clean air flow through a rotating slit. Aerosol, clean air and channel walls have the same rotational speed to minimize flow instabilities. A cross section of the centrifuge rotor is shown in fig. 1.

The cylindrical rotor is rotating around the axis of symmetry driven by a high speed router motor (stanley tools) via the shaft K. The clean air is produced internally by passing the aerosol through the ring-channel V. At the entrance in the deposition chamber M only the gas is left. The aerosol enters into the rotating slit S, which is perpendicular to the axis, flows outward and is superimposed on the laminar clean air flow at the entrance ring slit. The P's are deposited at the outer wall of M at the distance L from the aerosol inlet depending on their D_p*. The stream leaves the sampling head via the ringslit S2, the ring channel F and six flow-determining orifices G. To make any theoretical calculations one has to know the total flow in M. To measure it one has to make an air tight seal between the rotating and the stationary parts of the centrifuge - a very difficult problem. This has been solved by use of a teflon ring as shown in fig. 2. The flow measurement was made using a calibrated cylindrical glass tube connected air tight to the inlet and observing the movement of a soap film. If the flow limiting orifices were replaced by plugs, no flow did enter the centrifuge proving such that the system was sufficently air tight. One also has to know the rotational speed very accurately. The centrifuge head was supplied with ten small magnets. They produced an electrical pulse in a coil mounted on the stationary hood of the centrifuge while passing. An electronic counter permitted rapid and exact measurement of the rotational speed.*

If a multidispersed test aerosol (e.g. spray dried Dow latex aerosol including aggregates) is supplied to the instrument the foil shows a deposit in the form of lines, each line corresponding to another aggregate.A photograph of such a "spectrum" is shown in fig. 3. If the number concentration of P's on the foil is measured as a function of L, a concentration profile as shown in fig. 4 is obtained. Using different sized latex aerosols the instrument can be calibrated, i.e. the L vs D_p *curve can be established experimentally.*

On the other hand we know from the flow measurement the linear downward velocity of the air in the centrifuge, v. Since the flow profile in the channel does not influence L, as has been shown by Stöber[20]*, the L vs D relation can be calculated. The particle velocity* v_p *perpendicular to the axis of rotation is given from simple theory:*

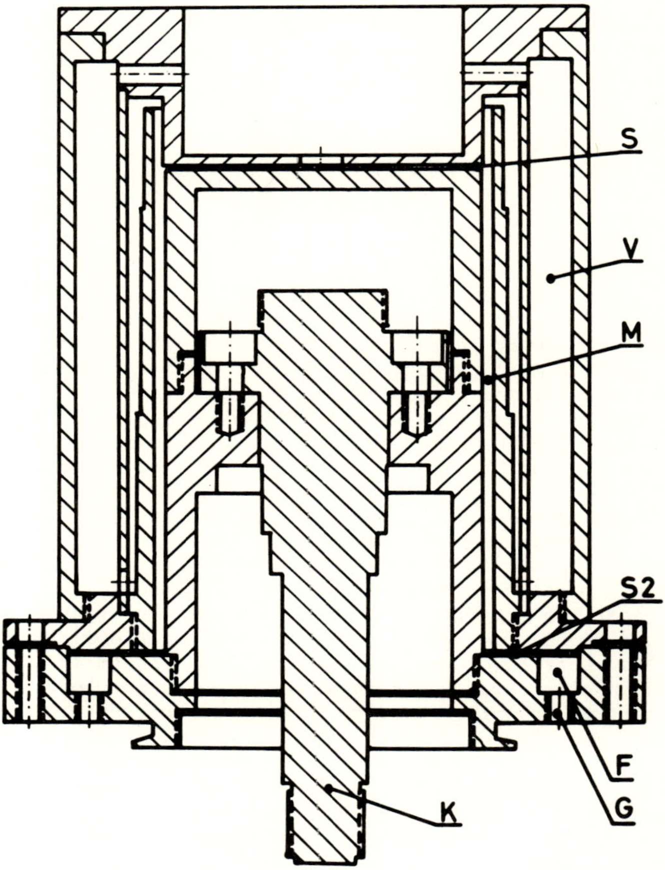

Fig. 1

Cross section of the rotor:

K shaft, V ring channels for producing clean air, M deposition chamber, S aerosol inlet, rotating slit (width 0,1 mm), S2 ringslit connecting M with channel E carrying the total flow through M, E outlet channel connecting S2 with the flow limiting orifices, G one of the six equally spaced flow limiting orifices.

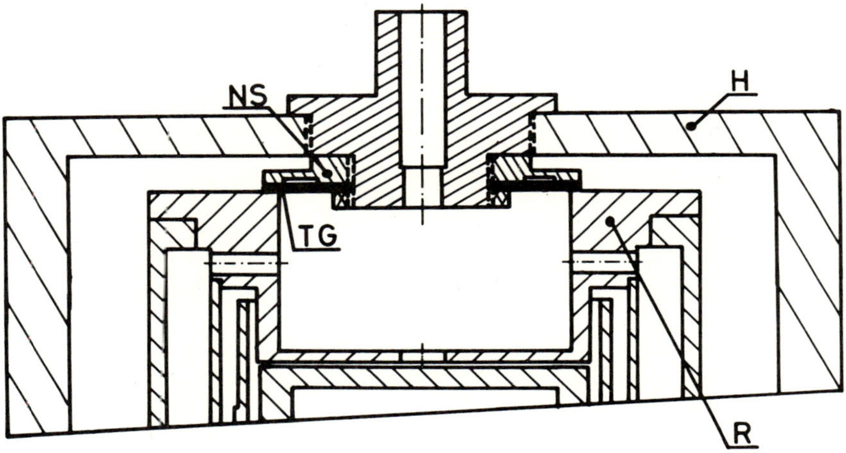

Fig. 2

cross section of the top of the rotor and the hood.

R rotor, H stationary hood, TG Teflon gasket facilitating the air tight seal between R and H, NS nylon spring pressing TG against R

Fig. 3

photograph of the deposit from a spray dried multidispersed latex aerosol

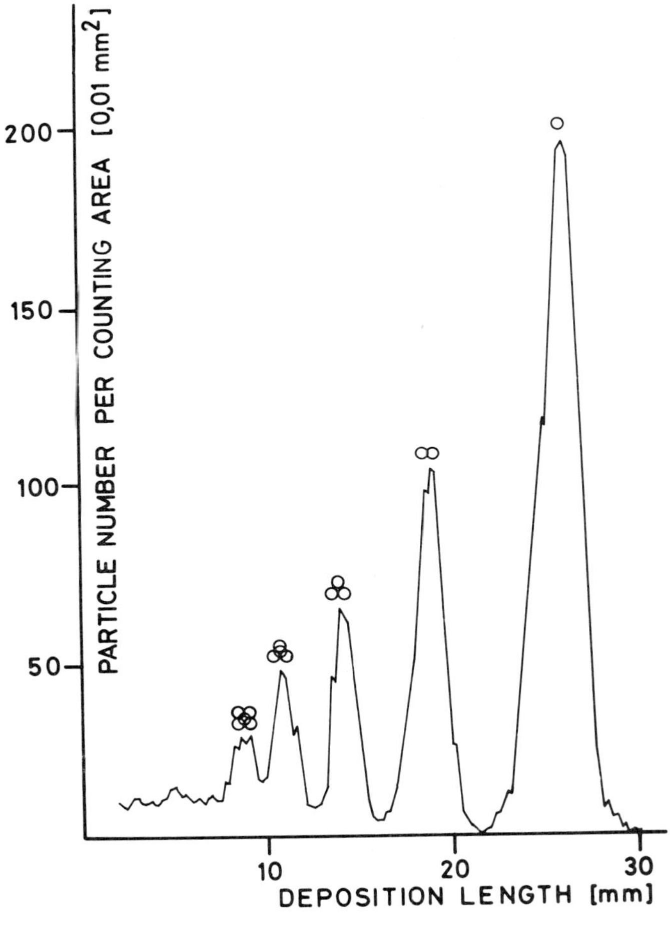

Fig. 4

concentration profiles across the lines shown in fig. 3
centrifuge operating conditions: 3000 RPM, Q = 31,2 ml/s
latex D = 0,557 μm

$$v_p = \frac{B\ 2\pi^2\ y_o\rho}{9\eta}\ N^2 D^2 \qquad (1)$$

with

N	*rotational speed of the centrifuge in rps*
D	*geometric diameter of a spherical P(cm)*
ρ	*density of P (1,o5)*
η	*viscosity of the air* ($0{,}183.1o^{-3}\,cm^{-1}s^{-1}$)
B	*correction factor for the Stokes law.*

$$B = 1 + 1{,}36\ \frac{\lambda}{D} + 0{,}70\ \frac{\lambda}{D}\ \exp(-3{,}68\ \frac{\lambda}{D}) \qquad (2)$$

λ	*mean free path of gas molecules* ($\lambda = 0{,}653.1o^{-5}\,cm$)
y	*channel height (o,3 cm)*
y_o	average channel (cylinder) radius (38.5 mm)

Assuming constant centrifugal force over y yields the residence time of the airborne P in the channel

$$t = \frac{y}{v_p} \qquad (3)$$

and

$$L = vt = y\ \frac{v}{v_p} \qquad (4)$$

Using all numerical constants and entering the volume flow rate through the channel Q(ml/s) one gets for

$$L = k\ Q/N^2 D^2 B\ \rho \quad \text{with } k=0{,}85325.1o^{-6} \qquad (5)$$

Equation (5) yields directly the inverse squared aerodynamic diameter L^*, *a normalised deposition length.*

$$L^* = \frac{1}{D_p^{*2}} = \frac{1}{D^2 B\rho} = \frac{L\ N^2}{k\ Q} \qquad (6)$$

The measured L-values and the operating conditions give L^* *and* L^* *in turn permits one to calculate* D_p^*.

The variation of the centrifugal force accross the channel alters the constant in equation 5 only slightly ($k'=0{,}85366.1o^{-6}$)*,so one does not need to considered it.*
Calculated values L_C, *for the different operating conditions and actually measured* L_M *values are given in table 2 together with the* L_M^* *as derived from the* L_M *values.*

The results are summarized in fig. 5 showing the good agreement between theory and experiment. However more important for the application of aerosol centrifuges is their high size resolution and the stability

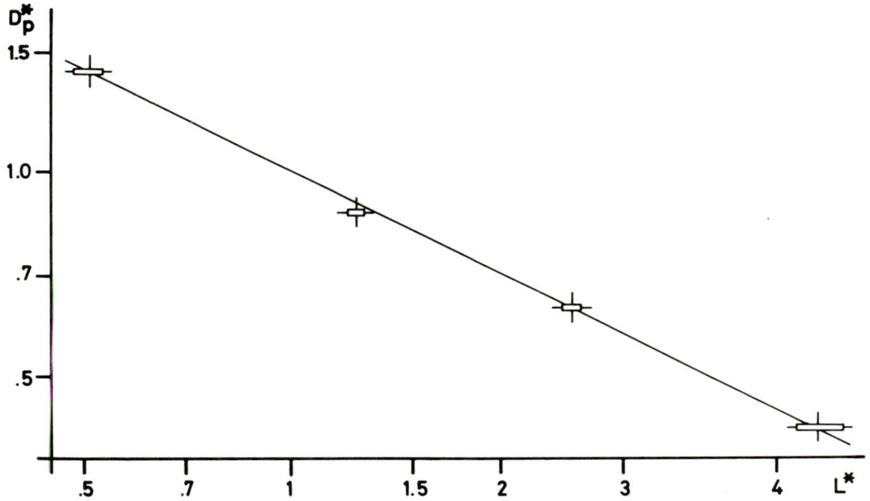

Fig. 5

average D_p^* values vs L^* for the different latex aerosols used. Solid line, calculated relation using equation 6. horizontal bar: range of measured L^* values, vertical mark: average L^* value, horizontal mark D_p^* calculated from manufactures date using ρ = 1,o5 and B from equation 2.

of operation , e.g. the calibration must remain constant during hours or even days of continuous operation. Since some warming up occurs, the independence of the calibration from this warming up effect has to be ensured experimentally. Fig. 6 shows the counted line profiles for cold and warmed up conditions; the coinciding curves prove that additional cooling is not necessary with our instrument.

The flow rate through the centrifuge has to be kept below a critical value, otherwise secondary parasite flows develop rapidly, first producing regular wiggles in the lines indicating a cellular structure of the parasite flow.
Still higher flow rates lead to turbulence in the entrance region and the line structure of the deposit disappears. All these effects can be avoided by restricting the aerosol flow by a limiting orifice.

Finally the modes of operation which ensure highest resolution are discussed. Obviously one may expect sharper lines as the superimposed aerosol sheet becomes thinner. Therefore the line width was measured for different aerosol flow limiting orifices, leaving the other operating parameter constant. Fig. 7 shows two line profiles as obtained with different aerosol flow limiting orifices.

The variances V_m *and standard deviation* σ_m *of the measured concentration profiles, the lines, were calculated directly from the counts using the equation*

$$\sigma_m^2 = V_m = \sum_{i=1}^{n} f_i (D_i^* - \bar{D}_G^*)^2 \Big/ \left(\sum_{i=1}^{n} f_i\right) - 1 \qquad (7)$$

with

$$\bar{D}_G^* = \frac{1}{\sum_{i=1}^{n} f_i} \sum_{i=1}^{n} f_i D_i^*$$

with

D_i^* *particle diameter of the i-th interval*

D_G^* *mean diameter of all P's*

f_i *number of P's in an area of* 10^{-2} mm^2 *located so that its L corresponds to* D_i^*

The dependence of the measured variance, V_m*, on the diameter of the aerosol flow limiting orifice is given in fig. 8*

Since the variance of "lines" produced in different experiments under constant conditions were determined separately it was also possible

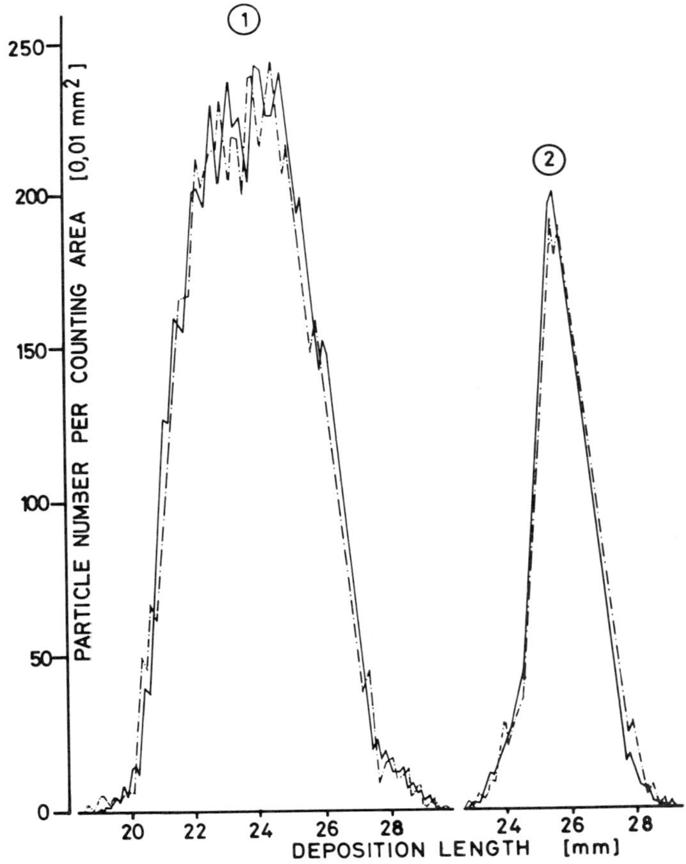

Fig. 6

line profiles produced in a cold (solid line) and a warmed up (dashed line) centrifuge.

1 6ooo RPM, Q = 1o9 ml/s,aerosol flow limiting orifice diam.2,o mm

2 5ooo RPM, Q = 74,2ml/s,aerosol flow limiting orifice diam.o,35mm

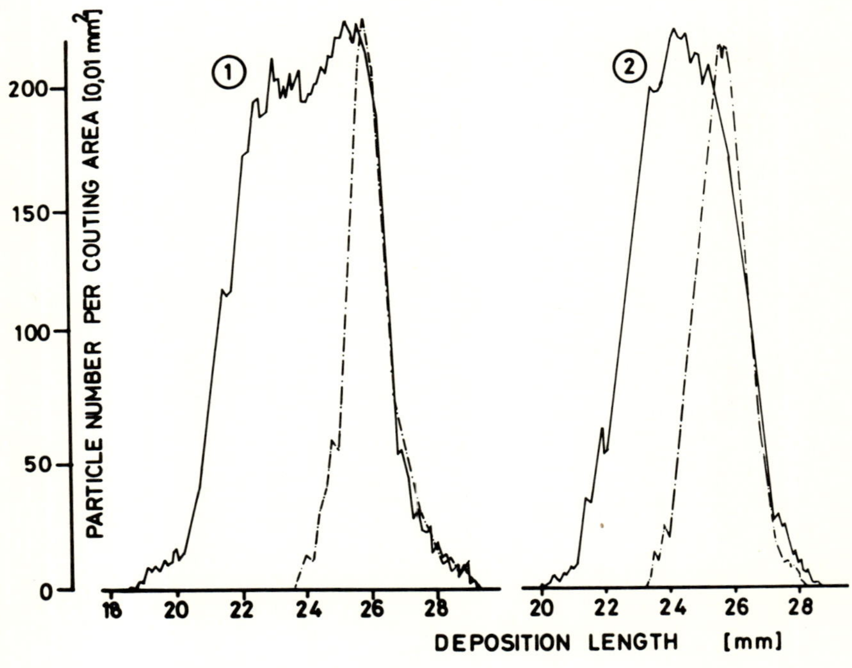

Fig. 7

two line profiles obtained with aerosol limiting orifices of diameters of 2,o mm solid line and o,35 dashed line
1 6ooo RPM, Q = 1o9 ml/s
2 5ooo RPM, Q = 74,2 ml/s

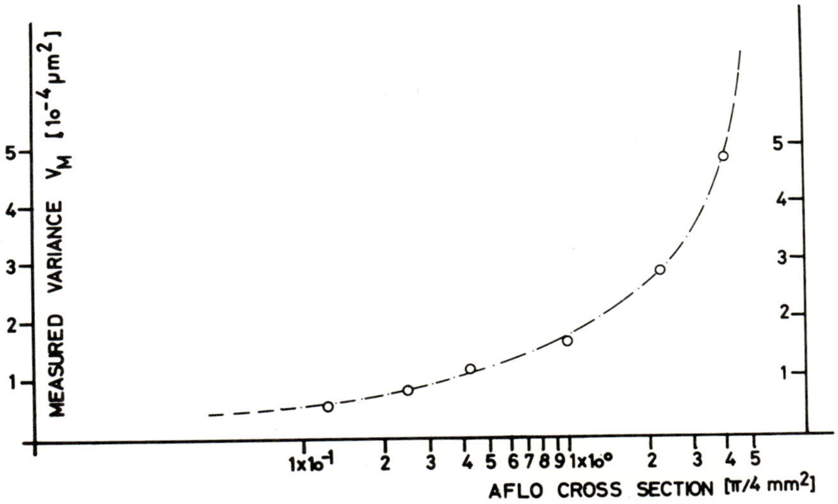

Fig. 8

measured variance V_m (absolut value)
vs. the diameter of the aerosol flow limiting orifices (AFLO). Size of circels corresponds approximately to the range of the experimentally derived variances.

to estimate the scatter of the variance values by calculating the variance of the variance. The results are given in table 3; V_m appears as a function of the inlet orifice of the aerosol flow. This measured variance V_m consists of both the variance V_A describing the width of the size distribution of the aerosol and the variance V_I describing the instrument influences.

$$V_m = V_A + V_I \tag{8}$$

Since V_A is not known with sufficient accuracy V_I cannot be calculated but V_m is certainly an upper limit of V_I. Comparing the V_m values with the V_A given by the manufacturer of the latex suspension (Dow Chemical) one finds V_A and the smallest measured V_m of about the same magnitude.

ACKNOWLEDGEMENT

The work reported was partly supported by the US-EPA-Grant R 8o1 983 and RO1 AP OO468.

Table 1

Correction factor for the Stokes law according to equation 2

D	B
1,3o5	1,o97
o,79o	1,156
o,557	1,214
o,357	1,316

Table 3

The measured variances V_m of the "lines" and the variances of these variances versus the diameters of the aerosol flow limiting orifices

diameter of the aerosol flow limiting orifice mm	V_m 10^4 μm^2	variance of V_m 10^4 μm^2
o,35	o,53	o,o9
o,5o	o,81	o,o6
o,65	1,15	o,o3
1,oo	1,58	o,12
1,5o	2,79	o,o5
2,oo	4,76	o,o8

operating condition: 3ooo RPM, Q = 49,5 ml/s

Latex:

manufactures data   *mean geometric diameter O,79O µm*
standard deviation O,OO44 µm

Table 2
Experimental results for 4 different latex aerosols and different operating conditions of the centrifuge.

RPM	Q	$D_g = 1.305\ \mu m$			$D_g = 0.790\ \mu m$			$D_g = 0.557\ \mu m$			$D_g = 0.357\ \mu m$		
$\left[\frac{1}{min}\right]$	$\left[\frac{ml}{sec}\right]$	$D_p^* = 1.409\ \mu m$			$D_p^* = 0.872\ \mu m$			$D_p^* = 0.630\ \mu m$			$D_p^* = 0.421\ \mu m$		
		$L_{th}^* = 0.507\ \mu m^{-2}$			$L_{th}^* = 1.341\ \mu m^{-2}$			$L_{th}^* = 2.517\ \mu m^{-2}$			$L_{th}^* = 5.646\ \mu m^{-2}$		
		L_{th}	L_m	L_m^*	L_{th}	L_m	L_m^*	L_{th}	L_m	L_m^*	L_{th}	L_m	L_m^*
3000	19.3	3.52	3.70	0.533	9.13	8.70	1.270	17.47	17.70	2.547	39.25	40.00	5.759
3000	31.5	5.69	5.90	0.526	14.75	13.75	1.255	28.28	28.20	2.509	64.06	60.20	5.311
3000	49.5	9.03	9.09	0.506	23.41	21.50	1.206	44.86	46.00	2.581	100.67	–	–
4800	35.8	2.55	–	–	6.61	6.20	1.232	12.68	12.80	2.540	28.44	31.20	6.199
4800	55.0	3.92	4.00	0.518	10.16	9.50	1.228	19.47	20.00	2.585	43.69	44.10	5.704
4800	75.0	5.35	5.10	0.484	13.86	13.50	1.279	26.55	26.10	2.474	59.85	63.50	6.023
6000	47.5	2.16	–	–	5.62	5.20	1.251	10.76	10.90	2.549	24.15	23.50	5.499
6000	71.5	3.26	3.30	0.541	8.45	8.10	1.259	16.20	15.70	2.438	36.35	36.00	5.597
6000	109.0	4.97	4.90	0.500	12.88	12.20	1.244	24.70	25.00	2.547	55.41	55.50	5.661
		$\bar{L}_m^* = 0.512 \pm 0.016\ \mu m^{-2}$			$\bar{L}_m^* = 1.240 \pm 0.025\ \mu m^{-2}$			$\bar{L}_m^* = 2.530 \pm 0.048\ \mu m^{-2}$			$\bar{L}_m^* = 5.719 \pm 0.282\ \mu m^{-2}$		
		$\bar{D}_p^* = 1.398 \pm 0.022\ \mu m$			$\bar{D}_p^* = 0.898 \pm 0.009\ \mu m$			$\bar{D}_p^* = 0.629 \pm 0.006\ \mu m$			$\bar{D}_p^* = 0.418 \pm 0.010\ \mu m$		

REFERENCES

1. *Sawyer, K.F., and Walton, W.H. The Conifuge - A size-separating sampling device for airborne particles. J. Sci. Instruments 27: 272, 195o*

2. *Goetz, A. An Instrument for the quantitative separation and size-classification of air-borne particulate matter down to o.2 micron. Geofisica Pura e Appl. Proc. II, 36, 1957a, 49*

3. *Goetz, A., Stevenson, H.J.R. and Preining, O. The Design and Performance of the Aerosol Spectrometer. A.P.C.A. Journal 1o: 378-383, October 196o.*

4. *Goetz, A. and Preining, O. Bestimmung der Größenverteilung eines Aerosols mittels des Goetz'schen Aerosolspektrometers. Acta Physica Austriaca, 14:292-3o4, 1961*

5. *Gerber, H.E. On the Performance of the Goetz Aerosol Spectrometer. Atm. Environment 5: 1oo9-1o31, 1971.*

6. *Keith, C.H. and Derrick, J.C. Measurement of the particle size distribution and concentration of cigarette smoke by the "conifuge". J. Coll. Science 15: 34o-356, 196o.*

7. *Kast, W. Neues Staubmeßgerät zur Schnellbestimmung der Staubkonzentration und der Kornverteilung. Staub, 21: 215-223, April 1961.*

8. *Stöber, W. and Zessack U. Zur Theorie einer konischen Aerosolzentrifuge. Staub, 24: 295-3o5, August 1964.*

9. *Stöber, W. und Zessack, U. Zur Messung von Aerosol-Teilchengrößenspektren mit Hilfe von Zentrifugalabscheidern. Zentralblatt für Biologische Aerosolforschung, 13: 1-19, Dezember 1966.*

1o. *Stöber, W. and Flachsbart,H. Aerosol Size Spectrometry with a Ring Slit Conifuge. Environmental Sci., 3: 641-651, July 1969.*

11. *Stöber, W. and Flachsbart, H. High Resolution Aerodynamic Size Spectrometry of Quasi-Monodisperse Latex Spheres with a Spiral Centrifuge. J. Aerosol Sci. 2: 1o3-116, 1971.*

12. *Stöber, W., Flachsbart, H. and Boose, Ch. Distribution Analyses of the Aerodynamic Size and the Mass of Aerosol Particles by Means of the Spiral Centrifuge in Comparison to other Aerosol Precipitators J. Coll. Sci. 39: 1o9-12o, April 1972*

13. Stöber, W. Design and Performance of a Size-Separating Aerosol Centrifuge Facilitating Size Spectrometry in the Submicron Range. In: Assessment of Airborne Radioactivity. Vienna, International Atomic Energy Agency, 1967.

14. Hauck, H. und Schedling, J.A. Über ein modifiziertes Modell einer Konifuge. Staub 28: 18-19, January 1968.

15. Hochrainer, D. and Brown, P.M. Sizing of Aerosol Particles by Centrifugation. Environmental Sci. and Technology, 3: 83o-835, September 1969.

16. Hochrainer, D. A New Centrifuge to Measure the Aerodynamic Diameter of Aerosol Particles in the Submicron Range. J. Coll. Interf. Sci., 36: 191-194, June 1971.

17. Berner, A. und Reichelt, H. Über Einlaßspaltsysteme in Konifugen Teil I: Das ROSL-System. Staub 29: 92-95, 1969

18. Burson, J.H., Keng, E.Y.H., and Orr, C.,Jr. Particle Dynamics in Centrifugal Fields. Powder Technology 1: 3o5-315, 1967.

19. Matteson, M.J., Boscoe, G.F. and Preining, O. Design Theory and Calibration of a Field Type Aerosol Spectrometer. Aerosol Sci. 5: 71-79, 1974.

2o. Stöber, W. and Boose, Ch. Developing Flow and Particle Deposition in Horizontal Elutriators and Semi-Dispersive Aerosol Centrifuges. Atmospheric Env. 7: 119-13o, 1973.

PART IV

AEROSOL MEASUREMENT AND ANALYSIS

METHODS FOR DETERMINATION OF AEROSOL PROPERTIES

RUPRECHT JAENICKE

MAX PLANCK-INSTITUT FÜR CHEMIE
D 65 MAINZ, GERMANY

ABSTRACT

In aerosol research, the measurement of the aerosol size distribution today is in a transition stage from first generation measurements and methods to the second generation. The first generation used instruments and methods which attempted to filter certain size fractions out of the aerosol. From the data obtained and an assumed rectangular filter function the size distribution was estimated. In the second generation, the size distribution is obtained by trial and error methods, applying any filter function in its exact form without falsifiing assumptions. Typical methods - as they are applied in Germany today - are summarized and additional methods for determination of the size distribution and physical and chemical properties of aerosols are surveyed.

Finally the problem of monitoring aerosols is discussed. It is recommended for monitoring purposes that instruments be selected with filter functions which simulate typical effects of the aerosol.

METHODS FOR DETERMINATION OF AEROSOL PROPERTIES

Ruprecht Jaenicke
Max Planck-Institut für Chemie, D 65 Mainz, Germany

INTRODUCTION

One of the most essential properties of aerosols is the size distribution: the particle number concentration as a function of particle size. It is known, that most observable effects of natural aerosols are influenced by the size distribution. This can be seen, for example, for visibility, water adsorption, mass deposition, cloud formation, and scattering of radiation. In general the effect E is caused by the aerosol size distribution $n^*(r)$ and some specific weighting functions w (r).

$$E = \int w(r) \cdot n^*(r) \, d\lg r \qquad (1)$$

Where r is the particle radius.
The integration over dlgr is prefered, because the size distribution is usually expressed as the differential distribution of the particle concentration N:

$$n^*(r) = \frac{dN}{d\lg r} \qquad (2)$$

In contrast to laboratory aerosols, the size distribution of atmospheric aerosols is not only a function of the particle size (given here as particle radius r), rather it is a function of time (this includes certain meteorological parameters like relative humidity) and location (including altitude).
If the effect E changes from place to place, with altitude, or with time, it is usually caused by a change of the size distribution $n^*(r)$. It, therefore, was one of the most important aims to measure the size distribution. All measurements M, however, can only be done the following way

$$M = \int m(r) \; n^*(r) \, d\lg r \qquad (3)$$

Each measurement or method exhibits a typical filter function m (r), so only some information of $n^*(r)$ is contained in each M. Until re-

cently, the measurement of $n^*(r)$ was accomplished through the suitable selection of m (r). m (r) was assumed to be a rectangular function. Typical examples are cascade impactors. This kind of measurements can be regarded as belonging to the first generation. Today we are in a transition stage to the second generation of measurements. In these measurements the m (r) are not selected as rectangular, instead they can have more or less any shape.
The evaluation of $n^*(r)$ is then done by variing $n^*(r)$ in (3) in a fitting procedure, until the measurement M is reproduced. These second generation measurements of the size distribution do not depend on falsifiing assumptions which had to be made in the first generation. This paper will review some of the new methods which are being used in Germany today.
Until now, we know only some of the weighting functions w (r) which relate effects of the aerosol with its size distribution. This paper will review some new methods for determing w (r), mostly as physical or chemical integral properties of aerosol samples.
Since it was realized that the measurement of the aerosol size distribution is too difficult and time consuming to be used for monitoring the atmospheric aerosol, single parameters have been selected for this purpose. From equation (3) one can see that such monitoring gives only limited information about the changes of the aerosol. It will be discussed how monitoring can be done.

AEROSOL SIZE DISTRIBUTION

AITKEN PARTICLES

The particle size distribution of atmospheric aerosols covers 5 orders of magnitude, from a radius of 0.001 μm to several 100 μm. This wide range can be observed in polluted aerosols as well as in clean aerosols. To measure any size distribution the entire radius range has to be subdivided into small fractions which can be covered by single methods. As a rule of thumb a single method can cover one order of magnitude in radius. This rule is valid for natural aerosols with steep size distributions, but in cases where the size distribution is comparatively constant, a wider fraction can be covered. Below a radius of 0.1 μm the knowledge of the size distribution in natural aerosols is very unsatisfactory. This is because only two methods are

available - diffusion and electrostatic precipitation - both of which can be regarded as poor filters, based on the considerations of equation (3).
Attempts with a method employing the charging of particles have shown success for particle radii greater than 0.01 μm and concentrations greater than 10^4 cm^{-3} (Willeke et al [29]).

Integrated Diffusion Analyzer

For the concentration range from 100 cm^{-3} to 10^6 cm^{-3} and the radius range of 0.001 μm to 0.1 μm the problem has been undertaken by Jaenicke[17]. An integrated diffusion analyzer in connection with an electrostatic precipitator was used. The measurement of the particle concentration is done with Condensation Nuclei Counters*). Whereas diffusion channels have been used in the past, the present method has some new features.
To avoid uncontrolled particle losses in connection pipes, the diffusion system was designed very compactly. The losses that do occur in the connecting pipes are integrated into the intented loss of particles in the system. The evaluation of the size distribution from measured values is done using an iterative fitting procedure of a minimum seach program. Thus it is no longer required to employ narrow-banded filter functions in equation (3). It is only necessary to know the weighting function w (r) whether analytical or stepwise. The size distribution $n^*(r)$ described by a number of parameters (whether as a set of normal distributions or as any other distribution) is varied until calculated values M_i agree to within the given accuracies with the measurements.
Since only a number of 7 diffusion channels is used the evaluated size distribution is given without fine structures. Size distributions in clean air have been evaluated as a superposition of two normal distributions. As opposed to classical ideas, these distributions showed two maxima, one at a radius of 0.1 μm the other close to 0.02 μm.

---

*) Similiar instruments have been recently used by A. Schultz, Heidelberg (personal communication) and Maigne et al [21].

LARGE PARTICLES

Spiral Centrifuge

For the measurement of particles greater than 0. 1 μm a spiral centrifuge has been developed (Stöber et al [27]) and will be presented in another paper of the symposium. Spiral centrifuges have been used in the past and the one developed by Goetz [9] being best known. The disadvantages have been discussed by Hochrainer [12]. The new development shows features avoiding these handicaps. For the best size resolution, the aerosol particles are spread over a large surface. The instrument thus is well adapted for polluted or laboratory aerosols with high particle concentration. In low concentration natural aerosols long sampling times are required and its use is limited due to the variability of the aerosol.

Laser Particle Counter

The size range covered by the spiral centrifuge is larger than that covered by optical particle counters. Optical particle counters have been used because of their capability to observe the particles in the airborne state. In addition optical particle counters have been regarded as absolute instruments in terms of particle concentration. Because of the cross-sensitivity, however, this certainly is not the case. (Jaenicke [14]).

An optical particle counter using small angle scattering of a laser beam has been developed by Gebhardt et al [7]. The laser beam provides excellent focussing and thus a sensitive volume as small as 0. 01 mm^3 can be obtained. The small sensitive volume makes it possible to count particle concentrations up to 10^4 cm^{-3} without dilution and only 10% loss due to coincidence. Following Jaenicke [14] the size distribution then can be measured in particle concentrations up to 1000 cm^{-3}. This is about a factor of 100 more than with other optical counters. Scattered light is measured between 1. 5° and 7. 5° in the forward scattering regime where the output signal becomes more or less independent of the index of refraction of the aerosol particles. This method has been used by earlier investigators, although it produces a certain ambiguity in the range 0. 5 μm to 1 μm. Thus the upper limit of the instrument is at a radius of 1 μm, while the lower limit of 0. 08 μm is given by the scattered light of the particle free sensitive volume. Gebhardt et al [7] give the resolution of the optical counter as 10-20%.

Sheet Air Impactor

Impactors are instruments which are used widely in aerosol physics and chemistry. The construction is usually very simple and thus the instrument is well suited for field measurements. In impactors, aerosol particles are physically sampled on a substrate and are thus available for studies of physical as well as chemical properties. Besides being used for sampling, the impactor can separate particles due to inertial characteristics. However, the size distributions derived from specially designed set ups remained comparatively crude, although sophisticated evaluations have been applied. (Jaenicke [15], Berner [3]). Zebel et al [31] have introduced a round jet impactor which seperates the particles on the substrate according to their aerodynamical behavior. This is achieved through the use of sheet air. The sheet air is released at the beginning of convergent part of the nozzle. The stream is thus focussed hydrodynamically and the aerosol particles are separated and deposited as rings with the largest - 2 µm - in the center and those with a radius of 0. 3 µm in a ring of 1. 4 mm diameter. The resolution of this instrument is not given, but can be estimated to be 20% or greater. As with most instruments with sheet air, the flow of aerosol to be measured is comparatively small - 3 cm^3/s - and thus the instrument is well suited for highly polluted or laboratory aerosols. It was tested first in a coal mine.

Rotating Disc Impactor

Another impactor for size distribution measurements of particles with radii greater 0. 1 µm has been developed by Schütz [33]. The sampling plate is shaped like a disc and rotates under a rectangular slit. This design produces high area densities in the center of the disc and low concentrations at its border. This impactor eliminates the uncertainty about the optimal sampling time for best evaluation, which usually is a handicap with impactors. A location with optimal area density always can be found somewhere on the disc. The impactor was designed for studies of mineral dust in clean air aerosols, so evaluation by electron microscopy can be applied.

Optical Instrument Combination

The instruments described above determine the aerosol size distribution. The results obtained for limited size ranges can be compiled to from the entire size distribution. For clean air this has been done by

Abel et al [1] and in polluted air Willeke et al [29] have published results. The resulting size distributions usually show some discrepancies in the overlapping ranges of different instruments, since different properties of the aerosol (optical, inertial, geometric) have been measured and transformed into radii. Attempts have been made to bring results of different instruments together into a single evaluation program in order to produce a consistent size distribution instead of a sum of adjecent distributions. (Jaenicke et al [19]). Recently Heintzenberg [11] has combined the results of 3 different instruments: an integrating nephelometer (Ahlquist et al [2]), an optical particle counter*) and a Concensation Nuclei Counter**). He expanded the fitting procedure described above to 3 instruments. A size distribution of 7 normal distributions (with given and unaltered average radii and standard deviations) is varied until calculated results of the instruments agree with the measured values. Heintzenberg [11] showed that the resulting size distribution depends on the initial values which must be assumed in the fitting procedure. This ambiguity however, remains within narrow limits, so useful information of the size distribution in the range 0.03 μm to 1 μm is obtained. The advantage of this method is, that all instruments employed monitor the aerosols continuously, thus the size distribution can be calculated for any given time.

VERTICAL AEROSOL PROFILES

Twilight Polarization

Similar optical methods have been applied by Steinhorst [26], however, he used them for remote measurements of atmospheric aerosols The depolarization of sky light by aerosol particles has been known for a long time. The Agung eruption in 1963 brought attention to the fact that the degree of polarization of red light can be used to investigate the stratospheric aerosol layer (Shah [24]). From measurements of the degree of polarization during twilight, Steinhorst [26] derived information about the particle size distribution in the aerosol layer and the

---

*) ROYCO 225, distributed by Royco, Menlo Park, Calif.

**) distributed by General Electric, Pittsfield, Mass

vertical profile of the large particle concentration. During twilight the altitude of the atmospheric layer illuminated by the sun depends on the depression of the sun. The information of polarization therefore originates predominately from a narrow layer at known altitude. Since all functions used are non linear, he applied a fitting procedure similar to Jaenicke [15] and Heintzenberg [11]. As his analysis shows, the information about the particle size distribution is rather vague, but the vertical profile of particle concentration can be determined comparatively accurately.

As opposed to lidar soundings, such measurements can only be performed in twilight, but the instrumental set-up remains comparatively small. The evaluation methods applied by Steinhorst [26] can be used to derive more information from lidar soundings than is done today.

The most recent vulcanic eruption by the volcano Fuego late in 1974 could be monitored by Steinhorst [26] as the stratospheric layer arrived over Europe.

From the reasons given above, one can consider the methods used by Heintzenberg [11], Steinhorst [26], Jaenicke [15, 17] to measure the size distribution of aerosols as of second generation type compared to typical size separation measurements.

PHYSICAL PROPERTIES OF AEROSOL PARTICLES

As equations (1) and (3) show, effects and measured properties of aerosols depend on the size distribution as well as physical and chemical properties of the aerosol. Measurements of physical and chemical properties depend mostly on the aerosol mass density which is small compared to that of trace gases and do not exceed 300 $\mu g/m^3$ as the average. Until recently this tiny mass forced such measurements to be restricted to informations on the total mass.

BULK DENSITY

The density of the aerosol particles is of importance for a number of processes and has influence on the collection efficiency of impactors as collecting devices. Usually the aerodynamic density, however, is needed. Measurements of optical properties and conclusions about the origin of aerosol particles need the knowledge of the bulk particle density. Efforts have been made to deduce this parameter from other measurements. (Hänel [10]).

Gaspycnometer

Recently Thudium [28] attempted to determine the bulk density from the two parameters defining it: Mass and Volume. The mass of a sample of atmospheric aerosol particles can be determined very accurately with commercially available microbalances. The determination of volumes of the order of 10 mm^3 remain difficult however. Thudium [28] developed a gaspycnometer for small volumes. The instrument measures the pressure reduction of air in a confined volume due to the volume of the sample. It was necessary to minimize all volumes used and measure with a very sensitive pressure sensor. Exact data are not published but the confined volume of the instrument is of the order of 2 cm^3. Volumes of 5 mm^3 could be determined within $\pm$ 10%. Impactorsamples in atmospheric polluted aerosols showed bulk densities of 2.1 g/cm^3 at 35% relative humidity. Respective values in continental clean air were 3.3 $g cm^{-3}$. Both values confirm the usual assumption of an aerodynamic density of 1 gcm^{-3} for irregularly shaped particles at mean atmospheric relative humidities.

MASS

Spiral centrifuge

The mass-distribution in aerosols was recently measured with a centrifuge from Stöber et al [27] by Mönig et al [22] with a newly developed direct method. They mounted quarze-crystalls at distinct locations on the foil usually used for precipitation of the particles. The change of oszillation of the quarze-crystalls due to mass precipitation was monitored. An accumulation-time of 30 h was estimated for atmospheric aerosols with 100 $\mu g/m^3$. The mass of 0.25 μm particles then can be determined with 20% accuracy.

Interferometer

The mass of single particles of 10^{-12} g (equiv. of 0.65 μm at 1 gcm^{-3} density) was measured by Holländer et al [13]. The system uses an aerosol particle beam brought into a vacuum chamber impacting single particles on a thin membrane. The impact causes small vibrations which are detectable because the membrane is part of an interferometer. The vibrations are converted into deviations of the interference pattern. This method has another advantage that optical properties can be analysed simultaneously. The method, however, is restricted to aerosols which remain stable in a vacuum.

LIGHT ABSORPTION

Sphere - Photometer

Fischer [6] has determined the absorption of light by aerosol particles. Comparisons between measured and calculated intensities of sky brightness indicated absorption by aerosol particles which had been neglected in earlier assumptions. Fischer [6] collected aerosol particles with impactors and measured the absorption in a sphere-photometer. He compared the total scattered light with the light of the incident beam to evaluate the absorption for wavelengths of 400 - 1000 nm and found the imaginary part of the complex index of refraction to be $n \cdot \kappa \sim 10^{-2}$. The aerosol samples showed approximately grey absorption. As was expected, the imaginary part of the complex index of refraction is a function of the water content of the sample.
This result effects our understanding of the energy budget of the cloud-free atmosphere. It also forces us to assume absorption of energy by clouddroplets, since the absorption of the cloud condensation nuclei can be larger than that of the water in the cloud droplets.

SCATTER OF LIGHT

Measurements and calculations in the field of atmospheric optics always assume the aerosol particles to be ideal spheres. This certainly is correct if the relative humidity is close to 100% and considerable amounts of water are attached to the particles. For lower humidities this certainly is incorrect. It was assumed, however, that the huge number of particles in the volumes under investigation are arranged randomly, so despite their irregular shape, ideal spheres could be assumed.

Experiments with Microwaves

Recently Zerrull [32] has conducted scattering experiments of microwaves (λ = 8 mm) with single particles of a size, that the results can be compared with Mie-scattering in the atmosphere. The dielectric constant was selected to be comparable with an index of refraction of 1. 5 - 0. 005 i similar to atmospheric particles. For test purposes ideal spheres were investigated and showed scattering properties as predicted by Mie-theorie. As nonspherical particles rough spheres, octaeders, and cubes were investigated. The results showed that randomly distributed non-spherical particles in a scattering volume cannot be treated as spheres.

For single particles, the scattering function of non-spherical particles did not show the typical spikes observed for spherical particles, as was to be expected. Randomly distributed non-spherical particles comparable to a polydisperse aerosol did not show the scattering function of a polydisperse aerosol of spheres. The forward scattering up to 45° of cubes agreed with that of spheres, but right angle scattering was a factor of 5 larger than expected and backward scattering was smaller. This result has large effects on all calculations in atmospheric optics. It also influences the interpretation of size distribution measurements with optical particle counters.

CHEMICAL PROPERTIES OF AEROSOL PARTICLES

In the discussion of the physical properties of aerosol particles, it could be seen that the investigations have been done on single particles and samples of particles, depending on the information desired. Analysis of a single chemical substance can usually be done on single particles, if the particle can be observed under the microscope. If a survey of chemical components is conducted, a sample of aerosolparticles is usually required.

SULPHATE

Single particle analysis

An example of the first kind of investigation is the work of Georgii et al[8]. They investigated $SO_4^{=}$ (sulphate) in aerosol particles with the well known barium sulphate reaction in a gelatine-layer. They increased the sensitivity of the method, so that "sulphate"-particles down to 0.26 μm radius could be detected. This radius is an equivalent value and describes the size of an ammonium sulphate ($(NH_4)_2SO_4$) sphere which contains the same amount of sulphate as the particle under investigation. Georgii et al [8] collected impactor samples at various locations and various altitudes in the troposphere and determined the size distribution of the sulphate particles. They found that about 50% of the sulphate mass can be associated with the Aitken-particles (< 0.1 μm)

ELEMENTS

X-ray fluoreszence of Aitken particles

Winkler [30] investigated the chemical composition of tiny masses of Aitken nuclei samples. The samples were obtained with absolute filters behind an impactor stage used as a precollection device. The impactor collected all particles greater 0.2 μm so only smaller particles were collected by the filter. The sample was analysed by X-ray fluoreszence for the elements S, Cl, K, Ca, V, Se and Br. The detection limit for S was 0.07 μg and 0.6 μg for Cl. The standard deviation of the analysis was around 15%. The results showed that sulpher compounds - expected as $(NH_4)_2SO_4$ - accounted for 90% of the mass, the overwhelming component in the investigated aerosols.

ORGANIC COMPOSITION

A very sensitive method to determine other soluble organic material in aerosol samples was developed by Ketseridis et al [20]. The method is based on a combination of well known procedures: Classical separation and thinlayer chromatography together with gas chromatography and mass spectrometry. The method is sensitive enough for the determination of all organic compounds in 100 mg of aerosol particles (with about 10% organic matter). In samples from various locations approximately 150 - 200 organic substances could be seen. Of these, up to 50 substances have been identified and quantitatively determined. On the average in clean air 1 $\mu g/m^3$ ether soluble organic material was found.

MONITORING OF AEROSOLS

With the concern today about air pollution, attention has been turned to monitoring and controlling atmospheric aerosols, which means making measurements as a function of time. For practical reasons, only single effects or properties can be monitored. A good example for this is the compilation of total particle concentration or total mass, done by governmental agencies as part of their air pollution survey. However, one has to ask what conclusions about the aerosol can be derived from the monitored parameter. As discussed in connection with equations (1) and (3), the measured parameter is always an integral over the size distribution and a certain weighting function. In Jaenicke [16] it was discussed that monitored parameters only contain information of limited particle size ranges of the size distribution. Under the assumption of a continental aerosol size distribution, the total particle number is affected by particles of 0. 001 μm to 0. 1 μm radius only. The sampled mass on the contrary is affected by particles from 0. 1 μm to 50 μm. The dry fall-out mass used as a very simple measure in a German monitoring network, gives information for particles of 10 μm and greater. The electrical conductivity of air - recently cited by Cobb et al [5] as a measure of air pollution - over ocean areas gives information for particles in the narrow range 0. 07 μm to 0. 1 μm only (the range 0. 01 to 0. 1 μm was assumed by Smic [25].
From this, we conclude, that the aerosol can only be meaningfully monitored by the simultaneous use of a number of different instruments. If the procedure recommended by Heintzenberg [11] is used, the size distribution at any time can be derived. Results of single monitoring instruments have to be discussed very carefully to draw conclusions about changes of the aerosol size distribution. Working with single monitoring instruments there is one possibility for application. Instruments can be selected which have filter functions similar to the weighting functions of aerosol effects. One example can be given. Breuer et al[4] describe a dust photometer working with scattered light. The filter function of the instrument is carefully selected, so the measured value is proportional to the volume of respirable dust retainable or deposi-

table in the aveoli of human lungs. The instrument is designed for aerosols with mass concentrations greater than 100 $\mu g/m^3$. Other instruments with filter functions corresponding to the weighting function of aerosol effects are already known (i. e. visibility, Ruppersberg 23). In this way, the aerosol and its effects are monitored and conclusions can be directly related to the influence of aerosols on man and his environment.

REFERENCES

(1) Abel, N., R. Jaenicke, C. Junge, H. Kanter, P. Rodriguez Garcia Prieto and W. Seiler. Luftchemische Studien am Observatorium Izana (Teneriffa). Met Rundschau. 22: 158-167, 1969.

(2) Ahlquist, N. C., and R.J. Charlson. Measurement of the wavelength dependence of atmospheric extinction due to scatter. Atm Env. 3: 551-564, 1969.

(3) Berner, A. Praktische Erfahrungen mit einem 20-Stufen-Impaktor. Staub. 32: 315-320, 1972.

(4) Breuer, H., J. Gebhardt, K. Robock and U. Teichert. Fotoelektrisches Meßgerät zur Bestimmung der Feinstaubkonzentration. Staub. 33: 182-185, 1973.

(5) Cobb, W. E., and H. J. Wells. The electrical conductivity of oceanic air and its correlation to global atmospheric pollution. J Atm Sci. 27: 814, 1970.

(6) Fischer, K. Bestimmung der Absorption von sichtbarer Strahlung durch Aerosolpartikeln. Beitr Phys Atm. 43: 244-254, 1970

(7) Gebhardt, J., J. Bol, W. Heinze and W. Letschert. Ein Teilchengrößenspektrometer für Aerosole unter Ausnutzung der Kleinwickel streuung der Teilchen in einem Laserstrahl. Staub. 30: 238-245, 1970.

(8) Georgii, H. W., D. Jost and W. Vitze. Konzentration und Größenverteilung des Sulfataerosols in der unteren und mittleren Troposphäre. Bericht des Institutes für Meteorologie und Geophysik der Universität Frankfurt/Main, May 1971.

(9) Goetz, A. An instrument for the quantitative separation and size classification of airborne particulate matter down to 0. 2 micron. Pageoph. 36: 49-69, 1957.

(10) Hänel, G. The real part of the mean complex refractive index and the mean densitiy of samples of atmospheric aerosol particles. Tellus. 20: 371-379, 1968.

(11) Heintzenberg, J., Über die Bestimmung der Größenverteilung luftgetragener Partikeln mit optischen Methoden. Dissertation Universität Mainz, 1974.

(12) Hochrainer, D.. On the reliability of measurements with the Goetz aerosol centrifuge. Atm Env. 6:699, 1972.

(13) Holländer, W., and J. Schörmann. Mass determination of single aerosol particles by optical interferometry. Atm Env. 8: 817-822, 1974.

(14) Jaenicke, R.. The optical counter: cross- sensitivity and coincidences. J Aer Sci. 3: 95-111, 1972.

(15) Jaenicke, R.. Der Doppelstufenimpaktor, eine weitere Anwendung des Impaktorprinzipes. Staub. 31: 229-236, 1972.

(16) Jaenicke, R.. Monitoring of aerosols by measurement of single parameters. Stockholm Tropospheric Aerosol Seminar, University of Stockholm, 29-31, 1973.

(17) Jaenicke, R.. Size distribution of condensation nuclei in the NE Trade-wind regime off the African coast. J Rech Atm. Submitted, 1974.

(18) Jaenicke, R.. Comments on "Size distribution of atmospheric particles" by C. N. Davies. J Aer Sci. Submitted 1975.

(19) Jaenicke, R., C. Junge and H. J. Kanter. Messungen der Aerosolgrößenverteilung über dem Atlantik. Meteor Forsch Erg. B7: 1-54, 1971.

(20) Ketseridis, G. and J. Hahn. Bestimmung der organischen Bestandteile von Aerosolpartikeln in Reinluft. Z Anal Chem. 273: 257-261, 1975.

(21) Maigne, J. P., P. Y. Turpin, G. Madelaine et J. Bricard. Nouvelle methode de determination de la granulometrie d'un aerosol au moyen d'une batterie de diffusion. J Aer Sci. 5: 339-355, 1974.

(22) Mönig, F. J., N. Schwarzer und W. Stöber. Bestimmung der Aero sol-Massenverteilung in einer Aerosol-Zentrifuge mit Hilfe von Schwingquarzen. Gesellschaft für Aerosolforschung, Jahresbericht 1973, p. 58-61.

(23) Ruppersberg, G. H. . Principes et Procédés de Mesure automatique de la Visibilité. Bull. de l'A. I. S. M. 31: 11-19, 1967.

(24) Shah, G. M. . Enhanced twilight glow caused by the volcanic eruption on Bali island in March and September 1963. Tellus. 21: 636-640, 1969.

(25) Report of the Study of Man's Impact on climate (SMIC). Inadvertent climate modification. Cambridge, MIT-Press, 1971, p. 209.

(26) Steinhorst, G. . Inversion gemessener Polarisationsgrade zur Bestimmung der stratosphärischen Aerosolkonzentration. Dissertation Universität Mainz, 1975.

(27) Stöber, W. , H. Flachsbart, and C. Boose. Distribution analysis of the aerodynamic size and the mass of aerosol particles by means of the spiral centrifuge in comparison to other aerosol precipitators. J Coll Interf Sci. 39: 109-120, 1972.

(28) Thudium, J. . Entwicklung eines Gaspyknometers zur Volumenmessung im Kubik-Millimeter-Bereich zwecks Dichtebestimmung an Proben atmosphärischer Aerosolteilchen. Diplomarbeit Universität Mainz, 1973.

(29) Willeke, K. , K. T. Whitby, W. E. Clark, and V. A. Marple. Size distribution of Denver aerosols - a comparison of two sites. Atm Env. 8: 609-633, 1974.

(30) Winkler, P. . Chemical analysis of Aitken particles ($< 0.2\ \mu m$ radius) over the Atlantic ocean. Geoph. Res. Let. 2: 45-48, 1975.

(31) Zebel, G. und D. Hochrainer. Zur Messung der Größenverteilung des Feinstaubes mit einem verbesserten Spektralimpaktor. Staub. 32: 91-95, 1972.

(32) Zerull, R. . Mikrowellenanalogieexperimente zur Lichtstreuung an Staubpartikeln. Forschungsbericht W 73-18, Bundesministerium für Forschung und Technologie, 1973.

(33) Schütz, L. . Mineralische Komponente im maritimen Aerosol. In: Arbeitsbericht des Sonderforschungsbereichs 73 "Atmosphärische Spurenstoffe", 1974, p. 82-85

AEROSOL MASS MEASUREMENT USING PIEZOELECTRIC CRYSTAL SENSORS

Dale A. Lundgren, Ph.D., University of Florida
Lawrence D. Carter, University of Florida
Captain Peter S. Daley, Ph.D., USAF

Piezoelectric crystals, when mechanically stressed, develop electrical charges on certain crystal surfaces. Quartz, as a piezoelectric material, vibrates at a very precise natural frequency, which can easily be determined to one part in ten million when the crystal is placed in an appropriate electronic oscillating circuit. When foreign material, such as particulate matter, is deposited onto the crystal active area, the vibrational frequency characteristics of the crystal change in a predictable and measurable manner. This phenomenon led to the development of two commercial instruments which were described in the literature of 1970. These two devices, one based upon particle deposition by impaction and one by electrostatic precipitation, were revolutionary because of their extreme sensitivity and rapid, near real time, response characteristics.

Many studies have been conducted and papers published on the use and performance of piezoelectric crystal sensors for aerosol mass concentration measurement. This paper carefully reviews these studies, defines the capabilities and limitations of the piezoelectric technique, and discusses recent developments and improvements made to commercially available instruments.

AEROSOL MASS MEASUREMENT USING PIEZOELECTRIC CRYSTAL SENSORS

Dale A. Lundgren, Ph.D., University of Florida
Lawrence D. Carter, University of Florida
Captain Peter S. Daley, Ph.D., USAF

INTRODUCTION

Piezoelectric crystal mass monitors are devices which measure the mass concentration of an aerosol by depositing particles from a sample stream onto a vibrating crystal sensor and determining the change in frequency caused by the additional mass. Since the sample gas volume is known, the aerosol's concentration can be calculated manually or electronically.

Current interest in piezoelectric crystals as aerosol mass concentration measuring devices is based on the properties of extreme sensitivity and rapid response which are inherent in the piezoelectric method. For example, in 1970 Leemhorst reported a first generation quartz crystal microbalance that measured the mass of single particles of 10^{-11} gram.[1] The quantitative output of mass concentration from crystal microbalances can be nearly real-time; electronic circuits generally average the output over a 1 or 10 second interval.[2] (The desire for representative average concentrations may dictate averaging readings over time periods of 1 to 10 minutes.)

Along with the potential for precision evident in piezoelectric mass monitors, several investigators have reported that these devices are capable of significant errors.[3-7] Research, recently conducted at the University of Florida, identified the sources of these errors and suggested realistic operating limits for piezoelectric aerosol monitors in order to reduce the effects of such error sources.[6,7]

This paper reviews the early developmental work of several researchers, two commercial first generation devices, analytical work at the University of Florida, and current work underway on second generation piezoelectric devices.

THEORETICAL CONSIDERATIONS

Due to the temperature and vibratory parameters associated with various cuts of quartz crystals, type AT crystals are generally accepted as most appropriate for mass monitors. Such a crystal vibrating in the fundamental thickness-shear mode has a natural frequency of

$$f = \frac{N}{b} \tag{1}$$

Where: N = frequency constant of AT-cut quartz, 0.166 MHz·cm
b = crystal thickness, cm.

Differentiating equation (1) shows how an incremental change in crystal thickness affects frequency.

$$df = -\frac{N}{b^2}db \tag{2}$$

Dividing equation (2) by equation (1) yields:

$$\frac{df}{f} = -\frac{db}{b} \tag{3}$$

By definition:

$$m = \rho Ab \tag{4}$$

Where: m = mass of vibrating quartz, g
ρ = density of vibrating quartz, 2.65 g/cm^3
A = Area of vibrating quartz, cm^2

Since A and ρ are constant,

$$dm = \rho Adb \tag{5}$$

so,

$$db = dm/\rho A \tag{6}$$

Substituting equation (6) into equation (3) yields:

$$\frac{df}{f} = - \frac{dm}{m} \tag{7}$$

Rearranging, we find:

$$\frac{df}{dm} = - \frac{f}{m} = - \frac{f}{\rho Ab} \tag{8}$$

Recalling $f = \frac{N}{b}$, we see:

$$\frac{df}{dm} = - \frac{f^2}{\rho AN} . \tag{9}$$

Introducing a new parameter, C_f, the layer sensitivity constant, such that

$$C_f = \frac{f^2}{\rho N} , \tag{10}$$

it follows that

$$\frac{df}{dm} = \frac{-C_f}{A} . \tag{11}$$

Replacing the differentials in equation (11) with finite deltas is acceptable if Δf and Δm are both small relative to f and m. Since C_f is a known quantity ($C_f = 2.27f^2$, for AT-cut quartz), we now have the following working relationship for the change in resonant frequency of a quartz crystal due to an incremental addition of mass to the vibrating area

$$\frac{\Delta f}{\Delta m} = \frac{-C_f}{A} . \tag{12}$$

It should be emphasized that the area in equation (12) is the vibrating crystal area and that practical piezoelectric devices limit this area to the electrode area. Daley has shown that the crystal and electrode can be designed to damp out both harmonics problems and vibration beyond the electrode (facilitating crystal mounting, in the process).[7] Hence the vibrating area in a properly designed crystal is the electrode area. It has also been demonstrated that the deposit area may be substituted for the vibrating area if the deposit area is greater than the vibrating area.[7]

If the deposit diameter is a significant fraction of the electrode diameter, then theoretical calculations are generally based on the assumption that the collected mass is evenly deposited over the deposit area, that the deposit area is a constant, and that the layer of deposit is thin enough to act as part of the electrode's surface. Essentially, the latter is the assumption of Sauerbrey who initially generated interest in piezoelectric crystals as mass monitors.[10]

However, when particles exceed a certain size or when the thickness of the layer is too great, the deposit begins to move relative to the electrode surface and the theory does not apply.

Sauerbrey determined that the vibrational amplitude and, therefore, the mass sensitivity varies across the active (vibrating) portion of the crystal surface.[10] This fact was verified by Daley, whose findings are presented schematically in Figure 1, together with the dimensions of a typical crystal.[6] This mass sensitivity distribution characteristic is of greatest importance for impaction devices in which the deposit area is small compared to the sensitive area.

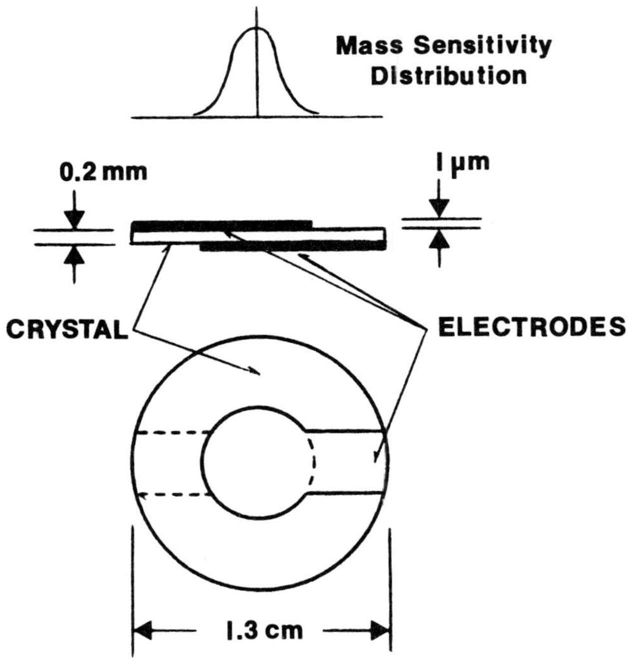

Figure 1. Size and mass sensitivity distribution of a typical crystal.

GENERAL STUDIES

Most of the work with crystal monitors conducted prior to 1970 focused on thin film measurement or gas concentration measurement. King has noted a list of references which deal with quartz crystal applications ranging from measuring dew points to detection of hydrocarbons and sulfur compounds.[11] He reports thickness gauges measuring deposited films of 0.1-30,000 Angstroms. He suggests personal monitors for measuring employee exposure to specific gases. These consist of a coated crystal carried by the employee all day and plugged into a circuit for frequency measurement before and after working hours. King developed CO_2, H_2O, and dust monitors to accompany the proposed 1975-1976 Viking Mars lander.[12]

Frechette and Fasching have investigated the use of crystals to monitor SO_2 concentrations.[13] A thin film of styrene-dimethylaminopropylmaleimide 1:1 copolymer adsorbs SO_2. Response was linear from 20-300 ppm and concentrations as low as 0.1 ppm were detectable.

Quite recently, Scheide and Taylor reported a quartz sensor capable of measuring the concentration of mercury in air from about 0.1 to over 3 ppb.[14]

The work of Warner and Stockbridge is significant for its analysis of temperature and crystal parameters as they relate to optimal design of the crystal.[15]

Pulker and Schadler have reported on factors affecting the accuracy of thin film measuring devices.[16] They also noted the higher mass sensitivity of the central portion of the electrode which, as mentioned earlier, Daley has since confirmed.[6]

WORKING MODELS OF PARTICULATE MASS MONITORS

The work of Chuan[8,17-20,26-28] (Celesco Industries, Inc.), Olin and Sem[21-23] (Thermo-Systems, Inc.), and Carpenter and Brenchly[3,24] (Purdue University) is of particular importance to piezoelectric mass sampling because these men constructed working models, two of which became commercially available.

A schematic of Chuan's quartz crystal microbalance is shown in Figure 2. Particles are collected by impaction on a sensor crystal. A reference crystal is exposed to the sample stream but does not collect particles. The difference between sensor crystal frequency and reference crystal frequency is converted to a voltage which is electronically differentiated to give a mass concentration readout. The original Celesco device had only one impaction jet. Because of the relationship between mass sensitivity and position on the electrode, a small misalignment of the crystal and the impactor nozzle could cause a significant error. This single nozzle design was abandoned in favor of a 4 nozzle design. The result was an increased mass response and a decreased sensitivity to deposit position. The sample rate is 1.5 ℓpm with 0.6 ℓpm flowing thru the impaction jets and the remainder being by-passed around the crystal.[28] When concentrations are low enough that particles impact at intervals greater than 0.01 second, Chuan reported the microbalance could detect a single particle mass of 10^{-11} gram.[28]

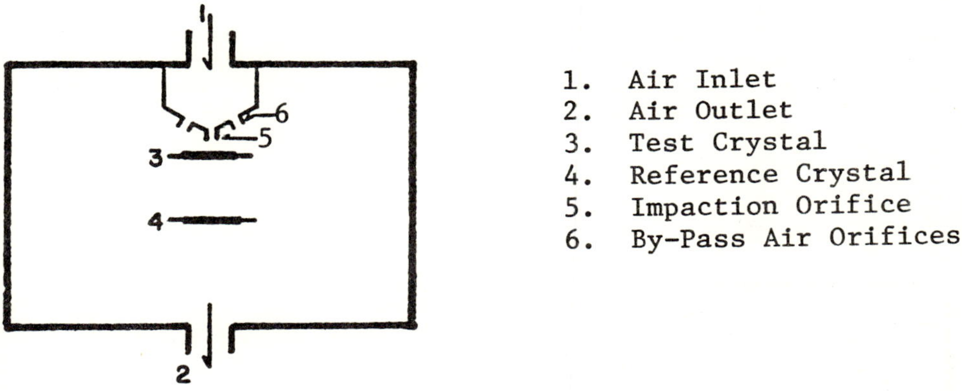

Figure 2. Schematic of the Celesco quartz crystal microbalance.

This impactor device was used to measure the mass concentration of effluents from a power plant, mineral wool cupola and electric furnace (all measurements downstream of collection devices).[8,17] Electron microscope analysis of the impacted particles showed a size range of

0.1-4 μm diameter for the power plant and slightly larger particles for the cupola and electric furnace. It should be noted that the 50% cut size for this impactor is about 0.55 μm.[7] It is not clear from the data if the upper size limit was due to capabilities of the impaction device or the control equipment. Nevertheless, parallel filter samples for the electric furnace showed comparable results.[17] Data from other tests showed collected particles as large as about 18 μm; however, very few exceed 10 μm diameter.[18,19,20]

Chuan also ran parallel tests of another aerosol with a Hi-Vol sampler and reported comparable results.[8] Unfortunately, size distribution and environmental conditions were not reported with these results.

Volatile particles are known to cause rapid negative then positive frequency shifts of the quartz crystal as particulate material is first sensed and then lost by evaporation.[18] Chuan reported in 1972 that the volatile particulate mass concentration could only be measured qualitatively by the quartz monitor.[20]

The piezoelectric mass monitor developed by Olin and Sem and sold by Thermo-Systems, Incorporated (TSI) differs from the Celesco device in several respects. Most importantly, the TSI unit collects particles by electrostatic precipitation and its deposit area is significantly larger than that of the Celesco impactor. A schematic of TSI's Model 3200 is shown in Figure 3. The sample flow rate is 1 ℓpm and the crystal's resonant frequency is about 5 MHz (versus 10 MHz for the Celesco device). TSI reports their device has a particle sensing range of 0.01-20 μm (diameter) and a concentration range of 1-200,000 μg/m^3.[2]

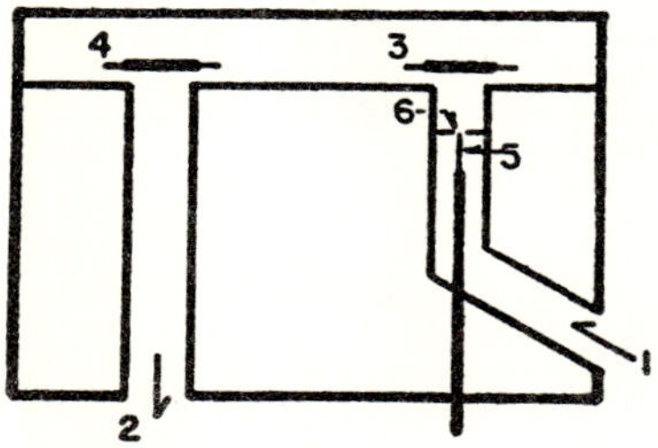

1. Air Inlet
2. Air Outlet
3. Test Crystal
4. Reference Crystal
5. Precipitator Electrode
6. Aerosol-Corona Contact Orifice

Figure 3. Schematic of the TSI piezoelectric mass monitor.

TSI indicates the frequency response to mass is linear until saturation occurs; i.e., until an additional mass deposit produces no additional frequency response.[2] The sensor crystal reportedly can operate 8 hours before cleaning is necessary, when sampling a 100 $\mu g/m^3$ aerosol.[2] The mass sensitivity of the instrument (change in frequency per unit mass) was reported to be 180 Hz/μg.

TSI reports their precipitator is essentially 100% efficient on tobacco smoke of 10,000-100,000 $\mu g/m^3$ concentration and that the majority of the mass of tobacco smoke is in the size range of 0.01-3.0 μm diameter. Other aerosols in that size range were also reported as collected with equal efficiency.[22] Data for larger particles have not been published by TSI.

A reported problem with the precipitator design is that the precipitator may "manufacture" aerosol by interaction of the corona with certain vapors (as much as 20 $\mu g/m^3$ of sampled air).[22] Daley did not find that this occurred. Another problem encountered with the precipitator design was inefficient collection of very dry particles, like sand. In one experiment, poor sensitivity was reported on particles of 10 μm diameter.[22] In TSI's commercial device an inlet air stream treatment apparatus was added to alleviate this problem. Olin acknowledges this treatment may aggravate humidity problems (discussed later).[23]

Carpenter designed, constructed and calibrated a 4-stage cascade impactor which used quartz crystals in each stage as mass sensors. In his calibration experiments, Carpenter noted that the frequency response ($\Delta f/\Delta m$) decreased as the mass loading on the crystal increased. Frequency changes associated with large mass loadings were not considered accurate measurements of true mass.[24]

Carpenter was unable to obtain accurate data from his impactor's first stage. He attributed this error to the fact that the impaction area exceeded the crystal electrode area. Another possible problem was the inability of a crystal to sense large particles (>20 μm) with high efficiency.

EVALUTION OF COMMERCIAL INSTRUMENTS

An in-depth study of piezoelectric mass concentration monitors was conducted at the University of Florida in 1973 and 1974.[7] In this study two commercially available quartz crystal mass monitors (TSI Model 3200 and Celesco Model 37A) were the subject of carefully controlled experiments to quantitatively determine the effects of temperature and humidity changes on mass measurement accuracy, linearity of response, particle collection characteristics, and mass sensitivity for a number of specific aerosols.

The effect of fluctuating temperature is minimized by the choice of an AT-cut quartz crystal with a cut angle of $35^{\circ}14' \pm 1'$. (This optimum cut angle may vary with crystal source). The $\pm 1'$ specification is within the accuracy of commercial crystal suppliers. Both the Celesco and TSI devices contained reference crystals to compensate for temperature (and other) effects. However, the reference crystals failed to correct for rapid temperature changes since these were effectively damped out as the air passed from test to reference crystal sensor. Daley found that the addition of metal tubing to the inlet line provided a heat sink capable of reducing temperature related errors to 5 $\mu g/m^3$.[7] This represents a temperature change rate of 0.3 and $1^{\circ}C$/minute in the Celesco and TSI units respectively (these are relatively large change rates which would not normally be encountered in ambient air sampling). The temperature behavior of beveled, lower frequency crystals was found to be significantly more stable than non-beveled crystals.

The Florida study found humidity affected the mass sensing ability in two ways. First, the crystals increased in mass as they adsorbed moisture from the air stream. Second, the accumulated particulate deposit increased in mass by adsorbing moisture. Figure 4 shows the change in crystal frequency with respect to humidity for the TSI unit without a reference crystal (Note: TSI supplies platinum electrode crystals).[7] It was possible to maintain almost "new" crystal response by cleaning the sensor crystal with detergent and distilled water. The precipitator corona in the TSI unit apparently activated its sensor crystal surface thereby increasing its hygroscopicity. Since the reference crystal was not activated, humidity change compensation was less effective than that theoretically obtainable.[7] The crystals of the Celesco unit were less sensitive to humidity changes; the hydrophobic crystal coating used to help retain impacted particles was partially responsible.

Error due to moisture adsorption by the aerosol deposit was found to be an order of magnitude greater than error due to crystal adsorption. This problem is basic to the method and not the device since the reference crystal clearly cannot compensate. Curve A in Figure 5 shows that for ambient aerosols at high relative humidities, significant changes in relative humidity can cause large errors.[25] Figure 6 indicates how use of a dirty crystal in an environment of changing humidity may cause large errors.[25] For an aerosol of constant concentration the piezoelectric unit indicated high concentration (steep slope) and no concentration (flat slope) as relative humidity increased and decreased, respectively. This error was due to water addition and removal from the deposited aerosol. Clearly, the rapidly

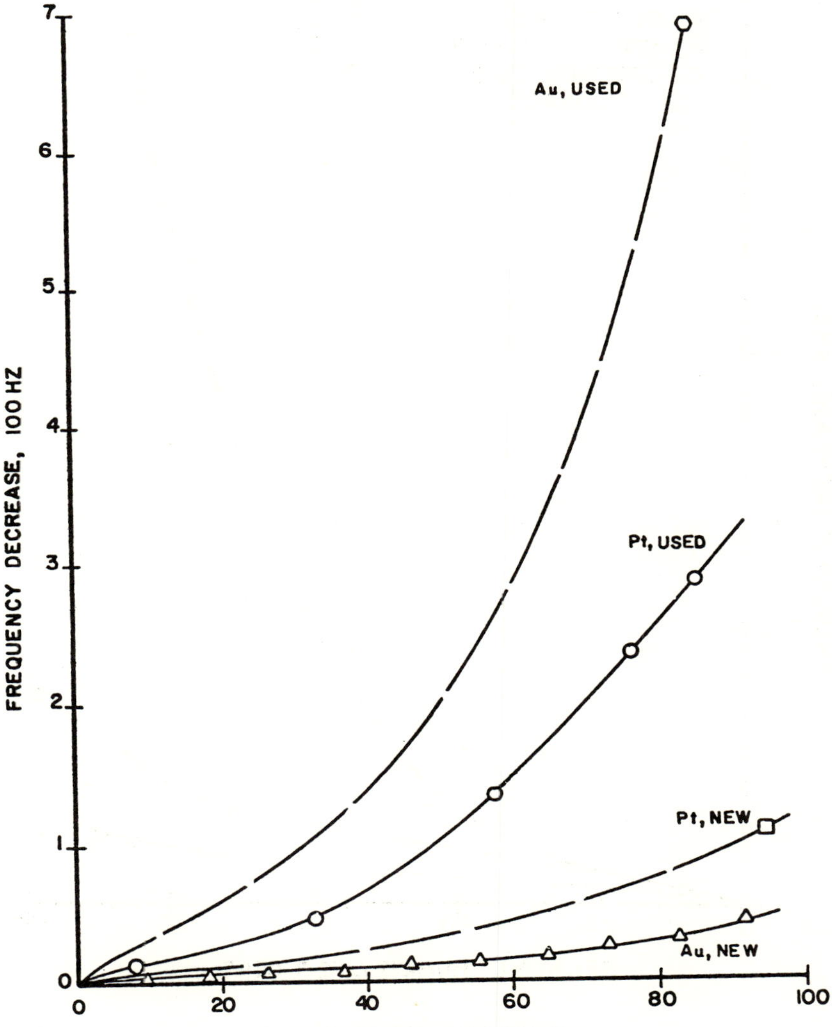

Figure 4. Humidity response of various 5 MHz crystals.

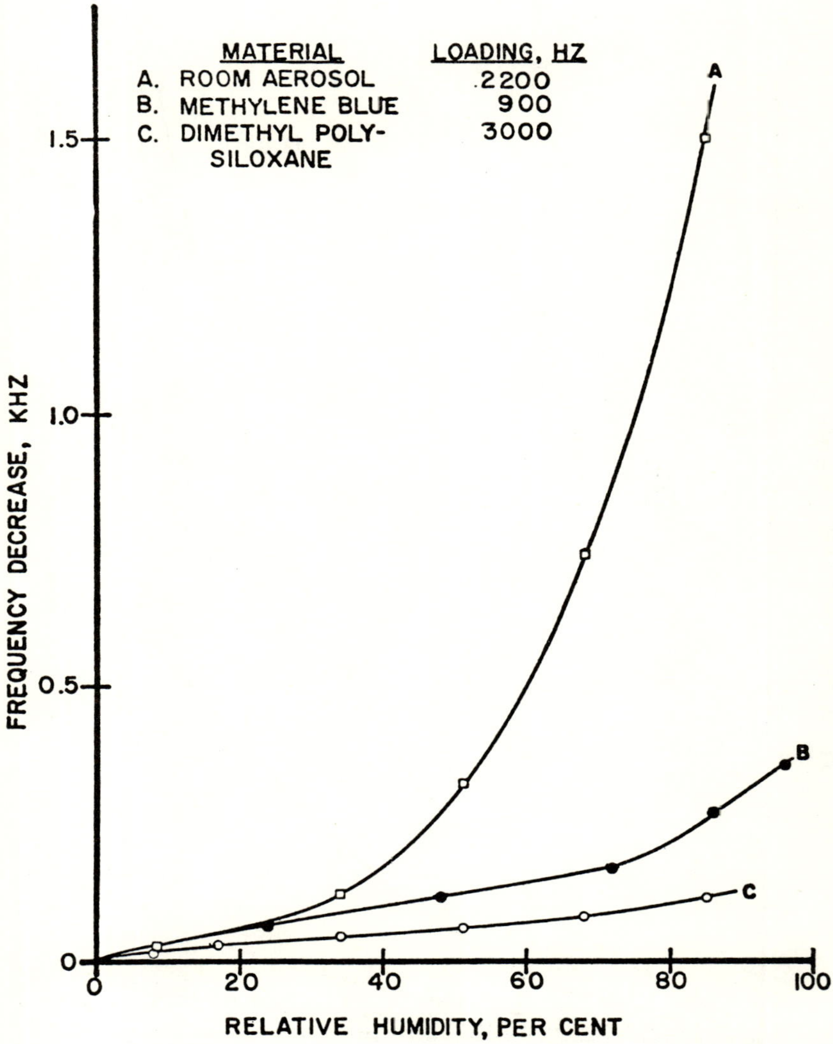

Figure 5. Humidity response of platinum electrode, 5 MHz crystals loaded with various materials.

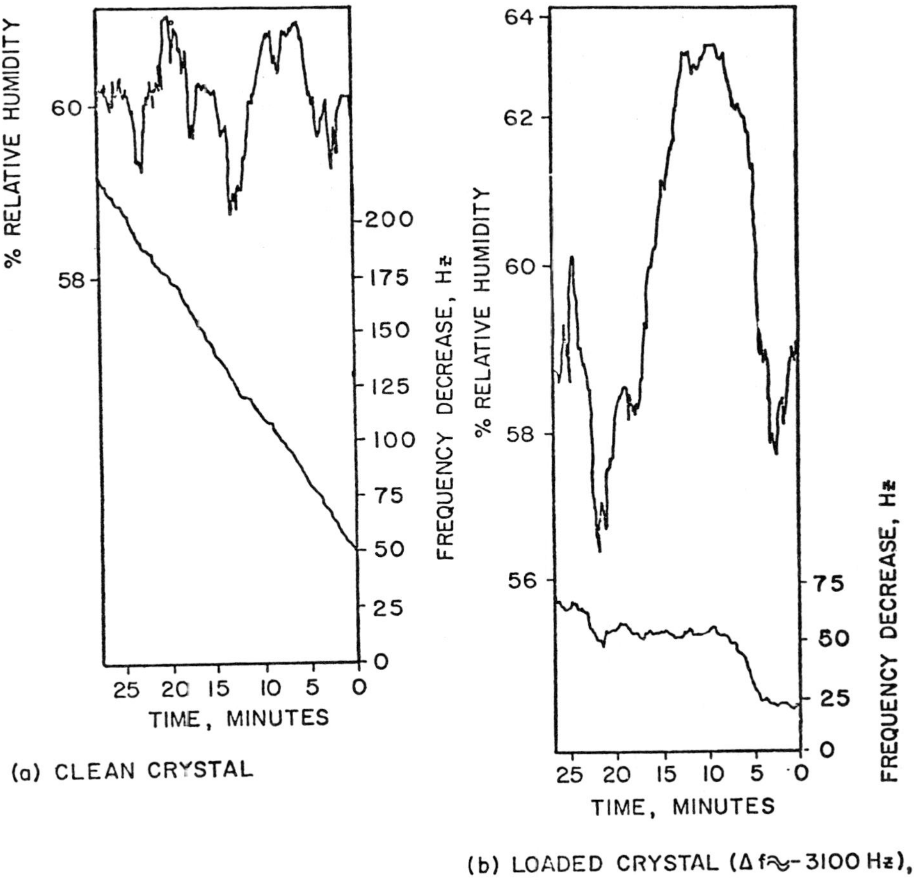

Figure 6. The effect of relative humidity fluctuations on a piezoelectric device used in ambient monitoring.

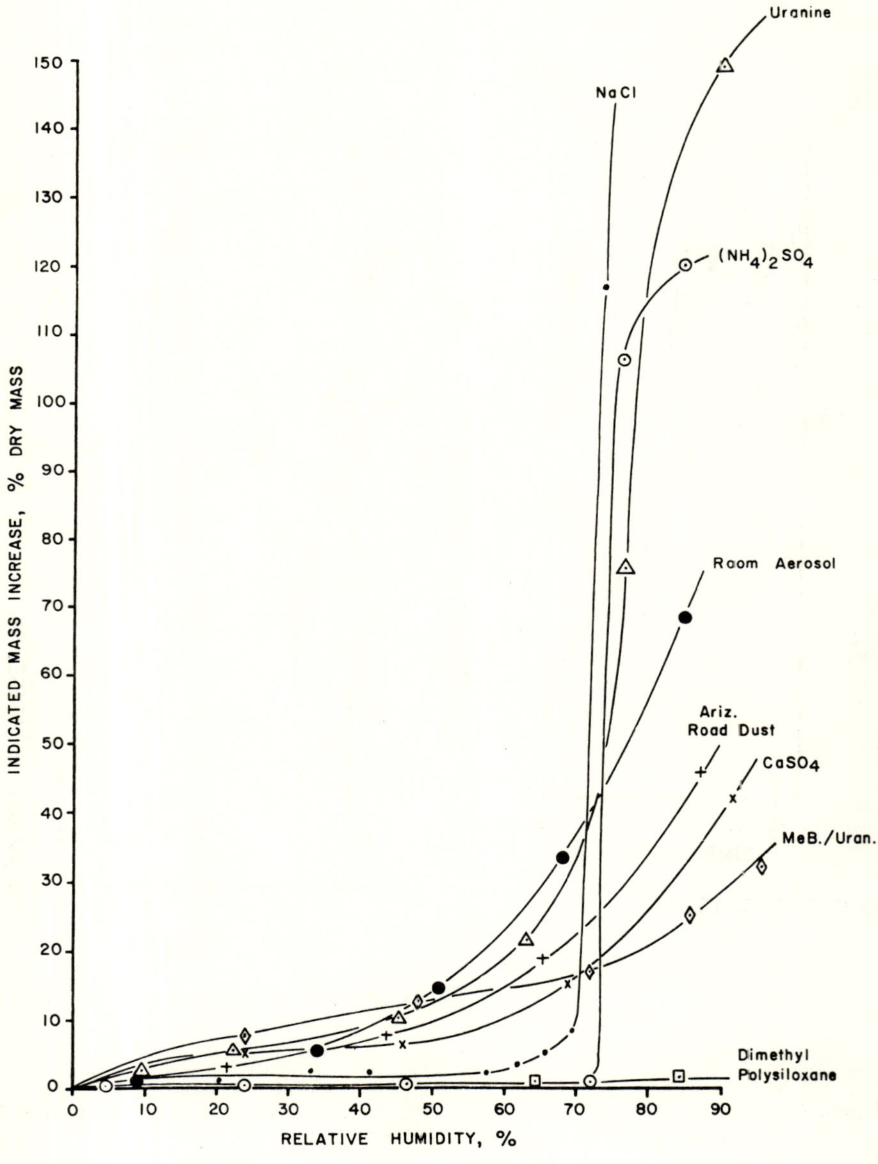

Figure 7. Hygroscopicity of various aerosol deposits.

changing relative humidity was responsible. Figure 7 shows the varying effect of humidity changes on indicated deposit mass.[6] Suggestions for reducing the effects of humidity are discussed later in this paper.

The linearity of response of the Celesco and TSI units was analyzed by using them to monitor a constant aerosol concentration. Variation of the readout from actual concentration by $\geq$20%, constituted nonlinear response.

Linearity was found to be a function of particle characteristics. For grease aerosols, frequency shifted 30 KHz before reaching nonlinearity on the TSI unit. However, for a methylene blue aerosol the shift was only 600 Hz. Figure 8 shows the effect of increased mass on linearity for a variety of deposits. Tables 1, 2 and 3 give the linear response limits of several different aerosols.[6] Note the absolute mass limits for the Celesco's impactor (with small deposit areas) are much lower than for TSI's precipitator, but the mass per unit area is greater. Daley reports that the sampling periods dictated by linearity considerations are 0.5 hour and 2 hours for the Celesco and TSI units, respectively, when sampling a 100 $\mu g/m^3$ ambient aerosol.[7] When linearity is lost it is necessary to cease sampling and clean the sensor crystal, a relatively simple operation.

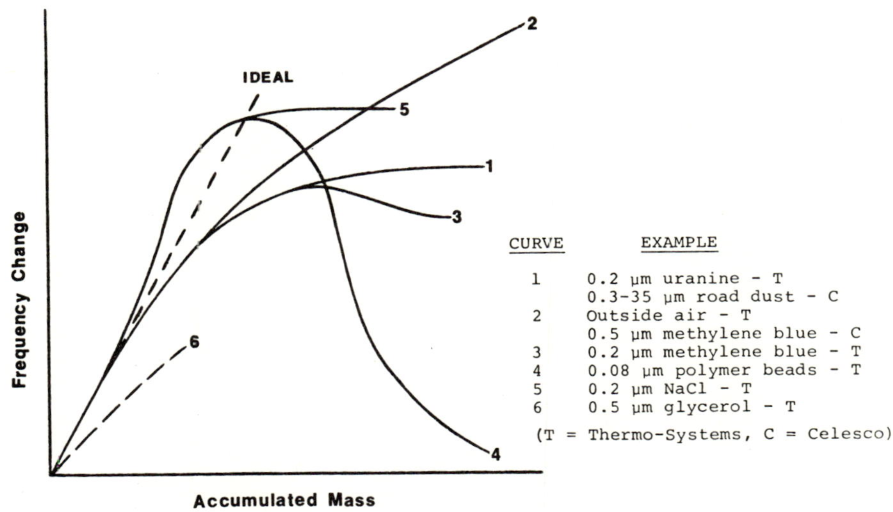

Figure 8. Typical crystal responses to various aerosols' deposits.

Table 1

TSI Mass Monitor, Polydisperse Aerosol Response Summary

Material	*Particle Size, μm	Deposit Diameter, mm	Crystal Sensitivity, Hz/μg	Linear Response Limit			Saturation	
				kHz	μg	$\frac{\mu g}{mm^2}$	kHz	$\frac{\mu g}{mm^2}$
Ariz. road dust	0.3-35	3.8	255	≈0.5	≈2.5	≈0.2	0.8	0.7
Methylene blue-uranine, 9:1, 22% RH, 1.5 mg/m^3	0.2	5.3	190	0.7	3.5	0.16	2.0	0.5
Ariz. road dust 27% RH	0.3-3	4.3	230	0.7	2.7	0.21	>3.0	>1.
$(NH_4)_2SO_4$, 27% RH	0.2	5.0	205	0.9	4.5	0.23	1.1	0.3
Laboratory air, ≈50% RH	-	5.	205	1.0	5.0	0.25	3.5	0.4
NaCl, 27% RH, 1.5 mg/m^3	0.2	4.3	230	1.0	4.5	0.30	3.5	0.9
Uranine, 45% RH	0.5	5.3	190	1.5	7.5	0.35	-	-
$CaSO_4 \cdot 1/2\ H_2O$, ≈50% RH	1-3	5.	205	1.5	7.2	0.5	2.5	1.3
Outside air, high visibility, 60% RH	-	5.	205	1.5	7.2	0.5	>4.	>1.5
Glycerol	0.5	6.4	155	3.5	23.	0.70	9.5	3.5
Dimethyl polysiloxane, viscosity = 10^5 cp	0.5	6.4	155	4.1	27.	0.83	9.5	4.
Methylene blue-uranine, 9:1, 22% RH, 0.2 mg/m^3	0.2	5.3	190	4.0	21.	0.97	20.	>>3.
NaCl, 27% RH, 0.15 mg/m^3	0.2	4.3	230	4.0	17.	1.5	4.3	1.6
Dimethyl polysiloxane viscosity = 10^5 cp	0.2	6.4	155	9.0	60.	1.8	9.5	2.0
Uranine, 75% RH	0.5	5.3	190	12.0	63.	3.0	>15.	>4.2

*Ranges estimated optically; 0.2, 0.5 μm aerosols were Collison generated, see text.

Table 2

TSI Mass Monitor, Monodisperse Aerosol Response Summary

Material	Particle Diameter, μm	Deposit Diameter, mm	Crystal Sensitivity, Hz/μg	Linear Response Limit				Saturation	
				KHz	μg	$\frac{\mu g}{mm^2}$	Monolayers	KHz	$\frac{\mu g}{mm^2}$
Polystyrene	0.08	5.0	205	1.6	8.8	0.40	8.0	4.0	1.2
Polystyrene	0.14	4.3	230	1.9	7.5	0.57	7.7	5.4	2.5
Polystyrene	0.41	4.3+	230	3.7	1.45	1.15	4.0	4.2	1.3
Polystyrene	0.65	4.5	220	3.8	15.3	1.10	3.0	4.5	1.2
Polystyrene	1.24	4.0	240	6.0	25.	2.0	2.6	17.	5.
Polyvinyl toluene	1.89	3.4	265	No data					
Styrene divinyl-benzene copolymer	7.8*	2.5	285	Not detected efficiently					
Paper mulberry pollen	12.	1.5	300	Not detected efficiently					
Ragweed pollen	20.	1.5	300	Not detected efficiently					
Glass beads	34.	Not collected							

*Approximately normally distributed with mass mean diameter of 7.8 μ and standard deviation of 1 μ.
+Not measured but determined from plot of particle diameter vs. deposit diameter for other sizes.

Table 3

Celesco Quartz Crystal Microbalance Aerosol Response Summary

Monodisperse Materials	Particle size, μm	Linear Response Limit				Saturation	
		Hz	μg	$\frac{\mu g}{mm^2}$	mono-layers	Hz	$\frac{\mu g}{mm^2}$
Polystyrene beads	0.41	Not Collected					
Polystyrene beads	0.65	120	0.043	.34	1.0	215	0.9
Polystyrene beads	0.83	75	0.024	.19	0.5	275	2
Polystyrene beads	1.24	25	0.009	.09	0.1	>60	>1
Polyvinyl toluene beads	1.89	30	0.011	.15	0.1	>300	>2
Polydisperse Materials							
Methylene blue-uranine, 9:1, 27% RH	0.5	75	0.027	.21	N.A.	>600	>3
Arizona road dust	0.3-35	-	-	-	N.A.	150	≈2
Uranine, 27% RH	0.5	100	0.036	.29	N.A.	>400	>7
Uranine, 60% RH (Crystal not coated)	0.5	1900	0.68	5.4	N.A.	>2000	>10

NOTE: Crystal sensitivity of 2800 Hz/μg used in calculation of this table. Deposit diameter of 0.4 mm for all materials except 1.24 and 1.89 μ beads which had 0.35 and 0.3 mm deposits, respectively.

Daley found particle collection characteristics to be an important consideration. Obviously, a particle which is not deposited on the crystal will not be weighed. Less obvious, but also important, is the need for particles to be uniformly deposited over the deposit area if the deposit size is significant compared to the crystal size.

The Celesco device collects particles by impaction. A 50% cut size diameter of 0.55 μm was calculated and empirically verified for unit density spheres.[7] This lower size limitation must be recognized by users of the Celesco unit.

The TSI unit was found to have high inlet passage losses (20%) of small particles (0.2 μm) because of precipitation by the field around the precipitator electrode shaft. The actual deposit is somewhat smaller than the active area resulting in a higher than expected mass sensitivity. These factors, therefore, compensate for each other; however, the user must be aware of and utilize these facts to obtain accurate results.

If the deposit area is a significant fraction of the active area, a change in deposit area causes a change in the sensor's mass sensitivity. The mass sensitivity of a crystal is the change in frequency per unit change in mass ($\Delta f/\Delta m$). Tables 1 and 2 show the change in deposit area and mass sensitivity observed for various aerosols in the TSI unit.[6] Using a single value for crystal sensitivity, therefore, results in error. This error can be reduced by actually measuring the deposit area and calculating mass sensitivity.

It should be noted that large particles (>5 μm) collected by the TSI unit interfered with the collection of smaller particles. Further, these large particles are not reliably measured, as is seen in the next paragraph. (Daley suggests eliminating them with an inlet cyclone.)[7]

Figure 9 shows the sensing ability of the two devices for spherical particles with diameters between 0.1 and 12 μm.[6] The sharp drop in sensing ability for larger particles is evident. (Qualitative evidence indicated that sensing ability for larger particles is improved in polydispersed aerosols.) Hence the large particle problems of "bounce off" and inlet loss are magnified by the inability of the crystals to detect large particles even when they are properly deposited.

The California Department of Health tested the TSI Model 3200 A (with auxiliary pumps and electronics) in parallel with high volume samplers and found poor agreement for ambient air monitoring.[5] Daley's analysis of these results indicated that lack of agreement resulted mainly from sampling beyond the linear limit of the quartz crystal device.[7]

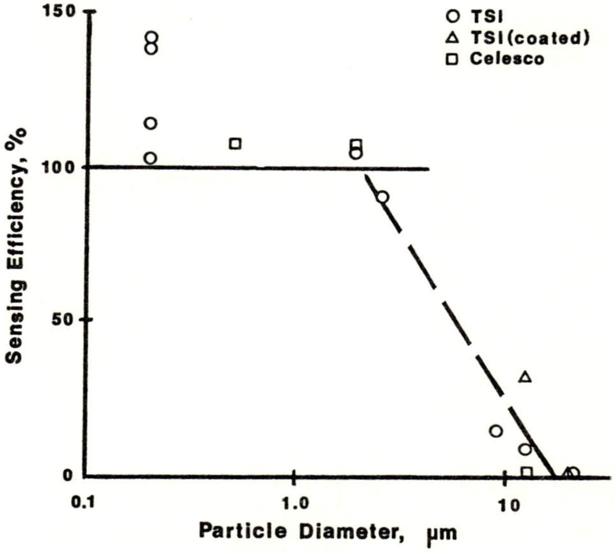

Figure 9. Particle sensing ability vs. particle size.

Tests by the EPA of a TSI device designed for measuring automotive exhaust had similar results. The TSI monitor mass determinations differed from filter measurements by ±30%.[4] The possible sources of error were several with the high moisture content of the gases most suspect. Temperature changes and the presence of volatile organics were other possibilities.

These three studies[4,5,7] may suggest that piezoelectric mass monitors are too error prone to be of significant value. However, this is not true. The latter two reports discussed above can be somewhat discounted because they show only that the piezoelectric monitor cannot do everything. Daley's work is more representative of reality. The devices do work, and they are quite accurate, if they are not taken beyond their ranges of linearity (short sample periods) or collection/ detection ability (particle size range), and if factors such as humidity changes and the presence of volatile aerosols are considered and minimized. These are a lot of if's, but they apply to first generation devices, which may simply be unsuited for many applications. We shall next consider second generation devices being built by Celesco and TSI which may alleviate some of the problems encountered with their predecessors.

RECENT DEVELOPMENTAL WORK

Two companies, Thermo-Systems, Inc. and Celesco Industries, Inc. have or are now developing second generation piezoelectric mass monitors. Celesco is developing a ten-stage impactor which will provide aerosol mass concentration and mass distribution data over a reported size range of approximately 0.05 to 100 μm diameter.[26] A six-stage abbreviated version of the above device has been built and tested on ambient California aerosol.[27] Performance of the unit's first two stages, with particle 50% cut sizes of 53 and 25 μm, may be suspect until the unit's ability to efficiently collect and sense large particles is convincingly demonstrated. Also, a low sampling rate coupled with a low concentration of large particles results in statistical problems associated with the large-particle collection frequency. However, discrete large particle impacts on a given stage may be counted if particle mass cannot be directly and accurately sensed because of poor particle adhesion. The multi-stage impactor with near real time aerosol mass distribution determination ability will certainly find many interesting applications.

TSI and a Japanese firm have developed an improved version of the original TSI mass monitor.[29] The new device continues to utilize an electrostatic precipitation collector. Basic information about the unit is still unpublished and proprietary, but the design appears to have eliminated several earlier problems. For example, the device is designed for a more specific purpose, measuring indoor particle concentrations, so the aerosol size distribution and concentration range is somewhat restricted. This restriction enables a more accurate mass sensitivity calculation. Large particles (>10 μm) which previously interfered with small particle collection, and which were not accurately detected anyway, are removed from the inlet air stream by an inlet impaction stage. The dust monitor's metal tubing inlet should dampen out error causing, short term temperature variations. Short sample periods (2 minutes) are recommended; hence, linearity problems should be resolved. A built-in crystal cleaner is included, and should prevent the need for frequent crystal changing. Other potential error sources include the crystal's ability to accurately sense particles in the 3 to 10 μm diameter size range, wall losses associated with particles in the 5 to 10 μm size range, the humidity fluctuation problem noted with earlier models and the precipitator's reported ability to "manufacture" particulate matter.

A possible new error source may be associated with the crystal wash solution leaving a volatile or hygroscopic residue, the effect of which may be amplified by short sampling periods. A new crystal mounting technique, epoxying the crystal's total perimeter to a base, departs from previous "clip mounts", but should not contribute to error if correctly designed.

In six parallel filter vs. piezoelectric instrument tests of a laboratory aerosol, a prototype dust monitor was reported to have an average error of about 10% as compared to a 0.8 μm pore size Nucleopore filter.[29] The aerosol size distribution was not stated but the aerosol concentrations were between 30 and 100 $\mu g/m^3$. Obviously it is desirable to conduct more extensive testing with various aerosols.

It should be noted that the model tested was a prototype for a possible commercial model and the effect of any design changes is of course unknown.

DIRECTIONS FOR FUTURE WORK

Piezoelectric crystal aerosol mass concentration monitors have tremendous potential in research and in field operations. Several improvements to early units have been made and were discussed earlier. A few other basic design modifications or suggestions, perhaps simpler in concept than execution, are set forth below.

With existing designs, the operator has only one means to compensate for high aerosol concentrations, i.e., reducing the sample time. (A noteworthy exception is the TSI accessory dilution system which offers a 50% or 80% aerosol dilution ability.) With high dust concentration, crystal response becomes nonlinear early in the sampling run. A dilution apparatus, perhaps with fixed dilution ratios of 2 to 1, 5 to 1, 20 to 1 and 100 to 1, would facilitate a large operating range for one measuring device so that the entire ambient air concentration range encountered could be covered.

Ambient air dilution with a temperature and humidity controlled air stream has several additional advantages such as reducing the effect of humidity variations in the sample gas stream. For example, diluting ambient air at 80% relative humidity (RH) with an equal volume of dry air will produce a mixture at 40% RH. Referring to Figures 4, 5 and 6 it becomes apparent that the lower the relative humidity, the less the tendency for water adsorption by the crystal or the deposited sample, therefore the lower the relative error resulting from humidity changes. Dilution can also be used to stabilize temperatures or to reduce the temperature of hot effluent gases.

Humidity effects could also be reduced by utilizing a crystal with a relatively flat temperature-frequency response at moderately high temperatures. This technique could be used to study or determine the volatility of aerosols by heating the sampled air stream to various temperatures (in order to drop the relative humidity) and then measuring the change in aerosol mass. TSI has utilized the principle of gas sample heating for a source sampling instrument.

Sampling of volatile aerosols, presently a rather difficult proposition, should be considered via the piezoelectric approach. Electronic circuitry which would filter the plus-minus signals of deposited and evaporated volatiles, and then sum the magnitude of plus-minus frequency changes, could measure total volatile mass associated with larger particles which were individually deposited on the sensing crystal.

Cascade impactors have several interesting possibilities. It may be possible to calibrate large particle impaction stages with inlet loss and sensing efficiency factors in order to accurately sense large particles (5-100 μm). If a lower size limit of 0.05 μm is obtainable (through low pressure impaction techniques or by other means) the staged impactor would effectively cover the particle range of primary interest to most aerosol investigators. The multi-stage impactor could further be optimized by adjusting the number of impaction jets per crystal for the anticipated mass loading for that stage. This would maximize sensitivity vs. sampling time but would require prior knowledge of the intended use of the monitor.

Apparently, the use of crystal sensors in other size selective particle collection devices, such as those involving centrifugal force or electrical force, has not been exploited. Crystal sensors also have potential use in thermal precipitators, diffusion batteries, and other special particle collection or sampling devices because of their extreme sensitivity. Although operational problems have made the first generation piezoelectric crystal mass monitors much less than universally acceptable, none of the problems encountered are insurmountable. Therefore, it must be concluded that the greatest potential use of crystal sensors remains to be found.

REFERENCES

1. Leemhorst, J.W. Direct Measurement of Particulate Mass and Humidity. Contamination Control. 4:11-14, July/August 1970.

2. Thermo-Systems, Inc. Particle Mass Monitor, Instruction Manual for Thermo-Systems Models 3200(A), 3200(B), 3205(A), 3205(B). St. Paul, Thermo-Systems, Inc.

3. Carpenter, T.E., and D.L. Brenchley. A Piezoelectric Impactor for Aerosol Monitoring. Am. Ind. Hyg. Assoc. J. 33:503-510, 1972.

4. Herling, R., W. Karches, and J. Wagman. A Comparison of Automotive Particle Mass Emissions Measurement Techniques. (Presented at Central States Meeting of the Combustion Institute, University of Michigan. Ann Arbor. March 23-29, 1971.) 23 p.

5. Imada, I., and P.K. Mueller. Evaluation of a Piezoelectric Quartz Crystal Microbalance for the Continuous Measurement of Aerosol. Air and Industrial Hygiene Laboratory, Calif. Dept. of Public Health, Berkeley, Calif. AIHL Report No. 114. 1971. 17 p.

6. Daley, P.S., and D.A. Lundgren. The Performance of Piezoelectric Crystal Sensors Used To Determine Aerosol Mass Concentrations. (Presented at American Industrial Hygiene Conference, Miami. May 12-17, 1974.) 41 p.

7. Daley, P.S. The Use of Piezoelectric Crystals In the Determination of Particulate Mass Concentrations In Air. Doctoral Dissertation, Gainesville, Univ. of Florida, 1974. 189 p.

8. Chuan, R.L. Application of an Oscillating Quartz Crystal To Measure the Mass of Suspended Particulate Matter. Celesco Industries, Inc., Costa Mesa, Calif. AT-159. 10 p.

9. Chuan, R.L. Particulate Mass Measurement by Piezoelectric Crystal. Nat'l. Bureau of Standards, Gaithersburg, MD. Special Publication 412. October 1974. 12 p.

10. Sauerbrey, G.Z. Verwendung von Schwingquarzen sur Wagung Junner Schichten und zur Microwagung. Zeits. Phys. 155:206-222, 1959.

11. King, W.H., Jr. Using Quartz Crystals as Sorption Detectors, Parts I and II. Research/Development. 20:29-33, April 1969, and 20:28-30, May 1969.

12. King, W.H., Jr. The State of the Art in Piezoelectric Sensors. In: Proc. of 25th Ann. Symposium on Freq. Control, U.S. Army Electronics Command. Ft. Monmouth, 1971. 7 p.

13. Frechette, M.W. and J.L. Fasching. Simple Piezoelectric Probe for Detection and Measurement of SO_2. E.S.&T. 7:1135-1137, December 1973.

14. Scheide, E.P., and J.K. Taylor. Piezoelectric Sensor for Mercury in Air. E.S.&T. 8:1097, December 1974.

15. Warner, A.W., and C.D. Stockbridge. Mass and Thermal Measurements With Resonating Crystaline Quartz. Vacuum Microbalance Techniques. 2:71-92, 1962.

16. Pulker, H.K., and W. Schadler. Factors Influencing the Accuracy of a Quartz Crystal as a Thickness Monitor for Thin-Film Deposition. Il Nuovo Cimento (Italy). 57B:19-24, September 11, 1968.

17. Chuan, R.L. Measurement of Particulate Pollutants in the Atmosphere. (Presented at Joint Conference on Sensing of Environmental Pollutants, Am. Inst. of Aeronautics and Astronautics, et al., Palo Alto, November 8-10, 1971.) 6 p.

18. Chuan, R.L. Electron Microscope Analysis of Particulate Sample From Rocket Launch at Cape Kennedy. Celesco Industries, Inc., Costa Mesa, Calif. AT-155. November 9, 1972. 3 p.

19. Chuan, R.L. Aerial Sampling and Post-Flight Analysis of Particulates Over the San Juan Basin at Four Corners. Costa Mesa, Celesco Industries, Inc., 1972. 8 p.

20. Chuan, R.L. Measurement of Vertical Temperature and Particulate Distributions Over the San Gabriel Valley. Celesco Industries, Inc., Costa Mesa, Calif. AT-154. November 1972. 35 p.

21. Olin, J.G., and G.J. Sem. Piezoelectric Microbalance for Monitoring the Mass Concentration of Suspended Particles. Atmospheric Environment. 5:653-668, 1971.

22. Olin, J.G., G.J. Sem, and D.L. Christenson. Piezoelectric-Electrostatic Aerosol Mass Concentration Monitor. Am. Ind. Hyg. Assoc. J. 32:209-220, April 1971.

23. Olin, J.G. Design and Operation of a Piezoelectric-Electrostatic Particle Microbalance for Automatic Monitoring of the Mass Concentrations of Air-Borne Particles, Paper 71-558. (Presented at Instrument Society of America International Conference and Exhibit. Chicago. October 4-7, 1971.) 10 p.

24. Carpenter, T.E. The Design, Construction, and Calibration of a Piezoelectric Cascade Impactor for Monitoring Aerosols. Master's Thesis, Lafayette, Purdue University, 1972. 104 p.

25. Daley, P.S. Real Time Aerosol Mass Concentration Measurement: Capabilities and Limitations of the Piezoelectric Microbalance Technique. (Presented at International Conference of Human Environment Conservation. Warsaw. February 1974.) 26 p.

26. Chuan, R.L. An Active Cascade Impactor for Real Time Sizing of Airborne Particulates. Celesco Industries, Inc., Costa Mesa, Calif. AT-149. 3 p.

27. Chuan, R.L. Recent Test Results With an Active Cascade System. Note accompanying correspondence to D.A. Lundgren. December 12, 1974.

28. Chuan, R.L. Application of an Oscillating Quartz Crystal to Measure the Mass of Suspended Particulate Matter. (Presented at 165th National Meeting of the American Chemical Society. Dallas. April 1973.) 37 p.

29. Tsurubayashi, K., and G.J. Sem. Personal Communication. Nihon Kagaku Kogyo Co., Ltd., Osaka, Japan. Thermo-Systems, Inc. St. Paul.

MEASUREMENTS OF AEROSOL OPTICAL PARAMETERS

A. P. WAGGONER
R. J. CHARLSON

UNIVERSITY OF WASHINGTON
SEATTLE, WASHINGTON 98195

ABSTRACT

An aerosol is most usefully described in terms of its integral properties, in particular those integral properties that directly describe the aerosol effect of interest. Atmospheric optical properties normally considered would include: visibility, scattering extinction coefficient, absorption extinction coefficient, hemispheric backscatter coefficient and the ratio of hemispheric backscatter to absorption coefficients.

These parameters can be calculated via Mie solutions if the aerosol particle size distribution and complex refractive index are known or assumed. These calculated values of integral aerosol properties will contain errors that arise from an inaccurate description of the aerosol particles. The calculations assume: (1) the particle size distribution is accurate, (2) the particles all have the same complex refractive index, (3) the complex refractive index is uniform within each particle and (4) the particles are spherical in shape. Most of these assumptions are poorly justified.

Techniques have been developed at the University of Washington for direct measurement of the integral aerosol optical properties listed in paragraph one. This paper describes these techniques and measured values of these aerosol optical parameters.

MEASUREMENTS OF AEROSOL OPTICAL PARAMETERS

A. P. WAGGONER
R. J. CHARLSON

UNIVERSITY OF WASHINGTON
SEATTLE, WASHINGTON 98195

I. INTRODUCTION

The aerosol is composed of particles that range in size from smaller than 0.01 μm to larger than 10 μm diameter. The particles are of various chemical compositions and each particle can be a mixture of substances or a single substance. The integral optical effect of the aerosol particles is dependent on all of these parameters. Atmospheric optical properties normally considered would include those of interest from a human impact standpoint, i.e., visibility and colored haze, and those of scientific interest, i.e., scattering and absorption extinction coefficients.

II. ATMOSPHERIC OPTICS AND VISIBILITY

It is convenient to define several parameters commonly used to describe atmospheric optics.

The extinction coefficient b_{ext} of a real atmosphere defines the change in intensity of light traversing a pathlength Δx by the Beer-Lambert law:

$$\frac{\Delta I}{I} = -b_{ext}\Delta x \qquad (1)$$

b_{ext} is the sum of two terms:

$$b_{ext} = b_{ext}\text{ (gases)} + b_{ext}\text{ (particles)}$$

$$b_{ext}\text{ (gases)} = b_{Rg} + b_{ag}\text{, where}$$

$b_{Rg}\Delta x$ is the fraction of incident light scattered into all directions by gas molecules in Δx.

$b_{ag}\Delta x$ is the fraction of incident light absorbed by gas molecules in Δx.

Our interest is in b_{ext} (particles) which can be broken down as follows:

$$b_{ext} \text{ (particles)} = b_{ap} + b_{sp} \tag{2}$$

where $b_{ap}\Delta x$ is the fraction of incident light absorbed by particles in Δx.

$b_{sp}\Delta x$ is the fraction of incident light scattered into all directions by particles in Δx.

The observer visibility, or visual range, is that distance at which a black object can be just discerned against the horizon. Koschmieder[1] showed that a turbid media, such as urban air, reduces the contrast (ratio of brightness of an object to the horizon brightness, minus one) of distant objects as given by

$$C = C_o\, e^{-b_{ext}\Delta x} \quad \text{(Middleton}^2\text{)} \tag{3}$$

where C_o and C are the contrast relative to the horizon of an object at zero distance and at distance x. A black object has a C_o of -1. Experiments have determined that typical observers can detect objects on the horizon with a visual contrast of 0.02 to 0.05. Assuming horizontal homogeneity of aerosol properties and illumination and a 0.02 detectable contrast, the visible range is

$$L_v = \frac{3.9}{b_{ext}} \tag{4}$$

For a contrast of 0.05,

$$L_v = \frac{3.0}{b_{ext}} \tag{5}$$

Usually the assumption is made that $b_{ext} = b_{sp}$.

b_{sp} can be calculated from known or assumed aerosol particle size distribution, concentration and refractive index, as discussed below.

III. PARTICLE OPTICS

The atmospheric aerosol is composed of particles that range in size from smaller than 0.01 μm to larger than 10 μm diameter. The particles are of various chemical compositions and each particle can be a mixture of substances or a single substance. The integral optical effect of the aerosol particles is dependent on all of these parameters.

The scattering coefficient of a single particle divided by the particle volume is shown in figure 1. The value of b_{sp} is the product of the curve in figure 1 times the particle volume distribution function. The aerosol particle volume per log radius interval usually is similar to that of figure 2, bimodal with the two volume modal diameters about 0.6 μm and 10 μm. The calculated optical scattering per log radius interval is predominately due to particle volume in the 0.1 μm to 1.0 diameter range, as shown in figure 2. <u>In all the measurements we have made, the particles in the 0.1 to 1.0 decade dominate scattering extinction in the visible spectrum</u> although there clearly are cases in fogs, rain, snow, clouds and dust storms in which large particles influence or dominate visible extinction.

The correlation of b_{sp}, measured with an MRI 1550 nephelometer, and 0.1 to 1.0 μm diameter particle volume, measured using an electrostatic mobility and single particle optical counters from Thermo Systems, was 0.95 at various locations in the Los Angeles basin. These measurements, shown in figure 3, are from the 1973 State of California Air Resources Board ACHEX[3] program.

The wavelength dependence of b_{sp} depends almost exclusively on particle size distribution[7]. If the wavelength dependence is described by a simple power law:

$$b_{sp}(\lambda) \propto \lambda^{-\alpha} \tag{6}$$

where α is an experimentally determined exponent, the usual range of α is 0.5 to 2.5 with occasional measured values of α in the -1 to 0 range at background sites. Rayleigh scattering always occurs simultaneously and has a wavelength dependence that is similar:

$$b_{Rg} \propto \lambda^{-4} \tag{7}$$

As a result, blue scattered light (against a dark background) or red transmitted light (from the sun or a bright white object) is no indication by itself of the presence of particles. Whether b_{sp} or b_{Rg} dominates is determined by the amount of particulate matter that is present. In remote, clean marine locations at sea level, Porch, *et al*[8] showed that $b_{sp} \leq b_{Rg}$ at 500 nm. In continental, low altitude sites, b_{sp} is usually larger than b_{Rg}, so that such hazes can often be assumed to be dominated by b_{sp}. However, clean arctic air intruding or air from aloft subsiding into mid continent cities occasionally produce $b_{sp}<b_{Rg}$. This sort of situation typically arises under post-cold frontal conditions in the midwest and results in unusually high visibility and clear air.

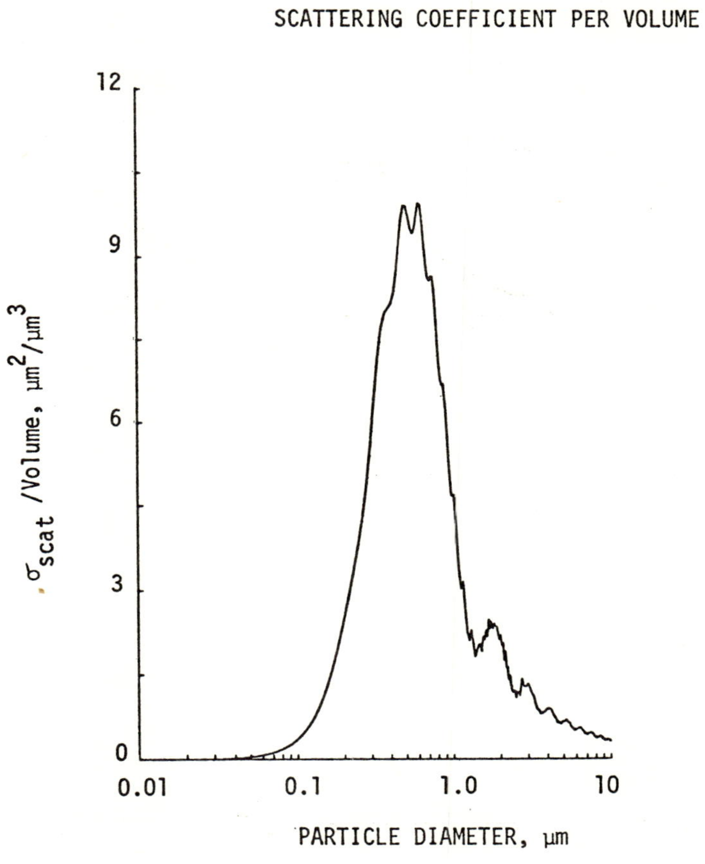

Figure 1. Scattering coefficient per particle divided by particle volume plotted as a function of diameter. The particles are assumed to be spheres of refractive index 1.50 illuminated by 550 nm light.

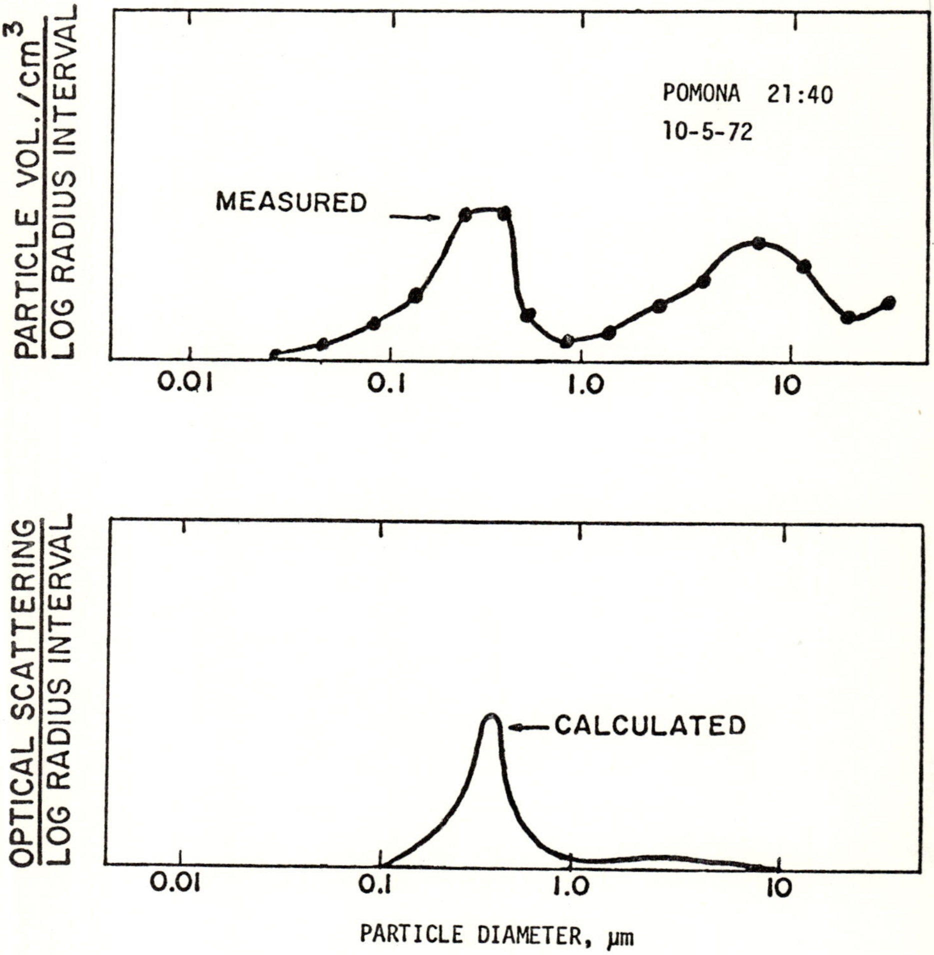

Figure 2. Top: Aerosol particle size distribution measured at Pomona during 1972 State of California Air Resources Board ACHEX program.

Bottom: Calculated optical scattering by particles, b_{sp}, for measured size distribution.

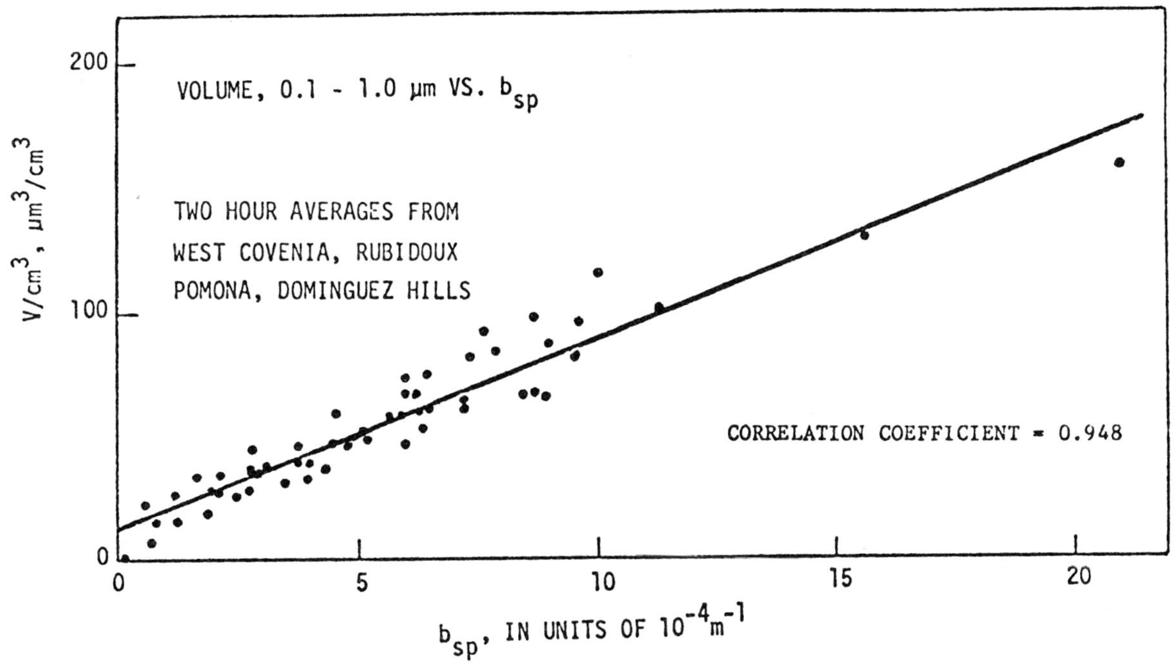

Fig. 3. Plot of measured aerosol particle volume including only those of 0.1 to 1.0μm diameter versus measured b_{sp}. Measurements were part of State of California Air Resources Board ACHEX program. Data was supplied by Dr. Clark of North American Rockwell.

Blue hazes - for example, in mountainous areas - may or may not be due to scattering by particles, depending on viewing conditions (e.g., dark or light background) and the distance from the observer to the background. If the product of b_{Rg} times distance to the background is much above two, the blue haze has a significant input due to b_{Rg}. On the other hand if the product of b_{sp} times distance is of this magnitude, then the haze is likely to be due to particles. Since b_{Rg}, 530 nm = $0.15 \times 10^{-4} m^{-1}$, if $b_{sp} \sim 0$ mountains should not appear to be behind a haze if they are within 10 km. or so. They will, however, appear hazy if the distance is much more than 100 km. due to the omnipresent scattering by gas molecules. Conversely, if such a distant mountain is not visible at all, $b_{sp} \gg b_{Rg}$ and the haze is due to particles.

When viewing bright objects (the sun and moon, sunlit snow capped peaks and cumulous clouds) hazes with $1 \leq \alpha \leq 2$ of sufficient optical depth cause the color to be reddened[2,9]. The color thus produced is remarkably similar to that observed through an optically thin layer of NO_2[10] so that the presence of color thus viewed is no proof of the existence of NO_2. To further complicate this issue, Husar[11] has shown that light scattered in the backward hemisphere calculated from typical measured size distribution is enriched in the red wavelengths also causing the haze itself to appear reddened. In forward scatter this same haze appears white. Charlson[12] showed that, in perhaps 20% of the measured cases during August, 1969 in Pasadena, CA, there was enough NO_2 to influence the coloration of white objects viewed through the haze and that the remaining cases particles dominated the wavelength dependence of total extinction (b_{ext}).

IV. MOLECULAR COMPOSITION

The particle interaction with water, biological effects and complex refractive index depend on the molecular composition. Therefore, it is important that the composition of various aerosol systems be classified, particularly insofar as this determines the imaginary part of the refractive index and hygroscopicity. Unfortunately, this is an area in which so far very little work has been done. Rasmussen[13] suggested that organic materials (terpenes) are a major source of atmospheric particles, but did not quantify their work adequately for application to optics. The reaction products of SO_2 with water and ammonia have been shown to play an important part in urban and rural aerosols by Junge[14] although he did not attempt to relate quantitatively the composition with optical effects. We have data from rural Missouri suggesting that continental aerosol optics is often dominated by H_2SO_4 and the products of its neutralization with NH_3[15,16].

The molecular nature of individual particles is a function of the source and removal mechanisms for these particles. The most important observable effects of composition on particle optics are the relationships of b_{sp} and relative humidity and the complex refractive index.

V. RELATIVE HUMIDITY EFFECTS

The humidity effects in aerosol optics fall into three categories:

RH $\leq$ 100%: particles between and above water cloud (including high RH hazes);

RH > 100%: unactivated particles in water clouds and fog;

RH > 100%: activated cloud droplets.

Our efforts have been limited to the first case and are discussed in the following paragraphs.

Since a large fraction of submicrometer particles are hygroscopic or deliquescent[14-18] the size distribution of an atmospheric aerosol and hence its optical or climatological properties, depend largely on relative humidities, even at RH < 50%.

First, light scattering always increases with humidity, although for relatively hygrophobic systems the increase may be very slight up to extremely high RH. While for most aerosols such as H_2SO_4 droplets the curve increases monotonically, definite inflection points due to deliquescent salts are seen at some locations indicating the dominance by rather pure inorganic substances such as $(NH_4)_2SO_4$ or sea salt (NaCl)[15,16,19].

A system has been designed and operated by this laboratory that (over a period of about 120 seconds) sweeps the relative humidity of air containing aerosol particles from 30% to 95%. Changes in particle diameter are detected as changes in the scattering coefficient of the aerosol particles[15,16,19].

In the midcontinent region 30 km southwest of St. Louis, this system detected $H_2SO_4/(NH_4)HSO_4/(NH_4)_2SO_4$ as dominate materials in the 0.1 to 1 μm decade of aerosol size. Injection of sub ppm concentrations of NH_3 converted the b_{sp}(RH) response characteristic of H_2SO_4 to that of $(NH_4)_2SO_4$. The $(NH_4)_2SO_4$ is detected by comparing the value of relative humidity at the deliquescence point for the unknown sample with that of laboratory generated $(NH_4)_2SO_4$ aerosol. 98% of the time either H_2SO_4 or $(NH_4)_2SO_4$ was the dominant substance in terms of optical effect[15,16].

VI. TECHNIQUES FOR MEASUREMENT OF RELEVANT OPTICAL PROPERTIES

In the past several years our efforts have been focused on design and testing of methods to measure aerosol otpical properties that directly determine aerosol radiative interactions. Methods for measurement of these relevant integral aerosol optical properties; namely, b_{sp}, b_{bsp}, $b_{sp}(RH)$, and b_{ap}, are described in the following sections.

A. b_{sp}

Consider a small volume of thickness dx illuminated by a parallel beam of wavelength λ and intensity $I_{o,\lambda}$. For unpolarized light, the intensity of light scattered into solid angle $d\Omega$ at scattering angle Θ is

$$\frac{dI_\lambda}{d\Omega}(\Theta)d = I_{o,\lambda}\beta_\lambda(\Theta)dx \tag{8}$$

A visibility meter using the operator's eye as a detector was devised by Buettell and Brewer[20] that geometrically performs the integration of $\beta_\lambda(\Theta)$ over solid angle to measure $b_{sp,\lambda}$[1].

The geometric errors of the instrument have been studied by Middleton[1], Ensor and Waggoner[23], Heintzenberg and Quenzel[24] and Rabinoff and Herman[25] and are estimated to be 10% or less for the aerosol particle size distributions normally found in the atmosphere.

Ahlquist and Charlson[21] increased the original instrument sensitivity by using a photomultiplier tube to detect scattered light from a xenon flash lamp. Ahlquist, et al.[22] improved the sensitivity, stability and dynamic range by substituting an incandescent lamp for the xenon flash lamp and detecting the scattered light using digital photon counting techniques. This insturment, called an integrating nephelometer, is shown in Figure 4. Modern versions of Buettell and Brewer's device have sufficient sensitivity to be calibrated in an absolute sense with b_{Rg}, the scattering coefficient of particle-free gases such as He, CO_2, CCl_2F_2.

The modern high sensitivity instrument is alternately filled with ambient and particle-free air and the difference in scattered light intensity is proportional to the scattering extinction coefficient due to aerosol particles, b_{sp}. The instrument calibrates itself by inserting a white object (end of a wire painted with Kodak optical white paint) into the scattering volume during a portion of the clean air cycle. Three digital counts of detected photons are accumulated in three memories under the operational conditions described.

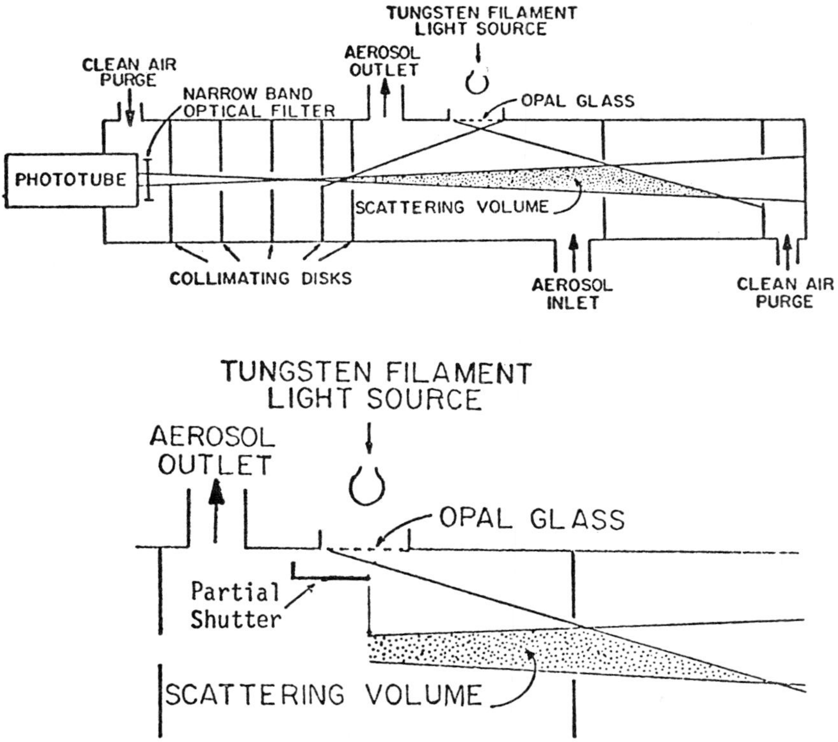

Figure 4. Diagram of nephelometer with enlarged view of the partial shutter. Without the shutter, the instrument integrates the particle scattering coefficient over ~ 7° to 170° to measure b_{sp}. With the shutter in place, the instrument integrates over ~ 90° to 170° to measure b_{bsp}.

Memory 1: Instrument Background plus air Rayleigh scattering, instrument filled with clean air.

Memory 2: Memory 1 plus white object brightness, clean air plus calibrate object.

Memory 3: Memory 1 plus scattering due to particles, instrument filled with ambient air.

Memory one is sub tracted from m emories two and three, giving digital counts proportional to the brightness of the particle scattering and calibrate object with instrument background and air Rayleigh scattering subtracted. The particle scattering count is divided by the calibrate object count. This sequence, repeated several times per hour, zeros and spans the nephelometer.

Two instruments of this type have been constructed and the span is stable within $\pm$ 2% over several months. Both instruments measure the particle scattering coefficient at several wavelengths determined by optical interference filters to measure the value of α, defined in equation 6 as

$$b_{sp}(\lambda) \propto \lambda^{-\alpha} \tag{6}$$

Measured values of b_{sp} in the atmosphere range from $3 \times 10^{-3} m^{-1}$ in polluted Los Angeles to $10^{-7} m^{-1}$ at Mauna Loa Observatory in Hawaii. High sensitivity, multiwavelength instruments have been purchased by Institute für Meteorologie, Mainz, Germany, Air Force Cambridge Research Lab and the National Oceanographic and Atmospheric Administration. Several hundred lower sensitivity, single wavelength instruments have been produced and are in regular use for both research and monitoring. The draft version of volume I of the ACHEX final report from Rockwell International to the Air Resources Board, State of California, recommends the integrating nephelometer for both long term monitoring and short term surveillance of aerosol properties.

B. b_{bsp}

An optically thin aerosol layer over a dark surface increases the albedo by scattering incident radiation backwards into space. The albedo per unit thickness of an aerosol layer illuminated by a zenith sun can be determined by integrating the aerosol volume scattering function over the backward hemisphere of scattering angle. A partial shutter, shown in Figure 4, can change the angle of integration of the nephelometer so that the scattered light intensity is proportional to the backward hemisphere scattering extinction coefficient b_{bsp} due to aerosol particles. b_{bsp} normally is in the range 0.1 to 0.2 times the aerosol scattering extinction coefficient b_{sp}.

C. b_{ap}

The two aerosol parameters needed in simple radiative climatic models are the particle backward hemisphere scattering coefficient, b_{bsp} and the particle absorption extinction coefficient, b_{ap}. There are a number of ways of measuring b_{ap}, and none is entirely satisfactory.

Long path extinction cannot be used because b_{ap} is $10^{-4}m^{-1}$ to $10^{-8}m^{-1}$ or smaller. Various techniques based on inverting angular scattering information have been used by Eiden[26] and Grams, et al.[27], etc., but these methods require precise knowledge of the aerosol size distribution, and contain errors of unknown size and magnitude, since the scattering by irregular particles is calculated using Mie formulae for spheres. The absorption coefficient of collected aerosol samples can be estimated with low precision from measurement of the transmission of KBr pellets containing dispersed aerosol[28]. Lindberg and Laude[29] measured aerosol absorption by measuring the decrease of diffuse reflectance of a white powder when a small amount of aerosol is dispersed in it.

All of the above methods, in our opinion, are poorly suited for measurements in background locations. Measurement of the angular dependence of the aerosol volume scattering function is difficult when molecular scattering dominates. The methods of Volz and Lindberg require collecting an aerosol sample over several days, scraping the sample off the collecting surface and dispersing the sample in another media. Any treatment of the sample that alters the aerosol size distribution will alter the optical absorption coefficient[30,31]. A different technique for measurement of b_{ap} has been developed in our laboratory that we believe is superior to those described above.

Atmospheric aerosol is collected by passing ambient air through a Nuclepore filter. The filter consists of a 10 μm thick film of polycarbonate plastic with 0.4 μm holes etched through it. The holes are etched along damage tracks from highly ionizing particles and are round and perpendicular to the surface of the film. Individual particles with a mean separation of several diameters are collected on the surface of the filter. The filter and the particles are placed in an optical system that illuminates the particles and the filter with a parallel beam of, in this case, green light and collects both direct transmitted and forward scattered light. The extinction or change in transmission between a clean filter and the filter plus aerosol is assumed to be the same as absorption by the same aerosol dispersed in a long column of air. Knowing the volume of air passed through the filter during collection of the aerosol, one can calculate the optical absorption coefficient due to particles, b_{ap}.

This Method has been checked for accuracy using laboratory aerosols of known (including zero) absorption coefficient and is described by Lin, et al.[32]. The disadvantages of the method center on errors introduced by sample alteration that may take place during collection, but the sample alteration is probably much less than in the techniques of Volz and Lindberg. The sample collection is simple and only requires 10 to 20 $\mu g/cm^2$ of aerosol on the filter.

VII. ATMOSPHERIC MEASUREMENTS AND DATA

A. b_{sp} AND VISIBILITY

As discussed in Section II, Koschmieder[1] related b_{ext} to the distance at which a black object is just visible when viewed against the horizon sky. The distance of visibility is given by

$$V = \frac{3.9}{b_{ext}} \qquad \text{(Middleton}^{2}\text{)} \tag{9}$$

assuming aerosol homogeneity, uniform illumination and a 0.02 detectable contrast. Commonly it is assumed that $b_{ext} = b_{scat}$, i.e., $b_{abs} = 0$. Measurements of b_{scat} and observer visibility show good agreement with the formula above.

Horvath and Noll[33] conducted a study in Seattle between total light scattering, b_{scat} measured with an integrating nephelometer, and prevailing visibility observed by two separate people. Their results were in good agreement with the theoretical expression of Koschmieder when data for RH >65% RH were excluded. Apparently the location of the nephelometer in a heated room caused reduced RH in the scattering measurements. In the cases where RH < 65%, the correlation between b_{scat} and prevailing observer visibility was 0.89 and 0.91 respectively with a coefficient in the Koschmieder expression of 3.5 $\pm$ 0.36 and 3.2 $\pm$ 0.25 respectively. This can be compared with the theoretical value of 3.9, indicating a slightly lower prevailing visibility than meteorological range. Since no ideal balck targets were used (only trees, buildings, etc.) these would have caused just such a deviation.

Samuels, *et al.*[4] conducted the most extensive tests to date of the relationship of prevailing visibility to light scattering and various mass concentration measures.

They conclude that b_{sp} as measured with the integrating nephelometer is a good predictor of prevailing visibility and that the regression analysis in agreement with Koschmieder's theory. These workers noted that there was a smaller observed prevailing visibility than that predicted from theory and b_{sp} measurement, which they suggested was due to non-ideal black visibility targets.

B. b_{sp} AND AEROSOL VOLUME OR MASS

As discussed in section III, b_{sp} should be and <u>is</u> highly correlated with particle volume in the 0.1 to 1.0 μm diameter range. Correlation of b_{sp} and total particle volume (or mass) is not expected unless the two modes of particle volume happen to be correlated. Thus we would not expect to find a particularly good correlation between b_{sp} and measured filterable mass concentration, for example, measured with the high volume air sampler.

It is somewhat surprising, in view of this, that the measured correlation coefficient between b_{sp} and total aerosol mass concentration is as high as the observed range between 0.5 and 0.9. While the former value is not impressive nor particularly useful, the latter is sufficiently high to allow inference of mass concentration from b_{sp}. Table I summarizes the various published correlations of b_{sp} and mass. Included in the table are correlation coefficients, r, and regression constants A and B.

The correlation coefficient of 0.9 in New York City must be due to either a correlation between the upper and lower volume (i.e., mass) modes or an absence of the upper mode. The location at the 16th floor of a Manhattan building suggests the latter since it was well removed from sources of wind blown dust and other mechanically produced particles.

In contrast, the low correlation coefficient in San Jose, CA, of 0.6 was obtained at a dusty athletic field, with the air intake at approximately 7 meters above the ground. In this case, the poor correlation was likely due to a large and variable fraction of the aerosol in the supermicrometer mode.

C. b_{sp} AND EXTINCTION

In a cooperative experiment with Dr. John Hall, Lowell Observatory, Flagstaff, AZ, b_{sp} and long path extinction were simultaneously measured at a number of wavelengths in the visible and ultraviolet.

The nephelometer measured b_{sp} at three wavelengths; 430 nm, 530 nm and 640 nm, and is described in section IV, A. Hall measured extinction by

Mass Sampling Method / Location (Reference)	2.5 cm dia. open face, glass fiber filter			2.5 cm dia. Nuclepore filter			High Volume Air Sample			Glass Fiber filter behind Lipmann-Harris Cyclone		
	r	A	B	r	A	B	r	A	B	r	A	B
Los Angeles (4)	0.83	-0.57	2.4	---	---	---	0.53	-0.09	3.3	0.83	0.33	3.7
Oakland, CA (4)	0.69	-0.40	1.3	---	---	---	0.86	-0.61	2.4	0.79	0.34	3.2
Sacramento, CA (4)	0.95	0.0	2.2	---	---	---	0.93	-0.56	2.8	0.98	0.13	4.4
New York, NY (5)	---	---	---	0.92	-0.33	3.0	---	---	---	---	---	---
San Jose, CA (5)	---	---	---	0.56	1.5	1.7	---	---	---	---	---	---
Seattle, WA (5)	0.83	-0.08	3.5	---	---	---	0.73	-0.26	3.6	---	---	---
Boston, MA (6)	---	---	---	---	---	---	0.86	0.15	2.0	---	---	---

Table I. Summary of light scattering-filterable particulate mass concentration studies.

The parameters are: r = linear correlation coefficient;

A and B defined by $b_{sp}(10^{-4}m^{-1}) = A + 10^{-2}B\ (\mu g/m^3)$.

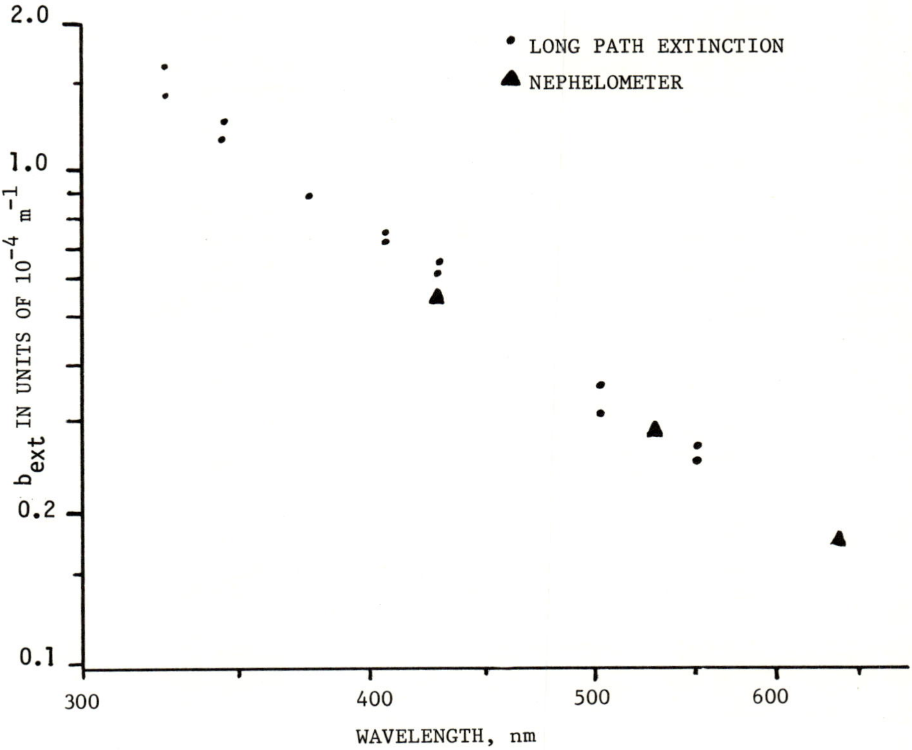

Figure 5. Measurments of $b_{ext}(\lambda)$ by long path extinction and by nephelometer. Data taken near Flagstaff Az. on 6 Nov. 1974.

measuring the brightness of a regulated incandescent light source at distances of 0.5 and 21 km. Extinction over the path was determined at eight wavelengths from 330 nm to 660 nm during the night of 6 November, 1974 at a rural site near Flagstaff. Good agreement was obtained between nephelometer and long path extinction measurements as shown in figure 5.

D. MEASUREMENTS OF SCATTERING PARAMETERS

Under support from the Environmental Protection Agency, National Science Foundation and the California Air Resources Board we have measured various aerosol scattering parameters in urban and rural locations in California, Colorado and Missouri. In all locations the incoming air was heated 5°C to 20°C above ambient to lower relative humidity of the sample. The measured parameters were:

$\underline{b_{sp}}$ - Scattering extinction coefficient of particles at 530 nm (Rayleigh at 530 nm = $0.15 \times 10^{-4} m^{-1}$)

$\underline{\alpha}$ - Wavelength dependence of b_{sp} parameterized as

$$b_{sp} = K\lambda^{-\alpha} \tag{10}$$

Two values of α were computed from Red-Green b_{sp} and Blue-Green b_{sp}. Red is 640 nm. Blue is 430 nm. Green is 530 nm.

Scat. ratio - Ratio of half sphere back scatter to b_{sp} from particles at 530 nm.

The sites were:

Richmond - Northeast corner of San Francisco Bay in vicinity of petro chemical plants.

Point Reyes - Coast Guard station on cliff 150 meters above the sea surface, 50 km NW of San Francisco.

Fresno - Central valley of California, urban agricultural site.

Hunter Liggett - Rural California site 20 km inland from ocean. Local elevation 400 m. Local vegetation consisted of dry grass and sparce trees.

Cal. Tec. - Site on campus in Pasadena in Los Angeles basin.

Pomona - Site at county fairgrounds in inland area of Los Angeles basin.

Washington Univ. - Campus site located in residential area of St. Louis, MO.

Tyson - Rural area 25 km WSW of St. Louis.

St. Louis Univ. - Campus site in industrial St. Louis.

Henderson - Site 10 km NE of Denver.

Trout Farm - Site 8 km N of Denver.

Table II lists the measured values at each site. For each measurement parameter, the range of that parameter containing 63% of the data is specified. For b_{sp}, the units are $10^{-4}m^{-1}$ and the range low to high contains 63% of data.

E. b_{ap} MEASUREMENTS

Using the technique described in section VI, C, measurements were made of b_{ap} at two locations NE of Denver and three sites near St. Louis during Fall of 1973. The measured values of the ratio of absorption to extinction are presented in figure 6. In Denver, the absorption to extinction ratio is very high, indicating that the aerosol heats and stabilizes the lower atmosphere. At the three Missouri sites the measured values are as one would expect - the rural area (Tyson) has a less absorbing aerosol than the industrial site (St. Louis University). Only the industrial MO site had absorption comparable to that measured outside Denver.

The probable chemical species that produces the absorption is graphitic carbon. Without chemical analysis for this material it is only possible to speculate about the nature of Denver's very absorbing aerosol. The absorption could result from:

(1) high graphitic carbon content.
(2) large concentrations of graphitic carbon particles smaller than 0.1 μm.
(3) lack of $(NH_4)_2SO_4$ as a major component of Denver aerosol when compared to that found in rural Missouri.

The ratio of absorption to filterable particulate mass can be used to estimate an imaginary refractive index for the aerosol if a size distribution and chemical uniformity are assumed. We believe the particles are not uniform chemically and prefer to report b_{ap} rather than n_2. With this warning, the average aerosol b_{ap} at Denver was $0.35 \times 10^{-4}m^{-1}$. The imaginary refractive index, n_2, given the stated assumptions was 0.035.

Location	b_{sp} (530nm)	b_{sp} low	b_{sp} high	αRG	αBG	Scat. Ratio	Start	End
	(Units	of	10^{-4} M^{-1})				Mo./Day/Hr.Yr.	
Richmond	0.4	0.2	1.4	0.8 ± 0.9	1.2 ± 1	- - - - - -	8/7/17/72	8/12/15/72
Point Reyes	0.12	0.04	0.4	0.25 ± 1.0	- - -	- - - - - -	8/15/11/72	8/25/6/72
Fresno	1.0	0.3	1.9	1.0 ± 0.4	1.7 ± 0.5	$18\pm5\%$	8/29/9/72	9/8/14/72
Hunter Liggett	0.4	0.2	0.8	1.4 ± 0.8	1.6 ± 0.6	- - - - - -	9/12/9/72	9/15/10/72
Cal. Tec.	1.5	0.8	3.0	1.5 ± 0.4	1.5 ± 0.3	$20\pm8\%$	9/20/10/72	10/2/8/72
Pomona	1.8	0.6	6.0	1.3 ± 0.4	1.3 ± 0.7	$16\pm8\%$	10/4/11/72	10/31/15/72
Wash. Univ.	1.58	1.12	2.24	1.47 ± 0.4	1.00 ± 0.4	$11\pm1\%$	8/25/2/73	8/31/9/73
Tyson	0.63	0.28	1.41	1.80 ± 0.4	1.25 ± 0.5	$14\pm4\%$	9/3/19/73	9/27/12/73
St. Louis Univ.	0.71	0.40	1.25	1.85 ± 0.3	1.25 ± 0.3	$14\pm2\%$	9/27/20/73	10/4/15/73
Henderson	0.31	0.08	1.25	1.65 ± 0.8	1.15 ± 0.6	$17\pm8\%$	11/10/--/73	11/14/--/73
Trout Farm	0.56	0.22	1.58	1.75 ± 0.7	1.30 ± 0.5	$18\pm6\%$	11/15/10/73	11/23/11/73

Table III. Listing of mean and variation including 63% of measurements for four scattering parameters in 11 locations. The parameters and sites are discussed in the text. Note that in all locations the sample air was heated 5° to 20°C above ambient temperature.

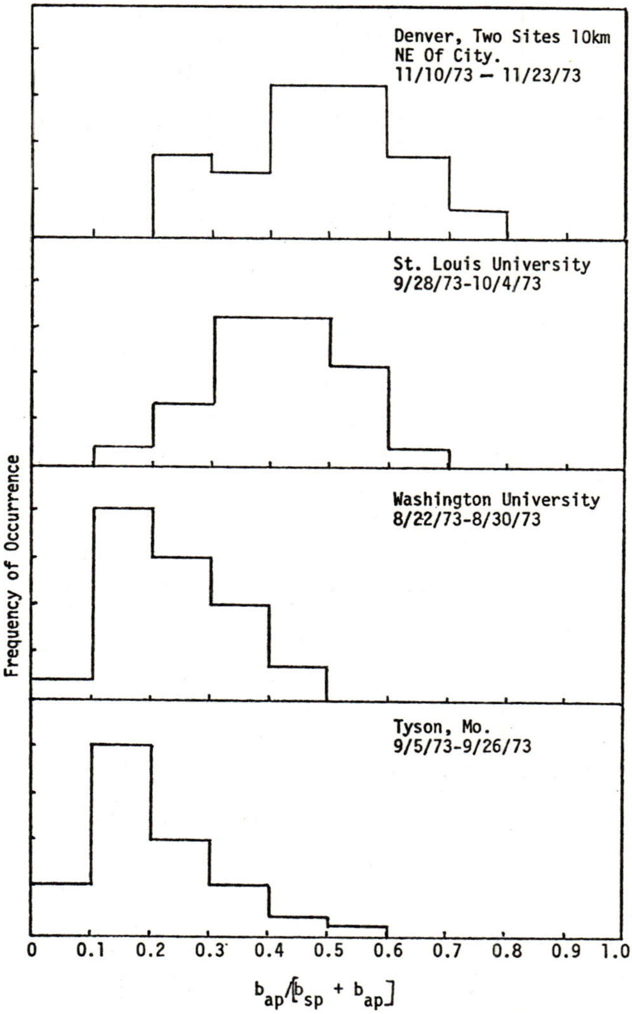

Figure 6. Ratio of absorption to extinction by particles.

VIII. CONCLUSIONS

Comparisons can be made between our measurements at Denver and other location. Deliquescent salts were not detected in the aerosol at Denver and the b_{sp}(RH) curves were at times quite hygrophobic. The aerosol is less water soluble in Denver than at other sites.

The aerosol had somewhat higher backscatter to b_{sp} ratio and much higher b_{ap}/b_{ext} than values of the same parameter at other locations. both measurements could be explained by a shift of the small particle mode to smaller particles. The absorbing character of Denver aerosol may enhance the brown or yellow color of distant white objects viewed through the urban plume.

ACKNOWLEDGEMENTS

This research has been supported by Environmental Protection Agency, National Science Foundation and California Air Resources Board funds.

REFERENCES

1. Koschmieder, H., Beitr. Phys. Freien Atm., 12, 33-55 & 171-181 (1924).
2. Middleton, W. E., Vision Through The Atmosphere, University of Toronto Press, Toronto, Canada (1968).
3. ACHEX, Aerosol Characterization Experiment of the State of California Air Resources Board. Prime contractor is Rockwell International Science Center.
4. Samuels, H. J. et al., "Visibility, Light Scattering and Mass Correlation of Particulate Matter," Report of California Air Resources Board (1973).
5. Charlson, R. J., et al., Atm. Env. 2, 455 (1968).
6. Simmons, W. A., et al., "Correlation of the Integration Nephelometer to High Volume Air Sampler," Mass. Dept. of Pub. Health (1970).
7. Thielke, et al., Aerosols and Atmospheric Chemistry, G. M. Hidy, editor, Academic Press, New York (1972).
8. Porch, W. M., Science, 170, 315 (1970).
9. Horvath, H., Atmospheric Environment, 5, 333 (1971).
10. Waggoner, A. P., et al., Applied Optics, 10, 957 (1971).
11. Husar, R. B., Private Communication (1974).

12. Charlson, R. J., et al., Aerosols and Atmospheric Chemistry, G. M. Hidy, editor, Academic Press, New York (1972).

13. Rasmussen, R. A., et al., PNAS 53, 1, 215 (1965).

14. Junge, C., J. Meteorology, 11, 323 (1954).

15. Charlson, R. J., et al., Science, 184, 156 (1974).

16. Charlson, R. J., et al., Atmospheric Environment, 8, 1257 (1974).

17. Winkler, P., Aerosol Science, 4, 373 (1973).

18. Hänel, G., Beitr. Z. Phys. Atm., 44, 137 (1971).

19. Covert, D. S., Ph.D. Thesis, University of Washington (1974).

20. Buettell, R. G., et al., J. Sci. Inst., 26, 357 (1949).

21. Ahlquist, N. C., et al., J.A.P.C.A., 17, 467 (1967).

22. Ahlquist, N. C., et al., Patent Application (1974).

23. Ensor, D., et al., Atmos. Env., 4, 48 (1970).

24. Quenzel, H., Atmos. Env., 9 (1975).

25. Rabinoff, R., et al., J.A.M., 12, 184 (1973).

26. Eiden, R., Applied Optics, 5, 569 (1966).

27. Grames, G. W., et al., J.A.M., 13, 459 (1974).

28. Volz, F. E., J.G.R., 77, 1017 (1972).

29. Lindberg, J. D., Applied Optics, 13, 1923 (1974).

30. Waggoner, A. P., et al., Applied Optics, 12 896 (1973).

31. Bergstrom, R. W., Beitr. Z. Phys. Atm., 46, 223 (1973).

32. Lin, C. I., Applied Optics, 12, 1356 (1973).

33. Howath, H., Atmos. Env., 3, 543 (1969).

A REVIEW OF ATMOSPHERIC PARTICULATE MASS MEASUREMENT VIA THE BETA ATTENUATION TECHNIQUE

E. S. Macias
Department of Chemistry
Washington University
St. Louis, Missouri 63130

and

R. B. Husar
Department of Mechanical Engineering
Washington University
St. Louis, Missouri 63130

ABSTRACT

The mass of atmospheric aerosols can be determined by using the attenuation of beta particles from radioactive sources. In this method atmospheric particulates are removed by filtration or impaction and the mass density of this deposit is determined from the decrease in the number of beta particles passing through the deposit. This paper reviews the theory of beta attenuation, the strengths and limitations of the method for mass measurement, the design of various instruments reported in the literature, and the characteristics of these instruments for both laboratory and field measurements. The criteria for selection of beta sources, electron detectors, and pulse processing electronics are discussed with an emphasis on state of the art instrumentation. Several aerosol collection systems are also described including a two stage on-line mass monitor with aerosol size separator (TWOMASS). This instrument independently analyzes the mass concentration of two particle size fractions. Calibration and field testing of this instrument coupled with aprticulate sulfate measurements for detailed aerosol characterization studies are also discussed.

A REVIEW OF ATMOSPHERIC PARTICULATE MASS MEASUREMENT VIA THE BETA ATTENUATION TECHNIQUE

E. S. Macias
Department of Chemistry
Washington University
St. Louis, Missouri 63130

and

R. B. Husar
Department of Mechanical Engineering
Washington University
St. Louis, Missouri 63130

INTRODUCTION

Air borne particulates, due to their effects on visibility, are the most obvious air pollutants. For this reason, particulate emissions were the first air pollutants to be controlled. In the past ten years, gaseous pollutants have received major attention from control agencies, but continuing deterioration of visibility in many urban areas, along with strong evidence of health effects on humans, and possible effects on climate, has led to a renewed concern for controlling atmospheric particulates.[1]

The present air quality standard for particulate matter is expressed in terms of the total aerosol mass in $\mu g/m^3$ as measured by the high-volume air filter sampler. Recently, there has been concern among researchers, as well as within control agencies, that total particulate mass or any other single parameter is an inadequate and to some extent misleading measure of adverse effects of atmospheric particulates. There is an increasing body of evidence which suggests that particulates found in the atmosphere that are less than 5 μm in size contribute significantly to the adverse effects of air pollution, and in fact constitute a large segment of the total air pollution problem. Accordingly the current thinking on a first step toward a proper characterization of atmospheric particulates is to divide the total aerosol population into two size classes, i.e., fine and coarse particulates. In studying the characteristics of the Los Angeles smog aerosol, Whitby *et al.*[2] discovered and substantiated by a variety of data, that most of the atmospheric aerosol volume is distributed bimodally; the lower mass mode is in the size range 0.1 to 1.0 μm, while the upper mode is over 5 μm as shown in Fig. 1. The saddle point between the two mass (or

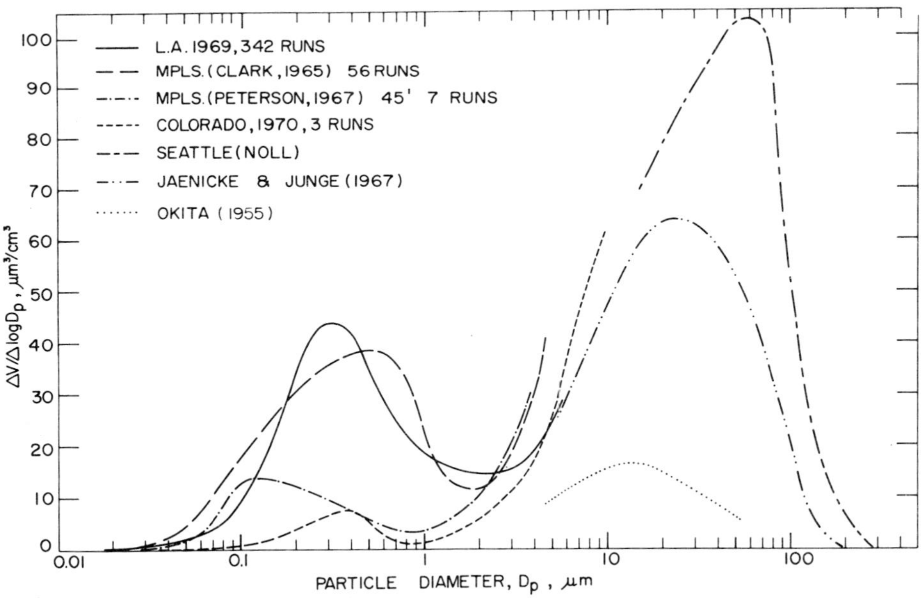

FIG. 1 - Typical atmospheric aerosol volume distribution data.

volume) modes is between 0.8 and 3.0 μm. The significance of bimodal aerosol mass spectra is that the two modes have distinctly different physical characteristics (particle shape, volatility) and chemical composition. Furthermore, the two modes are produced by different sources, and they are associated with different effects. These findings provide a scientific rational for the separate consideration of fine and coarse particulates.

It is anticipated that within the next few years these findings will be recognized by establishing new standards requiring the determination of aerosol mass and composition as a function of particle size. Independent determination of the mass concentration of only two size fractions, divided at about 1 to 3 μm will probably be adequate for monitoring purposes.

The high volume filter technique and other gravimetric weighing methods currently in use for atmospheric monitoring are inadequate for future monitoring purposes. Slow time response and tedious manual operation are the principal disadvantages of these methods. Other methods which are more easily automated have been proposed based on indirect measurements such as acoustic attenuation, pressure drop across a nozzle, pressure drop across a filter, unbalance of a centrifuge, acoustic particle counting, hot-wire anemometry, decrease in natural frequency of a vibrating band or wire, tape spot photometry, light scattering photometry or nepholometry, lidar, single particle light scattering, light transmission holography, electrostatic bounce, electrostatic probe-in-nozzle, electrostatic ion capture and electrostatic contact charging. However Sem, Borgos and Olin[3] point out that all of these techniques have serious limitations for monitoring of particulate mass emissions.

It has been known for some time that the attenuation of beta particles as they pass through an absorber can be used as a thickness gauge or mass monitor. In a recent comprehensive study of potential techniques for measurements in emissions from fossil fuel combustion sources, Sem *et al.*[4] identified beta attenuation as the most promising method for the sensing of particle mass concentration based on an evaluation of the basic sensing technique applied to a specific measurement. This conclusion was not a recommendation or criticism of presently available commercial beta instruments nor a recommendation of beta attenuation for use in other specific applications.[3]

This paper is concerned with application of the beta attenuation technique for automated atmospheric particulate mass measurements. Another promising technique for atmospheric mass measurements, the piezoelectric microbalance, has been discussed in a previous paper. A review of the physics of beta absorption and its use as a mass monitor is reviewed. The strengths and limitations of this method for use in monitoring atmospheric aerosols are also discussed.

A two stage on-line mass monitor with aerosol size separator (TWOMASS) employing the beta attenuation technique is described. The operating characteristics and calibrations are discussed as tested in St. Louis during the summer 1974 and spring 1975. The application of this instrument coupled with sulfur detection techniques for detailed characterization of atmospheric aerosols is also described.

ATMOSPHERIC MEASUREMENT VIA BETA ATTENUATION

PHYSICS OF BETA ABSORPTION

Although the absorption of beta particles has been used extensively for mass and thickness determinations, the physics of beta emission and interaction with matter is not obvious. The following is a brief discussion of the underlying physical principles which govern these processes.

BETA DECAY

In the process of β^- decay, a negative electron is emitted from the atomic nucleus and the nuclear charge changes from Z to Z + 1 in units of the electron charge. This process transmutes the beta-active element into the next heavier element in the periodic system. This nuclear emission of negative electrons is not to be confused with the emission of atomic orbital electrons as in internal conversion or Auger processes. In many cases beta decay does not populate the ground state of the daughter but rather populates an excited state which may de-excite by the emission of gamma rays or conversion electrons.

Beta decay can take place only if the mass of the daughter is less than that of the parent. The beta transition energy is given by the

$$\Delta E_{\beta^-} = [M(Z,A) = M(Z+1,A)]c^2 \tag{1}$$

where M(Z,A) is the atomic mass of a nuclide with atomic number Z and atomic weight A and c is the speed of light.

Beta particles are emitted with a continuous energy spectrum as shown in Fig. 2 rather than a single discrete energy equal to the transition energy. This apparent contradiction of conservation of energy and two-body kinematics is explained by a third massless particle, the anti-neutrino which carries off the missing energy. For example the beta decay of ^{14}C can be depicted as follows.

$$^{14}C \longrightarrow {}^{14}N + \beta^- + \bar{\nu} \tag{2}$$

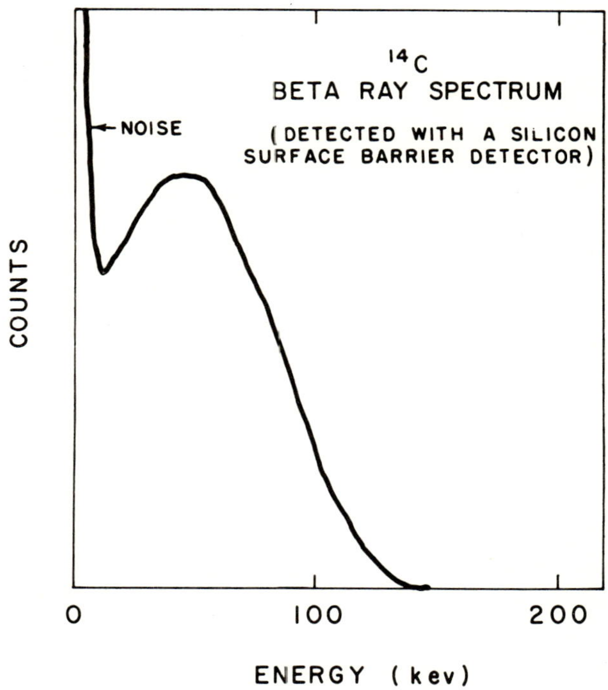

FIG. 2 - ^{14}C beta ray spectrum detected with a silicon surface barrier detector used in TWOMASS.

The shape of the beta spectrum is not the same in all cases. However, the average energy is at approximately one-third of the maximum value for most beta emitters.

INTERACTION OF BETA PARTICLES WITH MATTER

Beta particles interact with matter through elastic and inelastic scattering with atomic electrons and elastic nuclear scattering. For low energy electrons (E_β < 0.5 MeV) ineleastic scattering (ionization) with atomic electrons is the predominant mode of energy loss. The number of beta particles passing through an absorber decreases exponentially with absorber thickness, to a good approximation, given by the following

$$I = I_o e^{-\mu_m x} \quad (3)$$

where I_o is the beta intensity without an absorber, I is the intensity observed through an absorber of thickness x and μ_m is the mass absorption coefficient. The exponential form of the curve is fortuitous, since it also includes the effects of the continuous energy distribution of the beta particles and the scattering of the particles by the absorber.[5] The range R_β is the distance traversed by the most energetic particles emitted, and corresponds to the energy at the endpoint of the continuous spectrum. The exponential absorption of beta particles is not expected to hold for absorber thicknesses near the beta particle range.[5]

The absorber thickness, x, is usually given in units of mg/cm^2 with the actual thickness multiplied by the density. For low energy beta emitters, the mass absorption coefficient is nearly independent of the chemical composition of the absorber. This has been shown experimentally by a number of researchers.[6,7] This is because the absorption of electrons depends on the initial energy and the number of electrons with which they collide in passing through the absorber. The latter depends on the absorber electron density, i.e., the number of electrons per unit mass. Therefore the absorption of particles depends on the ratio of atomic number to the mass number (Z/A). Although this ratio decreases from light to heavy elements, the effect of this variation on μ_m for low energy beta emitters is not large. Most chemical compounds have Z/A ratios in the range 0.44-0.53. The absorption of high energy radiation (E_β > 1 MeV) includes a radiative contribution from Bremsstrahlung radiation (inelastic scattering with the nuclear Coulomb field). This effect causes large variations in the mass absorption coefficient with increasing atomic number[8] and is largest for heavy element absorbers. The ratio of energy loss due to ionization to that due to radiation is given approximately by

$$\frac{(dE/dx)\text{ioniz}}{(dE/dx)\text{rad}} = \frac{800}{EZ} \tag{4}$$

where E is the beta-particle energy in MeV and Z is the atomic number of the absorber.

MASS DETERMINATION USING THE ABSORPTION OF BETA PARTICLES

HISTORICAL SURVEY

The exponential absorption of beta particles and the implications of this absorption for thickness measurements of thin films have been known for some time.[8] The work of Crowther[8] and others has shown that the absorption of beta particles from sources with $E_\beta > 1$ MeV is strongly dependent on the elemental composition of the source.

Clapp and Bernstein[9] described a beta attenuation thickness gauge in 1950. They used a ^{90}Sr-^{90}Y source which has high beta transition energy (E_β = 2.16 MeV) and therefore did not have an element independent device. They used an ionization chamber to detect the beta particles but did not discuss a linear relationship between the logarithm of activity and absorber thickness. This device was intended to be calibrated for each particular absorber material. Peterson and Downing[10] built an early device which used the attenuation of low energy beta radiation for determining absorber mass. They use a ^{14}C source and a quartz fiber electroscope for detecting beta particles. It was found that the relation between the logarithm of the discharge time and the weight in mg/cm^2 of thin cellophane films and aluminum foils is almost linear.

Mandel[11] showed that ^{147}Pm (E_β = 0.23 MeV) was a more suitable beta source than the more commonly used (in 1954) ^{204}Tl (E_β = 0.74 MeV) for high sensitivity thickness measurements of very thin materials. Later Anders and Meinke[12] used a ^{147}Pm source and a Geiger counter for measuring thin films. Pate and Yaffe[13] reported measurements of the thickness of thin films using a ^{61}Ni source (E_β = 0.067 MeV) and a 2-π beta proportional counter.

The adaption of the beta absorption thickness measurement technique for use as a particulate mass monitor requires that particles be collected on a filter or other substrate. The mass determination is made by determining the beta intensity transmitted through the substrate before and after the particles are deposited. Mass concentrations are determined by also measuring the air volume from which the particles were collected.

An early application of the beta absorption technique for determining the mass concentration of atmospheric particulates was reported by Nader and Allen.[14] Several beta emitters, (^{14}C, ^{137}Cs, ^{204}Th) were studied; a ^{14}C source and a gas flow end-window proportional counter were chosen. Investigating several relative source-absorber-detector distances and source diameters, they found the optimal geometry with the absorber relatively close to a small diameter source. They also determined that the typical background due to radioactivity in the aerosol deposit is negligible compared to the beta source intensity. This work also demonstrated the feasibility of the beta absorption technique for use in automated tape samplers.

More recently many researchers have reported the use of beta attenuation mass monitors using a variety of experimental configurations.[6,7,15,24] The variations are mainly in the source, detector, aerosol sampling and electronics discussed in detail below.

BETA SOURCE

The important characteristics of a beta source for beta absorption atmospheric aerosol mass measurementsare long halflife (>1 yr), pure beta decay (no gamma decay), and low beta transitions energy (<1 MeV). The beta transition energy is directly proportional to the maximum thickness which can be pentrated and inversely proportional to the sensitivity of the measurement. At least 13 sources meet these criteria.[25] They are ^{3}He, ^{10}Be, ^{14}C, ^{26}Cl, ^{39}Ar, ^{63}Ni, ^{79}Se, ^{85}Kr, ^{87}Rb, ^{107}Pd, ^{147}Pr, ^{204}Tl, and ^{228}Ra. Of these nuclides, ^{14}C, ^{79}Se and ^{147}Pr have beta transition energies $\sim$0.2 MeV which optimize sensitivity and penetration of the absorber (6-12 mg/cm^2) in a typical mass monitor (due to filter paper, air, deposit detector window, etc.). ^{14}C is preferred because it has a longer halflife than ^{147}Pr (5730 yr vs. 2.6 yr) and is less expensive than ^{79}Se.

DETECTORS

A desireable beta detector for atmospheric mass measurements must have high stability, portability, and efficiency for detecting beta particles. Detectors in use for beta particle mass measurements which meet these requirements include gas flow proportional counters, geiger counters, scintillation detectors, and solid state detectors. However, Geiger counters and gas flow proportional counters have slow time response limiting the counting rate to about 1000 cps without large dead time. Scintillation and solid state detectors are much faster and allow the use of counting rates in excess of 10,000 cps. The efficiency for full energy events is somewhat less in solid state detectors than in the other detectors. However the efficiency for simply scoring an event as is needed in beta attenuation measurements is very high even with fairly thin detectors.[26] Silicon surface barrier solid state detectors have recently been made light tight, rugged and with very thin windows (40 $\mu gm/cm^2$) and appear to meet all the important criteria for beta attenuation mass monitors. The use of such a detector is described in a later section of this paper.

ELECTRONICS

The simplest electronic components needed for a beta detector include a preamplifier, amplifier, discriminator and counter. Geiger counters can be operated without a preamplifier. The NIM modular nuclear electronics operate from a single power supply located in a bin are recommended because of their flexability and reliability. This equipment is available from several commercial suppliers.

AEROSOL SAMPLING

Ideally, the physical and chemical characterization of atmospheric particulates should be performed 'in situ'. Unfortunately,'in situ' mass measurement of suspended particulates at present does not seem possible. Beta attenuation, for instance, requires the deposition of particulate matter onto an impactor, filter or electrostatic precipitator substrate placed between the beta source and the detector. In the past all three aerosol deposition mechanisms have been used by various investigators.

Filtration is the most convenient sampling method since it can be easily automated for quasi-continuous operation.[7,14,16,17,20,23] Such an instrument typically consists of a filter tape configuration similar to the commonly used tape sampler, A variation of this approach is the use of a casette-type system similar to those used in slide projectors. Inertial impaction as the mechanism has also been used.[17,23,27]

SAMPLING CONSIDERATIONS

The problems associated with aerosol sampling may be divided into the following categories.

Losses in the Inlet System

Inertial or gravitational deposition in the inlet system before particles reach the sensing area prevents the effective sampling of large atmospheric particulates, above 30-50 μm in diameter. In systems with impaction sampling[19,23] the large particle cutoff size is smaller because of the interference of the beta source or detector. Design criteria for minimizing the losses in the inlet system are discussed by

Fuchs.[28] In beta attenuation mass units that we are aware of, the upper size cutoff was taken for granted and no special effort was made to increase the cutoff size.

Filtration Efficiency and Bouncing

The physical retention efficiency of the aerosol substrate must be high for meaningful data interpretation. In the case of filtration, high (>98%) efficiency for any size particles may be attained by conventional glass fiber filters. Other criteria for filter choice such as low flow resistance, low tare weight, adequate mechanical strength, adequate capacity before clogging, inertness to water and chemicals, are generally much more difficult to satisfy.[17]

The use of impaction as the aerosol deposition mechanism imposes the problem of large particle bounce-off following impaction. Although the details of the bounce-off mechanism are not well understood it is generally accepted that it can be minimized by using a sticky substrate surface. Crystaline mineral particles, impacted on an uncoated teflon substrate has been shown to be the poorest combination.[29] Methods for bounce-off reduction include the use of sticky (greesy) or bushy glass fiber filters as the impaction substrates.

Sample Loss or Gain During and After Deposition

Atmospheric aerosols consist of solid or liquid particles. The large particle mode is typically composed of solid mineral material, while the mass in the small particle mode is contributed by liquid droplets consisting of an aqueous solution of salts and organic matter. Upon deposition, hygroscopic and deliquescent particles may gain or lose volatile matter in accordance with their own properties and the thermodynamic conditions of the surrounding air. Therefore, aerosol stability on a filter is an inherent problem, common to all measurement methods in which the detection is not performed 'in situ'. Presently, there is little quantitative information available on this subject, mainly because of experimental problems. One of the virtues of beta attenuation is that it can be used as a tool for the investigation of aerosol volatility at the moment of deposition. An example of such results is shown in Fig. 3. A fraction of a cigarette smoke puff was deposited in a beta mass monitor (described in the next section) and the mass concentration as a function of time was monitored in one second increments. As shown in Fig. 3 there is 30% loss in the deposited mass within about 20 seconds. Turning the sampling pump off immediately after deposition of the smoke puff, reveals that the smoke is stable on the filter with no air flow. Immediately after turning on the pump, a fraction of the deposit volatilized, indicating that the change in the pressure and/or flow conditions caused the loss of volatile compounds,

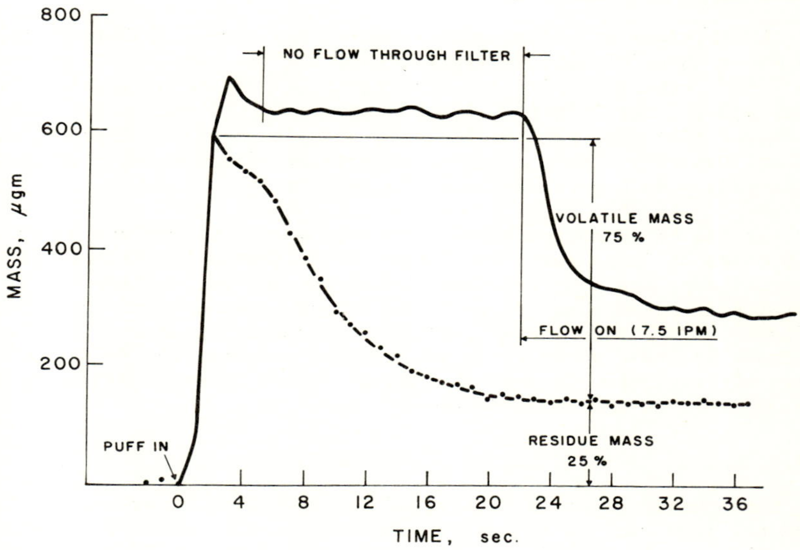

FIG. 3 Loss of volatile cigarette smoke mass after deposition on a glass fiber filter. The dashed line is for continuous pump operation. The solid line is for interrupted pump operation as indicated.

probably water and volatile organics. Unfortunately, such experiments can only be performed when aerosol concentration is high, i.e. the aerosol deposition rate is higher than the evaporation rate. Accordingly, the actual behavior of atmospheric aerosol upon deposition on a substrate can only be inferred from laboratory simulation experiments.

It is well established that a large fraction of atmospheric fine particulates is composed of hygroscopic or deliquescent particles.[13] Their mass, therefore, depends on the relative humidity of the ambient air. Although extensive studies have been carried out on the relative humidity (RH) dependence of the aerosol light scattering coefficient[31] we are only aware of one work in which the RH dependence of particle mass was studied in detail.[32] Here again beta attenuation provides a simple and convenient means of detecting the mass dependence on RH. A typical result on the RH dependence of ammonium sulfate aerosol as detected by beta attenuation is shown in Fig. 4.[33]

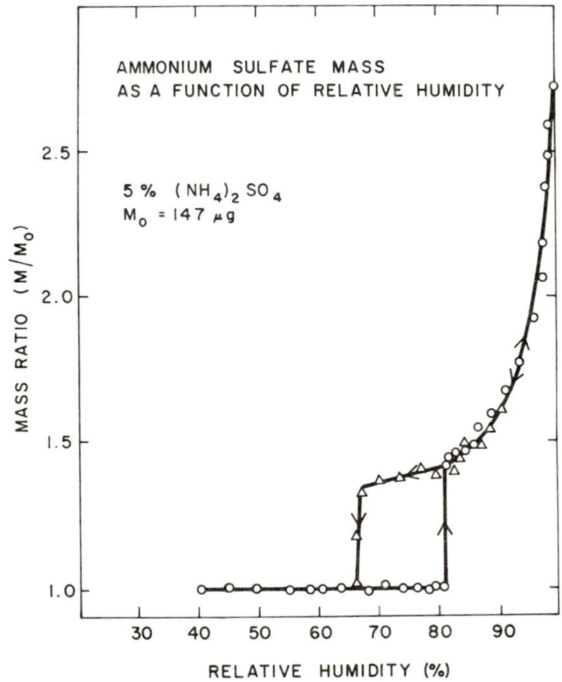

FIG. 4. Ammonium sulfate mass determined with a beta attenuation mass monitor as a function of relative humidity.

Recently, concern has been raised that particulate matter may form a filter media as a consequence of catalytic gas-particle conversion.[34] It has been speculated that the catalytic effect may be caused by the filter or by the deposited particulate matter itself.[35]

In summary, the following have been identifidd as essential for the proper use of the technique. A source with low beta transition must be used for mass measurements to be independent of chemical composition. A collimated source with small source-detector distance is optimal. The total abosrber mass including the aerosol deposit must be a small fraction of the range of the beta particle. The source must have a long lifetime (>1 yr) and decay without the emission gamma radiation. The pressure drop and flow rate should not change over the sampling period. Problems such as losses in the inlet system, filtration efficiency and bouncing, and aerosol stability on the collection substrate must be minimized in the sampler design.

TWOMASS

OBJECTIVES

The present section describes a two stage on-line mass monitor with aerosol size sampler (TWOMASS) which was designed to monitor atmospheric aerosols in the two size ranges described previously, namely fine particles with submicron diameter and coarse particles with supermicron diameter. This instrument was optimized for high sensitivity over the shortest possible measurement time. TWOMASS shown in Fig. 5, is fully automated so that continuous on-line determinations of particle mass are available every ten minutes over long periods of time (1-10 days). TWOMASS was developed for use in ground based ambient aerosol monitoring stations but it could be adapted for other uses.

INSTRUMENT DESIGN

Aerosol Collection System

The flow system separates particles into two size fractions. Coarse particles (diam >3 μm) are impacted on a glass fiber filter with a cellulose backing; submicron particles are collected on the same type of high-efficiency glass fiber filter tape. This system is shown schematically in Fig. 6. The single stage impactor head has a 3-mm diam inlet aperture with a 1-mm jet to plate distance. Various impaction materials have been tested, the glass fiber filter with thin cellulose backing was found to be most satisfactory because of its low mass density (3.2 mg/cm^2), efficient impaction properties, and uniform thickness. The impaction collection efficiency was designed to approximate the removal efficiency of the human respiratory system with 50% efficiency for 3 μm particle collection.

The second filter stage retains all remaining particles. An earlier study by Lilienfeld and Dulchinos[17] indicated that Pallflex E70/207W has high particle collection efficiency at high face velocities, low flow resistance, low tare weight, adequate mechanical strength, and adequate filter capacity before clogging. The mass density of this filter is 3.2 mg/cm^2 and the efficiency is over 99% for

FIG. 5 Photograph of TWOMASS.

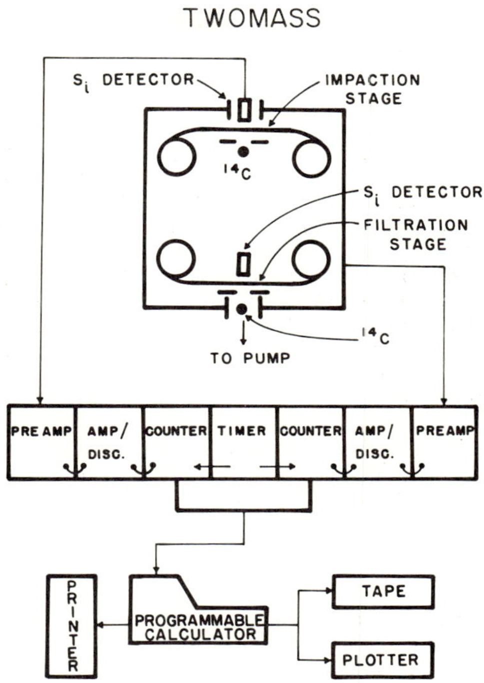

FIG. 6. Schematic diagram of TWOMASS including the data acquisition and analysis system.

both cigarette smoke and 0.312 μm diameter polystryrene latex spheres at the maximum face velocity obtainable, 6.4 m/sec. The flow rate through the TWOMASS was set at 12 ℓ/min using a GAST model 1531 rotary vane pump.

Beta Attenuation Mass Monitor

Both the impaction and filtration heads of TWOMASS were fitted with independent source-detector systems which were run simultaneously. The beta source chosen was 3 mCi of ^{14}C deposited as $BaCO_3$ mixed with Krylon spray adhesive in a 6-mm diameter brass disk covered with 800 $\mu g/cm^2$ of mylar film. This source, a pure low energy beta emitter with 5.7×10^3 year half life, E_{max} = 156 keV, and range R_β = 28.26 mg/cm^2 was chosen for the reasons outlined previously. The total mass density of all absorbers (air, detector window, filter deposited mass, etc.) in the system is about 6 mg/cm^2 or about 21% of the range of ^{14}C betas which is optimal for measurement sensitivity.[17] The cross sectional area of the collection spot is 32 mm^2 and 7 mm^2 in the filtration and impaction stages respectively.

The beta detector is a solid-state ruggedized silicon surface-barrier detector with 50-mm^2 surface area, 40-μgm aluminum window and noise width of <11 keV (manufactured by ORTEC, Inc.). This detector is light tight and rugged, thus it is suited for field operation. It was chosen because of its high count rate capability, low noise characteristics, low cost and simplicity of operation. As shown in Fig. 6 the output of each solid state detector is sent to an ultra low-noise (<5 keV) preamplifier with very fast rise time (<20 nsec) and then to a low-noise gaussian-shaping amplifier (timing constant 1.2 μsec) with a lower level discriminator set to remove noise. The energy spectrum of ^{14}C beta particles measured with this system is shown in Fig. 2.

The outputs of the two discriminators are sent to a Tectronix 31/53 calculator instrumentation system with two DC503 counters. This programmable calculator determines the mass with the following formula

$$M = \frac{A}{f\mu_m} \ell n(I_o/I) \quad (2)$$

where M is the mass concentration in units of μgm/m^3, A is the cross sectional area of the collector, f is the flow rate, μ_m is the mass attenuation coefficient in units of cm^2/μg, I_o is the count rate of the previous time interval and I is the count rate of the current interval. This mass concentration data is plotted with an x-y point plotter, printed, and stored on a magnetic tape casette.

On-line Operation

TWOMASS was designed for mass measurements over extended time periods with a minimum of operator maintanence. By measuring the beta intensity transmitted through the filter paper as the particulates are being deposited, mass concentrations are obtained at the end of each counting period. Furthermore, because the initial beta intensity, I, is always the intensity measured in the previous counting interval, there is no error introduced due to irregularities in the filter medium. This counting arrangement also minimizes the effect of electronic instability on the mass measurement. The system is automated with two motors moving both the filter and impaction tapes at programmed intervals. In order to avoid clogging the filter paper which can cause decrease in the flow rate, the tape is moved every two hours. In sampling tests of atmospheric aerosol with TWOMASS using 2 hour periods, the flow rate changed by less than 5%. A related effect, the change in the pressure below the filtration-stage filter tape with increasing filter loading, was also examined. This pressure decrease during on-line operation results in a decrease of the total absorber mass between the ^{14}C source and the beta detector. The geometry of the filtration stage in TWOMASS was re-designed to minimize this effect which was experimentally determined to be <6% of the mass of all aerosols studied. The automated tape movement is controlled by an automatic timer and therefore this interval can be adjusted to any experimental requirements. Note that with this arrangement the tape is not moved while particulates are being deposited or measured and thus, the tape movement does not effect the mass measurement.

OPERATING CHARACTERISTICS

Sensitivity

Emphasis in the TWOMASS design has been placed on low noise and fast timing characteristics in order to minimize instabilities in the counting system due to noise drift and to maximize the counting rate without large dead time. Typical counting rates are 10,000 cps with the unloaded filter tape in place. This high counting rate was possible with minimal dead time (<1%) because of the fast timing characteristics of the solid state detector and associated electronics. The total sensitivity of the instrument is a function of the stability of the system and the counting statistics. The sensitivity S, based only on counting statistics using the 2σ error limit given[25] as follows

$$S_{min(2\sigma)} = \frac{2A}{\mu_m f t \sqrt{C}} \qquad (3)$$

where t is the counting time and C is the detected count rate. With A, μ_m, f, and C fixed for a given instrumental design, the statistical sensitivity becomes a function of counting time only, i.e.

$$S_{min(2\sigma)} t = K_o \ . \qquad (4)$$

For TWOMASS with the conditions described above, this constant is K_o = 2000 μgmsec/m^3. From the above considerations it is clear that statistical sensitivity increases while stability decreases with the counting time. For sampling times of 600 sec the total sensitivity of TWOMASS is 4 μgm/m^3. The system has also been run with subsecond time intervals for laboratory experiments using high aerosol mass concentrations as described in the previous sections.

Calibration

The instrument was calibrated against a Cahn Microbalance with laboratory and St. Louis atmospheric aerosols. The mass of laboratory aerosol added to the filter was in the range of 1-7 mg/cm^2 as shown in Fig. 7. The mass of atmospheric aerosols was in the range of 0.2-1

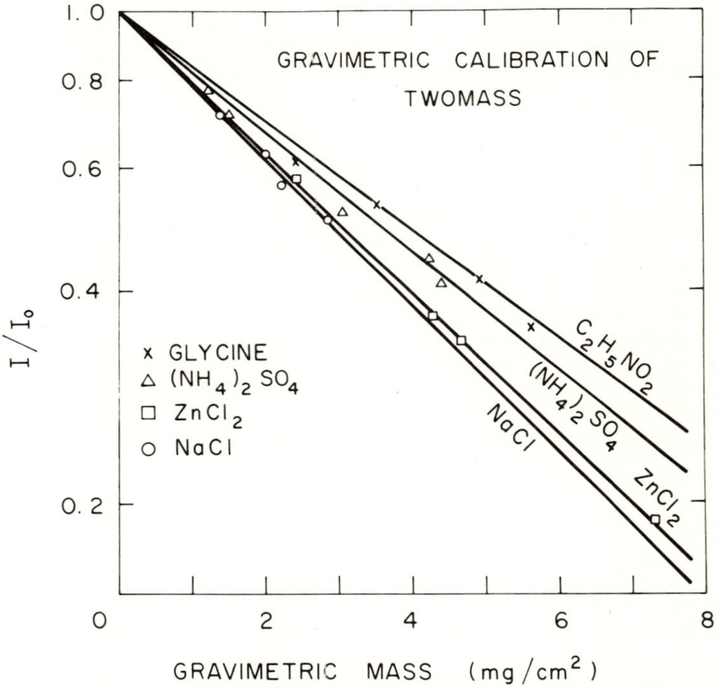

FIG. 7. Gravimetric calibration of TWOMASS using laboratory aerosols.

mg/cm^2 as shown in Fig. 8. These calibration data yield an exponential relationship between mass loading and beta attenuation (I/I_o) as expected from equation 3 within the accuracy of the measurements. Sem and Borgos[24] used an uncollimated ^{14}C source and various filter absorbers to simulate aerosol loading and found some deviation from the exponential absorption law particularly for high masses near the range of ^{14}C beta particles. The operation characteristics and source-detector configuration of TWOMASS are quite different from the Sem and Borgos arrangement e.g. use of a highly collimated ^{14}C source and relatively low total absorber mass, and a filter of uniform mass. Therefore the

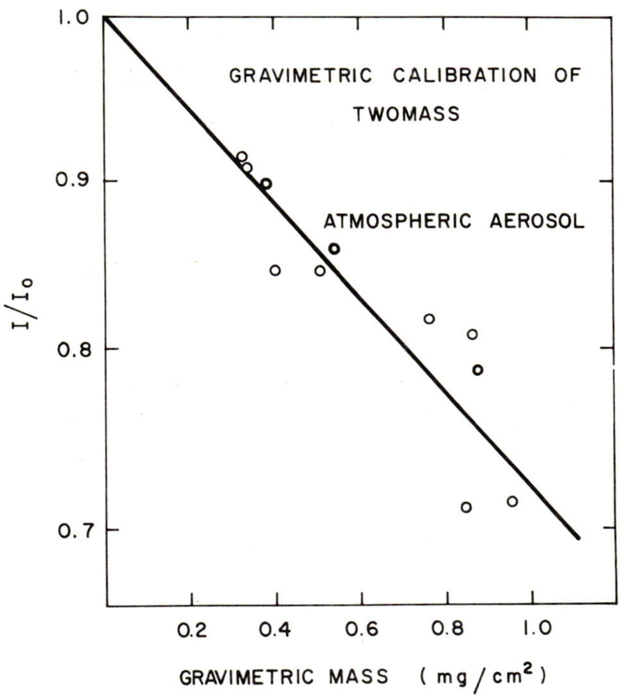

FIG. 8. Gravimetric calibration of TWOMASS using St. Louis atmospheric aerosol.

exponential calibration curve determined for TWOMASS is not inconsistent with their results.

As mentioned previously the mass attenuation coefficient, μ_m, is a weak function of the chemical form of the aerosol. The values of μ_m extracted from Figs. 7 and 8 are plotted versus the atomic number divided by atomic weight (Z/A) of the various aerosols is shown in Fig. 9. Because the Z/A ratio for the atmospheric aerosol is not known the μ_m value for the atmospheric aerosol is given in a shaded horizontal band. This data indicates that some variation in μ_m does exist over the Z/A range for most chemical compounds (0.44-0.53). Thus we suggest that determination of μ_m is necessary when using a beta gauge for studying a specific aerosol. However our data indicate that, at least for St. Louis, the atmospheric aerosol has a value of $\mu_m = 0.260 \pm 0.025$

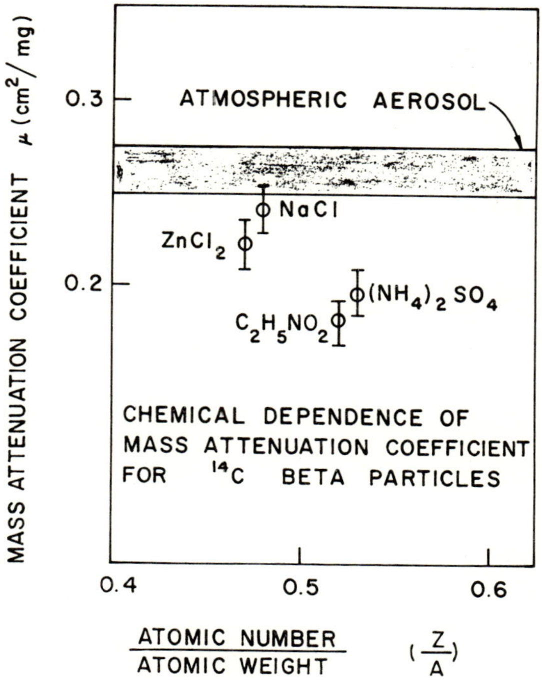

FIG. 9. Chemical dependence of mass attenuation for ^{14}C beta particles determined with TWOMASS.

cm^2/mg is satisfactory. It should be noted that this value is identical with the standard literature value of 0.2619 cm^2/mg.[5]

This calibration was carried out over a much wider mass range than required of TWOMASS when used for ambient monitoring. In a typical 3 hour period of aerosol collection in St. Louis the deposited particulate mass on either stage is no more than 1 mg/cm^2.

FIELD TESTING WITH ATMOSPHERIC AEROSOLS

TWOMASS was field tested for use in monitoring atmospheric aerosols during the August 1974 St. Louis Regional Air Pollution Study experiment. The instrument was run continuously determining mass in two size ranges in 1000 sec increments for 10 days in a manual mode. An example of the data, collected on 23 August 1974 is shown in Fig. 10. As can

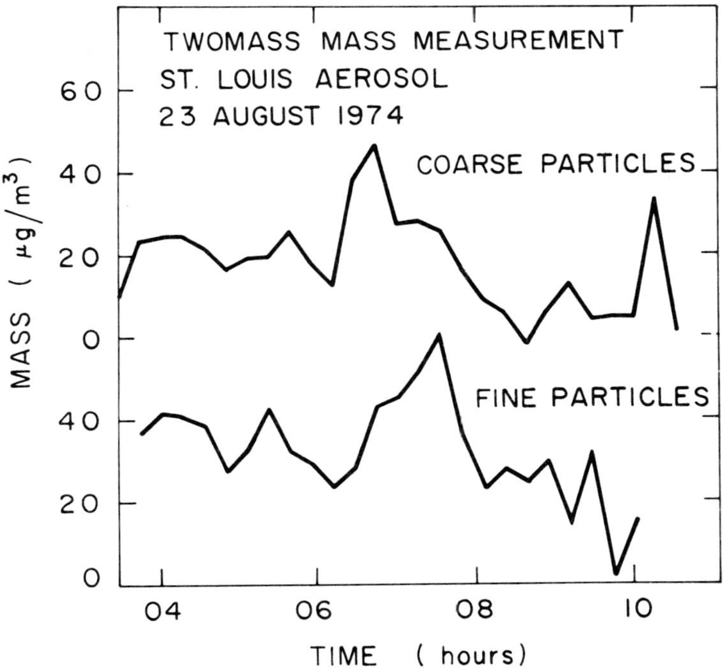

FIG. 10. On-line mass concentration data in two particle size ranges determined with TWOMASS during the August 1974 St. Louis RAPS experiment.

be seen from this data, both the fine and coarse particulate mass concentrations exhibit strong temporal variations; much more than expected from earlier low time response measurements with filters. In addition it is evident from these data that the mass concentrations of each size fraction are each roughly half of the total aerosol mass and at times vary independently. It is clear from these results that TWOMASS gives more detailed mass concentration data than previously available which should lead to a better understanding of atmospheric aerosols.

A more detailed characterization of the ambient aerosol using TWOMASS with 10 min sampling intervals and moving the filter tapes every two hours was carried out in April 1975 in St. Louis at Station number 112 of the Regional Air Monitoring System (RAMS). In these experiments the light scattering coefficient, (b_{scat}), sulfur content of the fine particle mass and relative humidity were also measured simultaneously. These parameters measured during a five day experiment are shown in Fig. 11. These data again exhibit a rather striking temporal variation of mass over the sampling period. Data from 18 April indicate the arrival of a polluted air mass, drastically changing both the mass concentration and light scattering coefficient within a one hour interval. A rain storm at 1600 on that day caused the sharp drop in fine particulate mass concentration.

It should be noted that such observations on this time scale were not possible with conventional gravimetric measurements or beta attenuation instruments. The fine particulate mass concentration, after the arrival of a clean air mass on 18 April dropped to an extremely low value (3-5 $\mu g/m^3$) as shown in Fig. 11. Even at these low concentrations, meaningful results can be obtained with TWOMASS operating with 10 min counting intervals.

Analysis of the data in Fig. 11 shows a strong correlation between fine particulate mass and light scattering coefficient with a correlation coefficient of 0.96 for two hour averages. The ratio b_{scat}/fine particulate mass extracted for these data was 0.2 gm/m^2.

The TWOMASS two hour filter deposits were analysed for sulfur using flash vaporization-flame photometric detection method.[36] The data indicate that sulfur in the fine particulates expressed as sulfate ($\mu g/m^3$) constitutes 15-50% of the fine particulate mass concentration. It is interesting to note, however, that this fraction exhibits substantial temporal variations.

The results of this study show that TWOMASS with high time resolution, ability to separate fine and coarse particulates, high sensitivity, and ability to analyse the chemical composition of the deposit, is a significant advance over existing atmospheric particulate mass instrumentation.

CONCLUSION AND RECOMMENDATIONS

Based upon the above discussion of the physics of beta absorption, beta source characterics, detection methods and sampling considerations, we conclude that the problems of precise determination of ambient aerosol mass concentration with high sensitivity are essentially solved. We recommend that future research effort be focused on a better understanding of the processes influencing the aerosol sampling.

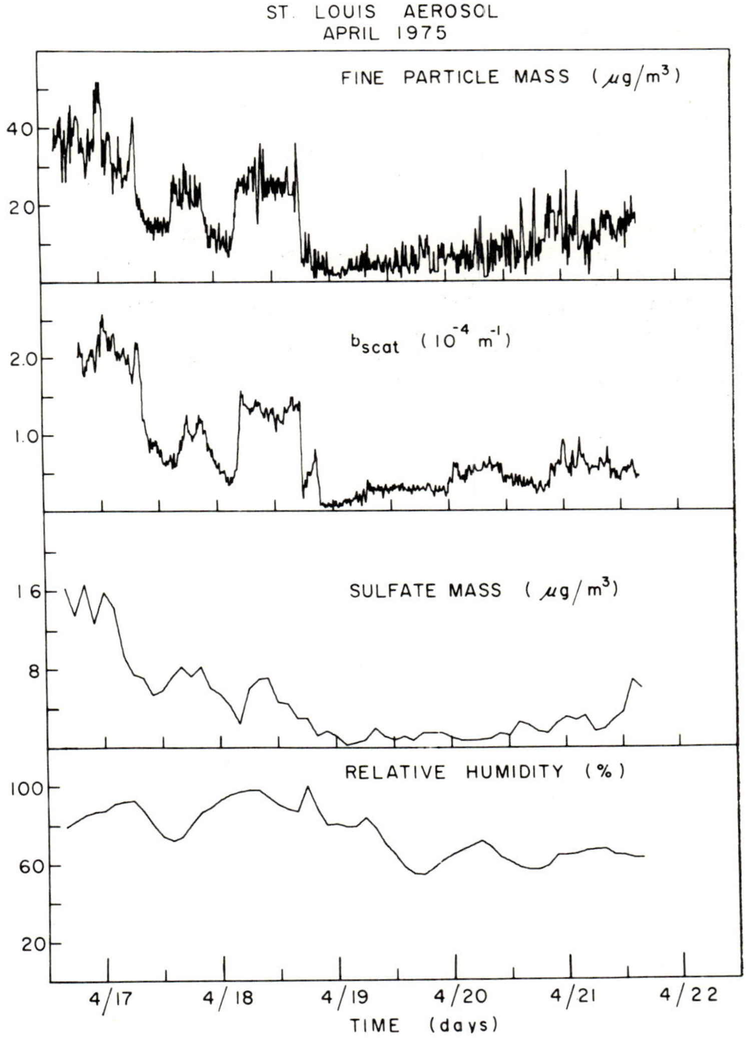

FIG. 11. Typical data for the characterization of the St. Louis aerosol by simultaneous determination of fine particle mass, light scattering coefficient, sulfate mass and relative humidity.

ACKNOWLEDGEMENTS

The authors gratefull acknowledge the assistance of Robert Fletcher, Larry Friedman, Roland Head, Connie Rutledge and Pamela Stubits in this work. Part of this work was supported by the U.S. Environmental Protection Agency, Research Triangle Park, North Carolina, Division of Chemistry and Physics, Field Methods Branch, under grant number R803115-01-0. Special thanks are due the Environmental Protection Agency and the St. Louis Regional Air Pollution Study for their cooperation in a portion of this work.

REFERENCES

1. Abatement of Particulate Emission from Stationary Sources. National Research Council, National Academy of Engineering, COPAC-5, 1972.

2. Whitby, K.T., R.B. Husar, and B.Y.H. Liu. The Aerosol Size Distribution of the Los Angeles Smog. J Colloid Interface Sci 39:117, 1972.

3. Sem, G.J., J.A. Borgos, and J.G. Olin. Monitoring Particulate Emmissions. Chem Eng Prog 67:83-89, October 1971.

4. Sem, G.J., J.A. Borgos, J.G. Olin, J.P. Pilney, B.Y.H. Liu, N. Barsic, K.T. Whitby, and F.D. Dorman. State of the Art 1971:Instrumentation for Measurement of Particulate Emissions from Combustion Sources. Thermo-System Inc., St. Paul, Minn. Prepared for the U.S. Environmental Protection Agency of 1971.

5. Kaplan, I. Nuclear Physics. Reading, Mass. Addison-Wesley Publishing Co., 1962. 770 P.

6. Dresia, H., P. Fischötter, and G. Felden. Kontinuierliches Messen des Staubgehaltes in Luft und Abgasen mit Betastrahlen. VDI-Z 106:1191, August 1964.

7. Husar, R.B. Atmospheric Particulate Mass Monitoring with a β Radiation Detector. Atm Env 8:183-188, 1974.

8. Crowther, J.A. On the Coefficient of Absorption of the β Rays From Uranium. Phil Mag 12:379-392, 1906.

9. Clapp, C.W. and S. Bernstein. Noncontacting Thickness Gauge Using Beta Rays. Elec Eng 69:308-10, April 1950.

10. Peterson, J.H. and J.R. Downing. Film Thickness Measurements for Infrared Spectrophotometry. J Opt Soc Am 41:862-862, November 1951.

11. Mandel, L. The β-Ray Absorption Spectrum of ^{147}Pm and its Application to Thickness Measurement. Brit J Appl Phys 5:287-9, August, 1954.

12. Anders, O.U. and W.W. Meinke. Beta Gauge for Localized Measurements on Thin Films Rev Sci Inst 27:416-17, 1956.

13. Pate, B.D. and L. Yaffe. A New Material and Techniques for the Fabrication and Measurement of Very Thin Films for use in 4π-βCounting. Can J Chem 33:15-23, 1955.

14. Nader, J.S. and D.R. Allen. A Mass Loading and Radioactivity Analyzer for Atmospheric Particulates. Am Ind Hyg Assoc J 21:300-307 (1960).

15. Izamilov, G.A. Measuring the Gravimetric Concentration of Dust in the Air Using β-Radiation. Ind Lab (Eng Tran of Zavodskaya Laboratoriya 27:40-43, January 1961.

16. Benarie, M. and A. Loverdo. Pesée automatique des polluants atmospheriques par jauge beta. Mesures, Regulation et Automatisme. 32:90-95, September 1967.

17. Lilienfeld, P. and J. Dulchinos. Vehicle Particulate Exhaust Mass Monitor. Final Report to Environmental Protection Agency, Durham, N.C. Contract No. 68-02-0209. March 1972. 41 p.

18. Horn, W. Process for Continuous Gravimetric Determination of the Concentration of Dustlike Emissions. Staub-Reinhalt 28:20-25, 1968.

19. Lilienfeld, P. Beta-Absorption-Impactor Aerosol Mass Monitor. Amer Ind Hyg Assoc J 31:722-29, November, 1970.

20. Dresia, H. and F. Spohr. Experience with the Radiometric Dust Measuring Unit "Beta Staubmeter." Stuab-Reinhalt 31:19-27, 1971.

21. Duke, C.R. and B.Y. Cho. Development of a Nucleonic Particulate Emission Gauge. Industrial Nucleonics Corp., Columbus, Ohio. Prepared as final report under U.S. Environmental Protection Agency Contract No. 68-02-0210, 1972.

22. Dresia, H. and R. Mucha. Registrierendas Radiometrisches Messerät zur Kombinierten Messung der Immissioner von Staub und Radioaktivität in Luft. Staub-Reinhalt 34:125-128, 1974.

23. Macias, E.S. and R.B. Husar. High Resolution On-Line Aerosol Mass Measurement by the Beta Attenuation Technique. In: Proceedings of the Second International Conference on Nuclear Methods in Environmental Research, Vogt, J.R. (ed.) USAEC, Oak Ridge, Tenn., 1975.

24. Sem, G.J., and J.A. Borgos. An Experimental Investigation of the Exponential Attenuation of Beta Radiation for Dust Measurement. Staub-Reinhalt 35:5-9, January 1975.

25. Lilienfeld, P. Beta Absorption Mass Monitoring of Particulates--A Review. Paper No. 71-1031, Joint Conference on Sensing of Environmental Pollutants. Palo Alto, California. November 1971. 30 p.

26. Instruction Manual for Surface Barrier Detectors. Ortec, Inc. Oak Ridge, Tenn. 1974.

27. Denzel, P. and W. Horn. Ein empfindliches Messverfahren zur Kontinuierlichen Bestimmung der gravimetrischen Konzentration von Staubförmigen Emissioner. Elektrotechnische Z PT. A., 87:311, 1969.

28. Fuchs, N.A. Mechanics of Aerosols. Pergamon Press, London, 1964.

29. Hidy, G.M., et al. Characterization of Aerosols in California. Interim Report for Phase 1. Science Center Rockwell International. Submitted to Air Resources Board, State of California, April 1973.

30. Junge, C.E. Air Chemistry and Radioactivity. Academic Press, N.Y., 1963.

31. Covert, D.S., R.J. Charlson, and N.C. Ahlquist. A Study of the Relationship of Chemical Composition and Humidity to Light Scattering by Aerosols. Submitted to J Appl Met, 1975.

32. Winkler, P. The Growth of Atmospheric Aerosol Particles as a Function of the Relative Humidity-II. An Improved Concept of Mixed Nuclei. Aerosol Sci 4:373-387, 1973.

33. Yamamoto, K. Measurement of Aerosol Mass as a Function of Relative Humidity. MS Thesis, Dept. of Mechanical and Aerospace Engineering, Washington University, St. Louis, Mo. 1975.

34. Novakov, T., S.G. Chang, and A.B. Harker. Sulfates as Pollution Particulates: Catalytic Formation on Carbon (Soot) Particles. Science 186:259-261, 18 October 1974.

35. Lee, R.E. Jr. and T. Wagman. A Sampling Anomaly in the Determination of Atmospheric Sulfate Concentration. Amer Ind Hyg Assoc J 27:266-271, 1966.

36. Husar, J.D., R.B. Husar, and P.K. Stubits. Determination of Submicrogram Amounts of Atmospheric Particulate Sulfur. Submitted to Anal Chem. May, 1975.

DETECTION OF ULTRA-FINE PARTICLES BY MEANS OF A CONTINUOUS FLUX CONDENSATION NUCLEI COUNTER

ABSTRACT

A new continuous flux condensation nuclei counter is described. Gases containing these nuclei go through an alcohol vapor saturator, then over-saturation required to generate condensation is produced in a Peltier cooling nozzle. Droplets cross a well lit measurement spacing of variable sizes so that the probability to emerge with a single droplet is equal to 0.95. The luminous flux scattered at 45° by each droplets is picked by a photomultiplier. A device enables to extend the field of use of this system to low pressures and low temperatures.

Thesecounter's operating characteristics are given and the study of particle detection as these were produced through photolysis of gaseous impurities in air within a pressurized environment varying from 760 to 60mm of Hg is discussed.

DETECTION OF ULTRA-FINE PARTICLES BY MEANS OF A CONTINUOUS FLUX CONDENSATION NUCLEI COUNTER

by Messrs. J. BRICARD, Professor, University of Paris VI, France

P. DELATTRE, PhD, Engineer, French Atomic Energy Commission, Fontenay-aux-Roses, France

G. MADELAINE, PhD Sciences, "

M. POURPRIX , PhD Sciences, "

\+ + + + + + + + + + + + + + +

In order to count very fine particles suspended in a carrier gas, we use artificial means wherein suspension is easily saturated with readily condensable vapor, then such vapor condensation is generated over these particles. Each give rise to a droplet and their sizes are adequate enough to allow for their detection. The concentration of droplets, that is of particles, can be derived either by measuring a luminous beam attenuation through a given thickness of the mist represented by all the droplets, or by measuring the luminous flux scattered by a given volume of said mist, once properly illuminated.

In general, when using presently marketed measuring instruments, the oversaturation required for vapor condensation is obtained by by lowering the temperature of the previously steam saturated gaseous media through adiabatic release, and in which the aerosol to study is suspended. The use of such particle counters results in two major difficulties : they do not allow for the detection of loosely concentrated aerosols, on the one hand, and they operate only by means of intermittend samplings of samplets in a media involved in a study, on the other hand, this resulting in media interferences.

So that both of these drawbacks are eliminated, we have devised a continuous flux counter with which we count particles whose radius is somewhere above 10^{-7} cm and whose concentration varies between 1 particle per cm^3 and 10^5 particles per cm^3.

I. INSTRUMENT DESCRIPTION

The particle-charged gas first goes through a saturator containing ethyl alcohol or n-butylic alcohol (FIGURE1). This saturator is a inner-lined cylindrical box, with spongy walls inhibited with liquid to obtain condensation. It is fitted with a 4mm Ø and 8 cm in length calibrated nozzle coolled by a Peltier effect module and a constant flow intake device. Condensation appears at nozzle outlet and generates the formation of droplet mists whose droplets radius is between 0 µ 5 and 1 µ. They can be seen with the naked eye. The observed droplet volume V which will now define is located at 2mm beyond the nozzle. The droplet velocity through it is even and

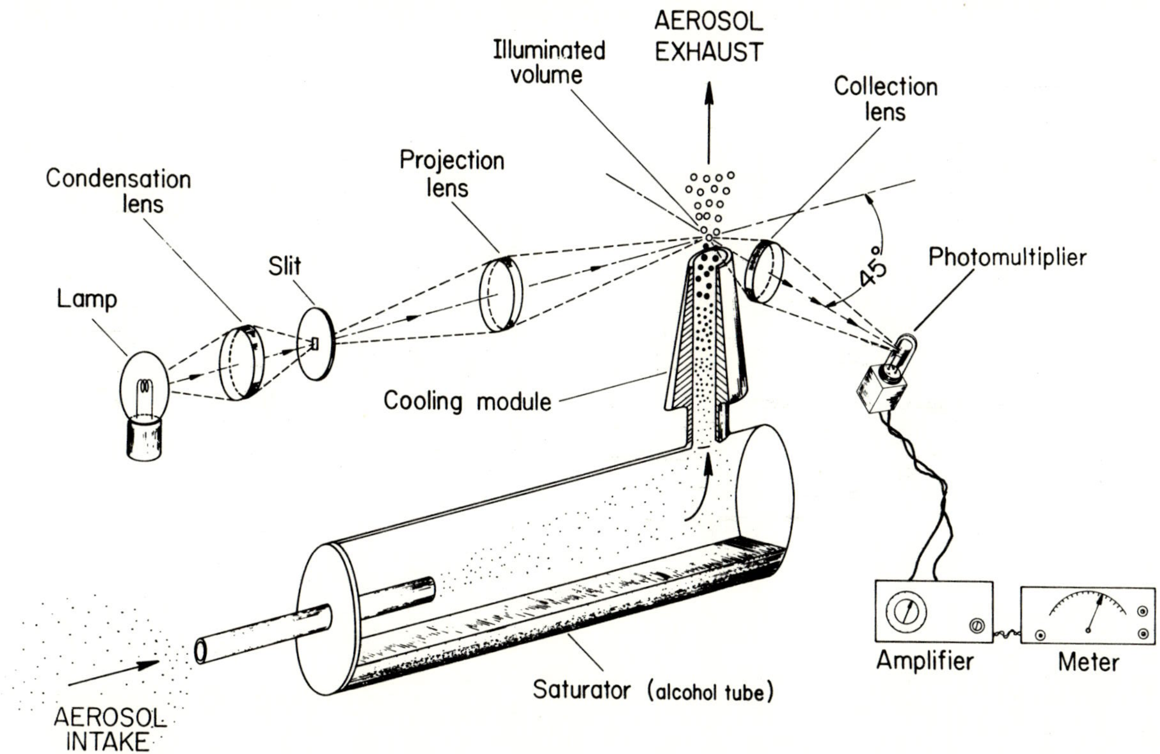

Figure 1. Schematic diagram of the Continuous Flux Condensation Nuclei Counter.

adequately reduced to be considered as evenly distributed. It is equal to 35 cm s^{-1}.

The measuring volume is defined in the following manner (FIGURE 2). A source of light S illuminates a diaphragm D_1 evenly. The latter is set against a condenser lens shown by L_2. The condenser L_2 creates the image of diaphragm D_1 at A in the measuring field of photometer whose axis is angular by 50° with that of the condenser. Lens L_3 in this photometer creates the image of D_2 at A. Behind D_2, a lens L_4 creates the image of lens L_3 on the photocathode of a photomultiplier. Under these conditions, the measuring volume has the shape of a parallelepiped defined in space by the images of diaphragms D_1 and D_2 ; it is evenly illuminated because D_1 is also illuminated evenly.

The particle carrying air current is directed at right angle to the plane of FIGURE 2. When a particle crosses the field, the light scattered within the photometer field is received by the photomultiplier. It delivers a current impulse that is proportional to the luminous flux scattered by the droplet, which can also be measured.

Knowing the number of impulses delivered during a given time interval, the straight surface of the measuring volume that is perpendicular to the air current charged with droplets and its specific velocity, we cans derive the droplet concentration.

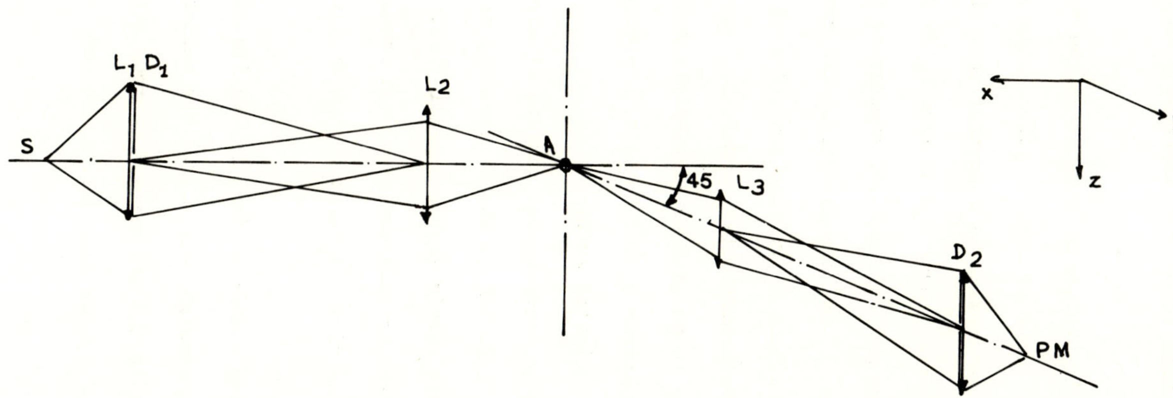

Figure 2. Schematic diagram of the optical system.

To effect an individual count of the droplets while mini-mizing the simultaneous presence of several particles in a single field, we can modify the sizes of diaphragms D_1 and D_2. Nine combinations are possible, so that, theoretically, it is possible to detect at least 1 particle per cm^3 and at most 10^6 particles per cm^3, with a probability of double presence of 5 %. The experimental checkout of such forecast is actually underway and should enable to eventually change the sizes of diaphragms in order to make sure that coincidences are of negligeable value.

By comparing the present counter findings and those of a General Electric counter, we realize that there is a linear relationship between concentrations given by both instruments when they do not exceed 10^5 particles per cm^3. Impulse amplitude enabled us to determine the size of droplets whose diameter exceeds 0.3 μ , and to check the operation of our instrument. With the photometer axis tilted by 50° over the collimating lens, this amplitude is about independent from the droplet index. Moreover, if the cooling system is shut down, the unit operates as a conventional granulometer for the study of particles with diameters over 0.3 μ.

We have designed a simplified version of this instrument to effect measurements in sity, when less accuracy is satisfactory. Diaphragm D_2 is then removed and the photomultiplier is replaced by a photo-transistor which directly gathers all the light diffused by all droplets.

II. OPERATING CHARACTERISTICS

The continuous flux counter operating characteristics have been defined as follows : an ultra fine aerosol was generated through radiolysis of airbone gaseous impurities, introduced, after filtration, into a previously dust-freed vessel under the action of α rays emitted by a source of P_o^{210} of 12 mC. The mean diameter of particles thus produced - practically monodispersed and determined by a Whitby analyser - is in the order of 0.1 μ. FIGURE 3 shows results obtained by varying the temperature difference ΔT between alcohol and the cold wall of the Peltier effect module. Of course, this difference in temperature is endowed with a qualitative significance only and relates particularly to the geometric configuration of the instrument.

Curve 1 represents the determination of homogeneous nucleation threshold and of the nucleation threshold over ions generated by a source of C_o^{60} of 10 mC placed near the counter. These thresholds were determined in the absence of any and all foreign particle in the counter. They correspond to these differences in temperature ΔT = 32° C and 30°C, respectively. The presence of ions lowers nucleation threshold, as we had expected.

Curve 2 shows the same curve obtained in the presence of a prior concentration of about 3000 particles per cm^3. We noted the

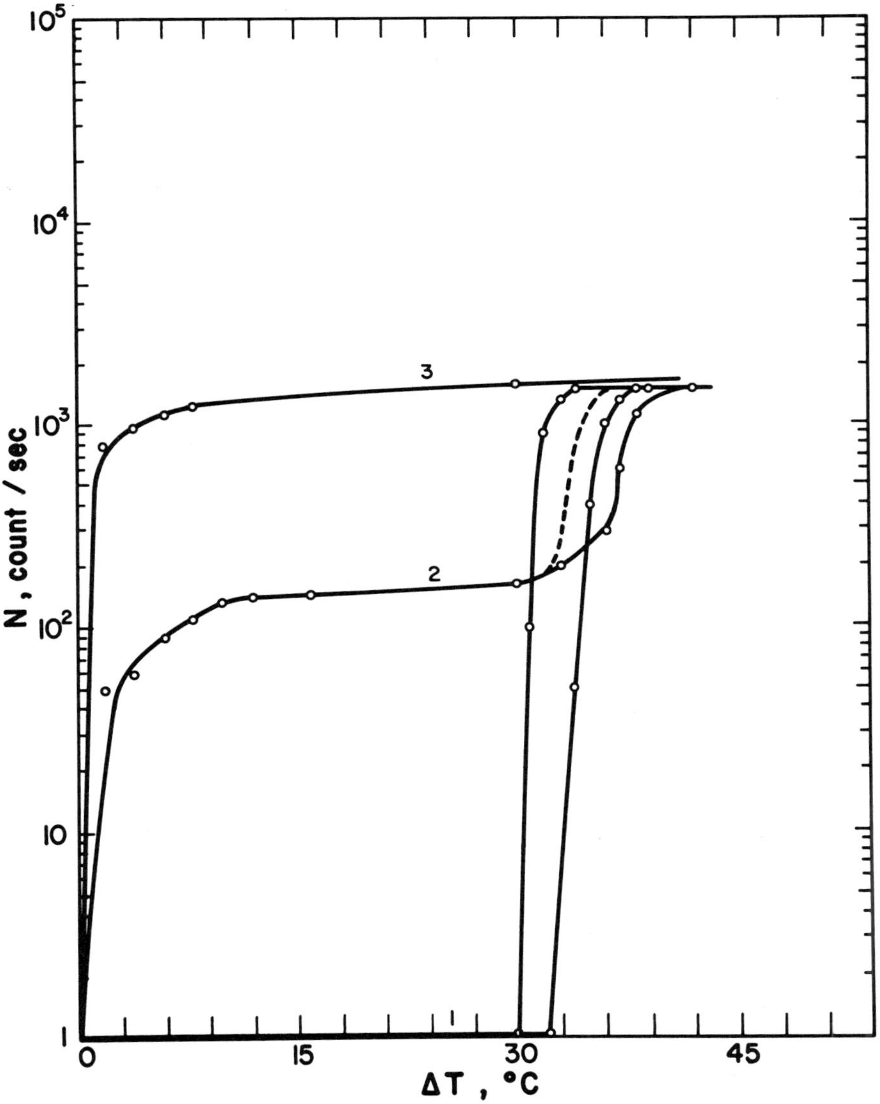

Figure 3. Nuclei concentration as a function of ΔT of counter in different conditions.

existence of a level, thereby indicating that we had detected actually all present particles, and as well a slight shift of nucleation threshold corresponding to the air cleansed of particles, or to ionized air, and this can be explained by the prior condensation of steam over present particles. We noted that homogeneous nucleation was obtained for a difference in temperature $\Delta T = 32°C$. This would correspond, were we to allow that vapor temperature was 0°C inside the module and that the air was previously saturated with steam in the saturator, to an oversaturation of 700% of ethyl alcohol. The theoretical value which corresponds to homogeneous nucleation (formation of 1 particle per cm^3 and by sec.), is of 230 %. This results would imply that either saturation was not reached in the saturator, or that condensation takes place on the walls of the cooled nozzle.

Curve 3, obtained with the same diaphragms system in the presence of a very high concentration of particles (about 10^6 per cm^3) still shows the existence of a counting top level ; however, it is not possible to place in evidence the nucleation over ions, nor homogeneous nucleation, and this can be explained by the catching of ions over particles and by vapor condensation available over them.

OPERATION OF THE LOW PRESSURE COUNTER

During our preliminary experiments which were intended to arrive at the development of a counter whose purpose was to effect condensation nuclei counts in the stratosphere, we have found that more and more important count losses were encountered in the proportion in which the carrier gas pressure decreased, in corroboration of findings already obtained by SCHLARB (3) and JUNGE (4) with relief counters.

In order to eliminate this drawback, we carried out a series of experiments using the following device (FIGURE 4) :

- a reactive vessel of 200 l. in capacity, in stainless steel, linked to a pumping system to reduce the pressure in this vessel.
- a continuous flux counter linked to the vessel by means of a 1 cm Ø hose and just a few cm in length, to reduce aerosol diffusion losses.

We worked at laboratory temperature. After emptying the reaction vessel, we filled it at atmospheric pressure p_o with air taken inside and charged with the atmospheric aerosol, hence with an aerosol generated by heating a platinum wire through the Joule effect. An experiment was also conducted by directly generating in the vessel a photolytic aerosol obtained by irradiation of air gaseous impurities.

Counting is done under such conditions. Supposing at C_o the concentration of particles. We then reduce pressure through pumping inside the vessel to the value p. Supposing at C'_p the pressure that we then measure. We see that the value C'_p is systematically

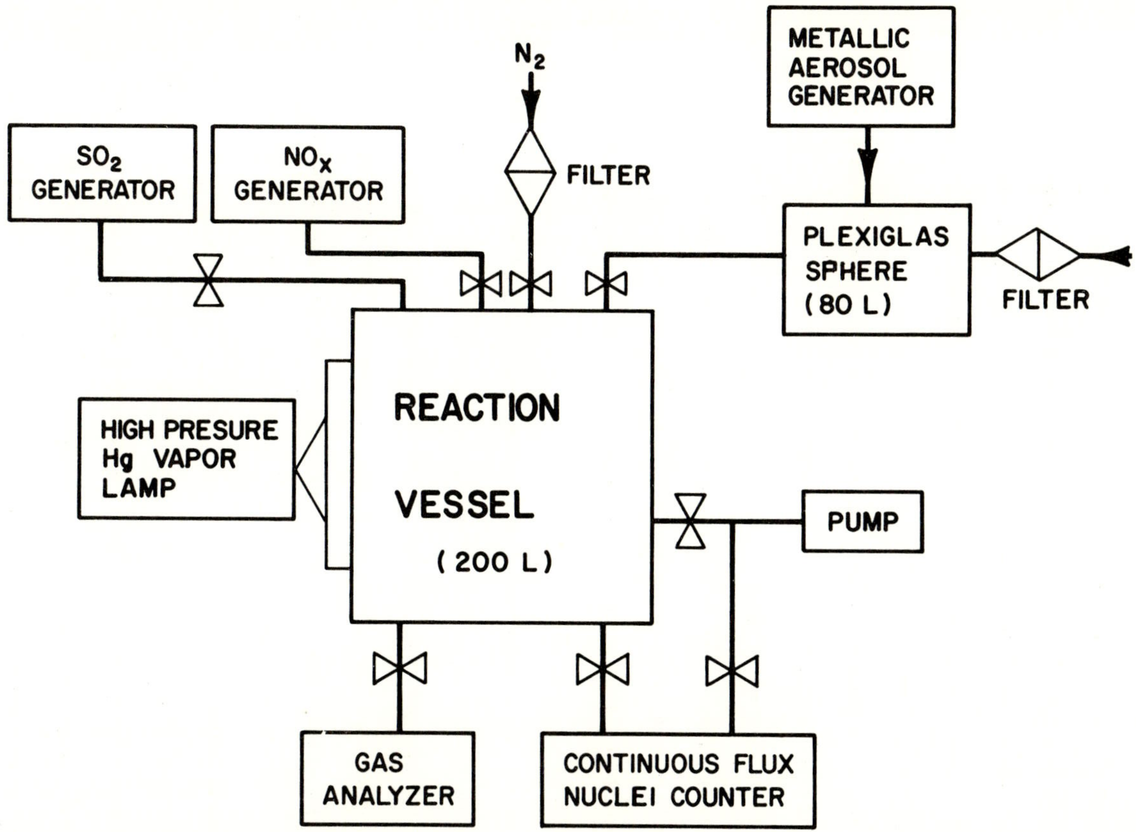

Figure 4. Schematic diagram of experimental set-up.

lower than the value $C_p = C_o \left(\frac{p}{p_o}\right)$ that we should obtain by allowing thăt aerosol particles followed the air flow during pumping. By bringing the pressure in this vassel to the value p_o by introducing filtered nitrogen, we found that the particle concentration, measured under these conditions, assumes the value C_p. Errors in counting which were discovered at low pressures cannot be attributed to evaporation, nor to losses resulting from diffusion over walls, nor again to aerosol coagulation. Therefore, we can conclude that a portion of aerosols, although existing at low pressure, is not counted under limited pressure.

FIGURE 5 show counting losses obtained with the three aerosols. They practically reach 100 % at a pressure of 60 torr. Curves 1, 2, and 3 correspond to atmospheric aerosol, platinum aerosol and photolytic aerosol, respectively. Counting loss is even more pronounced in the case of photolytic aerosol whose particles (100 Å Ø) are even smaller than those of platinum aerosol (600 Å Ø) and of atmospheric aerosol.

The measurement of signals delivered by the photomultiplier shows that their amplitude lessens with decreasing pressure, and that below a given value of it they cannot be differentiated from background noise. Thus, it would seem that at low pressure a certain number of droplets were not able to developp sufficiently and the light diffused by each became too weak to be detected.

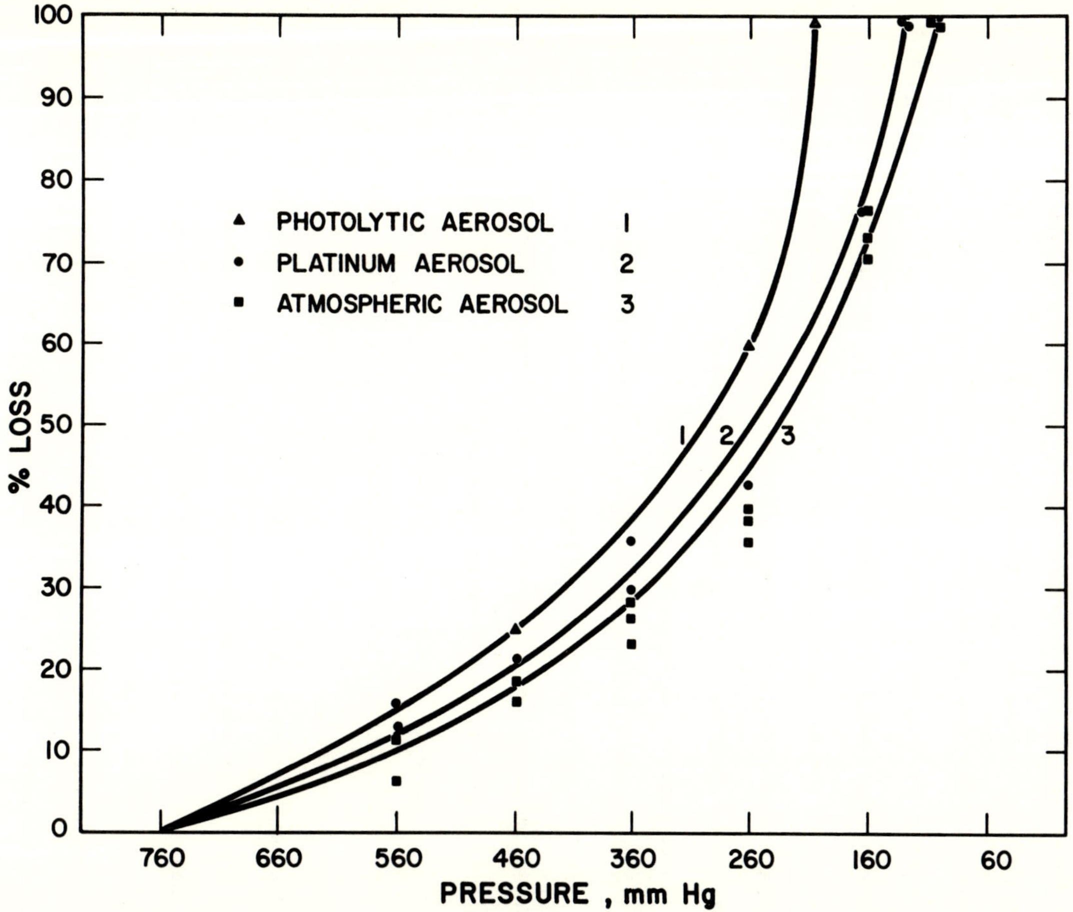

Figure 5. Loss of count as a function of the pressure.

We have compensated for the reduction in the velocity in which droplets are developing under low pressures by increasing the over-saturation in the counter by heating the alcohol in the saturator and by maintaining as constant the temperature of the cooling nozzle to a value of 0° C ± 1° C.

Thus, we are able to obtain a sufficiently high over-saturation at cooling nozzle outlet to cancel out the counting loss up to a pressure reaching about 60 torr.

A model specifically developed for the study of stratospheric particles, insensitive to ionic nucleation, has enabled us recently to carry out a primary test up to an altitude of 17 km. The analysis of measurments is presently underway and it should enable us to evaluate a concentration of at least 100 particules per cm^3 of air at that altitude.

BIBLIOGRAPHY

1) J. BRICARD, G.MADELAINE, P. REISS, P. Y. TURPIN,
C. R. Acad. of Sciences , PARIS, 275 Series B 387 (1972)

2) J. BRICARD, P. DELATTRE, G. LEBEAUPAIN & G. MADELAINE,
C. R. Acad. of Sciences, Paris, 278 Series B 191 (1974)

3) G. SCHLARB, Bioklim Beiblätter 7.86 (1940)

4) C. E. JUNGE, Journ. Meteor. 18.181 (1960).

+ + + +

ELECTRICAL MEASUREMENT OF AEROSOLS

by

Kenneth T. Whitby
Mechanical Engineering Department
University of Minnesota
Minneapolis, MN. 55455

ABSTRACT

This paper reviews the basic principles of electrical size distribution and concentration measuring methods. Electrical aerosol measuring instruments are discussed under three headings: charging, classification and detection.

Recent work by Liu and his students shows that the particle charging rate, based on the Boltzmann Law, can best describe bipolar and unipolar diffusion charging of aerosol particles, over the size range and charging conditions of most interest to electrical aerosol measurement. Working equations for calculating diffusion charge are presented and the latest experimental charging data are presented.

Several charged aerosol classifiers or mobility analyzers are described, including the latest differential mobility analyzers developed at the University of Minnesota.

Particle charge versus size curves for bipolar, diffusion and field charging are integrated with some of the latest aerosol size distribution models, in order to calculate the response characteristics of the different kinds of electrical instruments.

Three late model electrical aerosol instruments developed at the University of Minnesota are described. These are the Electrical Aerosol Analyzer, the Differential Mobility Analyzer, and an Electrical Aerosol Concentration Meter.

NOMENCLATURE

a - particle radius, cm
D_i - diffusion coefficient of ion, cm^2/sec
D_p - particle diameter, μm or cm
e - charge per electron - stat coul per charge
$f(r)$- particle size distribution function
$f_c(r)$- fraction of particles charged
$g(r)$- characteristic response function of the instrument
h - dimensionless parameter $K\left(\frac{k-1}{K\left(k+1\right)}\right)/(2akT)$
I - current, amperes or pico amperes.
J_c - charging rate in the continuum regime, number/sec
J_k - charging rate in the free molecular regime, number/sec
K - dielectric constant, dimensionless
k - Boltzmann's constant
$l(r)$- instrument sampling efficiency
m_i - mass of ion, g
m_j - mass of gas molecules, g
N - particle concentration, no/cm^3
N_i - ion concentration, no/cm^3
NT - total number concentration of particles, no/cm^3
n_p - number of unit charges per particle
n_{pf} - final particle charge, elementary charge unit
n_{pi} - initial particle charge, elementary charge unit
P - pressure, atmospheres
r_o - distance from center of particle where the image force of attraction is equal to the columbic force of repulsion
SG - geometric standard deviation
ST - total surface area concentration, $\mu m^2/cm^3$
$s(r)$- sampling efficiency of aerosol transport system
T - temperature - K°
t - time, sec.
U_i - mean velocity of ions, cm/sec
$\overline{U}_i$ - mean thermal speed of ions, cm/sec
V - voltage, volts
VT - total volume concentration, $\mu m^3/cm^3$
Z - function of charge parameter, $n_p e^2/akT$ (See Fig. 2, Gentry & Brock, 1967)
Z_i - electrical mobility of ions, cm^2/volt - sec
Z_p - particle electric mobility, cm^2/ volt - sec
α - fraction ion captured by the particle from the limiting sphere (See Table I, Fuchs, 1963)
$1/\alpha_i$- electrical aerosol analyzer sensitivity, pa/(10^6 part./cc)
β - a factor of the order of unity (see Eq. 3, Natanson, 1960)
β_i - constant in the equation $D_i = \beta_i \overline{U}_i \lambda_i$

δ - radius of limiting sphere, cm (see Eq. 5, Fuchs, 1963)
η - $=\delta/a$, dimensionless
λ - $=\pi^{1/2}(a/\lambda i)\left(\frac{m_j+m_i}{m_i}\right)$, dimensionless
λ_i - mean free path of ion, cm
λ_j - mean free path of gas molecule, cm
ρ - $=r_o/a$, dimensionless
ϕ - electrical potential, statvolt

ELECTRICAL MEASUREMENT OF AEROSOLS

by

Kenneth T. Whitby
Mechanical Engineering Department
University of Minnesota
Minneapolis, MN. 55455

INTRODUCTION

The high electric mobility of aerosol particles in an electric field makes it possible to separate and classify aerosol particles electrically. If the electric mobility is a single valued function of particle size, it is possible to use mobility classifiers as size analyzers. Also the net electrical charge carried by an aerosol can be measured by an electrometer and used as a measure of aerosol concentration.

Although these principles have been known for a long time it seems that Rohman[1] in 1923 was the first one to attempt the measurement of an aerosol mean size by electrical means.

The rapid advances in electronics during the past few decades combined with an increasing need for fine particle measurement, has resulted in a number of electrical measurement techniques and instruments being brought into general use.

It is the purpose of this paper to review the basic principles of the various electrical techniques in the light of the latest knowledge about electric aerosol charging and the size distributions of aerosols and then to discuss in more detail several of the latest techniques and instruments developed at the University of Minnesota for the classification and size distribution measurement of aerosols.

Emphasis will be on the developments of the last decade and on those techniques which have either demonstrated their usefulness or have the greatest potential for future development.

BASIC PRINCIPLES OF ELECTRIC AEROSOL MEASUREMENT

An electric aerosol measuring system can logically be broken down into three parts. These are:

* aerosol charging
* precipitation or classification by electric mobility
* aerosol detection

The following discussion will be organized in this way.

Particle Charging

There are three different ways of charging aerosols:

* electrification of particles which results from the contact and separation of a particle from a surface.
* the induced charge which results when a liquid drop separates from the bulk liquid in an electrical field.
* charging by coagulation of one charged particle with another. The charged particle may be an electron, a small ion or another charged particle.

Contact Electrification - Contact electrification is of concern to aerosol technology for two reasons. First, it produces enough unwanted and uncontrollable charge on the aerosols produced by dust generators so that all such generators should include some provisions for removing or reducing this charge. Secondly, contact electrification may also be used to charge aerosols for aerosol concentration measurement.

Because of its industrial and scientific importance there have been periodic conferences to review the state of knowledge of particle electrostatics. Harper[2] (1967) is a good review on static electricity. Static Electrification 1971, is a good recent conference review as is Moore[3] (1973).

There have been numerous instruments developed, a few of which have been developed for industrial dust measurement applications. Among these are the "Konitest", the deveopment of which is described by Prochizka[4] (1966) and the IKOR Air Quality Monitor[5] (1974).

John[6] in another paper in this symposium has reviewed the application of contact charging for particle monitoring. In most of these instruments the particle charge generated by collision with the walls of the plumbing leading up to the sensor or with special impact surfaces in the sensor, generates an electric charge which is then collected by a final impact of the particles with a probe connected to a current measuring circuit. Although the relationship between aerosol mass concentration and current may be linear and reasonably reproducible for a given material and given humidities, John[5] (1974) using the IKOR instrument found that the sensitivities varied over a range of more than 20 to 1. The insulators he tested had a mean sensitivity of 0.2 μcoul/g, semiconductors 0.62 μcoul/g and metallic conductors 2.25 μcoul/g.

The variable and low sensitivity of contact electrification has so far limited its use to the measurement of high concentrations of aerosol from industrial sources.

Guichard and Chauvelier[7] (1973) have described a novel instrument in which the aerosol is first charged by passage through a fluidized bed and then the charge measured by a second fluidized bed connected to a current measuring circuit. It is claimed that this instrument is less sensitive to variation in material properties.

Charging With Small Ions - The most important method for charging aerosol particles, prior to measurement, is by the coagulation of the particle with a charge carrier. The charge carrier may be an electron, a small ion or another charged particle. Because of the importance of this method of charging, it will be discussed in much more detail than other charging methods.

In the most common and useful approach, small ions generated by a radioactive source or by a corona discharge are allowed to charge the aerosol to be detected or collected.

Making small ions for aerosol charging is not difficult. However, making an appropriate and constant concentration and exposing the aerosol to it for the correct length of time is much more difficult. Before discussing some of the latest advances in methods of precision for aerosol measurement it is necessary to review the basic principles of particle charging by small ion clouds.

There are three basic kinds of small ion charging which may be identified, each of which results in a different magnitude of charge and a different relationship between particle electric mobility and size.

The first kind of charging exists when a particle is immersed in a bipolar mixture of small particles in the absence of an electric field. Under these conditions of bipolar diffusion charging, individual particles fluctuate in charge as the particle acquires + or - ions as a result of collisions of ions with the particle.

The second kind of charging exists when the ion mixture is unipolar, but there is no electric field. Under these conditions ions diffuse to the particle until the repelling force of the charge on the particle reduces the probability of acquiring further charge to a small value.

The third kind of charging occurs when an electric field is superimposed on the unipolar ion cloud. The additional velocity of the ions along the field lines which pass through the particle can increase the equilibrium charge for given ion concentrations and charging times above those for unipolar diffusion charging, especially for particles larger than a few tenths of a μm.

Apart from such operational problems as charging stability, uniformity of charging in different parts of the charger, etc., there are three characteristics of ion charging which affect usefulness of a given charging method for aerosol sizing by electric methods. These are:

1. The relationship between electric mobility, Z_p, and particle size, D_p. The average charge for the different kinds of charging are shown in Figure 1, and the resulting mobilities in Figure 2. Note that the Z_p vs. D_p curves are single valued for the Boltzmann bipolar diffusion and the unipolar diffusion but not for the field charging. Also, note that the slope of the unipolar diffusion curve is essentially zero above 1 μm. Thus, sizing methods using unipolar diffusion charging will have very poor resolution above about 0.6 μm at atmospheric pressure.

2. fraction charged. Particles which do not acquire a charge during their passage through a charger, cannot be influenced by subsequent electric fields and, therefore, cannot be measured electrically. Recently, Liu and Pui[13] have verified that Equation 1 is an accurate description of the fraction charged during unipolar diffusion charging.

$$f_c = 1 - \exp(- \frac{\pi}{4} D_p^2 \bar{u}_i N_i t) \qquad (1)$$

   In Figure 3 the fraction charged, f_c, calculated from Equation 1 is compared to the loss fraction for the University of Minnesota electrical aerosol analyzer. At an $Nt = 10^7$ the fraction charged decreases rapidly below 0.02 μm to a value of only 0.01 at 0.002 μm. From Figure 1 it is seen that for bipolar diffusion charging the fraction charged is only 0.003 at 0.01 μm. For any electrical instrument measuring aerosol concentration (neutral as well as charged particles), the indicated concentration in a given size range must be corrected by dividing by f_c. Since the error in the corrected concentration will be proportional to $1\Delta f_c$ when Δf_c is the uncertainty in f_c, errors in corrected concentrations become unacceptably large at values of $f_c < 0.03$. Thus, fraction charge in combination with aerosol losses, becomes the principal factor which limits the lower useful size of an instrument.

3. Discrete nature of electric charge. From Figure 1 it can be seen that the average charge goes from less than unity to more than unity at about 0.04 μm for unipolar diffusion charging and at about 0.3 μm for unipolar diffusion charging. The discreteness of the charge at these sizes limits the resolution that can be achieved. Husar[9] estimated this as corresponding to a charge $\sigma_g = 1.3$ for unipolar diffusion

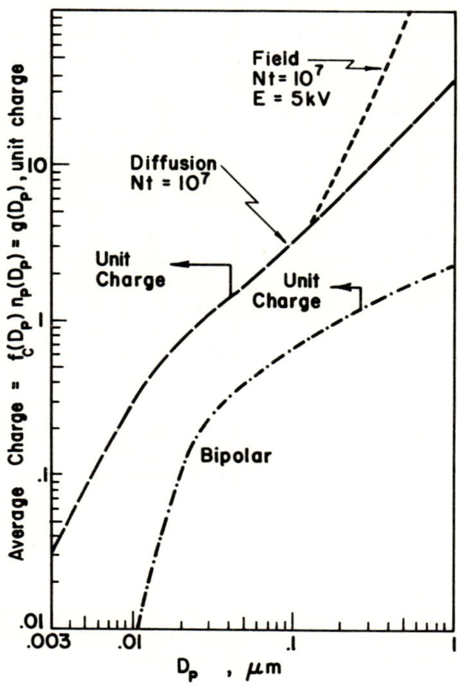

Figure 1 Average charge of particles charged in four different ways For sizes smaller than that at which the maximum charge is a unit charge, the curve for smaller sizes corresponds to the fraction charged.

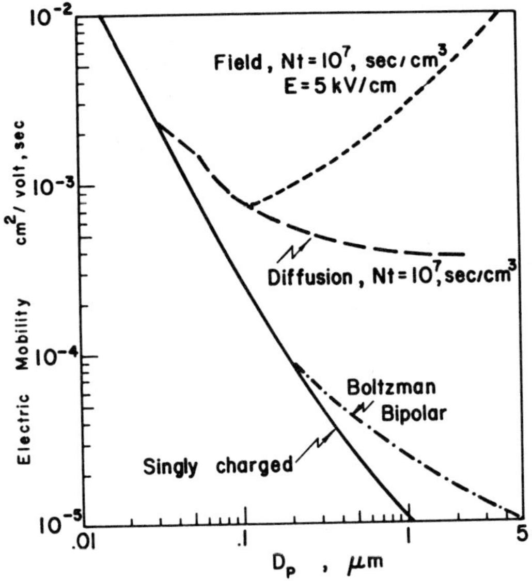

Figure 2. The electric mobility of particles charged in four different ways. Note that because of the minimum mobility at about 0.1 μm for the field charging, two sizes may have the same mobility.

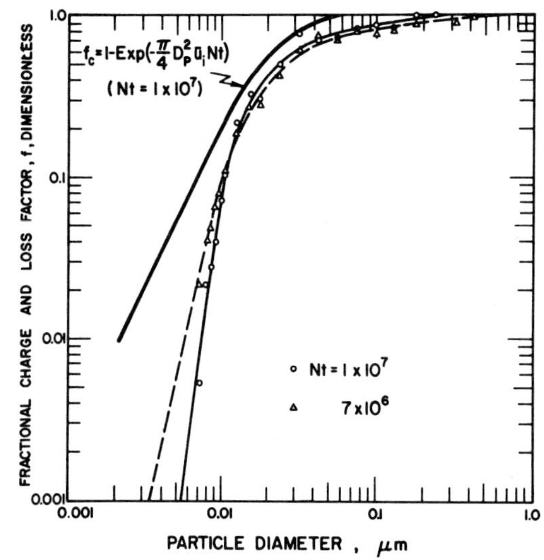

Figure 3. Comparison of the theoretical and measured charge and loss factors for the EAA. For sizes smaller than 0.01 μm space charge losses, mostly in the charger, reduce f_c to below the theoretical values. These losses effectively limit the lower sizing limit of the EAA to about 0.007 μm.

charging at 0.04 μm and similar calculations by Knutson[10] for bipolar diffusion charging give $\sigma_g = 1.3$ at 0.3 μm. Knutson[11] has developed a calculation procedure that corrects for this effect on the measured size distribution. Cantrell[12] has also developed a correction procedure for the electrical aerosol analyzer that includes a correction for discrete charge. Although these correction procedures have not been thoroughly evaluated yet they appear promising enough so that resolution of monodisperse aerosols with geometric standard deviations of less than 1.1 at the worst size may be possible. Of course if measurements are made using only the singly charged particles, then the resolution is as good as the resolution of the mobility analyzer. This requires, however, that the fraction of aerosol carrying unit charge be known accurately.

Diffusion Charging Theory

Since most existing and potential electrical methods for the measurement of aerosols utilize diffusion charging, the theory of this charging method will be examined in more detail than contact, induction or field charging.

In a recent thesis Pui[13], has examined various diffusion theories and compared them with experimental measurements. The various theories are listed in Table I along with brief comments on their range of application. Only those which have the most experimental confirmation will be discussed in detail.

Liu and Pui[14] found that the bipolar charge equilibrium reached by aerosol particles immersed in a bipolar ion atmosphere is best described by Boltzmann's Law, Equation 9, Table I, over the size range from 0.02 to 1 μm. for which they did experiments. Liu and Pui[15] also found that the $N_i t$ product required to reduce a highly charged aerosol particle to the Boltzmann equilibrium charge to be best described by Equations 8a and 8b, Table I. For a 1 μm particle $N_i t$ values on the order of 10^6 cm^3/sec are required to neutralize it with small ions.

For unipolar diffusion charging Pui[13] found Equation 10, Table I, agreed best with his experimental data over the size range of his experiments, 0.0072 to 5.04 μm. This modification of the White Equation, equation 1, Table I, contains the Liu-Bademosi[16] interpolation formula between the free molecular and continuum region. Why this relatively simple formula which contains no correction for image force should agree with the experimental data so much better than some of the more inclusive theories, is not known at this time.

Table I Summary of Diffusion Charging Theories[13]

	Theory		Comments and Range of Application
1.	White (1951)[18] confirmed by: Liu, Whitby & Yu (1967)[17] Gentry & Brock (1967)[18]	$n_p = \frac{akT}{e^2} \ln \left(1 + \frac{\pi a \bar{U}_i e^2 N_i t}{kT}\right)$	Applicable in free molecular regime. Without image force consideration.
2.	Keefe, Nolan & Rich (1959)[18], Keefe & Nolan[19] (1961), Keefe, Nolan & Scott[20] (1968)	a. Same as White (1951) b. $\frac{dn_p}{dt} = \pi a^2 \bar{U}_i N_i \left(1 + \frac{n_p e^2}{akT}\right)$ c. $n_p = \pi a^2 \bar{U}_i N_i \, t \left(1 + \sqrt{\frac{\pi e^2}{2akT}}\right)$ d. See numerical results in Figures 3 and 4 (Ref, 20)	Applicable in free molecular regime a. without image. Ion and particle have same sign of charge. b. without image. Ion and particle have opposite sign of charge. c. with image. For collision of uncharged particle with ion of either sign. d. with image. For initial particle charge $N_p > 0$.
3.	Natanson[21] (1960) confirmed by: Brock[22] (1970), Brock & Wu[23] (1970)	a. $\frac{dn_p}{dt} = \pi a^2 \rho^2 \bar{U}_i N_i \exp[-\phi_{r=a}/kT]$ b. $\frac{dn_p}{dt} = \frac{\pi}{4} a^2 \bar{U}_i N_i \left(\frac{\pi e^2}{2kTa}\right)^{1/2} \left[1 + 1.11 \left(\frac{a}{\lambda}\right)^{3/4} \left(\frac{e^2}{kT\lambda}\right)^{1/4}\right]$ c. $\frac{dn_p}{dt} = \frac{4\pi\rho^2 \bar{U}_i N_i \exp(2n_p / 1 + \frac{\rho}{\beta\lambda})}{1 + \frac{\rho^2 \bar{U}_i}{2Dn_p \beta\lambda} \left[\exp \frac{2n_p}{1 + \frac{\rho}{\beta\lambda}} - 1\right]}$	Applicable in free molecular regime. With image force consideration a. capture of a charge particle by similarly charged ion. b. capture of an ion by a neutral particle. c. capture of a charged particle by an opposite charge ion.
4.	Fuchs[24] (1947) Also, Pluvinage (1945[25], 1946[26]) Bricard[27] (1948)	a. $\frac{dn_p}{dt} = \frac{4\pi D_i a N_i}{\int_1^\infty \left(\frac{1}{x^2}\right) \exp\left(\frac{e^2}{akT}\left[\frac{n_p}{x} - \frac{K-1}{K+1} \frac{1}{2x^2(x^2-1)}\right]\right) dx}$ b. $\frac{dn_p}{dt} = \frac{4\pi n_p Z_i N_i e}{\exp(n_p e^2/akT) - 1}$	Applicable in continuum regime a. with image force consideration. b. without image force consideration.

5. Fuchs[28] (1963)	a. $\frac{dn_p}{dt} = \frac{\pi\alpha\delta^2\overline{U}_i N_i \exp[-e\phi(\delta)/kT]}{1 + \frac{\exp[-e\phi(\delta)kT]\alpha\overline{U}_i a\eta^2}{4D_i}\int_a \frac{1}{\eta}\exp[e\phi/kT]\,dx}$	Applicable in transition regime a. with image force consideration
	b. $\frac{dn_p}{dt} = \frac{N_i \pi a^2 \overline{U}_i \eta^2 \alpha \exp(-n_p e^2/akT\eta)}{1 + \frac{a}{4}\frac{\overline{U}_i \alpha\eta^2 akT}{D_i n_p e^2}\exp\left(-\frac{n_p e^2}{akT\eta}\right)\left[\exp\left(\frac{n_p e^2}{akT\eta}\right)-1\right]}$	b. without image force consideration
6. Gentry and Brock[18] (1967) Marlow & Brock[19] (1974)	a. $\frac{dn_p}{dt} = \frac{\pi a^2 \overline{U}_i N_i \exp(-n_p e^2/akT)}{1 + 0.217\left(\frac{m_j}{m_i}\right)^{1/2}\left(\frac{D_j}{D_i}\right)\frac{\overline{a}}{\lambda_j} Z}$	Applicable in transition regime a. without image force consideration
	b. $\frac{dn_p}{dt} = \frac{\pi a^2 \overline{U}_i N_i\,[1 + (\pi h)^{1/2}]}{1 + [\lambda G_{in}/1 + (\pi h)^{1/2}]}$	b. with image force consideration
7. Liu & Bademosi[30] (1971)	$\frac{dn_p}{dt} = J_c\left[1 + \frac{Jc}{Jk-1}\cdot\frac{1}{\left\{1 - \frac{0.13}{\beta_i}\left(\frac{a}{\lambda i}\right)\left(\frac{Jc}{Jk}\right)Z\right\} + 6.04\beta_i\frac{Jk}{Jc}}\right]^{-1}$	Applicable in transition regime J_c and J_k are the charging rate for theory 4 and 1, respectively.
8. Liu and Pui[14] (1974)	a. $N_i t = \frac{1}{4\pi Z_i e}\ \ln\frac{n_{pi}}{n_{pf}}$ b. $N_i t = \frac{kT}{2\pi a\overline{U}_i e^2}\left[\log\frac{\exp(n_{pi}e^2/akT)-1}{\exp(n_{pi}e^2/akT)+1} - \log\frac{\exp(n_{pf}e^2/akT)-1}{\exp(n_{pf}e^2/akT)+1}\right]$	Reduction of high initial charge N_{pi} to low final charge N_{pf} under bipolar charging a. applicable in continuum regime b. applicable in free molecular regime
9. Boltzman's Law: Keefe, Nolan & Rich[18] (1959). Also, Liu and Pui[15] (1974).	$\frac{N_{n_p}}{N_o} = \exp\left(-\frac{n_p^2 e^2}{2akT}\right)$	Stationary ratio obtained with bipolar charging. Applicable over entire size range.
10. Pui, D.Y.H.[13] (1975)	$\frac{dn_p}{dt} = \left(\frac{1}{\frac{1}{\pi a^2 \overline{U}_i} + \frac{1}{4\pi D_{ia}}}\right) N_i \exp\left(-\frac{n_p e^2}{2akT}\right)$	Applicable in entire size range. Agrees with all experimental data, including low pressure data.

The average number of charges per particle, n_p, calculated from Equation 10, Table I, is shown in Figure 4 for three $N_i t$ values in the range of practical interest. The average number of charges for values of n_p less than 1 have been calculated from the fraction charged equation. It can be seen that $N_i t$ values on the order of 10^7 cm^{-1} or higher should be used in order to obtain a sufficient fraction charged for sizes smaller than 0.02 μm and to obtain sufficient charge for the larger particles.

Pui[13] also found that the ions from the unipolar charger used appeared to have an electric mobility of about 1.4 cm^2 $volt^{-1}$ sec^{-1} corresponding to a molecular weight of 109. amu. These values have been used in the calculations for Figure 4.

Recently, Pouprix[31] (1973) has developed an arrangement in which the ions produced by alpha particles near their maximum range in air, can be used to produce a unipolar diffusion charge that is nearly identical to that for an $N_i t = 10^7$ as shown in Figure 1.

If aerosol particles immersed in a unipolar ion cloud are also subject to an electric field, particles larger than about 0.1 μm acquire additional charge above that acquired purely by diffusion. The magnitude of the charge acquired by field charging is a function of the dielectric constant of the particles and the electric field as well as the variables significant in diffusion charging.

The electric mobility of aerosols charged by bipolar, diffusion and field charging for a particular set of conditions is shown in Figure 2. For particles smaller than a 0.1 μm, the electric mobility is essentially independent of the method of charging because the particles either have a unit charge or no charge. For particles smaller than about 0.1 μm the principle charging mechanism is diffusion and an external electric field makes little difference. For particles larger than a few μm, the charge in an electric field is essentially proportional to D_p^2 with the result that the electric mobility, Z_p, is proportional to D_p.

Mobility Analyzers and Precipitators

Electrical aerosol concentration measurement requires some method of detecting the charged aerosol. In addition, electrical sizing methods require a precipitator or classifier by which the particles of different electric mobilities are separated before detection. The various approaches are shown in Figures 5 to 10. The characteristics and advantages and disadvantages of each are discussed below.

Condenser Ion Counter - Figure 5 - The condenser of this type counter

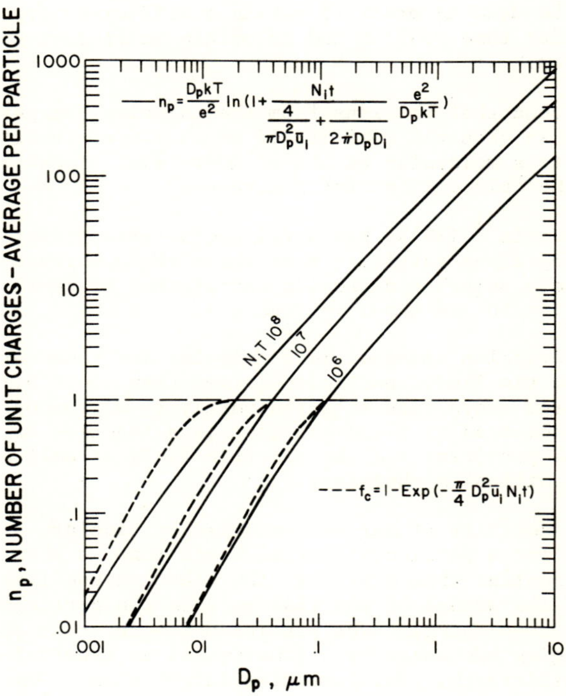

Figure 4. Number of unit charges calculated from Equation 10, Table I for diffusion charging at three different $N_i t$ values. The fraction charged f_c calculated from Equation 1 is also shown for comparison.

may consist either of parallel plates or of concentric cylinders. The current collected from one of the electrodes is measured as a function of applied potential difference by means of a pico ammeter. Although this instrument is deceptively simple, to obtain ΔI and hence the number of particles falling within a given mobility range requires that the current versus voltage curve be differentiated twice. This seriously limits its accuracy for size distribution measurement purposes. It has mostly been used for measuring small and intermediate ions of high mobility because having the pico ammeter connected to one electrode of the precipitator also limits the maximum voltage that can be applied to about 700 volts and hence limits the lowest Z_p that can be measured.

Israel[32] (1969) has made a comprehensive review of the design, application and application of the condenser ion counter. Mohnen[33] (1966) used a variation of the cylindrical condenser that is similar to that shown in Figure 9 in that the aerosol is introduced as a layer around a sheath of clean air. Measurement was from the collected deposit on the center electrode.

Parallel Plate Sheath Air Precipitator - Figure 6 - By introducing the charged aerosol as a thin sheet between layers of particle free air differential deposits may be obtained. Instruments of this type have been developed by MeGaw and Wells[34] (1964) and in a miniature form by Yeh, Raabe and Wood[35] (1973) and used for precision measurement of aerosol charge distribution. Its greatest disadvantage is the great amount of labor necessary to evaluate the deposits on the plate in order to obtain good size distribution data.

Denuder - Figure 7 - The Denuder developed by Rich et al.[36] (1959), uses a condensation nuclei counter as the detector at the end of a precipitator to measure the fraction of the aerosol charged. If the fraction charged is related in a unique manner to the particle size as it is for a Boltzmann equilibrium bipolar charge distribution, then the instrument can be used as a particle size distribution analyzer. However, Figure 7 shows that this technique is limited to the size range from about .02 to 1 μm. If the aerosol to be measured has a large number of particles smaller than .02 μm, then the change in nuclei count as the voltage is varied is too small to provide useful estimates of aerosol size distribution. This is often the case near sources of aerosol from relatively clean combustion. Nevertheless, this is a useful technique which because of its simplicity, has been used in atmospheric field studies, Fluger et al.[37] (1973).

Ion Capture Aerosol Concentration Measurement - Figure 8 - If essentially neutral aerosol particles are passed through a corona charger as shown in Figure 8, the aerosol particles will be charged by the flow of ions from the center to the outer electrode. The capture of the ions by the

Figure 5. Mobility Analyzers and Precipitators.

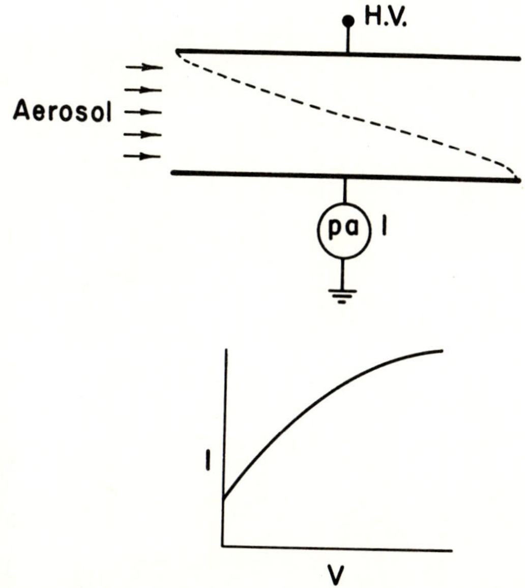

Figure 5. Condenser type mobility analyzer.

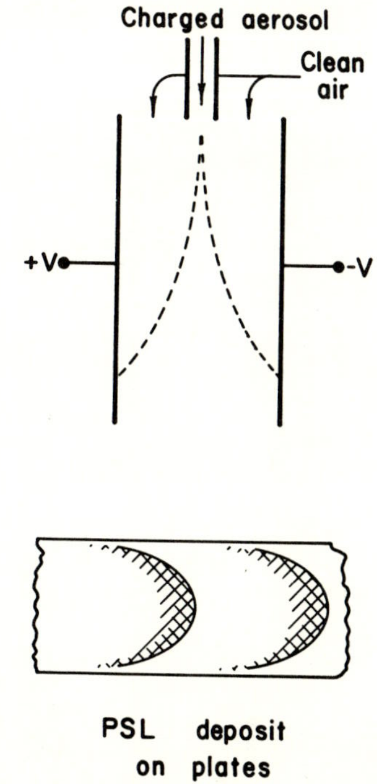

Figure 6. Parallel plate sheath air precipitator.

aerosol particles reduces the effective electrical mobility of the charges so that they are carried out of the device instead of reaching the outer electrode. This reduction of current is a measure of aerosol concentration. The exact relationship between I and the aerosol concentration depends on the method of charging, and the aerosol size distribution. The relationship between I and aerosol concentration (called sensitivity), will be discussed and illustrated later for several typical size distributions.

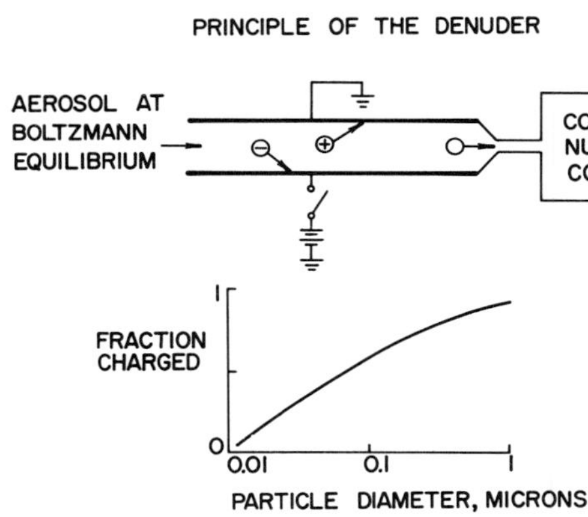

Figure 7. Denuder .

a suitable filter charge collector. Then I becomes an increasing function of the concentration, Mohnen and Holtz[38] (1968).

Whitby Mobility Analyzer - Figure 9- To overcome the limitations of the previously described types of mobility analyzers, Whitby et al.[39] (1966) developed the mobility analyzer shown in Figure 9. To eliminate the inevitable end effects of the parallel plate designs, axial symmetry is maintained. To obtain good resolution with a reasonably high aerosol to clean air ratios, the charged aerosol is introduced next to the outer wall. Charge particle collection for current measurement is accomplished by a separate carefully shielded and isolated filter connected to the picoammeter. Separation of the precipitator and current collection functions permits the use of precipitator voltages near breakdown and, therefore, the collection of particles with electrical mobilities less than 10^{-4} cm/sec per volt/cm. With good shielding, the precipitator voltage may be varied without shorting of the electrometer input. This permits completely automatic operation as in the latest commercial instruments, Liu, Whitby, and Pui[40] (1974).

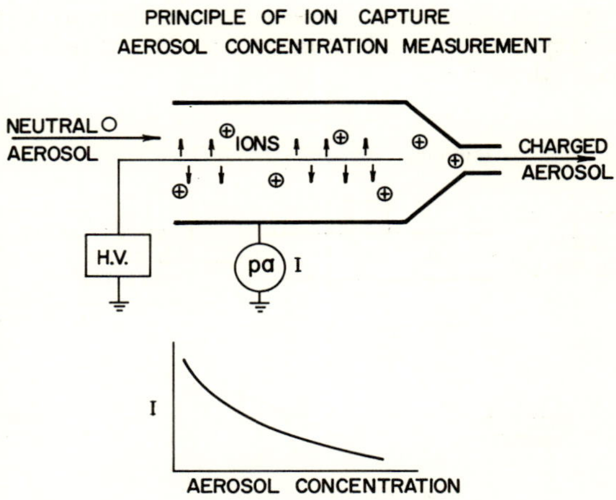

Figure 8. Ion capture detector of the type in which the current is reduced by the presence of particles.

With this appraoch, obtaining the current increment in a given particle size range requires differentiating the I vs. V curve once. This will be discussed later.

Differential Mobility Analyzer - Figure 10 - For high accuracy research on particle charge distribution Knudsen[10] (1971) and Knudsen and Whitby[41] (1975) developed a precision mobility analyzer that retains all of the desirable features of the Whitby mobility analyzer, but in addition is capable of classifying out and delivering a suspended, charged aerosol having less than 2% spread in electric mobility to either a current collector or to an external device. Liu and Pui[15] (1974 have further developed this analyzer into a precision monodisperse aerosol generator and have applied it to the calibration of condensation nuclei counters and other instruments. Its features and application will be discussed in more detail later.

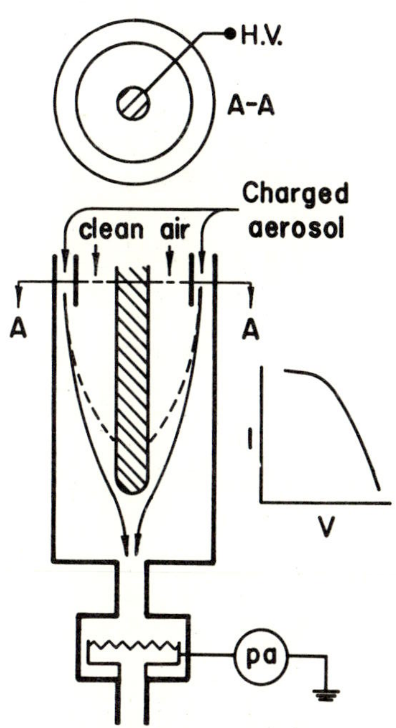

Figure 9. Whitby mobility analyzer arranged to measure charge carried by particles not collected by the precipitator.

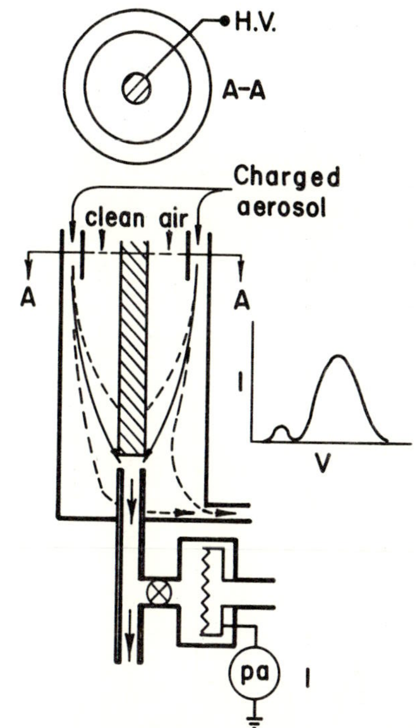

Figure 10. Differential mobility analyzer to separate particles having a narrow range of electric mobilities and deliver them as an aerosol.

Time of Flight Ion Mobility Analyzers - For the measurement of the electrical mobility of ions and particles having mobilities greater than about 0.1 cm^2/volt, sec, time of flight mobility analyzers have been developed which have two pairs of grids spaced a few cm apart. With the proper application of combined AC and DC fields to the grids, the ion mobility spectra can be measured with high accuracy and resolution. Typical of the instruments of this type used for atmospheric ion mobility analysis is that of Huertas et al[42] (1970).

Electrical Aerosol Instrument Response - The response of an aerosol measuring system is rather complicated. The response is generally a function of the characteristics of the instruments, the characteristics of the aerosol transport system and the characteristics of the aerosol. For an electrical aerosol measuring instrument the current contribution dI to the radius range, dr, is given by

$$dI = f(r)\ g(r)\ l(r)\ s(r)\ dr \tag{2}$$

where $f_c(r)$ is the current contribution of the total instrument output due to particles in the radius range dr, f(r) is the particle size distribution function, g(r) is a characteristic response function of the instrument l(r) is the instrument sampling efficiency (ratio of aerosol concentration in the instrument sensing volume to the true aerosol concentration at the inlet to the instrument), and s(r) is the sampling efficiency of the aerosol transport system bringing the aerosol from the sampling point to the instrument inlet. For electrical aerosol analyzers the instrument response function g(r) may be further expressed as:

$$g(r) = f_c(r)\ n_p(r) \tag{3}$$

where $f_c(r)$ is the fraction of particles charged and $n_p(r)$ is the number of unit charges per particle. The fraction charged $f_c(r)$ is given by Equation 1 and $D_p(r)$ depends on the method of charging, e.g. Figure 1.

The sampling efficiency of the aerosol transport system up to the inlet of the instrument determines s(r). Because instruments intended for atmospheric aerosol measurement respond primarily to particles smaller than 1 μm radius, s(r) may be assumed equal to 1 except for the particles smaller than 0.007 μm, where diffusion might be significant if very long sampling lines are used.

Instrument Response for Several Size Distributions - To illustrate the response characteristics of instruments which utilize bipolar diffusion or field charging, the average charge in Figure 1 has been integrated with four model size distributions who's parameters are tabulated in

Table II

Model size distributions used to calculate the response of electrical aerosol instruments

Model Aerosol	DGN	SGN	NT	ST	VT	DGN	SGN	NT	ST	VT	NT	ST	VT
1- Nuclei[a]	.017	1.74	1(6)	1677	10.2						1(6)	1677	10.2
2- Accumulation[b]						.096	2	1.3(4)	1000	53	1.3(4)	1000	53
3- Bimodal[c]	.017	1.74	2(5)	335	2	.133	1.75	3.3(4)	3500	171	2.3(5)	3835	173
4- Power law[d]											2.2(4)	2268	178

a) DGN and SGN choosen equal to average atmospheric values, Whitby[43] NT is in the range that will produce only a nuclei mode from relatively clean combustion.

b) DGN and SGN choosen approximately equal to average urban aerosol Whitby[43] NT is in the range of average urban values.

c) Distribution parameters obtained by fitting the Acetone smoke distribution shown in Figure 6.

d) Power law distribution in the size range 0.1 to 10 μm, $dN/dD_p = 0.4\ VT\ D_p^{-4}$. VT choosen to be about equal to VT for the bimodal distribution.

Table II. Whitby[43,44] has recently shown that atmospheric aerosol size distributions may be modeled by adding three log normal distributions, two of the modes being in the submicron size range. It has also been shown that the smallest of these submicron modes (names the nuclei mode) originates from the primary particles produced in combustion. The second submicron mode (named the accumulation mode) results from coagulation and condensation.

The submicron size distribution of combustion aerosols also show the same modes. This is illustrated in Figure 11, where the surface distributions of several combustion aerosols is shown. It has also been found that these combustion size distributions can be modeled well by two log normal distributions.

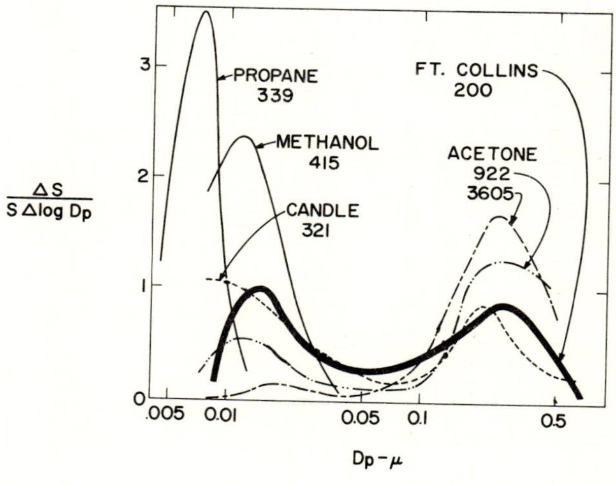

Figure 11. Normalized submicron surface area size distribution of several combustion aerosols produced by diffusion flames compared with an atmospheric aerosol size distribution observed in a relatively clean site in Ft. Collins Colo. The numbers are the surface area in $\mu m^2/cm^3$ for the aerosol as measured. Note that the nuclei modes and the accumulation modes for the combustion aerosol correspond to those observed for the atmospheric aerosol.

The nuclei mode distribution in Table II represents the kind of distribution that would probably result from relatively clean combustion aerosol that is diluted rather rapidly. It is not visible because it contains practically no particles in the light scattering range.

The accumulation mode distribution is representative of a well aged aerosol that is fairly dilute. It is highly visible because its surface mode corresponds well with the peak of the light scattering extinction curve.

The bimodal distribution was obtained by fitting an actual acetone smoke distribution from a diffusion flame that had undergone moderately rapid dilution. Because it is fresh, the nuclei mode is still apparent, but hetrogeneous coagulation of the nuclei mode with the accumulation mode has transferred most of the aerosol mass to the accumulation mode.

The power law model distribution has been included because it is a fair fit to the number distribution over the size range from about 0.1 to 10 μm.

The resulting sensitivities expressed as the pico amps. current that would flow if the charged aerosol was collected at 1 cm^3/sec at 10^6 particles/cm^3, 1000 $\mu m^2/cm^3$ and 100 $\mu m^3/cm^3$ respectively are given in Table III. The last two columns give average sensitivities and the % standard deviation of the sensitivities for the four distributions. A number of conclusions may be made from this table.

1. An electrical concentration instrument using diffusion or field charging would be from 7 to 8 times as sensitive as one using bipolar charging.

2. The variability of sensitivity with variations in size distribution is about the same for all three methods of charging, about 140% for number, 90% for surface area and 180% for volume.

Real smoke distributions would be the closest to the bimodal distribution so that the sensitivity figures for that distribution would be the best choice for predicting actual instrument performance.

The number distribution, dN/dlog Dp, and the number distribution multiplied by g(Dp) from Figure 1, is shown in Figure 12. It is seen that all of the charging methods weight the accumulation mode most heavily. This is especially true for diffusion and field charging. This would be desirable for a fire detector because the nuclei mode results mostly from clean efficient combustion whereas a heavy concentration of smoke would result in a large accumulation mode.

Table III

Current sensitivities for Bipolar, diffusion and Field charging for three model size distribution.

Sensitivity	1-Nuclei	2-Accumulation	3-Bimodal	4-Power law	Mean Sensitivity	% Std. Dev.
	MODEL DISTRIBTUION					
Bipolar						
pa/(10^6 particles/cm^3)[b]	.018	.114	.04	.57[a]	.19	140
pa/(10^3 $\mu m^2/cm^3$)	.011	.0015	.002	.005	.005	90
pa/(10^3 $\mu m^3/cm^3$)	.176	.003	.005	.007	.048	179
Diffusion (Not = 10^7 ion,sec/cm^3)						
pa/(10^6 particles)	.161	.68	.23	3.53[a]	1.15	139
pa/(10^3 $\mu m^2/cm^3$)	.078	.009	.013	.034	.034	94
pa/(10^2 $\mu m^3/cm^3$)	1.28	.017	.029	.043	.34	183
Field (Not = 10^7, E = 5 x 10^3 Volts/cm)						
pa/(10^6 particles)	.161	1.11	.316	5.7[a]	1.82	144
pa/(10^3 $\mu m^2/cm^3$)	.078	.014	.019	.05	.04	74
pa/(10^2 $\mu m^2/cm^3$)	1.28	.026	.043	.07	.36	174

a) These values are abnormally high because NT was set equal to the number of particles in the size range 0.1 - 10 μm for which the power law has been shown to fit real distribution, Whitby and Clark (1966).

b) currents in Pico amps/cm^3/sec.

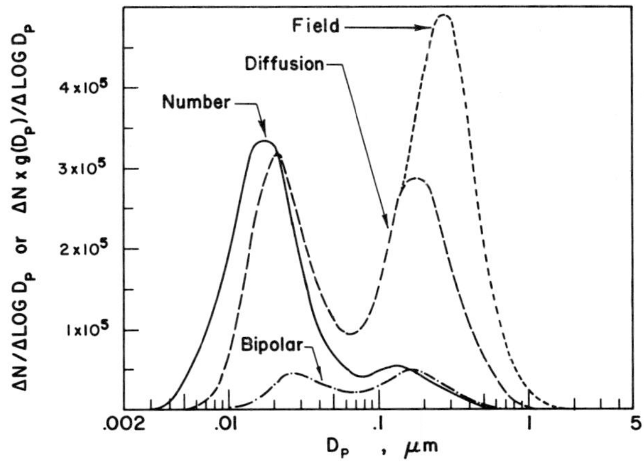

Figure 12. Typical bimodal smoke number - size distribution integrated with the average charge from Figure 1. Note that all three methods of charging weight the accumulation mode (largest size mode) the most heavily.

ELECTRICAL SIZE DISTRIBUTION MEASUREMENT AND CLASSIFICATION INSTRUMENTS

In the previous sections the basic principles of electrical size distribution and classification measurement instruments have been presented. In the sections below, instruments for size distribution and classification measurement developed at the University of Minnesota will be described and the performance of latest instruments discussed.

Differential Mobility Analyzer and Aerosol Generator - The principle of the differential mobility analyzer first constructed by Knudsen[10,41] and later developed into a high accuracy monodisperse aerosol generator by Liu and Pui[15], is shown in Figure 10. Details of its internal construction are shown in Figure 13.

From a careful theoretical and experimental study, Knudsen[10] was able to show that the electric mobility of the particles drawn through the slot could be calculated with an absolute accuracy of better than 2% and that the mobility spread of the aerosol was also less than 2%.

Thus, if the aerosol being classified consisted of spheres (oil drops for example), and if only singly charged particles were separated, then monodisperse aerosols of almost any material that could be prepared as an aerosol in the proper size range, could be prepared with size spreads of less than 2%. Furthermore, if the aerosol has a unit charge, measurement of the current carried by the aerosol is then a measure of the absolute concentration. Thus, this technique can be used as an absolute standard for both particle size and concentration.

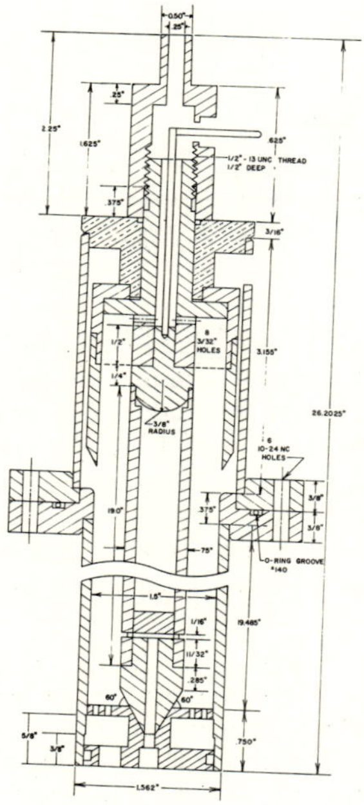

Figure 13. Internal details of differential mobility analyzer.

When used as an aerosol size and concentration standard the system consists of the following components:

1. Aerosol Generator. Among those used are the Collison atomizer aerosol generator [Liu and Pui[15]], condensation aerosol generator producing DOP aerosols, [Liu and Pui[45]], fluidized fluidized bed coal dust generator, Willeke, Lo, and Whitby[46]

Liu, Marple, Whitby, and Barsic[47] and Nichrome wire aerosols Liu, Pui, Hogan and Rich[48]. To be useful the generator must produce a large number of particles in the desired size range.

2. Aerosol Charger. The method of charging should produce the maximum number of singly charged particles at the size which it is desired to separate. For most purposes a Krypton 85 neutralizer containing about 5 mc and producing a Boltzmann bipolar charge distribution has been used. Occasionally the natural charge produced by the aerosol generator is satisfactory and under some conditions diffusion charging may be used. For larger size particles from a few tenths up to above a micron it is often impractical to eliminate the multiply charged particles. If bipolar charging is used it is possible to make corrections to the concentration Liu and Pui[15].

3. Mobility Classifier. Details of the mobility classifier are shown in Figure 13. It has been found that near perfect internal axial symmetry is essential to good classification. The concentricity between the two electrodes should be kept below 0.002 in. and the surfaces must be smooth and the edges free from burrs.

4. The standard operating condition of the mobility classifier is 2 ℓ/min of aerosol. Since the output is very stable, it is possible to measure the concentration by switching the charged aerosol stream to a filter collector connected to a sensitive and well calibrated electrometer.

5. The classifier is capable of generating monodisperse aerosols in the size range from about 0.01 to 0.5 μm depending on the method of charging. The output concentration depends on the number concentration in the feed aerosol in the size range of the classification. To obtain high output concentrations it is necessary to feed an aerosol that is initially fairly monodisperse with a mode near the classification size. The highest output concentrations obtained are about 6 x 10^5 per cm^3 obtained using a collision atomizer. Higher concentrations could probably be obtained with an optimized condensation generator.

Electrical Aerosol Analyzer - The electrical aerosol analyzer is a size distribution measuring instrument with in situ measurement capabilities over the 0.007 to 1 μm diameter range. The operating principle of the device is that of electrical charging and mobility analysis, a principle first described by Whitby and Clark[39]. Recent advances (Liu, Whitby, and Pui[40] and in Liu and Pui[45] in charger and mobility design and the use of all solid-state electronics have resulted in an improved instrument that is portable (about 30 Kg in weight) and considerably more versatile. Following is a brief description of this more recent device.

Figure 14. is a schematic diagram of the most recent commercial instrument, Thermo-Systems Inc., 2500 Cleveland Avenue North, St. Paul, MN 55113, showing its major components: the aerosol charger, the mobility analyzer, the current sensor, and associated electronic and flow controls. The instrument samples aerosols at the rate of 4 liters per minute with an additional 45 liters per minute of clean air needed to operate the mobility analyzer.

In the instrument the aerosol is first sampled into the charger where the particles are exposed to unipolar positive ions and become electrically charged. The most recent instrument differs from the one described by Liu, Whitby, and Pui[40] in having the charger located in the upper end of the same tube containing the mobility analyzer. Besides its mechanical advantages, close proximity of the charger and mobility analyzer reduces aerosol space charge losses.

The diffusion charging curve in Figure 2 shows the relationship between the electrical mobility and the size of the particles under different charging conditions. It is observed that there is montonic functional relationship between particle mobility and size. Using this relationship, the size distribution of the aerosol can be calculated from the mobility distribution measured by the mobility analyzer.

The mobility analyzer shown in Figure 14 is in the form of a cylindrical condenser with clean air and aerosol flowing down the tube in a laminar stream. The charged aerosol are deflected through the clean air core by the voltage applied on the center electrode. For a given voltage on the center rod, particles above a certain critical mobility and size are precipitated, while those with lower mobility and a larger particle size escape precipitation and are sensed by the electrometer current sensor. By changing the voltage on the center rod and measuring the corresponding electrometer current, the mobility and size distribution of the aerosol can be determined. The standard operating condition of the instrument provides a total of 11 voltage steps dividing the size range of the instrument into 10 equal geometrical intervals of four intervals per decade in particle size. The size interval boundaries are located at 0.0032, 0.0056, 0.01, 0.0178, 0.032, 0.056, 0.1, 0.178, 0.32, 0.56, and 1.0 μm. The complete voltage sequence can be scanned in about two minutes, thus allowing a complete size distribution analysis to be made in the same time period.

Calibration of the Electrical Aerosol Analyzer - The absolute calibration of the EAA for both aerosol concentration sensitivity for particle size and for size distribution resolution has been a difficult task because

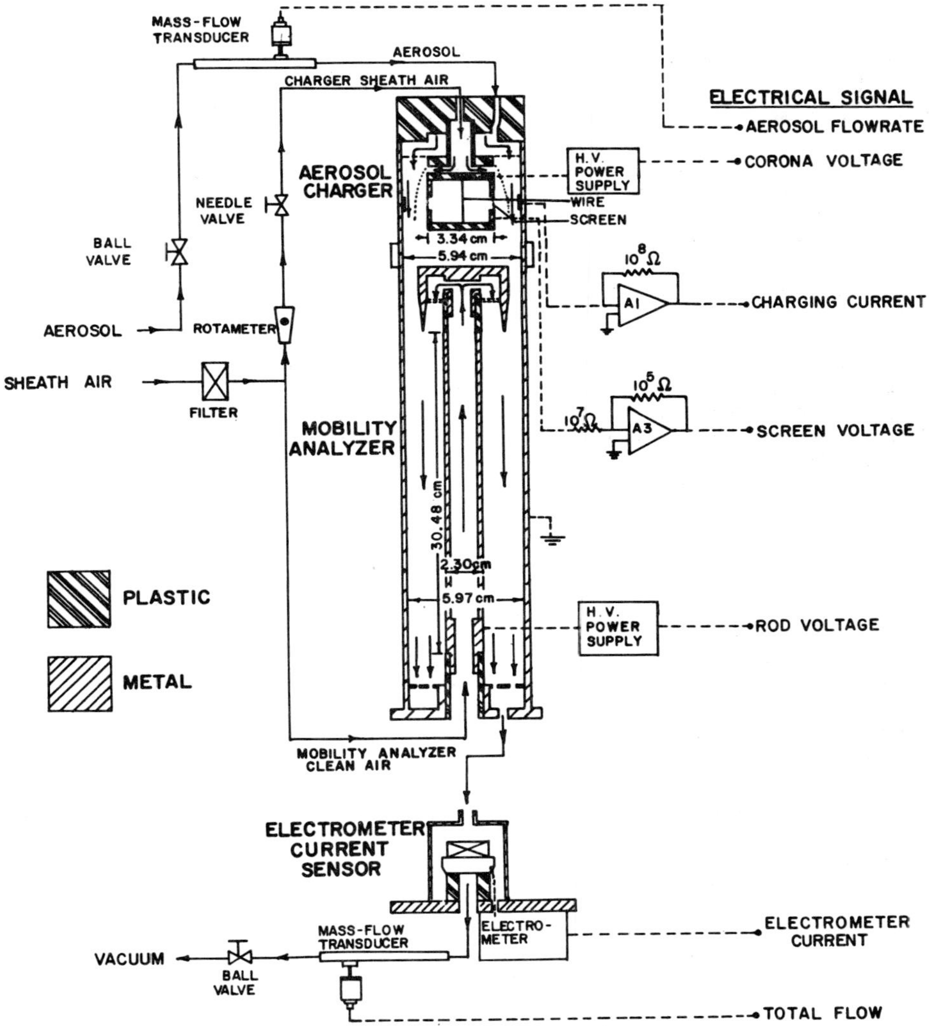

Figure 14. Schematic of the Minnesota - T.S.I. Model 3030 Electrical Aerosol Analyzer.

of the lack of suitable aerosol standards in the size range from 0.003 to 1 μm. However, the recent development by Liu and Pui[15] of the electrical classification generator described earlier in this paper has now made it possible to make such calibrations with acceptable accuracy over most of the size range of the aerosol analyzer, Liu and Pui[45]. The results to date from these calibration studies are summarized here.

1. Liu and Pui[45] studied the losses and mobility versus size curve at several charging conditions and concluded that $N_i t = 1 \times 10^7$ was the best for the commercial instrument. It was also found that although the commercial instrument had actually been originally designed for this Nt, the actual $N_i t$ was only 0.6×10^7. It is possible to adjust the commercial instruments to an $N_i t$ of 1×10^7 by means of a few simple component changes and major adjustments.

2. The current sensitivity in pico amps/10^6 particles per cm^3, for monodisperse aerosols of known size and known concentration generated with the monodisperse generator described earlier has been determined. These are shown at the top of Table IV. These range from 0.105 for 0.0075 μm to 285 for 0.75 μm particles. Also, from these data it is possible to calculate the fractional charge and loss factor for the instrument, Figure 3.

It can be seen by comparing the theoretical loss factor for sizes smaller than 0.02 μm in Figure 3 with actual loss factor, that the combined loss factor is much smaller than the fraction charged. Note that at 0.01 μm the fraction charged is 0.3, but that the combined factor is only 0.075 at the same size. Recent studies of the instrument indicate that most of this difference is caused by space charge and diffusion losses in the charger. If these losses are reduced so that the overall loss factor approaches that for fraction charged alone it should be possible to obtain useful size distribution measurements down to at least 0.003 μm. These results indicate that at the most desirable operating $N_i t = 10^7$, the loss factor is so great below about 0.006 μm that the lowest size range 0.0032 to 0.0056 μm, would only be usable under special conditions and that for most purposes the size range of the present correctly adjusted commercial instrument should be considered to be from 0.0056 to 1 μm.

3. Size calibration of the instrument was accomplished by monodisperse aerosols haveing sizes corresponding to the geometric mean diameter of each channel on the EAA and then measuring the fraction of the total current that was measured in each channel of the instrument. The results are shown in Table IV. The row at the top gives the actual size of the monodisperse aerosol, the second row gives the sensitivity, the main part of the table gives the fraction of current in each channel and the last two rows give the geometric mean channel size and geometric standard deviation of the response to each size, respectively. In making these cal-

Table IV
Response of the electrical aerosol analyzer to monodisperse aerosols for Nt = 1 x 10^7 (ions.cc)(sec.)

$\Delta N = \alpha \Delta I$		Dpi - Particle Diameter, µm	0.0042	0.0075	0.013	0.024	0.042	0.075	0.13	0.24	0.42	0.75
		Sensitivity, $1/\alpha_i$ - pa/(10^6part./cc)		0.105	2.40	6.00	11.5	22.5	41.5	81.0	150	285
Channel	Channel Boundary, µm	Collector-Rod Voltage, Volt	Channel Midpoint Dp_i - µm									
	0.0032	19										
1			0.0042	0	0	0	0	0	0	0	0	0
	0.0056	59										
2			0.0075	1	0	0	0	0	0	0	0	0
	0.0100	186										
3			0.0133	0	1	0.08	0	0	0	0	0	0
	0.0178	588										
4			0.0237	0	0	0.92	0.49	0	0	0	0	0
	0.0316	1,870										
5			0.0422	0	0	0	0.07	0.14	0	0	0	0
	0.0562	2,600										
6			0.075	0	0	0	0.44	0.65	0.42	0	0	0
	0.100	4,440										
7			0.133	0	0	0	0	0.17	0.49	0.22	0	0
	0.178	6,600										
8			0.237	0	0	0	0	0.02	0.09	0.48	0.31	0.11
	0.316	8,380										
9			0.422	0	0	0	0	0.02	0	0.20	0.36	0.28
	0.562	9,600										
10			0.750	0	0	0	0	0	0	0.06	0.18	0.25
	1.000	10,600										
										(0.04)	(0.15)	(0.36)
	1.78	(>10,600)										
DG				.0075	.0133	.023	.041	.081	.110	.269	.465	.692
SG			up to 1.78 µm	1	1	1.17	1.74	1.54	1.44	1.76	1.81	1.81

culations the responses shown in parenthesis were assumed to fall in the 1 to 1.78 μm size range.

The geometric means of the response versus the actual aerosol size is shown in Figure 15. It will be noted that in the size range from 0.0056 to 0.561 μm, the correlation between the actual and indicated sizes is quite good at these operating conditions. In the size interval 0.56 to 1 μm, 36% of the current that should be indicated in this channel is not.

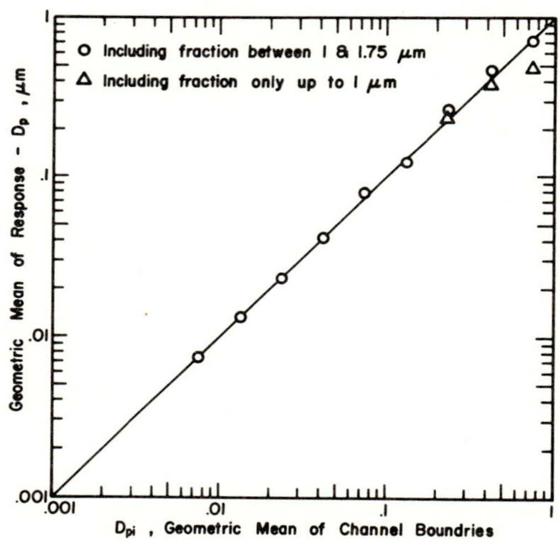

Figure 15. Comparison of the instrument indication with the response to monodisperse aerosols for the TSI model 3030 electrical aerosol analyzer. Data from Table IV.

4. The geometric standard deviations (SG), tabulated in the last row of Table IV, show that the resolution varies from that corresponding to less than one channel for the two smallest channels, to SG=1.81 for the two upper channels.

The sizing resolution is essentially a function of three factors: the inherent mobility resolution of the instrument, which corresponds to an SG < 1.1, the discrete nature of the electri charge and the slope of the Zp versus Dp curve, Figure 2. The Liu and Pui[45] studies indicate that mobility analyzer is functioning very close to theory with respect to its mobility resolution and with respect to what can be accomplished considering the discrete nature of particle charging. It appears that some improvement in the optimum Zp vs Dp curve is possible if the instrument were to be operated at a variable pressure, Liu, Marple, and Yazdani[49]. However, if the Nt were to be reduced to obtain more slope and better resolution at the larger sizes, then the fraction charged at the smaller sizes might decrease thereby reducing the instrument usefulness below 0.01 μm. In short it appears that the maximum size range over which such an instrument can be made to operate with useful resolution is about two orders of magnitude. As Whitby and Liu[50] have previously pointed out, this is probably the maximum range of sizing usefulness for any single measurement technique or instrument operation at a given set of conditions.

Table IV constitutes the calibration matrix of the instrument and it is possible to use these data to correct the indicated size distributions to obtain results much closer to the true distribution. Initial attempts by Cantrell[12] to do this have been reasonably successful and as soon as the techniques are more fully developed and sufficient examples have been evaluated, the results will be reported. Using the matrix inversion technique preliminary indications are that it should be possible to resolve an aerosol with SG > 1.4 for sizes up to 0.56 μm.

Most fine particle aerosol emissions and most atmospheric aerosols have SG's greater than two and Whitby[43] has shown that the size distribution modes occuring in the size range below 1 μm, have SG's on the order of two. Thus, the EAA appears to be able to resolve most of the significant aerosol size distribution structure found in smoke and atmospheric aerosols.

In the size range below 0.03 μm, where the particles are singly charged, by operating the EAA in the manual precipitator voltage control mode, it is possible to resolve size distributions with SG > 1.1.

It will be noted also from Table IV that SG = 1.74 at 0.042 μm. The resolution at this size is caused by the transition from singly charged to multiple charged aerosols. This is a fundamental problem that cannot be eliminated. From Figure 1, it can be seen that this transition would occur at about 0.2 μm for bipolar charging.

5. The precision of aerosol concentration measurements as a function of size is determined principally by two factors; short term noise fluctuations of the electronics and the longer term stability of the instrument and aerosol during the approximately two minutes required to complete a measurement cycle. Liu, Whitby and Pui[40] evaluated the precision due to electronic fluctuations by the following procedure.

In the University of Minnesota smog chamber, Clark and Whitby[51], 0.23 ppm of SO_2 was illuminated for 17 minutes, the lights turned off, and the aerosols allowed to coagulate until practically all of the particles smaller than a few hundredths of a μm had disappeared by coagulation with larger particles and the changes of aerosol surface and volume were small and linear with time. The ΔI's about the line of regression were calculated. By multiplication by the appropriate sensitivity, α, the standard deviation in ΔI can be translated into a standard deviation in particle number, surface area, and volume. These results are shown in Figure 16 as a function of particle size.

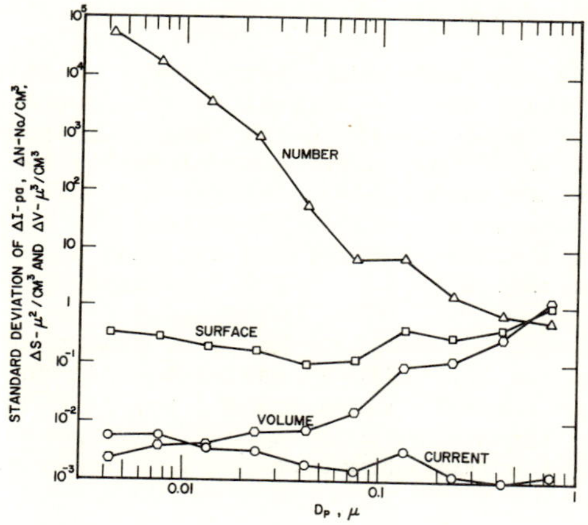

Figure 16. Variability in electrometer current, particle number, surface area, and volume due to inherent instrument variability.

It will be noted that the standard deviation of ΔI ($\sigma_{\Delta I}$), ranges from a maximum of about 0.01 pa at 0.004 μm to a low of about 0.001 pa at 0.4 μm. This is approximately equal to the electronic noise levels at the voltages used for measuring the particular sizes. For these measurements a Doric data acquisition system (Doric Scientific Corp., San Diego, California) was used having a filter with an effective averaging time of about 0.25 sec. By using computer averaging of the electrometer output as was done on the Air Resources Mobile Laboratory in California during the California aerosol characterization project, Whitby et al[52], the effect of the electrometer noise can be reduced even further to values on the order of 0.0005 pa.

From Figure 16, it will be noted that the effect of the variability in ΔI is quite different for particle number, surface, and volume. Plus, or minus one standard deviation in number at 0.0075 μm corresponds to 10^4 particles/cm^3. It is seen that the variability in number decreases sharply with size. the variability of surface is relatively constant, and the variability for volume increases with size. Thus, the instrument is able to measure surface area distributions with approximately the same precision over the full range of sizes. It has been found that the variability due to noise is only significant for the number distribution below 0.02 μm for number concentrations less than $10^3/cm^3$ and for the volume distribution above 0.5 μm. Therefore, for most field work, data from the instrument has only been used for calculating the volume for sizes smaller than 0.422 μm. An optical particle counter has been used to obtain the surface and volume distributions for larger sizes. When measuring atmospheric or smog chamber aerosols having a significant number of particles smaller than 0.02 μm, it is preferable to measure the total number concentration with a properly calibrated CNC instead of with the EAA.

The effects of stability of the instrument and aerosol during the measurement cycle is more difficult to predict and control. The current increments, ΔI, represent the first derivative of the I vs. V curve. Therefore, any changes in the aerosol concentration as a function of time tend to cause errors in the calculated ΔN's. In field studies where ambient aerosol concentrations can fluctuate significantly, a 20 to 50 liter buffer chamber has been used to damp the aerosol fluctuations. In such applications only the aerosol sampling stream need go through the buffer chamber. The clean air stream need not. Also, an automated bag sampling system has been used aboard an aircraft to grab a sample and hold it for the two minutes required by the EAA measurement cycle.

In addition to the above aerosol stability effect, other non-ideal current drift effects have sometimes been observed that so far are not completely understood. These seem to vary from instrument to instrument and are probably associated with insulator charging effects in the mobility analyzer. One manifestation of these effects is negative ΔI's for the smallest several size ranges when measuring aerosols which contain negligible numbers of particles in these size ranges. Use of gold-plated brass parts in the mobility analyzer instead of oxidized aluminum seems to reduce these effects.

Electrical Aerosol Analyzer Applications - The EAA was originally developed as an instrument to measure in situ atmospheric aerosol size distributions in the size range below that which could be measured by any other in situ instruments. Through development it has become smaller, lighter, more automatic, more reliable and its useful sizing range has been extended. Incorporated into instrument systems with one or more optical single particle counters (OPC) and condensation nuclei counters (CNC), it has been used in a number of large aerosol field studies, Whitby et al.[52], Willeke et al.[55], Sverdrup et al[56].

Figure 17 shows a sample size distribution obtained in St. Louis by a large mobile air pollution laboratory owned by the U.S. Environmental Protection Agency. In this laboratory, aerosol size distribution data from the EAA, and two optical particle counters is processed on line by a minicomputer and the data recorded. Although Figure 12 was produced later, plots similar to Figure 12 can now be produced on line as data is collected. Note the excellent agreement between the EAA and the Royco 220 OPC data in the size range from .561 to 1 μm where they overlap. These data were processed using the latest calibrations for both the EAA and the OPC's.

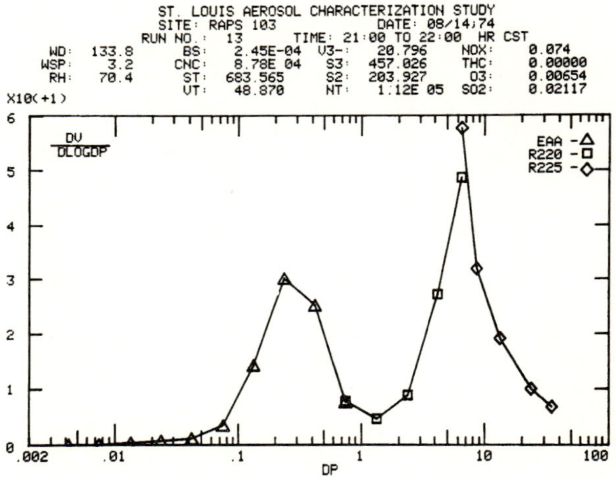

Figure 17. Volume-size distribution obtained with the large EPA mobile laboratory in St. Louis, No. on a site located in an area with industrial sources in all directions.

Within the last year the EAA along with an OPC have been used successfully in several aircraft[53]. For near urban use, where ambient aerosol fluctuations are too large to be tolerated, the aerosol is first forced into a 50 liter sampling bag by ram air pressure, the bag is closed, the measurement cycle made and the bag emptied.

This bag sampling system has also been used for the sizing of aerosol emissions from a Diesel Engine.[57]

Electrical Aerosol Size Distribution Measurement at Low Pressures - From Figure 2 it can be seen that if a sufficiently high NT is used to obtain a satisfactory fraction charged at 0.01 μm, the Zp vs Dp curve has only a very small slope above 0.5 μm. Liu, Marple and Yazdani[49] showed that if the pressure at which the mobility classification is carried out is reduced below 1 atmosphere, then the whole Zp vs Dp curve can be shifted to the right, the amount being determined by the product Dp P, where P is the pressure in atmospheres. For example, if P = 1/2, the mobility minimum would occur at 2 μm instead of 1 μm. This approach can also be used to correct the data from the standard EAA when it is operated at pressures below 1 atm. as in aircraft.

This approach is only useful for aerosol that are stable with respect to pressure. It is not satisfactory for atmospheric aerosol measurement because a substantial fraction of the total mass of these aerosols is usually water or other volatile matter. However, it has considerable potential for precision measurement of stable laboratory aerosols.

Electrical Aerosol Samplers - Quite a number of electrostatic aerosol samplers have been developed. These range from a wire and tube samplers that are in essence scaled down versions of large industrial electrostatic precipitators, to sophisticated samplers for special purposes. The chief advantages of an electric sampler are:

1. low pressure drop compared to a filter.

2. can sample particles to below 0.01 μm as compared to an impactor which is rarely useful below 0.3 μm.

Its chief disadvantages are:

1. complexity.

2. classification of the sample unless special care is taken to obtain non-classified deposits.

3. orientation of the particles by the electric field prior to deposition. This effect can be significant for very irregular particles such as chain aggregates of very small particles.

Limpann[58] has reviewed many of the electrical precipitators for aerosol sampling that were available up to 1971, including the historic wire and tube designs of Drinker and of Barnes and Penny. These samplers as well as later designs marketed by Mine Safety Appliance and Western Precipitation and Del Electronics were designed for low pressure drop, high volume sampling for mass determination. They are not readily used for obtaining samples for microscopy or chemical analysis.

The special requirements of samplers for electron microscopy has lead to the development of samplers which can obtain a satisfactory sample onto an electron microscope grid or stud.[59] This application requires that the sampling efficiency as a function of particle size be reasonably constant and be known if less than 100%.

Several point- to -plane samplers in which a corona needle opposite the collecting surface is used have been developed[60, 61, 62]. Baum[60] seems to be the first one to have developed this type of sampler while the developed at Rochester[61], seems to be the most widely used.

Reist[63] has evaluated the sampling efficiency and other characteristics of the point-to-plane precipitator and found that the efficiency was only a few percent and varied with size. Thus, this type of sampler although convenient because of its ability to concentrate a sample directly onto a small area, should be used with caution if more than rough estimates of the aerosol size distribution are to be made.

In an attempt to develop a sampler for both light and electron microscopy, which would not significantly classify the deposited aerosol, Liu Whitby, and Yu[64] and Liu and Verma[65] developed an electrostatic sampler in which the particles charged by a corona are precipitated in a pulsed electric field in such a way that classification is eliminated. Its principal problem however is that the sampled aerosol is spread over an area several orders of magnitude larger than for the point-to-plane samplers. This makes it less suitable for sampling low concentrations.

It cannot be emphasized too strongly that the biases of sampling methods must be known especially for sizes smaller than 0.01 and larger than 5 μm if accurate assessment of size distribution are to be made.

<u>Miscellaneous Electrical Aerosol Measuring Techniques</u> - Meyers et al.[66] has recently described a technique in which the ions produced when a particle strikes a hot surface, are detected in order to determine particle size, something about chemical composition and particle concentration. The ionized molecular fragments are analyzed in a continuous mass spectrometer.

ACKNOWLEDGEMENT

This paper was prepared with the support of EPA Research Grant No. R 800971 Aerosol Research Section, Division of Chemistry and Physics, Research Triangle Park, North Carolina.

REFERENCES

1. Rohmann, H., Method of Size Measurement for Suspended Particles, Z. Phys. 17, 253-265, 1923.

2. Harper, W. R, Contact and Frictional Electrification, W. R. Harper Co., Oxford (1967).

3. Moore, A. D. (Editor), Electrostatics and Its Applications, John Wiley & Sons, N.Y. (1973).

4. Prochazka, R., Staub, 26:22 (1966).

5. John, W., Investigation of Particulate Matter Monitoring Using Contact Electricity, EPA Grant No.Final Dept. (Sept., 1974).

6. John, W., Contact Electrification Applied to Particulate Matter-Monitoring, presented at the Fine Particle Symposium, University of Minnesota, May 27-30 (1975).

7. Guichard, J. C. and Chauvelier, Nouveau Procede de Mesuew en Continu de la Concentration en Poussieres de Particles Solids. Proceedings of the Conference on Aerosol in Physik, Medizin und Technik, Gesselschaft fur Aerosolforschung, e.v. Bad Soden, Germany (1973).

8. Liu, B. Y. H., and D. Y. H. Pui, Aerosol Science. 5:465 (1974a).

9. Husar, R. B., Recent Developments in 'In Situ' Size Spectrum Measurements of Submicron Aerosols, APRL Comm. #2, paper presented at the Conference on "Instrument Monitoring for Ambient Air", Boulder, Colorado, Aug. 14-16, (1973).

10. Knutson, E. O., The Distribution of Electric Charge Among the Particles of an Artifically Charged Aerosol, Ph.D. Thesis, University of Minnesota, Mechanical Engineering Department, Minneapolis, Mn. 55455(1971).

11. Knutson, E. O., Extended Electric Mobility Method for Measuring Aerosol Particle Size, presented at the Fine Particle Symposium, University of Minnesota, May 27-30 (1975).

12 Cantrell, B. C., unpublished procedures developed in the Particle Technology Laboratory, University of Minnesota.

13. Pui, Y. H., Experimental Study of Diffusion Charging of Aerosols, Ph.D. Thesis, University of Minnesota, Mechanical Engineering Department, Minneapolis, Mn. 55455 (1975).

14 Liu, B. Y. H. and D. Y. H. Pui, J. Colloid and Interface Sci., 49: 305 (1974b).

15 Liu, B. Y. H. and D. Y. H. Pui, J. Colloid and Interface Sci., 47: 155 (1974c).

16 Bademosi, F., Diffusion Charging and Related Transfer Processes in Knudsen Aerosols, Ph.D. Thesis, University of Minnesota, Mechanical Engineering Department, Minneapolis, Minnesota 55455 (June 1971).

17 Liu, B. Y. H., K. T. Whitby and H. H. S. Yu, J. Appl. Phys., 38:1592 (1967).

18 Keefe, D., P. J. Nolan and T. A. Rich, Proc. Royal Irish Academy, Section A, 60:27 (1959).

19 Keef, D., and P. J. Nolan, Geofis. Pura Appl. 50:155 (1961).

20 Keef, D. P. J. Nolan, and J. A. Scott, Proc. Roy. Irish Acad. 66A:2 (1968).

21 Natanson, G. L., Sov. Phys. Tech. Phys., 5:538 (1960).

22 Brock, J. R., J. Appl Phys., 41:843-844(1970).

23 Brock, J. R. and M. S. Wu, J. Colloid and Interface Sci., 33:473 (1970).

24 Fuchs, N. A., Investiza Acad. Nauk. USSR, Ser. Geogr. Geophys., 11: 341 (1947).

25 Pluvinage, P., Etude Theorique et Experimental de la Conductibilite Electrique duns les Nuages non Orageuz Thesis, Faculte des Sciences de l'Universite Paris (1945).

26 Pluvinage, P., Ann Geophys., 2:31 (1946).

27 Bricard, J., J. Geophys. Res. 54:39(1949).

28 Fuchs, N. A., Geofis. Pura Appl. 56:185-193(1963).

29 Marlow N.H., and J. R. Brock, J. Colloid Interface Sci.50:32-38(1971).

30 Liu, B. Y. H. and F. Bodemosi, Diffusion Charging of Kundson Aerosols. II. Theory, Publication No. 156, Particle Technology Lab., University of Minnesota.

31 Pourprix, M., UnNouveau Precipitateaur Electrostatique - Applications a L'etude de la Charge des Aerosols par Diffusion d'ions Bipolaires, Thesis, University of Paris IV (1973).

32 Israel, H., Atmospheric Electricity - Fundamentals, Conductivity, Ions, Vol I, Israel Program for Scientific Translations, Jerusalem 1970.

33 Mohnen, V. Investigation of the attachment of neutral and electrically charged emanation decay products to aerosols- Measurement of the aerosol size spectrum and of the attachement coeffocoemt pf meitra; decay products. Thesis to the Natural Science Faculty, Ludwig-Maximilians Universitat, Munich,(Sept. 1966).

34 Megaw, W. J. and A. C. Wells, A High Resolution Charge and Mobility Spectrometer for Radioactive Submicrometer Aerosols, J. Scientific Instruments (J. Phys. E.), Series 2, Vol. 2:1013-1016 (1969).

35 Yeh, H. C., O. G. Raabe, and B. E. Wood, Development of a Miniature Aerosol Charge Spectrometer, Report No. LF-46, Annual Report by the staff of the Inhalation Toxicology Research Institute, Lovelace Foundation, Dec. (1973).

36 Rich, T. A., L. W. Pollak and A. L. Metnieks, Estimation of Average Size of Sub-micron Particles from the Number of all and Uncharged Particles, Geofis. Pura. Appli., 44:233 (1959).

37 Flyger, H., K. Hansen, W. J. Megaw, and L. C. Cox, The background Level of the Summer Tropospheric Aerosol Over Greenland and the North Atlantic Ocean, J. Applied Meteor., 12:161 (1973).

38 Mohnen, V. A. and Holtz, J.A.P.C.A., 18:667 (1968).

39 Whitby, K. T. and W. E. Clark, Tellus, 18:573 (1966).

40 Liu, B. Y. H., K. T. Whitby and D. Y. H. Pui, J.A.P.C.A., 24:1067 (1974).

41 Knutson, E. O. and K. T. Whitby, Aerosol Classification by Electric Mobility: Apparatus, Theory and Applications, J. of Aerosol Science, 6:6 (1975).

42 Huertas, M. L., A. M. Marty, J. Fontan, L. Lanegrasse, and D. Blanc, A Method of Measuring the Mobility of Radioactive Ions, Rev. of Scientific Instruments, 41:1567-1569 (1970).

43 Whitby, K. T., Modeling of Multimodal Aerosol Distributions, Presented at the GAF meeting Bod Soden, Germany (Oct 17, 1974).

44. Whitby, K. T., Trimodal Log Normal Model for Atmospheric Aerosol Size Distributions, to be submitted to <u>J. of Aerosol Science</u> (1975).

45. Liu, B. Y. H., and D. Y. H. Pui, On the Performance of the Electrical Aerosol Analyzer," <u>J. Aerosol Sci</u>., 6:249(1975).

46 Willeke, K., C. S. K. Lo, and K. T. Whitby, <u>J. of Aerosol Science</u>, 5:465 (1974).

47 Liu, B. Y. H., V. A. Marple, K. T. Whitby, and N. J. Barsic, <u>American Industrial Hygiene Association Journal</u>, 8:443 (1974).

48 Liu, B. Y. H., D. Y. H. Pui, A. W. Hogan, and T. A. Rich, <u>J. of Applied Meteorology</u>, 14:46(1975).

49 Liu, B. Y. H., V. A. Marple, and H. Yazdani, <u>Env. Sci. and Tech</u>., 3: 381 (1969).

50 Whitby, K. T. and B. Y. H. Liu, Advances in Instrumentation and Techniques for Aerosol Generation and Measurement, presented at the Harold Heywood Memorial Symposium, Loughborough, England (Sept 17-18 1973).

51 Clark, W. E., and K. T. Whitby, Measurement of Aerosols Produced by the Photochemical Oxidation of SO_2 in Air, <u>J. of Colloid and Interface Sci</u>., 51:477 (1975).

52 Whitby, K. T., B. Y. H. Liu, R. B. Husar, and N. J. Barsic, "The Minnesota Aerosol Analyzing System Used in the Los Angeles Smog Project", <u>J.</u> Colloid and Interface Seicnec, 39:136-164 (1972).

53 Cantrell, B. C. and K. T. Whitby, Airborne Minnesota Aerosol Analyzing System, Program Report in preparation EPA Grant No. R800971 (1975).

54 Whitby, K. T., R. V. Husar, and B. Y. H. Liu, <u>J. Colloid and Interface Sci</u>., 39:211 (1972).

55 Willeke, K., K. T. Whitby, W. E. Clark, and V. A. Marple, <u>Atmospheric Environment</u>, 8:609 (1974).

56 Sverdrup, G. M., K. T. Whitby, and W. E. Clark, Characterization of California Aerosols, II. Aerosol Size Distribution Measurements in the Mojave Desert, <u>Atmospheric Environment</u>, 9:463 (1975).

57 Dolan, D. F., D. B. Kittelson, and K. T. Whitby, Measurement of Diesal Exhaust Particle Size Distribution, to be presented at the Winter ASME Meeting, Houston, Texas, Dec. 1975.

58 Lipmann, M., "Electrostatic Precipitators," Air Sampling Instruments, ACGIH, P.O. Box Box 1937, Cincinnati, Ohio 54201 (1972 ed.)

59 Mercer, T. T., Aerosol Technology in Hazard Evaluation, Academic Press, New York, N.Y. (1973).

60 Baum, J. W., "Electrical Precipitator for Aerosol Collection on Electron Microscope Grids," USAEC Rep. HW-39129 (1955).

61 Billings, C. E. and L. Silverman, "Aerosol Sampling for Electron Microscopy," J.A.P.C.A., 12:586 (1962).

62 Morrow, P. E. and T. T. Mercer, "A Point to Plant Electrostatic Precipitator for Particle Size Sampling," Amer. Ind. Hyg. Ass. J., 25:8 (1964).

63 Reist, P.C., "Size Distribution Sampling Errors Introduced by the Point to Plane Electrostatic Precipitator Sampling Device," Proc. A.E.C. Air Cleaning Conference, 9th CONF-66094, p. 613 (1966).

64 Liu, B. Y. H., K. T. Whitby, and H. S. Yu, J. Colloid Interface Sci., 23:367 (1967).

65 Liu, B. Y. H. and A. C. Verma, "A Pulse-Charging-Pulse Precipitating Electrostatic Aerosol Sampler", Analytical Chemistry, 40:843-847, (1968).

66 Myers, Richard L. and Wade L. Fite, "Electrical Detection of Airborne Particulates Using Surface Ionization Techniques," Environmental Science and Technology, 9:334-336 (1975).

67 White, H. J., "Particle Charging in Electrostatic Precipitation", AIEE Transactions, 70:1186-1191, 1951.

RECENT DEVELOPMENTS REGARDING THE USE OF A FLAME IONIZATION DETECTOR AS AN AEROSOL MONITOR

Lawrence (Tom) L. Altpeter, Jr. (North Star Division, MRI), J. P. Pilney (NS/MRI), L. W. Rust (NS/MRI), A. J. Senechal (Hennepin Technical Center), D. L. Overland (Digital Systems)

North Star Division, MRI
3100 38th. Ave. So.
Minneapolis, Minnesota 55406

ABSTRACT

High speed waveform studies were performed with the output signals of individual aerosol particles as they were consumed in the flame of a flame ionization detector. These output signals were shown to consist of modulated current pulses with mean amplitudes which varied with particle size from 10^{-9} to 10^{-6} amperes and with durations from a fraction of a millisecond to several milliseconds.

Aerosol measurements were performed with the FID centered in an aerosol test chamber. Sampling conditions of size, number concentration and composition were carefully controlled. Repeatability of the flame ionization detector was estimated to be 5 to 15 percent. Response of the detector was shown to be linear throughout the tested size range of 0.5 to 10 μm. Size limit of detection was 0.15 to 0.2 μm with evidence of ways to improve this limit significantly. Sensitivity varied with particle size as the 3/2 power. There was evidence of limited selectivity when the detector was used with parallel plate electrode geometry.

Considerations of the structured current pulses observed with the conventional FID led to design and testing of a novel FID. All features of the novel FID were identical to the conventional model with the exception of the collection electrode which has been segmented into a vertical array of horizontal strips. Each strip was a separate and individually addressable electrode. The detector was referred to as a segmented-plate FID, or SFID. Preliminary results with this device suggest that the current pulse observed with the conventional FID is a composite of two signals, one of which is the time-of-flight delayed current of the charge carriers from the flame.

RECENT DEVELOPMENTS REGARDING THE USE OF A FLAME IONIZATION DETECTOR AS AN AEROSOL MONITOR

Lawrence (Tom) L. Altpeter, Jr., (North Star Division/MRI), J. P. Pilney (NS/MRI), L. W. Rust (NS/MRI), A. J. Senechal (Hennepin Technical Center), D. L. Overland (Digital Systems)

INTRODUCTION

Although the flame ionization detector (FID) can detect single particles on a real time and continuous basis,[1-5] there has been relatively little interest in its use as an aerosol monitor for several reasons. First, the response of the FID varies with particle composition which means that accurate size information is not possible when the device is challenged with typical heterogeneous aerosols. Second, the coincidence of two or more particles in the flame cannot easily be deciphered so that the FID is count-rate limited. With appropriate sample-conditioning, the count-rate limitation can be minimized or eliminated. However, the variation of response with particle composition has remained as a drawback of the FID.

We wish to report the results of high speed waveform studies in which the output current pulses of the FID, produced by single particles, were examined on an expanded time base. If a high speed amplifier is connected to the collection electrode of an FID, then the output signal, produced by a single particle as it is consumed in the flame, can be observed as a pulse, 10^{-9} to 10^{-6} amperes in amplitude, and, from a fraction of a millisecond to several milliseconds in duration. If parallel plate electrode geometry is employed, this current pulse is observed to possess an irregular pattern of modulation. The modulation pattern is repeatable for a given aerosol species. The signal amplitude is repeatable for a given particle size. Further, during preliminary studies at this laboratory there was evidence that for certain aerosols the waveform of the current pulse might serve as an indicator of specie.

Because relatively little detailed data has been reported for the FID, a general performance study was undertaken to evaluate analytical parameters such as precision, response, sensitivity and size limit of detection. A variety of aerosols were examined under controlled conditions of size, number concentration and composition.

EXPERIMENTAL

APPARATUS

FID Detectors

The FID employed throughout this study is shown schematically in Figure 1. Significant features of this configuration included plane parallel plate electrodes, a pre-mixed stoichometric air-hydrogen flame, a ground-side amplifier, and removal of combustion products by evacuation. An exploded view of the actual configuration is shown in Figure 2. Significant dimensions and specifications are listed in Table 1. A polarizing voltage of 500 VDC was applied throughout most of this study. Hydrogen is premixed with the sample-laden air in a stainless steel tee below the burner. The stoichmetric air-hydrogen mixture burned at the orifice of the burner. Secondary air entered the enclosed space of the FID chamber at the base of the rear wall (not shown). A copper exhaust line, 0.6-cm OD (0.25 inches), lead to an aspirator (not shown). The latter was required to develop sample air flow and to direct hydrogen and secondary air into and through the FID.

In order to minimize electronic interferences from pickup, the burner was enclosed in a Faraday-like cage. The cage was fabricated from panels of copper-laminated epoxy board (Table 1) which were soldered together.

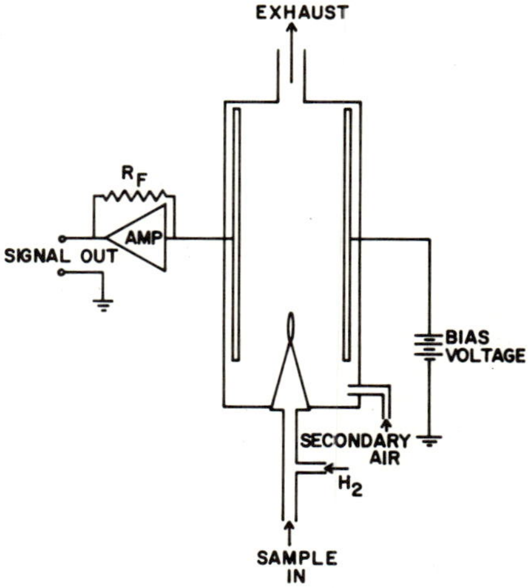

Figure 1. Schematic Illustration of Conventional FID Aerosol Monitor (Parallel Plate)

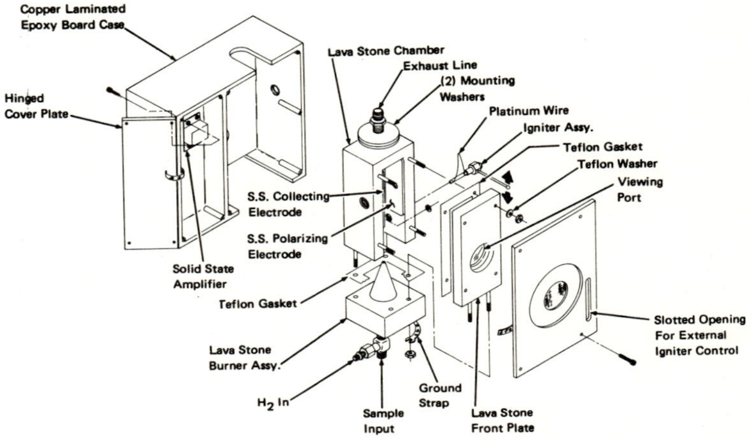

Figure 2. Exploded View of Experimental FID

Table 1. Dimensions and Specifications of an FID with Plane Parallel Electrode Geometry

1. Copper Laminated Epoxy Board Case--15.2 by 10.2 by 7.6 cm (6 by 4 by 3 inches); fabricated from panels of copper laminated epoxy board 0.32-cm (0.125-inch) thick; composed of two compartments sized to accept the lava stone FID chamber and the solid state amplifier with its associated discrete components.
2. Lava Stone FID Chamber (with burner assembly)--outer dimensions: 14.6 by 7 by 7 cm (5.75 by 2.75 by 2.75 inches); inner dimensions: 7.6 by 1.9 by 1.9 cm (3 by 0.75 by 0.75 inches).
3. Lava stone Burner Orifice--1 mm (0.04 inches); projects above lower edge of electrodes by 0.6 cm (0.24 inches).
4. Electrodes--plane parallel, stainless steel, 3.8 by 1.9 by 0.3 cm (1.75 by 0.75 by 0.125 inches); separated by 1.6 cm (0.625 inches).
5. Premix fitting--1/4-inch Swaglok Tee with one 1/8-inch NPT.
6. Igniter Assembly--platinum wire, 0.08 cm (0.030 inches), bent in a loop; attached to rotatable shaft which is sealed into FID housing, powered by 2.5 VAC stepdown transformer.
7. Applied Voltage--500 VDC.
8. Gaseous Flow Rates--100 cc/min hydrogen, 250 cc/min aerosol-laden air, 300 cc/min secondary air.

With the indicated structural features and 500 VDC applied, the rms background current from the FID was approximately 50 pA.

Segmented-Plate (SFID) Detectors

A simple functional sketch of the SFID is shown in Figure 3. Basically, the SFID is identical to the FID with the exception of the segmented collection electrode. The latter was designed to produce an electric field as nearly identical to that of the conventional FID as possible. Each segment is insulated from its neighbors and individually addressable. Those segments not in use are grounded. The area of each segment of the SFID collection electrode should be physically identical to the corresponding area of the conventional FID collection electrode. By means of the SFID, it should be possible to measure indirectly, the vertical spatial distribution of ion currents impinging on the surface of the collection electrode of the conventional FID.

An exploded view of the SFID developed at North Star is shown in Figure 4. The same basic three-pieced construction of the conventional FID is maintained (Figure 2). The burner housing and igniter assembly are identical to that of the conventional FID. The polarizing electrode assembly is also identical. A cut-away view above the rectangular viewing port of the main housing shows the location of the polarizing electrode. Partial details of the segmented collection electrode can be seen in the indicated blow-up on the right side of the figure.

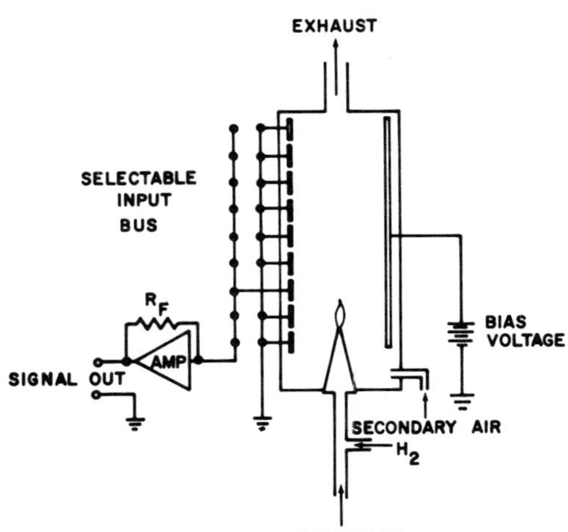

Figure 3. Schematic Illustration of SFID

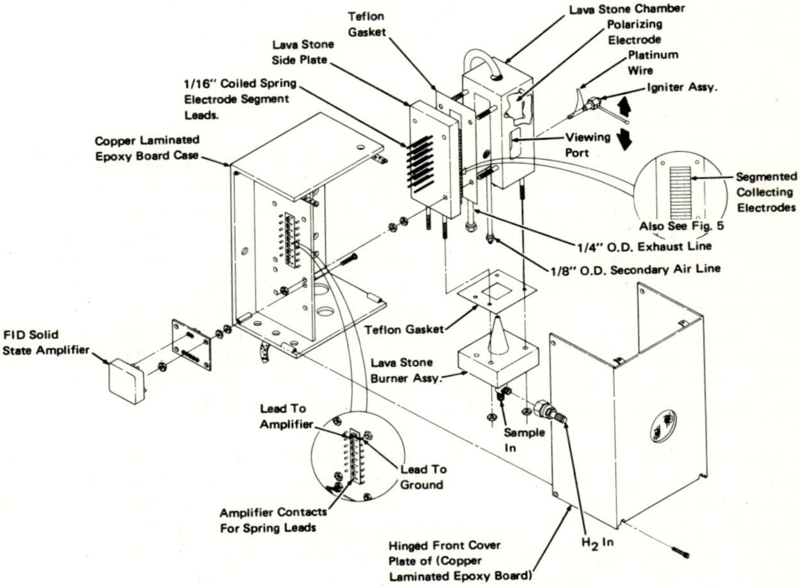

Figure 4. Exploded View of Experimental SFID

Experimental System

A schematic layout of the experimental system employed in this study is shown in Figure 5. The FID was centered within an atmospheric simulation chamber which had a volume of 3.68 m^3 (128 cubic foot). The chamber was made from plywood sheets, 2.5-cm thick (1 inch) and 1.22 m by 2.44 m (4 by 8 feet). The inner walls were treated with two coats of sealer and three coats of epoxy paint with sufficient time allowed for complete curing. The polarizing voltage for the FID was fed to the detector from an external power supply through a BNC cable. The output signal was carried through a BNC cable to an oscilloscope and Visicorder oscillograph, also outside the chamber. The oscilloscope provided a convenient means for visually monitoring the FID output, while the Visicorder provides a permanent record on command. Further details regarding equipment are listed in Table 2.

A 100 CFM exhaust blower drew air into the chamber. With such a flow rate, it was possible to fill or purge the chamber in a conveniently brief period of time. The exhaust blower permitted the chamber to be maintained at a slight negative pressure. Intake air passed through an

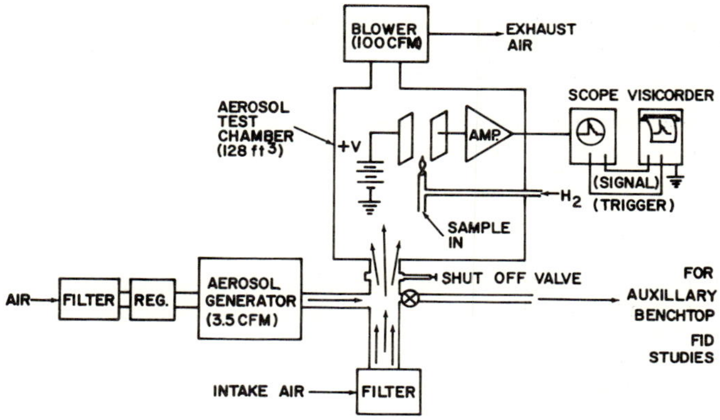

Figure 5. Schematic Layout of Experimental System for Evaluation of FID

Table 2. Experimental Equipment and Specifications

Aerosol Generator - Berglund-Liu Monodisperse Aerosol Generator, Model 3050, Thermo-Systems, Inc., equipped with 5, 10, and 20 μm orifices for liquid solution samples and 100 μm orifice for suspensions of large particle size.

Oscilloscope - Techtronix Type 535 operated typically at ordinate sensitivities of 5-200 mV/cm and time-base of 0.2-1 msec/cm.

Oscillograph - Honeywell Model 1806 Fiber Optics CRT Visicorder, operated with ordinate and abscissa scales identical to those of oscilloscope.

FID Power Supply - John Fluke Precision DC Supply, Model 301E, 0 to ± too VDC/300 mA.

FID Amplifier - Analog Devices, Model 48K Current Feedback Amplifier, 15 MHz bandwidth (at unity gain), open loop gain of 10^5 (500 Ω DC load), settling time of 500 ns to 0.01 percent (at unity gain), input impedance of 10^{11} Ω and 3.5 pf, powered with Analog Devices 15 VDC Power Supply.

Chamber Exhaust Blower - Dayton, Shaded Pole, 115 VAC, 126 W, rated at 100 CFM.

Chamber Intake Filter - Flanders, Model 7C10-L-N2C2 absolute filter rated at 600 DFM maximum.

Rotameters (H_2. air) - Brooks, tube size R-2-15D CO equipped with needle valves.

Rotameter (sample) - Brooks-Mite Model 2051, 2 SCFH (sample line normally open).

absolute filter. Aerosol samples were mixed with this clean air as it entered the chamber. If desired, a sampling probe was available for auxiliary benchtop studies of the FID.

Purge time for the aerosol chamber was anywhere from 1/2 to 4 hours, depending on previous state of the chamber.

AEROSOL SAMPLE PREPARATION

Atmospheric aerosols can be divided into three general categories: inorganic, organic, and biological. Sodium chloride was used extensively as an example of an inorganic aerosol. Monodisperse latex particles served as an organic aerosol. For biological aerosols a set of commonly occurring microorganisms were prepared. A complete list of the aerosols used during these studies appears in Table 3.

Table 3. Experimental Materials

Inorganic Aerosols	CsI, NaCl, $BaCl_2 \cdot 2H_2$), $NiCl_2$, $FeCl_3$, $UO_2(NO_3)_2 \cdot 6H_2O$, $Na_2WO_4 \cdot 2H_2O$, $Fe(NH_4)_2 \cdot 6H_2O$, AR grade quality.
Organic Aerosols	Latex, monodisperse, 0.8 µm, 3.4 µm, 9.5 µm, Coulter Diagnostics, Inc., Miami Springs, Florida.
Microbiological Aerosols	*Serratia marcescens*, *Proteus vulgaris*, *Azotobacter agilis*, *Bacillus cereus*, *Bacillus subtilis*, *Escherichia coli*, *Staphlococcus albus*, *Streptococcus lactis*, *Saccharomyces cerevisiae*, *Aspergillus niger*, obtained from private sources or from the American Type Culture Collection (12301 Parklawn Drive, Rockville, Maryland 20852).
Other Aerosols	Coal Dust (200 mesh), Arizona Road Dust, Stack simulants (alumina, silica).
FID Gases	Hydrogen, ultrapure (99.9995 percent, Air Products and Chemicals, Inc.), Air zero grade (less than 1 ppm total hydro-carbons) (Air Products and Chemicals, Inc.).

Inorganic and organic aerosols were prepared with deionized and distilled water which was doubly filtered through Wattman 40 paper and Millipore 0.45 μm paper. Test cells of biological aerosols were removed from their nutrient broth by centrifugation and then resuspended in tap water which was dechlorinated, sterilized and doubly filtered as described above. Biological aerosols were measured promptly to minimize the extent of lysis.

Aerosols were generated with a Berglund-Liu Model 3050 Monodisperse Aerosol Generator (Table 2).

The number concentration produced initially in the aerosol generator, $N^o(v)$, was readily calculated for a liquid solution as $N^o(v) - f/Q_g$, where f is the frequency of the orifice vibration which defined the rate of droplet formation, and Q_g is the flow rate of the dilution gas (air) which carried away the aerosol. For a suspension, the number concentration of the particles in the liquid phase was adjusted so that, statistically, the mean number of particles per emerging aerosol droplet was one or less. Knowing the liquid phase feed rate into the aerosol generator was sufficient to define the gas phase aerosol number concentration.

Aerosol particle size could be predicted for the case of a non-volatile solute in a reasonably volatile solvent such as water. Assuming a spherically-shaped particle and typical solute weight fractions of less than 1 percent, one can derive the relation:

$$d_s = \frac{6\theta Q_\ell}{\pi \rho_s f}$$

where d_s is the diameter of evaporated solute particle, θ is the original solute weight fraction; Q_ℓ is the liquid feed rate from a driven syringe through the orifice, and ρ_s is the solute density.

RESULTS

FID OUTPUT SIGNAL

A typical amplified signal is shown in Figure 6 for an organic or microbiological aerosol of approximately 1 μm in diameter. Brief studies with many materials showed that the waveform of Figure 6 was generally produced by organic and microbiological aerosols in the 0.5 to 5 μm range. A number of exceptions were noted among the tested materials such as NaCl, CsI, and certain of the larger biological aerosols. In the case of NaCl, only a single shoulder was observed behind the leading spike of the output waveform. For CsI, no shoulders were observed (see "Selectivity" section for further details).

FID RESPONSE, SENSITIVITY AND DETECTION LIMITS

The response of the FID to particles in the size range 1.5 to 10 μm was measured for selected inorganic, organic and biological aerosols. Solution and suspension samples were prepared as described above and dispersed from the aerosol generator. The FID was centered in the atmospheric simulation chamber. Approximately twenty output signals were recorded for each particle size of a given aerosol specie. Output

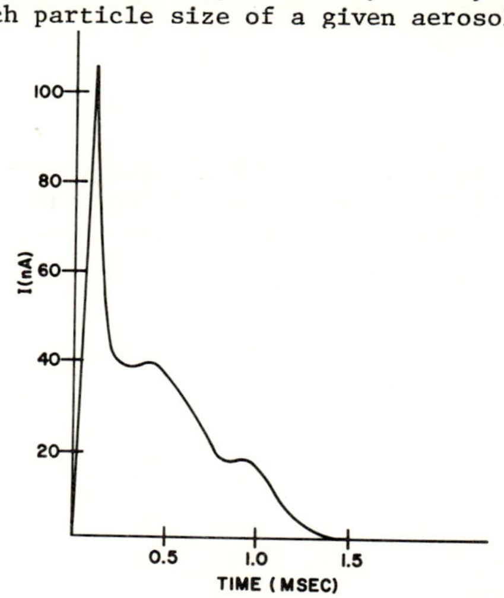

Figure 6. Typical Current Pulse Produced by a Particle of Approximately 1 μm as It Burns in the Flame of a Parallel Plate FID

signals were measured with a planimeter. The mean area under the curve was taken as a measure of detector response.

Sodium chloride served as an example of an inorganic aerosol. The results of two runs on separate days are shown in Figure 7 for a size range from 0.5 to 10 μm. The error bar at 4 μm is the result of six separate runs and is discussed below in the Precision section. The response for this aerosol appeared to be linear over the tested size range. A linear regression analysis of the data, transformed to a log-log format, yielded the equation:

$$\log_{10}y = 0.723 + 1.54 \log_{10}x \tag{2}$$

with a correlation coefficient of 0.996.

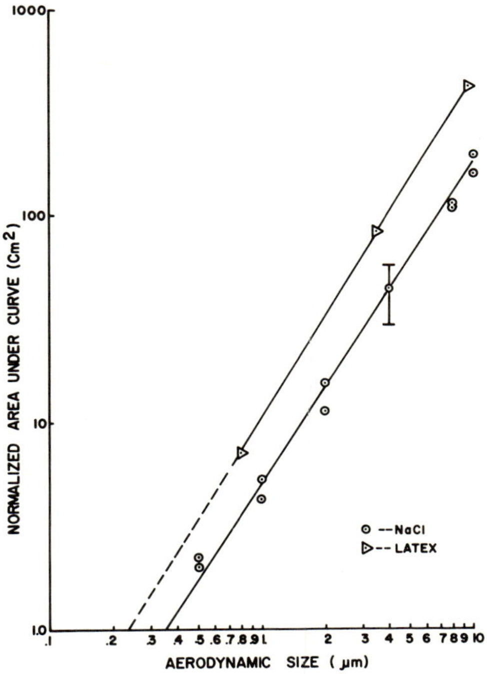

Figure 7. Response of Conventional FID (Parallel Plate) as a Function of Aerosol Particle Size

The results for latex particles, an organic aerosol, also appear in Figure 7. Monodisperse samples were dispersed from the aerosol generator in three sizes: 0.8 μm, 3.5 μm, and 9.5 μm. The response for this aerosol appeared, also, to be linear, though only three points were developed.

The data for latex and sodium chloride can be fit to straight lines which are slightly displaced and of very similar slope. A 2 σ noise limit was estimated as 0.44 cm^2. By extrapolation the lower limit of detection for NaCl and latex corresponded to be a particle size of 0.2 μm and 0.15 μm, respectively. The value for NaCl agreed well with that found by Crider and Strong.[4]

Several microbiological aerosols were selected to cover the size range 0.9 to 5.5 μm. These included *S. marcescens*, *P. vulgaris*, *B. cereus*, and *S. cerevisiae*. The results for microbiological aerosols are summarized in Table 4. The data points for these aerosols have not been included with those of NaCl and latex. Among six common, linearly transformable functions, a best least-squares fit was obtained with the exponential function:

$$y = 14.8e^{.04x} \tag{3}$$

While the difference in response for the microbiological aerosols is not completely understood, it should be recalled that these particles are composed of 70 to 80 percent water, to which the FID does not respond, and an equally uncertain percentage of non-aqueous mass. As shown in Table 4, the inherent variation in the mean volume of each specie is significant.

The slope of the response curve, or the sensitivity, was nearly 3/2 in the cases of the organic and inorganic aerosols (Equation 2). For small particles, in the limit, it might be expected that the sensitivity would relate to the cube of the particle size, *i.e.*, with volume, or mass. For larger particles, it might be expected that this correlation would shift toward the square of the diameter, *i.e.*, to the surface area.

Table 4. Summary of Microbiological Mass Response Data from the FID

Microbiological Aerosol	Mean Volume[7] (μm^3)	Mean Area Under The Curve (cm^2)	Relative Standard Deviation
S. marcescens	0.4 ± 0.4	15.1	15%
P. vulgaris	4.1 ± 3.3	19.0	18%
B. cereus	15.0 ± 7.7	23.0	20%
S. cerevisiae	90.1 ± 74.5	420.0	25%

Table 5. Summary of Experimental Conditions for Sodium Chloride Mass Response Study of the FID

Equipment Conditions	NaCl Aerosol Diameter					
Aerosol Generator	0.5 μm	1.0 μm	2.0 μm	4.0 μm	8.0 μm	10.0 μm
Weight Fraction (θ)	2.19×10^{-5}	1.75×10^{-4}	1.40×10^{-3}	4.09×10^{-3}	3.28×10^{-2}	6.40×10^{-2}
Liquid Feed Rate (Q, cc/min)	0.0774	0.0774	0.0774	0.212	0.212	0.212
Orifice Diameter (μm)	10	10	10	20	20	20
Vibrating Orifice Frequency (kHz)	200	200	200	200	200	200
Dispersion Air (SCFH)	4	4	4	4	4	4
Dilution Air (SCFH)	200	200	200	200	200	200
FID						
Hydrogen Flow (cc/min)	100	100	100	100	100	100
Secondary Air Flow (cc/min)	300	300	300	300	300	300
Sample Flow (cc/min)	250	250	250	250	250	250
Polarizing Voltage (VDC)	500	500	500	500	500	500
Readout						
Ordinate (mV/cm)	5	5	10	50	50	100
Abscissa (msec/cm)	0.2	0.2	0.2	0.2	0.2	0.2

A summary of experimental conditions appears in Table 5.

FID PRECISION

The 4 μm sodium chloride aerosol was selected for a precision study. The usual techniques of sample preparation and aerosol generation were employed. A set of at least 20 output waveforms was measured each day by planimeter for a total of six days. Precision data is listed in Table 6. This data was subjected to an analysis of variance to separate analytical and sampling errors. Relative standard deviations of 16.8 percent and 2 percent, respectively, were obtained. The small sampling error suggests that good control was achieved over the sample generation and sample transfer systems. Built into the relatively large analytical error (16.8 percent) were certain undefined sources of error which included electronic settings and planimeter data reduction (see Discussion). It would seem reasonable, therefore, to assume that a representative figure for the precision of the FID alone should lie in the range of 5 to 15 percent with the experimental arrangements employed in these studies. The 4 μm data point in Figure 7 is enclosed by 2 σ error bars.

FID SELECTIVITY

In the following section real FID output data is presented for a number of aerosols. This data, photographed from the original Xerox copy, consists of one or more oscillographic recordings, in which each recording corresponds to the pulse produced by a single particle. The oscilloscope was set to unblank and record only signals above the noise level; thus, leading edges of the waveforms are missing. Superimposed on the waveform data is a calibrated grid in which each horizontal line is separated from the next by one centimeter and each vertical hash mark is separated from its neighbors by one centimeter. Waveforms do not necessarily begin at the first vertical hash mark. The ordinate and abscissa for each waveform are mV/cm and sec/cm, respectively. Specific values are listed in the caption of each figure along with particle dimensions in parentheses.

The NaCl data (Figure 8) reveals a sharp leading peak with one shoulder. Several extraneous smaller peaks can be seen, occurring, perhaps, due to memory effects from the experimental system. Data for CsI (Figure 9) shows a single peak with no shoulders. An inorganic aerosol with a divalent metal ion, barium chloride, produces an output

Table 6. Analysis of Variance of FID Data from 4 μm NaCl Aerosol

Day	Mean Area Under the Curve (cm^2)	Variance	Relative Standard Deviations: Analytical	Relative Standard Deviations: Sampling
1	46.2	64.0	16.8%	2.2%
2	44.7	47.6		
3	45.5	49.0		
4	44.9	38.4		
5	50.0	100.0		
6	47.6	68.9		

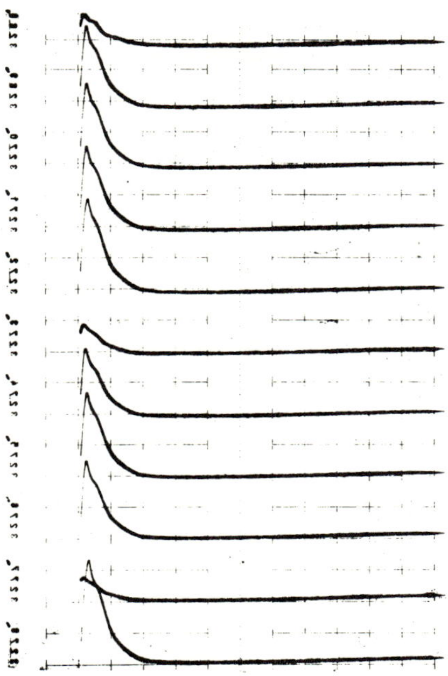

Figure 8. FID Output Signals from 4 μm NaCl Aerosol. Oscillograph Sensitivity 100 mV/cm. Time Base 1 msec/cm.

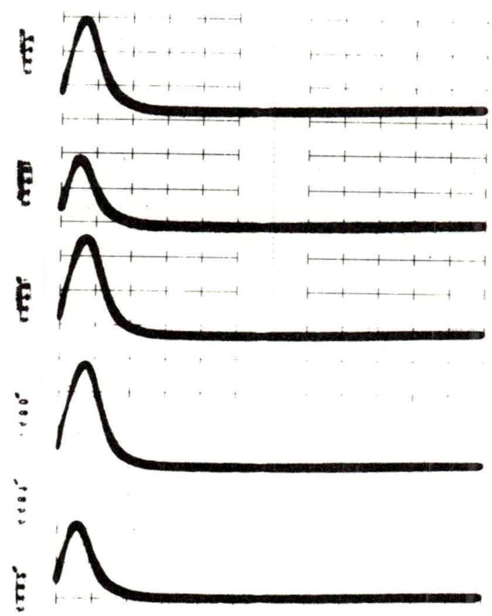

Figure 9. FID Output Signals from 4 μm CsI Aerosol. Oscillograph Sensitivity 200 mV/cm. Time Base 1 msec/cm.

waveform consisting of a leading peak followed by two shoulders (Figure 10). Other multivalent cation salts produced data like that of $BaCl_2$.

Data for latex, the only organic aerosol tested, reveals the more commonly observed type of waveform, namely that of a leading spike with two trailing shoulders (Figure 11).

All but two of the biological aerosols that were tested produced FID waveforms much like *Serratia marcescens* (Figure 12), *i.e.*, a leading spike followed by two shoulders. Unique waveforms were obtained from a yeast (*Saccaromyces cerevisiae*) and a mold (*Aspergillus niger*, as shown in Figures 13 and 14). Data for *Aspergillus niger* produced the largest signal of all species tested.

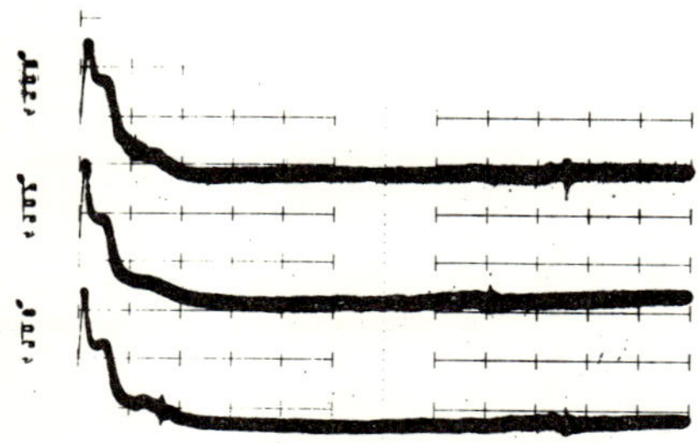

Figure 10. FID Output Signals from 4 μm $BaCl_2$ Aerosol. Oscillograph Sensitivity 20 mV/cm. Time Base 1 msec/cm.

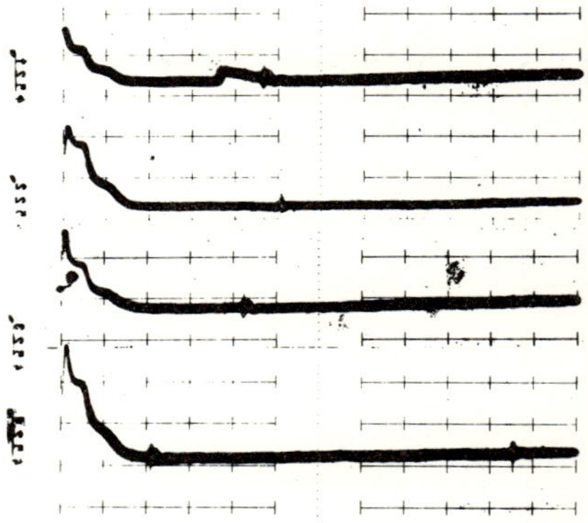

Figure 11. FID Output Signals from 0.8 μm Latex Aerosol. Oscillograph Sensitivity 20 mV/cm. Time Base 1 msec/cm.

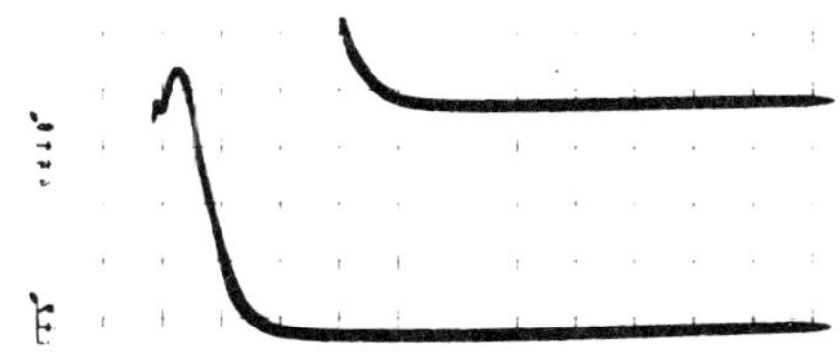

Figure 13. FID Output Signals from *S.c.* Aerosol (4 to 5 μm approximately). Oscillograph Sensitivity 200 mV/cm. Time Base 1 msec/cm.

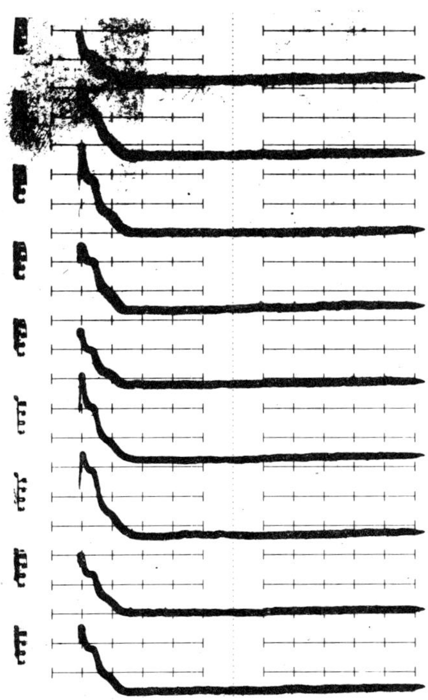

Figure 12. FID Output from *S.m.* Aerosol (0.5 to 1.0 μm by 1.0 μm. Oscillograph Sensitivity 20 mV/cm. Time Base 1 msec/cm.

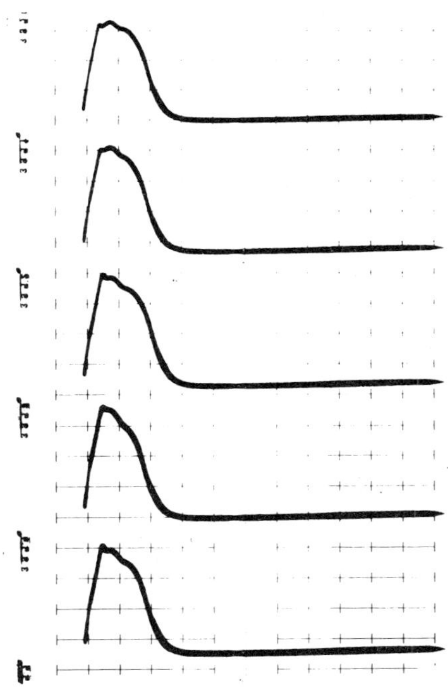

Figure 14. FID Output Signals from *A.n.* Aerosol. Oscillograph Sensitivity 500 mV/cm. Time Base 1 msec/cm.

SFID Signal

Typical data, gathered from the SFID are shown in Figure 15. The current pulses are recorded here as negative signals.* A series of waveforms are presented, beginning with electrode segment 1 of the SFID (the lowermost electrode of Figure 3) and running down the page with successive segments until the signal amplitude becomes trivial. These data were obtained by connecting the SFID amplifier to each electrode segment, separately, and recording a series of signals from individual aerosol particles of the same size and composition. During this period, all other electrode segments were grounded so as to maintain the electric field undisturbed. The hand-traced data of Figure 15 are a tableau of representative signals collected at each electrode. The abscissa (milliseconds/cm) and ordinate (millivolts/cm) are shown.

The signal, shown for a given segment of the SFID collection electrode, should be equivalent to that which would have been developed from the corresponding area of the collection electrode in the conventional FID. As such, each SFID signal was effectively a part of the total FID signal. At first glance, however, the SFID data did not appear at all related to the FID data. The SFID data was uniquely characterized by the presence of bipolar features in the waveforms. These signals predominated at the fringe electrodes furthest from the flame. They appeared to develop continuously from the start of the waveform. A closer inspection of the SFID data revealed the presence of a second signal component which was most apparent at those segments directly across from the flame. This component can be seen most clearly, in the case of the *Serratia marcescens* aerosol, at electrodes 2 to 4, as a trailing second peak. A similar signal can be observed for the 4 μm NaCl aerosol at electrodes 2 to 6--although not as clearly.

*An instrument malfunction prevented recording the data in conventional form as positive signals. Note that the ordinate is actually on millivolts--the tranduced readout unit of the oscillograph.

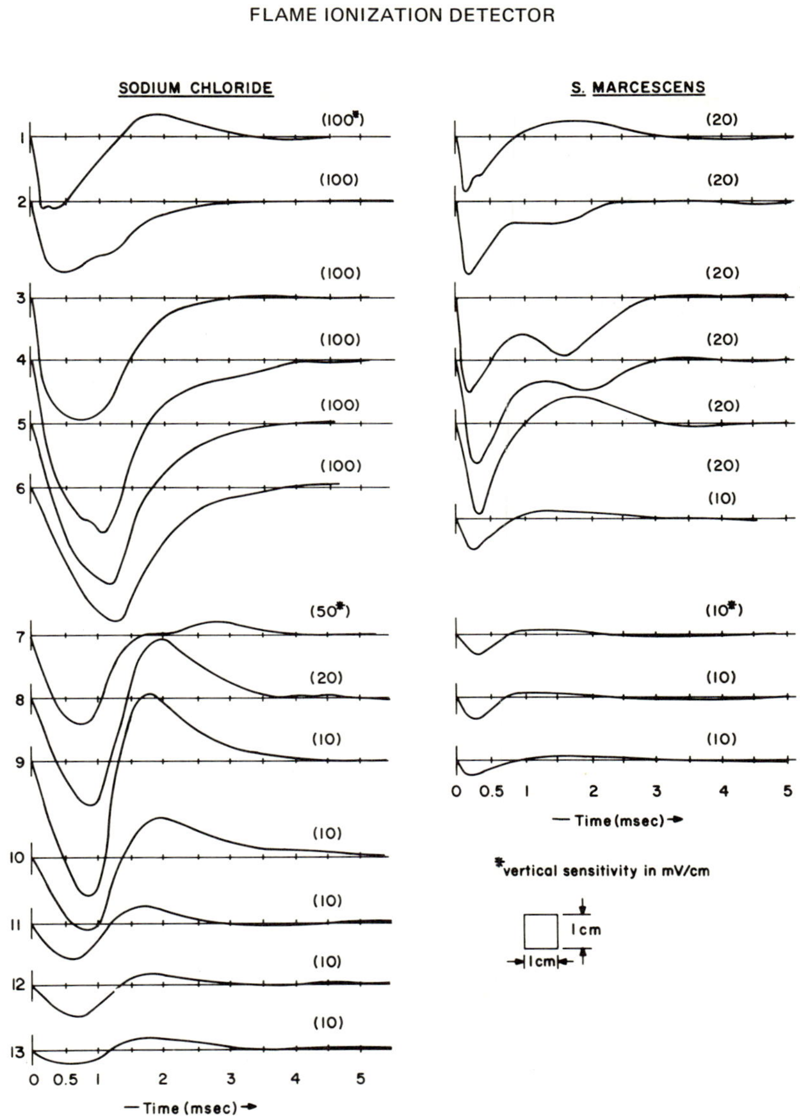

Figure 15. Tableau of Representative Output Signals from Collection Electrode Segments 1 to 13 of the SFID, as Produced by Individual particles from Separate Aerosols of NaCl and *S. marcescens* (hand traced from original data)

DISCUSSION

For organic and inorganic aerosols, the response of the FID was shown to be linear over the tested size range. The sensitivity appeared to be related with the particle diameter as the 3/2 power rather than the cube, as might have been expected. It has been reported that the combustion of hydrocarbon aerosols, in the 1 to 10 μm size range occurs in 1 to 100 microseconds.[5,8] The mean residence time in the flame of this study was approximately 500 microseconds. It may be that, with a larger flame, a cube relationship between particle size and sensitivity could be achieved.

The biological aerosols, tested in this study, displayed a different response. It is not clear what effect the presence of 70 to 80 percent water would have upon the particle integrity during the vaporization and excitation processes.

Remarks, above, regarding the relationship between particle size and sensitivity may also have a bearing on the precision of the FID. The significant error bars shown for some measurements in this study may reflect the incomplete production of chemiions brought about by the residence time limitation of the flame size. Nevertheless, the mean values reported should be reasonably accurate for the conditions employed.

Of particular interest during these studies was the selectivity of the FID, *i.e.*, the extent to which the waveform of the output current pulse could be uniquely related to composition. Of the tested electrode geometries, only from the parallel plate electrodes were signals obtained with discrete information which might serve potentially as a "fingerprint" for identification. These signals were composed, generally, of a leading spike with one or two trailing shoulders. Organic and biologic aerosols produce pulses with two shoulders while certain inorganic aerosols like NaCl of CsI produce pulses with marked diminuation or disappearance of one or both shoulders. Of possible practical significance were certain biological aerosols like yeast (*Saccharomyces cerevisiae*) and a mold (*Aspergillus niger*) which produced unique waveforms. Presence of these aerosols in the air might be detected by an FID equipped with a suitable pattern recognition system.

The electrodes of the FID, described above, form a parallel plate capacitor with the flame centered between the plates. As a particle is consumed in the flame and the resultant charge carriers are separated by the field, there occurs, effectively, a transient, localized change in the permittivity of the medium between the electrodes. The result of such an "upset" event is a transient fluctuation in the field and, also, in the steady state charge accumulated at each electrode, or plate. During, or following this process, the created charge carriers then arrive at their respective electrodes. The resultant output current pulse would be the algebraic sum of these two effects. If one takes a typical mobility of positive ions in ambient air as 1 cm^2/volt-sec,[9] and allows for the effect of the elevated temperatures within the FID enclosure upon the mean speed, then at 500 VDC and a flame-to-plate distance of approximately 7 to 8 mm, a time-of-flight of less than 3 milliseconds is reasonable. This is consistent with the duration of the FID pulse.

The net current from the positive electrode of the FID can be described, for the simple one dimensional case, by the expression:

$$I(t)^- = A\varepsilon_\circ \frac{\partial E(o,t)}{\partial t} - Aq^- b^- E(o,t)n(o,t) \qquad (4)$$

where $I(t)^-$ is the net output current at any time "t", "A" is the area of the collection electrode (cm^2), $\varepsilon_\circ$ is the permittivity (coul/volt-cm), E(o,t) is the electric field component normal to the positive electrode surface at time "t", b^- is the mobility of the negative charge carriers (cm^2/volt-sec) and n^- is the number concentration at the surface of the positive electrode (cm^{-3}). The first term refers to the current associated with the changing field at the electrode surface. The second term arises from the neutralization of the gaseous charge carriers as they arrive at the electrode surface. Consistent with the principle of electronuetrality, the net current at the collection electrode, $I(t)^+$, must be equal to that at the polarizing electrode, $I(t)^-$.

If one calculates the variation of the electric field with time at the surface of the electrode, then the first term in Equation 4 can be plotted as shown qualitatively in Figure 16.

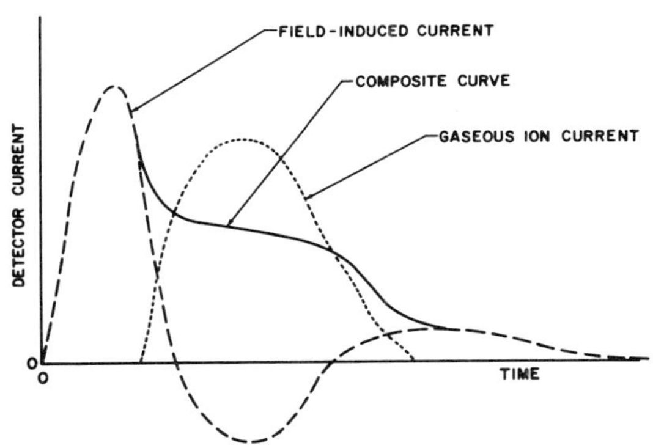

Figure 16. Sketch of Component Current Output Signals from a Parallel Plate FID

Also shown in Figure 16 is a qualitative sketch of the charge carrier current (dotted line), which is the second term in Equation 4. The arrival of the gaseous charge carriers at the positive plate would be delayed because of the time of flight factor.

Although qualitative, the presentation in Figure 16 is consistent with actual data from both the FID and the SFID. Thus, the composite curve of the overall current in Figure 16 can be compared with the FID data of Figure 6. The bipolar waveform of Figure 16 for the field-induced current can be compared with those observed in the SFID data of Figure 15. In the SFID data of Figure 15 these bipolar waveforms were observed most at the peripheral electrodes, *i.e.*, those above and below the flame. Many of the electrodes closest to the heart of the flame showed waveforms with a second peak. This is readily observed for *S.m.*, less so for NaCl. Consistant with the previous discussion, the second shoulder could be assigned to the gaseous charge carrier current. Thus, the field-induced current predominates at the peripheral electrodes while the ion current is restricted to the central set of electrodes. This is reasonable since one should expect the charge carrier current to be greatest at those electrode segments in a direct line of flight from the flame.

It should be noted that an enormous amount of data is produced by the SFID from a single particle. The sum total of all these data could, itself, serve as a fingerprint for particle composition.

BIBLIOGRAPHY

1. Ohline, R. W., "Hydrogen Flame Ionization for the Detection and Sizing of Organic Aerosols," *Anal. Chem., 37,* No. 1, 93 (1965).

2. Ohline, R. W., Thall, E., and Oey, P. H., "General Considerations Concerning Atmospheric Aerosol Monitoring with Hydrogen Flame Ionization Detector," *Anal. Chem, 41,* No. 2, 302 (1969).

3. Frostling, H., and Lindgren, P.H., "A Flame Ionization Instrument for the Detection of Organic Aerosols in Air," *J. Gas Chromatography, 4,* 243 (1966).

4. Crider, W. L., and Strong, A. A., "Flame Ionization-Pulse Aerosol Particle Analyzer (FIPAPA)," *Rev. Sci. Instrum., 38,* 1772 (1967).

5. Bolton, H. C., McWilliam, I. G., "Pulse Characteristics of the Flame Ionization Aerosol Particle Analyzer," *Anal. Chem., 44,* No. 9, 1575 (1972).

6. Berglund, R. N., and Liu, B. Y. H., "Generation of Monodisperse Aerosol Standards," *Env. Sci. Technology, 7,* No. 2, 147 (1973).

7. Breed, R. S., Murray, E. G. D., Smith, N. R., *Bergey's Manual of Determinative Bacteriology, 7th Edition,* Williams and Wilkins, Baltimore, (1957).

8. Essenhigh, R. H. and Fells, I., "Combustion of Liquid and Solid Aerosols," *Discuss. Faraday Soc., 30,* 208 (1960).

9. Von Engel, A., *Ionized Gases, 2nd Edition,* Oxford-at-the-Clarendon Press, (1965).

10. Altpeter, L. Jr., Senechal, A. J., Overland, D. L., *Performance of a Flame Ionization Detector as an Atmospheric Aerosol Monitor, II: The Multi-Electrode Flame Ionization Detector,* (SFID).

CONTACT ELECTRIFICATION APPLIED TO PARTICULATE MATTER-MONITORING

Walter John, Ph.D.

Air and Industrial Hygiene Laboratory
California Department of Health
Berkeley, California 94704

ABSTRACT

The theory of the charging of aerosol particles by contact electrification is reviewed, as well as the development over the past decade of monitors based on this phenomenon. Such monitors offer the advantage of continuous measurements over a wide range of concentrations. Moreover, the total electrical charge collected has been found to correlate well with the total mass determined gravimetrically. The lack of widespread use of these detectors may be due to some contradictory reports on their performance as well as uncertainty arising from the presently incomplete understanding of the operating principle.

The major factors which influence the instrumental response are discussed, including the recent finding by the author that the sensitivity is strongly dependent on the electrical resistivity of the material sampled. It is also found that the condition of the probe's surface has an important effect on the sensitivity. At present, theoretical understanding of the contact charging is semiquantitative for particles of metals but only rudimentary in the case of insulators.

CONTACT ELECTRIFICATION APPLIED TO PARTICULATE MATTER-MONITORING

Walter John, Ph.D.

Air and Industrial Hygiene Laboratory
California Department of Health
Berkeley, California 94704

INTRODUCTION

The transfer of charge between aerosol particles and a probe via contact electrification can be used as the basis of a monitor for particulate matter.[1-6] Important advantages of this type of monitor include the ability to make real time measurements with a response time of less than 10 seconds, sensitivity over a wide range of concentrations and an electrical signal convenient for data collection. Moreover, the simple probe can withstand the high temperatures encountered in-stack.[7] Most importantly, the electrical charge collected by the instrument has been found to correlate accurately with the mass determined gravimetrically.[4,7,8,9]

Despite more than a decade of development the contact electricity monitor has not seen widespread application. The principal reason appears to be uncertainty attributable to the lack of a quantitative theory for the contact electrification process.[10] While there is a qualitative theory for metal particles, there is no theory available for insulating materials. An experimental difficulty, as yet incompletely investigated, concerns changes in instrumental sensitivity depending on the condition of the surface of the probe.[2,9,11]

In the present paper the theory of the charging process will be discussed. The development of contact electricity monitors will be reviewed including data on the experimental characteristics. Finally, the present status of the monitor will be assessed.

THEORY OF CONTACT ELECTRIFICATION OF AEROSOL PARTICLES

Contact Charging

Contact electrification refers to electric charge transferred between two bodies as a result of pure contact with no sliding or rubbing.[11,12] By contrast, according to the strict definition, triboelectric charging involves rubbing. The latter may include the transfer of material and local heating.[11] In the case of the charging of aerosol particles there is little evidence on the relative importance of the various types of interactions. The present discussion will be limited to contact charging, which is better understood than the other charging mechanisms. It should be recognized that the other mechanisms may also be present.

For contact charging of large metal spheres good agreement between theory and experiment has been achieved by Harper.[11] When two metals touch, the difference in the contact potentials of the two metals will cause a flow of electrons to take place. The buildup of charge will continue until the resulting electrical potential equals the difference in the two contact potentials. Described in terms of energy levels, charge will flow to equalize the two Fermi levels (Figure 1). The resulting difference of potential of the two metals will be equal to the difference in the work functions of the metals.[11,13,14,15]

This simple picture of the charging is complicated by two additional effects.[11] First, some charge can flow by quantum-mechanical tunneling even when there is a small gap between the surfaces. Second, upon separation of the surfaces, the air can break down, resulting in an appreciable back flow of charge.

The theory for metal-metal contact can be generalized to include semiconductors.[11,15,16,17,18] Whereas the charges reside entirely on the surfaces of metals, for semiconductors there will also be a flow of charge from the interior. There will be an associated time constant given in seconds by $8.85 \cdot 10^{-14}\ K\rho$, where K is the dielectric constant and ρ the resistivity in ohm-cm. This implies that the amount of charge transferred will depend on the duration of contact. In the case of semiconductors having high resistivity there will be no appreciable flow of charge during a brief contact.

Insulators have an even higher resistivity so that there is no flow of charge from the bulk material. Surface states become the important source of charge for transfer. Such surface states are complicated, depending on the detailed physical and chemical condition of the surface. Adsorbed ions giving rise to electric double layers play an important role. According to Harper,[11] the charge transferred consists of ions rather than electrons.

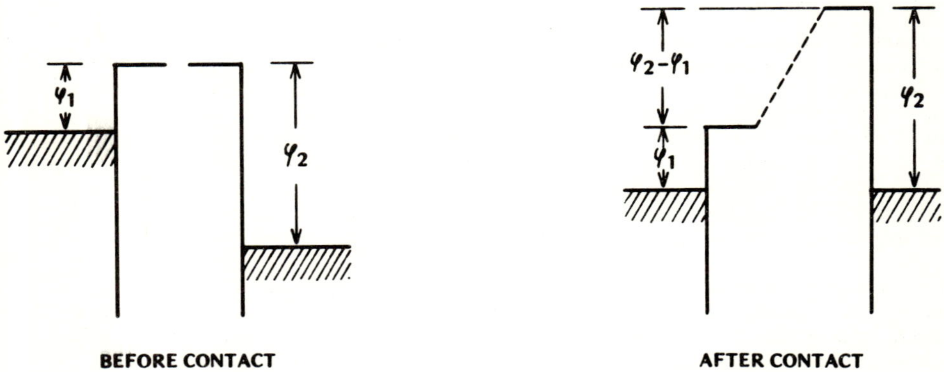

Figure 1. Energy level diagrams for two metals having work functions ϕ_1 and ϕ_2. After contact, the Fermi levels equilibrate resulting in a contact potential $\phi_2 - \phi_1$.

The foregoing discussion has concerned ideal surfaces; in actuality, the charging may be drastically altered by contamination of the surfaces or by the physical condition. The surfaces of metals will be inevitably oxidized. According to Harper[11] such oxide layers behave as metals with work functions of about 5.5eV. Most oxides are semiconductors; if the surface layer is not too thick, charge can be transported either by conduction or tunnelling. For insulators particularly, charging mechanisms other than by contact may have to be considered, such as electrolytic effects for moist surfaces and frictional effects resulting in transfer of material.

Charging of Aerosol Particles

Review of Cheng and Soo Theory - The charging of an aerosol particle is a dynamic process involving the impact of the particle with the surface of a probe. The theory developed by Cheng and Soo[6,19] considers the transfer of charge during a collision between two elastic spheres of radii a_1 and a_2. During the impact there will be a mean area of contact A_{21} lasting for Δt, both of these quantities depending on the kinematics of the collision and the elastic properties of the materials. While the spheres are in contact, the current density will be proportional to the voltage difference between the two spheres. If the current is integrated over the period of impact, the charge transferred is obtained:

$$Q_{21} = \frac{C_1C_2}{C_1 + C_2} (\varphi_2 - \varphi_1) [1 - e^{-\alpha\Delta t}]$$

where $C_{1,2}$ are the electrical capacities of the spheres
$\varphi_{1,2}$ are the work functions of the materials

and
$$\alpha = A_{21}h_{21} \cdot \frac{C_1 + C_2}{C_1C_2}$$

h_{21} the charge transfer coefficient, is given by

$$h_{21} = \frac{\sigma_1 \sigma_2}{d_1'\sigma_2 + d_2'\sigma_1}$$

with $\sigma_{1,2}$ the electrical conductivities and $d_{1,2}'$ the charge transfer lengths.

The transfer lengths are defined by the equations

$$J_{21} = \frac{\sigma_1}{d_1'} (V_c - V_1) = \frac{\sigma_2}{d_2'} (V_2 - V_c)$$

where J_{21} is the current density and V_c is the potential at the point of contact. Thus d_1' and d_2' are the effective distances in the material over which current is transported.

The following expressions for A_{21} and Δt are derived from Hertzian theory:

$$A_{21} = \frac{\pi a_1a_2}{a_1 + a_2} \cdot \alpha_1$$

$$\Delta t = \frac{2.94}{\Delta u_{21} \cos \theta} \cdot \alpha_1$$

where
$$\alpha_1 = \left[\frac{15}{16} \pi \left(\frac{1 + r^*}{2}\right) (\Delta u_{21})^2 \cos \theta \left(\frac{m_1m_2}{m_1+m_2}\right) \left(\frac{a_1+a_2}{a_1a_2}\right)^{1/2} (k_1+k_2)\right]^{2/5}$$

(note: α_1 is distinct from α)
with r^* the ratio of the rebound speed to the incoming speed
Δu_{21} is the velocity of approach
θ is the angle of impact
$m_{1,2}$ are the masses of the spheres

and $k_{1,2}$ are elastic parameters given by

$$k = \frac{1-\gamma^2}{\pi E}$$

and γ is Poisson's Ratio, E the modulus of elasticity.

Numerical Evaluation for Typical Parameters - It is instructive to evaluate numerically some of the above quantities using reasonable values of the parameters. Take sphere 1 to be the aerosol particle and sphere 2 to be the probe. The $a_1 << a_2$.

Using $k_1 = k_2 = 3.10^{-13}$ cm^2/dyne, $\theta = 0$, $\bar{\rho}_1 = 8g/cm^3$ (density), $a_1 = 10^{-5}$ cm, $\Delta u_{21} = 1.5 \cdot 10^3$cm/s, and $r^* = 1$,

we find that

$$\alpha_1 = 3 \cdot 10^{-7} \text{cm}$$
$$\Delta t = 6 \cdot 10^{-10} \text{s}$$
$$A_{21} \cong \pi a_1 \alpha_1 = 1 \cdot 10^{-11} \text{cm}^2$$

The short duration of the impact implies that the theory applies only to metals or good semiconductors ($\rho \lessapprox 10^3$ ohm·cm) since only then will there be appreciable current flow. The area of contact has a diameter roughly 1/20 of the particle diameter.

Q_{21} is the charge transferred per particle. The total charge transferred to or from the probe is given by

$$Q_T = N_1 \, Q_{21}$$

where N_1, the number of aerosol particles, is given in terms of the total mass of particles, M_1, by

$$N_1 = \frac{M_1}{\frac{4}{3} \pi a_1^3 \bar{\rho}_1}$$

For $\rho_1 < 10$ ohm·cm, $(1-e^{-\alpha\Delta t}) \cong 1$, and $Q_{21} = 4\pi\varepsilon_0 a_1 (\phi_2 - \phi_1)$. Then $Q_T \propto a_1^{-2}$. Thus the theory predicts that the total charge transferred to the probe is inversely proportional to the particle radius squared. Also, for $\rho_1 < 10$ ohm·cm, the charge transferred is predicted to be independent of the velocity of the particles relative to the probe. For $\rho_1 > 10$ ohm·cm, $Q_T \propto a_1^{-n}$, where $1 < n < 2$.

The magnitude of the charge transferred can also be evaluated. Consider metal particles with a radius of 0.05 μm. Assume that $\phi_2 - \phi_1 \cong$ 1 volt. Then $Q_{21} = 10^{-17}$ C/particle and Q_T = 2000 μC/g. This is three orders of magnitude greater than observed. The calculated charge could be brought into agreement with experiment by assuming an effective resistivity of $5 \cdot 10^4$ ohm·cm which is in the semiconductor range. The oxide layer on the particle surface could easily increase the effective resistance, not to mention contaminant layers on the surfaces of the particles as well as on the probe surface.[6]

While the theory is not quantitative, the model may still be useful in understanding the dependence of the charge on the physical parameters. However, the theory cannot as yet be regarded as verified by experiment. The above theory applies only to metals or good semiconductors. For materials with $\rho > 10^3$ ohm·cm, which includes most semiconductors and all insulators the charge transferred is that already on the surface in the area of contact. In the absence of detailed knowledge of the physical processes involved, it is not possible to carry out the analysis for the charging of insulating materials.

CONTACT ELECTRICITY MONITORING INSTRUMENTS

Review of Instrument Development

In monitoring instruments based on contact electrification, a flow of the aerosol is directed around or through an insulated probe. The transfer of charge from particle-probe collisions results in a current which is continuously monitored with an electrometer.

An early version of an instrument employing a spherical metal probe was described by Schütz.[1] A later version of his instrument, shown in Figure 2, utilized a probe-in-nozzle technique.[2]

Prochzaka[3,4] has described the development of the Konitest, an instrument which was produced commercially for a time. In the Konitest, the electrode is a tube of a semiconductor, steatite (magnesium hydrosilicate). The gas is introduced radially and the resulting helical path causes the particles to impinge on the walls of the tube. In a second version, the tube is shaped as a Venturi nozzle (see Figure 3).

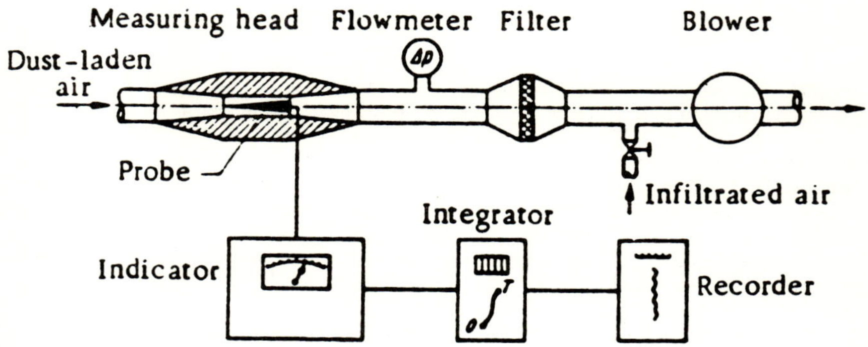

Figure 2. A dust monitor based on the contact electricity principle using a probe in a nozzle according to Schütz (Ref. 2).

Soo and his collaborators[6,20,21] constructed instruments using a spherical metal electrode and also a tubular electrode. In the USSR, Kisler[22] has described some contact electricity instruments including a monitor using a wire probe. The IKOR Air Quality Monitor was apparently independently developed and is now commercially available.[7] It utilizes a bullet-shaped Inconel probe in a pipe.

Characteristics of the Contact Electricity Monitor

A few experimental investigations have been made of the characteristics of the contact electricity monitor. Most were carried out under field rather than laboratory conditions. On some of the characteristics there are contradictory reports. Some of the findings on the important characteristics are reviewed below.

Electrical Current vs Mass Concentration - In most cases the instruments are to be used as an indirect measurement of the particulate mass concentration in the gas monitored. It is most important to establish the correlation between the electrical current from the probe and the mass concentration measured gravimetrically. Schütz[2] reported the following relationship for quartz dust:

$$I = ak^b$$

where I is the electrical current in amperes, k is the mass concentration in mg/m^3, the exponent b ranges between 1.26 and 1.30, and a is inversely proportional to the particle diameter.

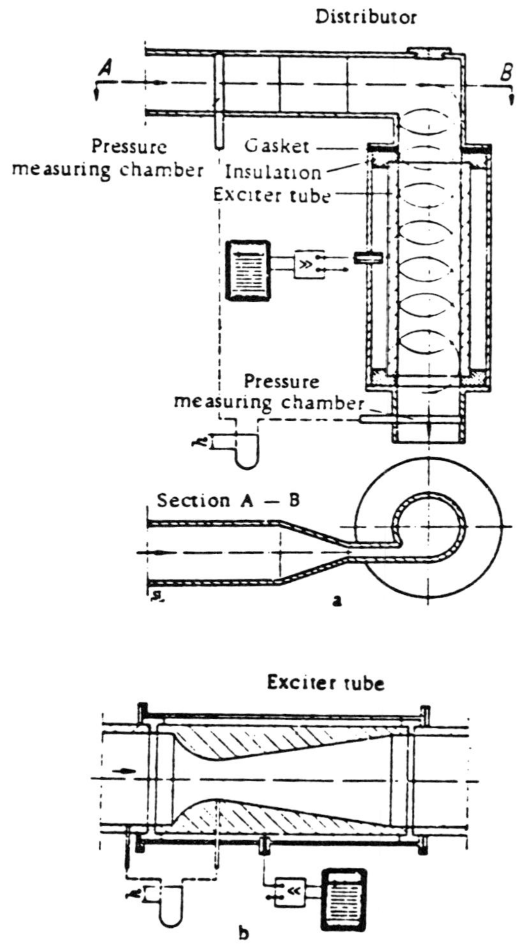

Figure 3. Two versions of the Konitest dust sampler are shown; in (a) the electrode is a cylindrical tubulence chamber, in (b) a Venturi nozzle is employed (from Prochazka, Ref. 4).

On the other hand, Prochazka[4] found the Konitest to give an accurately linear indication for concentrations from 0 to 3 g/m^3. Samples of his data are plotted in Figure 4. The dusts were sampled at various industrial sites. Ito, et al[23] also reported a linear relationship for a Konitest monitoring cement kiln exhaust at the exit of an electrostatic precipitator for concentrations up to 10 g/m^3. Schnitzler, et al[8,10] found a correlation coefficient of 0.92 between the gravimetric mass and the Konitest reading for effluent from a coal-fired plant. The performance of the Konitest compared very favorably to the best response obtained from optical transmissometers and beta radiation attenuation monitors at the same location.

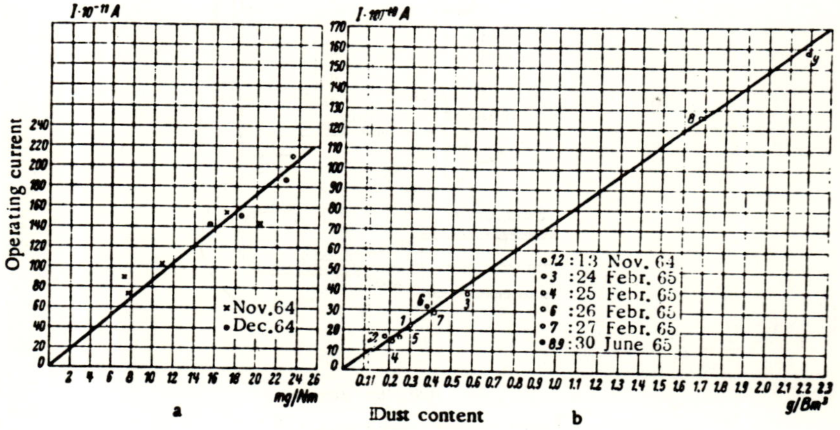

Figure 4. Calibration curves showing the Konitest current vs the mass concentration for (a) blast furnace gas and (b) flue gas, from Prochazka (Ref. 4).

Cheng and Soo[6,21] obtained plots of mass flow vs probe current using coal dust. The plots show small deviations from linearity, being concave upward.

Field tests of the IKOR Air Quality Monitor show that the integrated current (total charge) has a reproducible ratio to the gravimetric mass for aluminum oxide.[7] Good agreement was also obtained between the total charge and the mass concentration obtained with an EPA sampling train. These tests did not span a wide range of concentrations. The author has investigated the IKOR AQM using laboratory-generated aerosols.[9] The total charge was found to vary linearly with gravimetric mass for aluminum, a conductor, and aluminum oxide, an insulator (see Figure 5).

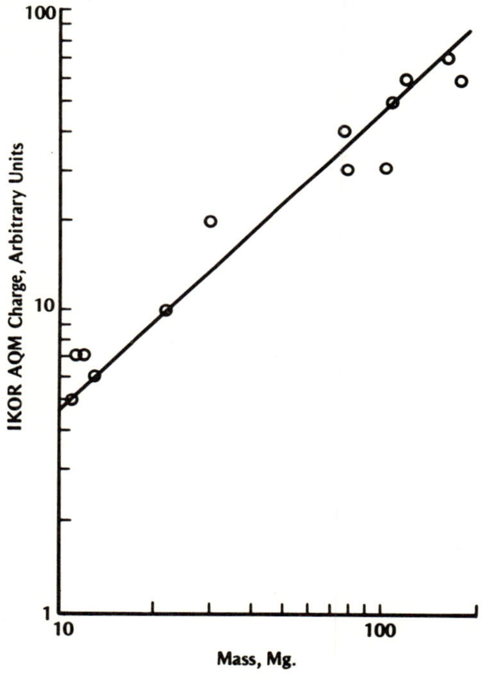

Figure 5. Charge vs mass of aluminum oxide particles measured with an IKOR monitoring instrument by the author (Ref. 9). The line is drawn at 45° corresponding to a linear dependence of charge on mass.

Composition of Dust Sampled - Although field tests have been conducted with a variety of dusts, there has been no reported correlation of the instrumental response to some characteristic of dust sampled. Schütz[2] concluded that the influence of the properties of the material was far less decisive than the particle size. Ito, et al[23] stated that the current obtained for a given mass concentration depended on the material, but that for cement dust the electrical resistance had little effect on the output. Most users of the contact electricity monitor have simply empirically calibrated the monitor for each type of dust sampled.

The author has measured the sensitivity (charge/mass) of the IKOR AQM for laboratory-generated aerosols of a number of substances.[9] The data are listed in Table 1 in order of increasing sensitivity. It is to be noted that this is also generally the order of increasing conductivity. In Figure 6, the sensitivities are plotted against the log resistivity of the materials.

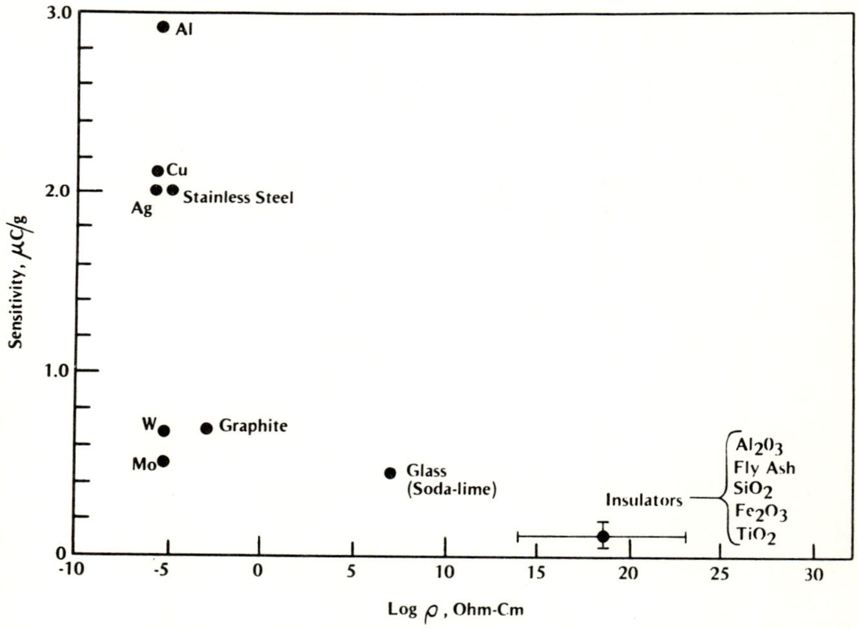

Figure 6. Sensitivity (charge/mass) of the IKOR AQM vs the log of the electrical resistivity of the material sampled, from the author's work (Ref. 9).

For the glass particles the resistivity was quoted by the supplier. The insulators have been lumped into a single point with a vertical bar covering the range of sensitivities. The horizontal bar covers the range nominally found for insulators; the individual resistivities of the samples used are not known. The correlation between sensitivity and resistivity is evident except for W and Mo. These two metals have resistivi-

TABLE 1

SENSITIVITY OF THE IKOR AQM

		μC/g
I.	Insulators	
	Titanium dioxide	0.051 ± 0.004*
	Red iron oxide	0.06 ± 0.01
	Silicon dioxide	0.09 ± 0.02
	Fly ash	0.148 ± 0.005
	Aluminum oxide	0.20 ± 0.01
	Glass beads 3-8μ	0.41 ± 0.03
	Glass beads 0.5-3μ	0.46 ± 0.01
II.	Intermediate conductors	
	Molybdenum**	0.51 ± 0.04
	Tungsten**	0.67 ± 0.13
	Carbon black	0.68 ± 0.09
III.	Metallic conductors	
	Stainless steel	2.0 ± 0.5
	Silver	2.0 ± 0.2
	Copper	2.1 ± 0.2
	Aluminum	2.9 ± 0.4

*standard deviation
**nonspherical metal particles

ties approximately three times that of the other metals used but the decrease in sensitivity was disproportionate. One possible factor was the nonspherical shape of the particles which could reduce the effective contact area. The other metal particles used in this study were spherical.

The ranking according to electrical resistivity can be understood in terms of the characteristic time for charge transport. For good insulators, this time is large so that only the area in the vicinity of the contact is involved in charge transfer. The area is larger for semiconductors and charge can also be transported from the interior. For metallic conductors the charge can probably be transported from the entire particle. The oxide layers on the metal particles are presumably thin enough so that they can be penetrated either by electron tunnelling or by conduction with a small time constant. The theory of Cheng and Soo[6] predicts a dependence of the sensitivity on the resistivity, however, the agreement with the experimental data is not quantitative.

These data provide a basis for predicting the sensitivity for other substances, at least for dry dusts. It is likely that they will fall within the range spanned by the table. The wide range in the sensitivities (factor of 60) implies that a small amount of an insulator would be difficult to detect in the presence of metals or good semiconductors.

Dependence of the Current on Particle Size - As pointed out above, Schütz found the constant a to be inversely proportional to particle diameter. Thus, higher sensitivity is predicted for small particles. On the other hand, Cheng and Soo[6] predict a current independent of particle size for one experimental configuration. They report that this was verified by Min. Ito, et al[23] report the Konitest to be little affected by particle size when sampling cement dust.

The author's experiments with the IKOR AQM showed that the sensitivity was not strongly dependent on particle size.[9] A polydisperse Al and a sized Al sample had nearly the same sensitivity and two sizes of glass beads gave similar results (Table 1).

On the other hand the IKOR AQM was found to barely respond to high levels of smoke from a filter cigarette. This shows there is an effective cut off in the response to very small particles. The experimental sensitivity depends not only on the contact charging parameters but also on the probability that the particles impact on the probe. Small particles tend to follow the streamlines around the probe. This smaller probability of impaction for small particles may offset the increased charging expected theoretically.

Effects of Temperature, Humidity, and Precharge - Schütz[2] stated that the contact electricity monitor should not be operated above 70°C due to the presence of thermal emf's. However, a modified IKOR AQM has been operated at 593°C[7].

Water layers on surfaces usually contain dissolved impurities. Some authors believe electrolytic ions to be important for contact charging.[12] However, Harper[11] does not, and tried unsuccessfully to obtain an experimental correlation between humidity and charging.

Schütz[2] reported no effect on his instruments' performance for relative humidity up to 99%. He did warn that droplets of water cause erroneous indications. Cheng and Soo[6] point out that the conductivity of mineral particles increases with the relative humidity. Ito, et al[23] state that humidity caused no problems in their tests of the Konitest. Kisler[22] states that the humidity of the gas should be below the dew point. The author found humidity up to 69% not to affect the IKOR AQM, but droplets caused the instrument to break down.[9]

The author did not observe any significant effect on the IKOR AQM produced by pre-existing charge on the aerosol particles.[9] Also, if precharging effects existed, humidity would be expected to affect the response. However, Kolar[24] suspected charge was interfering with his Konitest measurements on dust from an electrostatic precipitator.

Probe Material - With the exception of the Konitest, most instruments have used metal probes. According to Schütz,[2] the use of metal for the probe ensures that the charging will always be positive. It seems clear that this is a dubious assumption; in fact in field tests of the IKOR the sign of the current has been found to vary with material and the standard IKOR AQM is equipped with an automatic polarity-reversing circuit to maintain positive indication on the recorder.[7]

Dynamic Response - The author compared the response of the IKOR AQM to an opticle particle counter.[9] The time constant (1/e) of the IKOR AQM is approximately 10 seconds. Since the Climet optical analyzer is orders of magnitude faster, an R-C network was added to increase the Climet time constant to 4 seconds. Figure 7 shows the superimposed chart recordings of the two instruments for polydisperse aluminum oxide dust. Some of the differences in the records, particularly the more extreme highs and lows of the Climet trace, can be attributed to the 2.5 times shorter time constant of the Climet.

On the basis of many such runs, it was concluded that the correlation of the current records was satisfactory. There is one difference which should be mentioned. Sometimes at the very beginning of a run, the IKOR AQM was much less sensitive. The sensitivity increased by an order of magnitude in the first minute or so. This sometimes happened even if the instrument had been previously running on the same dust, sometimes even within the hour. This sensitizing effect will be discussed further below.

Surface Condition of the Probe - Direct evidence was obtained by the author that the condition of the surface of the probe of the IKOR instrument has a very important effect on the sensitivity.[9] It was mentioned above that there is sometimes a sensitizing effect at the beginning of a run. A long term change in sensitivity was also observed during a series of measurements, the sensitivity increasing by a factor of 5 to 10 over a period of about one month. Between runs with different substances, the probe was simply wiped clean with a dry cloth. On the other hand, the sensitivity was found to decrease by an order of magnitude when the probe was scrubbed with abrasive cleanser and detergent.

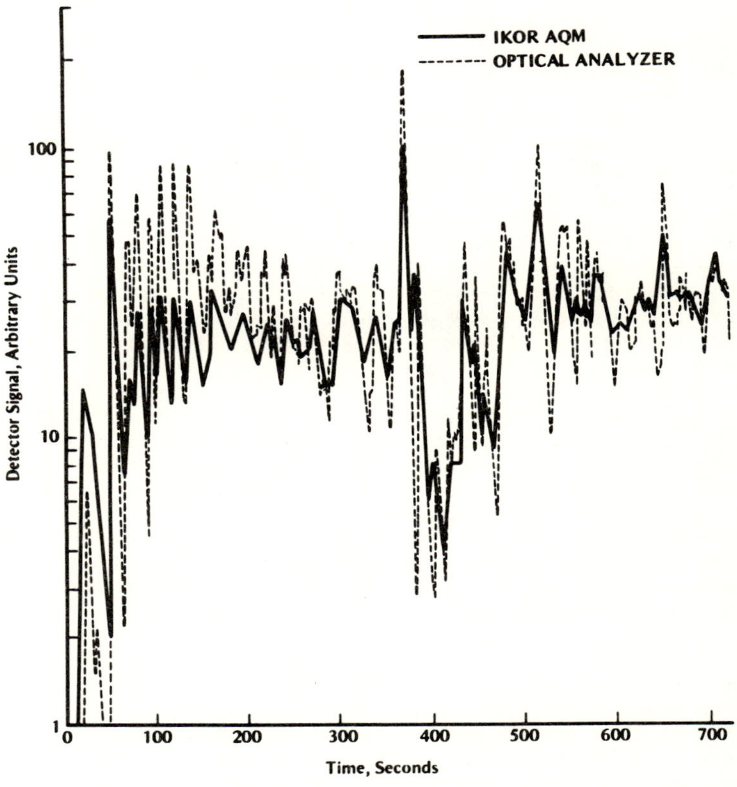

Figure 7. Dynamic response of the IKOR AQM compared to that of an optical particle counter for aluminum oxide particles, from the author's measurements (Ref. 9).

There would seem to be two possible explanations of the surface sensitizing effect. One would be a cleaning effect or exposure of fresh surface produced by the bombardment by particles, analogous to sandblasting. The other possibility is that a coating of particles builds up on the surface, altering the charging conditions, since particles would then encounter other particles rather than the surface itself.[5,25] At present the evidence is insufficient to allow one to decide between these two possible explanations.

CONTACT ELECTRIFICATION

SUMMARY

Experiments confirm that the total charge collected by the contact electricity monitor correlates well with the mass determined gravimetrically. No evidence has been found for disturbing effects of humidity below the dew point. The instrumental response cuts off for very small particles. The sensitivity for various materials depends on the electrical resistivity, ranging over a factor of 60 from metals to insulators. Also the sensitivity depends on the condition of the surface of the probe.

The theory of Cheng and Soo for particles of metals or good semiconductors overestimates the contact charge by several orders of magnitude but the discrepancy may be due to effects such as oxidation of the surfaces of real particles. No theory is available for the charging of insulators due to lack of understanding of the physical mechanism.

In view of the evidence reviewed here it would appear that the contact electricity monitor holds considerable promise for stationary source monitoring, expecially for non-sticky effluent of relatively constant composition. It is not suitable for ambient air because of insufficient sensitivity and ambiguity in the response to complicated, varying mixtures of materials.

Additional work is necessary to clarify the nature of the probe surface sensitization effect or to control it within limits. Better theoretical understanding of the charging mechanism is desirable but the complications imposed by actual particle surfaces are formidable.

ACKNOWLEDGEMENTS

The author's work on the contact electricity monitor was supported by Grant No. R-802726-01 from the U.S. Environmental Protection Agency and sponsored by California State College, Stanislaus.

I thank Mr. John Nader for his encouragement and helpful suggestions. Thanks are also due to Mr. Arnold H. Gruber of IKOR, Inc. for supplying valuable information.

REFERENCES

1. Schütz, A. Eine Anordnung zur Registrierenden Kontactelektrischen Staubmessung. Staub 24: No. 9, 359-363, 1964.

2. Schütz, A. A Recording Dust-Measuring Instrument Based on Electric Contact, with Logarithmic Indication. Staub 26: No. 5, 18-22, 1966.

3. Prochazka, R. Neueste Entwicklung des auf Kontaktelektrischer Basis Beruhenden Staubgehaltsmessgerates Konitest. Staub 24: No. 9, 353-359, 1964.

4. Prochazka, R. Recording Dust Measurement with the Konitest. Staub 26: No. 5, 22-28, 1966.

5. Schütz, A. Technical Dust Control Principles and Practice. Staub 26: No. 10, 1-8, 1966.

6. Cheng, L. and Soo, S. L. Charging of Dust Particles by Impact. J. Appl. Phys. 41: 585-591, 1970.

7. IKOR, Inc., unpublished reports, and Gruber, A. H., private communication.

8. Schnitzler, H. Messtand fur die Prufund und Kalibrierung von Registrierenden Staub-und Gasmessgeraten in einem Steinkohlengefeuerten Kraftwerk. SchrReihe Ver. Wass-Boden Lufthyg, Berlin-Dahlem, V. 33, Stuttgart, 1970.

9. John, W. Investigation of Particulate Matter Monitoring Using Contact Electrification. Environmental Protection Agency, Research Triangle Park, N.C., Technology Series Report Number EPA-650/2-75-043, February, 1975, 45p.

10. Sem, G. J., et al. Instrumentation for Measurement of Particulate Emissions from Combustion Sources. Thermo-Systems, Inc. Vol I: Particulate Mass, Report Number APTD-0733, April 1971 and NTIS PB 202 665.

11. Harper, W. R. Contact and Frictional Electrification. Oxford, Oxford U. Press, 1967. 369 p.

12. Loeb, L. B. Static Electrification. Berlin, Springer-Verlag, 1958, 240 p.

13. Hendricks, C. D. Charging Macroscopic Particles. In: Electrostatics and Its Applications, Moore, A. D. (ed.), New York, John Wiley & Sons, 1973, p. 64-67.

14. Harper, W. R. The Volta Effect as a Cause of Static Electrification. Proc. R. Soc. A 205: 83-103, 1951.

15. Vick, F. A. Theory of Contact Electrification. Brit. J. Appl. Phys. London, Suppl. No. 2: S1-S5, 1953.

16. Harper, W. R. How Do Solid Surfaces Become Charged? In: Static Electrification, Proceedings of the Conference organized by the Institute of Physics and the Physical Society, Static Electrification Group, London, May 1967. Inst. of Phys. and Phys. Soc. Conference Series Number 4. p. 3-10.

17. Krupp, H. Physical Models of the Static Electrification of Solids. In: Static Electrification, 1971, Proc. of the Third Conf. on Static Electrification organized by the Static Electrification Group of the Institute of Physics held in London, May 1971. Conf. Series No. 11, The Institute of Phys., London and Bristol. p. 1-16.

18. Soo, S. L. Fluid Dynamics of Multiphase Systems. Waltham, Mass., Blaisdell Publ. Co., 1967. p. 413-416.

19. Soo, S. L. Dynamics of Charged Suspensions. In: Topics in Current Aerosol Research, Vol. 2, International Reviews in Aerosol Physics and Chemistry. Oxford, Pergamon Press Ltd., 1971. p. 71-73, 90-93.

20. Soo, S. L., Stukel, J. J. and Hughes, J. M. Measurement of Mass Flow and Density of Aerosols in Transport. Envir. Sci. and Tech. 3: No. 4, 386-393, 1969.

21. Cheng, L., Tung, S. K. and Soo, S. L. Electrical Measurement of Flow Rate of Pulverized Coal Suspension. J. of Eng. for Power, Trans. of the ASME, 92A: No. 2, 135-145, 1970.

22. Kisler, S. YA. Monitoring of Concentration of Disperse Phase of Aerosol Flows by Electric Contact Method. Mekh. Avtomat. Proizvod., 26: No. 9, 27-28, 1972. (In Russian)

23. Ito, T., Saito, H. and Furuya, N. Continuous Dust Content Measurement Konitest. Proc. Japan Soc. of Air Poll., 13th, Nov. 1972. p. 247. (In Japanese)

24. Kolar, J. The Electrostatic Dust Measuring Device "Konitest" and its Functioning in a Remote Heating Plant. Tech., Ueberwach, Duesseldorf 10: No. 6, 188-190, 1969.

25. Soo, S. L. and Trezek, G. J. Turbulent Pipe Flow of Magnesia Particles in Air. Ind. and Eng. Chem. Fundamentals 5: No. 3, 388-392, 1966.

OPEN CAVITY LASER "ACTIVE" SCATTERING PARTICLE SPECTROMETRY FROM 0.05 TO 5 MICRONS

Robert G. Knollenberg
and
Robert Luehr

Particle Measuring Systems, Inc.
1855 South 57th Court
Boulder, Colorado 80301

ABSTRACT

An Active Scattering Aerosol Spectrometer is one that uses the active open cavity of a laser as the source of particle illumination. The interferometric aspects of the oscillating radiation illuminating the particles produces both forward and backward scattered radiation at all collecting angles. This, coupled with the fact that the collecting optics solid angle can be considerably greater than two steradians, results in a system with great particle sensitivity and reduced sensitivity to refractive index.

The particles are illuminated with a source of radiation many times greater than that possible with the hottest incandescent source known. This results in a system fully capable of sizing particles several hundred Angstroms diameter using photomultiplier detectors or 0.1 microns diameter using solid state silicon detectors. The fundamental limit of size detection is governed by particle evaporation rather than the noise component of the scattering from the molecular gas.

By using a beam splitter and double pulse height analysis, a system has been constructed which determines where a particle passes through a beam in the sampling volume, and reduces background light. The entire sampling volume has unrestricted flow precluding problems with evaporation of wet aerosols and the effects of plumbing walls.

OPEN CAVITY LASER

"ACTIVE" SCATTERING PARTICLE SPECTROMETRY

FROM 0.05 TO 5 MICRONS

Robert G. Knollenberg
and
Robert Luehr
Particle Measuring Systems, Inc.
1855 South 57th Court
Boulder, Colorado 80301

1. INTRODUCTION

Techniques that employ the active open cavity of a gas laser as a particle illumination source for light scattering measurements are what we designate as active scattering techniques. Such an instrument is particularly applicable to aerosol measurements. The open cavity laser was first used by Schleusener[1] as a particle illumination source. He performed particle extinction measurements to size particles of several tens of microns in size. Procter[2] has reported a similar instrument to measure the surface area of dusts. Schuster and Knollenberg[3] detailed the open cavity extinction process with particular attention given to the laser gain dynamics.

The open cavity extinction process is known to be quite insensitive to refractive index and particle shape, however, there are certain difficulties in achieving practical instrument designs to measure in the sub-micron range. The problem generally becomes one of resolving a very small loss in the presence of large signal background in the presence of noise.* A single frequency laser does have considerable

* Very few lasers are thermally quiet. Most exibit high frequency (100KHz-5MHz) components of 0.1 - 1% noise modulation in spite of current regulation. At low frequency these are attributed to axial mode competition with considerable FM. At high frequency is the characteristic anode associated modulation above 1MHz. The theoretical extinction size detection limit associated with the shot noise on the transmitted light is about 0.3u.

advantage in such applications.

Recognizing the lower limit of detectivity of the open cavity extinction process to be in the vicinity of 0.3u we turned to open cavity scattering or active scattering in 1971 to extend to smaller sizes. Some work was also done with interferometrically enhanced scattering wherein a standard laser was output coupled to a secondary total reflecting mirror. This method is still under study but has measurable disadvantages over active scattering. The first operational active scattering aerosol spectrometer (ASAS) was designed as an airborne instrument for measurements of the aerosol population in the urban plume of St. Louis as a part of Metromex (Metropolitan Meteorological Experiment). It was developed by Knollenberg and flown aboard the University of Chicago's Cloud Physics Laboratory aircraft. Figure 1 and 2 show some results of multilevel measurements using this instrument which operated multimode and had a useful size range of about 0.25 to 5 microns.

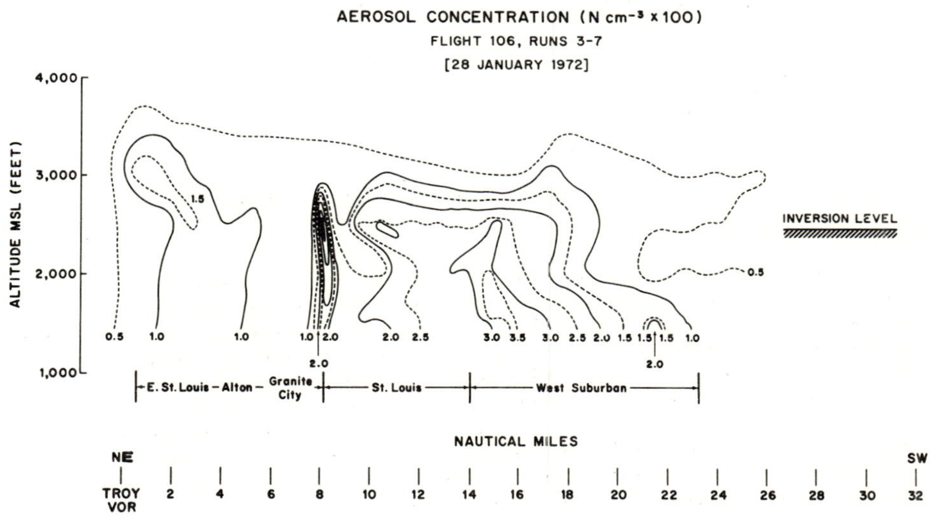

Figure 1. Vertical cross section of ASAS measured aerosol concentration in St. Louis urban plume. Note high density contours resulting from power plant plume.

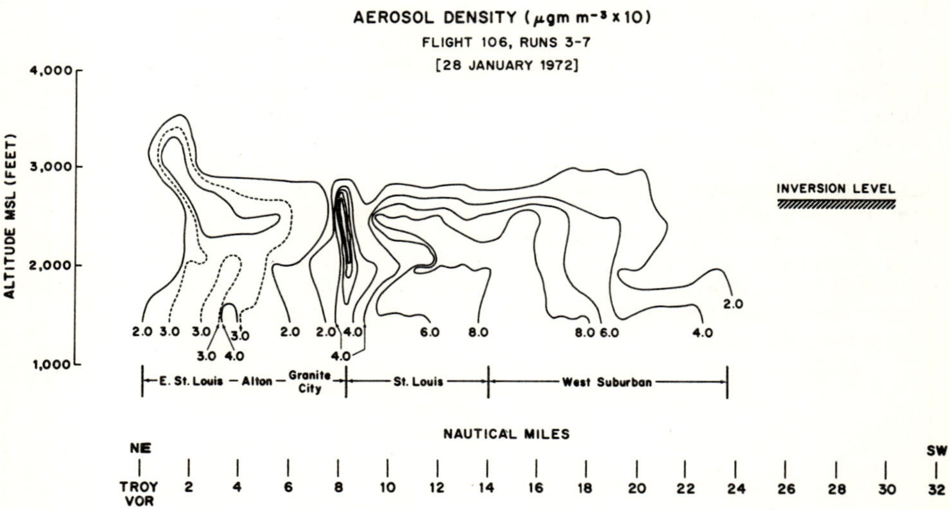

Figure 2. Aerosol density corresponding to Figure 1.

In early 1972, work was also initiated on ASAS instruments for NASA space related measurements in a hard vacuum (see Knollenberg[4] for a description of such a device). A number of active scattering aerosol spectrometer instruments have also been manufactured by PMS, Inc. Such an instrument is also described by Schehl, et.al[5].

The present work is a description of a significantly improved ASAS suitable for laboratory and ground based field measurements. An airborne fully "in situ" instrument is also under development. These instruments have ultimate sensitivities well below 0.1 microns and have several unique features to be described.

2. THE LASER CAVITY AS A PARTICLE ILLUMINATION SOURCE

There are perhaps five distinct advantages of using the laser cavity as a source of particle illumination rather than classical laser scattering approaches. These are according to decreasing importance.

1. Greatly increased illumination levels
2. The active interferometer reduces uncertainty due to refractive index and particle shape.
3. Available illumination reference exactly proportional to particle illumination.
4. Enhanced light collecting configurations with reduced stray light possible.
5. Available Doppler signals for velocimetry applications.

There is little doubt that our initial attraction to the active cavity was the available high energy density. As a source of illumination the laser cavity is unmatched in available intensity. The following exercise is worth following. Shown in the optical diagram of Figure 3 is the typical cavity configuration for ASAS production instruments. A hybrid laser tube is constructed with a 35cm high reflectivity (99.9+%) mirror sealed at one end of the plasma tube and a Brewster's window termination at the other. A second plane high reflectivity mirror completes the laser cavity. This results in a hemispherical cavity operating in TEM_{00} mode with a beam having a spot diameter of about 300u at the plane mirror and 800u at the curved mirror.

Now consider the fact that the manufacturer uses a 1% transmitting mirror (generally the curved mirror) in this same tube marketed as a 2mW device. It becomes obvious that in such a device, the radiation progressing towards this output mirror must be 100 times as great as the output or roughly 200mW. Now substitute a highly reflective mirror for the 1% transmitting mirror. Further intensity buildup is limited only by internal cavity losses. Measurements generally show an additional increase of a factor of 5 to 6 indicative of approximately 0.2% fixed losses at the mirrors and Brewster's window. Thus, we have achieved a source of approximately one watt of radiating power illuminating a particle from both directions. At the sample volume location the laser beam is approximately 500 microns diameter. Assuming a Gaussian intensity distribution, a brightness in excess 1000 W/cm^2 is computed. We have measured values as high as 4000 W/cm^2.

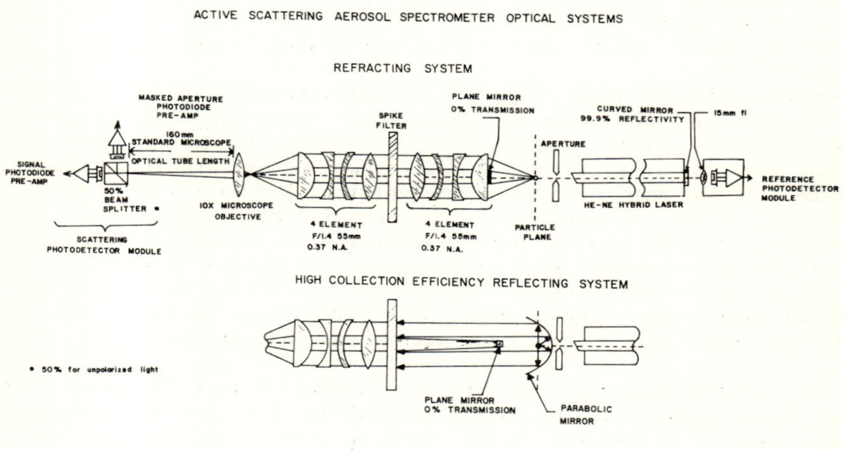

Figure 3. The refracting optical system is generally used. The high efficiency reflecting system is reserved for maximum resolution applications.

Probably the most intense incoherent source available is the PEK LABS 107/109 high pressure mercury arc lamp. It's corresponding brightness over it's entire visable spectrum approaches 250 W/cm^2. In any practical optical condensing system, it is nearly impossible to collect more than 20% of the light resulting in an illumination source more than one order of magnitude less intense than the laser just described. It is also no small matter that the laser as described is collimated to 0.5mr while the arc lamp would have a measured divergence of more than one radian.

It should be pointed out that a highly focussed laser beam can achieve even greater intensities than computed here. However, the resulting beam affords an impractically small sample volume. It should be apparent that the natural intrinsic beam width in the cavity is on the same order as the sample "view volume" dimensions of more standard optical counters. One might ask whether or not an argon or other more powerful laser might offer additional advantages. Possibly, however, the illumination levels of the He-Ne laser are already approaching potentially high enough levels to vaporize particles. For instance, a 0.05u diameter particle with an absorption cross section equivalent to it's geometric cross section and a heat of vaporization of water would vaporize in the above described beam for transit times longer than 300u sec.* In our instruments transit times are generally an order of magnitude faster.

Scattering within an open cavity laser is a quite distinct phenomenon from classical scattering. It is most closely approximated by scattering within a standing wave. Various experiments have contributed to a growing understanding of the phenomenon. Probably most important is the fact that the particle itself determines to a certain extent the intensity of radiation illuminating it. Specifically, it can be shown that when a particle interacts with the internal laser beam, that the extinction loss always exceeds the scattering loss for a non absorbing particle. This results in the non-linear response first reported by Schleusener[1] for extinction within the cavity. Schuster and Knollenberg[3] also predicted that the extinction process itself would not be expected to follow classical behavior showing rather suppressed resonance behavior due to interferometric cavity effects. More will be said in this regard in Section 5.

---

* The absorption cross section to mass ratio is the significant parameter here and typically has a maximum value in the vicinity of 500Å. The Qabs (efficiency factor) can be 10 times the Qsca at these sizes (see Van de Hulst[6] page 180).

A collimated laser beam is probably the most manageable light source available. The beam internal to the cavity has further advantages since it does not require dumping in a light trap and background light is more easily controlled. It is possible to design systems with nearly 4π steradian collecting solid angles (see the hemispherical mirror configuration of Schehl et.al for example).

The sealed mirror of the laser cavity is a protected surface whose transmission characteristics do not vary with time. The leakage out this mirror is thus directly proportional to the power illuminating the particles at all times. It's measurement affords a means of automatically compensating for changes in illumination level.

The internal laser cavity provides an internal standing wave of radiation with predicted Doppler phenomenon. Particles passing down the laser axis scatter forward and backscattered waves of frequencies differing in Doppler content. Such scattering pulses have high frequency ripple related to velocity. The modulation depth (index) of such pulses varies with size since the ratio of forward to backward scatter is also size dependent. The importance of the Doppler component may eventually lie in intrinsic measurement of flow rate or trajectory analysis. However, it is only easily observed in the submicron size range.

3. INSTRUMENT OPTICAL DESIGN CONSIDERATIONS

It was previously mentioned that a laser cavity allows enhanced designs for stray light shielding as well as collecting geometries. It should also be pointed out that the laser is itself a monochromatic source which allows spectral filtering. Spectral filtering is used in all ASAS optical designs. Two variations of optical systems are used in ASAS instruments. They differ only in the use of a parabolic mirror instead of refracting lenses in the primary light collecting optics. The reflective system is used for high efficiency light gathering for applications below 0.1u. The refracting system is a high resolution system which is used for particle trajectory analysis as well as light gathering. Because of it's unusual operating mode it will be described in detail.

Since we were certain that the light level would be adequate for 0.1u with refracting optics we chose to study designs that offered advantages in the sampling problem area. In this regards, one has generally only two choices. You either force the particle stream through a region of uniform intensity or by secondary measurements determine where each particle passes through the beam and reject or accept the measurement depending on whether it passes through the accepted part of the "view volume" which we will hereafter refer to as the sample volume. With regards to the laser beam in question we are dealing with a Gaussian intensity distribution which can be considered as sufficiently uniform in intensity only over it's central 20% of diameter. A monodispersed particle spectrum would then result in a Gaussian distribution of scattering pulse amplitudes without limiting the measurement to the center of the Gaussian. There are of course well known techniques involving hydrodynamic focusing coupled with sheath air flow for entraining particles into small laminar jets. However, under the best of conditions it is quite difficult to reduce the diameter to less than 100 microns and there is a comparable uncertainty in the positioning control of such a jet. In short the natural beam diameter is smaller than optimum. One could of course reverse the mirror configuration and double the beam diameter. This would also reduce the intensity of the beam by a factor of four. However, one of the primary goals was also to eliminate the plumbing and aerosol interaction with it. This requirement was the determining factor in the optical system design of ASAS instruments. The method used to define this volume requires a high resolution imaging system and the use of a masked beam splitter to derive two signals. The signal detectors in conjunction with double pulse height analysis provide a means of determining if a particle position is in the desired sample volume.

The optical system design involves a high resolution optical system wherein the particle trajectories at the object plane are magnified 10X or greater. (see Figure 3). The laser used is a 2mW He-Ne Hybrid tube tuned to TEM_{00} mode. The center of the sampling volume is about 35-40mm from the adjustable mirror surface where the beam is about 500 microns diameter. Particles passing through the laser beam in the sampling aperture scatter energy into the optics.

The collecting optics include 10 spherical refracting elements in 3 cells, an interference filter, and a beam splitter. The first lens cell contains four elements and is an F/1.4 55mm fl lens. The first element of this lens has the plane mirror of the laser cemented at it's center and has a central stop to exclude light collection over it's central aperture. This lens gives a 0.37 numerical aperture and collecting angles of 4° to 22° (backscatter collection at 176°-158° is also implied).

A dielectric interference filter is immediately behind the F/1.4 lens. This filter has 60% transmission at 6328A° and has a 100A° half width. There is collimated transmission by particle scattering at the object plane through the interference filter.

The second lens is identical to the first lens and reproduces the particle image at unity magnification. The third lens in the collecting optics is a standard microscope objective which forms a secondary image in the detector plane at 10X or higher magnification. The scattered energy is relayed from the object plane collected by the scattering photodetector module. A 50% beam splitter produces two image planes for two detectors. The reflected image prism face is masked with a 0.78mm diameter vertical slit to block central transmission. The transmitted laser beam is dumped at a central stop behind the mirror on the first lens element.

The relative size of the sample volume cross section with respect to the laser beam is depicted in Figure 4. The sample volume is seen to include only the region near the center of the laser beam. The sample volume cross section is noticeably diamond shaped. The center of this diamond cross section coincides with the object plane of the collecting optics. The points of this diamond cross section define the limiting depth-of-field of the ASAS. Both the width of the cross section and the depth-of-field vary inversely with the magnification used in the collecting optics. Why this is the case will now be explained.

The image plane forms at the two exit faces of the beam splitter in Figure 5. The light transmitted on axis through the beam splitter is the signal used to size the particles. The light reflected at 90° forms it's image on a circular aperture with a central opaque vertical slit. Shown in Figure 5 are the size and positions of images formed by particles at various positions in the illuminated volume. It is important to understand that the image size is only a function of it's axial displacement from the object plane. The image size is linearly related to the numerical aperture of the collecting optics and is given approximately by: Image Size = N. A. x Displacement from Object Plane.

It is apparent that only images that are near the object plane and the center of the sample volume form images with light concentrated on the opaque slit on the masked detector as illustrated in Figure 5. Thus, such images transmit little signal through the masked aperture.

By comparing the signals at the masked aperture with that from the unmasked aperture,it is possible to determine whether they had their origin within the diamond shaped sample volume cross section. Figure 6 depicts the size and position of images in the image plane originating on the line defining the diamond sample volume cross section. These images are seen to be fully enclosed within the envelope defined as the masked aperture slit width. It is apparent that if we used a very high gain or extremely sensitive detector on the masked aperture, that we would be able to see the edge of these images and the sample volume would be defined as shown if one rejected particles whose images extended beyond the mask.

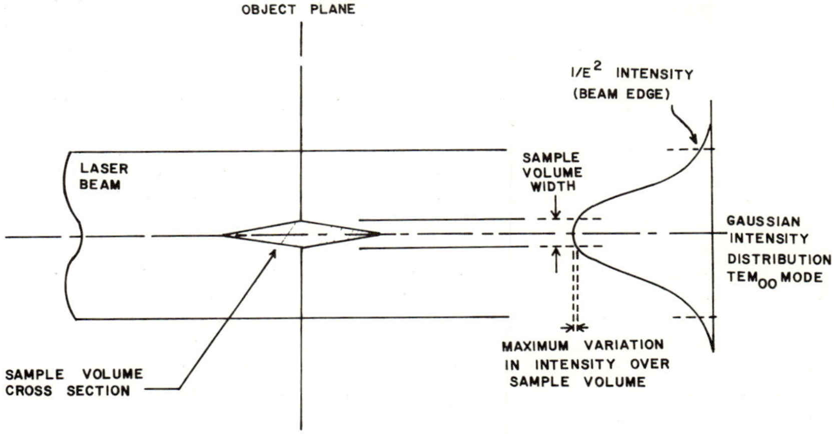

Figure 4. Relative size and position of sample volume cross section.

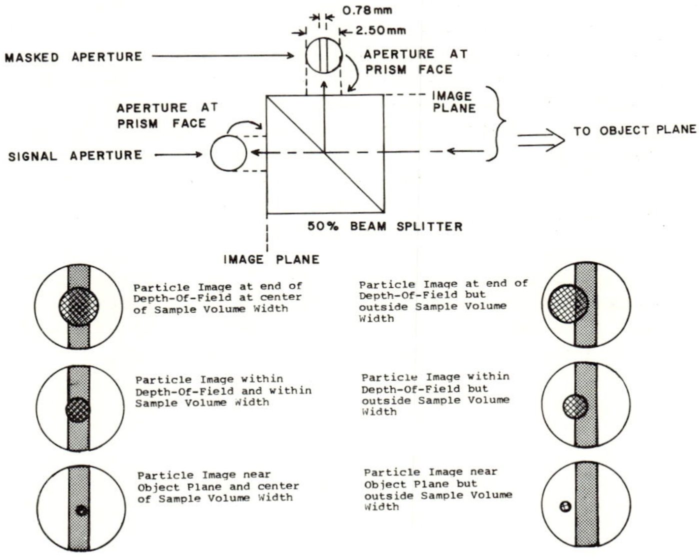

Figure 5. Image sizes and positions on masked aperture detector.

It is not practical to amplify signals beyond limits because of signal-to-noise ratio deterioration. Also because of the large range of signal levels encountered, it is more desirable to compare the amount of signal received by the masked aperture detector with the signal aperture detector to make more accurate measurements of particle image position. What is normally done in the ASAS is to use a gain ratio of masked aperture to signal aperture detector gain of 2X. Particles whose pulse amplitudes as seen by the masked aperture detector are greater than those seen by the signal aperture detector are rejected. This defines the volume and image relationships shown in Figure 7.

The gain ratio of 2X was selected as the best practical choice for several reasons. First of all, the gain ratio must be sufficient to reject images so diffuse that they fill the entire aperture. This also places limits on the percentage of the aperture that can be masked (not greater than 50% in the case of 2X gain ratio). Second, if one makes the gain ratio extremely high, there is less tolerance to slight misalignment and reflected light. Thirdly, low gain ratios do not sufficiently limit depth-of-field. The sample cross section is the product of the sample volume width and 1/2 the depth-of-field.

The parabolic mirror in the optical system used for high efficiency light collecting is not astronomical quality, hence, it does not lend itself to the rejection methods just described. The cost of a mirror of comparable resolution would be prohibiting. Since the application is limited to particles which may be adequately plumbed and the measurement restricted to a small axial segment of the laser beam it is not necessary to provide depth-of-field limiting. The measurements are restricted to the center of the Gaussian by transit time analysis of the pulses. Pulses of maximum width correspond to particles passing through the center of the beam and are accepted. Pulses of short duration correspond to trajectories through the weaker intensity regions and are rejected. A 4:1 rejection ratio guarantees particle trajectory acceptance only over the central beam region having intensity variations less than 10%. This rejection method has been used exclusively in the ASSP (Axially Scattering Spectrometer Probe) manufactured by PMS for airborne cloud droplet sizing.

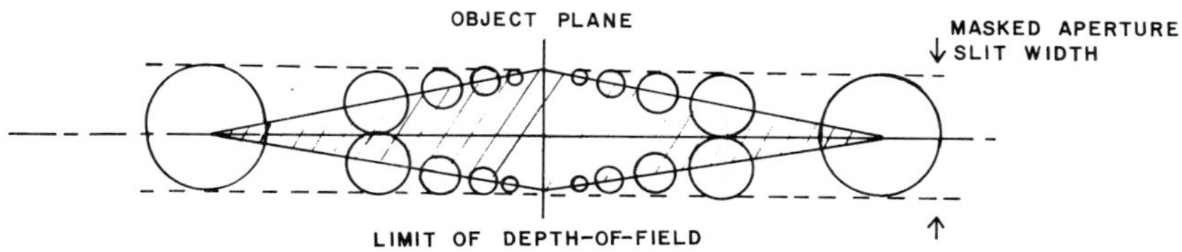

Figure 6. Image size as a function of sample volume position for the case of high gain ratio of masked-to-signal aperture.

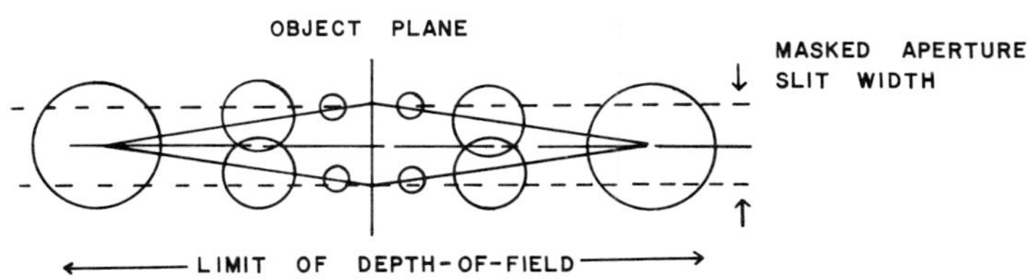

Figure 7. Image size as a function of sample volume position for the case of 2X gain ratio of masked-to-signal aperture.

4. NOISE ANALYSIS

One of the most important aspects of any instrument designed to measure transient phenomenon is the consideration of noise. The design of optical counters most often involves photomultiplier tubes. However, as was previously pointed out our baseline instrument design involves solid state photodiodes rather than photomultiplier detectors. It is a well known fact that the ultimate sensitivity of photomultipliers allows for light level measurements several orders of magnitude lower than with solid state photodiodes. This could be compensated by the fact that the ASAS instrument produces orders of magnitude greater light signal than conventional counters. However, our choice of the solid state photodiodes is primarily a result of noise analysis performed. It will be shown that the signal-to-noise ratio is in fact higher for solid state devices than for photomultiplier tubes for the bulk of our measurement range.

The appropriate starting point of such a calculation is to estimate the expected scattering signal as a function of particle size. For a 1000 watt/cm^2 illumination level a square micron of scattering cross section produces about 10uW of scattered energy. The percentage of light collected by the optical system can be as high as 50% for particles whose central diffraction lobe (Airy disc) just fills the collecting aperture. This occurs for a particle of about 1 to 2 microns diameter when the 0.37 numerical aperture is maintained. For larger and smaller particles it is substantially reduced although even our 4°- 22° system generally collects more than 15% of all the light scattered. A conservative estimate of 10% results in 1uW of scattering signal per square micron of scattering cross section. The conversion of the photon flux to photocurrent in the detector can be as high as 50% in solid state devices but is at best a few per cent in photomultiplier tube cathodes. In the case of the solid state detectors we are thus dealing with roughly 1uA of photocurrent for a particle slightly less than one micron in size. A typical response curve as measured from an instrument is shown in Figure 8. In terms of our noise analysis we will compare these with noise currents to define signal-to-noise ratios and noise equivalent particle size which defines the limit of detection. For a solid state detector the optimal detector configuration uses an operational amplifier in a current-to-voltage converter mode. The equivalent circuit is shown in Figure 9.

It is most convenient therefore to express the associated noise in the form of current noise referred to the amplifier input.

Three noise currents were calculated:

Shot noise $$I_{sn} = \sqrt{2\,eI_o \frac{1}{(\tau_F + \tau_A)}} \qquad (1)$$

Johnson noise $$I_{jn} = \sqrt{\frac{K_T}{R_T} \frac{1}{(\tau_F + \tau_A)}} \qquad (2)$$

Transister noise $$I_{tn} = \frac{V_T}{2\ R_F} \left(1 + \frac{C_O}{C_F}\right) \frac{1}{\sqrt{\tau_A}} \qquad (3)$$

Where:

$E = 1.6 \times 10^{-19}$ coulumbs

$K = 1.38 \times 10^{-23}$ eV deg $^{-1}$

T = Absolute temperature

R_F = Feedback resistance

C_F = Feedback capacitance

I_o = Sum of all detector leakage current, amplifier bias current and signal current

$\tau_F = R_F\ C_F$

$\tau_A = R_A\ C_A$

V_T = Input transister voltage noise

C_o = Sum of all input capacitance

The above equations represent a 2 pole networth transfer function analysis. The time constant $R_A\ C_A$ is that associated with the amplifier output, an output filter section or a second stage amplifier. The ratio of τ_F / τ_A for minimum transister noise is about 1.6.

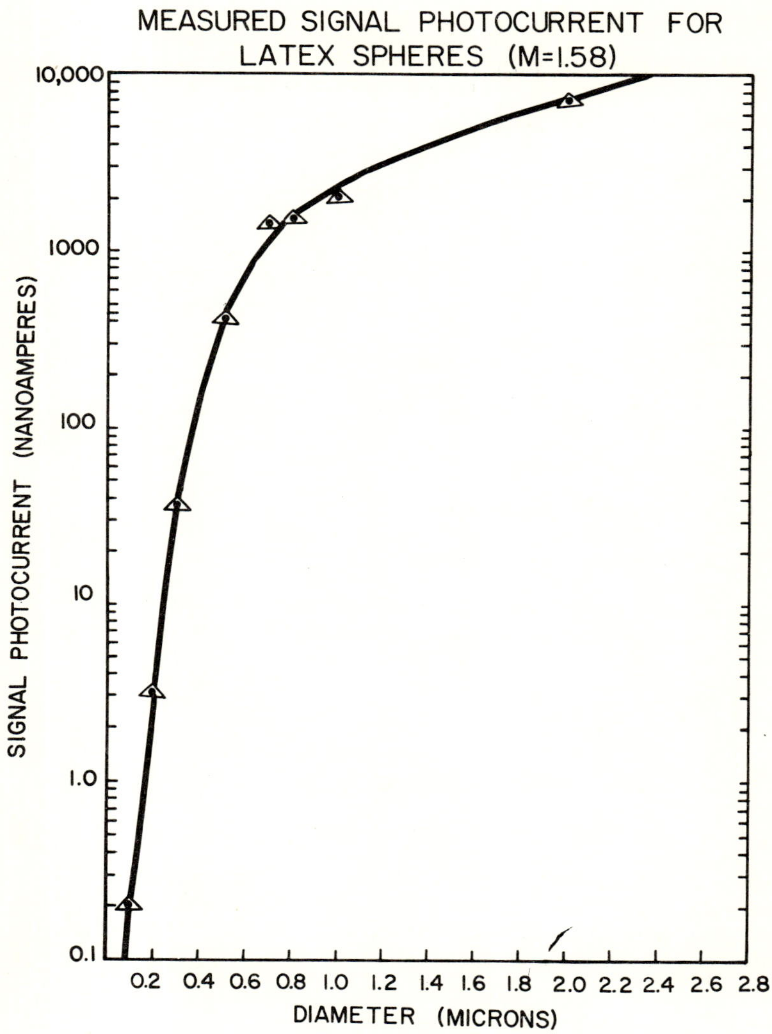

Figure 8. The measured photocurrents were with a laser generating approximately 2.5 W/cm^2 energy density.

SILICON PHOTODIODE-PREAMP NOISE

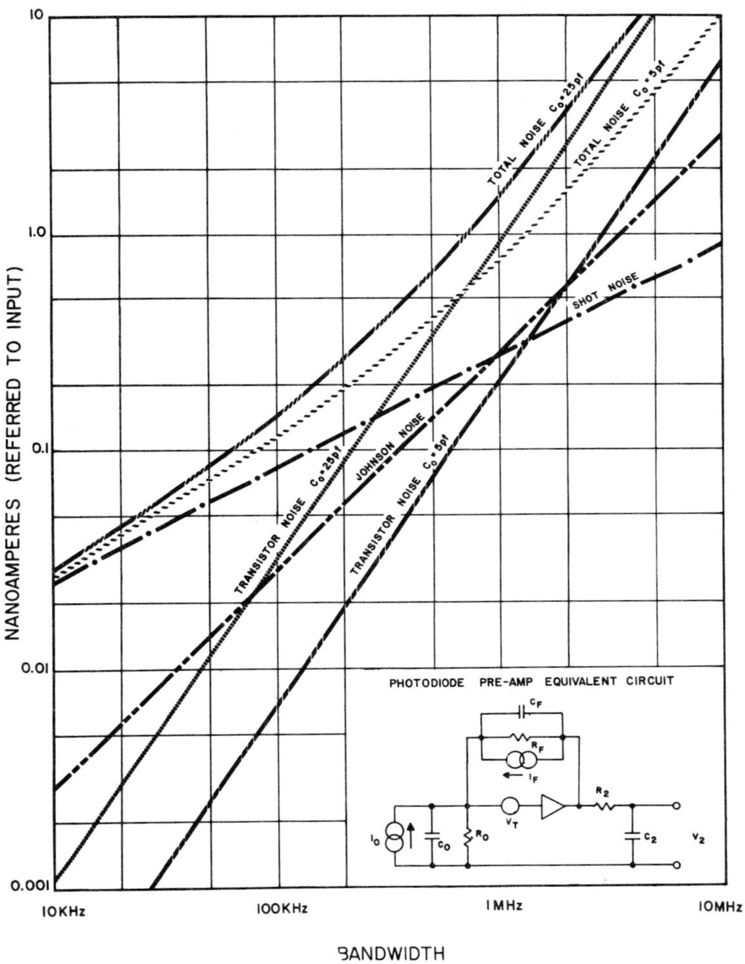

Figure 9. The above noise analysis represents a low noise input FET pre-amp. The following assumption was made:

$$\tau_1 = 1.6\,\tau_2 \text{ (condition for minimum transistor noise)}$$

$$V_T = 2.5 \text{ nV}\sqrt{\text{Hz}} \text{ (low noise input FET 2N5564)}$$

An important aspect of equation 3 is that the transistor noise is amplified by the ratio of Co/C. Clearly this noise would be minimal for Co=0. The actual minimum value one can obtain is about 5pf, while 25pf is measured in our system. Also the Johnson noise and transistor noise decrease with increasing Rf. For computational purposes, the minimum bandwidth requirement sets the value of R_F and the stray feedback capacitance (typically 0.5 pf) generally limits the maximum value of R_F. If the laser beam at the sample volume is 0.5mm diameter and the flow rate is approximately 5m/sec , the required amplifier rise time is 10u sec or an R_F C_F of 4u sec. Thus a maximum value of R_F would be 8 megohms. For computational purposes the bandwidth from 10KHz to 10MHz were used. The results are shown in Figure 9. The total noise curves represent the predicted minimum detectable signal that can be realized. There are two total noise curves which differ only in the total input capacitance. For most of our instrument designs the required bandwidth is in the neighborhood of 25KHz. It is quite clear from the above results shown that the detector electronics would be shot noise limited in the bandwidth range of interest. This may be a curious result to some who are accustomed to the notion that only a photomultiplier tube would be shot noise limited in this low current measurement range. This would in fact be true if the background light levels could be arbitrarily suppressed below the 100 nA baseline. However, since at 100 nA background we are governed by shot noise it should be clear that at even higher signal photocurrents,the overall signal-to-noise ratio,will be entirely governed by the quantum efficiency of the detector. Since most silicon photodiodes have quantum efficiencies of 50% versus 5% for the multialkalic red sensitive photomultipliers it's advantages are obvious. Figure 10 shows the resulting signal-to-noise ratios for the silicon photodiode and the photomultipliers of 0.2% and 0.5% quantum efficiency. Note the advantage of the photomulitplier exists only at low signal levels where background light is infinitesimally small.

Finally it must be stated that the above design is much more sophisticated than that required by a photomultiplier. The advantages of the photomultiplier lie in it's low noise amplifier qualities. The resultant anode current is at a high enough level where induced noise pick up is comparatively insignificant. The background light level can also be reduced to less than the 100 nA baseline by increased optical magnification. Under such conditions the photomultiplier has increased advantage. We use the photomultiplier for sizing below 0.1 u where further magnification is desirable for reducing sample volume size.

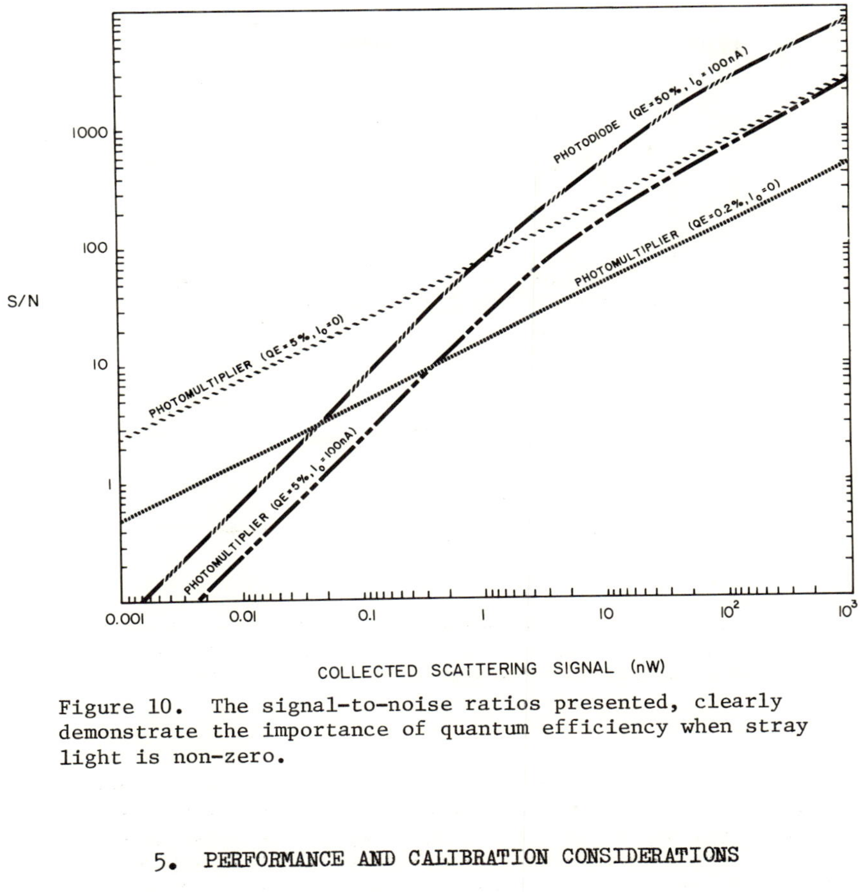

Figure 10. The signal-to-noise ratios presented, clearly demonstrate the importance of quantum efficiency when stray light is non-zero.

5. PERFORMANCE AND CALIBRATION CONSIDERATIONS

The exact calculation of the scattering response for an active laser cavity is closely analgous to electromagnetic scattering in a standing wave. Recent solutions to this problem will be illustrated. The solutions are not strictly correct because a particle traversing through the beam inside the cavity modifies the light field reducing the cavity "Q" and the laser gain. Schuster and Knollenberg[3] found effective extinction cross sections that were roughly a factor of 100 times the classical cross sections. More recent measurements in our laboratory of the relative extinction and scattering efficiency factors are shown in Figure 11. Here one sees that in the absence of absorption, that $Q_{sca} \neq Q_{ext}$ (Q = efficiency factor and $Q\pi r^2$ = cross section) but in fact differs by a factor of 10,000 in current instrument high "Q" cavities. The shape of the curves is not easily

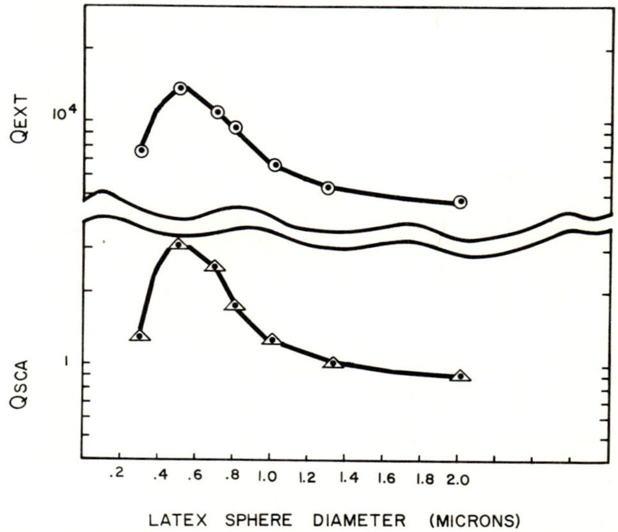

Figure 11. The above measurements are indicative of the role the cavity"Q" plays in the extinction process. The scattering efficiency factors are reasonably close to computed values.

explained but the decrease for larger size probably results from localized laser gain reduction in the near vicinity of the particle, (large particles can effectively burn holes in the laser beam). What is observed in the extinction process is not true extinction at all but largely cavity de-tuning and cannot be truly separated from it. The scattering phenomenon is however generally explained by scattering within a standing wave. It is only necessary to measure the light beam illumination and normalize the scattering signal by the amount of cavity detuning produced by extinction.

The theoretical calibration curves calculated for scattering in a standing wave are shown in Figure 12 and 13 for the two collecting geometries involving collecting angles of 4°- 22° and 5° - 90°. Eight response curves are shown for real refractive indices of m=1.3, m=1.4, m=1.5, m=1.6, m=1.7, m=1.8, m=1.9, m=2.0. The response function is an effective cross section for the particular collecting geometry. To calculate collected power,one simply multiplies the illumination level (W/cm ^{2}) by the effective cross section.

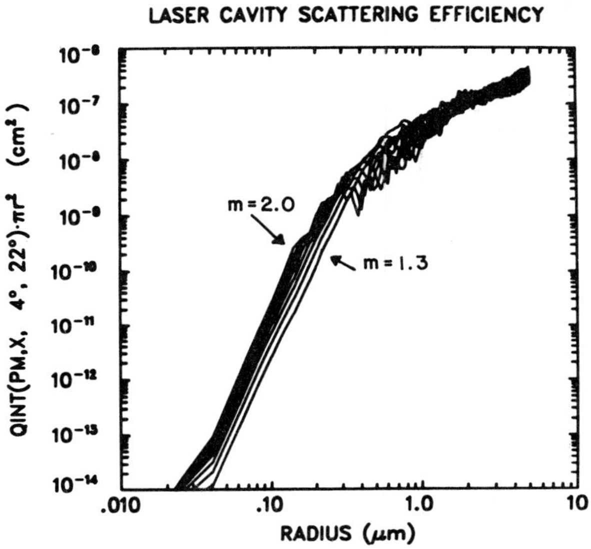

Figure 12. Theoretical instrument response for the 4°–22° collecting geometry for m = 1.3, m = 1.4, m = 1.5, m = 1.6, m = 1.7, m = 1.8, m = 1.9, m = 2.0

These results in general show finer structured resonance behavior than the computed response from superimposed MIE solutions for the forward and backward scattered components. They also show considerably less sensitivity to refractive index. The ASAS has less sensitivity to refractive index than most white light counters (see Cooke and Kerker[7] to compare). Outside of the reduced refractive index sensitivity in the 1 – 2u diameter range for the 5° – 90° collecting geometry the primary advantage of the hemispherical collecting geometry over the 4° – 22° collecting geometry is the significantly larger response at the small particle sizes.

Figure 14 shows the calculated sensitivity of the instrument to absorption for m = 1.5 and imaginary components of i = 0.5, 0.1, 0.08, 0.06, 0.04, 0.02 and 0.00. In general the greater the absorption the smaller the resonance. Experimental measurements have verified the predicted insensitivity of the ASAS to refractive index and to the accuracy we can measure result in monotonic calibration curves. The ASAS has less predicted sensitivity to complex refractive index than any white light instrument in the submicron range(compare with results of Cooke and Kerker[7]). Additional measurements on nonspherical particles show the hemispherical collecting geometry to be less sensitive to shape factor as one might predict.

Some general remarks on the associated resonance phenomenon associated with refractive index sensitivity are perhaps worthwhile. There are factors which also reduce the observed resonance. The cavity gain dynamics have already been mentioned. In other measurements of classical scattering we have found ourselves only able to reproduce the full theoretical resonance response if the beam was truly uniphasal which generally necessitated spatial filtering. The lack of spatial filtering or the presence of multimodes always suppresses the predicted resonances. These phenomenon may all be attributed to the lack of phase front coherence implicitly assumed in such calculations. In most cases the arguments would extend to scattering within a standing wave.

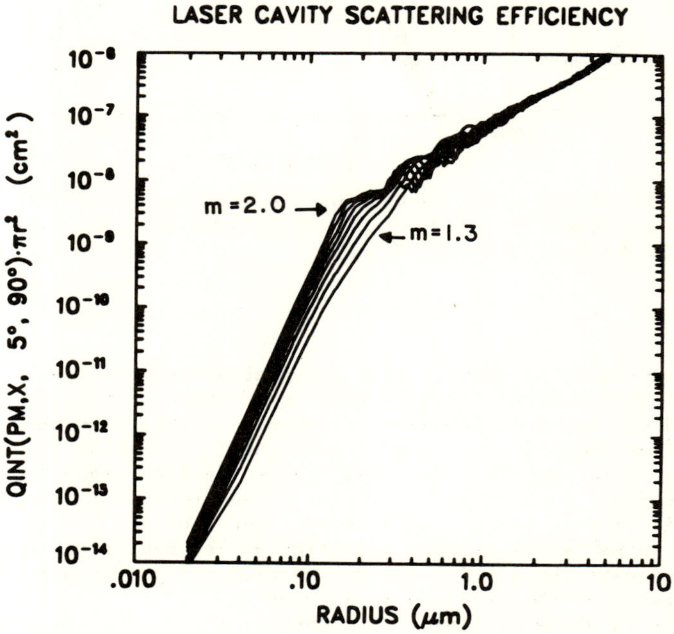

Figure 13. Theoretical instrument response for 5°- 90° collecting geometry for m = 1.3, m = 1.4, m = 1.5, m = 1.6, m = 1.7, m = 1.8, m = 1.9, m = 2.0.

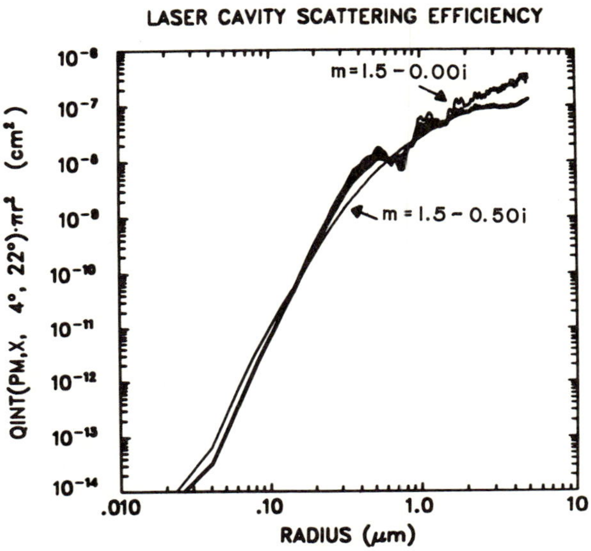

Figure 14. Theoretical instrument response for absorbing particles for 4°-22° for m=1.5, and imaginary components of i = 0.5, 0.1, 0.08, 0.06, 0.04, 0.02 and 0.00.

6. GENERAL INSTRUMENT DESCRIPTION

Our ASAS instruments are designed to size particles into fifteen linear size intervals, with a range change switch to provide up to four independent size ranges within a single system. Other features include trialkalia photomulitplier detectors for ultimate sensitivity and "free flow aspiration" for absolute "in situ" sampling. It is packaged in two separate enclosures housing the probe with it's optics and pre-amps and an electronics console (see Figure 15).

The laser used is a hybrid He-Ne (6328A$^{\circ}$) manufactured by Coherent Radiation of Palo Alto, California. The laser has optimized parameter selection for the ASAS. The optical system is much as described in Section 3. All glass optical elements are A-R coated for minimal light loss. Optical transmission is better than 90% without the filter.

The pulse height analyzer is capable of sizing pulses as short as 100n sec. Because it's reference voltage is derived from the source of illumination, the entire system has an effective automatic gain control (AGC). To accomodate the large dynamic range of the instrument, a programable amplifier is used to gain switch and provide several size ranges. Logarithmic amplification is provided in the tenth micron size range.

The data acquisition system within the electronics console has an active memory and is designed to decode, store and display particle size information. The memory capacity is 16 addresses. Fifteen of the addresses are used for particle size distribution storage. The remaining address is normally used for other housekeeping information such as selected size range and elapsed time. Each address has 16 bits storage capacity for 0-9999 counts BCD. A selectable digital display and a graphical CRT display are provided for real-time data monitoring requirements of precise particle counts or distribution functions.

The mechanical design of the probe section of ASAS instruments utilizes extruded aluminum sections for mechanical stability of optical bench quality. Laser and detector alignment is achieved with spring loaded #10-80 x-y screw adjustments. The probe section is approximately 36" long, 8" square, and weighs 25 pounds.

Two types of aspiration are used. The first type used primarily with the reflecting optical system involves a pair of diametrically opposed orifices of approximately 2mm inside diameter. A small vacuum pump buffered by a plenum chamber draws the aerosol sample through the laser beam. Since the sample volume is centered in the flow and constitutes a few percent of the flow cross section, wall effects are considerably reduced and sheath airflow is not required.

The second method of aspiration is unlike any used in conventional aerosol counters. As shown in Figure 16 the airflow is essentially unrestricted. A 10:1 accelerator produces a flow rate of 6.15 meters per second intersecting with a 3cm length of the laser beam. Again the sample volume is positioned at the center of flow. This system relies entirely upon optical definition of the sample volume. Problems associated with small bore tubing are eliminated. This makes the technique particularly useful for sizing volatile materials such as natural smog.

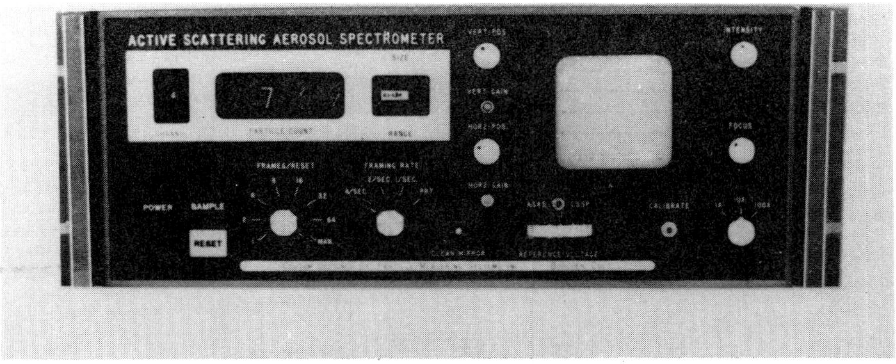

ELECTRONICS CONSOLE

PROBE

Figure 15. Photograph of ASAS and electronic console. Probe shown has free flow aspiration.

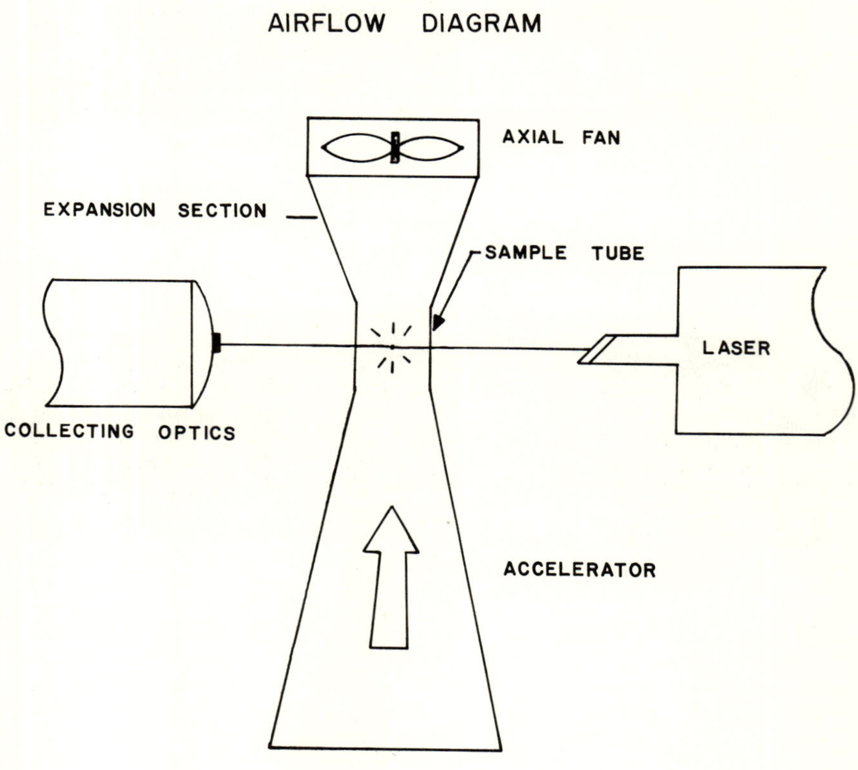

Figure 16. The free flow aspiration system depicted above has a minimum sample tube diameter of 3cm.

7. CONCLUSIONS

Active Scattering Aerosol Spectrometers are fully capable of particle size measurements down to 0.05 microns. As such they provide an important capability for studies in the submicron range. The extensive use of solid state detectors in this kind of system is a result of the high light levels developed and the design of preamplifiers pressing theoretical noise limits. However, photomultipliers are still required for maximum sensitivity.

Probably the single most important aspect of the ASAS as designed is the optical system and electronics which together allow for sample volume control without plumbing. Of further significance is the fact that the background light level can be reduced at will by increasing optical magnification. The ultimate sensitivity of photomultiplier tubes can be realized.

It is our opinion that the refractive index sensitivity of the ASAS is of the same order of magnitude as the sensitivity to non-spherical shape and need not be pursued further at this time. Of considerably more importance is the possibility of utilizing both extinction and scattering measurements to study the absorption properties of particles.

Some other aspects of the scattering phenomenon, are also worthy of further investigation. One that was observed early in our work was the Doppler modulated pulses produced by particle trajectories with axial components. For the collecting angles used in the ASAS, the Doppler modulation increases with decreasing size since the back-scattered radiation is in greater proportion to the forward scattered radiation. Thus the modulation depth is itself a size sensitive parameter. Also the frequency is a measure of particle velocity.

Other optical geometries are also useful in certain applications. Certainly collecting 2π steradians provides for a system more insensitive to particle shape. Ratio detection at certain paired angles can also be useful. However, for sizing speed sensitivity and size resolution the ASAS as currently designed would appear to be as satisfactory as most uses for such information demand.

ACKNOWLEDGMENTS

I would like to thank Jam Emming of Ball Brothers Research Corporation for the use of his amplifier circuit model. Also the solution for the electro magnetic scattering in a standing wave which was supplied by Ron Pinnick of the National Center for Atmospheric Research.

REFERENCES

1. S. A. Schleusener, J. Air Poll. Control Assoc. 19, 40 (1969).

2. T. D. Proctor, J. Sci. Instrum. 1, 631 (1968).

3. B. G. Schuster and R. G. Knollenberg, Detection and Sizing of Small Particles in an Open Cavity Gas Laser, Applied Optics, Vol. 11, page 1515, July 1972.

4. R. G. Knollenberg, An Active Scattering Aerosol Spectrometer, Atmospheric Technology, No. 2, June 1973.

5. R. Schehl, S. Ergun and A. Headrick, Size Spectrometry of Aerosols Using Light Scattering from the Cavity of a Gas Laser, Rev. Sci. Instrum., Vol. 44, No. 9, page 1193, September 1973.

6. H. C. Van De Hulst, Light Scattering by Small Particles (Wiley, New York, 1957).

7. D. D. Cooke and M. Kerker, Response Calculations for Light-Scattering Aerosol Particle Counters, Applied Optics, Vol. 14, No. 3, page 734, March 1975.

Single Particle Optical Counter: Principle and Application

Klaus Willeke and Benjamin Y. H. Liu
Particle Technology Laboratory
Mechanical Engineering Department
University of Minnesota, Minneapolis, Minnesota

ABSTRACT

Single particle optical counters (OPC) are used for in-situ measurements of aerosol concentrations and size distributions. This paper examines the optical configurations of several commercial OPC's and then the basic considerations for the theoretical calculation and experimental calibration of a counter's response to a particle of given size, shape and refractive index. Furthermore, detailed analyses are made of the influence on the counter's response by coincidence, resolution, statistical count accuracy, pulse processing and electrical noise.

Single Particle Optical Counter: Principle and Application

Klaus Willeke and Benjamin Y. H. Liu
Particle Technology Laboratory
Mechanical Engineering Department
University of Minnesota, Minneapolis, Minnesota

INTRODUCTION

Single particle optical counters[1-6] (OPC) have found widespread use in clean room monitoring, pollution research and laboratory aerosol studies because of their ability to make in-situ measurements[7-9] of the concentrations and size distributions of particles suspended in the air. A beam of light inside the instrument is focused onto a "view volume" through which the airborne particles pass one at a time. The amount of light scattered or absorbed from each individual particle is measured by a photosensitive detector. The signal amplitudes registered by this detector are then stored in the channels of a Multi-Channel Analyzer (MCA) from which the particle size distribution is determined. The range of particle sizes extends from about 0.5 μm to about 10 μm in diameter for most commercial OPC's.

Aerosol photometers, not discussed here, measure light scattering or extinction from a cloud of particulates, and do not resolve the individual particle sizes. These measurements will provide accurate concentration measurements provided the size distributions remain the same.

OPERATING PRINCIPLE

The light extinction system, shown in Figure 1a, will be used to exemplify the operating principle of the single particle optical counter. Light from a lamp, usually an incandescent lamp, sometimes a laser, is condensed onto an aperature with the filament of the light source focused in the plane of that aperture. The emanating light cone is condensed into the view volume where a sharp image of the defining aperture is formed. The aerosol flow is ducted to the light beam in this plane of focus. A view volume may then be defined as the region bounded by the cross-sectional area of the aerosol stream and the height of the aperture image. The light emanating from the view volume is collimated, condensed and focused onto a photosensitive detector, such as a photomultiplier or a photodiode.

In a single particle optical counter with an incandescent light source, a light extinction system, such as shown in Figure 1a, is only practical for particles larger than several micrometers in diameter. Since most particles in an ambient or industrial environment are smaller in size, this light extinction technique is rarely used for aerosols, but is frequently used for the characterization of particles in a liquid.

Most commercially available single particle optical counters measure the light scattered out of the incident beam rather than the light intensity reduction of that beam. The optical system of the common scattering instruments is schematically represented in Figures 1b and 1c. The illumination is, in general, as described in Figure 1a, the details of light collection and detection, however, may differ in how the scattered light is focused onto the photodiode or photomultiplier. Each commercial system has its own arrangement of illumination and acceptance angles. Specification of the following four angles (graphically represented in Figures 1b and 1c), fully describes the light interaction in the view volume: γ is the half angle of the illuminating cone, α, the light trap half angle, β, the collecting aperture half angle, and ψ, the inclination between illuminating and collecting cone axis.

The optical system in an OPC is sometimes characterized by the scattering angle limits. Referring to Figure 1c, scattering of light from illuminating angle γ to collecting angle β gives an upper limit of $\gamma + \beta$; scattering from γ to light trap angle α gives a lower limit $\alpha - \gamma$. For instance, an instrument with $\gamma = 15°$ and viewing angles $\alpha = 35°$ to $\beta = 90°$ has a scattering range of 20° to 105°.

When particles are illuminated by a monochromatic light source, such as a laser, the Mie scattering curve of scattered light flux shows periodic amplitude oscillations[10,11] with respect to particle size parameter $\alpha = \pi D_p/\lambda$, where D_p is the diameter of the particle (assumed spherical) and λ is the wavelength of illumination. The counter response may, therefore, not be a single valued function of particle size. Incandescent light usually washes out these oscillations and is, therefore, used in most commercial OPC's. When a powerful light source is used, e.g. a high intensity laser light, volatile particles may partially evaporate during the time they remain in the view volume.

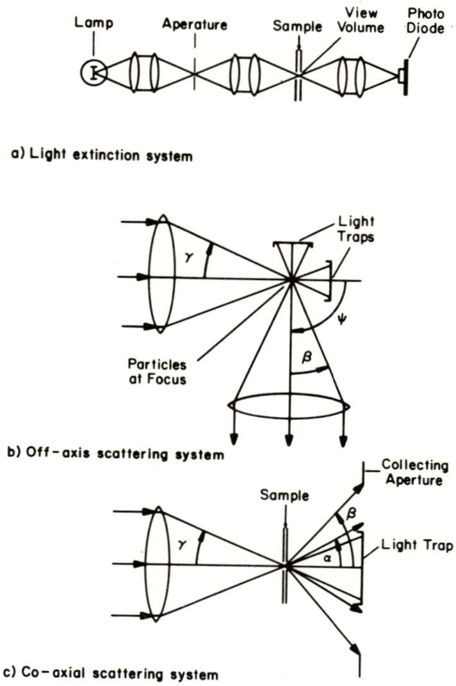

Figure 1 Geometries of illumination and collection in conventional single particle optical counters.

COMMERCIAL INSTRUMENTS

The significant characteristics of several commercial OPC's are listed in Table I. All of these instruments have incandescent light illumination.

The Royco 220 (Royco Instruments, 41 Jefferson Drive, Menlo Park, Calif. 94025) is a right angle scattering instrument with the axis of

Table I: Characteristics[a)] of Commercial Single Particle Optical Counters

Optical Counter	γ[c)] (degree)	α[d)] (degree)	β[e)] (degree)	v (mm^3)	Q_s ($(cm^3 min)^{-1}$)	Q_s (cfm)
Bausch & Lomb 40-1	13	33	53	0.5	170	0.006
Climet CI-201	15	35	90	0.5	7,080	0.25
Climet CI-250	12	18	28	0.4	472	0.017
Royco 218	5	11	30	0.25	283	0.01
Royco 220 (ψ = 90°)[b)]	24	-	24	2.63	2,830	0.1
Royco 245	5	16	25	4.0	28,300	1.0
Tech Ecology 200	5	8	20	0.46	283	0.01
Tech Ecology 208	5	10	20	2.5	2,830	0.1

[a)] Values given are approximate, for use as guidelines only

[b)] ψ = inclination between illuminating and collecting cone axis

[c)] γ = illuminating cone half angle

[d)] α = light trap half angle

[e)] β = collecting aperature half angle

the collecting aperture inclined at angle ψ = 90° to the axis of the illuminating cone (Fig. 1b). All other instruments, listed in Table I and schematically shown in Figure 1c, have a common illumination and collecting aperture axis, i.e. ψ = 0°, and collect the light scattered into a hollow cone.

The Royco 218 and 245, TechEcology 200 and 208 (TechEcology, Inc., 645 N. Mary Ave., Sunnyvale, Calif. 94086) and Climet CI-250 (Climet Instruments Co., 1620 W. Colton Ave., Redlands, Calif. 92373) collect light from the forward scattering lobe and may be termed forward or near forward scattering instruments. Their half angle of illumination is about 5° to 12°, their light trap half angle ranges from about 8° to 18°, and their collecting aperture half angle ranges from about 20° to 30°. The scattered light in these instruments is collimated, condensed and focused onto a photosensitive detector by lenses with a central dark stop. The light collection system is therefore much like a dark field microscope in which essentially all of the illuminating beam is rejected after passage through the view volume and the scattered light is focused onto a detector. The Climet CI-201 and the Bausch & Lomb 40-1A (820 Linden Ave., Rochester, N.Y. 14625) have viewing angles which extend from about 30° to as high as 90°. In the Bausch & Lomb instrument the view volume is located at the focal point of a parabolic mirror with a light trap at its center. The mirror reflects and collimates the scattered light onto a lens which condenses and focuses the light onto a photomultiplier. In the Climet CI-201 instrument, the use of lenses for light collection is avoided by locating the view volume at one of the two focal points of an elliptical mirror which reflects the scattered light to a photomultiplier at its other focal point.

The view volume, v, varies by about one decade from 0.25 to 4 mm^3 and the sampling rate Q_s varies by over two decades from 170 $cm^3\ min^{-1}$ (0.006 cfm) to 28,300 $cm^3\ min^{-1}$ (1 cfm) for the tabulated counters. Battery operated portable instruments are generally limited to pumps handling low flow rates.

A counter which utilizes the light extinction techniques is manufactured by HIAC (4719 West Brook Street, Montclair, Calif. 91763). The Company claims that the instrument sizes airborne particles from 150 μm to 5 μm, and from 60 μm to 2 μm, depending on the sensor used. In a counter[12,13] developed by Particle Measuring Systems (1855 So. 57th Court, Boulder, Colo. 80301) the particles are injected into the radiation field of an open cavity He - Ne gas laser. The light extinction by the particles perturbs the resonance in the cavity with a resulting power loss of the laser, i.e., the laser behaves as a nonlinear amplifier of perturbations in the resonant cavity. It is claimed that this counter, designated an Active Scattering Aerosol Spectrometer, sizes particles from 5 μm to 0.05 μm. Its cost, is on the order of $10,000. Conventional OPC's sell at $2,000 to $6,000. Peripheral equipment may increase this cost.

THEORETICAL RESPONSE CALCULATIONS

The response of single particle optical counters, which typically classify particle sizes from 0.5 μm to 10 μm in diameter, can be calculated through the use of the Mie - scattering theory[10-11,14-27]. This assumes that the particles are spherical and are illuminated by light of unit flux per unit beam cross-sectional area, and calculates the scattering intensity as a function of scattering angle θ (measured with respect to the incident ray), particle refractive index m and particle size parameter $\alpha = \pi D_p/\lambda$. The results of Mie theory calculations are usually presented as angular intensity functions i_1 (α,m,θ) and i_2 (α,m,θ), the components of scattered light polarized in and normal to the plane containing the directions of illumination and observation. The quantity $(\lambda^2/8\pi^2) \times (i_1+i_2)$ is then the flux per unit solid angle scattered in the direction θ by a particle of given size and refractive index. The intensity functions are represented by infinite series whose terms are functions of spherical Bessel and Legendre polynomials. The scattered angle θ is calculated with respect to the incident ray which is inclined at angle ϕ to the axis of illumination. The response calculation for each particular optical design, therefore, includes a geometrical factor $F(\theta,\phi)$. An additional factor $f(\lambda)$ accounts for the wavelength distribution of the emissive power of the light source and the spectral sensitivity of the photosensitive detector. The relative counter response R for spherical particles, in the absence of coincidence and cross sensitivity, is then

$$R(m,D_p) = \int\int\int(\lambda^2/8\pi^2)(i_1 + i_2)\ F(\theta,\phi)\ f(\lambda)\ d\lambda d\theta d\phi \quad (1)$$

As an example, the relative, theoretical amount of scatter[14] of one particular commercial counter is shown as a function of particle size in Figure 2. Calculations for other commercial counters are given elsewhere[10,11,14-19]. An OPC amplifies the photoelectric signal. If the amplification is linear, the counter response will closely follow the functional relationship given by these calculations.

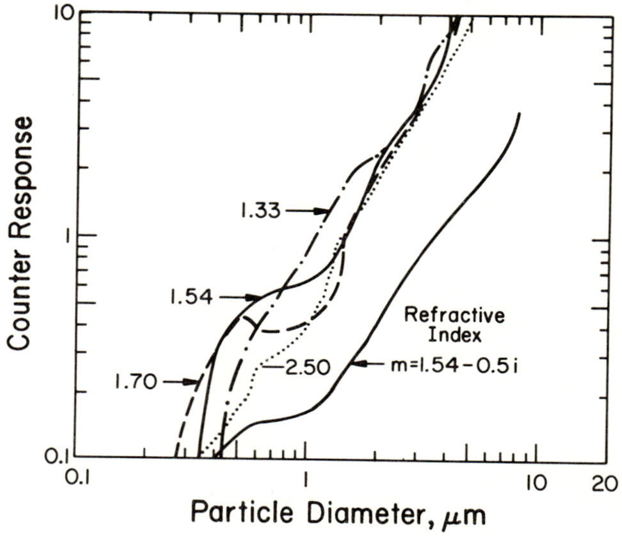

Figure 2 Theoretical response of the Bausch and Lomb 40-1A particle counter (adapted from D. D. Cooke and M. Kerker, 1975).

The counter response may also be represented as light flux received per unit particle projected area for unit luminance in the source (e.g. Ref. 16). The resulting curve shows the flux increasing by about 4 decades per decade increase of α when $\alpha << 1$ (Rayleigh or dipole scattering) The curve reaches a peak at about $\alpha = 2$ to 5, oscillates (Mie scattering) and settles to an approximately constant flux value with small or no oscillations when $\alpha > 50$, i.e. the flux per unit particle-projected area is approximately constant for large particles (geometric scattering). The rapid decrease of flux for $\alpha \lessdot 2$ imposes a lower size limit on conventional OPC's, ca 0.3 μm to 0.5 μm for white light illumination.

Stray light from optical elements such as lenses and light traps, and Rayleigh scattering of the air molecules results in background scattering intensities which should be lower than the scattered light obtained by passage of the smallest particle through the view volume. In a well-designed optical system, stray light scattering will be negligible with respect to Rayleigh scattering of the air molecules. The Rayleigh scattering intensity is given by the product of the mean illumination intensity, the number density of the air molecules, the scattering cross-section of a single air molecule, and the volume of scattering air in the viewing field.[11] The latter is generally larger than the view volume, i.e. sampled particles can scatter only within the view volume, but air molecules can scatter into the receiving aperture from the entire viewing field. OPC's designed to detect small particles should, therefore, have a very small illumination angle which can best be achieved through laser light illumination[11]. However, such illumination may result in the aforementioned non-single-valued response of the counter. Lowering the air pressure reduces the number concentration of the air molecules, and therefore, the Rayleigh scattering intensities from the air, but may evaporate liquid droplets or liquid coatings on solid particles, thus giving a particle size distribution which is different from that of the ambient air. In unusual circumstances, where the particles may be suspended in a gas such as Helium, the particle size limit can be lowered, because Helium atoms have a lower scattering cross-section than air molecules.[11] The lower particle size limit for special optical counters is of the order of 0.1 μm in diameter.

INSTRUMENT CALIBRATION AND MODIFICATION

Relative, theoretical calculations may be used to determine how monotonic the response of a counter will be with respect to particle diameter, and how much the response will depend on particle refractive index within the size range of interest. However, an experimental calibration[1-6,28-32] is always necessary to establish the actual voltage output of the counter as a function of particle size for a given aerosol.

As a first step, one needs to generate aerosols with a high degree of monodispersity. The least expensive technique is one in which polystyrene latex (PSL) or polyvinyl toluene latex suspensions (Dow Chemical Company, Midland, Michigan) are nebulized[33], dried, and sampled into the OPC. They are available in sizes from about 0.1 μm to about 3 μm, and have a refractive index of 1.6. They may be considered ideal because of their spherical, smooth shape, i.e. they fit the assumptions for the theoretical response calculations outlined above.

An alternate and more elaborate technique utilizes either the spinning disc (Sierra Instruments, P.O. Bx. 909, Village Sq., Carmel Valley, Ca. 93924) or the vibrating orifice monodisperse aerosol generator (Thermo-Systems Inc., 2500 Cleveland Avenue, St. Paul, Minnesota 55113). The latter[34] allows the calculation of particle diameter to an accuracy of about 2 percent using the experimental values of signal frequency, liquid flow rate and concentration of aerosol material in solution with a volatile substance. Aerosols generated with these instruments approximately cover the entire size range of conventional OPC's. The effect of particle refractive index may be studied by generating ideal particles of di-octyl phthalate (DOP, m = 1.49) and Cargille index-of-refraction liquids with m = 1.4, 1.5, and 1.6 (R. P. Cargille Laboratories Inc., Cedar Grove, New Jersey). The Cargille 1.7 refractive index liquid was found[2] to be unstable and sufficiently volatile at room temperature to cause significant uncertainties in the calculated particle diameter. An example[3] of an experimental calibration with DOP aerosols is shown in Figure 3. The instrument's theoretical response[13] is shown in Figure 2. Experimental calibration curves for other commercial optical particle counters are shown in Figures 4 through 6, and will be discussed in the next chapter. The effect of particle shape and light absorption may be studied by generating particles such as methylene blue, sodium chloride and india ink.[1]

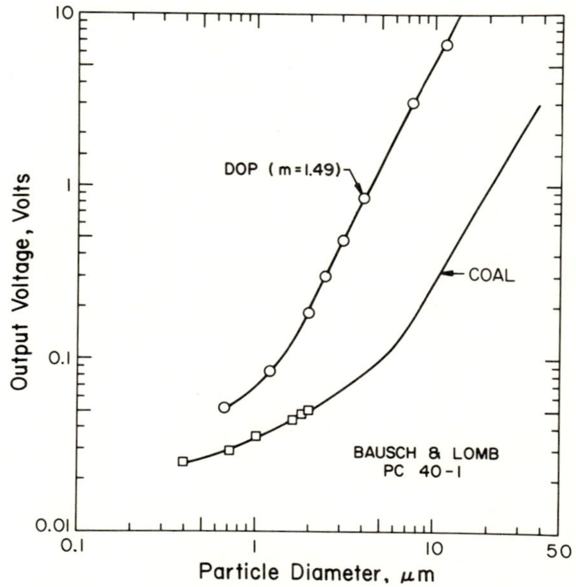

Figure 3. Experimental calibration curve for the Bausch & Lomb 40-1A particle counter.

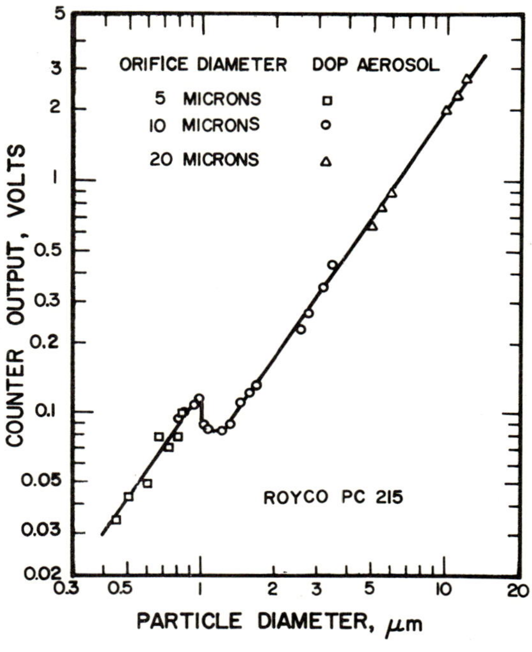

Figure 4 Vibrating orifice generated calibration curve for the Royco 215 particle counter. The optical system of the Royco PC 218 is similar to the older Royco PC 215.

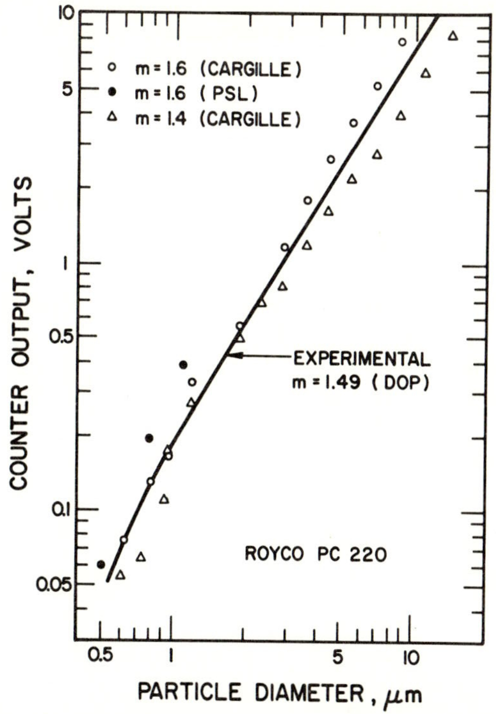

Figure 5 Experimental calibration curve for the Royco 220 particle counter.

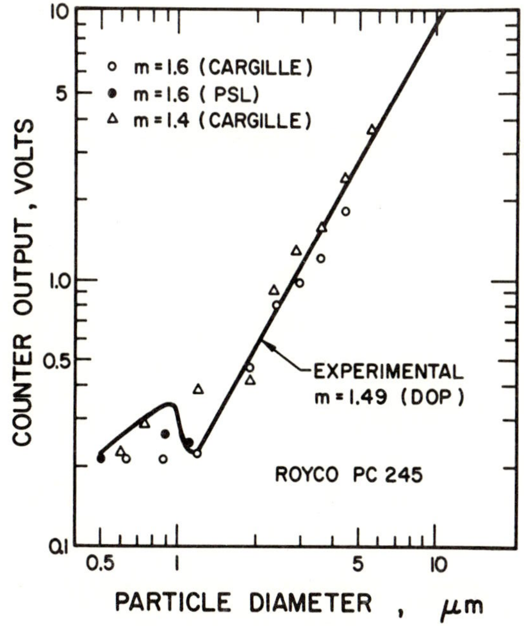

Figure 6 Experimental calibration curve for the Royco 245 particle counter.

The refractive index and shape of airborne particle found in environments such as mines and metallurgical processing may be quite different from these artificial aerosols. The main difficulty in calibrating OPC's for such dusts is the preparation of monodisperse calibrating aerosols. Conventional aerodynamic dispersion of powders such as coal does not give sufficiently monodisperse and de-agglomerated aerosols. It has been found that a fluidized bed[3,35,36] serves well in the de-agglomeration of powders. The airborne particle may then be classified to a known size by means of a differential mobility analyzer.[3] However, classification by electrical mobility is limited to submicron particles with the upper limit at about 2 to 3 μm. Unclassified larger particles may be sized by optical microscopy after passage through the OPC. However, the counter's response is spread over a range of channels for a particle of a given size so that the original size distribution of the dust must be deduced by numerical inversion.[3] This broadening effect limits the resolving power of the instrument as discussed in a later chapter.

Special attention should be paid to the uniform illumination of the view volume. The light source of which an image is formed in the view volume, should produce the most homogeneous radiation density possible. For instance, band-like tungsten filaments were found to be superior over spiral tungsten filaments[37].

A sheath-air inlet further improves the quality of the view volume. In a typical sheath-air inlet[2], shown in Figure 7, about 90 percent of the incoming aerosol flow is diverted through an external filter which separates the particles from the air stream. The filtered air is then reintroduced around the remaining 10 percent of the original aerosol flow which passes unimpeded through a straight, inner tube. The combined stream is then introduced into the optical view volume where the particles are counted. The flow is laminar and the aerosol stream flows as a central core within the clean-air sheath into the view volume.

A sheath air inlet increases the resolution of the counter by focusing the aerosol sample stream into a more uniformly illuminated portion of the view volume, and permits a higher aerosol concentration to be counted because of the decrease in size of that view volume. This also decreases the response time of the counter and prevents recirculation of aerosols into the view volume. The signal-to-noise ratio may be improved if the aerosol sample stream is focused into a portion of the view volume which has a higher mean intensity than the original view volume.

The measured aerosol concentration will be lower than the true aerosol concentration, if particles are lost in the OPC inlet. The ratio of

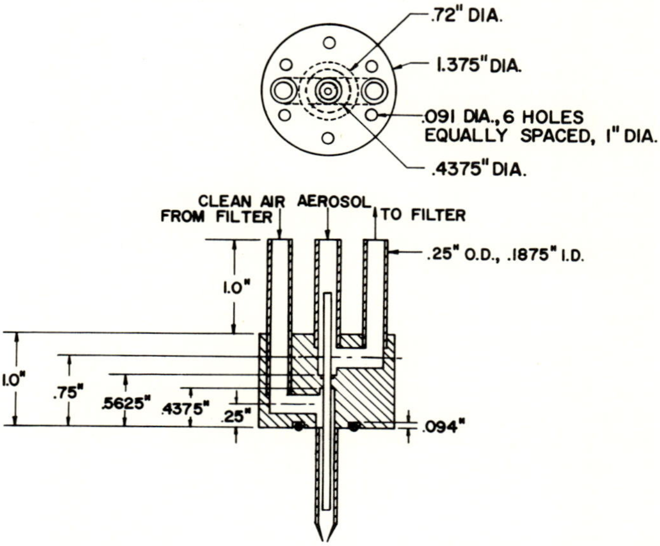

Figure 7 Sheath-air inlet for the Royco 220 particle counter.

these two concentrations defines the sampling efficiency. If particles of sizes larger than about 5 μm in diameter are to be sampled, the inlet tube should preferably be short without any bends. An example is given in Figure 8 which shows a conical inlet design[2] for the Royco 245 high-flow-rate counter. With this inlet the original efficiency for sampling from a calm air environment was improved to about 80% for 10 μm particles and about 43% for 30 μm particles as shown in Figure 9. One may conclude from Figure 9, that measured concentrations should always be modified by the sampling efficiency when particles larger than 5 to 10 μm in diameter are sampled.

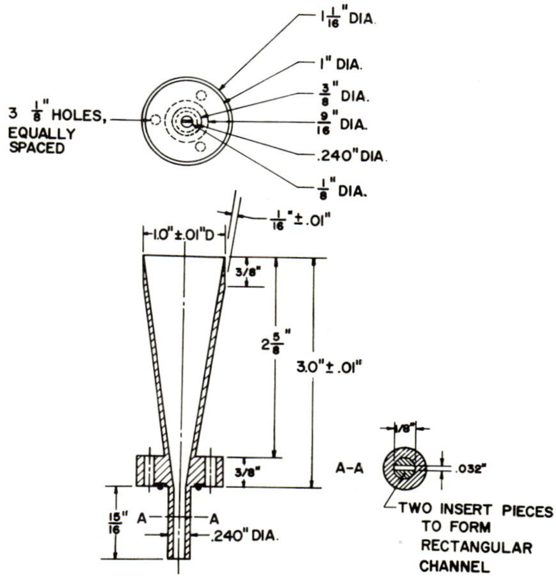

Figure 8 Conical inlet for the Royco 245 particle counter.

PARTICLE REFRACTIVE INDEX AND SHAPE

The chemical composition of natural aerosols may vary from particle to particle. The particle index of refraction, m, is therefore unknown. Many ambient aerosols contain a substantial quantity of water because of their hygroscopicity. The refractive index may therefore range from m = 1.33 (water) to, say, about m = 1.9. For practical purposes, OPC's are usually calibrated with ideal particles of refractive index 1.5 to 1.6. The size measured by an OPC is then[1] an "Equivalent latex sphere diameter" or an "Equivalent DOP sphere diameter" depending on the calibrating aerosol used. The theoretical response curve for the example given in Figure 2 (Bausch & Lomb 40-1A Counter) shows that the counter response is a strong function of refractive index, particularly for

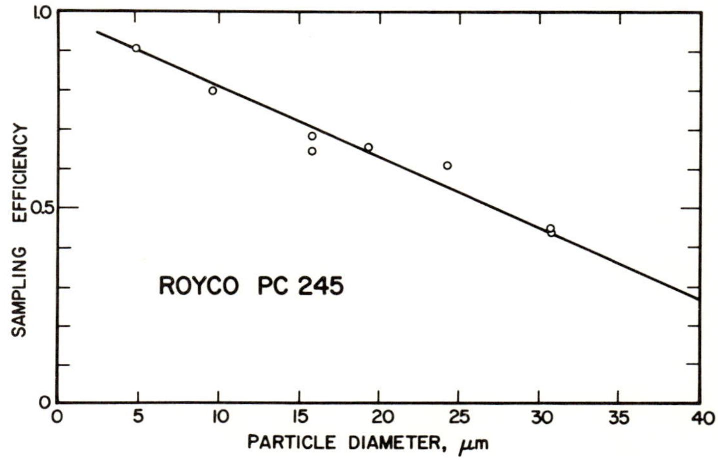

Figure 9 Sampling efficiency as a function of particle diameter for the modified conical inlet (Figure 8) of the Royco 245 high-flow-rate particle counter.

$D_p < 2$ μm. This dependence varies considerably from one type of counter to another because of the difference in optical configuration.

An experimental calibration[3] of the Bausch & Lomb 40-1A counter for DOP particles of refractive index m = 1.49 is shown in Figure 3. The counter output voltage is seen to monotonically increase with particle size. When calibrated with DOP particles, a similar monotonic increase was found for the Royco 220 particle counter (Figure 5), whereas the calibration curves for the Royco 215 (Figure 4) and the Royco 245 (Figure 6) show a dip for particle sizes of about 1 μm in diameter. No useful size distribution information can be obtained in the size range where counters display such a dip in the calibration curve. A counter operating with monochromatic laser light may have a multivalued curve over a large range of particle sizes.

Agreement between theoretical prediction and experimental calibration is generally very good[2]. After an experimental calibration curve has been established for a given refractive index, the theoretical response for this refractive index and for other refractive indices may be drawn onto the experimental plot. However, the reference size through which the theoretical response curve is dimensionalized to counter output voltage, should preferably be from a monotonic portion of the response curve where the response is least affected by variations in refractive index, i.e. $D_p > 2$ μm for the example in Figure 2.

Special aerosols may have refractive indices which greatly differ from those generally found in the ambient air. Figure 2 shows the relative theoretical counter response[14] to a particle of high refractive index, 2.5, and one to an absorbing particle with m = 1.54 - 0.5i. A typical light absorber is coal dust for which an experimental calibration curve[3] is shown in Figure 3. As seen, the output signal amplitude from the OPC ranges from a factor of 2 to 25 lower for coal than for ideal transparent DOP particles of the same size. One may define a size shift factor[1] as the ratio of the coal particle size to that of a DOP particle which generates the same output voltage from the OPC. This size shift increases monotonically from about 3 for small particles to about 5 for large particles in Figure 3. Submicron particles may have unity size shift; e.g., the curves for m = 1.54 and for m = 1.54 - 0.5i merge at a particle diameter of about 0.4 μm in the example of Figure 2.

The coal dust for the calibration curve (Fig. 3) was produced in a ball mill so that the particles were nodular in shape. However, coal dust particles produced by a cutting tool were found to have very irregular shapes[38], such as spears or platelets. This difference in shape does not necessarily change the calibration curve but will decrease the resolution of the counter because the scattered light received at a fixed angle from the optical axis will depend on the orientation of the particle in the view volume and will therefore have a larger range of intensities than one would expect from a spherical particle.

The effect of particle shape is expected to be less for a counter with a light collection aperture axis coaxial with rather than at some angle to the illumination axis, because the scattered light is integrated through 360° around the optical axis. A broad range of viewing angles ($\beta - \alpha$) for a coaxial system will further smooth out geometrical variations.

COINCIDENCE

Coincidence[37, 39, 40] refers to the simultaneous presence of more than one particle in the optical view volume. This determines the maximum particle concentration, which an OPC may measure accurately. In the case of coincidence, the indicated particle number concentration, N_i, is lower than the true concentration, N_t, and the shape of the size distribution is weighted towards larger particles because the pulse height analyzer registers two or more particles in the view volume as one particle of a larger size.

The statistical probability P, that no particle (p = o), one particle (p = 1) or more than one particle (p = 2, 3, 4, . . .) is present at the same time in the view volume v is given by the Poisson distribution

$$P_p = \frac{\mu^p e^{-\mu}}{p!} \tag{2}$$

where $\mu = N_t v$ (3)

is the average number of detectable particles in v. The probability of finding no particles in the view volume is then

$$P_o = e^{-\mu} = e^{-N_t v} \tag{4}$$

and the probability of finding one, two or more particles in v, i.e. the indicated number of particles in v is

$$(1 - P_o) = 1 - e^{-N_t v} = N_i v \tag{5}$$

so that the ratio of indicated to true particle concentration becomes

$$\frac{N_i}{N_t} = \frac{1 - e^{-N_t v}}{N_t v} \qquad (6)$$

This derivation, however, considers only the static situation of finding particles in the view volume[48]. One should also consider the dynamic situation of particles traversing the view volume. A particle will trigger a pulse only, if the view volume is not already occupied by particles in which case the indicated count is[37, 49-51]

$$\frac{N_i}{N_t} = e^{-N_t v} \qquad (7)$$

Equations (6) and (7) are plotted in Figure 10. For small values of $N_t v$, equation (6) reduces to $1 - N_t v/2$, and equation (7) to $1-N_t v$, i.e. twice the coincidence of equation (6). Since the view volume can usually not be defined accurately and the scattered light intensity of the particle entering the view volume may not immediately exceed the threshold value, either equation (6) or (7) can be used to predict the limiting number concentration for which coincidence may become significant. Equation (7) should be used for a conservative estimate.

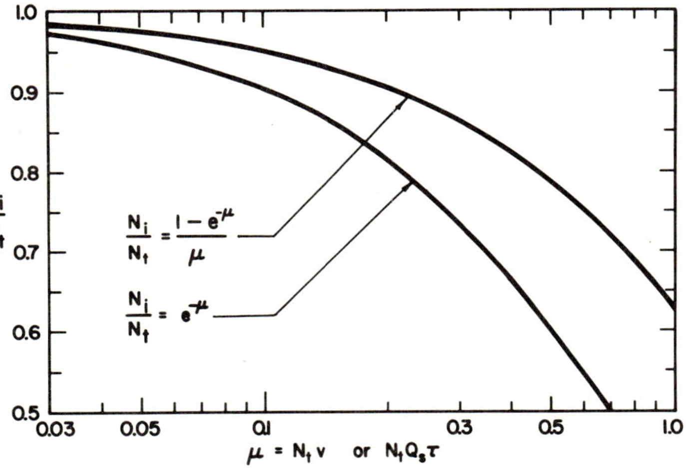

Figure 10 Ratio of indicated to true particle concentration as a function of average particle number concentration $N_t v$ in view volume v, or as a function of average particle count rate $N_t Q_s \tau$ during stretched pulse duration τ.

When several subcountable[41] particles are present in the view volume at the same time, the sum of the scattering intensities may result in a detectable light pulse so that an excess in counts will be registered by the multi-channel analyzer. This problem is usually alleviated by dilution of the aerosol concentration and by setting the discriminator voltage of the MCA to a sufficiently high value.

COUNT ACCURACY

The lowest particle concentration which an OPC may accurately measure is given by statistical considerations. The statistical probability of correct counting is given by the Poisson distribution. An inherent property of this distribution is the equality between the variance and the mean, i.e. the standard deviation is equal to the square root of the mean which equals the actual count to first approximation. The relative standard deviation, σ_r, defined as the ratio of standard deviation, σ, to particle count, n, is then:

$$\sigma_r = \frac{\sigma}{n} = \frac{n^{1/2}}{n} = n^{-1/2} \tag{8}$$

The particle count, is given by

$$n = N_t Q_s t \tag{9}$$

where Q_s is the volumetric sample flow rate and t is the time of sampling.

Equation (8) is plotted in Figure 11. For instance, a count of 100 particles has a relative standard deviation of 10%, and a count of 1000 particles has a relative standard deviation of about 3%. The counter is

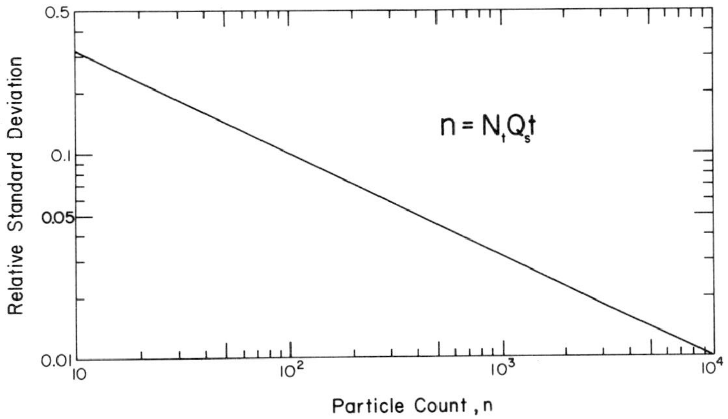

Figure 11 Particle count accuracy of OPC's.

thus limited by particle coincidence at high number concentrations and by particle count accuracy at low number concentrations. For normal sampling times, this limits most commercial counters to a number concentration range of about 3 to 4 decades. The number concentration of atmospheric aerosols is about 3 to 4 decades lower for particles 10 μm in diameter than for 0.5 μm particles, which corresponds to the nominal limits of conventional OPC's.

For ambient atmospheres, the number concentration for a given size range may vary within time as much as two decades, when sampled at surface locations, and three decades when vertical concentration profiles are measured[43], e.g., by use of an aircraft. One therefore needs to carefully observe the limits of operation of an OPC when sampling atmospheres with time-varying particle concentrations. A large size range is best sampled by two OPC's, one for sizes from, say, 0.5 μm to 5 μm, and the other for sizes from, say, 5 μm to 50 μm, with each OPC optimized for its respective size range.[44]

The count accuracy for a given counter may be improved by increased time of sampling, which results in a loss of time resolution. It may also be improved by registering a larger size subrange per channel of the multi-channel analyzer (MCA). This will, however, result in a decreased size resolution.

RESOLUTION

An MCA can, in principle, resolve very small size intervals and give detailed information on the structure of the size distribution curve. However, every optical counter has a limit of resolution. The term resolution[2,3,7] refers to the ability of the instrument to distinguish two monodisperse aerosols of different mean particle size. This is also referred to as channel cross-sensitivity.

The counter's response to monodisperse particles depends on the uniformity of the light intensity in the illuminated optical view volume, the uniformity of sensitivity of the photodetector surface, and on the baseline electronic and optical noise. The MCA therefore responds to monodisperse particles with pulses distributed over a range of channels.

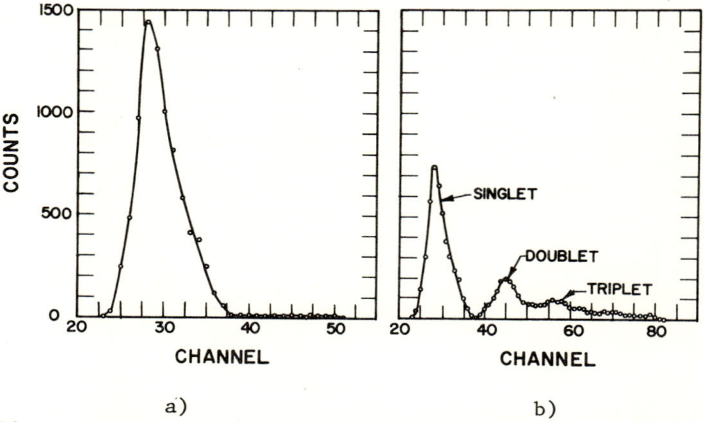

Figure 12 Optical particle counter output under good (a) and poor (b) dispersion conditions.

For example, Figure 12 shows some typical pulse height spectra from the Bausch & Lomb 40-1 counter when exposed to aerosols from the vibrating-orifice generator operated under good and poor dispersion conditions.

Good dispersion (Fig.12a) is indicated by a very pronounced peak in the displayed spectrum and by the near absence of pulses of a larger amplitude above the peak. Under poor dispersion conditions (Fig. 12b) additional peaks appear in the spectrum. These secondary peaks were caused by droplets formed from the collision and coalescence of the primary droplets during the dispersion process.

The shape of the counter signal, Fig 12a, is approximately Gaussian and may be characterized by the overall variance σ_o^2 (σ_o = standard deviation). The variance in signal from non-uniformity in illumination and detection, σ_s^2, and the variance in baseline electronic and optical noise, σ_b^2, are essentially independent of each other, so that

$$\sigma_o^2 = \sigma_s^2 + \sigma_b^2 \tag{10}$$

The relative standard deviation of the output pulse distribution for monodisperse aerosols is then[2]

$$\frac{\sigma_o}{V_o} = \left[\left(\frac{\sigma_s}{V_o}\right)^2 + \left(\frac{\sigma_b}{V_o}\right)^2\right]^{1/2} \tag{11}$$

where V_o is the pulse height at the peak of the distribution for monodisperse aerosols of particle diameter D_p.

The electronic noise is approximately constant and is equal to the counter output when no particles are passing through the optical view volume. For example, it is approximately equal to 3 mV for the Bausch and Lomb 40-1 counter[2]. Since the counter voltage output increases with particle size, approximately by one to two decades per decade of particle size, either electronic noise or Rayleigh scattering from the air molecules sets the lower limit of particle size which an OPC can detect, generally 0.3 μm to0.5μm in diameter. A discriminator is set at a voltage level exceeding the baseline noise signals.

The variance in signal σ_s^2 is found from measurements with large particles (large V_o) for which the ratio (σ_b/V_o) tends toward zero, and consequently (σ_o/V_o) → (σ_s/V_o). The ratio (σ_s/V_o) approaches 0.08 for large DOP test particles when measured with the Bausch and Lomb 40-1 counter[2]. These measurements also show that σ_s increased to a higher value than predicted by equation (10) when the particle size was reduced. A probable explanation of this discrepancy is that the smaller particles experience a greater lack of uniformity in illumination and detection than the large particles do, and that σ_s is actually a variable dependent upon particle size, rather than a constant, is as implied in equation (10). Also, small particles may scatter a finite number of photons, thus increasing the fluctuations in output signal.

The resolution of an OPC depends therefore on the ability of the counter to produce uniform pulses upon exposure to monodisperse aerosols. It also depends on the slope of the calibration curve, a steep one being the most desirable. Optical counters of the same model and produced by the same manufacturer may not have the same resolution due to normal manufacturing tolerances and differences in alignment of the optics.

When calibrating OPC's with non-ideal particles, such as monodisperse coal dust, the MCA may show a pulse height distribution with a spread which is considerably wider than the intrinsic properties of the instrument would dictate. The difference is attributed to the irregular shape of the monodisperse particles, and to a lesser extent, to refractive index variations of the particles[3]. This spreading effect further decreases the ability of the instrument to resolve small particle size differences. In general, the smallest variance is obtained with ideal aerosols, the worst with irregularly shaped particles, because the scattering intensity received in the collecting aperture from irregularly shaped particles, depends upon the orientation of the particle in the view volume. Absorbing particles may further increase the variance because the calibration curve of output voltage vs. particle size, e.g. Figure 3, has, in general, a lower slope than the one for non-absorbing particles.

PULSE PROCESSING

The photodetector in an optical counter, usually a photo-multiplier, receives a flux of photons when particles pass through the optical view volume. The photon flux is converted to a voltage signal and recorded by a multichannel analyzer which stores the signal in a channel assigned to a range of voltages and counts the number of pulses received in each channel. However, the output signal shape from an OPC is generally not compatible[45,46] with the input pulse shape requirement of an MCA.

In general, an MCA requires a constant height pulse of about 3 to 5μsec duration. A pulse converter is therefore used as an interface to convert the OPC signal to a compatible signal for the MCA. Some older MCA's had a low input resistance, which was incompatible with the output resistance of the OPC. Most of the newer MCA units, however, have an input impedance in excess of one megohm, which will not distort the output signal from the OPC[46].

The performance of the pulse converter will be illustrated by reference to a specific device, the University of Minnesota Model 170-1 peak detector, which was developed from the Royco 170-1 pulse converter. This device first removes low-frequency signals and d-c offset from the incoming signal. The incoming signal is generally Gaussian in shape and has a pulse width of a few to several hundred microseconds, depending on the OPC used. The signal is then linearly or logarithmically amplified. Logarithmic compression of the signal has the advantage of MCA signal storage in logarithmic voltage intervals which approximately corresponds to logarithmic particle size intervals. A baseline voltage offset is necessary to prevent the converter from being triggered by noise in the incoming signal.

When the amplified signal reaches its peak, the system stretches the pulse at that voltage level for a preset period of time at the end of which a "gating" pulse commands the MCA to read the signal amplitude. The signal voltage is held as a charge on a capacitor which is quickly discharged and reset to its baseline value after the MCA has read its amplitude.

When the pulse stretching technique is used, the pulse has to be stretched beyond the duration of the signal which introduces a dead time during which no additional signals are processed. The calculation of particle count loss is similar to the one for particle coincidence loss in the view volume. Referring to the Poission distribution, Equation (2), we have now

$$\mu = \dot{n}_t \tau = N_t Q_s \tau \tag{12}$$

where $\dot{n}_t$ is the true particle count rate (Equation 9) and τ is the time duration of the stretched pulse, i.e. μ is equal to the average number of counts made during pulse duration τ. The expression for the indicated particle count n_i is then similar to Equation (5)

$$\frac{n_i}{n_t} = e^{-N_t Q_s \tau} = \frac{N_i}{N_t} \tag{13}$$

which is plotted in Figure 10. As seen from Figure 10, a less than 5% loss in particle count approximately requires that

$$N_t Q_s \tau < 0.05 \tag{14}$$

Similarly, a less than 5% particle coincidence loss in the view volume requires

$$N_t v < 0.05 \tag{15}$$

The loss of signal count may be reduced by dilution of the aerosol concentration. However, dilution techniques may incur a loss of large particles.

An alternate technique of peak detection makes use of pulse differentation. With this technique, a comparator looks at the slope of the signal and gives a "gating" signal for amplitude reading by the MCA only when the differentiator output changes from positive to negative at the peak of the signal. However, fluctuations on the signal due to noise may produce local maxima so that the MCA may record more than one output pulse from a single input pulse.

A technique, reported by the University of North Carolina at Chapel Hill[46], eliminates the problem of dead time. It is similar to the stretching technique, but uses the comparator, which recognizes the peak of the pulse, to give a gating signal to the MCA at the beginning of the pulse stretching. The capacitor, which stretches the pulse at peak amplitude, is then discharged when the original signal has disappeared, rather than after a preselected time period.

CONCLUSIONS

Measurements by single particle optical counters are superior over manual sampling and optical microscope counting techniques when the particle size distribution and concentration of air environments is to be determined, because OPC's measure particles automatically and in situ. Detailed information may thus be provided on the time variations of these quantities, which cannot be achieved by microscope techniques with any degree of statistical counting accuracy without great expenditure in time. In addition, human error and bias is minimized, and possible errors resulting from sample collection onto a substrate, and from transfer of the sample to a proper measurement location are excluded.

Most OPC's can accurately count particles over a range of at least four decades of particle concentration. The limits are set by coincidence resulting from electronic dead time of simultaneous occurance of more than one particle in the view volume, when sampling high concentrations, and by particle count accuracy, when sampling low concentrations. However, any range of particle concentration can usually be measured by proper choice of particle counter(s), air dilution and sampling time.

The lower limit of particle size detected by conventional OPC's with white light illumination is approximately 0.3 μm in diameter. The upper limit is set by the sampling efficiency of the counter inlet tube. When two decades of particle size are to be measured, say from 0.5 μm to 50 μm, two OPC's should be operated, each one optimized for its respective size range.

The size indicated by the counter is equivalent to the size of the calibration aerosol, mostly an ideal (spherical and transparent) aerosol, such as polystyrene latex. The shape and refractive index of the measured particle may be different from that of the calibration aerosol so that its actual size may be larger or smaller than the one indicated by the OPC. However, this problem is not limited to optical counters alone. Practically all aerosol sizing instruments measure some equivalent sphere diameter.[47] The inertial impactor, the aerosol centrifuge, the elutriator and the cyclone measure an equivalent sphere diameter based on Stokes drag and particle mass - the so-called aerodynamic diameter. The electrical aerosol analyzer measures an equivalent sphere diameter based on electrical charge and mobility. The diffusion battery measures an equivalent sphere diameter based on the diffusion coefficient of the particles. This indicates that all particle size data need to be interpreted with caution and with due consideration for the method used for the measurement.

REFERENCES

1. Whitby, K. T., and R. A. Vomela. Response of Single Particle Optical Counters to Nonideal Particles. Environ. Sci. Technol. 1: 801-814, October 1967.

2. Liu, B. Y. H., R. N. Berglund, and J. K. Agarwal. Experimental Studies of Optical Particle Counters. Atmospheric Environment 8: 717-732, 1974.

3. Liu, B. Y. H., V. A. Marple, K. T. Whitby, and N. J. Barsic. Size Distribution Measurement of Airborne Coal Dust by Optical Particle Counters. Amer. Ind. Hyg. Assoc. J. 8:443-451, August, 1974.

4. Gucker, F. T., and D. G. Rose. A Photoelectronic Instrument for Counting and Sizing Aerosol Particles. Brit. J. Applied Physics: Supplement No. 3, 138-143, 1954.

5. Lieberman, A., and R. J. Allen. Theoretical and Experimental Light Scattering Data for a Near Forward System. (Presented at the Amer. Assoc. for Contamination Control), May 19-22, 1969.

6. Martens, A. W., and K. H. Fuss. An Optical Counter for Dust Particles. Staub-Reinhalt. Luft (English Transl.) 28:14-18, June 1968.

7. Whitby, K. T., B. Y. H. Liu, R. B. Husar, and N. J. Barsic. The Minnesota Aerosol-Analyzing System Used in the Los Angeles Smog Project. J. Colloid Interface Sci. 39:136-164, April 1972.

8. Willeke, K., K. T. Whitby, W. E. Clark, and V. A. Marple. Size Distributions of Denver Aerosols - A Comparison of Two Sites. Atmospheric Environment, 8:609-633, 1974.

9. Willeke, K. and K. T. Whitby. Atmospheric Aerosols: Size Distribution Interpretation. J. Air. Poll Control Assoc., 25:529-534, May, 1975.

10. Quenzel, H. Influence of Refractive Index on the Accuracy of Size Determination of Aerosol Particles with Light-Scattering Aerosol Counters. Applied Optics 8:165-169, January, 1969.

11. Gebhardt, J., J. Bol, W. Heinze, and W. Lebschert. A Particle-Size Spectrometer for Aerosols Utilizing Low-Angle Scattering of the Particles in a Laser Beam. Staub-Reinhalt. Luft (English Translation), 30:5-14, June 1970.

12. Schuster, B. G., and R. Knollenberg. Detection and Sizing of Small Particles in an Open Cavity Gas Laser. Applied Optics, 11: 1515-1520 July, 1972.

13. Knollenberg, R. G. Active Scattering Aerosol Spectrometry. In Aerosol Measurements, W. A. Cassatt and R. S. Maddock (eds), NBS Special Publ. No. 412, National Bureau of Standards (U.S.) October 1974, p. 57-64.

14. Cooke, D. D., and M. Kerker. Response Calculations for Light Scattering Aerosol Particle Counters. Applied Optics 14:734-739, March, 1975.

15. Martens, A. E., and D. D. Doonan. Comments on: Influence of Refractive Index on the Accuracy of Size Determination of Aerosol Particles with Light-Scattering Aerosol Counters. Applied Optics 9:1930-1931, August, 1970.

16. Hodkinson, J. R., The Optical Measurement of Aerosols. In; Aerosol Science, Davies, C. N. Academic Press, New York 1966, p. 287-357.

17. Hodkinson, J. R. and J. R. Greenfield. Response Calculations for Light Scattering Aerosol Counters. Applied Optics 4:1463-1474, November, 1965.

18. Hodkinson, J. R., and I. Greenleaves. Computations of Light-Scattering and Extinction by Spheres According to Defraction and Geometrical Optics, and some Comparisons with the Mie Theory. J. Optical Society of America 53:577-588, May 1963.

19. Gucker, F. T., and D. G. Ross. The Response Curves of Aerosol Particle Counters. (Presented at the Third National Air Pollution Symposium, Pasadena, Ca., April 18-20, 1955.)

20. Gucker, F. T. , R. L. Rowell, and G. Chui. A Study of Intensity Functions for Dielectric Spheres Calculated from the Mie Theory (Presented at the 1st Nat. Conf. on Aerosols, Liblice, Czechoslovakia, October 8-13, 1962).

21. Gucker, F. T. and J. Tuma. Influence of the Collecting Lens Aperture on the Light-Scattering Diagrams from Single Aerosol Particles. J. Colloid Interface Sci. 27:402-411, July 1968.

22. van der Hulst, H. C. Light Scattering by Small Particles. John Wiley, New York, 1957.

23. Mie, G. Beitraege zur Optik trueber Medien. Ann. Physik 25:377-445, 1908.

24. Kerker, M. The Scattering of Light and Other Electromagnetic Radiation, Academic Press, New York, 1969.

25. Cassatt, W. A., and R. S. Maddock (eds.) Aerosol Measurements (includes several articles on optical measurements.), NBS Special Publ. No. 412, National Bureau of Standards (U.S.), 1974.

26. Pinnick, R. G., and D. J. Hoffmann.Efficiency of Light-Scattering Aerosol Particle Counters. Applied Optics 12:2593-2597, November 1973.

27. Oeseburg, F. The Influence of the Aperture on the Optical System of Aerosol Particle Counters on the Response Curve. Aerosol Science 3:307-311, 1972.

28. Lundgren, D. A. and D. W. Cooper. Effect of Humidity on Light-Scattering Methods of Measuring Particle Concentrations, J. Air Poll. Control Assoc. 19:243-247, April 1969.

29. Rimberg D., and J. W. Thomas. Response of an Optical Counter to Monodisperse Aerosols. Atmospheric Environment 4:681-688, 1970.

30. Boden, A., and W. Coenen. Darstellung von Teilchengroessenvertlungen mittels eines einfachen Spektrometers als Zustaz zu Streulichtmessgeraeten. Staub-Reinhalt. Luft 30:483-487, November 1970.

31. Cooke, D. D., and M. Kerker. Particle Size Distribution of Colloidal Suspensions by Light Scattering Based upon Single Particle Counts - Polystyrene Latex. J. Colloid Interface Sci. 42:150-155, January, 1973.

32. Bakhanova, R. A., and L. V. Ivanchenko. The Calibration Curve of Photoelectric Counters and Computation of Particle Size Distribution when the Relationship between Particle Size and Electrical Pulse Amplitude is Ambiguous. Aerosol Science 4:485-490, 1973.

33. Whitby, K. T. and B. Y. H. Liu. Polystyrene Aerosols - Electrical Charge and Residue Size Distribution. Atmospheric Environment 2: 103-116, 1968.

34. Berglund, R. N., and B. Y. H. Liu. Generation of Monodisperse Aerosol Standards. Environ. Sci. Technol. 7: 147-153, 1972.

35. Guichard, J. C. and J. L. Magne. Application du lit fluidise en milieu gaseau a la generation des aerosols. Inst. Nat. Rech. Chim. Appl., Vert-le-Petit, France. Note interieure No. 51, 1967.

36. Willeke, K., S. K. Lo, and K. T. Whitby. Dispersion Characteristics of a Fluidized Bed. Aerosol Science 5:449-455, 1974.

37. Jaenicke, R., The Optical Particle Counter: Cross- Sensitivity and Coincidence. Aerosol Science 3:95-111, 1972.

38. Marple, V. A. Development, Calibration, and Application of Size Distribution Instruments at the University of Minnesota. In: Aerosol Measurements, Cassatt, W. A., and R. S. Maddock, (eds.) NBS, Special Publication No. 412, National Bureau of Standards (U.S.), 1974, p. 176.

39. Bader, H., H. R. Gordon, and O. B. Brown. Theory of Coincidence Counts and Simple Practical Methods of Coincidence Count Correction for Optical and Resistive Pulse Particle Counters. Rev. Sci. Instr. 43:1407-1412, 1972.

40. Nikiforova, N. K., and Y. S. Sedunov. An Investigation of Errors Arising in Photoelectric Counters - Aerosol Particle Analyzer. Aerosol Science. 3:441-453, 1972.

41. Whitby, K. T., and B. Y. H. Liu. Generation of Countable Pulses by High Concentrations of Subcountable Sized Particles in the Sensing Volume of Optical Counters. J. Colloid Interface Sci. 25:537-546, 1967.

42. Junge, C. E.. Air Chemistry and Radioactivity, Academic Press, New York, 1963.

43. Whitby, K. T., Dept. of Mechanical Eng., Univ. of Minnesota, private communication, 1975.

44. Whitby, K. T., W. E. Clark, V. A. Marple, G. M. Sverdrup, G. J. Sem, K. Willeke, B. Y. H. Liu, and D. Y. H. Pui. Characterization of California Aerosols. I. Size Distributions of Freeway Aerosol. Atmospheric Environment 9:463-482, May, 1975.

45. Graedel, T. E.. Channel Width Determination and Electronic Pulse Processing Losses in Optical Particle Counters. Aerosol Science 5:125-131, 1974.

46. Hemenway, D. R.. Pulse Processing from Particle Detectors. Ph.D. Thesis, Dept. Env. Sci. and Eng., Univ. of North Carolina at Chapel Hill, 1974.

47. Liu, B. Y. H., R. N. Berglund, and J. K. Agarwal. Author's Reply. Atmospheric Environment 8:1344, December 1974.

48. Wales, M., and J. N. Wilson. Theory of Coincidence in Coulter Particle Counters. RSI 32:1132-1136, 1961.

49. Princen, L. H., and W. F. Kwolek. Coincidence Corrections for Particle Size Determinations with the Coulter Counter. RSI 36:646-653, 1965.

50. Pisani, J.F., and G. H. Thomson. Coincidence Errors in Automatic Particle Counters. J. Phys. E. 4:359-361, 1971.

51. Mercer, T.T. Aerosol Technology in Hazard Evaluation, Academic Press, New York, 1973.

COMPARISON OF IMPACTION, CENTRIFUGAL SEPARATION AND ELECTRON MICROSCOPY FOR SIZING CIGARETTE SMOKE[1]

R. F. Phalen, Department of Community and Environmental Medicine, College of Medicine, University of California, Irvine, California 92664

W. C. Cannon, Biology Department, Battelle Northwest Laboratories, Richland, Washington 99352

D. Esparza, Aerosol Physics Department, Inhalation Toxicology Research Institute, Albuquerque, New Mexico 87108

ABSTRACT

Determination of the size distribution of aerosols composed of small liquid droplets involves special problems, especially when the particles are numerous per unit volume of carrier gas. Electron microscopy, useful in examining most solid aerosols, is not generally applicable to liquids. Using unfiltered smoke from research cigarettes as a test aerosol, three techniques for obtaining a size distribution were compared. The aerosol, having a mass median aerodynamic diameter of about 0.7 micrometers, was examined with a seven-stage cascade impactor, a centrifugal separator, and by rapid fixation of the aerosol with methyl 2-cyanoacrylate vapor followed by electrostatic precipitation and electron microscopy. The resulting size distributions were compared and found to be in fairly good agreement. Rapid fixation in the airborne state appears to be a useful technique for preparing certain liquid aerosols for electron microscopy.

COMPARISON OF IMPACTION, CENTRIFUGAL SEPARATION AND ELECTRON MICROSCOPY FOR SIZING CIGARETTE SMOKE[1]

R. F. Phalen, Department of Community and Environmental Medicine, College of Medicine, University of California, Irvine, California 92664

W. C. Cannon, Biology Department, Battelle Northwest Laboratories, Richland, Washington 99352

D. Esparza, Aerosol Physics Department, Inhalation Toxicology Research Institute, Albuquerque, New Mexico 87108

INTRODUCTION

The objective of this study was to compare three methods of obtaining size distributions of main-stream cigarette smoke. No attempt was made to reproduce cigarette smoke as it exists in the human smoking situation. Instead, laboratory methods of puffing, diluting, mixing and sampling were used so that the three sizing methods could be compared and so that other laboratories could repeat the studies. The methods for sizing were: a) use of a cascade impactor, b) use of a centrifugal, spiral aerosol separator, and c) coating of the smoke with methyl 2-cyanoacrylate, collecting with an electrostatic precipitator and sizing from electron micrographs. The latter method of sizing is not considered a standard laboratory technique.

Fresh cigarette smoke has several properties that make it difficult to characterize; it is chemically complex, initially highly concentrated and not in equilibrium with its surroundings with respect to moisture content, temperature, or vapor pressures of volatile components. In its fresh state it consists primarily of aqueous, spherical droplets with dissolved and suspended solid materials. Unless it is diluted, coagulation proceeds rapidly, and depending on the factors of dilution, the median size of a cloud may increase or decrease with time.

A great deal of research has been and is being done with respect to the health effects of cigarette smoke and there is a need for standardization of sizing methods. The methods compared in this study are all relatively straightforward and one or more probably could be made available to the majority of those investigating cigarette smoke.

MATERIALS AND METHODS

CIGARETTES

The Kentucky Reference IRI cigarette, stored at 60% RH and 20-23°C was used throughout.

SMOKING PARAMETERS

The smoking machine (Figure 1) was recommended by C.H. Keith (Celanese Fibers Company, Charlotte, N.C.). It consisted of a short glass cigarette holder connected to a 3-necked 500 ml glass flask fitted with a metal stirring paddle and a critical orifice through which a timed flow of air could be drawn. The cigarette holder was removed to permit sampling.

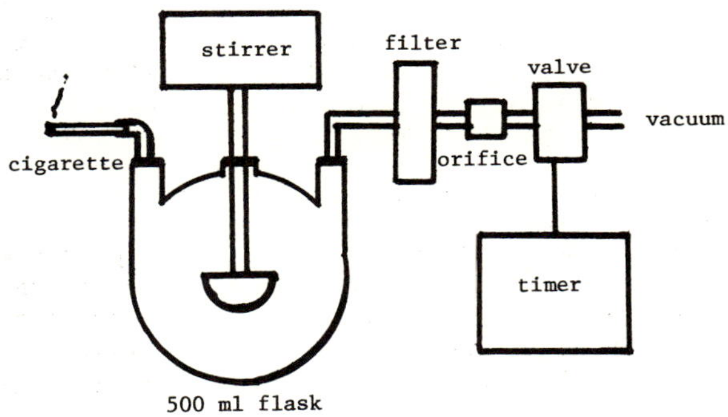

Figure 1. Machine used to generate cigarette smoke for comparison of three methods of sizing sub-micron droplets.

The smoking regime began with lighting the cigarette during a 35 ± .5 ml puff of 0.94 seconds duration. Two subsequent puffs of the same size were taken at 1 min. intervals. After the third puff the smoke was purged from the flask. The fourth puff, 1 min. after the previous puff, was sampled for 15 seconds, 5 seconds after it was drawn into the flask.

SAMPLING DEVICES AND METHODS

A seven stage cascade impactor of the type designed by T. T. Mercer[2] was used. When operated at 500 cc/min. the effective cut-off diameters of the impactor used were:

Stage 1	6.7 μm
2	4.3 μm
3	3.0 μm
4	2.3 μm
5	1.6 μm
6	1.0 μm
7	0.46μm
filter	0.0 μm

Impactor stages were covered with glass cover slips coated with a silicone-oil spray. After sampling, the deposit on each impactor stage was dissolved in a fixed volume of isopropyl alcohol and aliquots were assayed with a spectrophotometer at a wavelength of 350 nm. The relative distribution of material on each stage was fitted to a log-normal distribution and the mass median diameter and geometric standard deviation determined. Two cigarettes were used for each impactor run.

Smoke was drawn into a Lovelace Aerosol Particle Separator (LAPS) (P. Kotrappa and M.E. Light[3]) for 15 seconds at 310 cc/min. The device had been calibrated with monodisperse polystyrene spheres and found to collect particles from 0.54 μm to 5.1 μm along a 40 cm-long metal foil strip. After sampling, the strip was cut into 22 segments and assayed photometrically by the same method used with the impactor. The exit filter was also assayed. Six cigarettes were used for each run.

Samples for electron microscopy were collected using a point-to-plane electrostatic precipitator operated at 50 cc/min. for 15 seconds (P.E. Morrow and T. T. Mercer[4]). Before these sampling runs 3 drops of methyl-2-cyanocrylate (Polysciences, Inc., Warrington, Pa.) were placed in the 500 ml smoke chamber and the flask was heated to 70°C (this method is similar to that described by G.P. Morie, C.H. Sloan and V.G. Peck[5]). Samples were collected on carbon-coated diffraction grating replicas with line spacings of 0.833 μm. Sizing was performed using a Zeiss TGZ3 Analyzer (Carl Zeiss, Inc., N.Y., N.Y.) from photomicrographs with a magnification factor of about 30,000. An estimate of mass median aerodynamic diameter was made by assuming a density of unity and that the aerodynamic diameter of agglomerates containing n primary particles was equal to $n^{1/3}$ times the average primary particle diameter.

Impactor and spiral-centrifuge samples were taken at two locations, the Inhalation Toxicology Research Institute (ITRI), Albuquerque, New Mexico, and at Battelle Northwest Laboratories in Richland,Washington(BNW) using essentially identical smoking machines and sampling methods. The precipitator samples were taken only at ITRI.

RESULTS AND CONCLUSIONS

The agreement between all sampling methods and, in fact, all sample runs was within the experimental errors of a given technique. The table gives the results of each sampling run. In each case, the size distribution was approximately log-normal and the mass median and geometric standard deviation (σg) were used to describe the results.

run	method	average MMAD	σg	location
1,2	Impactor	0.74 μm	1.4	ITRI
3-7	Impactor	0.71 μm	1.4	BNW
8	LAPS	0.79 μm	1.3	ITRI
9-11	LAPS	0.72 μm	1.4	BNW
12	EM	0.57 μm	1.5	ITRI

The precipitator sample contained spherical particles, many of which were in cluster or chainlike agglomerates (Figure 2). Primary particles, sized from electron micrographs (n=191), were log-normally distributed with a count median diameter of 0.29 μm and a σg of 1.4. A background (no cigarette) precipitator sample yielded some spherical particles, negligible in quantity when compared to samples of cigarette smoke.

Since results from each of the three methods were in close agreement, it is concluded that any of the three techniques could be used for describing the size distributions of cigarette smoke except when extreme accuracy is required. Furthermore, the fixation by coating with methyl-2-cyanoacrylate may be of use in electron-microscopic studies of other aqueous aerosol particles.

ACKNOWLEDGEMENTS

This study was suggested by Dr. Charles H. Keith and performed with the invaluable assistance of Robert Yarwood.

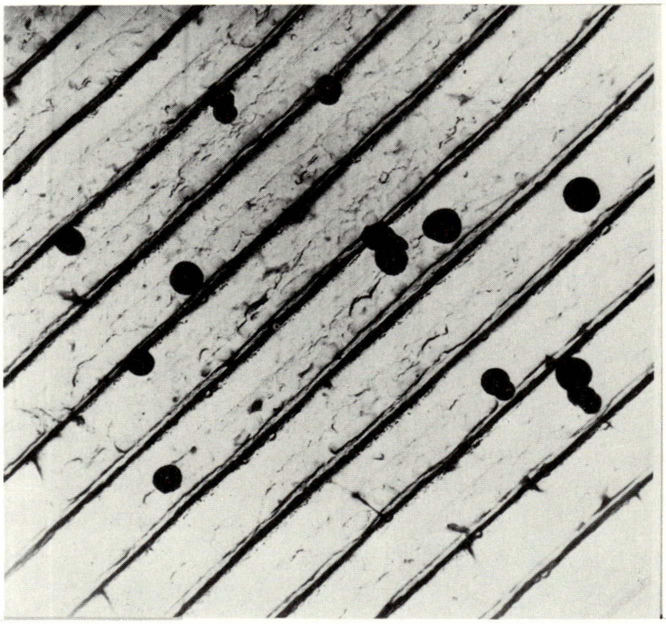

Figure 2. Electron micrograph of cigarette smoke particles fixed by coating with methyl-2-cyanoacrylate vapor. Diffraction-grating-replica spacing is 0.833 μm.

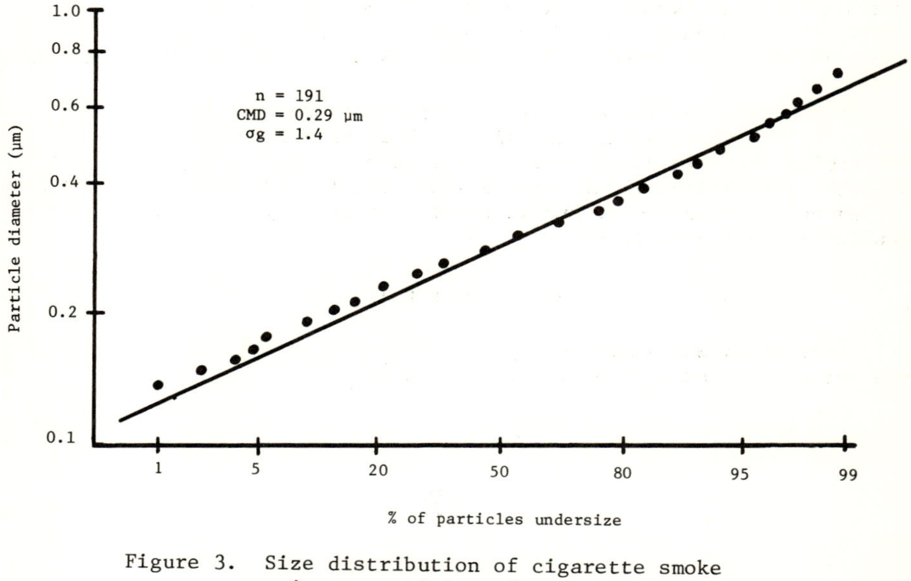

Figure 3. Size distribution of cigarette smoke primary particles after coating.

REFERENCES

1. Research conducted at the Inhalation Toxicology Research Institute of the Lovelace Foundation and at Battelle Northwest Laboratories.

2. Mercer, T.T., M.I. Tillery, and G.J. Newton. A Multi-stage, Low Flow Rate Cascade Impactor. Aerosol Sci. 1:9-15, 1970.

3. Kotrappa, P. and M.E. Light. Design and Performance of the Lovelace Aerosol Particle Separator. Rev. Sci. Instrmts. 43:1106-1112, 1972.

4. Morrow, P.E., and T. T. Mercer. A Point-to-Plane Electrostatic Precipitator for Particle Size Sampling. Am. Ind. Hyg. Assn. J. 25: 8-14, 1964.

5. Morie, G.P., C.H. Sloan and V.G. Peck. Study of Cigarette Smoke by Means of the Scanning Electron Microscope. Beitrage Z. Tabakforschung. 7:99-104, 1973.

EXTENDED ELECTRIC MOBILITY METHOD FOR MEASURING AEROSOL PARTICLE SIZE AND CONCENTRATION

Earl O. Knutson
IIT Research Institute
10 W. 35th Street
Chicago, Illinois 60616

ABSTRACT

An accurate method for determining the concentration and size distribution of aerosols has been developed, based on bipolar charging followed by differential electric mobility analysis. The size resolution is 10% or better throughout the size range 0.005 to 1.0 μm. The particles are kept airborne throughout the analysis, and are never subjected to heat or high vacuum. Thus, the particle size obtained by the method describes the particles in their natural state. Moderately volatile particles can be analyzed in the same way as solid particles.

The method depends critically on accurate expressions for particle mobility and charge distribution. The expressions used have been well established in the literature, at least for spherical particles above 0.01 μm diameter.

The method is applicable to both atmospheric and laboratory aerosols, provided that the number of particles larger than 1 μm is small compared to the number smaller. The lower limit of concentration depends on the type of aerosol sensor used in conjunction with the apparatus.

EXTENDED ELECTRIC MOBILITY METHOD FOR MEASURING AEROSOL PARTICLE SIZE AND CONCENTRATION

Earl O. Knutson
IIT Research Institute
10 W. 35th Street
Chicago, Illinois 60616

INTRODUCTION

The feasibility of measuring aerosol particle size by controlled electric charging and subsequent electric mobility analysis was first demonstrated by Rohmann[1] in 1923. Several variants of the method have since been described in the literature. The first convenient, practical, and accurate device was that of Whitby and Clark[2].

In a parallel but independent line of development, devices for the measurement of electric mobility spectra of atmospheric ions have been in use since about 1905. This led to the discovery and characterization of the small, intermediate, and large atmospheric ions. Junge[3] appears to have been the first to attempt deduction of the size and concentration of atmospheric particles from measured ion mobility spectra.

A common limitation of electric mobility-based methods for aerosol particle size measurement is poor size resolution in the range 0.1 to 1.0 μm diameter. Junge[3], whose method made use of the natural charge on atmospheric particles, restricted the application to sizes less than 0.1 μm to avoid multiple charged particles. Whitby and Clark's[2] method, which uses unipolar diffusion charging, is usable up to 1.0 μm, but has limited size resolution above 0.1 μm due to the relatively flat particle mobility versus size curve.

This paper describes an extended electric mobility method for aerosol size and concentration measurement. The method provides high resolution in the entire range 0.005 to 1.0 μm.

The method described here is similar to that suggested by Kudo and Takahashi[4] in that bipolar charging is used prior to the electric mobility analysis. The bipolar charger is a passive device employing a radioactive source. The mobility analyzer used in this work is a two-inlet, two-outlet flow coaxial cylinder design similar to that described by Hewitt[5]. The present method may be compared to the method of Whitby and Clark[2], which uses unipolar charging and a two-inlet, one-outlet mobility analyzer.

To date, the method described here has been used for approzimately 150 size distribution and concentration measurements on laboratory and atmospheric aerosols. For laboratory aerosols, 10% size resolution can be obtained throughout the size range 0.005-1.0 μm. For atmospheric aerosols, the size range coverage is 0.005-0.1 μm, since the concentration above 0.1 μm is usually too low to be accurately measured with the present sensors. A programmable desk calculator with plotter attachment is used to reduce the data and plot the resulting size distribution.

For those laboratory aerosols which are nearly monodisperse, the present apparatus may also be operated in the mode described previously by Knutson[6] to obtain highly accurate values of the mean and standard deviation of particle size.

APPARATUS

An overall sketch of the apparatus required for this method of particle size analysis is given in Figure 1. The main components are the aerosol charger, the mobility analyzer, and the aerosol sensor.

BIPOLAR CHARGER

The aerosol bipolar charger is shown in Figure 2. The aerosol is drawn through a 1.27 x 1.27 x 5.0 cm long channel. One wall of the channel is an aluminum-coated Mylar alpha-particle window. A foil containing 1.9 mCi ^{241}Am is cemented to the floor of a 0.48 cm deep recess behind the window. The radioactive area is 1.0 x 5.0 cm. Alpha particles penetrate the window and cross the channel, penetrating the aerosol flow at right angles. The ion concentration and production rate were determined by measuring the current between the Al-coated window and an electrode on the opposite channel wall, as a function of applied voltage. At low valtage, the resistance of the air gap was 3,170 MΩ. The saturation current (at 500 volts) was 63 nano-amps. These values imply an ion concentration of 1.74×10^{8} cc^{-1} and a production rate of 3.9×10^{11} ion-pairs sec^{-1}. The measured ion concentration checks well with the value 2.24×10^{8} cc^{-1} estimated as per Cooper and Reist[7].

The physics of aerosol charging by exposure to bipolar ions is known from extensive studies reported in the literature. The basic

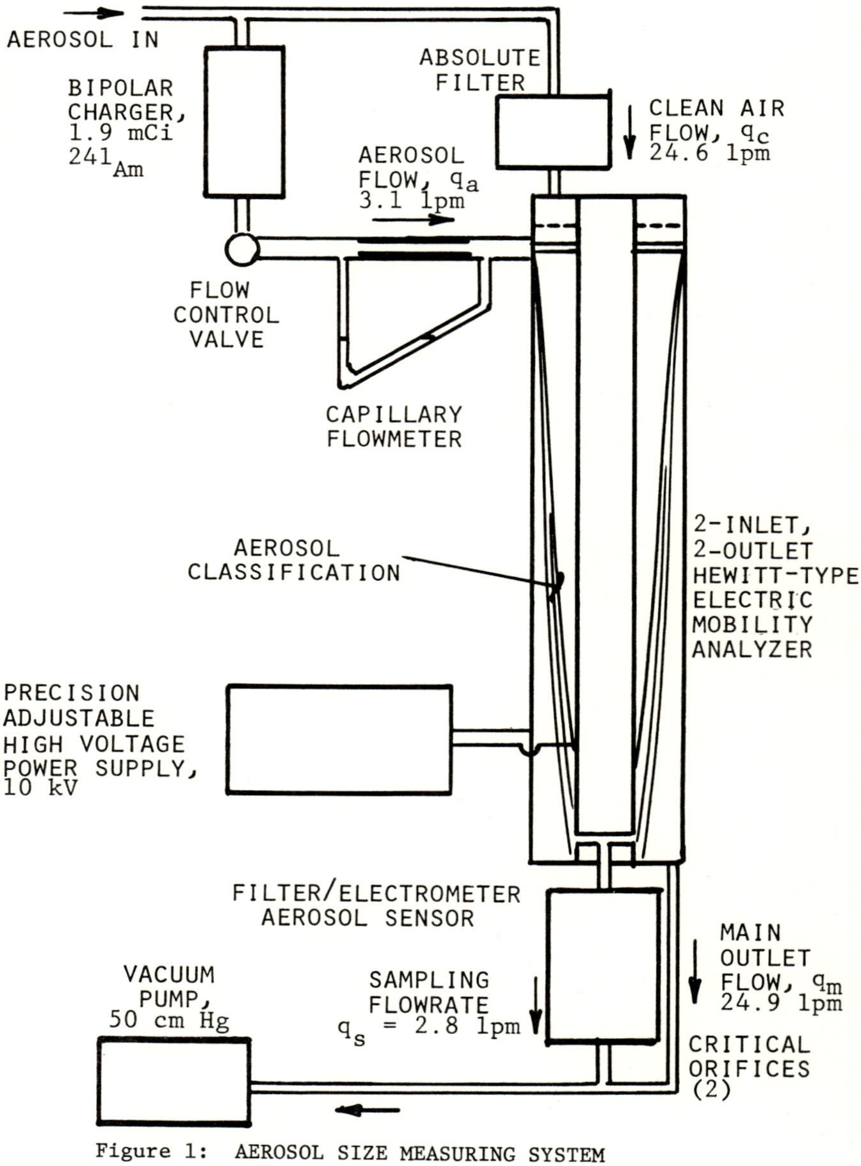

Figure 1: AEROSOL SIZE MEASURING SYSTEM

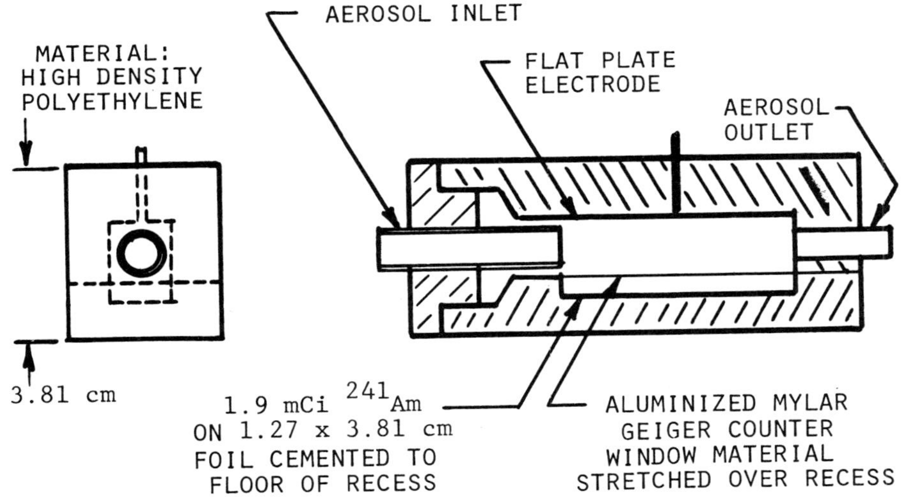

Figure 2

AEROSOL BIPOLAR CHARGER

theoretical equation was derived in various ways by Lissowski[8], Pluvinage[9], Gunn[10], Luchak[11], and Keefe, Nolan, and Rich[12]. This theory shows that, after sufficient exposure to bipolar ions, the particle charge reaches a stationary state which can be described to good approximation by

$$\Phi_\nu(x) = (2\pi a)^{-1/2}\exp\left[-\nu^2/2a\right] \qquad (1)$$

where

$\Phi_\nu(x)$ = the fractional number of particles of diameter x which carry ν units of charge ($\nu = 0, \pm 1, \pm 2, \ldots$)

a = $xkT/2e^2$, a dimensionless particle size

k = 1.38054×10^{-16} ergs/$^\circ$K, Boltzmann's constant

T = the absolute temperature, $^\circ$K

e = 4.803×10^{-10} statcoulomb

Experimental studies of bipolar charging of 0.4 to 4.0 μm diameter particles of various materials were conducted by Lissowski[8], Woessner and Gunn[13], and Gunn and Woessner[14]. Luchak[11] analyzed data by Kunkel and by Gillespie. These studies confirmed Equation 1 for slowly equilibrated aerosols, regardless of initial charge. For rapidly equilibrated aerosols, however, Gunn and Woessner[14] found that the charge distribution becomes assymetric, as described by the equation

$$\Phi_\nu(x) = (2\pi a)^{-1/2}\exp\left[-(\nu-a\sigma)^2/2a\right] \qquad (2)$$

where

$\sigma = \ln(\lambda_+/\lambda_-)$

λ_+, λ_- = the air conductivity due to positive and negative ions, respectively

This correction to Equation 1 has been found necessary for the charger in Figure 2. The best value of σ for the present charger is -0.1. The correction affects mainly particles above 0.5 μm diameter.

Takahasi and Kudo[15] have shown that Equation 1 is valid for 0.1 μm Pb and 0.14 μm Zn aerosols conditioned by exposure to a 2 mCi ^{241}Am alpha source similar to the one shown in Figure 2.

An extensive study of bipolar charging for 0.01 to 0.07 μm diameter hot nichrome wire particles was conducted by Pollak and Metnieks[16]. They found that Equation 1 is accurate in the size range 0.027 to 0.07 μm, but underestimates the number of charged particles for particle size below 0.027 μm. Their measurements corroborated earlier results by Nolan and Kennan[17]. The theoretical reasons for the discrepancy were identified by Fuchs[18], who presented improved calculations. Data and theory are shown in Figure 3.

The most comprehensive theory of small particle charging at present is that of Gentry and Brock[19], as amended by Brock[20] to include the image effect. The corresponding curve in Figure 3 was calculated by the simplified method of Gentry[21], assuming conducting particles and free molecular regime ion transport. Of the three theories shown, the Gentry curve best fits the data below 0.03 μm. In this range, Gentry's curve may be approximated by:

$$\Phi_\nu(x) = \begin{cases} \dfrac{1}{2 + (0.1/x)^{1.5444}} & \text{if } \nu = \pm 1 \\ 0 \quad \text{if} \quad \nu > 1 \end{cases} \tag{3}$$

where x is in μm.

Based on the above considerations, the following expression for $\Phi_\nu(x)$ was adopted in this work:

For $x > 0.03$ μm, Equation 2 with $\sigma = -0.1$ is used;

For $x < 0.03$ μm, Equation 3 is used.

Further data on bipolar charging have appeared recently. Liu and Pui[24] present data on NaCl and methylene blue particles with diameter from 0.02 to 0.2 μm and on dioctyl phthalate droplets 0.53 to 1.17 μm in diameter. The bipolar charger used was a 2 mCi ^{85}Kr beta-emitter. Several of their data points are shown in Figure 3.

Liu and Pui's data agreed well with Equation 1 down to the smallest particle size considered. It should not be inferred, however, that Equation 1 is valid for particle diameters less than 0.02 μm. For the present, Equation 3 is preferred over Equation 1 for particle size less than 0.03 μm due to its better theoretical basis and the supporting data of Pollak and Metnieks[16]. Equation 3 differs from Liu and Pui's data by at most 16%.

Liu and Pui[24] found no difference as to charging between NaCl and methylene blue.

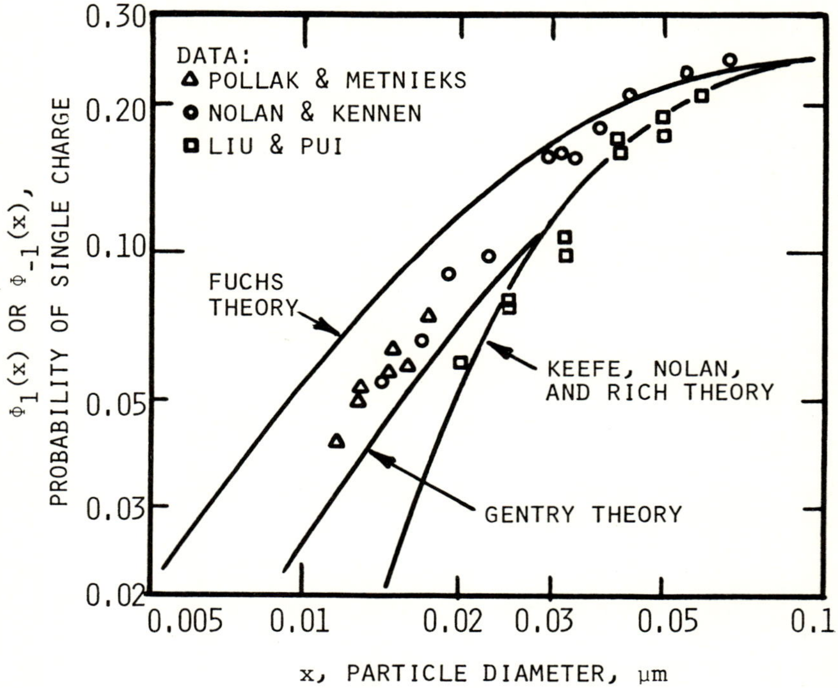

Figure 3

BIPOLAR CHARGING THEORY AND DATA

Kozima and Sekikawa[25] present data on bipolar charging of urban aerosols upon exposure to ^{210}Po alpha particles. Their results (not shown) lie somewhat above the theoretical curve of Fuchs[18].

MOBILITY ANALYZER

The differential electric mobility analyzer used in the present work is shown in Figure 4. It is a two-inlet flow, two-outlet flow coaxial cylinder design similar to those described by Hewitt[5], Knutson[6], and Lui and Piu[22].

The aerosol mechanics of the electric mobility analyzer have been discussed in some detail by Knutson[6]. The main result was an expression for the transfer function, Ω. This function is defined as the ratio of the number of particles carried out via the sampling outlet flow to the number entering via the aerosol flow. The function Ω is most easily presented in the form of a graph, as in Figure 5. It is seen that Ω is a function of: the particle mobility, K; the applied voltage, V; the mobility analyzer length, L; the radius ratio, r_2/r_1 and the four flow rates q_c, q_a, q_m, and q_s. For a given voltage, V, the half width, ΔK, of the mobility fraction extracted from the aerosol is

$$\Delta K = 1/2(q_a + q_s)/(2\pi\Delta V) \quad (4)$$

The centroid, K*, of this mobility band is

$$K^* = 1/2(q_c + q_m)/(2\pi\Delta V) \quad (5)$$

The resolution is therefore

$$\Delta K/K^* = (q_a + q_s)/(q_c + q_m) \quad (6)$$

AEROSOL SENSOR

The aerosol sensor (Figure 1) is used to determine the amount of aerosol emerging from the mobility analyzer sampling outlet at each mobility setting. Several devices can be used. The simplest is the collecting filter/electrometer arrangement indicated in Figure 1. This measures the electric current due to the stream of charged particles. Alternatives which have been used include a single particle optical counter (set to count all emerging particles within its size capability), a condensation nuclei counter, and an aerosol mass monitor. The latter sensor leads in a direct way to the aerosol size distribution by mass.

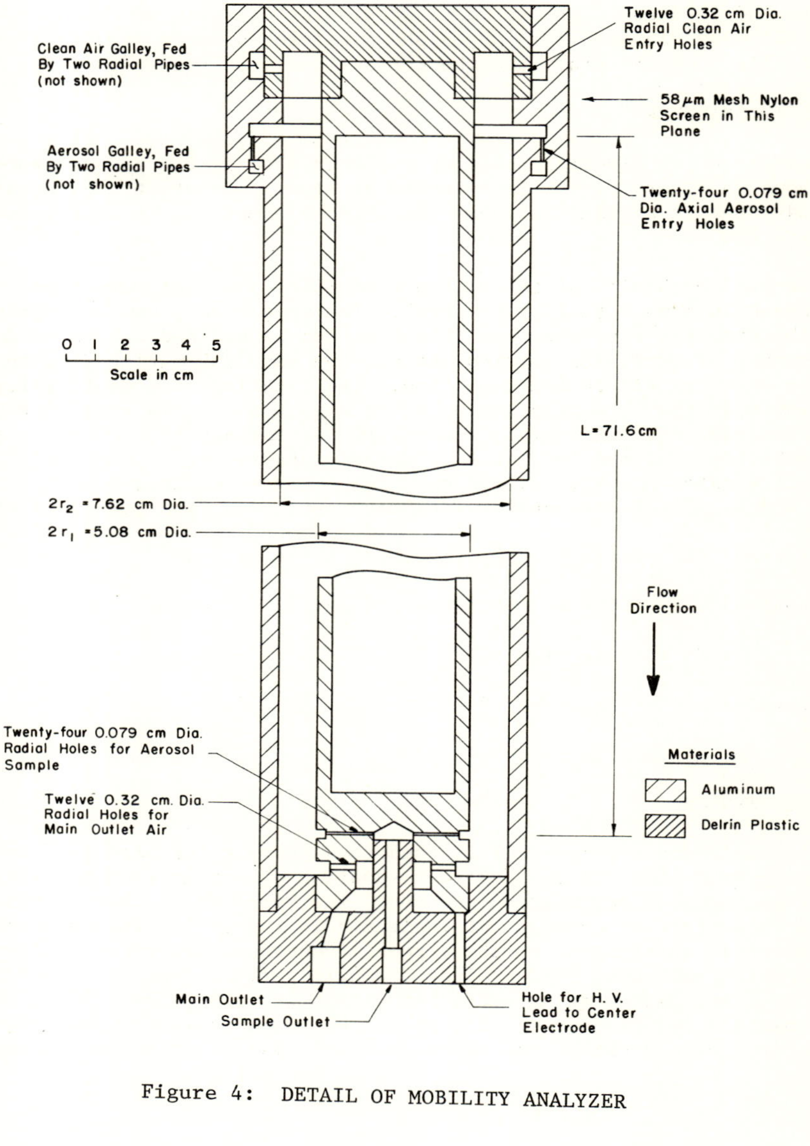

Figure 4: DETAIL OF MOBILITY ANALYZER

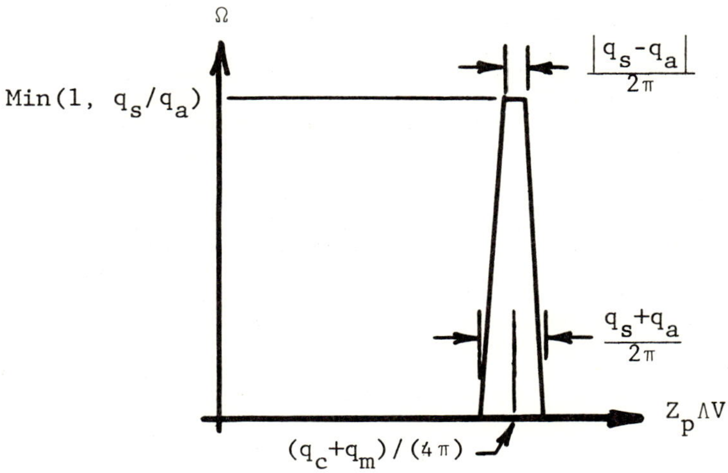

Figure 5

MOBILITY ANALYZER TRANSFER FUNCTION, Ω

q_a = aerosol flow rate

q_s = sampling flow rate

V = center rod voltage

L = distance between inlet and exit gaps

r_1, r_2 = inner and outer radius of annulus

q_c = clean air flow rate

q_m = main outlet flow rate

$\Lambda = L/\ln(r_2/r_1)$

OPERATION

The aerosol to be analyzed is drawn through the apparatus as shown in Figure 1. The bipolar charger establishes the charge distribution given by Equations 2 and 3. The mobility analyzer extracts mobility fractions selected by the applied voltage in accordance with the transfer function, Figure 5. The electric current carried by the extracted mobility fraction is measured by the aerosol sensor. Thus, the raw data consists of a table of current vs. voltage.

The mobility analysis process may be visualized with the aid of Figure 6. The sloping lines in figure 6 show the relationship between particle electric mobility and particle diameter for various particle electric charges. The horizontal lines labeled K_1, K_2, and K_3,..... represent a sequence of narrow mobility fractions. For reasons which will become clear, adjacent mobility fractions are chosen in the ratio 1:1.429. The vertical lines in Figure 6 represent the sequence particle sizes defined by the intersection of the mobility lines with the curve for mobility of singly charged particles.

Each mobility fraction consists of several discrete particle sizes. For example, the mobility fraction K_1 (2.042×10^{-4} cm^2/volt-sec.) in Figure 6 consists of particles of diameter 0.120, 0.186, 0.240, and 0.292 μm, carrying 1, 2, 3, and 4 units of charge, respectively. The first of these sizes is that labeled x_1 in Figure 6. The remaining three particle sizes nearly coincide with the sizes labeled x_3, x_4, and x_5 (0.183, 0.233, and 0.297 μm, respectively). This relationship is true also of the other mobility fractions: the mobility fraction K_i consists of particles whose sizes are given to close approximation by x_i, x_{i+2}, x_{i+3}, and x_{i+4}. This approximation, which is made possible by the choice of spacing for the mobility fractions, is used in the analysis which follows.

Particles of electric charge greater than four units are neglected in this treatment, since they are usually much less numerous than those considered above. An alternate procedure is available when greater accuracy is needed (see Appendix).

With the above approximations, the response, R_i, of the aerosol sensor at the ith voltage setting may be written as the sum of four terms:

$$R_i = q_a \sum_{\nu=1}^{4} W_{\nu j} \Phi_{\nu j} N_j w_{\nu j} \qquad (7)$$

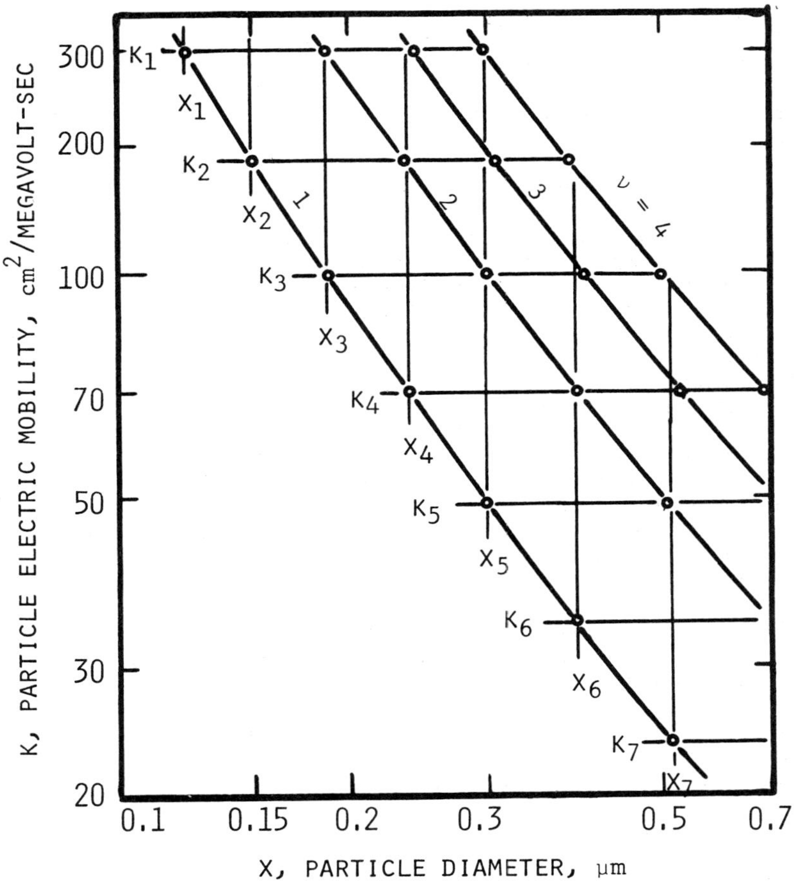

Figure 6

ILLUSTRATION OF THE SERIES OF
MOBILITY FRACTIONS AND PARTICLE SIZES

where q_a is the aerosol flow rate through the charger, N_j is the number concentration of aerosol particles in the size range (x, x + dx) centered at x_i, $W_{\nu j}$ is a weight factor giving the response of the sensor to a single particle of charge ν and size x_j, $w_{\nu j}$ is a weight factor related to the finite width of the mobility band, and j = i + 0, i + 2, i + 3, and i + 4 for ν = 1, 2, 3, and 4, respectively. A rigorous derivation of Equation 7, giving the weight factors, is given in the Appendix.

The aerosol analysis procedure consists of obtaining aerosol output readings R_1, R_2, R_3,... for the descending sequence of mobility fractions K_1, K_2, K_3,... A standard sequence of 25 voltage settings has been adopted for the operation, as shown in Table 1. The flow rates may be adjusted to suit the aerosol, subject to the restriction of high resolution, that is: $(q_a + q_s)/(q_c + q_m) \ll 1$. The sequence K_1, K_2, K_3,... is computed from the voltages using Equation 5. The corresponding particle sizes are computed from the Stokes-Cunningham mobility formula (Equation 25 of Wahi and Liu[23]). Particle size sequences at various flowrates are given in Table 1.

The voltage sequence may be entered at any point, but must be continued upward until the aerosol output drops to essentially zero, say at the nth step. The zero output indicates that the particle concentration for size x_n and above is negligibly small. Data taking is then stopped and Equation 7 is applied to the data to solve for the concentrations $N(x_1)$, $N(x_2)$,...$N(x_{n-1})$.

The solution of Equation 7 is accomplished by means of a programmable desk calculator. A plotting accessory automatically plots out the particle size distributions.

Aerosol losses in the analyzer and its piping due to diffusion and impaction have been estimated theoretically. Impaction losses appear to be inconsequential. Diffusion losses for the operating conditions shown in Figure 1 are estimated at 35% for 0.004 μm, 10% for 0.01 μm, and less for larger particles.

EXAMPLE MEASUREMENT RESULTS

Figure 7 shows three example aerosol size distributions obtained by the method just described. The plot format is dN/d(ln x) on a linear scale as a function of x on a logarithmic scale. The curves are not normalized. Thus, the area under the curve in each case is the total number concentration of the aerosol.

Table 1

COURSE VOLTAGE SEQUENCE AND CORRESPONDING PARTICLE SIZE FOR IITRI ELECTRIC MOBILITY AEROSOL ANALYZER

Voltage Setting	Particle Size (μm) at Various Flow Rates, $\bar{q}$ (lpm)				
	$\bar{q}$ = 10	15.0	20.0	25.0	30.0
1.902	0.0052	0.0043	0.0038	0.0034	0.0030
2.718	0.0062	0.0052	0.0045	0.0039	0.0037
3.884	0.0075	0.0061	0.0053	0.0048	0.0044
5,551	0.0089	0.0073	0.0065	0.0057	0.0051
7.932	0.0107	0.0087	0.0075	0.0067	0.0061
11.33	0.0128	0.0105	0.0090	0.0080	0.0074
16.20	0.0154	0.0125	0.0108	0.0097	0.0088
23.14	0.0.83	0.0149	0.0129	0.0116	0.0105
33.07	0.0221	0.0179	0.0155	0.0138	0.0128
47.26	0.0271	0.0216	0.0186	0.0166	0.0151
67.54	0.032	0.026	0.022	0.020	0.018
96.51	0.039	0.031	0.027	0.025	0.022
137.9	0.047	0.038	0.032	0.029	0.026
197.1	0.057	0.046	0.039	0.035	0.032
281.6	0.069	0.055	0.048	0.042	0.038
402.4	0.085	0.067	0.057	0.051	0.046
575.1	0.103	0.082	0.070	0.062	0.056
821.8	0.129	0.100	0.085	0.075	0.068
1,174.	0.159	0.124	0.105	0.092	0.087
1,678.	0.200	0.156	0.129	0.113	0.102
2,398.	0.25	0.19	0.16	0.14	0.13
3,427.	0.33	0.24	0.20	0.18	0.16
4,897.	0.43	0.32	0.26	0.22	0.20
6,998.	0.56	0.41	0.33	0.28	0.25
10,000.	0.76	0.54	0.43	0.37	0.32

$\bar{q} = 1/2(q_c + q_m)$

q_c = mobility analyzer clean airflow

q_m = mobility analyzer main outlet flow

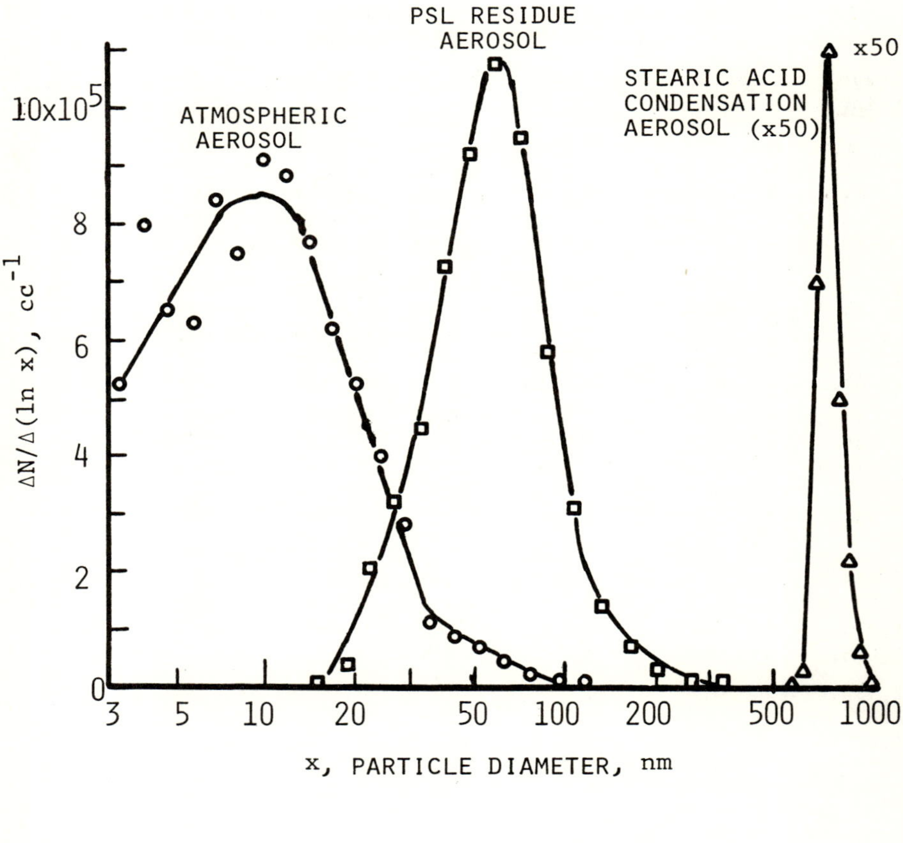

Figure 7

EXAMPLE AEROSOL SIZE DISTRIBUTIONS

The left curve in Figure 7 is an atmospheric aerosol sample taken in Chicago on February 12, 1974. The analyzer flow rates were as shown in Figure 1. The collecting filter/electrometer current measuring device indicated in Figure 1 was used as the aerosol sensor. The size distribution shows a weak mode at ~ 0.01 μm. The curve falls rapidly to the right of the mode, but continues leftward out of the limits of the analyzer. The area under the curve represents 1.59×10^6 particles/cc. This was among the highest concentration observed in 68 atmospheric aerosol samples in Chicago.

The total concentration for the atmospheric aerosol shown in Figure 7 was checked with a condensation nucleus counter (Gardner Associates, Schenectady, N.Y., Small Particle Counter, Type CN). The result was 3.4×10^5 particles/cc, or about 1/5 of the total number concentration indicated by the mobility analyzer. The reason for this difference has not been determined.

The center curve in Figure 7 was obtained when nebulizing an aqueous suspension of 0.557 μm polystyrene latex particles. The aerosol represented in Figure 7 consists of surfactant particles left when droplets not containing a latex particle evaporate. The latex particles themselves were too few in number to be detected accurately. The flow rates in this test were q_a = 1.4 lpm, q_c = 9.6 lpm, q_s = 1 lpm, and q_m = 10 lpm. The current sensor was used.

The number concentration of the surfactant residue particle aerosol, as determined by the area under the curve in Figure 7, was 1.17×10^6 cc^{-1} The geometric mean diameter and geometric standard deviation were 0.054 μm and 1.6, respectively. For comparison, the concentration of latex particles were measured separately and found to be 1,200 cc^{-1}.

The right curve in Figure 7 depicts a stearic acid aerosol generated by a high-volume condensation aerosol generator developed at IITRI. The generator output is 34 lpm, which was diluted to 7.1×10^4 lpm for the test shown in Figure 7. The analyzer flow rates for this test were q_a = 0.38 lpm, q_c = 7.07 lpm, q_s = 0.31 lpm, and q_m = 7.0 lpm. An optical single-particle counter (Royco, Inc., Menlo Park, California, Model 245) was used as the aerosol sensor in this test. The counter was set to count all particles above 0.3 μm diameter. The sampling flow rate (0.31 lpm) was diluted with filtered air to make up the 28.3 lpm flow required by the counter.

The number concentration computed from the stearic acid aerosol curve in Figure 7 is 4,600 cc^{-1}. The geometric mean diameter is 0.73 μm.

The geometric standard deviation of the stearic acid aerosol was too small to be accurately measured by the present technique. However, the analysis method described by Knutson[6], applicable to near-monodisperse aerosols, was applied to the data and yielded geometric mean diamether = 0.73 μm and geometric standard deviation = 1.12.

The number concentration of the stearic acid aerosol was checked by a microscope count of a glass slide sample taken with an electrostatic sampler (Thermo-Systems, Inc., St. Paul, Minnesota, Model 3100). The result was 2.03×10^4 particles/cc, 4.4 times the concentration determined by the mobility analyzer. The reason for this difference has not been identified.

SUMMARY

It has been demonstrated that the electric mobility method for measuring aerosol particle size and concentration can be extended to provide 10% size resolution throughout the range 0.005 to 1.0 μm. This is accomplished by using bipolar charging, a differential mobility analyzer, and a specially designed measurement procedure.

The extended electric mobility method has been applied to about 150 atmospheric or laboratory aerosols. The method was found to be convenient and moderately fast (20-30 min. for data collection). The size resolution has been adequately demonstrated with near-monodisperse aerosols. Further work is required, however, to determine the accuracy with which the device measures concentration.

ACKNOWLEDGEMENT

This work was supported by the U.S. Atomic Energy Commission through its Contract No. AT(11-1)-578.

REFERENCES

1. Rohmann, H. Method of Size Measurement for Suspended Particles. Z. Phys. 17: 253-265, 1923.

2. Whitby, K. T., and W. E. Clark. Electric Aerosol Particle Counting and Size Distribution Measuring System for the 0.015 to 1 μ Size Range. Tellus XV111,2: 573-586, 1966.

3. Junge, C. E., The Size Distribution and Aging of Natural Aerosols as Determined from Electrical and Optical Data on the Atmosphere. J. Meteor. 12: 13-25, February, 1955.

4. Kudo, A., and K. Takahashi. A Method of Determining Aerosol Particle Size Distribution Applying Boltzmann's Law. Atmos. Envir. 6: 543-549, August, 1972.

5. Hewitt, G. W. The Charging of Small Particles for Electrostatic Precipitation. Trans. Amer. Instit. Elec. Engrs. 76: 300-306, 1957.

6. Knutson, E. O., The Distribution of Electric Charge among the Particles of an Artificially Charged Aerosol. PhD thesis at University of Minnesota, 1971. 210 p.

see also Knutson, E. O., and K. T. Whitby. Aerosol Classification by Electric Mobility: Apparatus, Theory and Applications. To appear in J. Aerosol Sci. 6(6), 1975.

7. Cooper, D. W., and P. C. Reist. Neutralizing Charged Aerosols with Radioactive Sources. J. Colloid Interface Sci. 45: 17-26, 1973.

8. Lissowski, P. Das Laden von Aerosolteichen im ciner Bipolaren Ionenatmosphare. Act Phys. Chem. URSR. 13: 157-192, 1940.

9. Pluvinage, P. Etude Theorique et Experimentale de la Conductibilite Electrique dans les Nauges Non Orageux. Annales de Geophys. 2: 31-54, 160-178, 1946.

10. Gunn, R. The Statistical Electrification of Aerosols by Ionic Diffusion. J. Colloid Sci. 10: 107-119, 1955.

11. Luchak, G. The Theory of the Electric Charge Distribution of Monodispersed Lightly Charged Aerosols of Spherical Particles Coagulating in a Bipolar Ionized Atmosphere. J. Colloid Sci. 12: 144-160, 1957.

12. Keefe, D., P. J. Nolan, and T. A. Rich. Charge Equilibrium in Aerosols According to the Boltzmann Law. Proc. Roy. Irish Acad. 60 (Sect. A): 6-45, 1959.

13. Woessner, R. H., and R. Gunn. Measurements Related to the Fundamental Processes of Aerosol Electrification. J. Colloid Sci. 11: 69-75, 1956.

14. Gunn, R., and R. H. Woessner. Measurements of Systematic Electrification of Aerosols. J. Colloid Sci. 11: 254-259, 1956.

15. Takahashi, K., and A. Kudo. Electrical Charging of Aerosol Particles by Bipolar Ions in Flow Type Charging Vessels. J. Aerosol Sci. 4: 209-216, 1973.

16. Pollak, L. W., and A. L. Metnieks. On the Validity of Boltzmann's Distribution Law for the Charges of Aerosol Particles in Electrical Equilibrium. Geofis. Pura Appl. 53: 111-132, 1962.

17. Nolan, P. J., and E. L. Kennen. Condensation Nuclei from Hot Platinum: Size, Coagulation Coefficient and Charge Distribution. Proc. Roy. Irish Acad. (Dublin) 52A: 171-190, 1949.

18. Fuchs, N. A. On the Stationary Charge Distribution on Aerosol Particles in a Bipolar Ionic Atmosphere. Geofis. Pura Appl. 56: 185-193, 1963.

19. Gentry, J., and J. R. Brock. Unipolar Diffusion Charging of Small Aerosol Particles. Chem. Phys. 47: 64-69, 1967.

20. Brock, J. R. Aerosol Charging: The Role of the Image Force. Appl. Phys. 41: 843-844, 1970.

21. Gentry, J. W. Charging of Aerosol by Unipolar Diffusion of Ions. Aerosol Sci. 3: 65-76, 1972.

22. Liu, B. Y. H., and D. Y. H. Pui. A Submicron Aerosol Standard and the Primary, Absolute Calibration of the Condensation Nucleus Counter. Colloid Interface Sci. 47: 155-171, 1974.

23. Wahi, B. N., and B. Y. H. Liu. The Mobility of Polystyrene Latex Particles in the Transition and the Free Molecular Regimes. Colloid Interface Sci. 37: 374-381, 1971.

24. Liu, B. Y. H., and D. Y. H. Pui. Equilibrium Bipolar Charge Distribution of Aerosols. Colloid Interface Sci. 49: 305-312, 1974.

25. Kojima, H., and T. Sekikawa. An Attempt for Obtaining the Aerosol Size Distribution. Meteor. Soc. Japan. 51: 287-293, 1973.

APPENDIX
MATHEMATICAL DETAILS

In this appendix, a formal derivation is given for Equation 7 of the text, and methods of solving the equation are discussed.

The number concentration of particles with size in the range (x, x + dx) and charge ν after bipolar charging is given by the product $\Phi_\nu(x)N(x)$. The rate at which these particles enter the mobility analyzer is $q_a\Phi_\nu(x)N(x)$. The number of these particles extracted via the sampling outlet is obtained by multiplication with the transfer function, $\Omega(\nu\hat{K}\Lambda V)$, where $\hat{K}$ is the mobility of a singly charged particle of size x. The impact of each particle upon the aerosol sensor is given by a weight factor, $W_\nu(x)$, which may depend on ν, on x, or on neither or both. The response, R(V), of the aerosol sensor at the voltage setting V is obtained by summing over ν and integrating over x:

$$R(V) = q_a \sum_{\nu=1}^{\infty} \int_0^{\infty} W_\nu(x)\Phi_\nu(x)N(x)\Omega(\nu\hat{K}\Lambda V)dx \tag{8}$$

The ν-sum is taken over either the positive integers or the negative integers, as appropriate.

For each V and ν, the integral in Equation 8 may be evaluated approximately, as follows. Change variable from x to ξ where ξ is the argument of Ω:

$$\begin{aligned} \xi &= \nu\hat{K}(x)\Lambda V \\ d\xi &= \nu\Lambda V(d\hat{K}/dx)dx \end{aligned} \tag{9}$$

If the mobility analyzer is operated at high resolution (small $\Delta K/K^*$), the transfer function Ω (ξ) peaks sharply at $\xi = (q_c + q_m)/4\pi$ (See Figure 5). Hence, the terms $w_\nu(x)$, $\Phi_\nu(x)$, and dK/dx may be evaluated at x = x* and extracted from the integral, where x* is defined implicitly by

$$\hat{K}(x^*) = (q_c + q_m)/(4\pi\nu\Lambda V) \tag{10}$$

The integral that remains is $\int\Omega(\xi)d\xi$, the area under the curve in Figure 5 of the text. This area is $q_s/2\pi$. With these steps, Equation 8 becomes:

$$R(V) = q_a \sum_{\nu=1}^{\infty} W_\nu(x^*)\Phi_\nu(x^*)N(x^*)\frac{(dx/d\hat{K})^*}{\nu\Lambda V} q_s/2\pi$$

$$= \frac{2q_a q_s}{q_c+q_m} \sum_{\nu=1}^{\infty} W_\nu(x^*)\Phi_\nu(x^*)N(x^*)\hat{K}(x^*)(dx/d\hat{K})^* \qquad (11)$$

The last form is obtained using Equation 10.

In practice, the sensor response, R, is read at a discrete set of voltage settings, V_1, V_2, V_3,.., V_i,... . Thus x* depends on two indices, ν and i. A great simplification could be realized if x* could be made to depend on one index only. As indicated in Section 3 of this text, this can be accomplished by a proper choice of voltage settings.

Note that by Equation 10, x* depends only upon the product to ν and V. Consider the voltage sequence defined by $V_{i+1} = 1.429V_i$. This sequence has the property that

$$\begin{aligned} V_{i+2} &= 2.042V_i \cong 2V_i \\ V_{i=3} &= 2.918V_i \cong 3V_i \\ V_{i=4} &= 4.170V_i \cong 4V_i \end{aligned} \qquad (12)$$

Equations 10 and 12 show that the doubly charged particles at the ith voltage have nearly the same size, x*, as singly charged particles at the $(i + 2)^{th}$ voltage. Similar statements apply for the particles bearing 3 and 4 units of charge. This was demonstrated graphically in Figure 6. To good approximation, the reduction of x* to a single-index qualtity has been accomplished, at least for particle charges 1, 2, 3, and 4.

In many cases, Equation 11 can be truncated after four terms. Introducing index j,

$$j = \begin{cases} i & \text{if } \nu = 1 \\ i+2 & \text{if } \nu = 2 \\ i+3 & \text{if } \nu = 3 \\ i+4 & \text{if } \nu = 4 \end{cases} \qquad (13)$$

Equation 11 becomes

$$R_i = \frac{2q_a q_s}{q_c+q_m} \sum_{\nu=1}^{4} W_{\nu j}\Phi_{\nu j}N_j\hat{K}_j^*(dx/d\hat{K})_j^* \qquad (14)$$

where $R_i = R(V_i)$, $N_j = N(x_j^*)$, etc., and x_j^* denotes the root of Equation 10 for $\nu = 1$ and $V = V_j$.

Equation 14 is the formal version of the intuitive Equation 7. The weight factor $w_{\nu j}$ in Equation 7 is seen to have the value

$$w_{\nu j} = \frac{2q_s}{q_c + q_m} \hat{K}_j^*(dx/d\hat{K})_j^* \tag{15}$$

It is seen that $w_{\nu j}$ is independent of the index ν.

The weight factor $W_{\nu j}$ depends on the aerosol sensor used. If an optical single-particle counter is used, each particle contributes 1 count so that $W_{\nu j} = 1$. If a current sensor is used, as in Figure 1, $W_{\nu j}$ = the particle charge, νe. if a mass flow rate sensor is used $W_{\nu j}$ = the mass of a particle of size x_j.

Equation 14 may be treated as a fifth-order, inhomogeneous, linear difference equation. The coefficients, although not constant, are calculable from the sequence of particle sizes. Solution may be accomplished by applying Equation 14 recursively for $N_{n-1}, N_{n-2}, \ldots, N_1$ (in that order). The required starting values are $N_{n+4} = N_{n+3} = N_{n+2} = N_{n+1} = N_n = 0$. Alternatively, Equation 14 may be regarded as defining a linear system of equations for $N_1, N_2, \ldots, N_{n-1}$. The set of equations may be solved by standard methods, or by special methods which exploit the fact that the coefficient matrix has zeros below the main diagonal.

If the aerosol being analyzed contains appreciable numbers of particles larger than ~ 0.5 μm, a significant number of particles carries more than 4 units of charge. In this case, truncation of Equation 11 after 4 terms causes serious error in interpreting the data. This error is most serious when the current measurement is used to sense the output aerosol, because the weight factor $W_\nu(x)$ then increases with particle charge. An alternate procedure has been developed for use when appreciable numbers of particles exceed 0.5 μm diameter.

The alternate procedure makes use of the voltage sequence $v_{i=1} = 1.1053V_i$. For this sequence,

$$\begin{aligned}
V_{i+6} &= 2.015V_i \\
V_{i+10} &= 3.008V_i \\
V_{i+13} &= 4.062V_i \\
V_{i+15} &= 4.962V_i \\
V_{i+17} &= 6.062V_i
\end{aligned}$$

Hence, this voltage sequence permits incusion of the terms for particle charges 1 through 6 in Equation 11. This voltage sequence was used to obtain the stearic acid size distribution at the right of Figure 7.

RAPID MEASUREMENT OF PARTICULATE SIZE DISTRIBUTION IN THE ATMOSPHERE

R. L. Chuan
Celesco Industries
Costa Mesa, Calif.

ABSTRACT

A new device has been constructed and tested, which allows the rapid determination (in terms of minutes) of the distribution of particulate mass by size in the range 0.05 to 50 micrometers. The device is essentially an active cascade each stage of which consists of a quartz crystal microbalance. The mechanical and fluid-mechanical design of the system is entirely equivalent to a conventional cascade, except that the filter or impactor is replaced by a piezo-electric crystal whose resonant frequency decreases in proportion to accumulated mass, brought to the adhesive-coated crystal surface by aerodynamic impaction. The type of crystals and electronics employed offers a basic sensitivity of 7 x 10^8 Hz/gm and a frequency stability of 10^{-2} Hz/sec. which makes it possible to obtain sensible frequency changes in each of the 6-to 10-stage cascade in sampling time of 2 to 6 minutes at a sampling flow rate of 100 ml/min when the total particulate mass concentration is between 50 and 100 micrograms/m^3. Tests with the instrument have been conducted in both indoor and outdoor atmospheres, with total particulate mass concentration ranging from 25 to 200 $\mu g/m^3$. Some outdoor measurements during the passage of photo-chemical smog over the test site show the transition from a normal mono-modal to a bi-modal distribution.

RAPID MEASUREMENT OF PARTICULATE SIZE DISTRIBUTION IN THE ATMOSPHERE

R. L. Chuan, Celesco Industries Inc.

INTRODUCTION

The determination of the distribution of particulate mass according to size is of importance to the study of the dynamics of particulate generation and transport. The study of dynamical features necessarily requires the rapid measurement of size distribution. It is also desirable that a single means of measurement be applied over the complete size range of interest, for evident reasons of consistency in data interpretation. The technique of inertial separation of particulates by size is well developed and relatively free of theoretical and experimental uncertainties or ambiguities, and should lend itself well to particulate size determination if the means of particulate mass measurement can be made rapid. The conventional cascade impactors usually require the order of hours to collect sufficient samples for accurate weighing when sampling ambient atmosphere. With the development of the quartz crystal microbalance active impactor and its ability to measure in real time the mass of impacted particulates, as reported by Chuan[1,2], it is possible to achieve rapid determination of particulate size distribution by using a cascade of these active impactors. This paper describes the design and operation of such an active cascade system, and discusses some typical results obtained by sampling ambient air.

DESCRIPTION OF APPARATUS

The cascade is designed in the conventional way of pumping air through a series of chambers with increasing velocity as the air passes through successive stages. The flow rate to be sampled is determined by considerations of the area available for impacting particulates on the crystal (of the order of 0.2 cm^2), the sensitivity of the crystal (about 7×10^8 Hz/gm), the stability of the electronics and the concentration of particulates per size interval to be sampled. The basic response of the system expressed in terms of a frequency shift rate is

$$\frac{df}{dt} = 10^{-12} CVk \tag{1}$$

where C is the mass concentration in microgram/cubic meter ($\mu g/m^3$), V the flow rate in milliliter/minute (ml/min) and k the sensitivity in Hertz/gram (Hz/gm).

The electronics in each of the cascade stages consist of two oscillators and a mixer, all built into a hybrid chip. With the type of A-T cut matched crystals and the hybrid chip used the overall accuracy of the microbalance unit is typically about 0.02 Hz/min. A frequency shift of 1 Hz/min would then constitute a signal to noise ratio of 50; and is considered a criterion for designing the cascade system. With a sensitivity of 7×10^8 Hz/gm and a concentration of $10 \mu g/m^3$, one then arrives at a flow rate of 140 ml/min as that necessary to produce an adequate response.

In order to obtain a 50% cut-off for mass density 2 particulates of 0.05 micron diameter, the impaction jet velocity is to be set at about 3×10^4 cm/sec with a jet diameter of about 2×10^{-3} cm. The fabrication technique involved in making very small precision jets allows one to make an array of 37 jets within a circle of about 0.4 cm. This is within the working area of the crystal, and the total flow rate through the array of jets is about 100 ml/min, which is near the amount needed for adequate system response.

Once the final stage configuration is established the preceding stages can be configured to provide any size intervals one desires. The data to be shown later in this paper were produced by two cascade systems, one a 6-stage and the other a 10-stage. The 50% cut-off points for each cascade, calculated from actual, measured flows and pressure drops, are listed below:

Six-stage Cascade

Stage	50% cut-off
1	53 μm
2	25
3	6.6
4	1.4
5	0.32
6	0.06

Ten-stage Cascade

Stage	50% cut-off
1	77
2	36
3	17
4	7.9
5	3.7
6	1.7
7	0.87
8	0.40
9	0.17
10	0.06

The signal from each of the cascade stages is in the form of a beat frequency (between the sensing and reference crystals) in the kilohertz range. It is amplified and clipped to a square wave in the base of the cascade as shown in Fig. 1. The base also contains the power-supply for the stages and the corresponding amplifiers, as well as the pump and flow-meter for the cascade. Fig. 2 shows the internal details of a stage. A companion control unit, also seen in Fig. 1, contains a programmer and a printer to process the signals from the cascade. The beat-frequency from each stage is sequentially interrogated (for a 1-second gating time) and printed, requiring a total time of 2 seconds per stage (1 second to count and 1 second to print). After one complete scan of the stages the programmer remains in hold, to repeat the scan at one of three pre-set intervals — 2, 5, and 10 minutes. The change in frequency of each stage during the interval is then the frequency shift rate which can be converted to mass concentration by Eqn. (1).

The frequencies from the stages can also be fed to a computer which will read the frequencies, compute the concentration, and print out the concentration directly, as has been done by one user of a ten-stage system.

In order to compensate for occasional long-term drift in the frequencies, caused in part by drift in the electronics and partly by slow temperature changes which affect both the crystals and the electronics (but in no event greater than about 0.2 Hz/min), the instrument is usually operated continuously for several hours at a time and a filter is

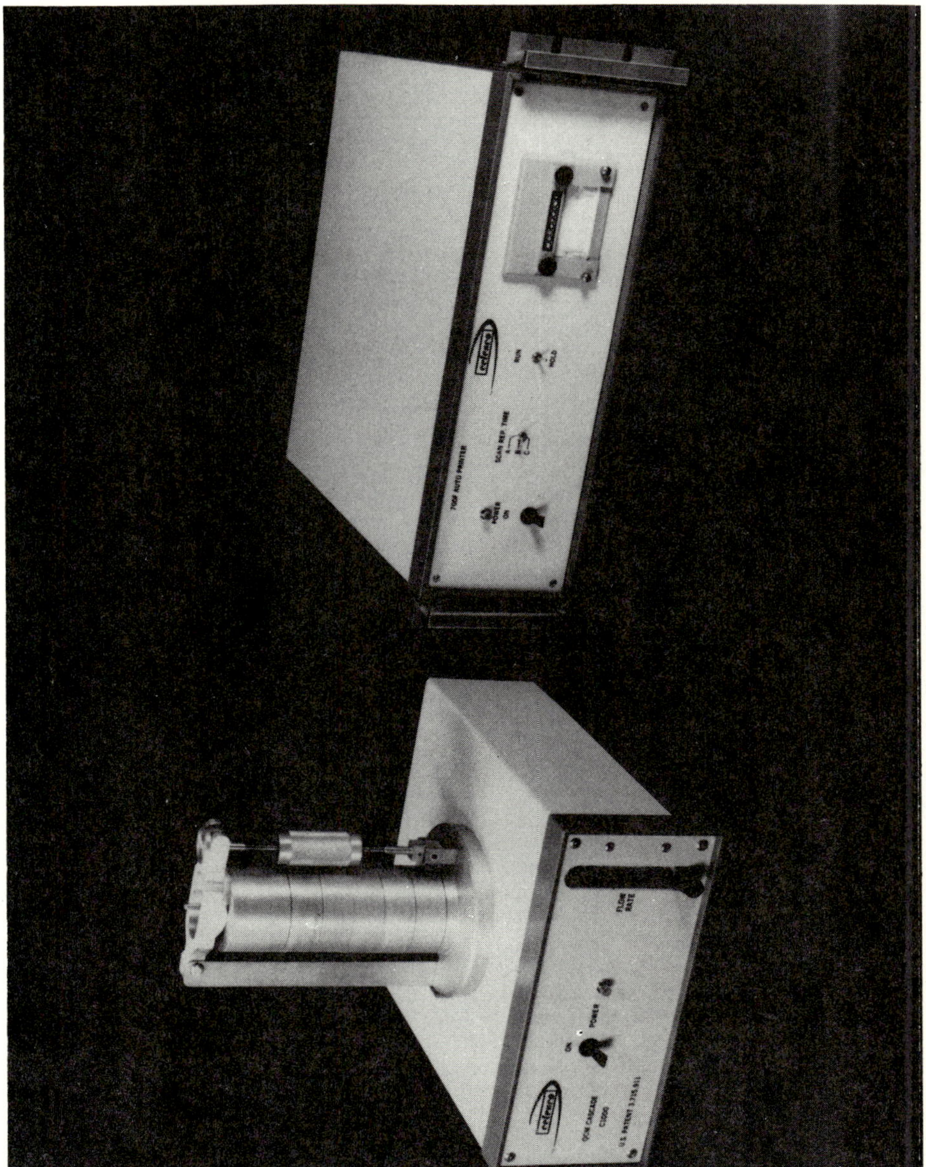

Fig. 1. Six Stage QCM Cascade.

Fig. 2. Interior Details of A QCM Stage.

used to obtain a tare correction before and after a short sampling run. It has been found from tests that at a total concentration of about 50 $\mu g/m^3$, five successive 2-minute scans are sufficient to produce good accuracy, with an 8- to 10-minute tare run with the filter before and after the sampling run if higher accuracy is desired. At total concentrations higher than 100 $\mu g/m^3$ a single 2-minute scan is sufficient.

RESULTS AND DISCUSSION

Some results of outdoor ambient air measurements are shown in Figs. 3 and 4. The measurements were made in Costa Mesa, California, which is located southeast of Los Angeles and about 7 miles from the coast. The location is generally free of the more severe smoggy conditions to the east of downtown Los Angeles, except during short periods (lasting a few days) when desert winds push the smog over and past the location out to sea, to return later when the wind shifts to an on-shore direction. The area is mainly light industrial and agricultural, with a major freeway and residential areas upwind.

Fig. 3 shows four distributions measured (with the 6-stage cascade) in the morning and afternoon of a day without any smog transported from Los Angeles. The time of day, brief weather description, the total particulate mass concentration and the mass mean diameter are shown on the figure.

The first distribution, taken in the early morning, appears to be log-normal, probably representing fairly clean condition (109 $\mu g/m^3$) left from the previous night, with a fairly small mass mean diameter of 0.19 μm. As the day warmed up and a light wind developed, the distribution showed a shift to the right, with the trend continuing into the early afternoon, as seen in the distributions taken at 1120 and 1345. The 1345 distribution showed a very large mass mean diameter of 1.5 μm, almost an order of magnitude larger than that of the early morning distribution. As a cooling trend developed in the late afternoon, accompanied probably by both locally generated and transported photochemical smog, a distribution suggestive of a bi-modal character developed, with a mass mean diameter of 0.50 μm.

Fig. 4 shows two distributions taken three hours apart on a day that saw a very visible and narrow band of smog that had been out at sea sweep over the measuring station around noon. The change from a bi-modal distribution with equal peaks in mid-morning to a bi-modal one with a much higher sub-micron peak is quite dramatic.

In measurement of indoor air, with presumably a large population of sub-micron particulates, a cascade with a stage between the last two (at 0.32 and 0.06 μm) of the 6-stage cascade was used. This was a 10-stage system with one stage at 0.17 μm. The result of sampling the air inside an instrumented van is shown in Fig. 5. The van had filtered and air-conditioned atmosphere; but smoking was allowed. The result was a very narrow distribution with a mass mean diameter of 0.15 μm and a total concentration of 282 $\mu g/m^3$. The filter system apparently allowed the very small aerosols from cigarette smoke to remain in the air.

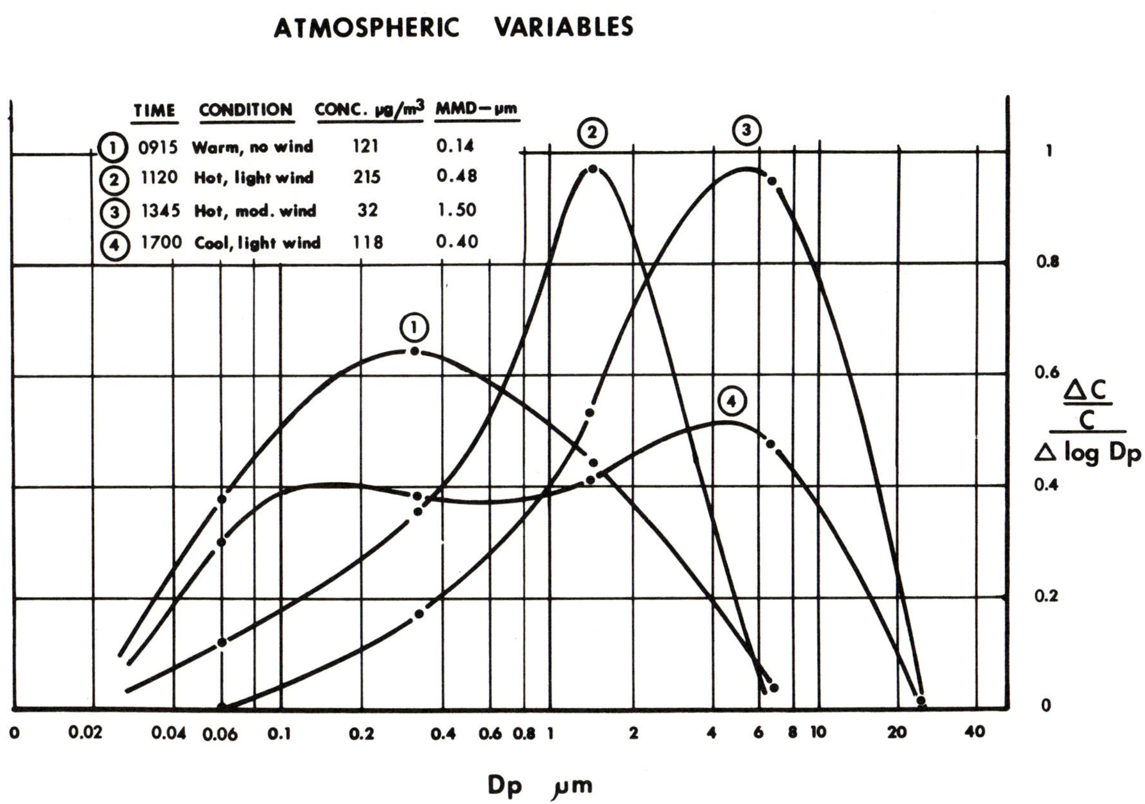

Fig. 3. Particulate Size Distributions in Outdoor Ambient Atmosphere.

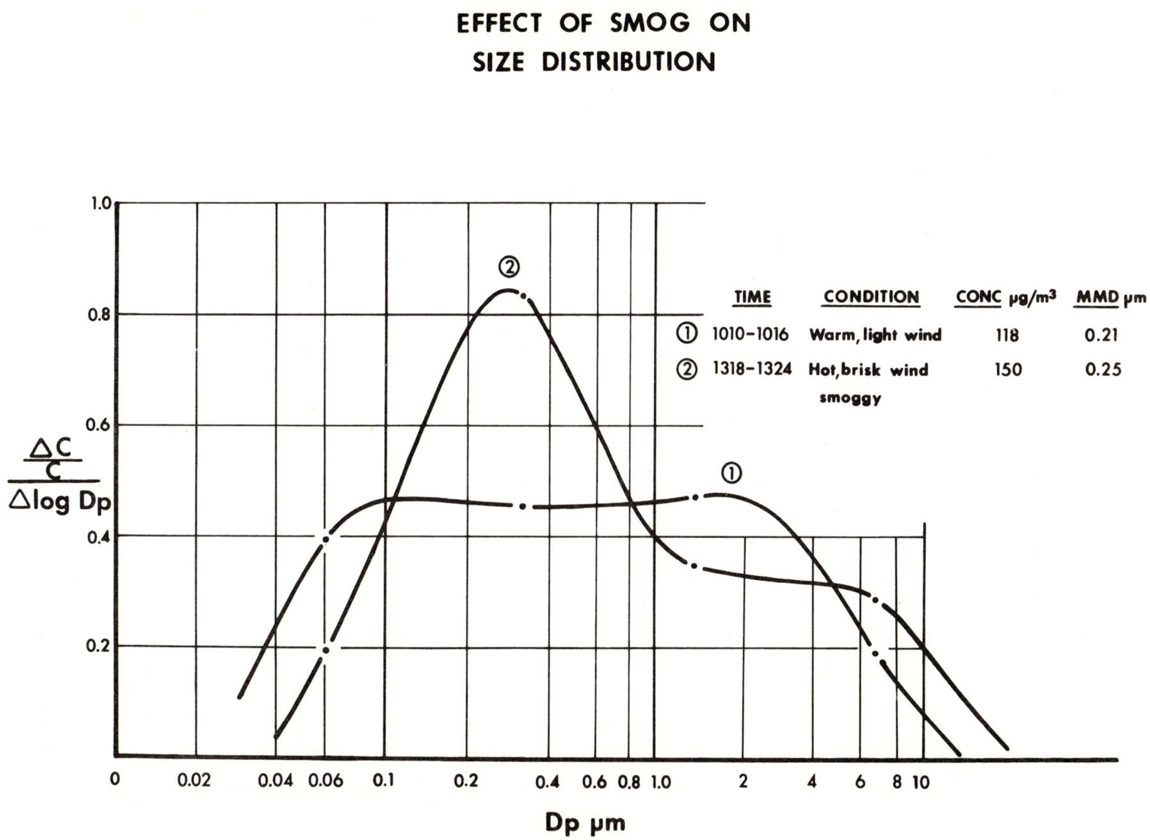

Fig. 4. Particulate Size Distributions in Smoggy Atmosphere.

INSTRUMENT VAN INTERIOR

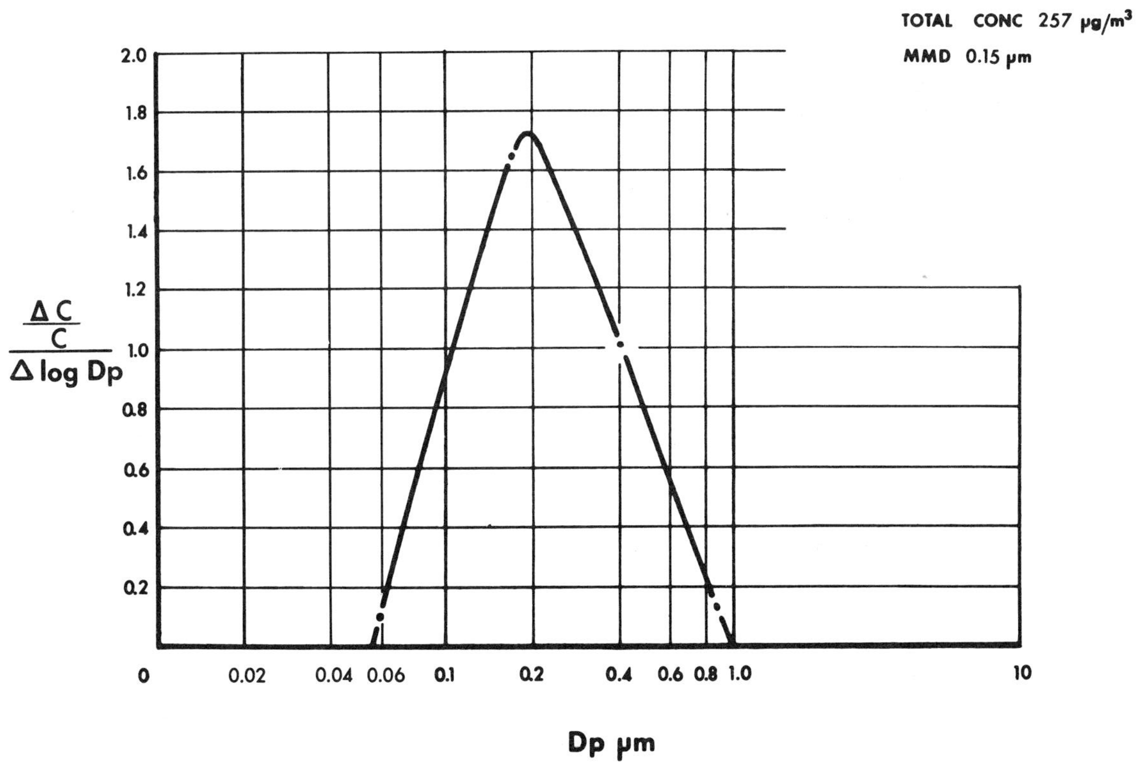

CONCLUDING REMARKS

The particulates impacted on the crystal remain there and eventually build up to such an extent (with many tens of layers of particles) that the response of the crystal to further accretion of mass becomes non-linear. The limits of linear response are established experimentally. In the case of the cascade it has been found that the final stage (with 0.05 μm cut-off) can accommodate a frequency change of 200 Hz before going non-linear. Other stages can accommodate more than 200 Hz. In the examples shown above the total frequency shift in the final stage per sampling run lasting long enough to yield accurate data was about 4 Hz. Thus, with a 200 Hz limit on total mass accumulation, 50 sampling runs can be made before it is necessary to change the sensing crystals. It is therefore possible, with one change of crystals (which requires about 10 minutes for a 10-stage cascade), to obtain four distributions per hour over a 24-hour period.

Preparations are under way to install a 10-stage cascade in an aircraft for plume measurements. With a plume concentration of several hundred $\mu g/m^3$ it will be possible to obtain a good sample in about 30 seconds, which is about the time required to traverse a plume at about one mile from the source. Other applications include the characterization of particulates in critical areas of a ship (such as the engine room) to develope fire warning criteria, and the monitoring of particulates at 1-hour intervals in a submarine to develope air quality data under prolonged cruise conditions.

REFERENCES

Chuan, R. L. An Instrument For the Direct Measurement of Particulate Mass. J. Aerosol Sci 1, No. 2, May 1970.

Chuan, R. L. Application of An Oscillating Quartz Crystal to Measure the Mass of Suspended Particulate Matter.
In: Analytical Methods Applied to Air Pollution Measurements, Stevens, R. K., and Herget, W. F. (ed.) Ann Arbor, Ann Arbor Science Publishers, Inc., 1974. p. 163-189.

IDENTIFICATION AND MEASUREMENT OF PARTICULATE TRANSPORT PROPERTIES

Donald L. Fenton

IIT Research Institute
10 West 35th Street
Chicago, Illinois 60616

ABSTRACT

The continuum approach to particulate flow in the wake region behind a circular cylinder is utilized to identify and experimentally evaluate the appropriate transport properties. The criteria of a dilute suspension is applied which allows separation of the gaseous phase equation of mean motion from the particulate equation of mean motion. This type of unrestricted turbulent flow is also found in jets and within boundary layers on the free side. Whereas the gaseous wake flow in the fully developed region can be characterized by a similar solution, the particulate flow cannot because of the nonlinearity of the gaseous-particulate transfer term. Neglect of this term in the governing equation of motion for particulate flow permits a solution analogous to the gaseous flow -- Gauss' function.

Measurements within the particulate wake include particulate velocity and particulate density (mass concentration) where the wake centerline defects persisted far downstream before fully developed conditions were established. Distributions of the particulate variables transverse to the wake became Gaussian at the downstream distance where the gaseous flow is fully developed. The experimental system consisted of a closed-loop air flow facility and the particulate material was composed of glass beads (count mean diameter = 11 microns). Knowledge of the particulate transport properties, especially their relation to the gaseous transport properties, is useful in calculations applied to particulate control devices.

NOMENCLATURE

b	half width of wake
C'	constant
D_H	hydraulic diameter of duct
d	diameter of circular cylinder
E_y	y-component of the electric field
F	time constant for momentum transfer between the particle and gas
$(N_{ev})_d$	electroviscous number based on d
$(N_m)_d$	gas-particle momentum number based on d
$(N_R)_d$	Reynolds number based on d
$(\frac{q}{m})_p$	charge-to-mass ratio
u	x-component of the mean gaseous velocity
$u_{1_{max}}$	maximum deviation in gaseous velocity from u_∞ within gaseous wake
$u_{p_{1_{max}}}$	maximum deviation in particulate velocity from $u_{p\infty}$ within particulate wake
v	y-component of gaseous velocity
x	free stream flow direction
x_o	geometrical origin of similarity of the wake
y	transverse flow coordinate

Greek

ε_{cp}	eddy diffusion coefficient for the particulate density
ε_{mg}	eddy diffusion coefficient for the gaseous momentum

ε_{mp}	eddy diffusion coefficient for the particulate phase momentum
η	similarity parameter
ξ	non-dimensional distance downstream from circular cylinder
ρ	gaseous density
ρ_p	particulate density
$\overline{\rho}_p$	density of the particulate material
$\rho_{p_{1_{max}}}$	maximum deviation in particulate density from $\rho_{p\infty}$ within particulate wake
ρ^*_∞	ratio of particulate density to gaseous density at free stream conditions
τ	shearing stress of the gaseous phase
τ_p	shear stress due to particulate collisions with the wall
Φ	volume fraction occupied by the particles

Subscripts

1	deviation from free stream conditions
∞	free stream condition
g	gaseous phase
p	particulate phase

IDENTIFICATION AND MEASUREMENT OF PARTICULATE TRANSPORT PROPERTIES

Donald L. Fenton

IIT Research Institute
10 West 35th Street
Chicago, Illinois 60616

INTRODUCTION

Many studies have been made concerning the behavior of dilute particulate suspensions in gaseous media. One motivation for this interest is the multitude of particulate control and instrumentation devices in use at present. In spite of this, however, little work has been expanded on the determination of the appropriate eddy diffusion transport coefficients. Determination of these transport coefficients for various flow systems and particulate materials would greatly improve the prediction of overall flow behavior. Also, the length of conduit required for the establishment of fully developed conditions could be more easily determined.

The continuum approach to particulate flow is utilized here where the ordinary plane two-dimensional wake is studied. This approach accounts for the separation of particulate and gaseous streamlines, relative slip between the particulate and gaseous phases, electrical field and charging effects, and interactions between particles and gas, mutual particles, and particles and gas with a solid boundary. For dilute suspensions, this continuum approach has been demonstrated to be valid[1,2]. Taking ρ_p as the particulate cloud density and $\overline{\rho}_p$ as the density of the particulate material, the volume fraction Φ is given as $\rho_p/\overline{\rho}_p$. Soo[2] indicates that Φ must be less than 0.05 for the existence of a dilute suspension (interparticle spacing > two particle diameters). Recently, successful application of the continuum approach has been made concerning conduit flows and other specialized geometries[3,4,5].

In the study reported here, the ordinary two-dimensional wake (generated by a circular cylinder) is investigated with emphasis on the transport coefficients. The particulate momentum and density eddy diffusion transport coefficients are compared to the gaseous momentum transport coefficient. The actual determination of the transport coefficients is based on the measured wake profile for the appropriate variable.

THEORETICAL BACKGROUND

The coordinate system used in conjunction with the conservation equations is given in Figure 1. The downstream distance along the wake axis

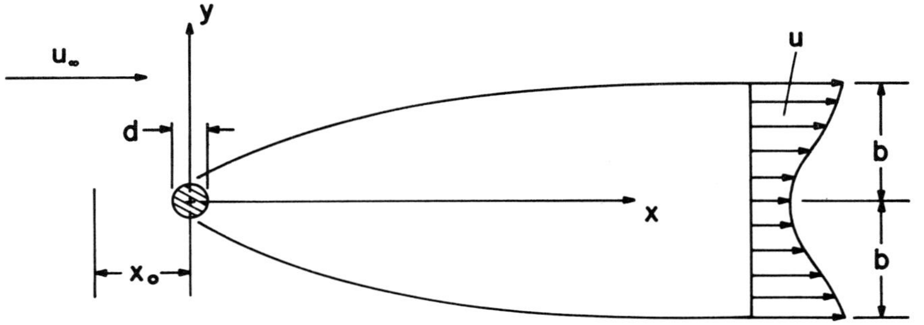

Figure 1. COORDINATE SYSTEM FOR THE PLANE TWO-DIMENSIONAL WAKE

is the x-coordinate and the upward vertical direction is the y-coordinate. The center of the circular cylinder is the origin of the coordinate system.

In light of the experimental conditions and following the continuum approach to particulate flow as given by Soo[1], the steady state continuity relations for the gaseous and particulate phases are

$$\frac{\partial u}{\partial x} + \frac{\partial v}{\partial y} = 0 \qquad (1)$$

$$\frac{\partial}{\partial x}(\rho_p u_p) + \frac{\partial}{\partial y}(\rho_p v_p) = 0 \qquad (2)$$

under the constraint of no phase transformations and no internal generation. The momentum equations for the gaseous and particulate phases in the x-direction are after simplification

$$\rho u \frac{\partial u}{\partial x} + \rho v \frac{\partial u}{\partial y} = -KF\rho_p(u-u_p) + \frac{\partial \tau}{\partial y} \qquad (3)$$

$$\rho_p u_p \frac{\partial u_p}{\partial x} + \rho_p v_p \frac{\partial u_p}{\partial y} = F\rho_p(u-u_p) + \rho_p E_x \left(\frac{q}{m}\right)_p + \frac{\partial \tau_p}{\partial y} \qquad (4)$$

The velocities u and v are to be interpreted as mean velocities for turbulent flow. The constitutive equation for τ has not been specified and therefore, the gaseous momentum equation is not restricted to laminar flow. Because the main flow velocity is large relative to the transverse velocity in wake flow, the y-component of fluid momentum is neglected. The pressure distribution in the main flow and throughout the wake was essentially uniform. The effectiveness parameter, K, in equation (4) incorporates the possibility of irreversible momentum transfer from the particulate phase to the gaseous phase.

With the assumption of a dilute suspension, equation (2) in combination with Fick's Law becomes

$$u \frac{\partial \rho_p}{\partial x} + v \frac{\partial \rho_p}{\partial y} = - \frac{\partial}{\partial y}\left[\left(\frac{q}{m}\right)_p \frac{E_y \rho_p}{F}\right] + \frac{\partial}{\partial y}\left[\varepsilon_{cp} \frac{\partial \rho_p}{\partial y}\right] \qquad (5)$$

where mass diffusion in the x-direction is taken to be much less than in the y-direction. Equation (5) includes diffusion by an electric field force and the transport coefficient, ε_{cp}, takes into account eddy diffusion of particulate density.

A similarity solution can be obtained for the gaseous phase and is reported in the literature[6]. The interparticle spacing is sufficient to prohibit the particulate phase from affecting the motion of the gaseous phase. Therefore, the effectiveness parameter is very nearly zero and gaseous motion is uncoupled from the particulate phase motion. Boundary conditions for the gaseous phase are as follows

$$y = 0 \qquad \frac{\partial u}{\partial y} = 0 \qquad v = 0 \tag{6}$$

$$y \to \pm \infty \qquad u = u_\infty \qquad v = 0 \tag{7}$$

In seeking a similarity solution for the wake profile, a similarity parameter is defined as

$$\eta = \frac{y}{d}\,\phi(\xi) \text{ and } \xi = \frac{x + x_o}{d} \tag{8}$$

where x_o is the distance from the origin of the coordinate system to the geometrical origin of similarity and d the diameter of the circular cylinder. An additional constraint on the gaseous motion is

$$\frac{\text{DRAG}}{\rho\, u_\infty^2} \int_{-\infty}^{+\infty} \frac{u_1}{u_\infty}\, dy \tag{9}$$

which also must be satisfied. In the process of identifying the transport coefficient, the following assumption is made

$$\tau_{xy} = \rho \varepsilon_{mg} \frac{\partial u_1}{\partial y} \tag{10}$$

where ε_{mg} is the coefficient of eddy viscosity for the gaseous phase. The resulting solution is

$$\frac{u_1}{u_{1_{max}}} = \exp\left(- \frac{u_\infty d}{2} \int_o^\eta \frac{1}{\varepsilon_{mg}}\, \eta d\eta \right) \tag{11}$$

where η equals $y/\sqrt{d(x + x_o)}$, ε_{mg} is taken as a scaler, and u_1 is the deviation between the velocity in the wake and the free stream velocity. The integral can be evaluated if ε_{mg} is prescribed and for simplicity is taken as a constant. Carrying out the integration yields

$$\frac{u}{u_{1_{max}}} = \exp\left(- \frac{u_{\infty}d}{4\varepsilon_{mg}} \eta^2\right) \tag{12}$$

which is the Gauss function.

An analogous situation exists -- similarity at sufficient downstream distance -- with the particulate phase if the following postulation is made[7]

$$\frac{u}{u_{p\infty}} - 1 = C_p \phi_p(\xi) f_p(\eta) \tag{13}$$

where ξ and η are defined in like manner to the gaseous phase and C_p is a constant. A condition of quasi-similarity for the particulate phase can then be formulated

$$\phi_p(\xi) f_p(\eta) \propto \frac{u}{u_{p\infty}} \tag{14}$$

With the above postulation, the solution for u_p proceeds as for the gaseous phase. The result is

$$\frac{u_{p_1}}{u_{p_{1_{max}}}} = \exp\left(- \frac{u_{p\infty}d}{4\varepsilon_{mp}} \eta^2\right) \tag{15}$$

where the integration is performed under the condition that ε_{mp} is a constant scaler.

The transport of particulate density in the wake region can also be obtained. With the assumption that the transport of ρ_p is analogous to the transport of gaseous momentum,

$$- v\rho_{p_1} = \varepsilon_{cp} \frac{\partial \rho_{p_1}}{\partial y} \tag{16}$$

and taking ε_{cp} as an unprescribed scaler, similar wake profiles of ρ_p should exist. If ε_{cp} is taken as a constant, Gauss' function is again obtained

$$\frac{\rho_{p_1}}{\rho_{p_{1_{max}}}} = \exp\left(-\frac{u_\infty d}{4\varepsilon_{cp}}\eta^2\right) \tag{17}$$

where ρ_{p_1} is the mean deviation between the wake region and free stream particulate density. Also, if the ratio $\varepsilon_{mp}/\varepsilon_{cp}$ is constant throughout the wake region, then

$$\frac{\rho_{p_1}}{\rho_{p_{1_{max}}}} = \left(\frac{u_{p_1}}{u_{p_{1_{max}}}}\right)^{\frac{\varepsilon_{mp}}{\varepsilon_{cp}}} \tag{18}$$

EXPERIMENTAL APPARATUS AND MEASUREMENT RESULTS

A recirculation-type wind tunnel 12 inches square was used to generate the turbulent suspension. The upstream distance from the test section was over 40 hydraulic diameters to insure fully developed gaseous flow upon generation of the wake. The Reynolds number based on the hydraulic diameter varied between 5.6×10^4 and 6.4×10^5 for the tests.

A circular cylinder placed transverse to the flow generated the two-dimensional wake. Circular cylinders were used because this geometry occurred most often in the literature to generate an ordinary symmetrical wake. Comparison directly to the gaseous flow literature is facilitated. To eliminate effects of periodic flow, the Reynolds number based on cylinder diameter was greater than 1,000.

In the determination of gaseous flow, a modified Prandtl pitot-static tube was employed. The gaseous phase requires only the measurement of velocity for characterization. Figure 2 gives typical results for the gaseous wake at various distances downstream. Similarity is observed to exist for downstream distances greater than $x/d = 40$. The curve drawn on the profile is a result of selecting the appropriate ε_{mg} which best satisfies the experimental data. The central portion of the gaseous wake is therefore seen to behave as predicted by the gradient-type diffusion concept of transport. Agreement at the outer wake boundary is not good and is due to the reduced coefficient of eddy diffusion near the boundary. Also important is the coalescence of the wake profiles for all Reynolds numbers in the region of similarity which, of course, indicates no peculiarities in the flow about the circular cylinder.

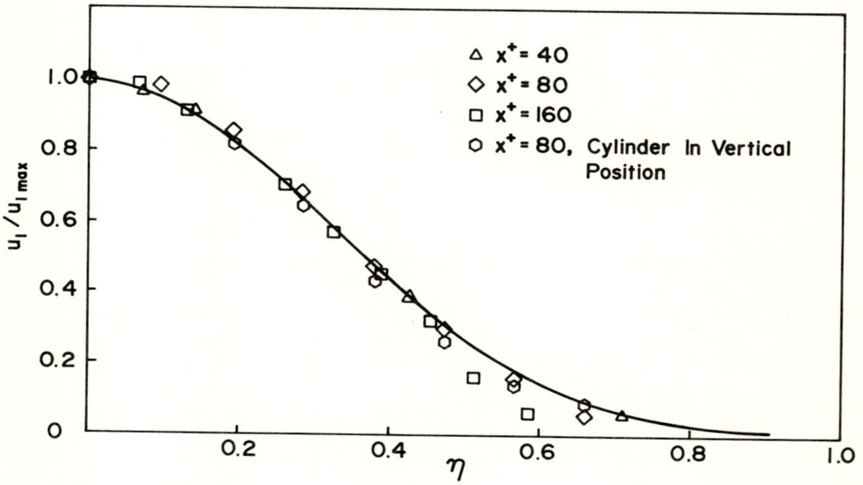

Figure 2. GASEOUS WAKE PROFILE AT $(N_R)_d = 3.42 \times 10^3$

Three quantities must be measured to fully characterize mean motion of the particulate phase: ρ_p, $\rho_p u_p$, and $(q/m)_p$. The electrostatic term was neglected in the solution of equation (4). Experimental justification (measurement of charge-to-mass ratio) is provided by the dimensionless form of the equation of motion[2] and typical values for the dimensionless groups are

$$(N_{ev})_d \sim 10^{-5} \text{ and } (N_R)_d \sim 10^3$$

In the dimensionless particulate diffusion equation, electrostatic effects can also be neglected because

$$(N_{ev})_d^2 \, (N_m)_d \sim 10^{-11} \text{ and } (N_R)_d \sim 10^3$$

With the measurement techniques utilized[7,8], the particulate velocity was not measured directly but through combination of the particulate mass flux and particulate density measurements. Essentially, the local mass flux divided by the particulate density at the same location yields the particulate velocity. Consequently, no information is obtained concerning the velocity of different particle sizes.

The particulate material was Ballentine Blast Beads (specific gravity = 2.65) that were bimodally distributed with the modes at 3 and 35 μm. Microscopic investigation determined the count mean diameter to be 11 μm and showed the particles to be 95% spheres and 5% platelets. Figure 3 indicates the duct centerline particle size distribution for the experimental conditions.

Figure 4 shows the particulate velocity defect for conditions corresponding to the gaseous momentum data already given. In all the data obtained, η serves as an effective similarity parameter. This is not surprising because the gas imparts momentum to the particles through drag effects and therefore the particulate momentum distribution should be approximately the same shape. The curve drawn through the data is the apparent coefficient of eddy diffusion for particulate momentum. Also, note that for the dilute suspensions employed, the distribution of particulate momentum is not significantly altered by the free stream particulate density. The two curves drawn reveal the actual influence of the free stream particulate density on the particulate momentum wake profile.

Particulate density defects within the wake region were also distributed according to Gauss' function and again, η served as an effective similarity parameter. Figure 5 indicates data obtained corresponding to

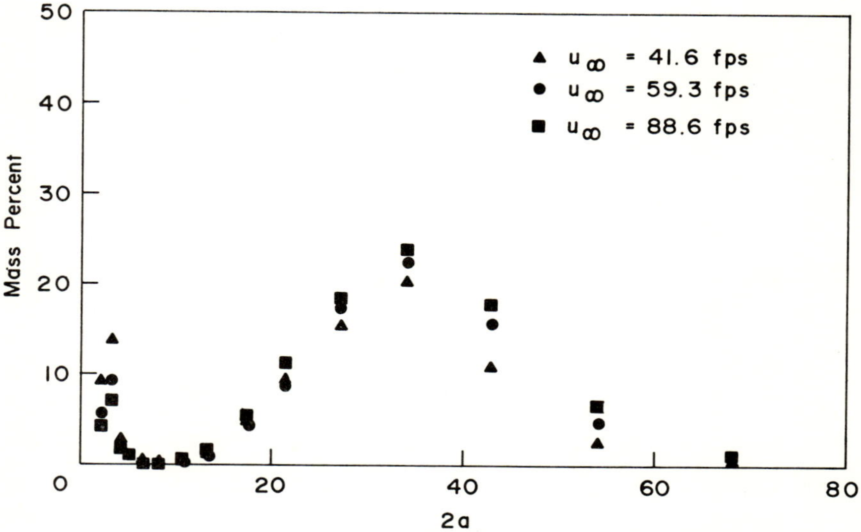

Figure 3. PARTICLE SIZE DISTRIBUTION AT DUCT CENTERLINE

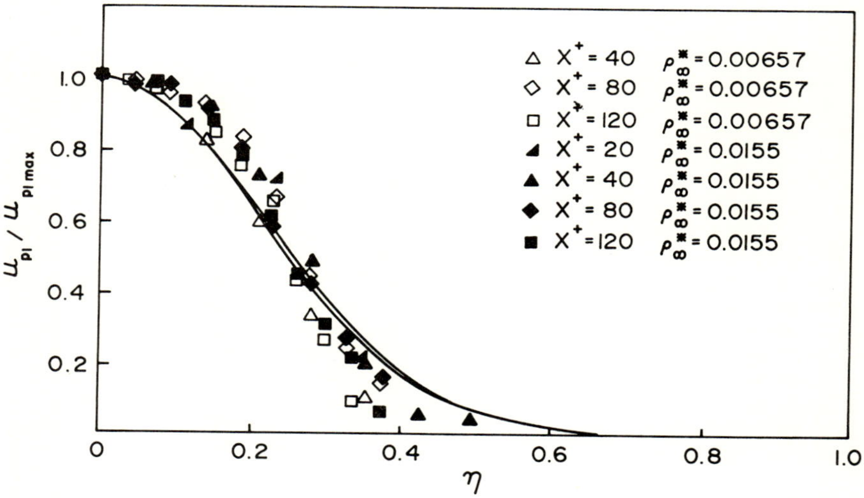

Figure 4. PARTICULATE VELOCITY WAKE PROFILE AT $(N_R)_d = 3.42 \times 10^3$

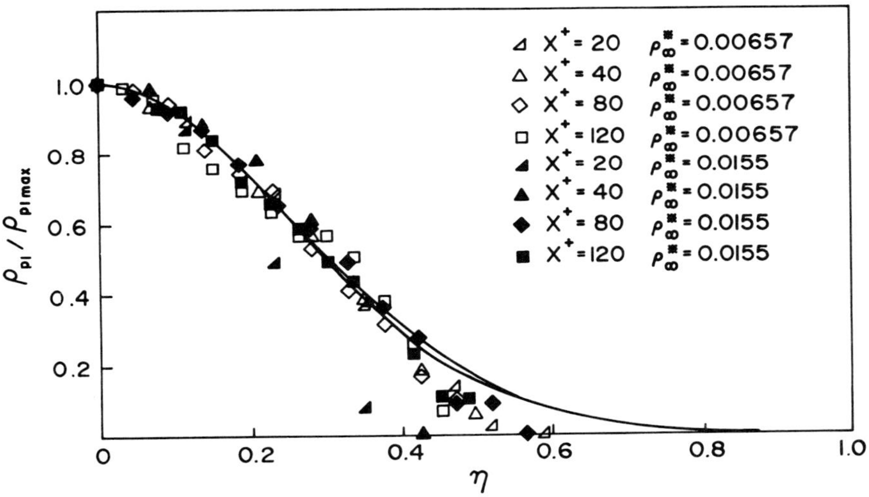

Figure 5. PARTICULATE DENSITY WAKE PROFILE AT $(N_R)_d = 3.42 \times 10^3$

gaseous momentum in Figure 2. Also, the free stream particulate density slightly influences the distribution of the particulate density defect. The curves drawn indicate the magnitude of this influence on the particulate density wake profile.

DISCUSSION OF RESULTS

Table 1 gives results for the eddy diffusion transport coefficients obtained for all the experimental conditions. Free stream values of the gaseous eddy diffusion coefficient, ε_m, are normalized and constant over the experimental conditions. The values for ε_m were obtained from measured duct velocity profiles. The eddy diffusion coefficient for gaseous momentum is also normalized and varies with the free stream velocity.

Table 1. EDDY TRANSPORT COEFFICIENTS

$(N_R)_d$	u_∞ (m/s)	d (cm)	ρ^*_∞	$\frac{2\varepsilon_m}{u_\infty D_H}$	$\frac{\varepsilon_{mg}}{u_\infty d}$	$\frac{\varepsilon_{mp}}{\varepsilon_{mg}}$	$\frac{\varepsilon_{cp}}{\varepsilon_{mg}}$	$\frac{\varepsilon_{mp}}{\varepsilon_{cp}}$
2.40×10^3	12.7	0.318	0.0141	1.41×10^{-4}	0.0481	0.489	0.632	0.773
3.42×10^3	18.1	0.318	0.00657	1.41×10^{-4}	0.0481	0.469	0.675	0.695
3.42×10^3	18.1	0.318	0.0155	1.41×10^{-4}	0.0481	0.489	0.698	0.700
6.93×10^3	18.3	0.635	0.00657	1.41×10^{-4}	0.0418	0.608	0.804	0.755
6.93×10^3	18.3	0.635	0.0157	1.41×10^{-4}	0.0418	0.653	0.830	0.786
5.12×10^3	27.0	0.318	0.0106	1.41×10^{-4}	0.0443	0.551	0.686	0.803
5.12×10^3	27.0	0.318	0.0146	1.41×10^{-4}	0.0443	--	0.747	--
5.12×10^3	27.0	0.318	0.0232	1.41×10^{-4}	0.0443	0.616	0.783	0.787

The two particulate transport coefficients are normalized by the gaseous momentum eddy diffusion coefficient for wake flow. The particulate momentum eddy diffusion coefficient is roughly half the eddy diffusion coefficient for gaseous flow. Values are also given for the apparent coefficient of eddy diffusion for particulate density in Table 1. The diffusion of particulate density is seen to occur more readily than particulate momentum for the same wake structure. Also, the ratio of eddy diffusivities for the particulate momentum and density for the various experimental conditions are calculated. Within experimental error, the ratio $\varepsilon_{mp}/\varepsilon_{cp}$ is constant (about 0.76) which justifies equation (18) determined from the integral forms of particulate momentum and density defect distributions.

Most of the symmetrical wake studies reported use circular cylinders for wake generation because at the downstream distances where similarity is established, object geometries are not significant. This turbulent wake type of unrestricted shear flow can also be found in jets and within boundary layers on the free side. Although the specific results of this work are unique to the particulate suspension investigated, significant information is yielded concerning the relative magnitude of the three transport coefficients.

REFERENCES

1. Soo, S. L., Fluid Dynamics of Multiphase Systems, First Edition, Blaisdell Pub., Mass., 1967.

2. Soo, S. L., Dynamics of Charged Suspensions, International Reviews in Aerosol Physics and Chemistry, V.2 (ed. Hidy, G. M., and Brock, J.), Pergamon Press, Oxford, 1971.

3. Soo, S. L., and Tung, S. K., Appl. Sci. Res., 24:83-97, June 1971.

4. Stukel, J. J., and Soo, S. L., Powder Technol., 2:278-289, 1969.

5. Ramadan, O. E., and Soo, S. L., Physics of Fluids, 12:1943-1945, September 1969.

6. Hinze, J. O., Turbulence, McGraw-Hill Book Company, New York, 1959.

7. Fenton, D. L., Ph.D. Thesis, Dept. of Mech. Engr., Univ. of Ill., 1974.

8. Soo, S. L., Stukel, J. J., and Hughes, J. M., Environmental Science and Technology, 3:386-393, 1969.

OPTICAL AEROSOL SIZE SPECTROMETRY BELOW AND ABOVE THE WAVELENGTH OF LIGHT - A COMPARISON

J. Gebhart, J. Heyder, C. Roth, W. Stahlhofen

Gesellschaft für Strahlen- und Umweltforschung m.b.H.
6 Frankfurt/Main, Paul-Ehrlich-Str.21, Fed. Rep. Germany

ABSTRACT

A critical review of the abundant literature on scattering of light on small particles revealed that different light scattering devices should be used in order to adjust their response to the light scattering properties of the particles. Below the wavelength of light laser light illumination and a mean scattering angle in the nearer forward direction are preferable. Such an instrument is the LASS (laser aerosol size spectrometer)with a sizing range between 0.05 and 0.7 µm. Above the wavelength of light white light illumination and low angle scattering is more advantageous. The LASI (low angle scattering instrument) which has a sizing range between 0.7 and about 6 µm serves for this purpose. The performance of these instruments with respect to particle diameter, refractive index and shape has been checked with monodisperse polystyrene spheres, condensation droplets and solid Fe_2O_3-particles as well as agglomerates of polystyrene spheres. It turned out that the response of the LASI is a function of the projected area of the particles independent of the optical properties and shape of the particles. The LASS measures directly the volume equivalent diameter of a particle regardless of its shape. However, the response of the LASS depends on the refractive index of the particle material. The response curves for different refractive indices differ more or less by a constant factor.

OPTICAL AEROSOL SIZE SPECTROMETRY BELOW AND ABOVE THE WAVELENGTH OF LIGHT - A COMPARISON

J. Gebhart, J. Heyder, C. Roth, W. Stahlhofen

Gesellschaft für Strahlen- und Umweltforschung m.b.H.
6 Frankfurt/Main, Paul-Ehrlich-Str.21, Fed. Rep. Germany

1. INTRODUCTION

Particle sizing by means of light scattering on single particles is known since more than 25 years. Meanwhile the subject has been steadily developed and since about 10 years optical particle counters using white light illumination are commercially available. After the invention of the laser principle several attempts have been made to replace the white light illumination of scattering devices by coherent and monochromatic laser light illumination. Although all instruments measure particle size distributions of aerosols they have been regarded as optical particle counters. Since the sensitivity as well as the resolution power of light scattering devices have been improved considerably by using laser light illumination, recent developments utilizing lasers have been regarded as opticle size spectrometers. Their performances will be reviewed in this paper. It will be also part of this paper to discuss the advantages and disadvantages of both types of illumination for particle size analysis.

To adjust light scattering devices to the light scattering properties of small particles it has been proved advisable to use two different optical arrangements because of the different optical behaviour of particles smaller or larger than the wavelength of light. For the size range below the wavelength a laser aerosol size spectrometer (LASS) has been developed in this laboratory. Above the wavelength a low angle scattering instrument (LASI) with white light illumination has been applied. The performance of both instruments with respect to particle diameter, refractive index and shape will be discussed based on experimental results.

2. WHITE LIGHT OR LASER LIGHT ILLUMINATION?

The outstanding characteristics of laser light are the high stability of the output, the degree of spatial and temporal coherence and the high flux density.

The high stability renders unnecessary a reference light source.

The spatial coherence or the low beam divergence leads to a high intensity at the focal point of an optical system and therefore the need of parabolic mirrors and complicated lens systems is eliminated. In the case of focal plane illumination the intensity of a typical 5 mW He-Ne-laser radiation is 4 to 5 orders of magnitude greater than the intensity from the most intense incoherent light source. Therefore the sizing range of usual optical particle counters can be extended by utilizing lasers. Since due to its spatial coherence a laser beam can be focussed to a very tiny, diffraction limited spot the dimensions of the sensing volume can be kept very small. For this reason aerosols of high concentrations (up to 10^6 cm^{-3}) can be analysed with negligible coincidence losses.

The temporal coherence or monochromacy on the other hand is advantageous for instruments analysing the angular dependence of the scattered light, but is a drawback for particle counters, since in the size range above the wavelength the response curves of monochromatic counters exhibit oscillations. These fluctuations can be smoothed over by applying white light and a large collecting aperture, however the smoothening of white light is more effective. The application of only a large collecting aperture is not sufficient since for monochromatic light, even for an aperture of 4π, oscillations occur as can be seen from the extinction coefficient of transparent spheres (Hodkinson[1], Quenzel[2]).

It has to be mentioned in this connection that the disadvantage of an oscillating response curve can be partly overcome by a new evaluation method. By this method a differential size distribution spectrum of a polydisperse aerosol of spherical particles and known refractive index is recorded with the instrument. The peaks in the spectrum are then related to the turning points of the theoretical calibration curve and act as internal size references (Schöck[3] Bakhanova and Ivanchenko[4]). But the resolution power of

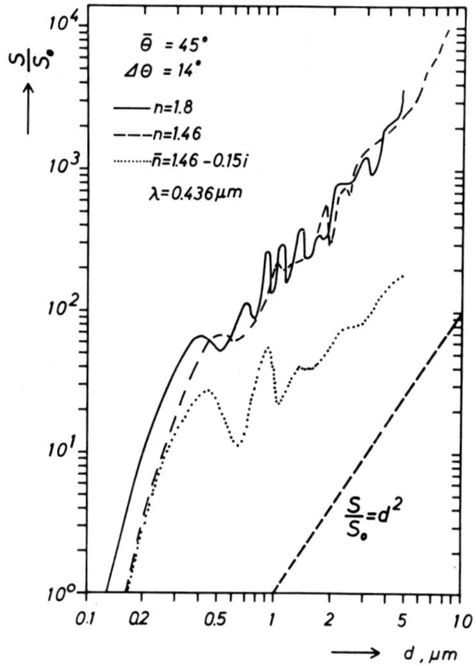

Fig.1. Relative response curves of optical particle counters for monochromatic light
$S_o = S(d=0.13\mu m; n=1.8)$
(according to Broßmann[7])

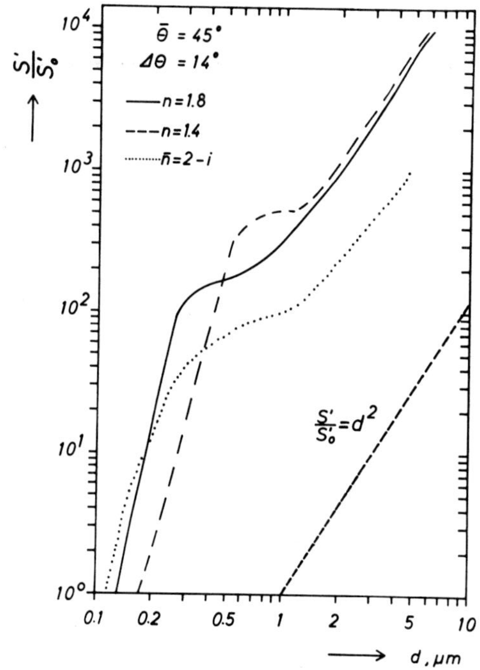

Fig.2. Relative response curves of optical particle counters for white light
$S_o = S(0.13\mu m; n=1.8)$
(according to Hodkinson and Greenfield[6])

this method is poor. If growing droplets are illuminated with laser light and the scattered light is observed under a fixed angle the response of the instrument as function of time shows also characteristic fluctuations which can be correlated with particle size. By this way the growth rate of a droplet due to condensation can be determined (Wagner[5]).

The influence of monochromatic and white light upon the calibration curve of an optical counter is demonstrated by the following graphes. Detailed calculations in this field have been made by Hodkinson and Greenfield[6], Quenzel[2], Broßmann[7] and Oeseberg[8]. In Fig. 1 and 2 the partial scattering cross-section as function of the particle size for monochromatic and white light and a mean scattering angle of 45° is shown. For particle sizes larger than the wavelength light scattering can be considered as a surface effect and outside the diffraction lobe the mean scattered light flux depends on the square of the particle diameter. For monochromatic light the curves show the typical oscillations. In the range below the wavelength a d^6-dependence is observed and the curves are straight lines even for monochromatic light. Fig. 2 shows the smoothening effect due to white light above the wavelength. In Fig. 3 the relative response curve for a receiver aperture between 2.5° and 5.5° is plotted against the particle diameter, d. For monochromatic light and transparent spheres the response curve exhibits periodical fluctuations with 5 maxima between 1 and 6 µm, which can be smoothed over by white light. (see chapter 6).

CONCLUSIONS: It has been shown that for size analysis of particles bigger than the wavelength white light illumination is advisable in order to avoid ambiguous response of the instrument. For size analysis of particles smaller than the wavelength either laser or white light illumination can be used. However, laser light illumination is preferable in order to achieve a higher sensitivity and resolution power of the instrument.

3. OPTICAL INSTRUMENTS USING WHITE LIGHT

A critical discussion of the performance of optical instruments using white light exceeds the limits of this paper Therefore, the reader is refered to the following papers describing the instruments: Gucker et al.[9], Zinky[10], Sinclair[11], Martens and Fuss[12], Turpin[13]. Because most of the instruments are nowadays commercially available their performance has been thoroughly studied in several aerosol laboratories: Whitby and Vomela[14], Jaenicke[15], Williams and Hedley[16], Liu et al. [17].

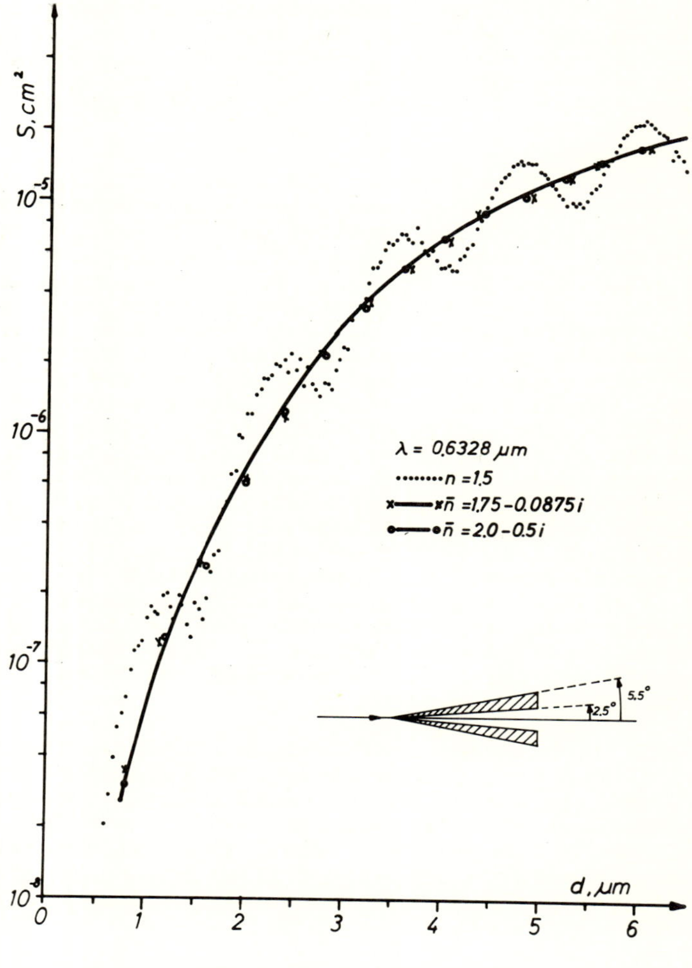

Fig. 3. Theoretical response curves of the LASI
solid line : transparent spheres
broken line: absorbing spheres

4. OPTICAL INSTRUMENTS USING LASER LIGHT ILLUMINATION

According to the special properties of laser light the optical setups can be devided into three groups:

i. Instruments with the sensing volume inside the optical cavity of the laser.

ii. Instruments detecting the angular intensity function of scattered light.

iii. Instruments with a fixed mean scattering angle and collecting aperture including the solid angle 4π.

ad i:
Schleusener[18] suggested recently a new approach to the detection and sizing of aerosol particles. In this method, each particle passing through the laser cavity,creates a cavity disturbance that results in a pulse shaped decrease in the laser output. Individual particles can be counted up to 10^4 per second and classified according to their light extinction. The light extinction is amplified by about a factor of 100 inside the laser resonator. Lateron Schehl[19] tried to minimize the sensing volume by passing the particles through the focus of one of the resonator mirrors and to increase the sensitivity by measuring the scattered light instead of the extinction. But nevertheless resolution and measurable size range is poor compared to the other optical counters. Although it is not mentioned the response of the device must be disturbed by particles passing not the sensing volume.

ad ii:
Instruments determining the whole scattering diagram of a particle with a small collecting aperture (Phillips and Wyatt[20], Wyatt and Phillips[21], Cooke and Kerker[22], Gucker et al.[23], Moser[24], Maier[25]) produce more information about the properties of the particle. These devices are useful for the exact determination of the refractive index or the diameter of a particle with an absolute accuracy comparable with that of electron microscopy. They are applicable for spherical particles in the size range from about 0,3 to 10 µm. The step by step detection of the angular dependence of the scattered intensity, however, is time consuming. Furthermore the immense amount of information has to be processed by a computer to obtain the best fitting of the experimental diagrams with the theoretical predictions. The most skillful work on this subject has been done by Gucker et al.[23]. Their high speed photometer detects the scattering diagram of single aerosol particles from 7° to 353° in as little as 14.4 ms. Although this instrument

has a high speed, the concentration of the aerosol must be lower than 10^2 cm^{-3} in order to prevent coincidences. All other devices have a much slower recording system.

ad iii:
Optical particle counters with a fixed mean scattering angle and collecting aperture can be further divided into two types. The first type of instruments detects with a moderate collecting aperture the light scattered in the forward direction. The optical design is realized with dark field illumination for the low angle scattering instruments and with an asymmetrically arranged collecting aperture for bigger mean scattering angles.

Although these instruments have quite different designs, they all attempt to minimize the sensing volume. For the low angle instruments the coherent radiation is absolutely necessary for the exact separation of illuminating and scattered light. Useful measurements can be made at angles as small as 1.5^o.

In the second type of instruments a maximum quantity of scattered light from all directions is collected in order to extend the sizing range of the instruments to smaller particles. If the collecting aperture is enlarged, however, the background noise - due to light scattering on gas molecules and on structural parts of the instruments - increases in the same way as the radiant flux through the aperture due to light scattering on the particle. An enlargement of the collecting aperture, therefore, does not result in a lower detection limit.

Low angle light scattering instruments with small sensing volumes and fairly high resolution powers are described by Kaye[26] and Gebhart et al.[27]. The instrument of Gebhart et al. can be used for particle size analysis between 0.15 and 1.2 μm. For larger particles the response curve becomes ambiguous. The testing procedure of the instrument of Kaye is still incomplete.

Another optical particle counter with dark field illumination having a collecting aperture from 7° to 12° has been developed by Hill[28]. This high resolving instrument is designed for high concentrated aerosols, but limited to a size range between 0.15 and 1 μm. Instruments detecting the light scattered in the forward direction by means of an asymmetrically arranged receiver aperture have been reported by Bol et al.[29] and Pinnick et al.[30]. The latter publication gives not much information about the actual performance of the device since the proof of the light scattering theory was given preferance.

The counter described by Bol et al. [29] was designed for highly concentrated aerosols in the size range below the wavelength of light; but the lower detection limit of this instrument is already met for 0.15 μm particles.

The most popular instrument which detects the light scattered in nearly the solid angle 4π is the Jacobi-counter (Jacobi et al.[31]) using a lucite cylinder as light collecting device. A copy of this instrument is described by Walsh[32]. These counters have the discussed disadvantages of non-monotonous response curves for particles in the micron size range. The smallest detectable particles are about 0,3 μm. Because the laser beam is not focussed the sensing volume is relatively large and the measurable concentrations are low, but nevertheless the resolving power of the instrument is quite excellent. Lioy et al.[33] tried to increase the sensitivity of their particle counter by collecting the scattered light from all directions by an ellipsoidal mirror and achieved a lower detection limit of about 0,2 μm.

5. RESPONSE TO PARTICLES OF DIFFERENT OPTICAL PROPERTIES AND SHAPES

5.1. Definitions

To analyse the influence of the optical properties and the shape of a particle upon the response of a light scattering device it is useful to introduce the definitions (Heyder et al.[34]) given in Fig. 4. In these diagrams the partial scattering cross-section, S, is drawn against the geometrical diameter, d, of the calibration spheres. If the aerosol to be investigated consists of spherical particles with the same refractive index, n_c, as the calibration spheres the geometrical diameter, d, is measured. For a spherical particle with a refractive index, $n \neq n_c$, an equivalent light scattering diameter, d_{sce}, is determined. For a non-spherical particle with refractive index, n_c, the light scattering diameter, d_{sc}, is obtained. The influence of the particle shape may be described by an optical shape factor, w, also defined in Fig. 4.

5.2. Particles larger than the wavelength

For particles larger than the wavelength light scattering

becomes more and more a surface effect. The incident light entering the particle is scattered due to refraction. This component dominantes in the angular range up to about 100° except within the forward scattering lobe of diffraction. For particles of absorbing materials the part of light entering the particle is absorbed. Instruments which mainly collect the light scattered due to refraction, therefore, show great differences between transparent and absorbing materials. This can be seen from Fig. 1 and Fig. 2. An experimental verification of this behaviour using commercially available particle counters has been given by Whitby

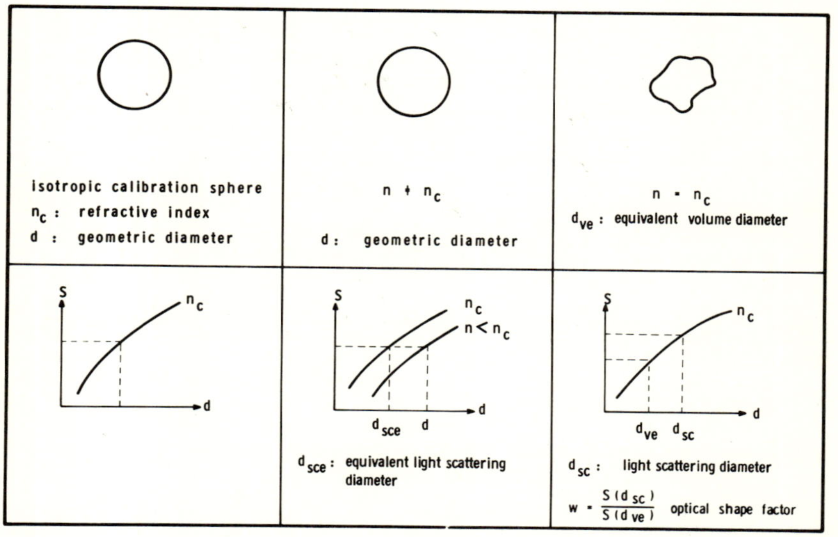

Fig.4. Definitions used in optical particle size spectrometry

and Vomela[14] for particles of india ink and by Liu et al.[35] for monodisperse coal particles.

Within the forward scattering lobe of diffraction the scattered light intensity is a function of the projected area of the particle independant of its optical properties and its shape. This can be seen theoretically in Fig. 3 and will be confirmed experimentally (see chapter 6). The oscillations of the response curve for transparent spheres are caused by interferences of light diffracted and refracted on the particle. They can be smoothed over by white light. For opaqué materials the refracted component is absorbed inside the particle and no interference occurs (see Fig. 3).

5.3. Particles smaller than the wavelength

For particles smaller than the wavelength light scattering becomes more and more a volume effect. Calibration curves for different real parts of refractive index tend to run parallel in a double logarithmic plot (Fig. 5). That means, that these curves differ more or less only by a constant factor which is a function of the real part of the refractive index. For absorbing materials with a complex refractive index, $\bar{n} = n - ik$, additionally the influence of the k-value is reduced considerably. As can be seen from Fig. 6 all curves of the same real part of refractive index cross over for particle diameters below the wavelength.

As long as light scattering is a volume effect the scattering cross-section of a particle becomes mainly a function of its volume and less of its shape. This can be shown theoretically by applying the light scattering calculations to spheroids which are smaller than the wavelength (van de Hulst[36]). Consequently, a light-scattering device below the wavelength measures the equivalent volume diameter, d_{ve}, of a non-spherical particle (see Fig. 4). An experimental verification of this fact will be given in chapter 7.

6. MEASUREMENTS WITH THE LASI ABOVE THE WAVELENGTH

For particle size analysis in the size range above the wavelength a low angle scattering instrument (LASI) using the white light of a mercury lamp has been developed (Gebhart and Roth[37]). The optical setup of the instrument is shown in Fig. 7. The lens, L_1, forms an image of the mercury arc, Q, on the slit, D_1, of cross-section (0.04 x 0.20) mm^2. D_1 is imaged into the sensing volume by the lense, L_2; the magnification is 5 so that the beam cross-section in the sensing volume is about (0.2 x 1) mm^2. By means of a hole in the mirror, M, the primary beam is seperated from the light scattered by the particles. The primary beam vanishes in a light trap, T. The scattered light is sampled by the lens, L_3, which forms an image of the particle onto the stop, D_4. Behind this stop the scattered light reaches the photomultiplier, PM, where the light flashes are converted into electrical pulses. The electrical pulses are linearly amplified and then classified in

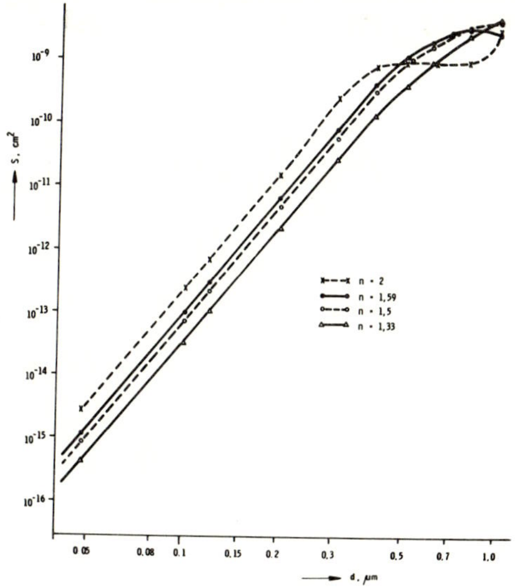

Fig.5. Theoretical response curves of optical particle counters for transparent spheres and monochromatic light (Λ=.6328μm) (angular intensity functions according to Maguire[48] and Lowan[49])

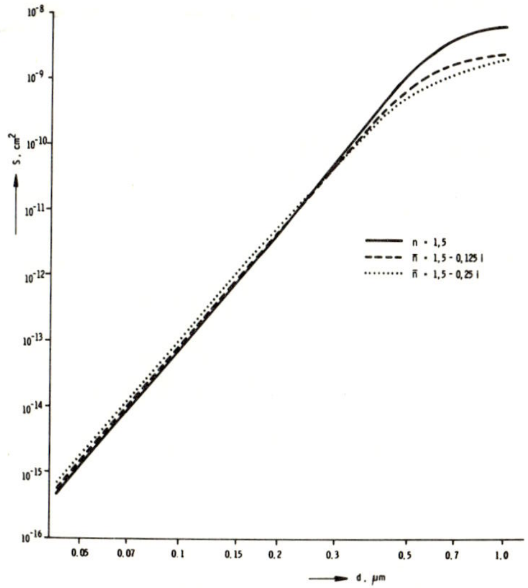

Fig.6. Theoretical response curves of optical particle counters for absorbing spheres and monochromatic light (Λ=.6328μm) (angular intensity functions according to Olaf und Robock[50])

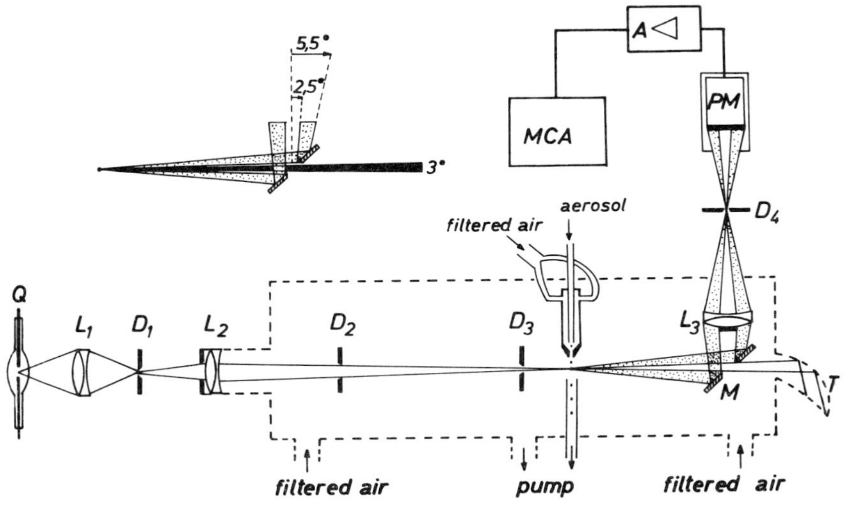

Fig. 7. Set-up of the LASI

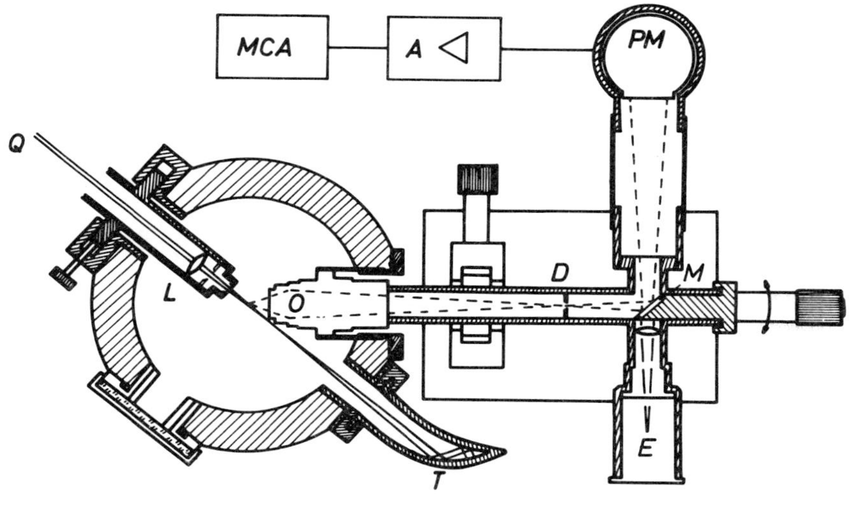

Fig. 9. Set-up of the LASS

a multi channel analyser, MCA. The primary beam has a semi-angle of 1.5^{o} and the aperture of the receiver is extended from 2.5 to 5.5^{o}. The whole primary beam passes a tank with filtered air. By means of a constant pressure drop between the tank and the aerosol inlet particles are sucked into the sensing volume. The aerosol tube has an inner diameter of0.500 mm. The sampling rate is about 100 $cm^3\ min^{-1}$. By use of a clean air jacket the aerosol stream is focused to a filament of about 100 μm. The pulse duration is a few μs and the effective sensing volume about $10^{-6}\ cm^3$. Thus concentrations of up to 10^5 particles per cm^3 are allowed with counting rates of more than 10^4 particles per second.

The experimental response curve of the instrument in the size range between the lower detection limit of about 0.7μm and the upper limit of about 6 μm is shown in Fig.8. Three kinds of monodisperse aerosols with different optical properties have been used for the calibration. To find the geometrical diameters of the particles independent methods have been applied. The size of polystyrene spheres was checked by electron microscopy (Porstendörfer and Heyder[38]). For the aerosol of di-2-ethylhexyl-sebacate (DES) produced in a Sinclair-La Mer generator (Stahlhofen et al.[39]) size measurements of the droplets were carried out in a sedimentation cell (Stahlhofen et al.[40]). Fe_2O_3-particles were produced by nebulizing a colloidal solution of Fe_2O_3 (Albert et al.[41]) by means of a spinning top(May[42]). The diameter of the dried solid particles was determined by electron microscopy. In spite of the different optical properties of these particles all measured points lie almost on one curve. No oscillations of the experimental calibration curve are observed. The partial scattering cross-section, S, depends on the third to fourth power of the particle diameter, d. These results confirm the theoretical predictions that light scattering within the forward lobe of diffraction is nearly independent of the optical constants of the particle material.

The response of the instrument to non-spherical particles has been investigated using aggregates of uniform polystyrene spheres. Let d_{scj} be the diameter of a sphere with refractive index, n_c, which has the same partial scattering cross-section, S, as an agglomerate formed by j uniform spheres with diameter, d_1, and refractive index, n_c. Then a relative light scattering diameter, F_j, of the aggregates can be defined as (Heyder and Porstendörfer[43]):

$$F_j = \frac{d_{scj}}{d_1}$$

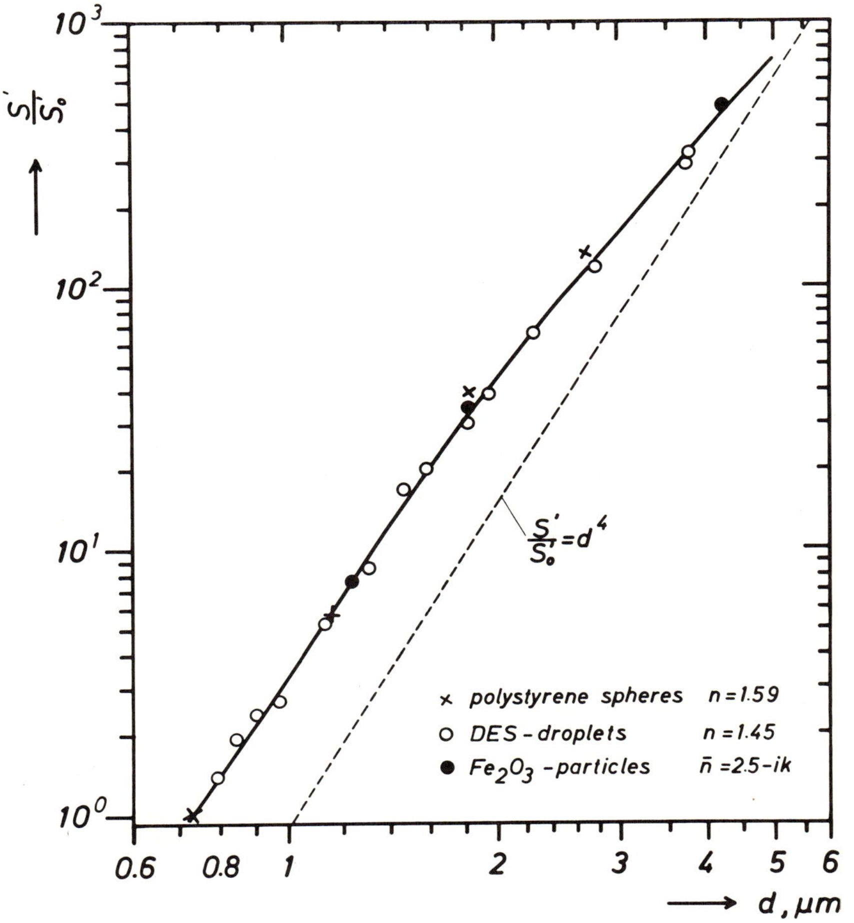

Fig.8. Experimental response curves of the LASI for particles of three different refractive indices

Some relative light scattering diameters measured for different diameters, d_1, of the single spheres are listed in Table 1. As can be seen, $F_j = j^{1/2}$, so that d_{scj} is identical with the diameter of the projected area of a non-spherical particle. In other words, as long as a compact

d_1 (μm)	F_2	F_3
1.158	1.39	1.69
1.83	1.41	1.75
average values	1.40	1.72
$F_j = j^{1/2}$	1.41	1.73

Table 1: Relative light scattering diameters, F_j, of aggregates of uniform polystyrene spheres measured with the LASI.

particle larger than the wavelength is considered and the scattered light is collected within the forward lobe of diffraction the response is a function of the projected area of the particle regardless of its shape. Outside the diffraction lobe, however, light scattering is effected by reflection and refraction at the particle surface. Then specular reflections and internal reflexes can produce a light scattering distribution which deviates considerably from the scattering diagram of a sphere of equal projected area (Broßmann[7], Zerull[44]).

Finally it should be mentioned that the response of a projected area instrument depends on the orientation of the particle in the sensing volume. It can be shown that even for randomly oriented agglomerates of spheres those orientations are more frequent which exhibit projected areas being multiples of the cross-section of a single sphere. Therefore separate peaks can be recorded for the different agglomerates.

7. MEASUREMENTS WITH THE LASS BELOW THE WAVELENGTH

In the submicron size range (below the wavelength of light) the laser aerosol size spectrometer (LASS) is used for particle size analysis (Heyder et al.[45], Roth and Gebhart[46]). The optical arrangement of the spectrometer is shown in Fig. 9 as cross-section through the plane of observation. The spectrometer is supplied with two light sources, Q, a helium-neon laser ($\lambda = 0.6328$ μm) with an output of 7 mW and an argon ion laser ($\lambda = 0.5145$ µm) with an output of 1.2 W. The laser light is focused by an astigmatic system of lenses, L, into the sensing volume. The focus has an elliptical cross-section of (0.02 x 0.12) mm^2.

The light scattered by the particle is collected by microscope objective, O, under a mean scattering angle of 40° and is passed via a mirror, M, to a red sensitive photomultiplier, PM. By turning the mirror, M, the scattered light is either reflected on the photomultiplier cathode or into an eyepiece, E, for observation. The sensing volume is imaged by the objective, O, into a plane of exchangeable stops, D. Behind the sensing volume the primary beam vanishes in alight trap, T. The aerosol nozzle is directed perpendicular to the laser beam and to the plane of observation. The width of the aerosol stream having an original diameter of 0.2 mm is reduced by aerodynamic focusing to a small filament of about 0.03 mm width. In this way a sensing volume of only $1.5 \cdot 10^{-8}$ cm^3 is obtained, so that aerosols with concentrations up to 10^6 cm^{-3} can be measured without coincidence losses. Under these conditions a sampling rate of about 4 $cm^3 min^{-1}$ is obtained. The output pulses of the photomultiplier (~ 0.25 µs) are linearly amplified and then accepted by a multi-channel analyser, MCA.

Calibration curves, S(d), of the instrument for two refractive indices are given in Fig. 10, where in a semi-logarithmic diagram S is plotted against the dimensionless size parameter, $\alpha = \pi d/\lambda$. The corresponding diameters of the particles for the two laser wavelengths are also drawn on the abscissa. The diameters of the polystyrene spheres with refractive index, $n_c=1.590$, have been checked by electron microscopy (Porstendörfer and Heyder[38]). The experimental values and the theoretical curve derived from light scattering theory summarized by Kerker[47] were matched for d = 0.206 µm. It turned out that the experimental points and the theoretical curve are almost in line. Where small deviations occured the experimental results have been considered more reliable. Calibration curves for aerosols of refractive indices, $n \neq n_c$, differ from the matched experimental one by a factor $A(n,n_c,d)$ which can be derived from light scattering theory. The curves in Fig. 10 increase monotoneously up to particle diameters of 0.7 µm and differ below 0.5 µm ($\alpha < 2.5$) only by constant factor $A(n,n_c)$.

To investigate the response of the LASS to non-spherical particles agglomerates of uniform polystyrene spheres with

d_1 (μm)	F_2	F_3	F_4	F_5	F_6
0.100	1.254	1.446	1.610	1.708	--
0.151	1.285	1.457	1.616	1.702	1.841
0.206	1.233	1.417	1.588	1.718	1.830
0.318	1.267	1.478	--	--	--
average values	1.259	1.449	1.604	1.709	1.836
$F_j = j^{1/3}$	1.260	1.442	1.588	1.710	1.818

Table 2: Relative light scattering diameters, F_j, of aggregates of j uniform polystyrene spheres measured with the LASS.

diameter, d_1, have been used. Due to the high resolution power of the instrument aggregates up to six spheres can be distinguished. By means of the relative light scattering diameter, F_j, defined in chapter 6 the results are summarized in Table 2. As can be seen, $F_j = j^{1/3}$, so that d_{scj} is identical with the equivalent volume diameter, d_{ve}, of a particle regardless of its shape. In other words the optical shape factor of a non-spherical particle smaller than the wavelength of light is 1. These experimental findings are valid within the experimental error of the method which is less than 2 %.

8. HIGH RESOLUTION OPTICAL SIZE SPECTROMETRY

In order to evaluate the resolution power of the spectrometers LASI and LASS monodisperse polystyrene latex aerosols of known size distributions are used. The spectra stored in the multichannel analyser are converted into particle diameter distributions by means of the calibration curves. Since the size distributions of the polystyrene aerosols can be approximated by Gaussian distribution functions, the modal diameter, $\hat{d}$, and the standard deviation, σ_{op}, are used to characterize the distributions;

σ_{op} depends on the standard deviation, σ_{el}, of the polystyrene latex aerosols (evaluated by electron microscopy) and, additionally, on the variance of the spectrometers.

	$\hat{d}_{el}$ (µm)	σ_{el} (µm)	σ_{op} (µm)	$\sigma_{op}/\hat{d}_{el}$
LASS	0.073	0.0050	0.0040	0.055
	0.100	0.0040	0.0035	0.035
	0.151	0.0030	0.0032	0.021
	0.206	0.0040	0.0035	0.017
	0.318	0.0050	0.0037	0.012
	0.488	0.0070	0.0065	0.013
LASI	1.158	0.021	0.043	0.037
	1.83	0.028	0.040	0.022
	2.68	--	0.050	0.018

Table 3: Resolution power of the LASS and the LASI measured with monodisperse polystyrene latex aerosols ($\hat{d}_{el}$ = modal diameter of the polystyrene latex spheres measured with an electron microscope).

The comparison of σ_{op} and σ_{el} is shown in Table 3. These values agree very well as far as the LASS is concerned. Hence, the variance of this instrument is negligible. The variance of the LASI is not negligible probably primarily due to the short-time stability of the mercury arc lamp which is less than the stability of the laser.

To achieve a high resolution power an optical size spectrometer has to be designed in such a way that the partial scattering cross-section of a particle depends strongly on particle size. Furthermore, the resolution can still be improved by applying a light illumination of high short-time stability and aerodynamic focusing of the aerosol stream.

The latter results in a constant aerosol flow and homogeneous illumination of the particles. However, a high resolution optical size spectrometer has a limited dynamic range for instantaneous size detection. Therefore, the size distributions of polydisperse aerosols have to be measured stepwise by varying the amplification of the electrical pulses.

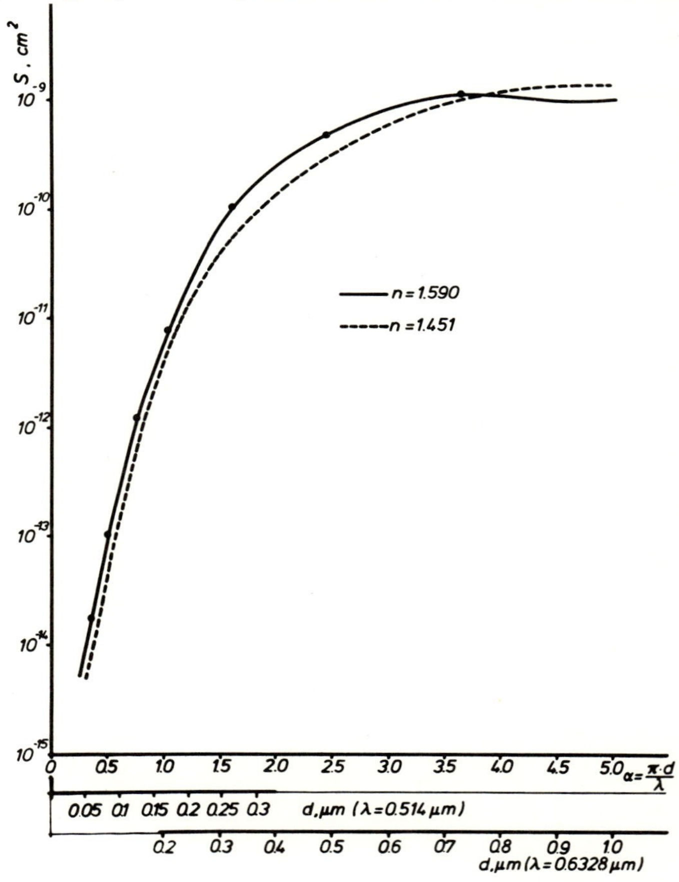

Fig. 10. Experimental response curves of the LASS for transparent spheres of different refractive index

REFERENCES

1) Hodkinson,J.R.: Dust measurements by light scattering and absorption, Ph.D.Thesis, University of London 1962

2) Quenzel,H.: Appl.Optics 8:165,1969

3) Schöck,W.: Kolloquium Aerosolmeßtechnik, Aachen 1975

4) Bakhanova,R.A. and Ivanchenko,L.V.: Aerosol Sci.4:485, 1973

5) Wagner,P.: Untersuchung des Tröpfchenwachstums in einer schnellen Expansionsnebelkammer, Ph.D.Thesis, Universität Wien 1974

6) Hodkinson,J.R. and Greenfield,J.R.: Appl.Optics 4:1463, 1965

7) Broßmann,R.: Die Lichtstreuung an kleinen Teilchen als Grundlage einer Teilchengrößenbestimmung, Ph.D.Thesis, Universität Karlsruhe 1966

8) Oeseburg,F.: Aerosol Sci. 3:307,1972

9) Gucker,F.T. and Rose,M.A.: Brit. J. Appl. Physics, Supplement No 3:138,1953

10) Zinky,W.R.: J. Air Pollution Control Assoc. 12:578,1962

11) Sinclair,D.:J. Air Pollution Control Assoc. 17:105,1967

12) Martens,A.E. and Fuss,K.H.: Staub-Reinhaltung Luft 28: 229,1968

13) Turpin,P.Y.: Revue d'Optique 46:309,1967

14) Whitby,K.T. and Vomela,R.A.: Env.Sci.Techn. 1:801,1967

15) Jaenicke,R.: Aerosol Sci. 3:95,1972

16) Williams,I. and Hedley,A.B.: Aerosol Sci. 3:363,1972

17) Liu,B.Y.H.;Berglund,R.N. and Argawal,J.K.: Atm. Env. 8: 717,1974

18) Schleusener,S.A. and Read,A.A.: Rev.Sci.Instr. 38:1152 1967

19) Schehl,R.;Ergun,S. and Headrick,A.: Rev.Sci. Instr. 44: 1193,1973

20) Phillips,D.T.;Wyatt,P.J. and Berkman,R.M.: J.Coll. Interface Sci. 34:159,1970

21) Wyatt,P.T. and Phillips,D.T.: J.Coll. Interface Sci. 39:125,1972

22) Cooke,D.D. and Kerker,M.: J.Coll. Interface Sci. 42: 150,1973

23) Gucker,F.T.;Tuma,J.;Lin,H.-M.;Huang,C.-M.;Ems,S.C. and Marshall,T.R.: Aerosol Sci. 4:389,1973

24) Moser,H.O.: Appl.Optics 13:173,1974

25) Meyer,U.: Bestimmung von Teilchengrößen mit einer Streulichtmethode, M.Sc.Thesis, TH Darmstadt 1975

26) Kaye,W.: J.Coll.Interface Sci. 44:384,1973

27) Gebhart,J,;Bol,J.;Heinze,W. and Letschert,W.: Staub-Reinhaltung Luft 30:238,1970

28) Hill,C.A.: Conference on Aerosol Physics, Southhampton 1973

29) Bol,J.;Gebhart,J.;Heinze,W.;Petersen,W.D. and Wurzbacher, G.: Staub-Reinhaltung Luft 30:475,1970

30) Pinnick,R.G.;Rosen,J.M. and Hofmann,D.J.: Appl.Optics 12:37,1973

31) Jacobi,W.;Eichler,J. and Stolterfoht,N.: Staub-Reinhaltung Luft 28:314,1968

32) Walsh,M.: private Mitteilung, 1973

33) Lioy,P.J.;Rimberg,D. and Hanghey,F.J.: Aerosol Sci 6: in press,1975

34) Heyder,J.;Gebhart,J. and Stahlhofen,W.: Water, Air and Soil Pollution 3:567,1974

35) Liu,B.Y.H.;Marple,V.A.;Whitby,K.T. and Barsic,N.J.: Am. Ind. Hyg. Assoc. J. 74:443,1974

36) van de Hulst,H.C.: Light scattering by small particles, John Wiley,New York,1957

37) Gebhart,J. and Roth,C.: 1st Annual Meeting of Gesellschaft für Aerosolforschung (Bad Soden/Ts. Rossertstr.11, Germany) 1973

38) Porstendörfer,J. and Heyder,J.: Aerosol Sci. 3:141,1972

39) Stahlhofen,W.;Gebhart,J.;Heyder,J. and Roth,C.: Aerosol Sci. 6,in press,1975

40) Stahlhofen,W.;Armbruster,L.;Gebhart,J. and Grein,E.: Atm. Env., in press,1975

41) Albert,R.E.;Petrow,H.G.;Salam,A.S. and Spiegelman,J.R.: Helath Physics 1/,933,1964

42) May,K.R.: J.Appl. Physics 20:932,1949

43) Heyder,J. and Porstendörfer,J.: Aerosol Sci. 5:387,1974

44) Zerull,R.: Mikrowellenanalogieexperimente zur Lichtstreuung an Staubpartikeln, Ph.D.Thesis,Univ.Bochum 1973

45) Heyder,J.;Roth,C. and Stahlhofen,W.: Aerosol Sci. 2:341 1971

46) Roth,C. and Gebhart,J.: 1st Annual Meeting of Gesellschaft für Aerosolforschung (Bad Soden/Ts. Rossertstr.11,Germany) 1973

47) Kerker,M.: The scattering of light, Academic Press, New York-London, 1969

48) Maguire,B.A.: Aerosol Sci. 2:417,1971

49) Lowan,A.N.: Nat.Bur.Standard, Appl. Math.Series 4, Tables of scattering functions for spherical particles

50) Olaf,J. and Robock,K.: Staub-Reinhaltung Luft 21:495, 1961

AUTHOR INDEX

AUTHOR INDEX

M

Q

R

AUTHOR INDEX

SUBJECT INDEX

SUBJECT INDEX